KU-030-315

Compendium of Organic Synthetic Methods

Compendium of Organic Synthetic Methods

Volume 12

MICHAEL B. SMITH
Department of Chemistry
The University of Connecticut
Storrs, Connecticut

A JOHN WILEY & SONS, INC., PUBLICATION

Published by John Wiley & Sons, Inc., Hoboken, New Jersey.
Published simultaneously in Canada.

Library of Congress Cataloging Card Number: 71-162800

ISBN 978-0-471-44530-2

Printed in the United States of America.

10 9 8 7 6 5 4 3 2 1

CONTENTS

PREFACE

Since the original volume in this series by Ian and Shuyen Harrison, the goal of the *Compendium of Organic Synthetic Methods* has been to facilitate the search for functional group transformations in the original literature of Organic chemistry. In Volume 2, difunctional compounds were added, and this compilation was continued by Louis Hegedus and Leroy Wade for Volume 3 of the series. Professor Wade became the author for Volume 4 and continued with Volume 5. I began editing the series with Volume 6, where I introduced an author index for the first time and added a new chapter (Chapter 15, Oxides). Volume 7 introduced Sections 378 (Oxides-Alkynes) through Section 390 (Oxides-Oxides). The *Compendium* is a handy desktop reference that remains a valuable tool to the working organic chemist, allowing a quick check of the literature. Even in the era of powerful computer searching, the *Compendium* allows one to "browse" for new reactions and transformations that may be of interest in a rapid and logical manner. The body of Organic chemistry literature is very large and the *Compendium* is a focused and highly representative survey of the literature, and is offered in that context.

Compendium of Organic Synthetic Methods, Volume 12 contains both functional group transformations and carbon-carbon bond-forming reactions from the literature appearing in the years 2002, 2003, and 2004. The classification schemes used for Volumes 6-11 have been continued. Difunctional compounds appear in Chapter 16. The experienced user of the *Compendium* will require no special instructions for the use of Volume 12. Author citations and the Author Index have been continued as in Volumes 6-11.

Every effort has been made to keep the manuscript error-free. Where there are errors, I take full responsibility. If there are questions or comments, the reader is encouraged to contact me directly at the address, phone, fax, or email given below.

As I have throughout my writing career, I thank my wife, Sarah, and my son, Steven, for their encouragement and support during this work. I also thank Dr. Darla Henderson, Michael Forster, Jonathan Rose, Lauren Hilger, Rebekah Amos, and Angioline Loredo of Wiley for their help in the publication of this volume.

MICHAEL B. SMITH

Department of Chemistry, University of Connecticut
55 N. Eagleville Road
Storrs, Connecticut 06269-3060

Voice phone: (860)-486-2881
Fax: (860)-486-2981
Email: michael.smith@uconn.edu

Storrs, Connecticut
December 2008

ABBREVIATIONS

Ac	Acetyl	$C(=O)CH_3$
acac	Acetylacetonate	
AIBN	*azobis*-isobutyronitrile	
aq.	Aqueous	
[9-BBN boryl structure]	9-Borabicyclo[3.3.1]nonylboryl	
9-BBN	9-Borabicyclo[3.3.1]nonane	
BER	Borohydride exchange resin	
BINAP	*2R,3S*-2,2'-*bis*-(diphenylphosphino)-1,1'-binaphthyl	
Bmim	1-butyl-3-methylimidazolium	
Bn	benzyl	
BOC	*tert*-Butoxycarbonyl	$C(=O)Ot$-Bu
bpy (Bipy)	2,2'-Bipyridyl	
Bu	*n*-Butyl	$-CH_2CH_2CH_2CH_3$
Bz	Benzoyl	
°C	Temperature in degrees Celsius	
CAM	Carboxamidomethyl	
CAN	Ceric ammonium nitrate	$(NH)_2Ce(NO_3)_6$
c-	Cyclo-	
cat.	Catalytic	
Cbz	Carbobenzyloxy	$C(=O)OCH_2Ph$
Chirald	2*S*,3*R*-(+)-4-Dimethylamino-1,2-diphenyl-3-methylbutan-2-ol	
COD	1,5-Cyclooctadienyl	
COT	1,3,5-Cyclooctatrienyl	
Cp	Cyclopentadienyl	
CSA	Camphorsulfonic acid	
CTAB	Cetyltrimethylammonium bromide	$C_{16}H_{33}NMe_3^+ Br^-$
Cy (c-C_6H_{11})	Cyclohexyl	[cyclohexyl structure]
DABCO	1,4-Diazabicylco[2.2.2]octane	
dba	Dibenzylidene acetone	
DBE	1,2-Dibromoethane	$BrCH_2CH_2Br$
DBN	1,5-Diazabicyclo[4.3.0]non-5-ene	
DBU	1,8-Diazabicyclo[5.4.0]undec-7-ene	
DCC	1,3-Dicyclohexylcarbodiimide	c-C_6H_{11}-N=C=N-c-C_6H_{11}

DCE	1,2-Dichloroethane	$ClCH_2CH_2Cl$
DCM	Dichloromethane	CH_2Cl_2
DDQ	2,3-Dichloro-5,6-dicyano-1,4-benzoquinone	
% de	% Diastereomeric excess	
DEA	Diethylamine	$HN(CH_2CH_3)_2$
DEAD	Diethylazodicarboxylate	$EtO_2C\text{-}N{=}NCO_2Et$
Dibal-H	Diisobutylaluminum hydride	$(Me_2CHCH_2)_2AlH$
Diphos (dppe)	1,2-*bis*-(Diphenylphosphino)ethane	$Ph_2PCH_2CH_2PPh_2$
Diphos-4 (dppb)	1,4-*bis*-(Diphenylphosphino)butane	$Ph_2P(CH_2)_4PPh_2$
DMA	Dimethylacetamide	
DMAP	4-Dimethylaminopyridine	
DME	Dimethoxyethane	$MeOCH_2CH_2OMe$
DMF	*N,N′*-Dimethylformamide	$HC(=O)N(CH_3)_2$
dmp	*bis*-[1,3-Di(*p*-methoxyphenyl)-1,3-propanedionato]	
dpm	Dipivaloylmethanato	
dppb	1,4-*bis*-(Diphenylphosphino)butane	$Ph_2P(CH_2)_4PPh_2$
dppe	1,2-*bis*-(Diphenylphosphino)ethane	$Ph_2PCH_2CH_2PPh_2$
dppf	*bis*-(Diphenylphosphino)ferrocene	
dppp	1,3-*bis*-(Diphenylphosphino)propane	$Ph_2P(CH_2)_3PPh_2$
dvb	Divinylbenzene	
e^-	Electrolysis	
% ee	% Enantiomeric excess	
EDA	Ethylenediamine	$H_2NCH_2CH_2NH_2$
EDTA	Ethylenediaminetetraacetic acid	
EE	1-Ethoxyethoxy	EtO(Me)CHO–
Et	Ethyl	$\text{-}CH_2CH_3$
FMN	Flavin mononucleotide	
fod	*tris*-(6,6,7,7,8,8,8)-Heptafluoro-2,2-dimethyl-3,5-octanedionate	
Fp	Cyclopentadienyl-*bis*-carbonyl iron	
FVP	Flash Vacuum Pyrolysis	
h	Hour (hours)	
hν	Irradiation with light	
1,5-HD	1,5-Hexadienyl	
HMPA	Hexamethylphosphoramide	$(Me_2N)_3P{=}O$
HMPT	Hexamethylphosphorus triamide	$(Me_2N)_3P$
iPr	Isopropyl	$\text{-}CH(CH_3)_2$
LICA (LIPCA)	Lithium cyclohexylisopropylamide	
LDA	Lithium diisopropylamide	$LiN(iPr)_2$
LHMDS	Lithium hexamethyldisilazide	$LiN(SiMe_2)_2$
LTMP	Lithium 2,2,6,6-tetramethylpiperidide	
MABR	Methylaluminum *bis*-(4-bromo-2,6-di-*tert*-butylphenoxide)	
MAD	*bis*-(2,6-di-*tert*-butyl-4-methylphenoxy)methyl aluminum	
mCPBA	*meta*-Chloroperoxybenzoic acid	
Me	Methyl	$\text{-}CH_3$
MEM	β-Methoxyethoxymethyl	$MeOCH_2CH_2OCH_2\text{-}$

Mes	Mesityl	2,4,6-tri-Me-C_6H_2
MOM	Methoxymethyl	$MeOCH_2$-
Ms	Methanesulfonyl	CH_3SO_2-
MS	Molecular Sieves (3 or 4 Å)	
MTM	Methylthiomethyl	CH_3SCH_2-
NAD	Nicotinamide adenine dinucleotide	
NADP	Sodium triphosphopyridine nucleotide	
Napth	Naphthyl ($C_{10}H_8$)	
NBD	Norbornadiene	
NBS	*N*-Bromosuccinimide	
NCS	*N*-Chlorosuccinimide	
Ni(R)	Raney nickel	
NIS	*N*-Iodosuccinimide	
NMP	N-Methyl-2-pyrrolidinone	
Oxone	2 $KHSO_5$•$KHSO_4$•K_2SO_4	
 ●	Polymeric backbone	
PCC	Pyridinium chlorochromate	
PDC	Pyridinium dichromate	
PEG	Polyethylene glycol	
Ph	Phenyl	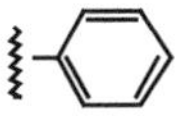
PhH	Benzene	
PhMe	Toluene	
Phth	Phthaloyl	
pic	2-Pyridinecarboxylate	
Pip	Piperidino	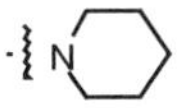
PMP	4-Methoxyphenyl	
Pr	*n*-Propyl	$-CH_2CH_2CH_3$
Py	Pyridine	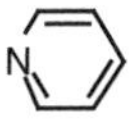
quant.	Quantitative yield	
Red-Al		$[(MeOCH_2CH_2O)_2AlH_2]Na$
s-Bu	*sec*-Butyl	$CH_3CH_2CH(CH_3)$
s-BuLi	*sec*-Butyllithium	$CH_3CH_2CH(Li)CH_3$
Siamyl	Diisoamyl	$(CH_3)_2CHCH(CH_3)$-
TADDOL	α,α,α',α'-tetraaryl-4,5-dimethoxy-1,3-dioxolane	
TASF	*tris*-(Diethylamino)sulfonium difluorotrimethyl silicate	
TBAF	Tetrabutylammonium fluoride	n-$Bu_4N^+ F^-$
TBDMS	*tert*-Butyldimethylsilyl	t-$BuMe_2Si$
TBDPS	*tert*-Butyldiphenylsilyl	t-$BuPh_2Si$
TBHP (*t*-BuOOH)	*tert*-Butylhydroperoxide	Me_3COOH
t-Bu	*tert*-Butyl	$-C(CH_3)_3$
TEBA	Triethylbenzylammonium	$Bn(Et)_3N^+$
TEMPO	Tetramethylpiperdinyloxy free radical	

TFA	Trifluoroacetic acid	CF_3COOH
TFAA	Trifluoroacetic anhydride	$(CF_3CO)_2O$
Tf (OTf)	Triflate	$-SO_2CF_3(-OSO_2CF_3)$
THF	Tetrahydrofuran	
THP	Tetrahydropyran	
TMEDA	Tetramethylethylenediamine	$Me_2NCH_2CH_2NMe_2$
TMG	1,1,3,3-Tetramethylguanidine	
TMP	2,2,6,6-Tetramethylpiperidine	
TMS	Trimethylsilyl	$-Si(CH_3)_3$
TPAP	tetra-*n*-Propylammonium perruthenate	
Tol	Tolyl	$4-C_6H_4CH_3$
Tr	Trityl	$-CPh_3$
TRIS	Triisopropylphenylsulfonyl	
Ts(Tos)	Tosyl = *p*-Toluenesulfonyl	$4-MeC_6H_4SO_2$
X_c	Chiral auxiliary	

INDEX, MONOFUNCTIONAL COMPOUNDS

Sections—**heavy type**
Pages—light type

PREPARATION OF →

FROM ↓

	Alkynes	Carboxylic acid derivatives	Alcohols, phenols	Aldehydes	Alkyls, methylenes, aryls	Amides	Amines	Esters	Ethers, epxoides	Halides, sulfonates	Hydrides (RH)	Ketones	Nitriles	Alkenes	Oxides
Alkynes	**1** 1	**16** 9	**31** 17	**46** 51	**61** 67	**76** 151	**91** 177	**106** 209	**121** 227			**166** 259		**196** 289	
Carboxylic acid derivatives	**2** 4	**17** 9	**32** 17	**47** 51	**62** 69	**77** 151		**107** 209	**122** 227			**167** 259		**197** 295	**212** 307
Alcohols, phenols		**18** 10	**33** 18	**48** 51	**63** 70	**78** 153	**93** 178	**108** 212	**123** 227	**138** 245	**153** 253	**168** 261	**183** 283	**198** 295	**213** 307
Aldehydes	**4** 5	**19** 10	**34** 18	**49** 56	**64** 71	**79** 153	**94** 179	**109** 216	**124** 230	**139** 246	**154** 254	**169** 265	**184** 283	**199** 297	
Alkyls, methylenes, aryls		**20** 11	**35** 31		**65** 72	**80** 155		**110** 217	**125** 231	**140** 247	**155** 254	**170** 266		**200** 298	**215** 308
Amides			**36** 32			**81** 155	**96** 181	**111** 217				**171** 267	**186** 284		**216** 308
Amines					**67** 74	**82** 162	**97** 182	**112** 217				**172** 267			**217** 308
Esters		**23** 11	**38** 32	**53** 57	**68** 74	**83** 165	**98** 191	**113** 217	**128** 232	**143** 247	**158** 254	**173** 268		**203** 299	**218** 309
Ethers, epxoides		**24** 12	**39** 33	**54** 57	**69** 78	**84** 166	**99** 191	**114** 220	**129** 232	**144** 247		**174** 269		**204** 299	**219** 309
Halides, sulfonates	**10** 6	**25** 13	**40** 36	**55** 58	**70** 79	**85** 166	**100** 192	**115** 221	**130** 233	**145** 247	**160** 255	**175** 270	**190** 285	**205** 300	**220** 311
Hydrides (RH)		**26** 13	**41** 36	**56** 59	**71** 96	**86** 168	**101** 196	**116** 222	**131** 236	**146** 248		**176** 271	**191** 287		**221** 312
Ketones		**27** 13	**42** 37		**72** 99	**87** 169	**102** 196	**117** 223			**162** 256	**177** 273		**207** 301	
Nitriles	**13** 8		**43** 45	**58** 59	**73** 99	**88** 170	**103** 198	**118** 224				**178** 276	**193** 287		
Alkenes	**14** 8	**29** 14	**44** 45	**59** 59	**74** 100	**89** 171	**104** 200	**119** 225	**134** **237**	**149** 252		**179** 276		**209** 302	**224** 314
Miscellaneous	**15** 8	**30** 15	**45** 46	**60** 60	**75** 148	**90** 173	**105** 202	**120** 225	**135** 242	**150** 252		**180** 278	**195** 287	**210** 304	**225** 315

PROTECTION

	Sect.	Pg.
Carboxylic acids	30A	15
Alcohols, thiols	45A	46
Aldehydes	60A	61
Amides	90A	176
Amines	105A	206
Ketones	180A	279

Blanks in the table correspond to sections for which no additional examples were found in the literature

INDEX, DIFUNCTIONAL COMPOUNDS

Sections—**heavy type**
Pages—light type

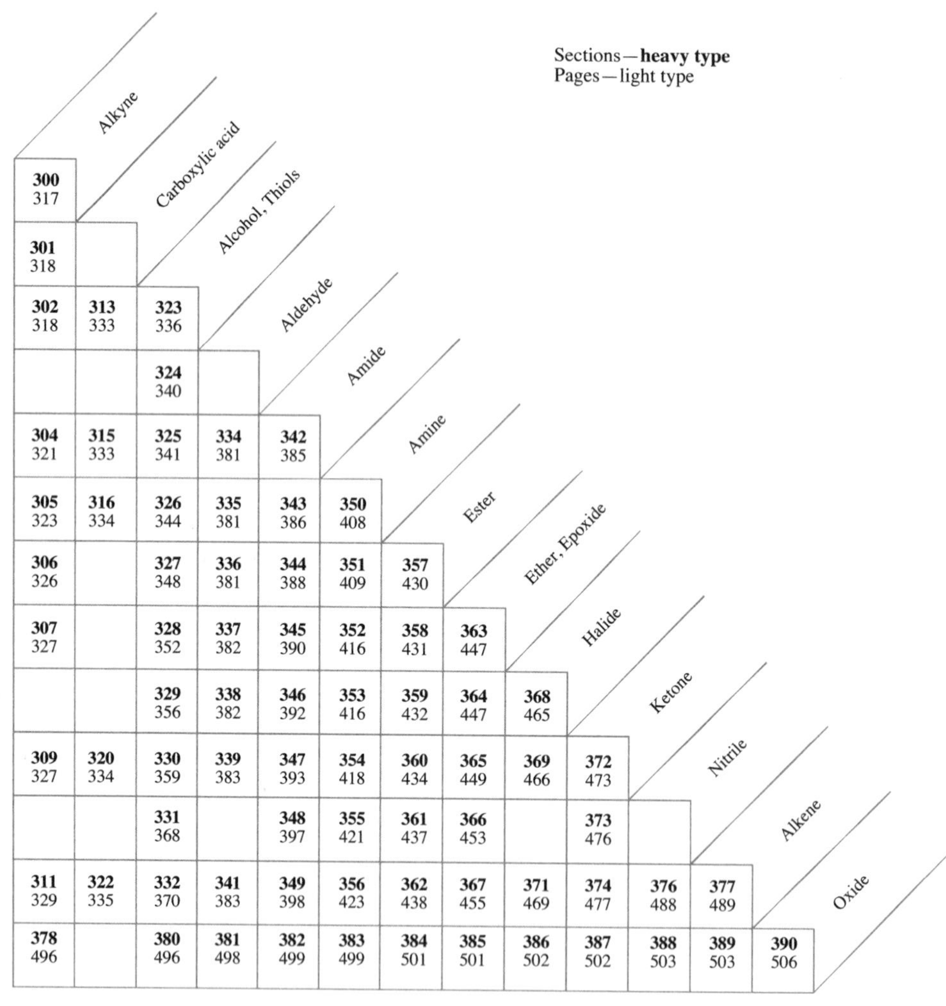

	Alkyne	Carboxylic acid	Alcohol, Thiols	Aldehyde	Amide	Amine	Ester	Ether, Epoxide	Halide	Ketone	Nitrile	Alkene	Oxide
Alkyne	**300** 317												
Carboxylic acid	**301** 318												
Alcohol, Thiols	**302** 318	**313** 333	**323** 336										
Aldehyde			**324** 340										
Amide	**304** 321	**315** 333	**325** 341	**334** 381	**342** 385								
Amine	**305** 323	**316** 334	**326** 344	**335** 381	**343** 386	**350** 408							
Ester	**306** 326		**327** 348	**336** 381	**344** 388	**351** 409	**357** 430						
Ether, Epoxide	**307** 327		**328** 352	**337** 382	**345** 390	**352** 416	**358** 431	**363** 447					
Halide			**329** 356	**338** 382	**346** 392	**353** 416	**359** 432	**364** 447	**368** 465				
Ketone	**309** 327	**320** 334	**330** 359	**339** 383	**347** 393	**354** 418	**360** 434	**365** 449	**369** 466	**372** 473			
Nitrile			**331** 368		**348** 397	**355** 421	**361** 437	**366** 453		**373** 476			
Alkene	**311** 329	**322** 335	**332** 370	**341** 383	**349** 398	**356** 423	**362** 438	**367** 455	**371** 469	**374** 477	**376** 488	**377** 489	
Oxide	**378** 496		**380** 496	**381** 498	**382** 499	**383** 499	**384** 501	**385** 501	**386** 502	**387** 502	**388** 503	**389** 503	**390** 506

Blanks in the table correspond to sections for which no additional examples were found in the literature

INTRODUCTION

Relationship Between Volume 12 and Previous Volumes. *Compendium of Organic Synthetic Methods, Volume 12* presents about 2500 examples of published reactions for the preparation of monofunctional compounds, updating the more than 15,500 examples found in Volumes 1-11. Volume 12 contains about 1500 examples of reactions that prepare difunctional compounds with various functional groups. Reviews have long been a feature of this series, and Volume 12 adds 97 pertinent reviews in the various sections.

Chapters 1-14 present the same functional group transformations as found in Volumes 1-11, as does Chapter 15, introduced in Volume 6. Difunctional compounds appear in Chapter 16, as in Volumes 6-11. The sections on oxides as part of difunctional compounds, introduced in Volume 7, continues in Chapter 16 of Volumes 8-11 with Sections 378 (Oxides-Alkynes) through Section 390 (Oxides-Oxides).

Following Chapter 16 is a complete alphabetical listing of all authors (last name, initials). The authors for each citation appear *below* the reaction. The principal author is indicated by underlining (i.e., Kwon, T.W.; Smith, M.B.), as done previously in Volumes 7-11.

Classification and Organization of Reactions Forming Monofunctional Compounds. Chemical transformations are classified according to the reacting functional group of the starting material and the functional group formed. Those reactions that give products with the same functional group form a chapter. The reactions in each chapter are further classified into sections on the basis of the functional group of the starting material. Within each section, an effort has been made to put similar reactions together when possible. Review articles are collected at the end of each appropriate section.

The classification is unaffected by allylic, vinylic, or acetylenic unsaturation appearing in both starting material and product, or by increases or decreases in the length of carbon chains; for example, the reactions *t*-BuOH → *t*-BuCOOH, $PhCH_2OH$ → PhCOOH, and $PhCH{=}CHCH_2OH$ → PhCH=CHCOOH would all be considered as preparations of carboxylic acids from alcohols. Sections 74D (Alkyls from Alkenes: Conjugate Reductions) and 74E (Alkyls from Alkenes: Conjugate Alkylations) contain the reactions conjugate reduction or conjugate alkylation of unsaturated ketones, aldehydes, esters, acids, and nitriles, respectively.

The terms hydrides, alkyls, and aryls classify compounds containing reacting hydrogens, alkyl groups, and aryl groups, respectively; for example, RCH_2-H → RCH_2COOH (carboxylic acids from hydrides), RMe → RCOOH (carboxylic acids from alkyls), RPh → RCOOH (carboxylic acids from aryls). Note the distinction between R_2CO → R_2CH_2 (Methylenes from Ketones) and RCOR' → RH (hydrides from ketones). Alkylations involving additions across double bonds are given in Section 74 (Alkyls, Methylenes, and Aryls from Alkenes).

The following examples illustrate the classification of some potentially confusing cases:

RCH=CHCOOH	→	$RCH{=}CH_2$	Hydrides from carboxylic acids
$RCH{=}CH_2$	→	RCH=CHCOOH	Carboxylic acids from hydrides
ArH	→	ArCOOH	Carboxylic acids from hydrides

ArH	→	ArOAc	Esters from hydrides
RCHO	→	RH	Hydrides from aldehydes
RCH=CHCHO	→	$RCH=CH_2$	Hydrides from aldehydes
RCHO	→	RCH_2	Alkyls from aldehydes
R_2CH_2	→	R_2CO	Ketones from methylenes
RCH_2COR	→	R_2CHCOR	Ketones from ketones
$RCH=CH_2$ →		RCH_2CH3	Alkyls from alkenes (Hydrogenation of Alkenes)
RBr + HC≡CH	→	RC≡CR	Acetylenes from halides; also acetylenes from acetylenes
ROH + RCOOH	→	RCOOR	Esters from alcohols; also esters from carboxylic acids
RCH=CHCHO	→	RCH_2CH_2CHO	Alkyls from alkenes (Conjugate Reduction)
RCH=CHCN	→	RCH_2CH_2CN	Alkyls from alkenes (Conjugate Reduction)

How to Use the Book to Locate Examples of the Preparation of Protection of Monofunctional Compounds. Examples of the preparation of one functional group from another are given in the monofunctional index on p. xiii, which lists the corresponding section and page. Sections that contain examples of the reactions of a functional group are given in the horizontal rows of this index. Section 1 gives examples of the reactions of acetylenes that form new acetylenes; Section 16 gives reactions of acetylenes that form carboxylic acids; and Section 31 gives reactions of acetylenes that form alcohols.

Examples of alkylation, dealkylation, homologation, isomerization, and transposition are given in Sections 1, 17, 33, and so on, lying close to a diagonal of the index. These sections correspond to such topics as the preparation of acetylenes from acetylenes; carboxylic acids from carboxylic acids; and alcohols, thiols, and phenols from alcohols, thiols, and phenols. Alkylations that involve conjugate additions across a double bond are given in Section 74E (Alkyls, Methylenes, and Aryls from Alkenes).

Examples of name reactions may be found by first considering the nature of the starting material and product. The Wittig reaction, for instance, is given in Section 199 (Alkenes from Aldehydes) and in Section 207 (Alkenes from Ketones). The aldol condensation may be found in the chapters on difunctional compounds in Section 324 (Alcohol, Thiol-Aldehyde) and in Section 330 (Alcohol, Thiol-Ketone). Examples of the synthetically important alkene metathesis reaction are provided primarily in Section 209 (Alkenes from Alkenes).

Examples of the protection of acetylenes, carboxylic acids, alcohols, phenols, aldehydes, amides, amines, esters, ketones, and alkenes are also presented. Sections (designated with an A: 15A, 30A, etc.) are labeled "Protection of" and are located at the end of pertinent chapters.

Some pairs of functional groups, such as alcohol, ester; carboxylic acid, ester; amine, amide; and carboxylic acid, amide, can be interconverted by simple reactions. When a member of these groups is the desired product or starting material, the other member should also be consulted in the text.

The original literature must be used to determine the generality of reactions, although this is occasionally stated in the citation. This is done only in cases where such generality is stated clearly in the original citation. A reaction given in this book for a primary aliphatic substrate may also be applicable to tertiary or aromatic compounds. This book provides very limited experimental conditions or precautions and the reader is referred to the original literature before attempting a reaction. **In *no* instance should a citation in this book be taken as a complete experimental procedure. Failure to refer to the original literature prior to beginning laboratory work could be hazardous.** The original papers usually yield a further set of references to previous work. Papers that appear after those publications can usually be found by consulting *Chemical Abstracts* and the *Science Citation Index.*

Classification and Organization of Reactions Forming Difunctional Compounds. This chapter considers all possible difunctional compounds formed from the groups acetylene, carboxylic acid, alcohol, thiol, aldehyde, amide, amine, ester, ether, epoxide, thioether, halide, ketone, nitrile, and alkene. Reactions that form difunctional compounds are classified into sections on the basis of two functional groups in the product that are pertinent to the reaction. The relative positions of the groups do not affect the classification. Thus preparations of 1,2-aminoalcohols, 1,3-aminoalcohols, and 1,4-aminoalcohols are included in a single section (Section 326, Alcohol, Thiol-Amine). Difunctional compounds that have an oxide as the second group may be found in the appropriate section (Sections 278-290). The nitroketone product of oxidation of a nitroalcohol is given in Section 384 (Ketone-Oxide). Conversion of an oxide (such as nitro or a sulfone moiety) to another functional group is generally given in the Miscellaneous section of the sections concerning monofunctional compounds. Conversion of a nitroalkane to an amine, for example, is given in Section 105 (Amines from Miscellaneous Compounds). The following examples illustrate applications of this classification system:

Difunctional Product	*Section Title*
$RC\equiv C{-}C\equiv CR$	Acetylene-Acetylene
RCH(OH)COOH	Carboxylic acid-Alcohol
RCH=CHOMe	Ether-Alkene
$RCHF_2$	Halide-Halide
$RCH(Br)CH_2F$	Halide-Halide
$RCH(OAc)CH_2OH$	Alcohol-Ester
$RCH(OH)CO_2Me$	Alcohol-Ester
$RCH{=}CHCH_2CO_2Me$	Ester-Alkene
RCH=CHOAc	Ester-Alkene
$RCH(OMe)CH_2SO_2CH_2CH_2OH$	Alcohol-Ether
$RSO_2CH_2CH_2OH$	Alcohol-Oxide

How to Use the Book to Locate Examples of the Preparation of Difunctional Compounds. The difunctional index on p. xiv gives the section and page corresponding to each difunctional product. Thus, Section 327 (Alcohol, Thiol-Ester) contains examples of the preparation of hydroxyesters; Section 323 (Alcohol, Thiol-Alcohol, Thiol) contains examples of the preparation of diols.

Some preparations of alkene and acetylenic compounds from alkene and acetylenic starting materials can, in principle, be classified in either the monofunctional or difunctional sections; for example, the transformation RCH=CHBr → RCH=CHCOOH could be considered as preparing carboxylic acids from halides (Section 25, monofunctional compounds) or as preparing a carboxylic acid-alkene (Section 322, difunctional compounds). The choice usually depends on the focus of the particular paper where this reaction appeared. In such cases both sections should be consulted.

Reactions applicable to both aldehyde and ketone starting materials are in many cases illustrated by an example that uses only one of them. Similarly, many citations for reactions found in the Aldehyde-X sections include examples that could be placed in the Ketone-X section. Again, the choice is dictated by the original publication in which the reaction appeared.

Many literature preparations of difunctional compounds are extensions of the methods applicable to monofunctional compounds. As an example, the reaction RCl ROH might be used for the preparation of diols from an appropriate dichloro compound. Such methods are difficult to categorize and may be found in either the monofunctional or difunctional sections, depending on the focus of the original paper.

The user should bear in mind that the pairs of functional groups alcohol, ester; carboxylic acids, ester; amine, amide; and carboxylic acid, amide can be interconverted by simple reactions. Compounds of the type $RCH(OAc)CH_2OAc$ (Ester-Ester) would thus be of interest to anyone preparing the diol $RCH(OH)CH_2OH$ (Alcohol-Alcohol).

CHAPTER 1

PREPARATION OF ALKYNES

SECTION 1: ALKYNES FROM ALKYNES

Br–C$_6$H$_4$–CN → (Bu–C≡C–BF$_3$K; 9% PdCl$_2$(dppf)•CH$_2$Cl$_2$, reflux; 3 Cs$_2$CO$_3$, H$_2$O, THF, 12 h) → Bu–C≡C–C$_6$H$_4$–CN 98%

Molander, G.A.; Katona, B.W.; Machrouchi, F. *J. Org. Chem.* **2002**, *67*, 8416.

Me–C$_6$H$_4$–Br → (HC≡C–CH$_2$CH$_2$–OH, 5% Pd/C; 10% PPh$_3$, 12 *i*-Pr$_2$NH, reflux; dimethylacetamide-H$_2$O) → Me–C$_6$H$_4$–C≡C–CH$_2$CH$_2$–HO 94%

Novák, Z.; Szabó, A.; Répási, J.; Kotschy, A. *J. Org. Chem.* **2003**, *68*, 3327.

HC≡C–Ph → (Ac–C$_6$H$_4$–ONf; 2 PMHS, 5 CsF, 4 h; 5% PdCl$_2$(PPh$_3$)$_2$, NMP, 5% CuCl) → Ac–C$_6$H$_4$–C≡C–Ph 96%

PMHS = poly(methylhydrosiloxane)

Gallagher, W.P.; Maleczka Jr. R.E. *J. Org. Chem.* **2003**, *68*, 6775.

C$_5$H$_{11}$–C≡C–SiMe$_2$OH → (PhI, 2 TMSOK, DME, rt; 2.5% PdCl$_2$(PPh$_3$)$_2$, 3 h) → C$_5$H$_{11}$–C≡C–Ph 93%

Denmark, S.E.; Tymonko, S.A. *J. Org. Chem.* **2003**, *68*, 9151.

Me–C$_6$H$_4$–I → (HC≡C–Ph, [bmim]PF$_6$; 5% PdCl$_2$(PPh$_3$)$_2$, 60°C; *i*-Pr$_2$NH, 2 h) → Me–C$_6$H$_4$–C≡C–Ph

Cu free Sonogashira

Fukuyama, T.; Shinmen, M.; Nishitani, S.; Sato, M.; Ryu, I. *Org. Lett.* **2002**, *4*, 1691.

2 (1,2-difluorobenzene, F, F) → (0.02M, dark, 60°C, 18 h; 6% PdCl$_2$(PPh$_3$)$_2$, 10% CuI; 5 THS-C≡CH, 6 DBU, 40% H$_2$O) → (2-F-C$_6$H$_4$)–C≡C–(C$_6$H$_4$-2-F) 93%

Mio, M.J.; Kopel, L.C.; Braun, J.B.; Gadzikwa, T.L.; Hull, K.L.; Brisbois, R.G.; Markworth, C.J.; Grieco, P.A. *Org. Lett.* **2002**, *4*, 3199.

5% $Mo(CO)_6$, PhCl
2-fluorophenol , 3 h
reflux
83%

Grela, K.; Ignatowska, J. *Org. Lett.* ***2002**, 4*, 3747.

, CuI , MeCN
cat $PdCl_2(PPh_3)_2$, TEA , 8 h
$H_2 + N_2$, reflux
89%

Elangovan, A.; Wang, Y.-H.; Ho, T.-I. *Org. Lett.* ***2003**, 5*, 1841.

, NaOH , PEG
H_2O , 170°C
91%

Leadbeater, N.E.; Marco, M.T.; Tominack, B.J. *Org. Lett.* ***2003**, 5*, 3919.
Leadbeater, N.E.; Tominack, B.J. *Tetrahedron Lett.* ***2003**, 44*, 8653.

Ph-Br → Ph-C≡C-H , 10% P(*t*-Bu)$_3$, 2.5% [(ayl)PdCl]$_2$ / MeCN , rt , 2 piperidine → Ph-C≡C-Ph 71%

Soheili, A.; Albaneze-Walker, J.; Murry, J.A.; Dormer, P.G.; Hughes, D.L. *Org. Lett.* ***2003**, 5*, 4191.

Ph—≡ → Et_2N-TMS , 1.65 $ZnCl_2$ / dioxane , 100°C , 1.5h → Ph—≡—$SiMe_3$ 95%

Andreev, A.A.; Konshin, V.V.; Komarov, N.V.; Rubin, M.; Brouwer, C.; Gevorgyan, V. *Org. Lett.* ***2004**, 6*, 421.

Ph—≡ + Br—C$_6$H$_4$—OMe → $Pd(OAc)_2$/phosphine , 8h / 65°C , THF , K_2CO_3 → Ph—≡—C$_6$H$_4$—OMe
copper and amine free

Cheng, J.; Sun, Y.; Wang, F.; Guo, M.; Xu, J.-H.; Pan, Y.; Zhang, Z. *J. Org. Chem.* ***2004**, 69*, 5428.

Ph-C≡CH , 2% $Pd(OAc)_2$, rt
1.5 Bu_4NOAc , DMF
73% (28:72 *E:Z*)

Urgaonkar, S.; Verkade, J.G. *J. Org. Chem.* ***2004**, 69*, 5752.

Ph—≡—Li → BuBr , PPh_3 / $Pd_2(dba)_3$ → Ph—≡—Bu 88%

Yang, L.-M.; Huang, L.-F.; Luh, T.-Y. *Org. Lett.* ***2004**, 6*, 1461.

Ph-I + Ph—≡ → *i*-Pr_2NEt , H_2O , 70°C / cat $Pd(PPh_3)_4$, CuI → Ph—≡—Ph 92%

Bhattacharya, S.; Sengupta, S. *Tetrahedron Lett.* ***2004**, 45*, 8733.

Ph—≡ → (1. 0.05 $InBr_3$, piperidine, rt; 2. 4-iodotoluene, 5% $PdCl_2(PPh_3)_2$, piperidine, rt) → Ph—≡—C_6H_4—Me 95%

Sakai, N.; Annaka, K.; Konakahara, T. *Org. Lett.* **2004**, *6*, 1527.

4-BrC$_6$H$_4$COMe → ($PhC{\equiv}CH$, CuI, K_2CO_3, 130°C; cat $[Pd(C_3H_5)Cl]_2$–2 $F_c(P)_2(t\text{-Bu})(Pirr)$) → Ph—≡—$C_6H_4$—COMe 94%

Hierso, J.-C.; Fihri, A.; Amardeil, R.; Meunier, P.; Doucet, H.; Santelli, M.; Ivanov, V.V. *Org. Lett.* **2004**, *6*, 3473.

C_5H_{11}—≡—$AlMe_2$ → (PhI, 2.5% $Pd(dba)_3$•$CHCl_3$, 4.5 h; 5% dppf, 20°C, heptane/DME) → C_5H_{11}—≡—Ph quant

Wang, B.; Benin, M.; Micouin, L. *Org. Lett.* **2004**, *6*, 3481.

MeO—C_6H_4—≡—Me → (10% 2-CF_3-phenol (OH), toluene, 30°C; Mo carbene catalyst) → MeO—C_6H_4—≡—C_6H_4—OMe 83% with removal of 2-butyne

Zhang, W.; Kraft, S.; Moore, J.S. *J. Am. Chem. Soc.* **2004**, *126*, 329.

C_6H_{13}—≡ → ($PhB(OH)_2$, 5% $PdCl_2(dppf)$, DCM; 2 Ag_2O, 5 K_2CO_3, rt) → C_6H_{13}—≡—Ph 90%

Zou, G.; Zhu, J.; Tang, J. *Tetrahedron Lett.* **2003**, *44*, 8709.

t-Bu—C_6H_4—I → (MeO—C_6H_4—≡; cat CuI, Cs_2CO_3, NMP, microwaves, 4h) → *t*-Bu—C_6H_4—≡—C_6H_4—OMe 76%

He, H.; Wu, Y.-J. *Tetrahedron Lett.* **2004**, *45*, 3237.

Ph—≡ → (Cl—C_6H_4—COMe; cat $[Pd(C_3H_5Cl)]_2$/tetraphosphine ligand, DMF, K_2CO_3, 140°C) → Ph—≡—C_6H_4—COMe 82%

Feuerstein, M.; Doucet, H.; Santelli, M. *Tetrahedron Lett.* **2004**, *45*, 8443.

Ph—≡ → (PhCl, $ZnCl_2$, $NaHO_3$, dioxane, 1d; cat phosphinito Pd-pincer complex, 160°C) → Ph—≡—Ph 64%

Eberhard, M.R.; Wang, Z.; Jensen, C.M. *Chem. Commun.* **2002**, 818.

MeCO—C_6H_4—Cl → ($PhC{\equiv}CH$, toluene, 100°C, 4h; $Na_2(PCl)_4$, $PBn(1\text{-Ad})_2$-HPr) → MeCO—C_6H_4—≡—Ph 90%

Köllhofer, A.; Pullmann, T.; Plenio, H. *Angew. Chem. Int. Ed.* **2003**, *42*, 1056.

Cl–C_6H_4–CN → (t-BuC≡CH, Cs_2CO_3, 70°C, 9h; cat $PdCl_2(MeCN)_2$/biaryl phosphine) → t-Bu–C≡C–C_6H_4–CN 89%

Gelman, D.; Buchwald, S.L. *Angew. Chem. Int. Ed.* **2003**, *42*, 5993.

Ph–C≡CH → (Ph-I, 5% CuI, doxane, piperidine; [polymer]-$N(CH_2PPh_2)_2$, 60°C, overnight) → Ph–C≡C–Ph 99%

Gonthier, E.; Breinbauer, R. *Synlett* **2003**, 999.

Ph–C≡CH → (PhI, CuI-PPh_3, microwaves; K_2CO_3, DMF, 10 min) → Ph–C≡C–Ph 91%

Wang, J.-X.; Liu, Z.; Hu, Y.; Wei, B.; Kang, L. *Synth. Commun.* **2002**, *32*, 1937.

O_2N–C_6H_4–C≡C–$B(Oi\text{-}Pr)_3$ → (2.5% $Pt(PPh_3)_4$, PhI, DMF; 5% CuI, DMF) → O_2N–C_6H_4–C≡C–Ph

Oh, C.H.; Reddy, V.R. *Synlett* **2004**, 2091.

CH_3–C≡C–$(CH_2)_9$–C≡C–CH_3 → ($Mo(CO)_6$, toluene, 110°C; 4-CF_3-C_6H_4OH) → cyclic alkyne C≡C $(CH_2)_{10}$ 42%

Hellbach, B.; Gleitr, R.; Rominger, F. *Synthesis* **2003**, 2535.

4-CH_3–C_6H_4–I → (C_8H_{17}–C≡CH, CuI, PPh_3/KF-Al_2O_3; nanosize Ni powder, microwaves, no solvent) → 4-CH_3–C_6H_4–C≡C–C_8H_{17} 85%

Wang, M.; Li, P.; Wang, L. *Synth. Commun.* **2004**, *34*, 2803.

Ph-I + HC≡C-$(CH_2)_4CO_2H$ → (polymer-supported Pd-phosphine complex; 3 CsOH, water, copper-free, 60°C, 12 h) → Ph-C≡C-$(CH_2)_4CO_2H$ 74%

Uozumi, Y.; Kobayashi, Y. *Heterocycles* **2003**. *59*, 71.

Ph–C≡C–H → (1. BuLi; 2. $GaClL_3$; 3. BEt_3, O_2, ICH_2CO_2Bn) → Ph–C≡C–CH_2CO_2Bn 90%

Usugi, S.-i.; Yorimitsu, H.; Shinokubo, H.; Oshima, K. *Bull. Chem. Soc. Jpn.* **2002**, *75*, 2687.

SECTION 2: ALKYNES FROM ACID DERIVATIVES

PhC(O)Cl → (1. $BtCH_2TMS$; 2. Tf_2O; 3. 2 NaOH; 4. $C_5H_{11}Li$) → Ph–C≡C–C_5H_{11} 76%

Katritzky, A.R.; Abdel-Fathah, A.A.A.; Wang, M. *J. Org. Chem.* **2002**, *67*, 7526.

SECTION 3: ALKYNES FROM ALCOHOLS AND THIOLS

NO ADDITIONAL EXAMPLES

SECTION 4: ALKYNES FROM ALDEHYDES

ROMP
CHO
N_2
K_2CO_3 , MeOH , 4d
91%

Barrett, A.G.M.; Hopkins, B.T.; Love, A.C.; Tadeschi, L. *Org. Lett.* ***2004***, *6*, 835.

$C_{11}H_{23}CHO$, *p*-TsN_3 , MeCN / K_2CO_3 , MeOH , rt → $C_{11}H_{23}$—≡ 89%

Roth, G.J.; Liepold, B.; Müller, S.G.; Bestmann, H.J. *Synthesis* ***2004***, 59.

SECTION 5: ALKYNES FROM ALKYLS, METHYLENES AND ARYLS

NO ADDITIONAL EXAMPLES

SECTION 6: ALKYNES FROM AMIDES

NO ADDITIONAL EXAMPLES

SECTION 7: ALKYNES FROM AMINES

NO ADDITIONAL EXAMPLES

SECTION 8: ALKYNES FROM ESTERS

NO ADDITIONAL EXAMPLES

SECTION 9: ALKYNES FROM ETHERS, EPOXIDES, AND THIOETHERS

NO ADDITIONAL EXAMPLES

SECTION 10: ALKYNES FROM HALIDES AND SULFONATES

Br — $NaAl(C{=}CSiMe_3)$, THF / cat $PdCl_2(PPh_3)_2$, 12 h → $SiMe_3$ 83%

Gelman, D.; Tsvelikhovsky, D.; Molander, G.A.; Blum, J. *J. Org. Chem.* ***2002***, *67*, 6287.

I, TMS, O, Si, Me, Me — 1. BEt_3, toluene 2. TBAF 3. O_2N, Cl, O, O_2N → NO_2, O, O, NO_2 83%

Sukeda, M.; Ichikawa, S.; Matsuda, A.; Shuto, S. *J. Org. Chem.* ***2003***, *68*, 3465.

O, O, Br, Br — ≡—TMS / cat $Pd(PPh_3)_2Cl_2$, CuI dioxane, NEt_3, 0.5 h → O, O, Br, TMS 83%

Lee, J.-H.; Park, J.-S.; Cho, C.-G. *Org. Lett.* ***2002***, *4*, 1171.

Ph—≡ — PhI, 2% CuI, NEt_3, 4h / 2.5% thioxodiphosphapropene → Ph—≡—Ph no yield

Liang, H.; Ito, S.; Yoshifuji, M. *Org. Lett.* ***2004***, *6*, 425.

PhI — 1. ≡—CMe_2OH, cat $PdCl_2(PPh_3)_2$, cat CuI / i-Pr_2NH 2. PhI, KOH, 110°C → Ph—≡—Ph 84%

Novák, Z.; Nemes, P.; Kotschy, A. *Org. Lett.* ***2004***, *6*, 4917.

PhBr — Cl, Br, $Pd(dppb)(Oc)_2$, KOH / 18-crown-6, toluene, 100°C, 42h → Ph—≡—Ph 47%

Abele, E.; Abele, R.; Arsenyan, P.; Kukevics, E. *Tetrahedron Lett.* ***2003***, *44*, 3911.

I — ≡—Ph, aq dioxane, 2 K_2CO_3 / 5% $Ni(PPh_3)_2Cl_2$ → Ph quant

Beletskaya, I.P.; Latyshev, G.V.; Tsvetkov, A.V.; Lukashev, N.V. *Tetrahedron Lett.* ***2003***, *44*, 5011.

PhBr — PhC≡CH, NEt_3, DMF/H_2O, 80°C, 3h / cat $[Pd(NH_3)_4]$ NaY zeolite → Ph—≡—Ph quant

Djakovitch, L.; Rollet, P. *Tetrahedron Lett.* ***2004***, *45*, 1367.

O, Br — HC≡CCH_2OH, CuI, DMF, 100°C / cat $[Pd(C_3H_5)Cl]_2$Tedicyp, K_2CO_3 → O, OH 94%

Feuerstein, M.; Doucet, H.; Santelli, M. *Tetrahedron Lett.* ***2004***, *45*, 1603.

I — 1. PhC≡CH, Ni(activated powder) 2. CuI, PPh_3, i-PrOH, KOH → Ph 98%

Wang, L.; Li, P.; Zhang, Y. *Chem. Commun.* ***2004***, 514.

4-bromoacetophenone → 4-(phenylethynyl)acetophenone: PhC≡CH , K_2CO_3 , 140°C; cat $[Pd(C_3H_5)Cl]_2$/tetraporphine ligand — quant

Feuerstein, M.; Berthiol, F.; Doucet, H.; Santelli, M. *Org. Biomol. Chem.* ***2003***, *1*, 2235.

3-bromopyridine → 3-(phenylethynyl)pyridine: PhC≡CH , pyrrolidine , TBAB , 140°C; Pd-phosphinous acid catalyst , H_2O — 80%

Wolf, C.; Lerebours, R. *Org. Biomol. Chem.* ***2004***, *2*, 2161.

Ph-I → Ph-C≡C-Ph: PhC≡CH , Pd complex , NEt_3 , rt , 30 min; copper free — quant

Méry, D.; Heuzé, K.; Astruc, D. *Chem. Commun.* ***2003***, 1934.

2-iodotoluene → 2-(phenylethynyl)toluene: PhC≡CH , piperidine , 120°C , 1h; cat bis(imidazole)PdClMe , bmim PF_6 — 42%

Park, S.B.; Alper, H. *Chem. Commun.* ***2004***, 1306.

2-iodoanisole (OMe, I) → 2-(phenylethynyl)anisole (OMe, Ph): PhCCH , K_2CO_3 , DMF , 100°C; CuI , N,N-dimethylglycine , 36h — 82%

Ma, D.; Liu, F. *Chem. Commun.* ***2004***, 1934.

2-bromonaphthalene → 2-(phenylethynyl)naphthalene: Ph—≡ , TBAB , H_2O , 15 min; Na_2CO_3 , 175°C , microwaves — 83%

Appukkuthan, P.; Dehaen, W.; van der Eycken, E. *Eur. J. Org. Chem.* ***2003***, 4713.

I—C₆H₄—OMe → Ph—≡—C₆H₄—OMe: Ph—≡—H , 6 h; cat Pd/CuI , 0.5M NH_3 , rt , THF — 96%

Mori, A.; Ahmed, M.S.M.; Sekiguchi, A.; Masui, K.; Koike, T. *Chem. Lett.* ***2002***, *31*, 756.

SECTION 11: ALKYNES FROM HYDRIDES

For examples of the reaction RC≡CH → RC≡C-C≡CR[1], see Section 300 (Alkyne-Alkyne).

NO ADDITIONAL EXAMPLES

SECTION 12: ALKYNES FROM KETONES

NO ADDITIONAL EXAMPLES

SECTION 13: ALKYNES FROM NITRILES

PhCN —— 1. Ph—≡—ZnBr , cat $Ni(PMe_3)_2Cl_2$, THF, reflux , 41h; 2. aq Na citrate ——→ Ph—≡—Ph 79%

Penney, J.M.; Miller, J.A. *Tetrahedron Lett.* ***2004***, *45*, 4989.

SECTION 14: ALKYNES FROM ALKENES

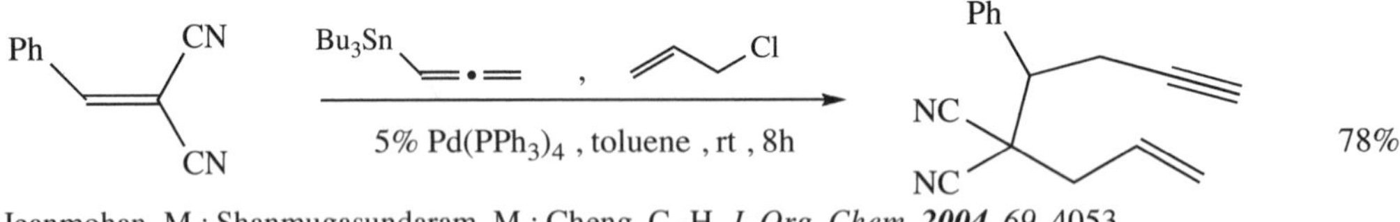

Jeanmohan, M.; Shanmugasundaram, M.; Cheng, C.-H. *J. Org. Chem.* ***2004***, *69*, 4053.

SECTION 15: ALKYNES FROM MISCELLANEOUS COMPOUNDS

NO ADDITIONAL EXAMPLES

REVIEW:

"One Century of Aryne Chemistry"
Wenk, H.H.; Winkler, M.; Sander, W. *Angew. Chem. Int. Ed.* ***2003***, *42*, 502.

SECTION 15A: PROTECTION OF ALKYNES

NO ADDITIONAL EXAMPLES

CHAPTER 2

PREPARATION OF ACID DERIVATIVES

SECTION 16: ACID DERIVATIVES FROM ALKYNES

$$C_6H_{13}-\!\!\equiv\!\!-C_6H_{13} \xrightarrow[9.9\ NaHCO_3\ ,\ MeCN/H_2O/EtOAc]{3.3\ Oxone\ ,\ 3\%\ RuO_2\ ,\ rt\ ,\ 1\ h} C_6H_{13}-CO_2H \quad 99\%$$

Yang, D.; Chen, F.; Dong, Z.-M. *J. Org. Chem.* ***2004***, *69*, 2221.

SECTION 17: ACID DERIVATIVES FROM ACID DERIVATIVES

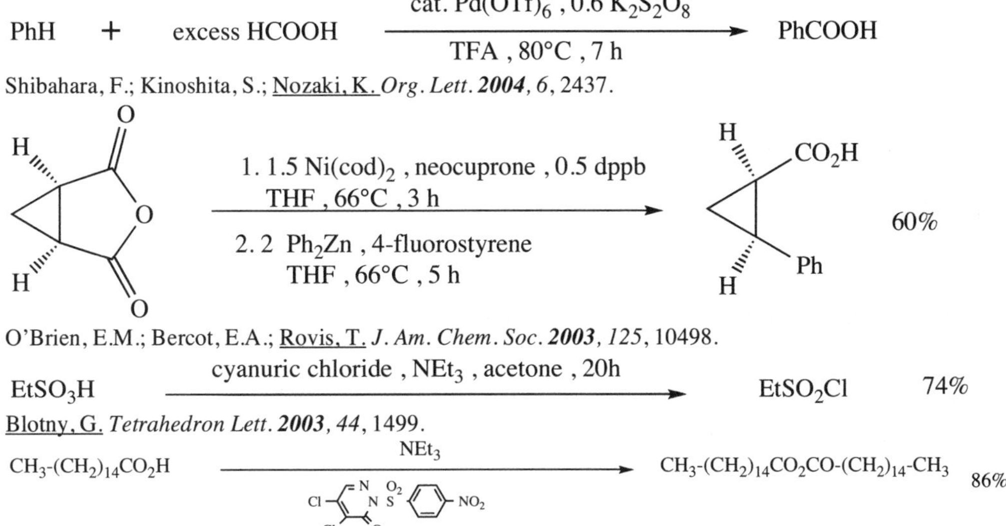

Shibahara, F.; Kinoshita, S.; Nozaki, K. *Org. Lett.* ***2004***, *6*, 2437.

O'Brien, E.M.; Bercot, E.A.; Rovis, T. *J. Am. Chem. Soc.* ***2003***, *125*, 10498.

Blotny, G. *Tetrahedron Lett.* ***2003***, *44*, 1499.

Kim, J.-J.; Park, Y.-D.; Lee, W.S.; Cho, S.-D. *Synthesis* ***2003***, 1517.

REVIEW:

"Asymmetric Alcoholysis of Cyclic Anhydrides"
Chen, Y.; McDaid, P.; Deng, L. *Chem. Rev.* ***2003***, *103*, 2965.

SECTION 18: ACID DERIVATIVES FROM ALCOHOLS AND THIOLS

1% TEMPO, NaBr, acetone, rt; trichloroisocyanuric acid, aq $NaHCO_3$, 3 h — quant

De Lucca, L.; Giacomelli, G.; Masala, S.; Purcheddu, A. *J. Org. Chem.* ***2003***, *68*, 4999.

1. $HC(CO_2Et)_3$, PMe_3, DEAD, toluene, –53°C → rt
2. NaOH, MeOH, reflux
3. AcOH, reflux

82x74% (90% ee)

Hillier, M.C.; Desrosiers, J.-N.; Marcoux, J.-F.; Grabowski, E.J.J. *Org. Lett.* ***2004***, *6*, 573.

$Ru\text{-}Co(OH)_2\text{-}C_4O_2$, O_2, rt, benzotrifluoride — 82%

Ji, H.; Mizugaki, T.; Ebitani, K.; Kaneda, K. *Tetrahedron Lett.* ***2002***, *43*, 7179.

$PhCH_2OH$ —(Co salen catalyst, 30% H_2O_2, 6h, MeCN, 80°C)→ $PhCO_2H$ 76%

Das, S.; Punniyamurthy, T. *Tetrahedron Lett.* ***2003***, *44*, 6033.

10% PS-TEMPO; PS-NMe_3 ClO_2; PS-NMe_3 H_2PO_4 — 95%

Yasuda, K.; Ley, S.V. *J. Chem. Soc. Perkin Trans. 1* ***2002***, 1024.

$C_9H_{19}CH_2OH$ —(1. 2.1 IBX, DMSO, rt, 30 min; 2. 2-hydroxypyridine, rt, 3d)→ $C_9H_{19}CO_2H$ 91%

IBX = 1-hydroxy-1,2-benzoidodxol-3(1H)-one-1-oxide

Mazitschek, R.; Mülbaier, M.; Giannis, A. *Angew. Chem. Int. Ed.* ***2002***, *41*, 4059.

SECTION 19: ACID DERIVATIVES FROM ALDEHYDES

$C_5H_{11}\text{-}CHO$ —(Oxone, DMF, rt, 3 h)→ $C_5H_{11}\text{-}CO_2H$ 97%

Travis, B.R.; Sivakumar, M.; Hollist, G.O.; Borhan, B. *Org. Lett.* ***2003***, *5*, 2573.

PhCHO + (AcO)₃I-benziodoxolone —(NaN_3, DCM, 0°C, 1.5h)→ $N_3C(O)Ph$ 89%

Bose, D.S.; Reddy, A.V.N. *Tetrahedron Lett.* ***2003***, *44*, 3543.

PhCHO —(cat $H_5[Y_5(\mu_4\text{-}O)(\mu_3\text{-}O)_4\text{-}(\mu\text{-}\eta^2\text{-}Ph_2(acac)_4)(\mu^2\text{-}Ph_2(acac)_6]$, DCM, 12h, air)→ $PhCO_2H$ >99%

Roesky, P.W.; Canseco-Melchor, G.; Zulys, A. *Chem. Commun.* ***2004***, 738.

OHC, S–S, CHO (bis(1-formylcyclohexyl) disulfide) → $NaClO_2$, NaH_2PO_4 / H_2O , DMSO → HO_2C, S–S, CO_2H 89%

Fang, X.; Bandarage, U.P.; Wang, T.; Schroeder, J.D.; Garvey, D.S. *Synlett* ***2003***, 489.

Ph–CH₂CH₂–CHO → (polymer)–$I(OAc)_2$, 0.5 TEMPO / acetone, H_2O , rt , 1 d → Ph–CH₂CH₂–CO_2H 91%

Tashino, Y.; Togo, H. *Synlett* ***2004***, 2010.

SECTION 20: ACID DERIVATIVES FROM ALKYLS, METHYLENES, AND ARYLS

NC–C₆H₄–Me → (N,N',N''-trihydroxyisocyanuric acid) / O_2 , 0.5% $Co(OAc)_2$, AcOH , 6 h → NC–C₆H₄–CO_2H >99%

Hirai, N.; Sawatari, N.; Nakamura, N.; Sakaguchi, S.; Ishii, Y. *J. Org. Chem.* ***2003***, *68*, 6585.

Ph–CH_3 → urea-H_2O_2 , microwaves , 150°C / 180 sec → Ph–CO_2H 75%

Paul, S.; Nanda, P.; Gupta, R. *Synlett* ***2004***, 531.

t-Bu–C₆H₄–Me → NBS , EtOAc , hv , aq NaOH , 12 h → *t*-Bu–C₆H₄–CO_2H 85%

Itoh, A.; Kodama, T.; Hashimoto, S.; Masaki, Y. *Synthesis* ***2003***, 2289.

$PhSiMe_3$ → CO_2 , $AlBr_3$, PhH / rt , 3 h → PhCOOH 20% 50% with *p*-$TolSiMe_3$

Hattori, T.; Suzuki, Y.; Miyano, S. *Chem. Lett.* ***2003***, *32*, 454.

SECTION 21: ACID DERIVATIVES FROM AMIDES

NO ADDITIONAL EXAMPLES

SECTION 22: ACID DERIVATIVES FROM AMINES

NO ADDITIONAL EXAMPLES

SECTION 23: ACID DERIVATIVES FROM ESTERS

TBDPSO(C=O)–C$_6$H$_4$–Cl → MeCN , THF/MeCN, 50°C , 1 h, 4 eq Me_2NC(=NH)NMe_2 → HO(C=O)–C$_6$H$_4$–Cl 91%

Oyama, K.-i.; Kondo, T. *Org. Lett.* **2003**, *5*, 209.

$[Re(CO)_3Cl]$, neat , 160°C , 3 h quant

Davies, T.J.; Jones, R.V.H.; Lindsell, W.E.; Miln, C.; Preston, P.N. *Tetrahedron Lett.* **2002**, *43*, 487.

MegBr , TMSCl , THF; $CuBr{\bullet}SMe_2$, –70°C 88%

Nelson, S.G.; Wan, Z.; Stan, M.A. *J. Org. Chem.* **2002**, *67*, 4680.

0.02 TMSOTf , DCM; rt , 1 h 96%

Nishizawa, M.; Yamamoto, H.; Seo, K.; Imagawa, H.; Sugihara, T. *Org. Lett.* **2002**, *4*, 1947.

$NaHSO_4{\bullet}SiO_2$, 5h PhCOOH 95%

Ramesh, C.; Mahender, G.; Ravindranath, N.; Das, B. *Tetrahedron Lett.* **2003**, *44*, 1465.

$CeCl_3{\bullet}7\ H_2O$, NaI , MeCN; reflux , 3.5 h 88%

Yadav, J.S.; Reddy, B.V.S.; Rao, C.V.; Chand, P.K.; Prasad, A.R. *Synlett* **2002**, 137.

$NiBr_2{\bullet}bpy$, e^- , CO_2 , aq TBAF; DMF 65%

Senboku, H.; Kanaya, H.; Tokuda, M. *Synlett* **2002**, 140.

Other reactions useful for the hydrolysis of esters may be found in Section 30A (Protection of Carboxylic Acid Derivatives).

SECTION 24: ACID DERIVATIVES FROM ETHERS, EPOXIDES, AND THIOETHERS

LiBr , DMF/MeCN/H_2O; 91°C , overnight no yield

Badham, N.F.; Medelson, W.L.; Allen, A.; Diederich, A.M.; Eggleston, D.S.; Filan, J.J.; Freyer, A.J.; Killmer Jr. L.B.; Kowalski, C.J.; Liu, L.; Novak, V.J.; Vogt, F.G.; Webb, K.S.; Yang, J. *J. Org. Chem.* **2002**, *67*, 5440.

SECTION 25: ACID DERIVATIVES FROM HALIDES AND SULFONATES

$PhCH_2Cl$ —— e^-, Ag cathode, MeCCN, aq Et_4NClO_4 ——→ $PhCO_2H$ 94%

Isse, A.A.; Gennaro, A. *Chem. Commun.* **2002**, 2798.

MeO–C6H4–I —— DMF, 80°C, 2 Ac_2O, 3 HCO_2Li, 1d; 3 LiCl, 2 $EtN(i\text{-}Pr)_2$, 2.5% $Pd_2(dba)_3$ ——→ MeO–C6H4–CO_2H 74%

Cacchi, S.; Fabrizi, G.; Goggiamani, A. *Org. Lett.* **2003**, *5*, 4269.

SECTION 26: ACID DERIVATIVES FROM HYDRIDES

Ph-H —— silica sulfuric acid, DCE, 80°C ——→ $PhSO_3H$ 80%

Hajipour, A.R.; Mirjalili, B.B.F.; Zarei, A.; Khazdooz, L.; Ruoho, A.E. *Tetrahedron Lett.* **2004**, *45*, 6607.

naphthalene —— 3 mPa CO_2, $AlCl_3$, PhH; 40°C, 1 d ——→ 1-naphthoic acid (CO_2H) 43%

Suzuki, Y.; Hattori, T.; Okuzawa, T.; Miyano, S. *Chem. Lett.* **2002**, 102.

3-aminophenoxide (H_2N, $O^-\ Na^+$) —— CO_2, K_2CO_3, 200°C ——→ H_2N, OH, CO_2^- 80%

Rahim, M.A.; Matsui, Y.; Matusyama, T.; Kosugi, Y. *Bull. Chem. Soc. Jpn.* **2003**, *76*, 2191.

SECTION 27: ACID DERIVATIVES FROM KETONES

4-methylphenyl ketone (O, N-OH) —— 1. NaOMe, MeOH; 2. $^1O_2$ ——→ 4-methylbenzoic acid (CO_2H) + methyl 4-methylbenzoate (CO_2Me) (1 : 2.8) 92%

Öcal, N.; Yano, L.M.; Erden, I. *Tetrahedron Lett.* **2003**, *44*, 6947.

Ph–C(O)–CH_3 —— cat $Mn(NO_3)_2$, cat $Co(NO_3)_2$, AcOH; 6h, O_2, 100°C ——→ $PhCO_2H$ 95%

Miniscui, F.; Recupero, F.; Fontana, F.; Biøsrvik, H.-R.; Liguori, L. *Synlett* **2002**, 610.

Cl–C6H4–C(O)CH3 → 10% 1,3-dinitrobenzene , Na_2CO_3 ; 1.5 H_2O_2 , *t*-BuOH/*t*-BuOK ; 80°C , 5h → Cl–C6H4–CO2H (OH) 73%

Bjørsvik, H.-R.; Merinero, J.A.V.; Liguori, L. *Tetrahedron Lett.* ***2004***, *45*, 8615.

3,4-(OMe)2C6H3–C(O)CH3 → 5% 1,3-dinitrobenzene , *t*-BuOH , O_2 ; *t*-BuOK , 80°C , 5h → 3,4-(OMe)2C6H3–CO_2H 78%

Bjørsvik, H.-R.; Liguori, L.; González, R.R.; Merinero, J.A.V. *Tetrahedron Lett.* ***2002***, *43*, 4985.

REVIEW:

"Homologation of Ketones into Carboxylic Acids"
Badham, N.F. *Tetrahedron* ***2004***, *60*, 11.

SECTION 28: ACID DERIVATIVES FROM NITRILES

NO ADDITIONAL EXAMPLES

SECTION 29: ACID DERIVATIVES FROM ALKENES

HO–CH2–$(CH_2)_7$–CH=CH2 → OsO_4 , Oxone , DMF ; rt , 3 h → HO–CH2–$(CH_2)_7$–CO_2H 88%

Travis, B.R.; Narayam, R.S.; Borhan, B. *J. Am. Chem. Soc.* ***2002***, *124*, 3824.

CH2=CH–Ph → CO , H_2O , MCM-41-Pd catalyst , LiCl ; TsOH → CH3CH(CO_2H)Ph 97%

Mukhopadhyay, K.; Sarkar, B.R.; Chaudhari, R.V. *J. Am. Chem. Soc.* ***2002***, *124*, 9692.

4-CH3C6H4–CH=CH2 → mesoporous silica FSM-16 , hν ; 0.3 I_2 , *i*-Pr_2O , rt , 2d → 4-CH3C6H4–CO_2H 81%

Itoh, A.; Kodama, T.; Masaki, Y.; Inagaki, S. *Synlett* ***2002***, 522.

SECTION 30: ACID DERIVATIVES FROM MISCELLANEOUS COMPOUNDS

1. $CHCl_2OMe$, LiO*t*-Bu
2. NaOH , H_2O , 1h

→ $C_8H_{17}CO_2H$ 83%

Soderquist, J.A.; Martinez, J.; Oyola, Y.; Kock, I. *Tetrahedron Lett.* ***2004***, *45*, 5541.

SECTION 30A: PROTECTION OF CARBOXYLIC ACID DERIVATIVES

NO ADDITIONAL EXAMPLES

Other reactions useful for the protection of carboxylic acids are included in Section 107 (Esters from Carboxylic Acid Derivatives) and Section 23 (Carboxylic Acid Derivatives from Esters).

CHAPTER 3

PREPARATION OF ALCOHOLS

SECTION 31: ALCOHOLS AND THIOLS FROM ALKYNES

Ph—≡ → 1. Bu_3SnH, toluene; 2. O_3; 3. BF_3•SMe_2 → Ph–CH(OH)–CH_2OH 61%

Gómez, A.M.; Company, M.D.; Valverde, S.; López, J.C. *Org. Lett.* ***2002***, *4*, 383.

SECTION 32: ALCOHOLS AND THIOLS FROM ACID DERIVATIVES

PhCOCl → 2 CH_2=$CHCH_2$Cl, Zn/TMSCl, THF → Ph(HO)C(CH_2CH=CH_2)$_2$ 92%

Ishino, Y.; Mihara, M.; Kageyama, M. *Tetrahedron Lett* ***2002***, *43*, 6601.

3,4,5-F$_3C_6H_2$–B(OH)$_2$ + $C_8H_{17}CO_2H$ → $NaBH_4$, THF, Na_2SO_4; rt, 10h → $C_8H_{17}CH_2OH$ 90%

Tale, R.H.; Patil, K.M.; Daurkar, S.E. *Tetrahedron Lett.* ***2003***, *44*, 3427.

3-hydroxy-4-nitrobenzoic acid (HO_2C, OH, NO_2) → BH_3•THF, BF_3•OEt_2 (1.25:1); THF, rt, 18 h → 3-hydroxy-4-nitrobenzyl alcohol (HOH_2C, OH, NO_2) 91%

Chen, M.H.; Kesten, E.I.S.; Magano, J.; Rodriguez, D.; Sexton, K.E.; Zhang, J.; Lee, H.T. *Org. Prep. Proceed. Int.* ***2002***, *34*, 665.

SECTION 33: ALCOHOLS AND THIOLS FROM ALCOHOL AND THIOLS

NO ADDITIONAL EXAMPLES

SECTION 34: ALCOHOLS AND THIOLS FROM ALDEHYDES

PhCHO → (H_2O_2 , H_3BO_3 , [BMI] PF_6 , cat H_2SO_4 ; BMI = 1-butyl-3-methylimidazolium) → PhOH 89%

Zambrano, J.L.; Dorta, R. *Synlett* ***2003***, 1545.

The following reaction types are included in this section:
A. Reductions of Aldehydes to Alcohols
B. Alkylation of Aldehydes, Forming Alcohols

SECTION 34A: REDUCTIONS OF ALDEHYDES TO ALCOHOLS

PhCHO → ($NaBH_4$, wet THF , 20 min, rt ; poly(*N*-(2-aminoethyl)acylamido)triethylammonium chloride) → $PhCH_2OH$ 95%

Tamami, B.; Maahdavi, H. *Tetrahedron* ***2003***, *59*, 821.

Ph–CH₂CH₂–CHO → (Ar_3SnH , BF_3•OEt_2 , DCM ; –40°C) → Ph–CH₂CH₂CH₂–OH 83%

Ar = 2,6-diphenylbenzyl

selective for aliphatic RCHO in the presence of conjugated RCHO

Sasaki, K.; Komatsu, N.; Shirakawa, S.; Maruoka, K. *Synlett* ***2002***, 575.

PhCHO + HCHO → (KF-Al_2O_3 , 5 min ; microwaves) → $PhCH_2OH$ + HCO_2H

Reddy, B.V.S.; Srinivas, R.; Yadav, J.S.; Ramalingam, T. *Synth. Commun.* ***2002***, *32*, 219.

PhCHO → (FeS•NH_4Cl , MeOH , H_2O ; reflux , 3 h) → $PhCH_2OH$ 78%

Desai, D.G.; Swami, S.S.; Nandudikar, R.S. *Synth. Commun.* ***2002***, *32*, 931.

(citral) CHO → ($NaBH_4$, I_2 , THF , 15°C) → CH_2OH 94%

Singh, J.; Kaur, I.; Kaur, J.; Bhalla, A.; Kad, G.L. *Synth. Commun.* ***2003***, *33*, 191.

O_2N–C₆H₄–CHO → (alumina-DMAP , MeOH ; microwaves , reflux) → O_2N–C₆H₄–CH_2OH 95%

Pradhan, P.K.; Jaisankar, P.; Pal, B.; Dey, S.; Giri, V.S. *Synth. Commun.* ***2004***, *34*, 2863.

SECTION 34B: ALKYLATION OF ALDEHYDES, FORMING ALCOHOLS

ASYMMETRIC ALKYLATIONS

PhCHO → (Et_2Zn, toluene, rt; 5% chiral sulfinylferrocene ligand) → Ph–CH(OH)–Et

Priego, J.; Mancheño, O.G.; Cabrera, S.; Carretero, J.C. *J. Org. Chem.* ***2002***, *67*, 1346.

Ph(CH$_2$)$_3$CHO → (2-MeOC$_6$H$_4$COCH$_2$OH; 10% chiral complex, THF; 9% KHMDS, 20% H_2O; –40°C, 35 h) → (76 (95% ee) : 24) 69% (74% ee)

Yoshikawa, N.; Suzuki, T.; Shibasaki, M. *J. Org. Chem.* ***2002***, *67*, 2556.

PhCHO → (Et_2Zn, 10% chiral amino acid ligand; hexane, 4°C, 1 d) → Ph–CH(OH)–Et 73% (70% ee)

Richmond, M.I.; Seto, C.T. *J. Org. Chem.* ***2003***, *68*, 7509.

PhCHO → (3 Et_2Zn, chiral N-acyl ethylenediamine ligand) → Ph–CH(OH)–Et 84% (68% ee)

Sprout, C.M.; Seto, C.T. *J. Org. Chem.* ***2003***, *68*, 7788.

PhCHO → (Cl_3SiCH$_2$CH=CH$_2$, Bu_4NI, DCM, –60°C; cat chiral 2,2'-bipyridyl N-oxide, *i*-Pr_2NEt) → Ph–CH(OH)–CH$_2$CH=CH$_2$ 85% (88% ee)

Malkov, A.V.; Orsini, M.; Pernazza, D.; Muir, K.W.; Langer, V.; Meghani, P.; Kocovsky, P. *Org. Lett.* ***2002***, *4*, 1047.

PhCHO → (Et_2Zn, hexanes; azole ligand) → Ph–CH(OH)–Et 91% (86% ee)

Wipf, P.; Wang, X. *Org. Lett.* ***2002***, *4*, 1197.

Ph–CH=CH–CH$_2$Cl → (PhCHO, $(SnMee_3)_2$, 14 h; 40°C, 5% PPh_3; 5% $[Pd(\eta^3C_3H_5)Cl]_2$) → Ph–CH=CH–CH(OH)–CH(Ph)–CH=CH$_2$ 67% (10:1 dr)

Wallner, O.A.; Szabó, K.J. *Org. Lett.* ***2002***, *4*, 1563.

PhCHO → (Et_2Zn, chiral amino alcohol, rt / hexane-toluene, 3 h) → OH, Ph, Et 58% (99% ee)

Nugent, W.A. *Org. Lett.* ***2002**, 4*, 2133.

$C_6H_{11}CHO$ → (Ph, NEt_3, $Zn(OTf)_2$ / Ph, NMe_2, HO, H_2O) → HO, C_6H_{11}, Ph 94% (97% ee)

Boyall, D.; Frantz, D.E.; Carreira, E.M. *Org. Lett.* ***2002**, 4*, 2605.

Ph, MeHN, N → (1. Bu_2Mg / 2. PhCHO) → OH, Ph, Bu 76% (89:11 *R:S*)

Yong, K.H.; Taylor, N.J.; Chong, J.M. *Org. Lett.* ***2002**, 4*, 3553.

PhCHO → (2 Et_2Zn, toluene, rt, 1 d / chiral aminomethyl hydroxy binaphthyls) → OH, Ph, Et 89% (94% ee)

Ko, D.-H.; Kim, K.H.; Ha, D.-C. *Org. Lett.* ***2002**, 4*, 3759.

→ (1. 2.5% $Pt(dba)_2$, 2.5% PCy_3 / 2. $C_7H_{15}CHO$ / 3. NaOH, H_2O_2) → OH, C_7H_{15}, Me, HO 68% (70% ee)

Morgen, J.B.; Morken, J.P. *Org. Lett.* ***2003**, 5*, 2573.

PhCHO → (Et_2Zn, toluene, rt / 0.5% N-Boc-diselenide) → OH, Ph 91% (95% ee)

Braga, A.L.; Lüdtke, D.S.; Paixao, M.W. *Org. Lett.* ***2003**, 5*, 2635.

PhCHO → (Et_2Zn, silica-immobilized chiral alcohol / BuLi) → OH, Ph, Et 99% (77% ee)

Fraile, J.M.; Mayoral, J.A.; Serrano, J.; Pericàs, M.A.; Solà, L.; Castellnou, D. *Org. Lett.* ***2003**, 5*, 4333.

PhCHO → (1.8 Et_2Zn, 1.2 $Ti(Oi\text{-}Pr)_4$, DCM, 2h / 3% N-triflated amino alcohol, –25°C, DCM) → OH, Ph, Et 99% (99% ee)

Kang, S.-W.; Ko, D.-H.; Kim, K.H.; Ha, D.-C. *Org. Lett.* ***2003**, 5*, 4517.

C_3H_7MgBr → (1. Cp_2ZrCl_2, THF, diisopropyl ketone / 2. PhCHO, toluene, –78°C) → OH, Ph, C_3H_7 98% (91:9 *anti:syn*)

Fujita, K.; Yorimitsu, H.; Shinokubo, H.; Oshima, K. *J. Org. Chem.* ***2004**, 69*, 3302.

PhCHO → (1. chiral Cr/Mn - allyl bromide complex, 2 TMSCl / 2. TBAF) → Ph, OH 93% (90% ee)

Inoue, M.; Suzuki, T.; Nakada, M. *J. Am. Chem. Soc.* ***2003**, 125*, 1140.

Ph CHO → (Si(chiral complex); DCM, 0°C, 20 h) → Ph, OH, 81% (98% ee)

Hackman, B.M.; Lombardi, P.J.; Leighton, J.L. *Org. Lett.* **2004**, *6*, 4375.

Ph, O, Si, Cl, Me, N, H → (PhCHO, toluene, –10°C; 2 h) → OH, Ph, 80% (81% ee)

Kinnaird, J.W.A.; Ng, P.Y.; Kubota, K.; Wang, X.; Leighton, J.L. *J. Am. Chem. Soc.* **2002**, *124*, 7920.

PhCHO → (Bpin, toluene, –78°C; cat $AlCl_3$-BINOL) → OH, Ph, 92% (39% ee)

Ishiyama, T.; Ahiko, T.-a.; Miyaura, N. *J. Am. Chem. Soc.* **2002**, *124*, 12414.

Ph, O, B, O → (PhCHO, 10% $Sc(OTf)_3$; DCM, –78°C) → OH, Ph, 85% (92% ee)

Lachance, H.; Lu, X.; Gravel, M.; Hall, D.G. *J. Am. Chem. Soc.* **2003**, *125*, 10160.

Cl—C_6H_4—CHO → (1% sulfonamide-amino alcohol; Et_2Zn, hexane, rt) → Cl—C_6H_4—C(HO)(Et)(H), 92% (89% ee)

Mao, J.; Wan, B.; Wang, R.; Wu, F.; Lu, S. *J. Org. Chem.* **2004**, *69*, 9123.

PhCHO → (20% Ph, CO_2H, OH, Et_2Zn, DCM, $Ti(Oi\text{-}Pr)_4$; rt) → Ph, Et, OH, 88% (77% ee)

Bauer, T.; Tarasiuk, J. *Tetrahedron Lett.* **2002**, *43*, 687.

PhCHO → ($SnBu_3$, 10% S-BINOL-$Zr(Ot\text{-}Bu)_4$, 1.5 h; MS 4Å, –20°C, toluene, pivalonitrile) → OH, Ph, 90% (90% ee)

Kurosu, M.; Lorca, M. *Tetrahedron Lett.* **2002**, *43*, 1765.

PhCHO → (Et_2Zn, toluene, 5% chiral dendritic polymer; rt) → OH, 99% (75% ee)

Hu, Q.-S.; Sun, C.; Monaghan, C.E. *Tetrahedron Lett.* **2002**, *43*, 927.

PhCHO → (Et_2Zn, chiral Os complex, toluene; $Ti(Oi\text{-}Pr)_4$, –30°C, 1d) → OH, Ph, Et, 87% (87% ee)

Muñiz, K. *Tetrahedron Lett.* **2003**, *44*, 3547.

PhCHO → (Et_2Zn, chiral azonia catalyst, toluene; 0°C, 2d) → OH, Ph, 61% (>99% ee)

Braga, A.L.; Milani, P.; Paixao, M.W.; Zeni, G.; Rodrigues, O.E.D.; Alves, E.F. *Chem. Commun.* **2004**, 2488.

Ph CHO → (HO, O, MeO-aryl ketone) 2% Et_2Zn , 1% BINOL derivative THF , −38°C , 20 h → Ph OH O OMe OH 94% (89:11 *syn:anti*) (91% ee; 89% ee)

Kumagai, N.; Matsunaga, S.; Kinoshita, T.; Harada, S.; Okada, S.; Sakamoto, S.; Yamaguchi, K.; Shibasaki, M. *J. Am. Chem. Soc.* ***2003***, *125*, 2169.

PhCHO → Et_2Zn , hexane , 0°C , chiral amino alcohol → OH Ph Et 85% (95% ee)

Le Goanvic, D.; Holler, H.; Pale, P. *Tetrahedron: Asymmetry* ***2002***, *13*, 119.

PhCHO → chiral napthylazepine ligand , Et_2Zn , THF / NEt_3 , reflux , 70 min → OH Ph Et 99% (94% ee)

Superchi, S.; Giorgio, E.; Scafato, P.; Rosini, C. *Tetrahedron: Asymmetry* ***2002***, *13*, 1385.

PhCHO → chiral bis(hydroxy)sulfonamide ligand / Et_2Zn , Ti(O*i*-Pr)$_4$, toluene , 0°C , 8h → OH Ph Et >95% (90% ee)

Yus, M.; Ramón, D.J.; Prieto, O. *Tetrahedron: Asymmetry* ***2002***, *13*, 1573.

PhCHO → chiral ferrocenyl amino alcohol , Et_2Zn / toluene , rt , 17h → OH Ph Et 97% (90% ee)

Bastin, S.; Ginj, M.; Brocard, J.; Pélinski, L.; Novogrocki, G. *Tetrahedron: Asymmetry* ***2003***, *14*, 1701.

PhCHO → Et_2 , chiral amino alcohol , toluene , rt → OH Ph Et 68% (81% ee)

Joshi, S.N.; Malhotra, S.V. *Tetrahedron: Asymmetry* ***2003***, *14*, 1763.

PhCHO → Et_2Zn , toluene , chiral diamine → OH Ph Et quant (70% ee)

Fonseca, M.H.; Eibler, E.; Zabel, M.; König, B. *Tetrahedron: Asymmetry* ***2003***, *14*, 1989.

PhCHO → 1. Et_2Zn , toluene , chiral amino alchol / 2. aq NH_4Cl → OH Ph Et 76% (65% ee)

Ionescu, R.D.; Blom, A.; Frejd, T. *Tetrahedron: Asymmetry* ***2003***, *14*, 2369.

PhCHO → Et_2Zn , toluene , chiral diamine , 0°C → OH Ph Et 94% (96% ee)

Lesma, G.; Danieli, B.; Passarella, D.; Sacchetti, A.; Silvani, A. *Tetrahedron: Asymmetry* ***2003***, *14*, 2453.

PhCHO → Et_2Zn , toluene-hexane , 0°C , 1d / ferrocenyl oxazoline ligand → OH Ph Et 86% (87% ee)

Fu, B.; Du, D.-M.; Wang, J. *Tetrahedron: Asymmetry* ***2004***, *15*, 119.

PhCHO → (Et_2Zn, toluene, 25°C; paracyclophane salen complex) → PhCH(OH)Et 88% (94% ee)

Danilova, T.I.; Rozenberg, V.I.; Starikova, Z.A.; Bräse, S. *Tetrahedron: Asymmetry* ***2004**, 15*, 223.

PhCHO → (Et_2Zn, chiral amino alcohol, rt; toluene, 1d) → PhCH(OH)Et 97% (94% ee)

Scarpi, D.; Lo Galbo, F.; Occhiato, E.G.; Guarna, A. *Tetrahedron: Asymmetry* ***2004**, 15*, 1319.

PhCHO → ($(CH_2{=}CH)_4Sn$, DMF, rt, 1d; L-asartic acid) → PhCH(OH)CH$_2$CH=CH$_2$ 88% (33% ee)

Yanigisawa, A.; Nakamura, Y.; Arai, T. *Tetrahedron: Asymmetry* ***2004**, 15*, 1909.

PhCHO → (Et_2Zn, toluene, 30h; chiral pentacycloundecane ligand) → PhCH(OH)Et 84% (90% ee)

Boyle, G.A.; Govender, T.; Kruger, H.G.; Mequire, G.E.M. *Tetrahedron: Asymmetry* ***2004**, 15*, 2661.

PhCHO → (Et_2Zn, chiral relay pyrazole ligand, 0°C; toluene) → PhCH(OH)Et 99% (89% ee)

Sibi, M.P.; Stanley, L.M. *Tetrahedron: Asymmetry* ***2004**, 15*, 3353.

PhCHO → (Et_2Zn, chiral amino thiol, 12h; toluene) → PhCH(OH)Et quant (99.6% ee)

Tseng, S.-L.; Yang, T.-K. *Tetrahedron: Asymmetry* ***2004**, 15*, 3375.

PhCHO → (Et_2Zn, Ti(O*i*-Pr)$_4$, chiral bipyridyl diol; toluene, rt, 30h) → PhCH(OH)Et 67% (91% ee)

Chen, Y.-J.; Lin, R.-X.; Chen, C. *Tetrahedron: Asymmetry* ***2004**, 15*, 3561.

PhCHO → (cat chiral ferrocenyl aziridine alcohol; Et_2Zn, toluene, rt, 2d) → PhCH(OH)Et 98% (90% ee)

Wang, M.-C.; Liu, L.-T.; Zhang, J.-S.; Shi, Y.-Y.; Wang, D.-K. *Tetrahedron: Asymmetry* ***2004**, 15*, 3853.

PhCHO → (cat bis(BINOL)-Ti(O*i*-Pr)$_4$, DCM, 0°C; Et_2Zn) → PhCH(OH)Et 98% (86% ee)

Hsarada, T.; Kanda, K.; Hiraoka, Y.; Marutani, Y.; Nakatsugawa, M. *Tetrahedron: Asymmetry* ***2004**, 15*, 3879.

TBSO-dienyl oxazolidinone → ($C_5H_{11}CHO$, $TiCl_4$; DCM, –78°C) → C_5H_{11}CH(OH)–CH$_2$–CH=C(CH$_3$)–C(O)–N(oxazolidinone) 97% (42:1 ds)

Shirokawa, S.-i.; Kamiyama, M.; Nakamura, T.; Okada, M.; Nakazaki, A.; Hosokawa, S.; Kobayashi, S. *J. Am. Chem. Soc.* ***2004**, 126*, 13604.

PhBr → (DCM, −10°C, 20h; chiral cyclohexanediamine-derived allylsilane reagent, N-4-ClC_6H_4, Si, Cl) → 1-phenylbut-3-en-1-ol (Ph, OH) 69% (98%ee)

Kubota, K.; Leighton, J.L. *Angew. Chem. Int. Ed.* ***2003***, *42*, 946.

PhCHO → (allyl-$SiCl_3$, quinoline type N-oxide catalyst; DCM, *i*-Pr_2NEt, 2h) → Ph, OH homoallylic alcohol 60% (87%ee)

Malkov, A.V.; Dufková, L.; Farrugga, L.; Kovovsky, P. *Angew. Chem. Int. Ed.* ***2003***, *42*, 3674.

PhCHO → (bis(sulfonamide) ligand, Et_2Zn, toluene; 1.5 $Ti(Oi\text{-}Pr)_2$, −35°C) → Ph, Et, OH 93% (96%ee)

Lake, F.; Moberg, C. *Eur. J. Org. Chem.* ***2002***, 3179.

PhCHO → (Et_2Zn, toluene, 20°C; paracycloquinoino ligand) → Ph, OH, Et 99% (64% ee)

Ruzziconi, R.; Pieratti, O.; Ricci, G.; Vinci, D. *Synlett* ***2002***, 747.

1-naphthaldehyde (CHO) → (Et_2Zn, 5% imidizolidine ligand; toluene, rt) → 1-(1-naphthyl)propan-1-ol (HO, Et) 94% ee

Casey, M.; Smyth, M.P. *Synlett* ***2003***, 102.

cyclohexenone (O) → (1. cat $Cu(MeCN)_4\,ClO_4$, thiophosphoramide ligand; 2. toluene, Et_2Zn, 20°C) → 3-ethylcyclohexanone (O, Et) 90% (94% ee)

Shi, M.; Wang, C.-J.; Zhang, W. *Chem. Eur. J.* ***2004***, *10*, 5507.

PhCHO + 2-methylallyl-$SnBu_3$ → (Rh catalyst, MS 4Å; DCM, 25°C) → Ph, OH, methylene product 95% (93% ee)

Motoyama, Y.; Nishiyama, H. *Synlett* ***2003***, 1883.

Ph–CH2CH2–CHO → (allyl-$SnBu_3$, DCM, −20°C; 10% Ti-S-BINOL, 5% $(ArBO)_3$) → Ph, OH homoallylic alcohol 93% (94% ee)

Xia, G.; Shibatomi, K.; Yamamoto, H. *Synlett* ***2004***, 2437.

PhCHO → ($PhB(OH)_2$, $ZnEt_2$, 10% ferrocene derivative; dimethyl polyethylene glycol) → Ph, Ph, OH 93% (96% ee)

Rudolph, J.; Schmidt, F.; Bolm, C. *Synthesis* ***2004***, 840.

PhCHO → Et_2Zn , chiral Zn complex / 20% $Sc(Oi\text{-}Pr)_3$, hexane → 1-phenylpropan-1-ol (OH, Ph) 58% (89% ee R)

Shiina, I.; Konishi, K.; Kuramoto, Y.-A. *Chem. Lett.* ***2002***, 164.

PhCHO → 1. ClMg-allyl , THF / 2. water or no water / 3. RCHO → homoallylic alcohol (OH, Ph)

PhCHO	THF (80%)	THF-water (85%)
BnCHO	–	76%
c-hex-CHO	73%	80%

Fukuma, T.; Lock, S.; Miyoshi, N.; Wada, M. *Chem. Lett.* ***2002***, *31*, 376.

PhCHO → allyl-$SiCl_3$, Bu_4NI , DCM , 1d / cat terpene-bipyridine N-oxide , –90°C → homoallylic alcohol (OH, Ph) 67% (92% ee)

Malkov, A.V.; Bell, M.; Orsini, M.; Pernazza, D.; Massa, A.; Herrmann, P.; Meghani, P.; Kocovsky, P. *J. Org. Chem.* ***2003***, *68*, 9659.

NON-ASYMMETRIC ALKYLATIONS

O_2N-C6H4-CHO → allyl-Cl , $(Me_3Sn)_2$, THF / cat $[(\eta^3\text{-}C_3H_5)PdCl]_2$ → O_2N-C6H4-CH(OH)-allyl 86%

Wallner, O.A.; Szabó, K.J. *J. Org. Chem.* ***2003***, *68*, 2934.

PhCHO → C_5H_{11}-C(OTMS)-$SiMe_3$, TMSOTf , –78°C / allyl-$SiMe_3$, toluene → ether ($SiMe_3$, O, C_5H_{11}, Ph) 62% (3:1 dr)

Huckins, J.R.; Rychnovsky, S.D. *J. Org. Chem.* ***2003***, *68*, 10135.

PhCHO → allyl-$SnBu_3$, $CeCl_3$•7 H_2O / 10% NaI , MeCN , 2 0.1N HCl , 27 h → homoallylic alcohol (OH, Ph) 95%

Bartoli, G.; Bosco, M.; Giuliani, A.; Marcantoni, E.; Palmieri, A.; Petrini, M.; Sambri, L. *J. Org. Chem.* ***2004***, *69*, 1290.

$C_7H_{15}CHO$ → allyl-$SnBu_3$, 2% activated $Ca(OTf)_3$ / $PhCO_2H$, MeCN → homoallylic alcohol (OH, C_7H_{15}) 75%

Aspinall, H.C.; Bissett, J.S.; Greeves, N.; Levin, D. *Tetrahedron Lett.* ***2002***, *43*, 319.

allyl-OH → PhCHO , InI , 5% $Ni(acac)_2$ / PPh_3 , DCM , rt , 5h → homoallylic alcohol (OH, Ph) 95%

Hirashita, T.; Kambe, S.; Tsuji, H.; Omori, H.; Araki, S. *J. Org. Chem.* ***2004***, *69*, 5054.

silane (O, Si, O, O) → cyclohexane-CHO / PhH , 40°C , 120 h → diol (OH, OH) 71% (86:11:3:1 dr)

Wang, X.; Meng, Q.; Nation, A.J.; Leighton, J.L. *J. Am. Chem. Soc.* ***2002***, *124*, 10672.

PhCHO → (Ph-CH=CH-CH$_2$Br , Sn , H_2O; rt , 1 d) → 80%

Tan, K.-T.; Cheng, S.-S.; Cheng, H.-S.; Loh, T.-P. *J. Am. Chem. Soc.* **2003**, *125*, 2958.

→ ($(HCHO)_n$, BF_3•OEt_2, MS 4Å , DCM; −5°C → −10°C) → 72%

Okachi, T.; Fujimoto, K.; Onaka, M. *Org. Lett.* **2002**, *4*, 1667.

PhCHO → (allyl bromide , H_2O , 1.8 h; nano-Sn) → 95%

Wang, Z.; Zha, Z.; Zhou, C. *Org. Lett.* **2002**, *4*, 1683.

→ (cyclohexadiene , $Ni(acac)_2$; $ZnCl_2$, THF , rt , 2 h) → 62% (72:28 *anti:syn*)

Loh, T.-P.; Song, H.-Y.; Zhou, Y. *Org. Lett*, **2002**, *4*, 2715.

BF_3^- K^+ → (*t*-BuCHO , DCM/H_2O; 10% Bu_4NI , rt , 15 min) → 96%

Thadani, A.N.; Batey, R.A. *Org. Lett.* **2002**, *4*, 3827.

CHO → (allyl chloride; $SnCl_2$, $TiCl_3$, H_2O) → 92%

Tan, X.-H.; Shen, B.; Deng, W.; Zhao, H.; Liu, L.; Guo, Q.-X. *Org. Lett.* **2003**, *5*, 1833.

PhCHO → (allyl-$SiMe_3$, cat. I_2 , MeCN , 0°C; 50 sec) → 90%

Yadav, J.S.; Chand, P.K.; Anjaneyulu, S. *Tetrahedron Lett.* **2002**, *43*, 3783.

C_3H_7-CH=CH-CHO → (allyl-$SnBu_3$, MeCN , 1h; Selectfluor) → 93%

Liu, J.; Wong, C.-H. *Tetrahedron Lett.* **2002**, *43*, 3915.

Br-allyl → (PhCHO , Ga , H_2O) → 76%

Wang, Z.; Yuan, S.; Li, C.-J. *Tetrahedron Lett.* **2002**, *43*, 5097.

PhCHO → ($MeNO_2$, $ZnEt_2$, 20% $NH_2CH_2CH_2OH$; toluene , 0°C , 2d) → 88%

Klein, G.; Pandiaraju, S.; Reiser, O. *Tetrahedron Lett.* **2002**, *43*, 7503.

Et_3SiH , 10 atm CO , ether
1% $Rh(acac)(CO)_2$, 70°C

CHO
Ts N
Ts N
$SiEt_3$
OH
74%

Kang, S.-K.; Hong, Y.-T.; Lee, J.-H.; Kim, W.-Y.; Lee, I.; Yu, C.-M. *Org. Lett.* ***2003***, *5*, 2813.

1. 1.2 $BnOSiMe_3$, 5% $FeCl_3$
DCM , rt , 2 h
2. $MeNO_2$, $SiMe_3$
rt , 1 h

Ph CHO
OBn
Ph
80%

Watahiki, T.; Akabane, Y.; Mori, S.; Oriyama, T. *Org. Lett.* ***2003***, *5*, 3045.

I , Zn/Cu , cat InCl
0.07M $Na_2Cr_2O_7$

NC CHO
OH
NC
71%

Keh, C.C.K.; Wei, C.; Li, C.-J. *J. Am. Chem. Soc.* ***2003***, *125*, 4062.

PhCHO , $In(OTf)_3$, DCM
rt

OH OH
OH
Ph
63%

Cheng, H.-S.; Loh, T.-P. *J. Am. Chem. Soc.* ***2003***, *125*, 4990.

I
Bu , Cp_2ZrCl_2 , *i*-Pr_2NEt
10% $CrCl_2/NiCl_2(dppp)$, Mn , LiCl
chiral sulfonamide, MeCN

Ph CHO
Ph Bu
OH
92%

Namba, K.; Kishi, Y. *Org. Lett.* ***2004***, *6*, 5031.

MgBr ,THF , –30°C
cat $Yb(OTf)_3$

O
Ph
Ph
OH
97%

Likhar, P.R.; Kumar, M.P.; Bandyopadhyay, A.K. *Tetrahedron Lett.* ***2002***, *43*, 3333.

1. Br , Sn , ultrasound
2. H_2O

PhCHO
OH
Ph
98%

Andrews, P.C.; Peatt, A.C.; Raston, C.L. *Tetrahedron Lett.* ***2002***, *43*, 7541.

Br , Bi , NH_4Cl , H_2O , 8h

PhCHO
OH
Ph
98%

Miyamoto, H.; Daikawa, N.; Tanaka, K. *Tetrahedron Lett.* ***2003***, *44*, 6963.

3 *t*-Bu-P4 base , 1.6 PhCHO
ZnI_2 , THF , –78°C → rt

CN
Br
HO CN
Ph
Br
86%

Imahori, T.; Kondo, Y. *J. Am. Chem. Soc.* ***2003***, *125*, 8082.

PhCHO —[allyl-$SiCl_3$, i-Pr_2NEt, DCM, 7h; 3 Tol-S(=O)Me]→ Ph–CH(OH)–$CH_2CH=CH_2$ 61%

Massa, A.; Malkov, A.V.; Kocovsky, P.; Scettri, A. *Tetrahedron Lett.* ***2003**, 44*, 7179.

PhCHO —[$(CH_2=CJCH_2)_4Sn$, wet SiO_2; rt, 1h]→ Ph–CH(OH)–$CH_2CH=CH_2$ 95%

Jin, Y.Z.; Yasuda, N.; Faruno, H.; Inanaga, J. *Tetrahedron Lett.* ***2003**, 44*, 8765.

PhCHO —[allyl-Br, Fe, NaF/H_2O, 96h]→ Ph–CH(OH)–$CH_2CH=CH_2$ 88%

Chan, T.C.; Lau, C.P.; Chan, T.H. *Tetrahedron Lett.* ***2004**, 45*, 4189.

EtCHO —[allyl-Br, $SnCl_2/PdCl_2$, H_2O, 16h]→ Et–CH(OH)–$CH_2CH=CH_2$ 92%

Tan, X.-H.; Hou, Y.-Q.; Shen, B.; Liu, L.; Guo, Q.-X. *Tetrahedron Lett.* ***2004**, 45*, 5525.

PhCHO —[allyl-Br, H_2O, Zn, $CdSO_4$, 96°C, 1.5h]→ Ph–CH(OH)–$CH_2CH=CH_2$ 85%

Zhou, C.; Zhou, Y.; Jiang, J.; Xie, Z.; Wang, Z.; Zhang, J.; Wu, J.; Yin, H. *Tetrahedron Lett.* ***2004**, 45*, 5537.

PhCHO —[allyl-Br, 2 Cp_2TiCl, THF, rt, 1.5h]→ Ph–CH(OH)–$CH_2CH=CH_2$ 94%

Jana, S.; Guin, C.; Roy, S.C. *Tetrahedron Lett.* ***2004**, 45*, 6575.

PhCHO —[1. $(c\text{-}C_6H_{11})_2B\text{-}Cl$, O_2; 2. H_2O]→ Ph–CH(OH)–cyclohexyl 90%

Kabalka, G.W.; Wu, Z.; Ju, Y. *Tetrahedron* ***2002**, 58*, 3243.

pyridine-3-CHO —[allyl-Br, Zn, THF; rt]→ 3-pyridyl–CH(OH)–$CH_2CH=CH_2$ 97%

Felpin, F.-X.; Bertrand, M.-J.; Lebreton, J. *Tetrahedron* ***2002**, 58*, 7381.

Cl–C_6H_4–CHO —[allyl-OH, cat $[RhCl(cod)]_2$; THF, H_2O, 50°C, 11h]→ Cl–C_6H_4–CH(OH)–$CH_2CH=CH_2$ 99%

Masuyama, Y.; Kaneko, Y.; Kurusu, Y. *Tetrahedron Lett.* ***2004**, 45*, 8969.

$(CH_3)_2CHCH_2$CHO —[Br-allyl, $SnCl_2/PdCl_2$; H_2O, rt, 1d]→ $(CH_3)_2CHCH_2$–CH(OH)–$CH_2CH=CH_2$ 99%

Tan, X.-H.; Hou, Y.-Q.; Huang, C.; Liu, L.; Guo, Q.-X. *Tetrahedron* ***2004**, 60*, 6129.

PhCHO → (allyl bromide, Bi, ball mill, rt) → 1-phenylbut-3-en-1-ol (OH, Ph) 81%

Wada, S.; Hayashi, N.; Suzuki, H. *Org. Biomol. Chem.* ***2003***, *1*, 2160.

PhCHO → (allyl bromide, Bi, 5 KF, H_2O; 12h, rt) → 1-phenylbut-3-en-1-ol (OH, Ph) 99%

Smith, K.; Lock, S.; El-Hiti, G.A.; Wada, M.; Miyoshi, N. *Org. Biomol. Chem.* ***2004***, *2*, 935.

PhCHO → (allyl-Si(*i*-Pr)$_3$, 5% CdF_2, aq THF, 30°C; 5% 2,2',4',2"-terpyridine, 9h) → 1-phenylbut-3-en-1-ol (OH, Ph) 93%

Aoyama, N.; Hamada, T.; Manabe, K.; Kobayashi, S. *Chem. Commun.* ***2003***, 676.

$C_9H_{19}CHO$ → (allyl chloride, Cp_2TiCl_2, THF, rt) → C_9H_{19}CH(OH)CH$_2$CH=CH$_2$ 90%

Rosales, A.; Oller-López, J.L.; Justicia, J.; Gansauer, A.; Oltra, J.E.; Cuerva, J.M. *Chem. Commun.* ***2004***, 2628.

3-chlorobenzaldehyde (CHO, Cl) → (3 Et_2Zn, $TiCl_4$, DCM, 2.5 h; glucosamine ligand) → 1-(3-chlorophenyl)propan-1-ol (OH, Et, Cl) 85%

Bauer, T.; Tarasiuk, J.; Pasniczek, K. *Tetrahedron: Asymmetry* ***2002***, *13*, 77.

PhCHO → (allyl bromide, 0.1$CrCl_2$, 0.2 NEt_3, 3 Mn; 1.5 TMSCl, salen catalyst, THF, 5°C) → 1-phenylbut-3-en-1-ol (OH, Ph) 72%

Berkessel, A.; Menche, D.; Sklorz, C.A.; Schröder, M.; Paterson, I. *Angew. Chem. Int. Ed.* ***2003***, *42*, 1032.

cyclohexanecarboxaldehyde (CHO) → (Ph-CH=CH-CH$_2$OH, 3.6 Et_2Zn, 0.2 PBu_3; 10% $Pd(OAc)_2$, toluene, 30h) → 1-cyclohexyl-2-phenylbut-3-en-1-ol (OH, Ph) 79%

Kimura, M.; Shimizu, M.; Shibata, K.; Tazoe, M.; Tamura, Y. *Angew. Chem. Int. Ed.* ***2003***, *42*, 3392.

acetophenone (O, Ph) → ($Sn(CH_2CH{=}CH_2)_4$, cat biaryl ligand; cat $Sn(CH_2CH{=}CH_2)_3$, H_2O, toluene) → 2-phenylpent-4-en-2-ol (HO, Ph) >98%

Cunningham, A.; Woodward, S. *Synlett* ***2002***, 43.

cyclohexanone (O) → (1. allyl-TMS, TFS-H, DCM; 2. TBAF, THF) → 1-allylcyclohexan-1-ol (OH) 80%

Cossy, J.; Lutz, F.; Alauze, C.; Meyer, C. *Synlett* ***2002***, 45.

PhCHO → (MeO–C₆H₄–SiEt(OH)₂; cat $[Ru(OH)(cod)_2]$, 70°C , THF) → 59%

Fujii, T.; Koike, T.; Mori, A.; Osakada, K. *Synlett* ***2002***, 298.

PhCH₂CH₂CHO → (allyl-$SnBu_3$, aq EtOH; 20% $Cd(ClO_4)_2$/diamine ligand) → 91%

Kobayashi, S.; Aoyama, N.; Manabe, K. *Synlett* ***2002***, 483.

I(CH₂)₅CO₂Me → (1. Zn , TMSCl; 2. PhCHO , NMP , 2 TMSCl) → 70%

Ito, T.; Ishino, Y.; Mizuno, T.; Ishikawa, A.; Kobayashi, J.-i. *Synlett* ***2002***, 2116.

PhCHO → ($Sm(I_2)$, *i*-PrOH) → $PhCH_2OH$ 96%

Fukuzawa, S.-i.; Nakano, N.; Saitoh, T. *Eur. J. Org. Chem.* ***2004***, 2863.

PhCHO → (allyl-Br , Zn , NH_4OAc; THF , 0°C) → 85%

Chen, C.; Dai, W.-C.; Chang, H.-G. *Org. Prep. Proceed. Int.* ***2002**, 34,* 507.

PhCHO → (allyl-Br , 10% $CrCl_2$(thf) , 2 Mn; collidine•HCl , THF , rt , 7 h) → 60%

Shaughnessy, K.H.; Huang, R. *Synth. Commun.* ***2002**, 32,* 1923.

PhCHO → (allyl-$SnBu_3$, $InCl_3$, 40°C , 16 h; bmim Cl) → 70%

Lu, J.; Ji, S.-J.; Qian, R.; Chen, J.-P.; Liu, Y.; Loh, T.-P. *Synlett* ***2004***, 534.

PhCHO → (allyl-Br , Bi* , H_2O , 1 h; * = nanometer size) → quant

Xu, X.; Zha, Z.; Miao, Q.; Wang, Z. *Synlett* ***2004***, 1171.

PhCHO → ($PhSi(OEt)_3$, cat $[Rh(OH)(cod)]_2$/cod; aq NaOH , dioxane) → 79%

Murata, M.; Shimazaki, R.; Ishikura, M.; Watanabe, S.; Masuda, Y. *Synthesis* ***2002***, 717.

PhCH(CH₃)CHO → (allyl-$SnBu_3$, $NbCl_5$, ether , –78°C) → 63% (3:1 *syn:anti*)

Andrade, C.K.Z.; Azevedo, N.R.; Oliveira, G.R. *Synthesis* ***2002***, 928.

PhCHO → (allyl-OH , In , $InCl_3$, $Pd(PPh_3)_4$; THF , H_2O , rt) → 94%

Jang, T.-S.; Keum, G.; Kang, S.B.; Chung, B.Y.; Kim, Y. *Synthesis* ***2003***, 775.

MeO–C6H4–CHO → (allyl OAc, cat $CoBr_2$; Zn, MeCN, 5 h) → MeO–C6H4–CH(OH)CH2CH=CH2 83%

Gomes, P.; Gosmini, C.; Périchon, J. *Synthesis* ***2003***, 1909.

PhCHO → (allyl $GaCl_2$, THF, hexane) → PhCH(OH)CH2CH=CH2 94%

Tsuji, T.; Usagi, S.-i.; Yorimitsu, H.; Shinokubo, H.; Matsubara, S.; Oshima, K. *Chem. Lett.* ***2002***, 2.

2-pyridyl-Si(Me)(Me)-allyl → (PhCHO, 10% AgOAc, 15 h; toluene, 100°C) → PhCH(OH)CH2CH=CH2 85%

Itami, K.; Kamei, T.; Mineno, M.; Yoshida, J. *Chem. Lett.* ***2002****, 31,* 1084

PhCHO → (1. Sr, THF, –15°C; 2. MeI) → PhCH(OH)Me 88%

Miyoshi, N,; Kamiura, K.; Oka, H.; Kita,A.; Kuwata,R.; Ikehara, D.; Wada, M. *Bull. Chem. Soc. Jpn.* ***2004****, 77,* 341.

REVIEWS:

"Asymmetric Alkynylzinc Additions to Aldehydes and Ketones"
Pu, L. *Tetrahedron* ***2003****, 59,* 9873.

"Catalytic Enantioselective Addition of Allylic Organometallic Reagents to Aldehydes and Ketones"
Denmark, S.E.; Fu, J. *Chem. Rev.* ***2003****, 103,* 2763.

"Asymmetric Transition Metal Catalyzed Allylic Alkylations: Applications in Total Synthesis"
Trost, B.M.; Crawley, M.L. *Chem. Rev.* ***2003****, 103,* 2921.

"Titanium-Catalyzed Enantioselective Additions of Allenes with a Nucleophilic Functionality Connected to the Carbon Atom"
Walsh, P.J. *Acc. Chem. Res.* ***2003****, 36,* 739.

SECTION 35: ALCOHOLS AND THIOLS FROM ALKYLS, METHYLENES, AND ARYLS

2-(bromomethyl)-5-vinyltetrahydrofuran → (1. [Cp_2ZrCl_2/BuLi/–78°C/DME]; DME, –78°C → 0°C, 2 h; 2. aq HCl) → (2-hydroxy-5-vinyltetrahydrofuran) 67% (20:1 *cis:trans*)

Williams, D.R.; Donnell, A.F.; Kammler, D.C. *Heterocycles* ***2004****, 62,* 297.

For reactions of the type RH → ROH (R = alkyl or aryl), see Section 41 (Alcohols and Thiols from Hydrides).

SECTION 36: ALCOHOLS AND THIOLS FROM AMIDES

$PhC(O)NH_2$ → 1. Boc_2O, DMAP; 2. $NaBH_4$ → $PhCH_2OH$ 77%

Ragnarsson, U.; Grehn, L.; Monteiro, L.S.; Maia, H.L.S. *Synlett* ***2003***, 2386.

SECTION 37: ALCOHOLS AND THIOLS FROM AMINES

NO ADDITIONAL EXAMPLES

SECTION 38: ALCOHOLS AND THIOLS FROM ESTERS

$Ph(CH_2)_2CO_2Me$ → 1. Ru catalyst, $HSiMe_2Et$, dioxane, 20°C; 2. H_3O^+ → $Ph(CH_2)_3OH$ 97%

Matsubara, K.; Iura, T.; Maki, T.; Nagashima, H. *J. Org. Chem.* ***2002***, *67*, 4985.

γ-Butyrolactone → 1. 2 allyl bromide, THF, 20°C, mischmetal/SmI_2 (cat); 2. H_3O^+ → $HO(CH_2)_3C(OH)(CH_2CH=CH_2)_2$ 72%

Lannou, M.-I.; Hélion, F.; Namy, J.-L. *Tetrahedron Lett.* ***2002***, *43*, 8007.

PhOAc → cat Amberlyst-15, MeOH, 2.5h, rt → PhOH 95%

Das, B.; Banerjee, J.; Ramy, R.; Pal, R.; Ravindranath, N.; Ramesh, C. *Tetrahedron Lett.* ***2003***, *44*, 5465.

$PhCO_2Me$ → 6 $NaBH_4$, MeOH, THF, 70°C, 4h → $PhCH_2OH$ 90%

Beechat, N.; da Costa, J.C.S.; Mendonca, J. de S.; Sanos, P.; de Oliveira, M.; DeSouza, M.V.N. *Tetrahedron Lett.* ***2004***, *45*, 6021.

$TfO-C_6H_4-CO_2Me$ → 1. 10% aq Et_4NOH, dioxane, rt, 1h; 2. 1M HCl → $HO-C_6H_4-CO_2Me$ 98%

Ohgiya, T.; Nishiyama, S. *Tetrahedron Lett.* ***2004***, *45*, 6317.

PhOAc → NH_4OAc, aq MeOH → PhOH 98%

Ramesh, C.; Mahender, G.; Ravindranath, N.; Das, B. *Tetrahedron* ***2003***, *59*, 1049.

N-Boc-tetrahydropyridin-2-yl OSO_2CF_3 → 6 PhCHO, 6 $CrCl_2$, 2% $NiCl_2$, rt, 15 h, DMF-THF, sonication → N-Boc-tetrahydropyridin-2-yl-CH(OH)Ph 60%

Easton, L.P.; Dake, G.R. *Can. J. Chem.* ***2004***, *82*, 139.

SECTION 39: ALCOHOLS AND THIOLS FROM ETHERS, EPOXIDES, AND THIOETHERS

i-PrMgBr , $CuBr \cdot SMe_2$

THF , –20°C , 1 h addition

Cl Cl OH 76% (22.1:1 *Z:E*)

Taber, D.F.; Mitten, J.V. *J. Org. Chem.* ***2002***, *67*, 3847.

InBr$_3$, DCM , 0°C , 4 h

Ph OH 70% (99% ee)

Bandini, M.; Cozzi, P.G.; Melchiorre, P. *J. Org. Chem.* ***2002***, *67*, 5386.

1. 2.3 B(Ipc)$_2$, 7.5% Sc(OTf)$_3$

THF , –78°C , 6 h

2. NaOH , H_2O_2 , 14 h , rt

Ph OH 88% (96% ee)

Lautens, M.; Maddess, M.; Sauer, E.L.O.; Ouellet, S.G. *Org. Lett.* ***2002***, *4*, 83.

PhO

$CeCl_3 \cdot 7\ H_2O$, NaN_3

MeCN , H_2O , reflux , 3 h

PhO OH N_3 99%

Sabitha, G.; Babu, R.S.; Rajkumar, M.; Yadav, J.S. *Org. Lett.* ***2002***, *4*, 343.

B–Ph , 0.5 [5 M aq Cs_2CO_3]

2.5 % [Rh(cod)Cl]$_2$, THF , rt

5% chiral PPt-P(*t*-Bu)$_2$

Ph OH 78% (92% ee)

Lautens, M.; Dockendorgg, C.; Fagnou, K.; Malicki, A. *Org. Lett.* ***2002***, *4*, 1311.

PhMe$_2$Si Ph

$BF_3 \cdot OEt_2$, DCM

0°C , 30 min

OH Ph + OH Ph (80 : 20) 75%

Barbero, A.; Castreño, P.; Pulido, F.J. *Org. Lett.* ***2003***, *5*, 4045.

SmI$_2$, H_2O , *i*-PrNH$_2$

THF , 1 min

OH OH >99%

Dahlén, A.; Sundgren, A.; Lahmann, M.; Oscarson, S.; Hilmersson, G. *Org. Lett.* ***2003***, *5*, 4085.

Ph

5% [Pd° En Cat] , 4 eq HCO_2H

4 NEt$_3$, EtOAc , 23°C

Ph OH 84%

Ley, S.V.; Mitchell, C.; Pears, D.; Ramarao, C.; Yu, J.-Q.; Zhou, W. *Org. Lett.* ***2003***, *5*, 4665.

OC_3H_7 [bmim] BF_4 , 115°C , 13h 47% HBr OH 95%

Boovanahalli, S.K.; Kim, D.W.; Chi, D.Y. *J. Org. Chem.* **2004**, *69*, 3340.

10 $PhNH_2$, 30°C , 2h 0.1% Pd catalyst OH + NHPh 94%

Murakami, H.; Minami, T.; Ozawa, F. *J. Org. Chem.* **2004**, *69*, 4482.

O BBr_3 , DCM , rt , 5 min OH 72%

with Me_2CHOMe , BBr_3 , DCM , rt , 35 min = 99%

Punna, S.; Neunier, S.; Finn, M.G. *Org. Lett.* **2044**, *6*, 2777.

CF_2 O Ph 2 eq BuLi , THF , 0°C OH F_2C Bu Ph 51% (73:27 *E:Z*)

Ueki, H.; Chiba, T.; Yamazaki, T.; Kitazume, T. *J. Org. Chem.* **2004**, *69*, 7616.

O Bu_2Zn , cat Pd(Tol-BINAL)Cl_2 DCM , rt OH Bu 84% (95% ee)

Lautens, M.; Hiebert, S. *J. Am. Chem. Soc.* **2004**, *126*, 1437.

$CO_2CH_2CH{=}CMe_2$ O $ZrCl_4$, $NaBH_4$ DCM , 1.5h $CO_2CH_2CH{=}CMe_2$ OH 96%

Babu, K.S.; Raju, B.C.; Srinivas, P.V.; Rao, J.M. *Tetrahedron Lett.* **2003**, *44*, 2525.

Me_2HC—C6H4—OMe $SiCL_4$, LiI , BF_3 , 70°C toluene/MeCN , 10h Me_2HC—C6H4—OH 89%

Zewge, D.; King, A.; Weissman, S.; Tschaen, D. *Tetrahedron Lett.* **2004**, *45*, 3729.

O Ph In , THF , rt , 3h OH Ph 75%

Hirashita, T.; Mitsui, K.; Hayashi, Y.; Araki, S. *Tetrahedron Lett.* **2004**, *45*, 9189.

O_2N—C6H4—O 10% KOH , 10% Pd/C , MeOH 9 h O_2N—C6H4—OH 94%

Ishizaki, M.; Yamada, M.; Watanabe, S.-i.; Hoshino, O.; Nishitani, K.; Hayashida, M.; Tanaka, A.; Hara, H. *Tetrahedron* **2004**, *60*, 7973.

cat nano Pd-6 , THF , InCl , rt , 1d

Ph OH 74%

Jiang, N.; Hu, Q.; Reid, C.S.; Lu, Y.; Li, C.-J. *Chem. Commun.* **2003**, 2318.

BuLi , sparteine , ether

–78°C → rt

Bu HO OH 46% (34%ee)

Hodgson, D.M.; Maxwell, C.R.; Miles, T.J.; Paruch, E.; Stent, M.A.H.; Malthews, I.R.; Wilson, F.X.; Witherington, J. *Angew. Chem. Int. Ed.* **2002**, *41*, 4313.

$Bu_{34}N\text{-}OSO_2\text{-}O\text{-}O\text{-}SO_2\text{-}O\text{-}NBu_4$

I_2 , aq MeCN

OH 93%

Yang, S.G.; Park, M.Y.; Kim, Y.H. *Synlett* **2002**, 492.

DDQ , DCM/H_2O

rt , 1.5 h

OH 72%

Vatèle, J.-M. *Synlett* **2002**, 507.

$PhCH_2OTr$ — $NaHSO_4$, SiO_2 , DCE-MeOH , rt → $PhCH_2OH$ 97%

Das, B.; Mashendeer, G.; Kumar, V.S.; Chowdhury, N. *Tetrahedron Lett.* **2004**, *45*, 6709.

NaN_3 , Oxone , aq MeCN , 2 h

OH N_3 90%

Sabitha, G.; Babu, R.S.; Reddy, M.S.K.; Yadav, J.S. *Synthesis* **2002**, 2254.

Ph — NaN_3 , wet *t*-BuOH , reflux / 0.2 M $LiBF_4$, 10 min

N_3 Ph OH + OH Ph N_3

(90 : 10) 96%

Kazemi, F.; Kiasat, A.R.; Ebrahimi, S. *Synth. Commun.* **2003**, *33*, 999.
Kazemi, F.; Kiasat, A.R.; Ebrahimi, S. *Synth. Commun.* **2004**, *34*, 999.

p-TsOH , DCM

rt , 1.5 h

OH 96%

Babu, K.S.; Raju, B.C.; Rao, S.A.S.; Kumar, S.P.; Rao, J.M. *Chem. Lett.* **2003**, *32*, 704.

$C_6H_{13}CH(OCPh_3)Me$ — 1. 10% $LiC_{10}H_8$, THF; 2. H_2O → $C_6H_{13}CH(OH)Me$ 62%

Yus, M.; Behloul, C.; Guijarro, D. *Synthesis* **2003**, 2179.

Additional examples of ether cleavages may be found in Section 45A (Protection of Alcohols and Thiols).

SECTION 40: ALCOHOLS AND THIOLS FROM HALIDES AND SULFONATES

2-(3-bromopropyl)naphthalene — H_2O , [bmim] BF_4 , 20 h; dioxane , 110°C → 2-(3-hydroxypropyl)naphthalene 95%

Kim, D.W.; Hong, D.J.; Seo, J.W.; Kim, H.S.; Kim, H.K.; Song, C.E.; Chi, D.Y. *J. Org. Chem.* **2004**, *69*, 3186.

$O_2N-C_6H_4-I$ — 2 $HC{\equiv}CCH_2OH$, DMSO; 2 KO*t*-Bu , microwaves → $O_2N-C_6H_4-OH$ 70%

Levin, J.I.; Du, M.T. *Synth. Commun.* **2002**, *32*, 1401.

PhBr — NaOH , microwaves , Montmorillonite K-10; $AgNO_3$, 5 sec → PhOH 78%

Hashemi, M.M.; Akhbari, M. *Synth. Commun.* **2004**, *34*, 2783.

$H_3C-CH(Cl)-C_{10}H_{21}$ — 2M NaOH , H_2O , 5 mPa; 250°C , 2 h → $H_3C-CH(OH)-C_{10}H_{21}$ 66% (86% ee) + $H_3C-CH{=}CH-C_{10}H_{21}$ 34%

Yamasaki, Y.; Hirayama, T.; Oshima, K.; Matsubara, S. *Chem. Lett.* **2004**, *33*, 864.

SECTION 41: ALCOHOLS AND THIOLS FROM HYDRIDES

N-benzylpiperidine (N—Bn) — *Sphingomonas* sp HXN-200; 5 h → 4-hydroxy-N-benzylpiperidine (HO—, N—Bn) 98%

Chang, D.; Feiten, H.-J.; Engesser, K.-H.; Van Beilen, J.; Witholt, B.; Li, Z. *Org. Lett.* **2002**, *4*, 1859.

1,3-dibromobenzene — 1. 2 $B(pinayl)_2$, 2% (Ind)Ir(cod), 2% dmpe , 150°C; 2. aq Oxone , acetone → 3,5-dibromophenol 87%

Maleczka Jr. R.E.; Shi, F.; Holmes, D.; Smith III, M.R. *J. Am. Chem. Soc.* **2003**, *125*, 7792.

$PhCH_3$ — cat Mo(VI)peroxo complex , H_2O , MeCN; reflux , 12h → $PhCH_2OH$ 37%

Das, S.; Bhowmick, T.; Punniyamurthy, T.; Dey, D.; Nath, J.; Choudhuri, M.K. *Tetrahedron Lett.* **2003**, *44*, 4915.

SECTION 42: ALCOHOLS AND THIOLS FROM KETONES

The following reaction types are included in this section:
A. Reductions of Ketones to Alcohols
B. Alkylations of Ketones, forming Alcohols

Coupling of ketones to give diols may be found in Section 323 (Alcohol → Alcohol).

SECTION 42A: REDUCTION OF KETONES TO ALCOHOLS

ASYMMETRIC REDUCTION

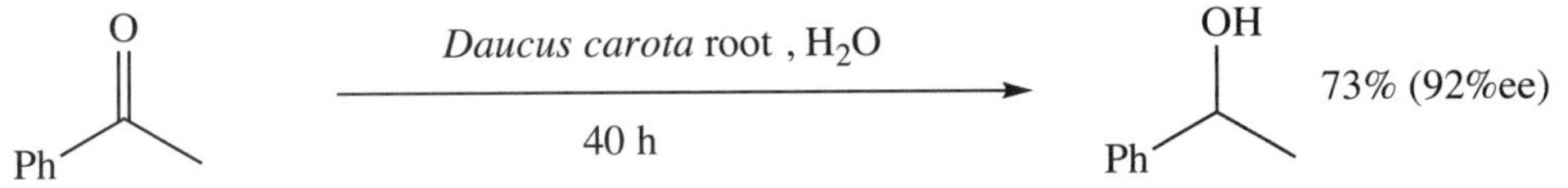

Daucus carota root , H_2O, 40 h — 73% (92%ee)

Yadav, J.S.; Nanda, S.; Reddy, P.T.; Rao, A.B. *J. Org. Chem.* **2002**, *67*, 3900.

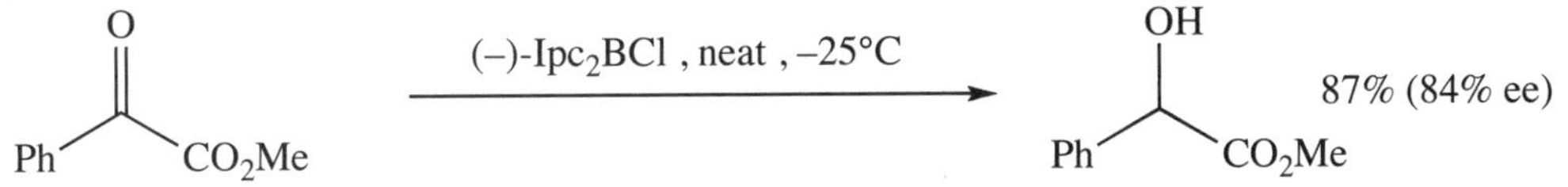

(–)-Ipc_2BCl , neat , –25°C — 87% (84% ee)

Ramachandran, P.V.; Pitre, S.; Brown, H.C. *J. Org. Chem.* **2002**, *67*, 5315.

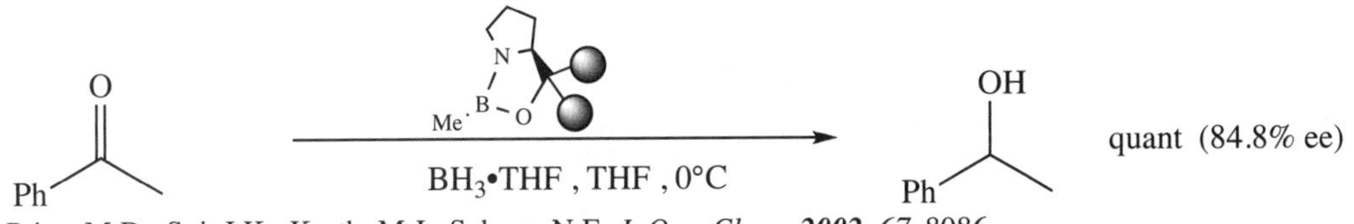

BH_3•THF , THF , 0°C — quant (84.8% ee)

Price, M.D.; Sui, J.K.; Kurth, M.J.; Schore, N.E. *J. Org. Chem.* **2002**, *67*, 8086.

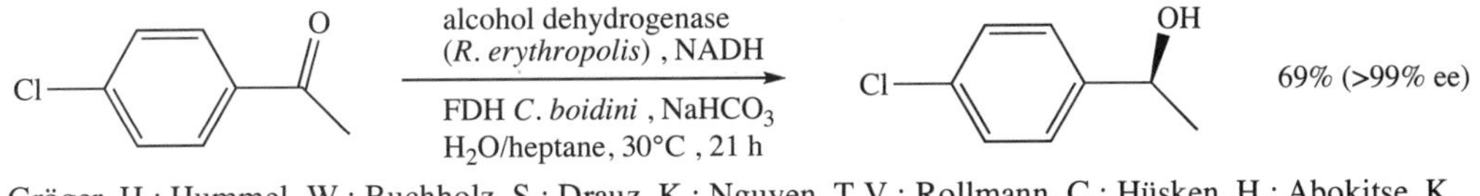

alcohol dehydrogenase (*R. erythropolis*) , NADH; FDH *C. boidini* , $NaHCO_3$; H_2O/heptane, 30°C , 21 h — 69% (>99% ee)

Gröger, H.; Hummel, W.; Buchholz, S.; Drauz, K.; Nguyen, T.V.; Rollmann, C.; Hüsken, H.; Abokitse, K. *Org. Lett.* **2003**, *5*, 173.

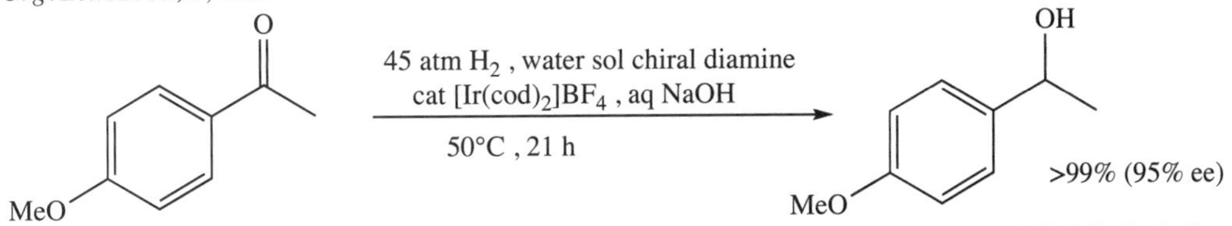

45 atm H_2 , water sol chiral diamine cat $[Ir(cod)_2]BF_4$, aq NaOH; 50°C , 21 h — >99% (95% ee)

Maillet, C.; Praveen, T.; Janvier, P.; Minguet, S.; Evain, M.; Saluzzo, C.; Tommasino, M.; Bujoli, B. *J. Org. Chem.* **2002**, *67*, 8191.

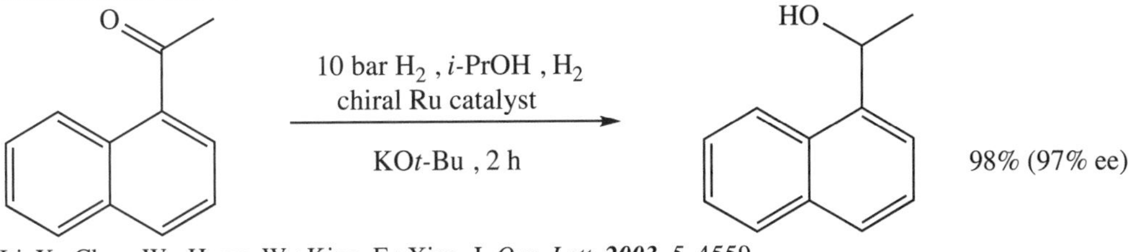

10 bar H_2 , *i*-PrOH , H_2 chiral Ru catalyst; KO*t*-Bu , 2 h — 98% (97% ee)

Li, X.; Chen, W.; Hems, W.; King, F.; Xiao, J. *Org. Lett.* **2003**, *5*, 4559.

silica immobilized amino sulfonamide ligands

$[RuCl_2(p\text{-cymene})]_2$, HCO_2H/NEt_3

>99% (97% ee)

Liu, P.N.; Gu, P.M.; Wang, F.; Tu, Y.Q. *Org. Lett.* ***2004***, *6*, 169.

10 HCO_2Na , SDS , 30°C

NHPh

97% (95.8% ee)

Rhyoo, H.Y.; Park, H.-J.; Suh, W.H.; Chung, Y.K. *Tetrahedron Lett.* ***2002***, *43*, 269.

9 atm H_2 , *i*-PrOH , *t*-BuOK
Ru-BINAP/1,4-diamine catalyst
25°C

quant (98% ee)

Ohkuma, T.; Hattori, T.; Ooka, H.; Inoue, T.; Noyori, R. *Org. Lett.* ***2004***, *6*, 2681.

0.1

1.2 BH_3•SMe_2 , THF

>99% (84% ee)

Gilmore, N.J.; Jones, S.; Muldowney, M.P. *Org. Lett.* ***2004***, *6*, 2805.

cat CuCl , cat NaO*t*-Bu , toluene

cat bis(phosphine) , 5 Ph_3SiH , –78°C

98% (94% ee)

Lipshutz, B.H.; Noson, K.; Chrisman, W.; Lower, A. *J. Am. Chem. Soc.* ***2003***, *125*, 8779.

H_2 , Rh catalyst

96% (94% ee)

Raja, R.; Thomas, J.M.; Jones, M.D.; Johnson, B.F.G.; Vaushan, D.E.W. *J. Am. Chem. Soc.* ***2003***, *125*, 14982.

100 bar H_2 , EtOH , 80°C , 72 h

chiral catalyst , chiral bis(phosphine)

quant (>99% ee)

Lei, A.; Wu, S.; He, M.; Zhang, X. *J. Am. Chem. Soc.* ***2004***, *126*, 1626.

chiral bis(amino diol) , *i*-PrOH

2 $SmI(\eta^8\text{-}C_8H_8)(thf)$, 1 d

>99% (>99% ee)

Ohno, K.; Kataoka, Y.; Mashima, K. *Org. Lett.* ***2004***, *6*, 4695.

Ph(C=O)CH3 → [LiBH4, THF, rt, 0.5 h; 2 O_2N-C6H4-B(tartaric acid ester, CO_2H, CO_2H)] → PhCH(OH)CH3 quant (99% ee)

Suri, J.T.; Vu, T.; Hernandez, A.; Congdon, J.; Singaram, B. *Tetrahedron Lett.* **2002**, *43*, 3649.

PhC(O)CH3 → [1. toluene, 5% (Ph, Ph oxazaborolidine, N·B Me), *t*-Bu(TMS)NH•BH_3, 20°C; 2. MeOH 3. H_2O] → PhCH(OH)CH3 84% (97% ee)

Huertas, R.E.; Corella, J.A.; Soderquist, J.A. *Tetrahedron Lett.* **2003**, *44*, 4435.

PhC(O)CH3 → [1. Ph_2SiH_2, 3% $Cu(OAc)_2$•H_2O, S-BINAP; toluene, 0°C 2. TBAF] → PhCH(OH)CH3 94% (79% ee)

Lee, D.-W.; Yun, J. *Tetrahedron Lett.* **2004**, *45*, 5415.

PhC(O)CH3 → [cat $IrHCl(cod)_2$, diamine-phosphine ligand; *i*-PrOH, 7.5h, rt] → PhCH(OH)CH3 92% (79% ee)

Chen, J.-s.; Li, Y.-y.; Dong, Z.-r.; Li, B.-z.; Gao, J.-x. *Tetrahedron Lett.* **2004**, *45*, 8415.

PhC(O)Me → [cat chiral lactam-alcohol, BH_3•THF; reflux] → PhCH(OH)Me 97% (97% ee)

Kawanami, Y.; Murao, S.; Ohga, T.; Kobayashi, N. *Tetrahedron* **2003**, *59*, 8411.

2-F-C6H4C(O)CH3 → [polymer supported amine sulfonamide ligand, cat $[RuCl_2(p\text{-cymene})]_2$, H_2O, HCO_2Na; Bu_4NBr, 58h] → 2-F-C6H4CH(OH)CH3 91% (90% ee)

Liu, P.N.; Dengt, J.G.; Tu, Y.Q. *Chem. Commun.* **2004**, 2070.

PhC(O)CH3 → [tetrahydroquinolilyl oxazoline ligand, KOH; cat $[Ru(p\text{-cymene})Cl_2]_2$, *i*-PrOH, –20°C, 2d] → PhCH(OH)CH3 78% (78% ee)

Zhou, Y.-B.; Tang, F.-Y.; Xu, H.-D.; Wu, X.-Y.; Ma, J.-A.; Zhou, Q.-L. *Tetrahedron: Asymmetry* **2002**, *13*, 469.

2-hexanone → [dry cells *Geotrichium cardedoni*; NAD^+, *i*-PrOH, buffer] → 2-hexanol 88% (>99% ee)

Matsuda, T.; Nakajima, Y.; Harada, T.; Nakamura, K. *Tetrahedron: Asymmetry* **2002**, *13*, 971.

PhC(O)CH3 → [H_2, cat $[It(cod)_2]$ BF_4, chiral diamine; 50°C, 15h] → PhCH(OH)CH3 71% (63% ee)

Ferrand, A.; Bruno, M.; Tommasino, M.L.; Lemaire, M. *Tetrahedron: Asymmetry* **2002**, *13*, 1379.

PhC(O)CHF_2 → [*Synechococcus elongatus* PC7942, 4d] → PhCH(OH)CHF_2 >99% (70% ee)

Nakamura, K.; Yamanaka, R. *Tetrahedron: Asymmetry* **2002**, *13*, 2529.

KOH , *i*-PrOH , chiral ferrocenyl amino alcohol

cat $[RuCl(p\text{-cymene})]_2$

86% (68% ee)

Patti, A.; Pedotti, S. *Tetrahedron: Asymmetry* ***2003***, *14*, 597.

PhCHO

BH_3-SMe_2 , THF , reflux

chiral amino alcohol

84% (97% ee)

Zhang, Y.-X.; Du, D.-M.; Chen, X.; Liu, S.-F.; Hua, W.-T. *Tetrahedron: Asymmetry* ***2004***, *15*, 177.

Diplogelasinospora grovesii IMI 171018

28°C , 72h

99% (99% ee)

Carballeira, J.D.; Álvarez, E.; Campillo, M.; Pardo, L.; Sinisterra, J.V. *Tetrahedron: Asymmetry* ***2004***, *15*, 951.

Synechococcus sp. PCC 7942

99% (98% ee)

Shimoda, K.; Kubota, N.; Hamada, H.; Kaji, M.; Hirata, T. *Tetrahedron: Asymmetry* ***2004***, *15*, 1677.

i-PrOH , BINOL/$AlMe_3$, toluene

rt , 16h

99% (80% ee)

Campbell, E.J.; Zhou, H.; Nguyen, S.T. *Angew. Chem. Int. Ed.* ***2002***, *41*, 1020.

Ph_2SiH_2 , Rh catalyst , 1.2% $AgBF_4$

DCM , rt

95% (79%ee)

Gade, L.H.; César, V.; Bellemin-Laponnaz, S. *Angew. Chem. Int. Ed.* ***2004***, *43*, 1014.

cat $[RuCl_2(C_6H_6)]_2$-cyclodextrin

HCO_2Na

90% (77%ee)

Schlatter, A.; Kundu, M.K.; Woggon, W.-D. *Angew. Chem. Int. Ed.* ***2004***, *43*, 6731.

cat $[RuCl_2(p\text{-cymene})]_2$, TsOPEN

HCO_2H-NEt_3 , H_2O , 40°C , 30 min

98% (97% ee)

Wu, X.; Li, X.; Hems, W.; King, F.; Xiao, J. *Org. Biomol. Chem.* ***2004***, *2*, 1818.

NON-ASYMMETRIC REDUCTION

chiral Ru cat , NEt_3

HCOOH , 50°C

(92 : 8) 94%

Eustache, F.; Dalko, P.I.; Cossy, J. *Org. Lett.* ***2002***, *4*, 1263.

Ph ketone → alcohol: Bu_2SnH_2 , 1% $Pd(PPh_3)_4$, 25°C / toluene , 3 h — >95%

Kamiya, I.; Ogawa, A. *Tetrahedron Lett.* ***2002***, *43*, 1701.

Cyclohexanone → cyclohexanol: 1 atm H_2 , Cu/SiO_2 , 90°C / heptane , 3h — quant

Ravasio, N.; Psaro, R.; Zaccheria, F. *Tetrahedron Lett.* ***2002***, *43*, 3943.

6.25 H_2O , 5 NEt_3 , THF / 2.5 SmI_2 — >99%

Dahlén, A.; Hilmersson, G. *Tetrahedron Lett.* ***2002***, *43*, 7197.

Ph(C=O)Ph → Ph(CHOH)Ph: Al , NaOH , aq MeOH , 2h — 85%

Bhar, S.; Guha, S. *Tetrahedron Lett.* ***2004***, *45*, 3775.

bis(carbene) Ru(III) complex , KOH/*i*-PrOH / reflux , 10h — >98%

Albrecht, M; Crabtree, R.H.; Mata, J.; Peris, E. *Chem. Commun.* ***2002***, 32.

SmI_2 , THF , H_2O , TMEDA / < 10 sec — >99%

Dahlén, A.; Hilmersson, G. *Chem. Eur. J.* ***2003***, *9*, 1123.

activated Al_2O_3 , KOH , 13 min / *i*-PrOH , microwaves — 90%

Kazemi, F.; Kiasat, A.R. *Synth. Commun.* ***2002***, *32*, 2255.

O*i*-Pr → O*i*-Pr: Ru-TSDPEN catalyst / *i*-PrOH , 1 h — 94%

Guo, M.; Li, D.; Sun, Y.; Zhang, Z. *Synlett* ***2004***, 741.

Cyclohexanone → cyclohexanol: 1.5 $CuCl_2$•2 H_2O , 6 h / cat Li-DTBB — 93%

Alonso, F.; Vitale, C.; Radivoy, G.; Yus, M. *Synthesis* ***2003***, 443.

REVIEW:

"Recent Developments in Asymmetric Reduction of Ketones with Biocatalysts"
Nakamura, K.; Yamanaka, R.; Matsuda, T.; Harada, T. *Tetrahedron: Asymmetry* ***2003***, *14*, 2659.

SECTION 42B: ALKYLATION OF KETONES, FORMING ALCOHOLS

Aldol reactions are listed in Section 330 (Alcohol, Thiol-Ketone)

ASYMMETRIC ALKYLATION

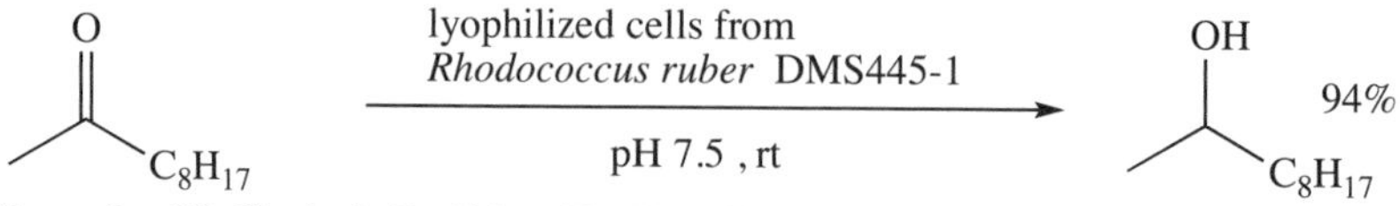

O
C_8H_{17}
lyophilized cells from *Rhodococcus ruber* DMS445-1
pH 7.5 , rt
OH
C_8H_{17}
94% conversion (>99% ee)

Stampfer, W.; Kosjeck, B.; Faber, K.; Kroutil, W. *J. Org. Chem.* ***2003***, *68*, 402.

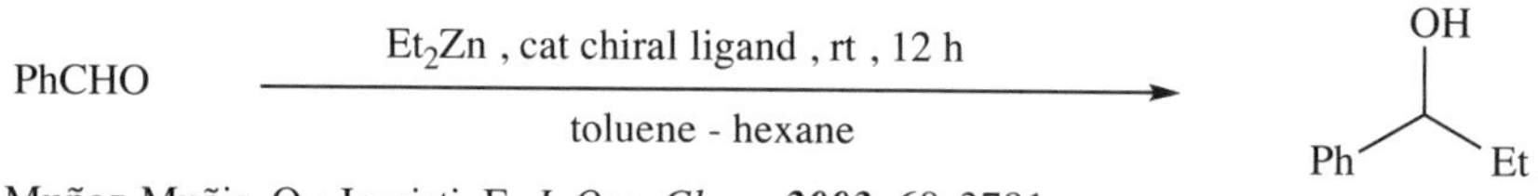

PhCHO
Et_2Zn , cat chiral ligand , rt , 12 h
toluene - hexane
OH
Ph Et
89% (79% ee)

Muñoz-Muñiz, O.; Juaristi, E. *J. Org. Chem.* ***2003***, *68*, 3781.

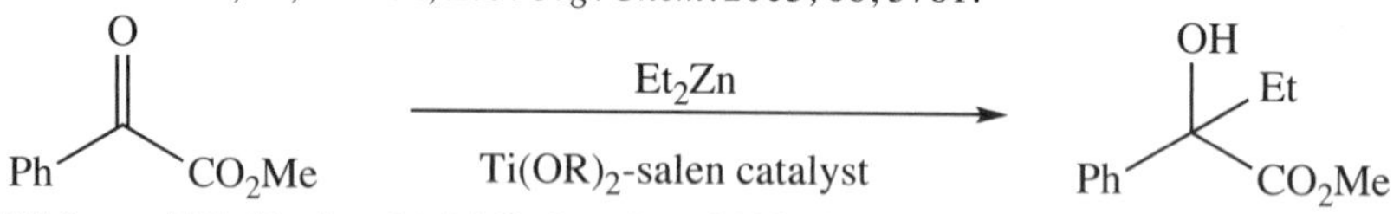

O
Ph CO_2Me
Et_2Zn
$Ti(OR)_2$-salen catalyst
OH
Et
Ph CO_2Me
99% (56% ee)

DiMauro, E.F.; Kozlowski, M.C. *Org. Lett*, ***2002***, *4*, 3781.

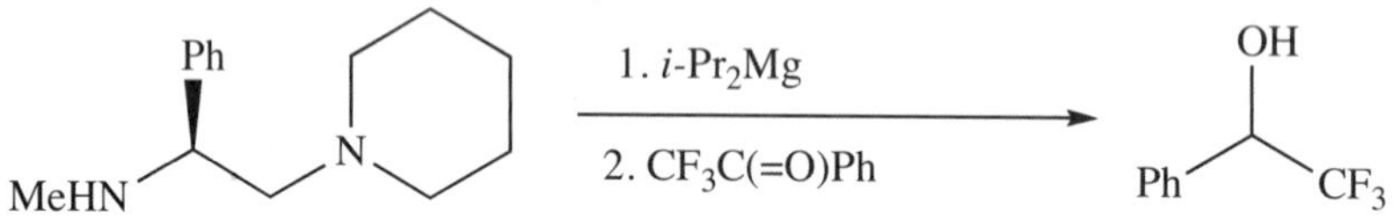

Ph
MeHN N
1. *i*-Pr_2Mg
2. $CF_3C(=O)Ph$
OH
Ph CF_3
93% (71:29 *R*:*S*)

Yong, K.H.; Chong, J.M. *Org. Lett*, ***2002***, *4*, 4139.

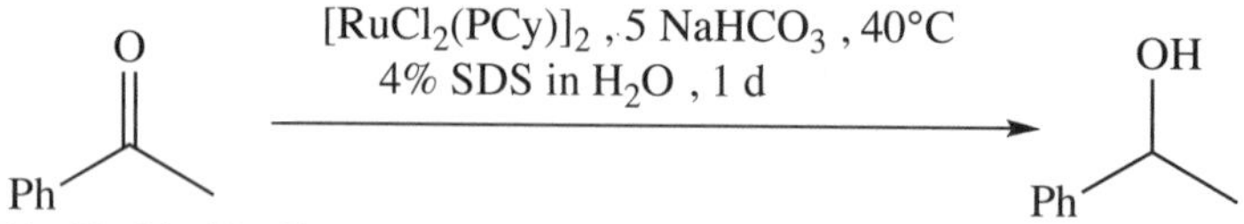

O
Ph
$[RuCl_2(PCy)]_2$, 5 $NaHCO_3$, 40°C
4% SDS in H_2O , 1 d
OH
Ph
>99% (95% ee)

Ma, Y.; Liu, H.; Chen, L.; Cui, X.; Zhu, J.; Deng, J. *Org. Lett.* ***2003***, *5*, 2103.

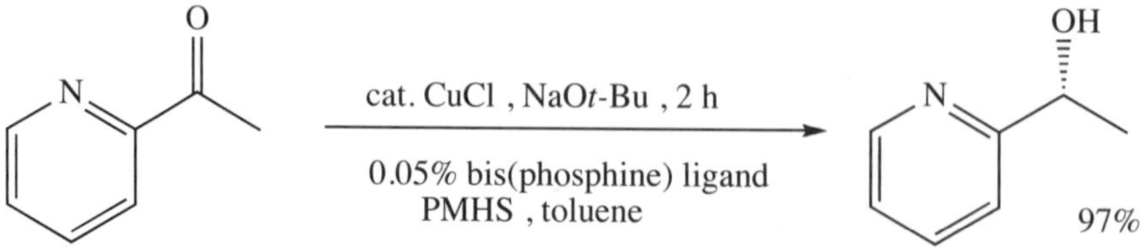

O
N
cat. CuCl , NaO*t*-Bu , 2 h
0.05% bis(phosphine) ligand
PMHS , toluene
OH
N
97% (90% ee)

Lipshutz, B.H.; Lower, A.; Noson, K. *Org. Lett*, ***2002***, *4*, 4045.

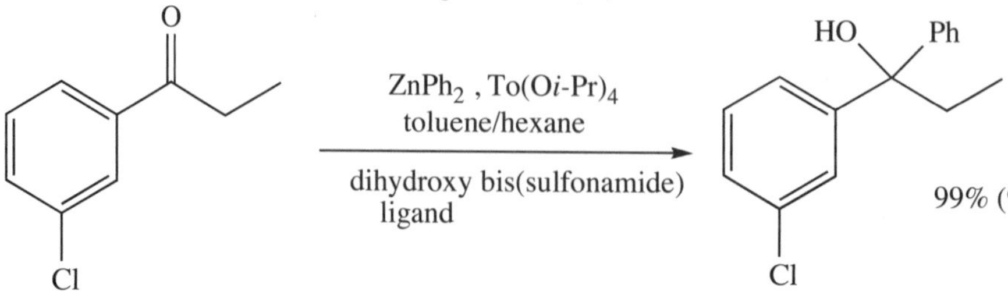

O
Cl
$ZnPh_2$, To(O*i*-Pr)$_4$
toluene/hexane
dihydroxy bis(sulfonamide) ligand
HO Ph
Cl
99% (92% ee)

García, C.; Walsh, P.J. *Org. Lett.* ***2003***, *5*, 3641.

B-allyl-3,3-binaphthyl boronate, toluene, $-78°C \rightarrow -40°C$, 2d — 88% (96:7 *R:S*)

Wu, T.R.; Shen, L.; Chong, J.M. *Org. Lett.* ***2004**, 6*, 2701.

1. Et_2Zn, $Ti(Oi\text{-}Pr)_4$, chiral diamine, hexane/toluene 2. aq NH_4Cl — 71% (96% ee)

García, C.; LaRochelle, L.K.; Walsh, P.J. *J. Am. Chem. Soc.* ***2002**, 124*, 10970.

, 3% $CuF_2{\bullet}3\ H_2O$, $-40°C$; 6% *i*-PrBUPHOS, 45% $La(Oi\text{-}r)_3$, DMF — 94% (82% ee)

Wada, R.; Oisaki, K.; Kanai, M.; Shibasaki, M. *J. Am. Chem. Soc.* ***2004**, 126*, 8910.

$Sn(CH_2CH{=}CH_2)_4$, DCM, rt; $Ti(Oi\text{-}Pr)_4$, BINOL, *i*-PrOH — 82% (96% ee)

Kim, J.G.; Waltz, K.M.; Garćia, I.F.; Kwiatkwski, D.; Walsh, P.J. *J. Am. Chem. Soc.* ***2004**, 126*, 12580.

2 Et_2Zn, $TiOi\text{-}Pr)_4$, toluene, 25°C, 8h; camphorsulfonamido cylcohexenes — 80% (98% ee)

Yus, M.; Ramón, D.J.; Prieto, O. *Tetrahedron: Asymmetry* ***2002**, 13*, 2291.

cat Cu-pybx derivative, Ph_2Zn, 1d; 1.1 $Ti(Oi\text{-}Pr)_4$, toluene, 24°C — 98% (92% ee)

Preito, O.; Ramón, D.J.; Yus, M. *Tetrahedron: Asymmetry* ***2003**, 14*, 1955.

1.5 $(CH_2{=}CH)_4Sn$, DCM, rt; cat BINOL/$Ti(Oi\text{-}Pr)_4$/*i*-PrOH — 82% (96%ee)

Waltz, K.M.; Gavenonis, J.; Walsh, P.J. *Angew. Chem. Int. Ed.* ***2002**, 41*, 3697.

3 PhC≡CH, 3 Me_2Zn, 20% salen; toluene, rt — 53% (57% ee)

Cozzi, P.G. *Angew. Chem. Int. Ed.* ***2003**, 42*, 2895.

Me_2Zn, toluene-hexane, $-20°C$, 1d; chiral phosphonate catalyst — 72% (82%ee)

Funabashi, K.; Jachmann, M.; Kanai, M.; Shibasaki, M. *Angew. Chem. Int. Ed.* ***2003**, 42*, 5489.

NON-ASYMMETRIC ALKYLATION

1. catecholborane , THF , 3 h
2. MeOH
89%

Huddleston, R.R.; Cauble, D.F.; Krische, M.J. *J. Org. Chem.* ***2003***, *68*, 11.

$BrCH_2CO_2Me$, cat Zn dust
I_2 , dioxane , ultrasound
20°C , 35 min
CO_2Me
quant

Ross, N.A.; Bartsch, R.A. *J. Org. Chem.* ***2003***, *68*, 360.

OAc , e^- , 2,2'-bipy , DMF
$Bu_4N\ BF_4$, $FeBr_2$
78%

Durandetti, M.; Meignein, C.; Périchon, J. *J. Org. Chem.* ***2003***, *68*, 3121.

MeO_2C
SmI_2 , *i*-PrOH
18 HMPA , THF
MeO_2C
84% (2:1 *syn:anti*)

Ohno, H.; Okumura, M.; Maeda, S.; Iwasaki, H.; Wakayama, R.; Tanaka, T. *J. Org. Chem.* ***2003***, *68*, 7722.

H_2=C=CHOMe , SmI_2 , THF
18 eq HMPA , 2 eq *t*-BuOH , 16 h
OMe
85% (60:40 *E:Z*)

Hölemann, A,; Reissig, H.-U. *Org. Lett.* ***2003***, *5*, 1463.

Bu-Cl , NdI_2 , THF
Bu
97%

Evans, W.J.; Workman, P.S.; Allen, N.T. *Org. Lett.* ***2003***, *5*, 2041.

Bu_3Sn Ph
acetophenone , $SnCl_2$, MeCN
rt , 3 h
82% (92:8 *anti:syn*)

Yasuda, M.; Hirata, K.; Nishino, M.; Yamamoto, A.; Baba, A. *J. Am. Chem. Soc.* ***2002***, *124*, 13442.

Ph CHO
1. O_2, UV , TPP , L-proline , DMF , –5°C
2. $NaBH_4$
91% (74:26 R:S)

Córdova, A.; Sundén, H.; Engqvist, M.; Ibrahem, I.; Casas, J. *J. Am. Chem. Soc.* ***2004***, *126*, 8914.

EtMgBr , 5% $InCl_3$, THF , 0°C
no $InCl_3$ (15:85) 99%
(45 : 55) 86%

Kelly, B.G.; Gilheany, D.G. *Tetrahedron Lett.* ***2002***, *43*, 887.

Ph–CH=CH–Ph —[acetone, Mg/TMSCl, DMF]→ 2-methyl-3,4-diphenylbutan-2-ol (HO, Ph, Ph) 70%

Yamamoto, Y.; Kawano, S.; Maekawa, H.; Nishiguchi, I. *Synlett* **2004**, 30.

PhC(O)Me —[Bu_3Sn-allyl; CAN, rt, 1 h]→ Ph(Me)C(OH)CH₂CH=CH₂ 86%

also with aldehydes

Yadav, J.S.; Reddy, B.V.S.; Krishna, A.D.; Sadasiv, K.; Chary, Ch.J. *Chem Lett.* **2003**, *32*, 248.

REVIEW:

"Catalytic Enantioselective Addition of Allylic Organometallic Reagents to Aldehydes and Ketones" Denmark, S.E.; Fu, J. *Chem. Rev.* **2003**, *103*, 2763.

SECTION 43: ALCOHOLS AND THIOLS FROM NITRILES

Ph(CH₂)₃O–CH₂–C₆H₄–CN —[2.4 Et_3GeNa, 1,4-dioxane]→ Ph(CH₂)₃O–CH₂–C₆H₄–OH 99%

Yokoyama, Y.; Takizawa, S.; Nanjo, M.; Mochida, K. *Chem. Lett.* **2002**, *31*, 1032.

SECTION 44: ALCOHOLS AND THIOLS FROM ALKENES

Ph–CH=CH₂ —[1. $AcNMe_2$, 2 eq catecholborane, THF, 0°C; 2. MeOH; 3. 2.5 NEt_3; 4. O_2, rt, 12 h]→ PhCH₂CH₂OH + PhCH(OH)CH₃ (12 : 1) 80%

Cadot, C.; Dalko, P.I.; Cossy, J.; Ollivier, C.; Chuard, R.; Renaud, P. *J. Org. Chem.* **2002**, *67*, 7193.

Ph–CH=CH–Ph —[1. $HSiCl_2$, Pd cat, phosphine ligand; 2. 6 KF, 6 $KHCO_3$, MeOH-THF; 3. aq $Na_2S_2O_3$]→ PhCH(OH)CH₂Ph 87% (99% ee)

Jensen, J.F.; Svendsen, B.Y.; la Cour, T.V.; Pedersen, H.L.; Johannsen, M. *J. Am. Chem. Soc.* **2002**, *124*, 4558.

Ph–CH=CH₂ —[1. $NaIO_4$, 1.75 h, 0°C, H_2O, $RuCl_3$•xH_2O, EtOAc/MeCN; 2. $NaBH_4$]→ PhCH₂OH 79%

Sharma, P.K.; Nielsen, P. *J. Org. Chem.* **2004**, *69*, 5742.

CH₂=CH–CN —[Et_3SiH, cat NHPI, AcOEt, 60°C, 10h; cat $Co(OAc)_2$/$Co(accac)_3$]→ Et_3SiCH₂CH(OH)CN 67%

Tayama, O.; Iwahama, T.; Sakaguchi, S.; Ishii, Y. *Eur. J. Org. Chem.* **2003**, 2286.

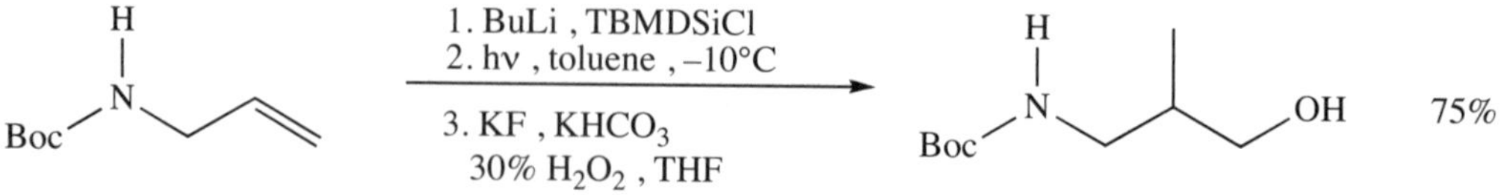

1. BuLi , TBMDSiCl
2. hν , toluene , –10°C
3. KF , $KHCO_3$
30% H_2O_2 , THF

75%

Blaszykowski, C.; Dhimane, A.-L.; Fensterbank, L.; Malacria, M. *Org. Lett.* ***2003***, *5*, 1341.

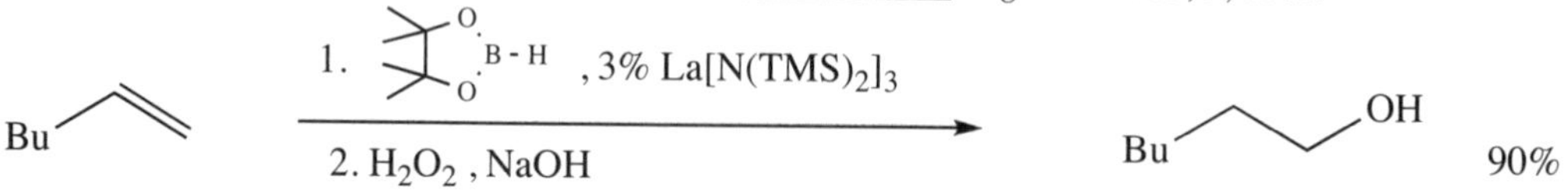

1. , 3% $La[N(TMS)_2]_3$
2. H_2O_2 , NaOH

90%

Horino, Y.; Livinghouse, T.; Stan, M. *Synlett* ***2004***, 2639.

CO_2 , 3 H_2 , [bmim] Cl
toluene , Ru catalyst

84% of alcohols

Tominaga, K.; Sasai, Y. *Chem. Lett.* ***2004***, *33*, 14.

SECTION 45: ALCOHOLS AND THIOLS FROM MISCELLANEOUS COMPOUNDS

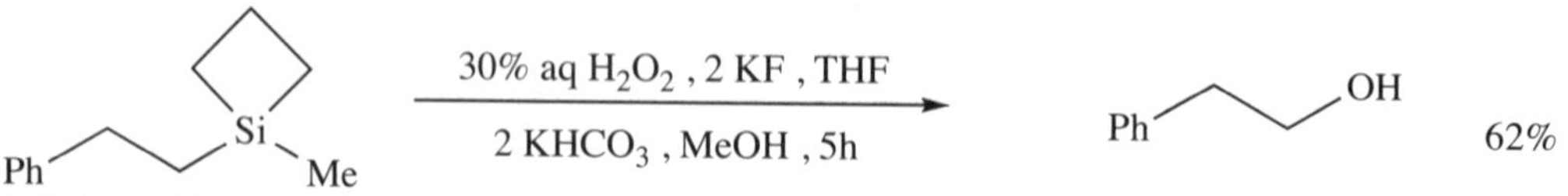

30% aq H_2O_2 , 2 KF , THF
2 $KHCO_3$, MeOH , 5h

62%

Suderhaus, J.D.; Lam, H.; Dudley, G.B. *Org. Lett.* ***2003***, *5*, 4571.

$C_8H_{17}S\text{-}SO_3Na$ → (THF , H_2O) → $C_8H_{17}S\text{-}H$ 78%

Zhan, Z.-P.; Lang, K.; Liu, F.; Hu, L.-m. *Synth. Commun.* ***2004***, *34*, 3203.

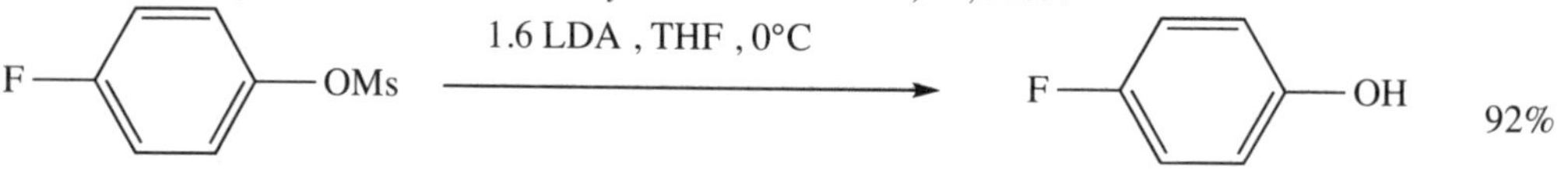

1.6 LDA , THF , 0°C

92%

Ritter, T.; Stanek, K.; Larrosa, I.; Carreira, E.M. *Org. Lett.* ***2004***, *6*, 1513.

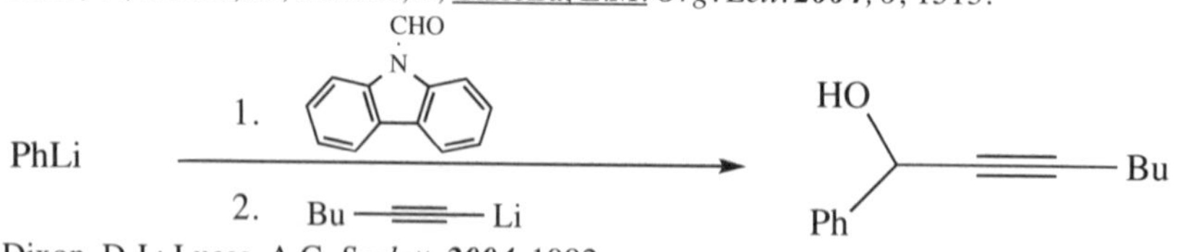

PhLi
1. (N-formylcarbazole)
2. Bu—≡—Li

Dixon, D.J.; Lucas, A.C. *Synlett* ***2004***, 1092.

SECTION 45A: PROTECTION OF ALCOHOLS AND THIOLS

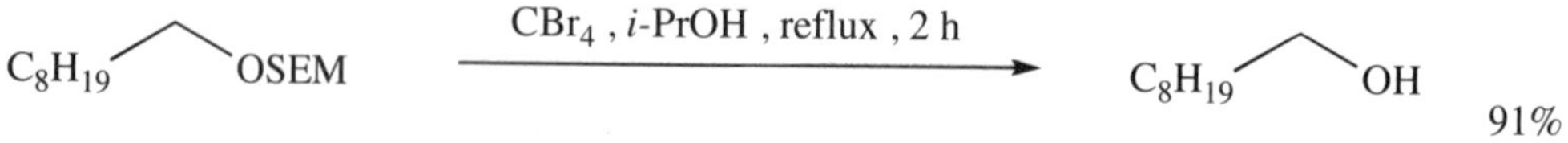

CBr$_4$, *i*-PrOH , reflux , 2 h

91%

Chen, M.-Y.; Lee, A.S.-Y. *J. Org. Chem.* ***2002***, *67*, 1384.

Ph~~O–(4-methoxyphenyl) →[20% $ZrCl_4$, MeCN , rt; 30 min] Ph~~OH 86%

Sharma, G.V.M.; Reddy, Ch.G.; Krishna, P.R. *J. Org. Chem.* **2003**, *68*, 4574.

PhO-MOM →[$NaHSO_4$, SiO_2 , DCM; rt , 1.5 h] Ph-OH 94%

Ramesh, C.; Ravindranath, N.; Das, B. *J. Org. Chem.* **2003**, *68*, 7101.

C_6H_{13}-epoxide →[$SnCl_2$, acetone] C_6H_{13}-dioxolane (2,2-dimethyl) 95%

Vyvyan, J.R.; Meyer, J.A.; Meyer, K.D. *J. Org. Chem.* **2003**, *68*, 9144.

PhOH →[Tf_2NPh , 3 K_2CO_3 , THF; microwaves , 6 min , 120°C] PhOTf 69%

Bengtson, A.; Hallberg, A.; Larhed, M. *Org. Lett.* **2002**, *4*, 1231.

$C_{10}H_{21}OTES$ →[MeOH , 10% Pd/C , rt , 1 h] $C_{10}H_{21}OH$ >95%

Rotulo-Sims, D.; Prunet, J. *J. Org. Chem.* **2002**, *67*, 4701.

4-methylphenol (OH) →[Tf_2O , aq K_3PO_4 , toluene] 4-methylphenyl OTf 81%

Frantz, D.E.; Weaver, D.G.; Carey, J.P.; Kress, M.H.; Dolling, U.H. *Org. Lett.* **2002**, *4*, 4717.

AcO~$(CH_2)_8$~OTr →[0.01 TESOTf , 80% AcOH; THF , rt] AcO~$(CH_2)_8$~OH 98%

Imagawa, H.; Tsuchihashi, T.; Singh, R.K.; Yamamoto, H.; Sugihara, T.; Nishizawa, M. *Org. Lett.* **2003**, *5*, 153.

Ph~~O~allyl →[cat $CpRu(MeCN)_3$ PF_6 , MeOH; quinoline-2-CO_2H , 30°C] Ph~~OH >99%

Tanaka, S.; Saburi, H.; Ishibashi, Y.; Kitamura, M. *Org. Lett.* **2004**, *6*, 1873.

DHP + ~~~~~~OH , 1% $AlCL_3$•6 H_2O ⇄ ~~~~~~OTHP 95%

Namboodiri, V.V.; Varma, R.S. *Tetrahedron Lett.* **2002**, *43*, 1143.

$PhCH_2OH$ →[1.6 DHP , 0.6 LiOTf , DCE , reflux; 2.5h] $PhCH_2OTHP$ 96%

Karimi, B.; Maleki, J. *Tetrahedron Lett.* **2002**, *43*, 5353.

$C_8H_{17}OTBS$ →[10% $Ce(OTf)_4$, MeCN , 0.8h , rt] $C_8H_{17}OH$ 95%

Bartoli, G.; Cupone, G.; Dalpozzo, R.; De Nino, A.; Maiuolo, L.; Procopio, A.; Sambri, L.; Tagarelli, A. *Tetrahedron Lett.* **2002**, *43*, 5945.

Ph~~~OTES →[$ZnBr_2$, H_2O , DCM] Ph~~~OH 87%

Crouch, R.D.; Polizzi, J.M.; Cleiman, R.A.; Yi, J.; Romany, C.A. *Tetrahedron Lett.* **2002**, *43*, 7151.

$PhOSiMe_2t$-Bu → (Et_3N-O , MeOH , 1h) → PhOH 98%

Zubaidha, P.K.; Bhosale, S.V.; Hashimi, A.M. *Tetrahedron Lett.* ***2002***, *43*, 7277.

$PhCH_2OH$ → (dihydropyran , cat $In(OTf)_3$, DCM; 0°C , 30 min) → $PhCH_2OTHP$ 85%

Mineno, T. *Tetrahedron Lett.* ***2002***, *43*, 7975.

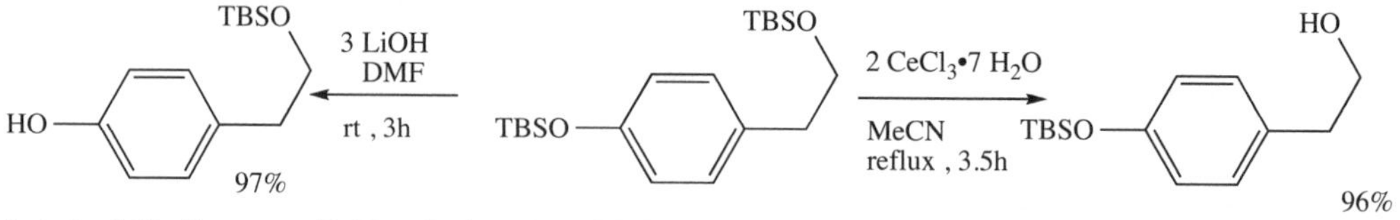

TBSO / TBSO (bis-TBS ether) → (3 LiOH, DMF, rt , 3h) → HO / TBSO 97%; → (2 $CeCl_3$•7 H_2O, MeCN, reflux , 3.5h) → TBSO / HO 96%

Ankala, S.V.; Fenteany, G. *Tetrahedron Lett.* ***2002***, *43*, 4729.

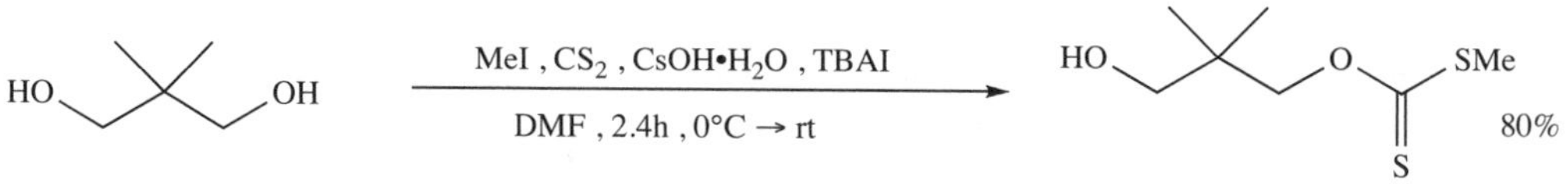

HO / OH → (MeI , CS_2 , $CsOH•H_2O$, TBAI; DMF , 2.4h , 0°C → rt) → HO / O / SMe / S 80%

Nagle, A.S.; Salvatore, R.N.; Cross, R.M.; Kapxhiu, E.A.; Sahab, S.; Yoon, C.H.; Jung, K.W. *Tetrahedron Lett.* ***2003***, *44*, 5695.

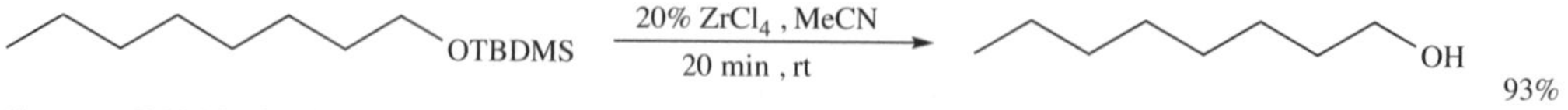

OTBDMS → (20% $ZrCl_4$, MeCN; 20 min , rt) → OH 93%

Sharma, G.V.M.; Srinivas, B.; Krishna, P.R. *Tetrahedron Lett.* ***2003***, *44*, 4689.

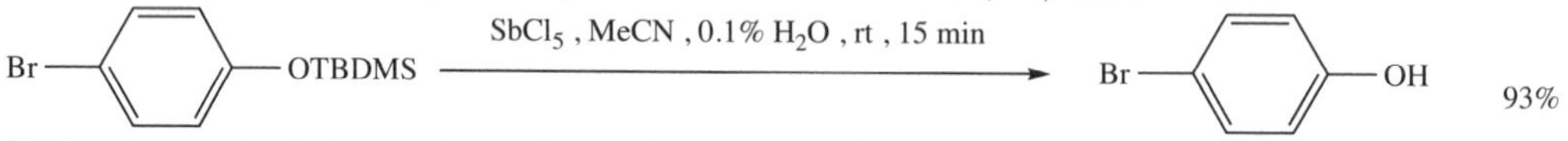

Br—OTBDMS → ($SbCl_5$, MeCN , 0.1% H_2O , rt , 15 min) → Br—OH 93%

Glória, P.M.C.; Prabhakar, S.; Lobo, A.M.; Gomes, M.J.S. *Tetrahedron Lett.* ***2003***, *44*, 8819.

$PhCH_2OH$ → (polystyrene supported $AlCl_3$, DHP) → $PhCH_2OTHP$ 97%

Tamami, B.; Borujeny, K.P. *Tetrahedron Lett.* ***2004***, *45*, 715.

Ph OTES → ($BiOCl_4$•x H_2O , DCM) → Ph OH 79%

Crouch, R.D.; Romany, C.A.; Kreshock, A.C.; Menconi, K.A.; Zile, J.L. *Tetrahedron Lett.* ***2004***, *45*, 1279.

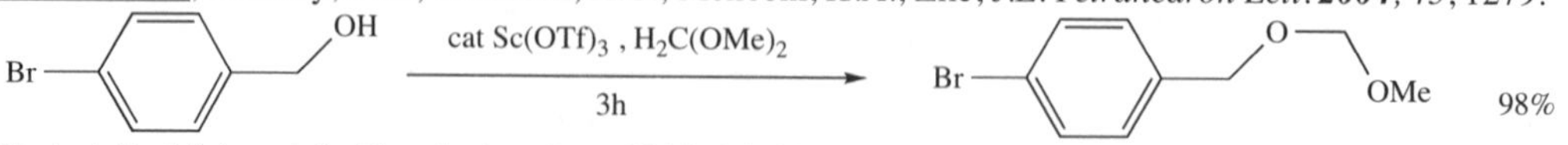

Br / OH → (cat $Sc(OTf)_3$, $H_2C(OMe)_2$; 3h) → Br / O / OMe 98%

Karimi, B.; Ma'mani, L. *Tetrahedron Lett.* ***2003***, *44*, 6051.

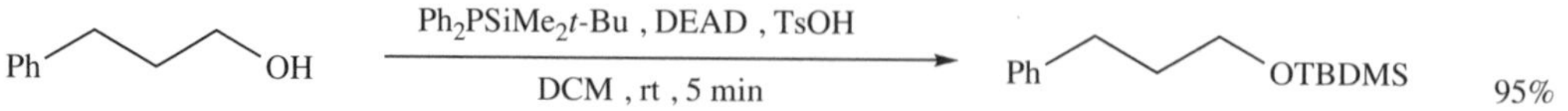

Ph OH → (Ph_2PSiMe_2t-Bu , DEAD , TsOH; DCM , rt , 5 min) → Ph OTBDMS 95%

Hayashi, M.; Matsuura, Y.; Watanabe, Y. *Tetrahedron Lett.* ***2004***, *45*, 1409.

97%

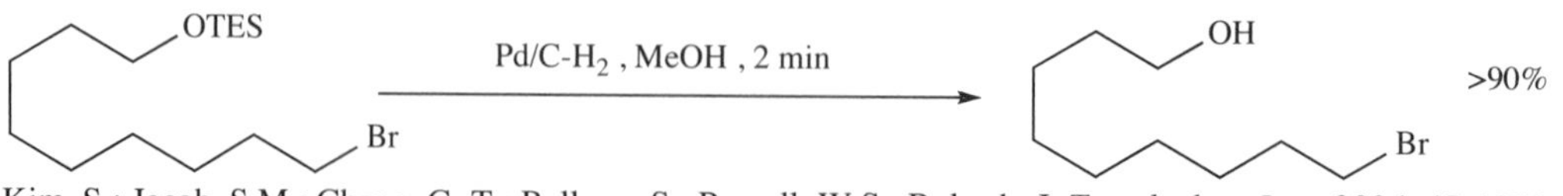

OTES / Br → (Pd/C-H_2 , MeOH , 2 min) → OH / Br >90%

Kim, S.; Jacob, S.M.; Chang, C.-T.; Bellone, S.; Powell, W.S.; Rokach, J. *Tetrahedron Lett.* ***2004***, *45*, 1973.

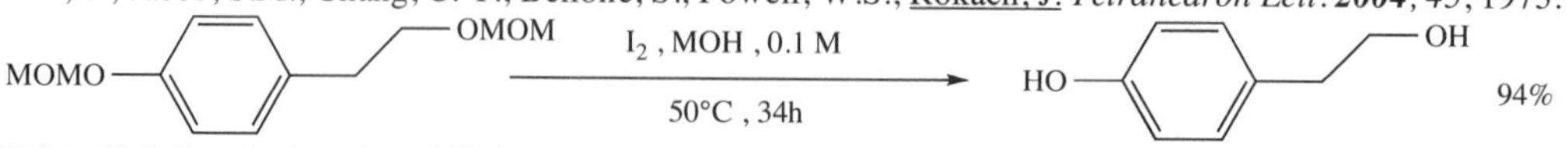

MOMO / OMOM → (I_2 , MOH , 0.1 M; 50°C , 34h) → HO / OH 94%

Keith, J.M. *Tetrahedron Lett.* ***2004***, *45*, 2739.

$PhCH_2CH_2OH$ —[DHP, cat $PdCl_2(MeCN)_2$, THF, rt, 1h]→ $PhCH_2CH_2OTHP$ 91%

Wang, Y.-G.; Wu, X.-X.; Jiang, Z.-Y. *Tetrahedron Lett.* **2004**, *45*, 2973.

OH / Ph —[Cl, O, OMe; 0.5% TsOH]→ O, OMe / Ph 84%

Watanabe, Y.; Ikemoto, T. *Tetrahedron Lett.* **2004**, *45*, 5795.

Ph, OMOM —[$HOCH_2CH_2OH$, 140°C, 1h]→ Ph, OH 92%

Miyake, H.; Tsumura, T.; Sasaki, M. *Tetrahedron Lett.* **2004**, *45*, 7213.

$PhCH_2OH$ —[10% $ZrCl_4$, DMFA]→ $PhCH_2OMOM$ 97%

$PhCH_2OMOM$ —[50% $ZrCl_4$, *i*-PrOH, reflux]→ $PhCH_2OH$ 97%

Sharma, G.V.M.; Reddy, K.L.; Lakshmi, .S.; Krishna, P.R. *Tetrahedron Lett.* **2004**, *45*, 9229.

Ph, OTBDMS —[10% Pd/C, H_2, rt, 1d]→ Ph, OH quant in MeCN

in MeOH, hydrogenation without deprotection

Ikawa, T.; Hattori, K.; Sajiki, H.; Hirota, K. *Tetrahedron* **2004**, *60*, 6901.

$CH_3(CH_2)_{12}CH_2OBn$ —[10 K, 2 *t*-$BuNH_2$; 2 *t*-BuOH, cat. 18-crown-6]→ $CH_3(CH_2)_{12}CH_2OH$ 98%

Shi, L.; Xia, W.J.; Zhang, F.M.; Tu, Y.Q. *Synlett* **2002**, 1505.

OH —[0.1 $Bi(O_2SCF_3)_3{\bullet}4\,H_2O$; O, 3.5 h]→ OTHP 87%

Stephens, J.R.; Butler, P.L.; Clow, C.H.; Oswald, M.C.; Smith, R.C.; Mohan, R.S. *Eur. J. Org. Chem.* **2003**, 3827.

Ph, OH —[0.01 Me_2S^+-Br Br^-, DHP, 5 min, rt]→ Ph, OTHP 85%

Ph, OTHP —[0.01 Me_2S^+-Br Br^-, DHP-MeOH, rt]→ Ph, OH 97%

Khan, A.T.; Mondal, E.; Borah, B.M.; Ghosh, S. *Eur. J. Org. Chem.* **2003**, 4113.

$(CH_2)_4$, OTBS —[cat AcCl, MeOH]→ $(CH_2)_4$, OH 98%

Khan, A.T.; Mondal, E. *Synlett* **2003**, 694.

Ph, OAc —[10 NaOH, THF, Bu_4NHSO_4]→ Ph, OH 93%

Crouch, R.D.; Burger, J.S.; Zletek, K.A.; Cadwallader, A.B.; Bedison, J.E.; Smielewska, M.M. *Synlett* **2003**, 991.

HO, OH —[SiO_2 supported BF_3, EtOAc; rt, 9 h]→ HO, OAc 81%

Das, B.; Venkataiah, B.; Madhusudhan, P. *Synth. Commun.* **2002**, *32*, 249.

OH —[dihydropyran, 3% $LiPF_6$, 15°C; hexane, 20 min, 2.5 h]→ OTHP 81%

Hamada, N.; Sato, T. *Synlett* **2004**, 1802.

Ph(CH$_2$)$_3$OTr $\xrightarrow{\text{BiCl}_3\text{ , MeCN , rt , 7 min}}$ Ph(CH$_2$)$_3$OH 92%

Sabitha, G.; Reddy, E.V.; Swapna, R.; Reddy, N.M.; Yadav, J.S. *Synlett* ***2004***, 1276.

$PhCH_2OTHP$ $\xrightarrow[\text{3 min}]{\text{cat Bi(OTf)}_3\text{ , MeOH , reflux}}$ $PhCH_2OH$ 99%

Mohammadpoor-Baltork, I.; Kharamesh, B.; Kolagar, S. *Synth. Commun.* ***2002***, *32*, 1633.

$PhCH_2OH$ $\xrightarrow[\text{no solvent}]{\text{dihydropyran , cat ZrO}_2\text{/SO}_4^{-2}}$ $PhCH_2OTHP$ 94%

Reddy, B.V.; Sreekanth, P.M. *Synth. Commun.* ***2002***, *32*, 3561.

Ph(CH$_2$)$_2$OH $\xrightarrow[\text{microwaves}]{\text{PhI(OAc)}_2\text{ , THF , 90°C}}$ Ph(CH$_2$)$_2$O–(2-tetrahydrofuranyl) 63%

French, A.N.; Cole, J.; Wirth, T. *Synlett* ***2004***, 2291.

OTBDMS (decalin, H, H) $\xrightarrow{\text{Me}_2\text{S}^+\text{–Br , MeOH}}$ OH (decalin, H, H) 99%

Rani, S.; Babu, J.L.; Vankar, Y.D. *Synth. Commun.* ***2003***, *33*, 4043.

BuOH $\xrightarrow[\text{imidazole , 30 min}]{t\text{-BuMe}_2\text{SiCl , bmim PF}_6\text{ , rt}}$ BuOSiMe$_2$*t*-Bu 94%

Xu, Z.-Y.; Xu, D.-Q.; Liu, B.-Y.; Luo, S.-P. *Synth. Commun.* ***2003***, *33*, 4143.

2,3-dihydrofuran $\xrightarrow[\text{PhCl , hv , 1 d , rt; Ru(salen) complex}]{\text{PhCH(Me)OH}}$ PhCH(Me)O–(2-tetrahydrofuranyl) quant (98% de)

Nagano, H.; Katsuki, T. *Chem Lett.* ***2002***, *31*, 782.

Cl–C$_6$H$_4$–OSiMe$_2$*t*-Bu $\xrightarrow[\text{0.3 h}]{\text{1.5 solid KOH , EtOH}}$ Cl–C$_6$H$_4$–OH 99%

Jiang, Z.-Y.; Wang, Y.-G. *Chem Lett.* ***2003***, *32*, 568

$PhCH_2OH$ $\xrightarrow[\text{rt , 1 h}]{\text{1.8 dihydropyran , 0.2\% Sc(OTf)}_3\text{ , AcOEt}}$ $PhCH_2OTHP$ 95%

Watahiki, T.; Kikumoto, H.; Matsuzaki, M.; Suzuki, T.; Oriyama, T. *Bull. Chem. Soc. Jpn.* ***2002***, *75*, 367.

CHAPTER 4

PREPARATION OF ALDEHYDES

SECTION 46: ALDEHYDES FROM ALKYNES

$HC{\equiv}C{-}C_7H_{15}$ $\xrightarrow[\text{2\% rhodium complex}]{5\ H_2O\text{ , acetone , 70°C , 3 h}}$ $C_7H_{15}CH_2CHO$ quant

Grotjahn, D.B.; Lev, D.A. *J. Am. Chem. Soc.* ***2004***, *126*, 12232.

SECTION 47: ALDEHYDES FROM ACID DERIVATIVES

2-methylbenzoic acid (CO_2H) $\xrightarrow[\text{cat }Pd(OAc)_4\text{ , }PCy_3]{NaH_2PO_2\text{ , }K_3PO_4\text{ , }(t\text{-}BuCO_2)_2O}$ 2-methylbenzaldehyde (CHO) 70%

Gooßen, L.J.; Ghosh, K. *Chem. Commun.* ***2002***, 836.

SECTION 48: ALDEHYDES FROM ALCOHOLS AND THIOLS

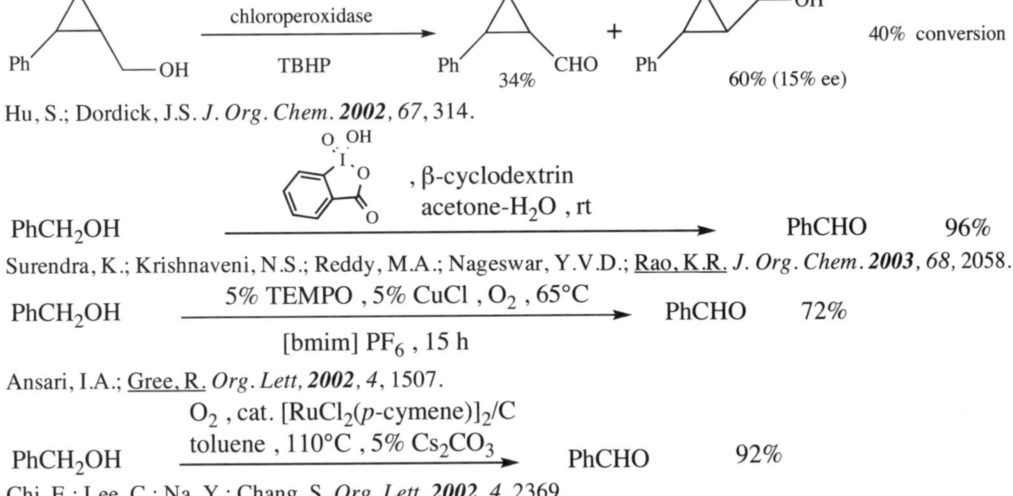

Hu, S.; Dordick, J.S. *J. Org. Chem.* ***2002***, *67*, 314.

Surendra, K.; Krishnaveni, N.S.; Reddy, M.A.; Nageswar, Y.V.D.; Rao, K.R. *J. Org. Chem.* ***2003***, *68*, 2058.

Ansari, I.A.; Gree, R. *Org. Lett*, ***2002***, *4*, 1507.

Chi, E.; Lee, C.; Na, Y.; Chang, S. *Org. Lett.* ***2002***, *4*, 2369.

OH — Ir complex, 2-butanol, reflux, 16 h → CHO 92%

Suzuki, T.; Morita, K.; Tsuchida, M.; Hiroi, K. *J. Org. Chem.* **2003**, *68*, 1601.

$PhCH_2OH$ + 3 (IBX) — 1. EtOAc, 80°C; 2. filter → PhCHO quant

More, J.D.; Finney, N.S. *Org. Lett.* **2002**, *4*, 3001.

O_2N–C_6H_4–CH_2OH — SIBX, THF, 60°C, 1 h → O_2N–C_6H_4–CHO 87%

SIBX = stabilized iodoxybenzoic acid

Ozanne, A.; Pouységu, L.; Depernet, D.; François, B.; Quideau, S. *Org. Lett.* **2003**, *5*, 2903.

$PhCH_2OH$ + (CO_2H, I, O, O reagent) — (bmim) Cl, H_2O, rt → PhCHO 99%

Liu, Z.; Chen, Z.-C.; Zheng, Q.-C. *Org. Lett.* **2003**, *5*, 3321.

Me–$(CH_2)_8$–CH$_2$CH$_2$OH — PEG-TEMPO → Me–$(CH_2)_8$–CH$_2$CHO >99%

Ferreira, P.; Phillips, E.; Rippon, D.; Tsang, S.C.; Hayes, W. *J. Org. Chem.* **2004**, *69*, 6851.

$PhCH_2OH$ — Pd-HAP-O, $PhCF_3$, O_2, 9°C → PhCHO 99%

Mori, K.; Yamaguchi, K.; Hara, T.; Mizugaki, T.; Ebitani, K.; Kaneda, K. *J. Am. Chem. Soc.* **2002**, *124*, 11572.

$PhCH_2OH$ — RuO_2-zeolite catalyst → PhCHO quant

Zhan, B.-Z.; White, M.A.; Sham, T.-K.; Pincock, J.A.; Doucet, R.J.; Rao, K.V.R.; Robertson, K.N.; Cameron, T.S. *J. Am. Chem. Soc.* **2003**, *125*, 2195.

$PhCH_2OH$ — TEMPO, Br_2, DCM, $NaNO_2$, 80°C; air → PhCHO 95%

Liu, R.; Liang, X.; Dong, C.; Bu, X. *J. Am. Chem. Soc.* **2004**, *126*, 4112.

$PhCH_2OH$ — (polymer)-$MoO_2(acac)_2/O_2$, toluene, 100°C, 12 h → PhCHO 86%

Velusamy, S.; Ahamed, M.; Punniyamurthy, T. *Org. Lett.* **2004**, *6*, 4821.

(E)-crotyl alcohol (OH) + (IBX bearing CO_2H, water soluble) — H_2O, rt, 18 h → crotonaldehyde (CHO) 86%

Thottumkara, A.P.; Vinod, T.K. *Tetrahedron Lett.* **2002**, *43*, 569.

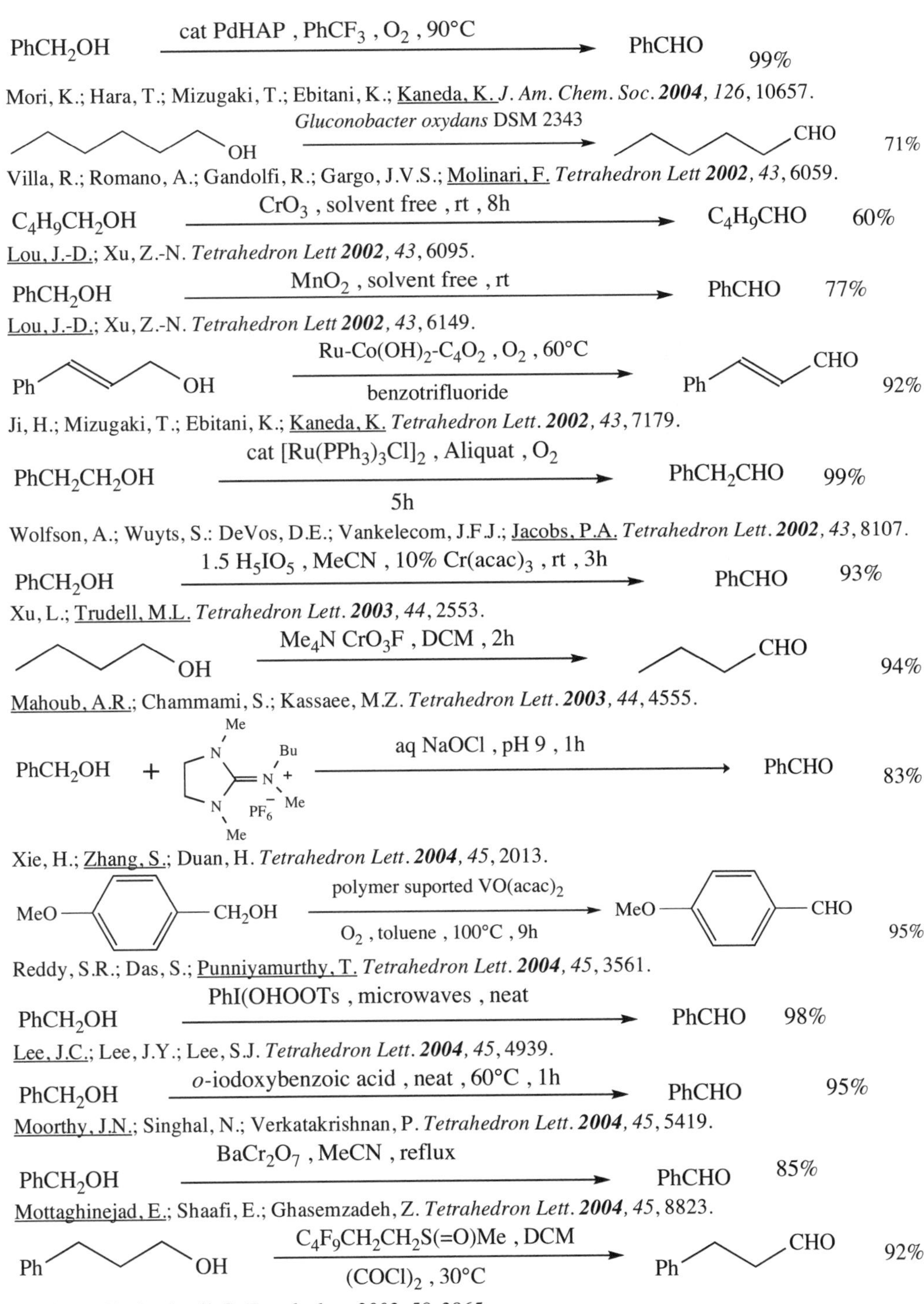
PhCH2OH
cat PdHAP , PhCF3 , O2 , 90°C
PhCHO
99%
Mori, K.; Hara, T.; Mizugaki, T.; Ebitani, K.; Kaneda, K. J. Am. Chem. Soc. 2004, 126, 10657.
OH
Gluconobacter oxydans DSM 2343
CHO
71%
Villa, R.; Romano, A.; Gandolfi, R.; Gargo, J.V.S.; Molinari, F. Tetrahedron Lett 2002, 43, 6059.
C4H9CH2OH
CrO3 , solvent free , rt , 8h
C4H9CHO
60%
Lou, J.-D.; Xu, Z.-N. Tetrahedron Lett 2002, 43, 6095.
PhCH2OH
MnO2 , solvent free , rt
PhCHO
77%
Lou, J.-D.; Xu, Z.-N. Tetrahedron Lett 2002, 43, 6149.
Ph
OH
Ru-Co(OH)2-C4O2 , O2 , 60°C
benzotrifluoride
Ph
CHO
92%
Ji, H.; Mizugaki, T.; Ebitani, K.; Kaneda, K. Tetrahedron Lett. 2002, 43, 7179.
PhCH2CH2OH
cat [Ru(PPh3)3Cl]2 , Aliquat , O2
5h
PhCH2CHO
99%
Wolfson, A.; Wuyts, S.: DeVos, D.E.; Vankelecom, J.F.J.; Jacobs, P.A. Tetrahedron Lett. 2002, 43, 8107.
PhCH2OH
1.5 H5IO5 , MeCN , 10% Cr(acac)3 , rt , 3h
PhCHO
93%
Xu, L.; Trudell, M.L. Tetrahedron Lett. 2003, 44, 2553.
OH
Me4N CrO3F , DCM , 2h
CHO
94%
Mahoub, A.R.; Chammami, S.; Kassaee, M.Z. Tetrahedron Lett. 2003, 44, 4555.
PhCH2OH +
Me
N
Bu
N +
Me
N
PF6−
Me
aq NaOCl , pH 9 , 1h
PhCHO
83%
Xie, H.; Zhang, S.; Duan, H. Tetrahedron Lett. 2004, 45, 2013.
MeO
CH2OH
polymer suported VO(acac)2
O2 , toluene , 100°C , 9h
MeO
CHO
95%
Reddy, S.R.; Das, S.; Punniyamurthy, T. Tetrahedron Lett. 2004, 45, 3561.
PhCH2OH
PhI(OHOOTs , microwaves , neat
PhCHO
98%
Lee, J.C.; Lee, J.Y.; Lee, S.J. Tetrahedron Lett. 2004, 45, 4939.
PhCH2OH
o-iodoxybenzoic acid , neat , 60°C , 1h
PhCHO
95%
Moorthy, J.N.; Singhal, N.; Verkatakrishnan, P. Tetrahedron Lett. 2004, 45, 5419.
PhCH2OH
BaCr2O7 , MeCN , reflux
PhCHO
85%
Mottaghinejad, E.; Shaafi, E.; Ghasemzadeh, Z. Tetrahedron Lett. 2004, 45, 8823.
Ph
OH
C4F9CH2CH2S(=O)Me , DCM
(COCl)2 , 30°C
Ph
CHO
92%
Crich, D.; Neelamkavil, S. Tetrahedron 2002, 58, 3865.

$O_2N-C_6H_4-CH_2OH$ —[Me_2S, TEMPO, PhCl, 90°C / $C_8F_{17}Br$, $CuBr{\bullet}SMe_2$, 1.5h]→ $O_2N-C_6H_4-CHO$ 93%

Ragagnin, G.; Betzemeieer, B.; Quici, S.; Knochel, P. *Tetrahedron* **2002**, *58*, 3985.

$C_9H_{19}CH_2OH$ —[Magtriere, microwaves, 20 min]→ $C_9H_{19}CHO$ 90%

Bogdal, D.; Lukasiewicz, M.; Pielichowski, J.; Mikciak, A.; Bednarz, Sz. *Tetrahedron* **2003**, *59*, 649.

$Cl-C_6H_4-CH_2OH$ —[5% PhSNH*t*-Bu, NCS, 0°C / 10 K_2CO_3, MS 4Å, DCM, 1h]→ $Cl-C_6H_4-CHO$ 93%

Matsuo, J.-i.; Iida, D.; Yamanaka, H.; Mukaiyama, T. *Tetrahedron* **2003**, *59*, 6739.

$C_7H_{15}CH_2OH$ —[cat $Ru_3(CO)_{12}$, tolane, 130°C, 15h]→ $C_7H_{15}CHO$ 80%

Meijer, R.H.; Lighart, G.B.W.L.; Meuldijk, J.; Vekemans, J.A.J.M.; Hulshot, L.A.; Mills, A.M.; Kooijman, H.; Spek, A.L. *Tetrahedron* **2004**, *60*, 1065.

$PhCH_2OH$ —[O_2, N-hydroxyphthalimide, MeCN, mcpba / cat $Co(OAc)_2$, rt]→ PhCHO 92%

Miniscui, F.; Punta, C.; Recupero, F.; Fontana, F.; Pedulli, G.F. *Chem. Commun.* **2002**, 688.

$PhCH_2OH$ —[5% $Pd(OAc)_2$, NEt_3, DCE, O_2, 25°C]→ PhCHO 84%

Schultz, M.J.; Park, C.C.; Sigman, M.S. *Chem. Commun.* **2002**, 3034.

1-pentanol (OH) —[1% $AuSiO_2$]→ CHO 63%

Biella, S.; Rossi, M. *Chem. Commun.* **2003**, 378.

$PhCH_2OH$ —[cat $CuBr_2$(bipy)-TEMPO, air, 25°C / cat *t*-BuOK, $MeCN/H_2O$]→ PhCHO quant

Gamez, P.; Arends, I.W.C.E.; Reedijk, J.; Sheldon, R.A. *Chem. Commun.* **2003**, 2414.

$Br-C_6H_4-CH_2OH$ —[(*o*-Tol)$_3BCl_2$, DBU, rt / toluene, 30 min]→ $Br-C_6H_4-CHO$ 94%

Mantano, Y.; Nomura, H. *Angew. Chem. Int. Ed.* **2002**, *41*, 3028.

$PhCH_2OH$ —[cat Ru/Al_2O_3, O_2, $PhCF_3$, 356 K, 1h]→ PhCHO 99%

Yamaguchi, K.; Mizuno, N. *Angew. Chem. Int. Ed.* **2002**, *41*, 4538.

$C_9H_{19}CH_2OH$ —[5% CuCl, phenanthroline, 5% DBAB, C_6H_5F / 5% *t*-BuOK, O_2, 10% DMAP, 70-80°C]→ $C_9H_{19}CHO$ 51%

Markó, I.E.; Gautier, A.A.; Dumeunier, R.; Doda, K.; Philippart, F.; Brown, S.M.; Urch, C.J. *Angew. Chem. Int. Ed.* **2004**, *43*, 1588.

Ph–CH=CH–CH$_2$OH —[H_5IO_6, TEMPO, DCM / rt, 6h]→ Ph–CH=CH–CHO 96%

Kim, S.S.; Nhru, K. *Synlett* **2002**, 616.

$PhCH_2OH$ —[48% HBr, DMSO, 100°C, 3 h]→ PhCHO 95%

Li, C.; Xu, Y.; Lu, M.; Zhao, Z.; Liu, L.; Zhao, Z.; Cui, Y.; Zheng, P.; Ji, X.; Gao, G. *Synlett* **2002**, 2041.

$PhCH_2OH$ + [IBX: 1-hydroxy-1,2-benziodoxol-3(1H)-one 1-oxide] —bmim Cl , rt , 15 min→ PhCHO 85%

Karthikeuan, G.; Perumal, P.T. *Synlett* **2003**, 2249.

$PhCH_2OH$ —Oxone , NaBr , 3 h / $MeCN/H_2O$→ PhCHO 87%

Koo, B.-S.; Lee, C.K.; Lee, K.-J. *Synth. Commun.* **2002**, *32*, 2115.

Ph–CH=CH–CH₂OH —*i*-PrOH , KOH , reflux , 2 h , 356 K / C,O-ZrO_2 catalyst→ Ph–CH=CH–CHO 80%

Sonovane, S.U.; Jayaram, R.V. *Synlett* **2004**, 146.

$PhCH_2OH$ —0.2 TEMPO , acetone / (polymer)–$I(OAc)_2$→ PhCHO 90%

Sakuraathani, K.; Togo, H. *Synthesis* **2003**, 21.

$PhCH_2OH$ —silica gel , $Co(NO_3)_2$, 9 min / microwaves→ PhCHO 91%

can also prepare ketones

Kiasat, A.R.; Kazemi, F.; Rafati, M. *Synth. Commun.* **2003**, *33*, 601.

$PhCH_2OH$ —$KMnO_4$, bmim BF_4 , rt , 1 h→ PhCHO 90%

Kumar, A.; Jain, N.; Chauhan, S.M.S. *Synth. Commun.* **2004**, *34*, 2835.

$PhCH_2OH$ —2 NCS , DCM , 2 *i*-Pr_2NEt , 0°C , 1 h / 20% (polymer)–O–C_6H_4–C(Cl)=N-*t*-Bu→ PhCHO 94%

Matsuo, J.-i.; Kawana, A.; Pudhom, K.; Mukaiyama, T. *Chem. Lett.* **2002**, 250.

Tr–CH₂CH₂CH₂–OH —5% 2-NO_2-C_6H_4-S-NH*t*-Bu / 1.1 Br_2 , K_2CO_3 , MS 4Å / DCM , rt , 1 h→ Tr–CH₂CH₂–CHO 90%

also for oxidation of 2° alcohols

Matsuo, J.; Kawana, A.; Yamanaka, H.; Mukaiyama, T. *Chem Lett.* **2003**, *32*, 182.

Cl–C_6H_4–CH_2OH —NaOCl , KBr-H_2O , 8 h / β-cyclodextrin→ Cl–C_6H_4–CHO 94%

Surendra, K.; Krishnaveni, N.S.; Rama Rao, K. *Can. J. Chem.* **2004**, *82*, 1230.

C_6H_5–CH_2OH —aluminosilicate KSF/0 , 1.5 h→ C_6H_5–CHO 94%

Farkas, J.; Békássy,S.; Madarász, J.; Figueras, F. *New J. Chem.* **2002**, *26*, 750.

$PhCH_2OH$ —1.5 Ph(Cl)S=N*t*-Bu , 2 DBU , –78°C / DCM , 30 min→ PhCHO 98%

Matsuo, J.-i.; Iida, D.; Tatani, K.; Mukaiyama, T. *Bull. Chem. Soc. Jpn.* **2002**, *75*, 223.

REVIEWS:

"Palladium Catalyzed Oxidation of Primary and Secondary Alcohols"
Murart, J. *Tetrahedron* ***2003**, 59*, 5789.

"Recent Developments in the Aerobic Oxidation of Alcohols"
Zhan, B.-Z.; Thompson, A. *Tetrahedron* ***2004**, 60*, 2917.

"Oxidation of Alcohols with Molecular Oxygen on Solid Catalysts"
Mallat, T.; Baiker, A. *Chem. Rev.* ***2004**, 104*, 3037.

"Green, Catalytic Oxidations of Alcohols"
Sheldon, R.A.; Arends, I.W.C.E.; ten Brink, G.-J.; Dijksman, A. *Acc. Chem. Res.* ***2002**, 35*, 774.

"Preparation of Tetramethylpiperidine-1-oxammonium Salts and Their Use as Oxidants"
Merbouh, N.; Bobbitt, J.M.; Brückner, C. *Org. Prep. Proceed. Int.* ***2004**, 36*, 1.

SECTION 49: ALDEHYDES FROM ALDEHYDES

Conjugate reductions and Michael alkylations of conjugated aldehydes are listed in Section 74 (Alkyls, Methylenes, and Aryls from Alkenes).

$C_6H_{13}CHO$ —[PhBr, 5% $PdOAc)_2$-$P(t\text{-}Bu)_3$; Cs_2CO_3, dioxane, 2h]→ $C_6H_{13}CH(Ph)CHO$ 77%

Terrao, Y.; Fukuoka, Y.; Satoh, T.; Miura, M.; Nomura, M. *Tetrahedron Lett.* ***2002**, 43*, 101.

$(EtO_2C)_2C(CH_2CHO)(CH_2CH_2CH_2CH_2I)$ —[(S)-2-methylproline (CO_2H, Me), NEt_3, –30°C; $CHCl_3$, 1 d]→ diethyl 2-formylcyclopentane-1,1-dicarboxylate (OHC, EtO_2C, EtO_2C) 92% (95% ee)

Vignola, N.; List, B. *J. Am. Chem. Soc.* ***2004**, 126*, 450.

Related Methods:

Aldehydes from Ketones (Section 57)
Ketones from Ketones (Section 177)
Also via: Alkene-Aldehydes (Section 341)

SECTION 50: ALDEHYDES FROM ALKYLS, METHYLENES AND ARYLS

NO ADDITIONAL EXAMPLES

SECTION 51: ALDEHYDES FROM AMIDES

NO ADDITIONAL EXAMPLES

SECTION 52: ALDEHYDES FROM AMINES

NO ADDITIONAL EXAMPLES

Related Method: Ketones from Amines (Section 172)

SECTION 53: ALDEHYDES FROM ESTERS

MeO–C$_6$H$_4$–CH$_2$CH$_2$C(O)SEt → (5% Pd/C, 3 Et_3SiH, acetone, rt, 10 min) MeO–C$_6$H$_4$–CH$_2$CH$_2$CHO 90%

Miyazaki, T.; Han-ya, Y.; Tokuyama, H.; Fukuyama, T. *Synlett* ***2004***, 477.

MeO_2C–naphthalene–CO_2Me → (2-methoxyethoxy Al hydride, 2.5 NMP, toluene) OHC–naphthalene–CHO 84%

Hagiya, K.; Mitsui, S.; Taguchi, H. *Synthesis* ***2003***, 823.

SECTION 54: ALDEHYDES FROM ETHERS, EPOXIDES, AND THIOETHERS

(vinyl, PMPO, OSEM, OMe tetrahydrofuran) → ("Cp_2Zr", BF_3•OEt_2) (vinyl, OH, PMPO, OSEM cyclobutane)

Paquette, L.A.; Cunière, N. *Org. Lett.* ***2002***, *4*, 1927.

(N-acryloyl oxazolidinone, Ph) → (PhCHO, Et_2Al-I, DCM, 0°C) (Ph, OH, I, Ph aldol product) 91% (>85% de)

Pal, M.; Parasuraman, K.; Yeleswarapu, K.R. *Org. Lett.* ***2003***, *5*, 349.

$PhCH_2OSiMe_3$ → (cat. $Co(CO_2C_9H_{19})_2$, cat NHPI, rt, 1 atm O_2, MeCN) PhCHO 92%

NHPI = *N*-hydroxyphthalimide

Karimi, B.; Rajabi, J. *Org. Lett.* ***2004***, *6*, 2841.

(Ph, N, N, Ph aziridinyl hydrazone ether, OMe) → (cat $Rh_2(OAc)_4$, DCE, 130°C) (H, O, Ph, OMe aldehyde) 82%

May, J.A.; Stolz, B.M. *J. Am. Chem. Soc.* ***2002***, *124*, 12426.

OSiPh$_3$ Cr(tpp)OTf , DCE , 83°C OSiPh$_3$ CHO 98%

Suda, K.; Kikkawa, T.; Nakajima, S.-i.; Takanami, T. *J. Am. Chem. Soc.* **2004**, *126*, 9554.

Ph O $(CH_2{=}CHCH_2)_4Sn$, cat $Bi(OTf)_3$, DCM rt , 0.5h Ph OH 87%

Yadav, J.S.; Reddy, B.V.S.; Satheesh, G. *Tetrahedron Lett.* **2003**, *44*, 6501.

Ph O 1% $IrCl_3{\bullet}x\ H_2O$, THF , 50°C Ph CHO quant

Karamé, I.; Tommasino, M.L.; Lomaire, M. *Tetrahedron Lett.* **2003**, *44*, 7687.

BnO O 0.2 $ZrCl_4$, 0.2 NaI , MeCN reflux , 1.5 h BnO OH 91%

Sharma, G.V.M.; Reddy, Ch.G.; Krishna, P.R. *Synlett* **2003**, 1728.

O 1% $Er(OTf)_3$, DCM , rt reflux CHO >99%

Procopio, A.; Dalpozzo, R.; DeNino, A.; Nardi, M.; Sindona, G.; Tagarelli, A. *Synlett* **2004**, 2633.

$PhCH_2OTBDMS$ β-cyclodextrin , NBS , H_2O PhCHO 80%

Reddy, M.S.; Narender, M.; Nageswar, Y.V.D.; Rao, K.R. *Synthesis* **2004**, 714.

Related Method: Ketones from Ethers, Epoxides, and Thioethers (Section 174)

SECTION 55: ALDEHYDES FROM HALIDES AND SULFONATES

Br N I C_3H_7CHO , *i*-PrMgCl , THF other electrophiles can be used Br N C_3H_7 OH 69%

Song, J.J.; Yee, N.K.; Tan, Z.; Xu, J.; Kapadia, S.R.; Senanayake, C.H. *Org. Lett.* **2004**, *6*, 4905.

$PhCHBr_2$ DMSO , 100°C , 2h PhCHO 95%

Li, W.; Li, J.; DeVincentis, D.; Masour, T.S. *Tetrahedron Lett.* **2004**, *45*, 1071.

I $NaHCO_2^-$/DMF , CO , 90°C , 8h a silica supported Pd catalyst CHO 56%

Cai, M.-Z.; Zhao, H.; Zhou, J.; Song, C.-S. *Synth. Commun.* **2002**, *32*, 923.

$Ph_3\overset{+}{P}$–(cyclohexene)–$\overset{+}{P}Ph_3$ S_2O_8, MeCN , H_2O

$PhCH_2Cl$ → (reflux , 123 min) PhCHO 97%

Badri, R.; Soleymani, M. *Synth. Commun.* **2003**, *33*, 1325.

SECTION 56: ALDEHYDES FROM HYDRIDES

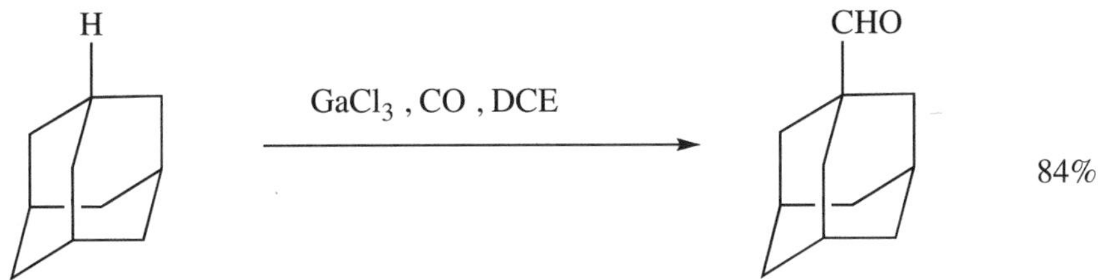

Oshita, M.; Chatani, N. *Org. Lett.* **2004**, *6*, 4323.

SECTION 57: ALDEHYDES FROM KETONES

NO ADDITIONAL EXAMPLES

SECTION 58: ALDEHYDES FROM NITRILES

PhCN → 1. PtO_2 , HCOOH , H_2O , 3 h; 2. H_3O^+ → PhCHO 94%

Xi, F.; Kamal, F.; Schenerman, M.A. *Tetrahedron Lett.* **2002**, *43*, 1395.

SECTION 59: ALDEHYDES FROM ALKENES

(OTBS alkene) → 0.02 OsO_4 , 4 $NaIO_4$; 2 eq 2,6-lutidine , 20 h; aq dioxane → (CHO, OTBS) 99%

Yu, W.; Mei, Y.; Kang, Y.; Hua, Z.; Jin, Z. *Org. Lett.* **2004**, *6*, 3217.

Br(alkene) → Rh catalyst , toluene , 70°C; 10 bar CO/H_2 → Br(CHO) no yield

Breit, B.; Seiche, W. *J. Am. Chem. Soc.* **2003**, *125*, 6608.

Ph(alkene) → CO/H_2(100 psi) , DCM , 22 h; Rh-dendrimer catalyst , 25°C → Ph(branched CHO) + Ph(linear CHO) (36 : 1) >99%

Li, S.-M.; Alper, H. *J. Am. Chem. Soc.* **2003**, *125*, 13126.

Related Method: Ketones from Alkenes (Section 179)

SECTION 60: ALDEHYDES FROM MISCELLANEOUS COMPOUNDS

MeO–C$_6$H$_4$–CH=N–OH → (BAcIM BF_4 (imidazolium salt); SiO_2, 72h) → MeO–C$_6$H$_4$–CH_2CHO 89%

Li, D.; Shi, F.; Guo, S.; Deng, Y. *Tetrahedron Lett.* ***2004**, 45*, 265.

MeO–C$_6$H$_4$–CH=N-OH → (DMIm BF_4, 96h; SiO_2, H_2O) → MeO–C$_6$H$_4$–CHO 99%

Li, D.; Shi, F.; Deng, Y. *Tetrahedron Lett.* ***2004**, 45*, 6791.

Ph–(2-oxazoline) → (1. $NiCl_2$, $NaBH_4$; 2. H_3O^+) → PhCHO 90%

Babu, M.S.; Rai, K.M.L. *Tetrahedron Lett.* ***2004**, 45*, 7969.

Cl–C$_6$H$_4$–CH=N-OH → ($(NH_4)_2Cr_2O_7$, $ZrCl_4$, 80°C; wet SiO_2, 10 min) → Cl–C$_6$H$_4$–CHO 80%

Shirini, F.; Zolfigol, M.A.; Pourhabib, A. *Synth. Commun.* ***2002**, 32*, 2837.

MeO–C$_6$H$_4$–CH=N-OH → ($K_5CoW_{12}O_{40}$•3 H_2O, MeCN; rt, 1 h) → MeO–C$_6$H$_4$–CHO 95%

Bose, D.S.; Reddy, A.V.N.; Das, A.P.R. *Synthesis* ***2003***, 1883.

Ph(H)C=N-OH → ($KMnO_4$, Al_2O_3, 50°C m 40 min) → Ph(H)C=O 78%

Imanzadeh, G.H.; Hajipour, A.R.; Mallakpour, S.E. *Synth. Commun.* ***2003**, 33*, 735.

Cl–C$_6$H$_4$–CH=N-OH → (30% H_2O_2, acetone, 5 h; 10% Na tungstate) → Cl–C$_6$H$_4$–CHO 98%

Manjula, A.; Reddy, G.N.; Rao, B.C. *Synth. Commun.* ***2003**, 33*, 3455.

PhCH=N-OH + (IBX: 1-hydroxy-1,2-benziodoxol-3(1H)-one 1-oxide) → (H_2O, β-cyclodextrin, rt) → PhCHO 92%

Krishnaveni, N.S.; Surendrak, K.; Nageswar, Y.V.D.; Rao, K.R. *Synthesis* ***2003***, 1968.

REVIEWS:

"Recent Synthetic Development in the Nitro to Carbonyl Conversion (Nef Reaction)"
Ballini, R.; Petrini, M. *Tetrahedron* ***2004**, 60*, 1017.

"Recent Advances in Rhodium-Catalyzed Asymmetric Hydroformylation Using Phosphite"
Diéguez, M.; Pàmies, O.; Claver, C. *Tetrahedron: Asymmetry* ***2004**, 15*, 2113.

"The Pummerer Reaction: Methodology and Strategy for the Synthesis of Heterocyclic Compounds"
Bur, S.K.; Padwa, A. *Chem. Rev.* ***2004***, *104*, 2401.

SECTION 60A: PROTECTION OF ALDEHYDES

Ph—CH(OMe)$_2$ —[$Bi(OTf)_3 \bullet xH_2O$, 1 h, rt / THF, H_2O]→ Ph—CHO 84%

Carrigan, M.D.; Sarapa, D.; Smith, R.C.; Wieland, L.C.; Mohan, R.S. *J. Org. Chem.* ***2002***, *67*, 1027.

Ph–(1,3-dithiolane) —[SiO_2–Cl, DMSO, DCM, rt, 40 min]→ PhCHO 94%

Firouzabadi, H.; Iranpoor, N.; Hazarkhani, H.; Karimi, B. *J. Org. Chem.* ***2002***, *67*, 2572.

also with ketones

PhC(O)H —[$(MeO)_3CH$, MeOH, 1 h / $Bi(OTf)_3 \bullet 4\ H_2O$]→ PhCH(OMe)(OMe) 82%

Leonard, N.M.; Oswald, M.C.; Freiberg, D.A.; Nattier, B.A.; Smith, R.C.; Mohan, R.S. *J. Org. Chem.* ***2002***, *67*, 5202.

PhCHO —[ethylene glycol, MeCN / $Bu_4N^+\ Br_3^-$, 0.08 h]→ Ph–(1,3-dioxolane) 92%

Gopinathi, R.; Haque, S.K.J.; Patel, B.K. *J. Org. Chem.* ***2002***, *67*, 5842.

PhCH)OMe)$_2$ —[β-cyclodextrin–H_2O, 50°C, 8 h]→ PhCHO 80%

Krishnaveni, N.S.; Surendra, K.; Reddy, M.A.; Nageswar, Y.V.D.; Rao, K.R. *J. Org. Chem.* ***2003***, *68*, 2018.

$CH_3(CH_2)_8CH(OM)e_2$ —[1. 2 Me_3SiOTf, 3 2,6-lutidine, DCM, 0°C; 2. H_2O]→ $CH_3(CH_2)_8CHO$ 83%

Fujioka, H.; Sawama, Y.; Murata, N.; Kitsu, T.; Kubo, O.; Matsuda, S.; Kita, Y. *J. Am. Chem. Soc.* ***2004***, *126*, 11800.

$CH_3(CH_2)_3CHO$ —[2 EtSH, DCM, 50 min / 4% $Sc(OTf)_3$]→ $CH_3(CH_2)_3CH(SEt)_2$ 90%

Kamal, A.; Chouhan, G. *Tetrahedron Lett.* ***2002***, *43*, 1347.

PhCH=N–NH–C(O)NH$_2$ —[$Mg(HSO_4)_2$, wet SiO_2, rt, 30 min; also with oximes and hydrazones]→ PhCHO 94%

Shivini, F.; Zolfigl, M.A.; Mallapour, B.; Mallakpour, S.E.; Hajipour, A.R.; Baltork, I.M. *Tetrahedron Lett.* ***2002***, *43*, 1555.

PhCH(OAc)$_2$ —[$Zr(MePO_3)_{1.2}(O_3PC_6H_4SO_3H)_{0.8}$, dioxane, 50°C]→ PhCHO 87%

PhCHO —[$Zr(MePO_3)_{1.2}(O_3PC_6H_4SO_3H)_{0.8}$, Ac_2O, rt]→ PhCH(OAc)$_2$ 84%

Curini, M.; Epifano, F.; Marcotullio, M.C.; Rosati, O.; Nocchetti, M. *Tetrahedron Lett.* ***2002***, *43*, 2709.

MeO—C_6H_4—CHO (85%) ⇌ [$HOCH_2CH_2SH$, TBATB , DCM; reverse: cat TMATB , DCM] MeO—C_6H_4—(1,3-oxathiolan-2-yl) 65%

Mondal, E.; Sahu, P.R.; Bose, G.; Khan, A.T. *Tetrahedron Lett.* ***2002***, *43*, 2843.

PhCHO → [$HSCH_2CH_2SH$, cat $NiCl_2$ / DCM-MeOH , rt] Ph—(1,3-dithiolan-2-yl) 96%

Khan, A.T.; Mondal, E.; Sahu, P.R.; Islam, S. *Tetrahedron Lett.* ***2003***, *44*, 919.

PhCHO → [cat $Bi(NO_3)_3$, THF , MeOH , 6h] $PhCH(OMe)_2$ 75%

also with ketones

Srivastava, N.; Dasgupta, S.K.; Banik, B.K. *Tetrahedron Lett* ***2003***, *44*, 1191.

PhCHO → [10% $MoO_2(acac)_2$, MeCN , $HSCH_2CH_2OH$ / 4h , rt] Ph—(1,3-oxathiolan-2-yl) 86%

Rana, K.K.; Guin, C.; Jana, S.; Roy, S.C. *Tetrahedron Lett.* ***2003***, *44*, 8597.

PhCHO → [$HSCH_2CH_2CH_2SH$, 5% $CoCl_2$, MeCN / reflux , rt] Ph—(1,3-dithian-2-yl) 91%

De, S.K. *Tetrahedron Lett.* ***2004***, *45*, 1035.

PhCHO → [$HSCH_2CH_2CH_2SH$, 5% $Yb(OTf)_3$, MeCN / rt , 45 min] Ph—(1,3-dithian-2-yl) 89%

De, S.K.; *Tetrahedron Lett.* ***2004***, *45*, 2339.

PhCHO → [$CoCl_2$, MeOH , reflux , 2h] $PhCH(OMe)_2$ 78%

Velusamy, S.; Punniyamurthy, T. *Tetrahedron Lett.* ***2004***, *45*, 4917.

PhCHO → [MeOH , cat $RuCl_3$, 5h] Ph—CH(OMe)(OMe) 85%

De, S.K.; Gibbs, R.A. *Tetrahedron Lett.* ***2004***, *45*, 8141.

Ph—C_6H_4—CHO → [Ac_2O , cat $ZrCl_4$, neat , rt / 30 min] Ph—C_6H_4—CH(OAc)(OAc) 90%

Smitha, G.; Reddy, Ch.S. *Tetrahedron* ***2003***, *59*, 9571.

MeO—C_6H_4—CHO ⇌ [$HOCH_2CH_2SH$, cat $HClO_4$, DCM , 0°C; reverse: H_2MoO_4•H_2O , H_2O_2 , NH_4Br , $HClO_4$ (92%)] MeO—C_6H_4—(1,3-oxathiolan-2-yl) 68%

Mondal, E.; Sahu, P.R.; Khan, A.T. *Synlett* ***2002***, 463.

PhCHO → [$HSCH_2CH_2SH$, 15% NBS , DCM / 40 min] Ph—(1,3-dithiolan-2-yl) 80%

Kamal, A.; Chouhan, G. *Synlett* ***2002***, 474.

PhCHO $\xrightarrow{Ac_2O\text{ , MeCN , cat }Bi(NO_3)_3\bullet 5\ H_2O\text{ , 1.5h}}$ $PhCH(OAc)_2$ 87%

Aggen, D.H.; Arnold, J.N.; Hayes, P.D.; Smoter, N.J.; Mohan, R.S. *Tetrahedron* ***2004***, *60*, 3675.

PhCHO $\xrightarrow{Ac_2O\text{ , }LiBF_4\text{ , MeCN , rt , 4.5 h}}$ $PhCH(OAc)_2$ 90%

Yadav, J.S.; Reddy, B.V.S.; Venugopal, C.; Ramalingam, V.T. *Synlett* ***2002***, 604.

$PhCH(OMe)_2$ $\xrightarrow{\text{EtSH , cat }InCl_3\text{ , DCE}}$ $PhCH(SEt)_2$ 91%

Ranu, B.C.; Das, A.; Samanta, S. *Synlett* ***2002***, 727.

PhCHO $\xrightarrow{\text{CAN , }Ac_2O\text{ , rt}}$ $PhCH(OAC)_2$ 95%

Roy, S.C.; Banerjee, B. *Synlett* ***2002***, 1677.

AcO–C6H4–(1,3-dithian-2-yl) $\xrightarrow[0°C \rightarrow rt]{NaNO_2\text{ , AcCl , }H_2O\text{ , DCM}}$ AcO–C6H4–CHO 85%

Khan, A.T.; Mandal, E.; Sahu, P.R. *Synlett* ***2003***, 377.

2-(1-naphthyl)-1,3-dithiane → 1-naphthaldehyde (CHO) 83%

1. Ph–S(=O)–N(piperidine), DCM, –60°C
2. Tf_2O , –60°C
3. THF/H_2O , –60°C → rt

Crich, D.; Picione, J. *Synlett* ***2003***, 1257.

PhCH=N-OH $\xrightarrow{\text{pyridinium }CrO_3F\text{ , wet alumina , 20 min}}$ Ph-CHO 90%

Ganguly, N.C.; De, P.; Sukai, A.K.; De, S. *Synth. Commun.* ***2002***, *32*, 1.

5-methylfuran-2-CH=N—OH $\xrightarrow[\text{microwaves , 30 sec}]{N\text{-Bn-1,4-azabicycloammonium dichromate}}$ 5-methylfuran-2-CHO 89%

Hajipour, A.R.; Mallakpour, S.E.; Khoee, S. *Synth. Commun.* ***2002***, *32*, 9.

O_2N–C6H4–CH=N-OH $\xrightarrow[\text{15 min}]{HIO_3\text{ , wet }SiO_2\text{ , steam bath}}$ O_2N–C6H4–CHO 90%

Shirini, F.; Zolfigol, M.A.; Hzadbar, M.R. *Synth. Commun.* ***2002***, *32*, 315.

PhCHO $\xrightarrow[\text{3 min , rt}]{1\%\ WCl_6\text{ , }Ac_2O\text{ , solvent free}}$ $PhCH(OAc)_2$ 98%

Karimi, B.; Ebrahimian, G.-R.; Seradj, H. *Synth. Commun.* ***2002***, *32*, 669.

PhCHO $\xrightarrow[\text{2.5 h , rt}]{HSCH_2CH_2SH\text{ , 10\% }InCl_3\text{ , DCM}}$ Ph–(1,3-dithiolan-2-yl) 90%

also with ketones

Yadav, J.S.; Reddy, B.V.S.; Pandey, S.K. *Synth. Commun.* ***2002***, *32*, 715.

PhCHO $\xrightarrow[\text{30 min}]{Ac_2O\text{ , 10 }InCl_3\text{ , DCM , rt}}$ $PhCH(OAc)_2$ 95%

Yadav, J.S.; Reddy, B.V.S.; Srinivas, Ch. *Synth. Commun.* ***2002***, *32*, 1175.
Yadav, J.S.; Reddy, B.V.S.; Srinivas, Ch. *Synth. Commun.* ***2002***, *32*, 2169.

PhCH=N-OH → (Cr-MCM-41 / SiO_2, microwaves) → PhCH=O 98%

Nagarapu, L.; Ravirala, N.; Akkewar, D. *Synth. Commun.* ***2002**, 32,* 2195.

PhCHO + HO–CH₂CH₂–SH → (cat CAN, MeCN, rt, 10 h) → 2-phenyl-1,3-oxathiolane 81%

Maiti, G.; Roy, S.C. *Synth. Commun.* ***2002**, 32,* 2269.

PhCHO → ($HC(OMe)_3$, 10% $LiBF_4$, MeOH) → $PhCH(OMe)_2$ quant

Hamada, N.; Kazahaya, K.; Shimizu, H.; Sato, T. *Synlett* ***2004***, 1074.

C_3H_7CHO → (Ac_2O, $FeCl_3$, SiO_2, microwaves, 3 min) → $C_3H_7CH(OAc)_2$ 78%

Wang, C.; Li, M. *Synth. Commun.* ***2002**, 32,* 3469.

PhCHO → (HS–CH₂CH₂CH₂–SH, 10% $LiBF_4$, 25°C, neat) → 2-phenyl-1,3-dithiane quant

Kazahaya, K.; Tsuji, S.; Sato, T. *Synlett* ***2004**,* 1640.

PhCHO → (Ac_2O, 1% $InBr_3$, neat, rt) → $PhCH(OAc)_2$

Yin, L.; Zhang, Z.-H.; Wang, Y.-M.; Pang, M.-L. *Synlett* ***2004**,* 1727.

c-C_6H_{11}CHO → ($HOCH_2CH_2SH$, 0.5 PPA/SiO_2) → 2-cyclohexyl-1,3-oxathiolane 96%

Aoyama, T.; Takido, T.; Kodomari, M. *Synlett* ***2004**,* 2307.

PhCHO → ($HOCH_2CH_2SH$, cat $Yb(OTf)_3$, bmim PF_6, rt) → 2-phenyl-1,3-oxathiolane 95%

Kumar, A.; Jain, N.; Rana, S.; Chauhan, S.M.S. *Synlett* ***2004**,* 2785.

PhCHO → ($HSCH_2CH_2CH_2SH$, 2.5 min, cat $H_3PW_{12}O_{40}$) → 2-phenyl-1,3-dithiane 98%

Firouzabadi, H.; Iranpoor, N.; Amani, K. *Synthesis* ***2002**,* 59.

2-phenyl-1,3-dithiolane + IBX (O, OH) → (β-cyclodextrin, H_2O, rt, 1 h) → PhCHO 90%

Krishnaveni, N.S.; Surendra, K.; Nagswar, Y.V.D.; Rao, K.R. *Synthesis* ***2003***, 2295.

PhCHO → ($HSCH_2CH_2CH_2SH$, $Al(OTf)_3$, rt, no solvent) → 2-phenyl-1,3-dithiane 97%

Firouzabadi, H.; Iranpoor, N.; Kohmarch, G. *Synth. Commun.* ***2003**, 33,* 167.

PhCHO $\xrightarrow{\text{Aberlyst 15, } Ac_2O\text{, DCM, rt}}$ $PhCH(OAc)_2$ 95%

Reddy, A.V.; Racinder, K.; Reddy, V.L.N.; Ravikanth, V.; Venkatewarlu, Y. *Synth. Commun.* **2003**, *33*, 1531.

MeO–C₆H₄–(1,3-dithiolan-2-yl) $\xrightarrow[\text{reflux, 3 h}]{FeCl_3\text{, KI, MeOH}}$ MeO–C₆H₄–CHO 90%

Chavan, S.P.; Soni, P.B.; Kale, R.R.; Pasupathy, K. *Synth. Commun.* **2003**, *33*, 879.

PhCHO $\xrightarrow[\text{DCE, heat, 1.2 h}]{HSCH_2CH_2SH\text{, } SiO_2\bullet AlCl_3}$ Ph–(1,3-dithiolane) 95%

Tamami, B.; Borujeny, K.P. *Synth. Commun.* **2003**, *33*, 4253.

PhCHO $\xrightarrow[\text{DCM, rt}]{HOCH_2CH_2SH\text{, cat } Sc(OTf)_3}$ Ph–(1,3-oxathiolane) 90%

Karimi, B.; Ma'mani, L. *Synthesis* **2003**, 2503.

PhCHO $\xrightarrow{NaHSO_4\bullet SiO_2\text{, EtSH}}$ $PhCH(SEt)_2$ 98%

Das, B.; Ramu, R.; Reddy, M.R.; Mahender, G. *Synthesis* **2004**, 250.

PhCHO $\xrightarrow[4\%\text{ aq } Zn(BF_4)_2]{HSCH_2CH_2SH\text{, DCM, rt}}$ Ph–(1,3-dithiolane) 90%

Islam, S.; Majee, A.; Mandal, T.; Khan, A.T. *Synth. Commun.* **2004**, *34*, 2911.

$PhCH(OAc)_2$ $\xrightarrow[POCl_3\text{-Montmorillonite}]{HSCH_2CH_2SH\text{, DCM}}$ Ph–(1,3-dithiolane) 96%

Jin, T.-S.; Sun, G.; Li, Y.-W.; Li, T.-S. *Synth. Commun.* **2004**, *34*, 4105.

PhCHO $\xrightarrow{CH(OEt)_3\text{, 10\% } NH_2SO_3H\text{, 1 h}}$ $PhCH(OEt)_2$ 97%

Gong, W.; Wang, B.; Gu, Y.; Yan, L.; Yang, L.; Suo, J. *Synth. Commun.* **2004**, *34*, 4243.

PhCHO $\xrightarrow[\text{rt, 6 h}]{HSCH_2CH_2SH\text{, cat } Lu(OTf)_3}$ Ph–(1,3-dithiolane) 83%

De, S.K. *Synth. Commun.* **2004**, *34*, 4401.

PhCHO $\xrightarrow[\text{rt, 3 h}]{Ac_2O\text{, } Zn(BF_4)_2\text{, neat}}$ $PhCH(OAc)_2$ 95%

Ranu, B.C.; Dutta, J.; Das, A. *Chem Lett.* **2003**, *32*, 366.

REVIEW:

"Regeneration of Carbonyl Compounds from Oximes, Hydrazones, Semicarbazones, Acetals, 1,1-Diacetates, 1,3-Dithiolanes, 1,3-Dithianes, and 1,3-Oxathiolanes"
Khoee, S.; Ruoho, A.E. *Org. Prep. Proceed. Int.* **2003**, *35*, 527.

CHAPTER 5

PREPARATION OF ALKYLS, METHYLENES, AND ARYLS

This chapter lists the conversion of functional groups into methyl, ethyl, propyl, etc. as well as methylene (CH_2), phenyl, etc.

SECTION 61: ALKYLS, METHYLENES, AND ARYLS FROM ALKYNES

t-Bu
N
Ph—≡—$SiMe_3$
5% $Pd(OAc)_2$, 10% PPh_3
100°C , 21 h
I
N
Ph
85%

Roesch, K.R.; Larock, R.C. *J. Org. Chem.* ***2002***, *67*, 86.

Ph
5% CuI , DMA–NEt_3 , 80°C
3 h
O
O
Ph
85%

Kel'in, A.V.; Gevorgyan, V. *J. Org. Chem.* ***2002***, *67*, 95.

CN
Ph–C≡C–Ph , Bu_4NCl , DMF
cat $Pd(OAc)_2$, 100°C , 2 d
I
NH_2
Ph
Ph
83%

Tian, Q.; Pletnev, A.A.; Larock, R.C. *J. Org. Chem.* ***2003***, *68*, 339.

$SiMe_3$
Me_3Si—≡—CO_2t-Bu
1. C_6H_{13}—≡ ,Ti(O*i*-Pr)$_4$
2 *i*-PrMgCl
2. ≡—\ Br , –50°C → rt
t-BuO_2C
52%
C_6H_{13}

Tanaka, R.; Nakano, Y.; Suzuki, D.; Urabe, H.; Sato, F. *J. Am. Chem. Soc.* ***2002***, *124*, 9682.

C_3H_7—≡ , DCE , 80°C
3% $AuCl_3$, 1.5 h
(92 : 8) 91%

Asao, N.; Takahashi, K.; Lee, S.; Kasahara, T.; Yamamoto, Y. *J. Am. Chem. Soc.* ***2002**, 124*, 12650.

C_3H_7—≡—C_3H_7 , o-xylene
cat $[IrCl(cod)]_2$, $(t\text{-}Bu)_3$, 25 h
reflux
91%

Yasukawa, T.; Satoh, T.; Miura, M.; Nomura, M. *J. Am. Chem. Soc.* ***2002**, 124*, 12680.

2.5% $[RhCl(cod)]_2$, 16% TPPTs
aq EtOH , 3 h
79%

Kinoshita, H.; Shinokubo, H.; Oshima, K. *J. Am. Chem. Soc.* ***2003**, 125*, 7784.

10% aq NaOH , PhH , N_2H_4
sonication , 40°C
Te , 2 $NaBH_4$, Aliquat 464 , 8 h
65%

Landis, C.A.; Payne, M.M.; Eaton, D.L.; Anthony, J.E. *J. Am. Chem. Soc.* ***2004**, 126*, 1338.

$(i\text{-}PrO)_2BC{\equiv}CBu$, BuC≡CH , rt
DCE , 5% Cp*RuCl(cod)
73%

Yamamoto, Y.; Ishii, J.-i.; Nishiyama, H.; Itoh, K. *J. Am. Chem. Soc.* ***2004**, 126*, 3712.

PPh_3 , EtOAc , sealed tube , 110°C
72%

Jung, C.-K.; Wang, J.-C.; Krische, M.J. *J. Am. Chem. Soc.* ***2004**, 126*, 4118.

$(OC)_5Cr$ carbene (OMe, cyclohexenyl)
THF , 55°C
70%

Jiang, M.X.-W.; Rawat, M.; Wulff, W.D. *J. Am. Chem. Soc.* ***2004**, 126*, 5970.

(1-)-naphth; O; (1-)-naphth → $MeOH_2C$—≡—CH_2OMe; cat $Ir(cod)_2$ + 2 MeDUPHOS; xylene, 100°C, 1 h → (1-)-naphth; O; OMe; OMe; (1-)-naphth 76% (99.5% ee)

Shibata, T.; Shibata, T.; Fujimoto, T.; Yokota, K.; Takagi, K. *J. Am. Chem. Soc.* ***2004***, *126*, 8382.

OTf → 1. Me_3Al, Cp_2ZrCl_2, DCE; 2. 10% Pd[(*p*-MeOPh)$_2$Cl$_2$; 20% P(*p*-OMeOPh), 0.1 NEt_3; 6 DABCO, MeCN, 100°C → 67%

Fillion, E.; Carson, R.J.; Trépanier, V.E.; Goll, J.M.; Remorova, A.A. *J. Am. Chem. Soc.* ***2004***, *126*, 15354.

Ph—≡—Ph → $Pd(OAc)_2$, NaOMe, MeOH, 25°C, 2d → Ph, Ph 92%

Wei, L.-L.; Wei, L.-M.; Pan, W.-B.; Leou, S.-P.; Wu, M.-J. *Tetrahedron Lett.* ***2003***, *44*, 1979.

O_2N; N-OH → HO_2CCHO, rt, 1.2h → O_2N; CHO 94%

Chavan, S.P.; Soni, P. *Tetrahedron Lett.* ***2004***, *45*, 3161.

MeO; N; Bn; CO_2Et → 10 *t*-BuOK, *t*-BuOH, THF; 90°C, 15 min → OH; Me; N; Bn; CO_2H

Hirayama, M.; Choshi, T.; Kumemura, T.; Tohyama, S.; Nobuhiro, J.; Hibino, S. *Heterocycles* ***2004***, *63*, 1765.

SECTION 62: ALKYLS, METHYLENES, AND ARYLS FROM ACID DERIVATIVES

H; N → $MeCO_2H$, microwaves; $ZnCl_2$, solvent free, 5 min → N; Me 79%

Koshima, H.; Kutsunai, K. *Heterocycles* ***2002***. *57*, 1299.

NH_2; NH_2 → N; CO_2H; PPA, microwaves, 3 min → H; N; N; N 96%

Yu, H.; Kawanishi, H.; Koshima, H. *Heterocycles* ***2003***, *60*, 1457.

SECTION 63: ALKYLS, METHYLENES, AND ARYLS FROM ALCOHOLS AND THIOLS

Ph/CH=CH/CH2OH → (p-tolyl-B(OH)$_2$, 10% RhCl-xH$_2$O•Cu(OAc)$_2$; [bmim]PF$_6$, 50°C, 2 h) → PhCH=CHCH$_2$(p-tolyl) 72%

Kabalka, G.W.; Dong, G.; Venkataiah, B. *Org. Lett.* ***2003***, *5*, 893.

PhC(=O)CH(CH$_3$)CO$_2$Et → (PhCH=CHCH$_2$OH, H$_2$O, reflux; 10% 1-AdCO$_2$H, 5 h; 5% Pd(PPh$_3$)$_4$) → PhC(=O)C(CH$_3$)(CO$_2$Et)CH$_2$CH=CHPh 90%

Manabe, K.; Kobayashi, S. *Org. Lett.* ***2003***, *5*, 3241.

PhCH(OH)C≡CH → (2-methylfuran, Ru catalyst, DCE; NH$_4$BF$_4$, 60°C, 1 h) → 2-methyl-5-(1-phenylprop-2-ynyl)furan 85%

Nishibayashi, Y.; Yoshikawa, M.; Inada, Y.; Hidai, M.; Uemura, S. *J. Am. Chem. Soc.* ***2002***, *124*, 11846.

PhCH$_2$OH → (PhMe, 0.1 Cl$_2$Si(OTf)$_2$, 50°C, 4.5h) → PhCH$_2$Ph(Me) 50% (41:59 *o:m+p*)

Shiina, I.; Suzuki, M. *Tetrahedron Lett* ***2002***, *43*, 6391.

cyclohex-2-en-1-ol → (CH$_2$(CO$_2$Et)$_2$, cat Pd(OAc)$_2$/PPh$_3$; BEt$_3$, 0.5 NaH, THF, 25°C, 4h) → cyclohex-2-enyl-CH(CO$_2$Et)$_2$ 74%

Kimura, M.; Mukai, R.; Tanigawa, N.; Tanaka, S.; Tamaru, Y. *Tetrahedron* ***2003***, *59*, 7767.

PhCH=CHCH$_2$OH → (PhB(OH)$_2$, cat Pd(PPh$_3$)$_4$; sealed tube, 80°C, THF, 3h) → PhCH=CHCH$_2$Ph 61%

Tsukamoto, H.; Sato, M.; Kondo, Y. *Chem. Commun.* ***2004***, 1200.

PhCH(OH)C≡CBu → (2-methylfuran, DCE, 60°C, 1h; 5% diruthenium complex) → 2-methyl-5-(CH$_2$C≡CBu)furan 91%

Nishibayashi, Y.; Inada, Y.; Yoshikawa, M.; Hidai, M.; Uemura, S. *Angew. Chem. Int. Ed.* ***2003***, *42*, 1495.

Ph$_2$CHOH → (cat InCl$_3$, DCM, 10 min, rt; CH$_2$=CH$_2$SiMe$_2$Cl) → Ph$_2$CHCH$_2$CH=CH$_2$ 80%

Yasuda, M.; Saito, T.; Ueba, M.; Baba, A. *Angew. Chem. Int. Ed.* ***2004***, *43*, 1414.

p-MeC$_6$H$_4$B(OH)$_2$ → (HOCH$_2$CH=CH$_2$, toluene, 2h; cat P(OPh)$_3$, Pd$_2$dba$_3$•CHCl$_3$) → p-MeC$_6$H$_4$CH$_2$CH=CH$_2$ 89%

Kayaki, Y.; Koda, T.; Ikariya, T. *Eur. J. Org. Chem.* ***2004***, 4989.

PhCH$_2$OH → (PhH, Ce$_2$(SO$_4$)$_3$, 125°C, overnight) → PhCH$_2$Ph 85%

Li, J.-H.; Liu, W.-J.; Yin, D.-L. *Synth. Commun.* ***2004***, *34*, 3161.

OH
NO_2
EtNHPh
210-215°C
O
N
CH_3
27%

Nishioka, H.; Ohmori, Y.; Iba, Y.; Tsuda, E.; Hrayama, T. *Heterocycles* ***2004**, 64*, 193.

SECTION 64: ALKYLS, METHYLENES, AND ARYLS FROM ALDEHYDES

N
H
Cl—C_6H_4—CHO , $In(OTf)_3$
bmim PF_6 , 15 min
Cl
HN
NH
95%

Ji, S.-J.; Zhou, M.-F.; Gu, D.-G.; Wang, S.-Y.; Loh, T.-P. *Synlett* ***2003***, 2077.

BnOH
1. BuLi , Ph_2PCl
2. MeI
3. $M(CH_2)_5MgBr$, THF , rt, 1 h
Bn —$(CH_2)_5Me$
82%

Shintou, T.; Kikuchi, W.; Mukaiyama, T. *Chem Lett*. ***2003**, 32*, 676.

NH_2
NH_2
PhCHO , 10% $Sc(OTf)_3$, THF
O_2 , rt , 44 h
N
N
H
Ph
97%

Nagata, K.; Itoh, T.; Ishikawa, H.; Ohsawa, A. *Heterocycles* ***2003**. 61*, 93.

CHO
NO_2
In , NH_4Cl , EtOH , H_2O
N
90%

Banik, B.K.; Banik, I.; Samajdar, S.; Wilson, M. *Heterocycles* ***2004**, 63*, 283.

OH
Ph
Ph
O
PhCHO , NH_4OAc , SiO_2
microwaves , 20 min
Ph
N
Ph
N
H
Ph
70%

Xu, Y.; Wan, F.; Salehi, H.; Deng, W.; Guo, Q.-X. *Heterocycles* ***2004**, 63*, 1613.

SH, NH_2 — 1.5 PhCHO , 2% $Sc(OTf)_3$ / O_2 ,THF , rt , 2 h → S, N, Ph 98%

Itoh, T.; Nagata, K.; Ishikawa, H.; Ohsawa, A. *Heterocycles* **2004**, *63*, 2769.

Related Method: Alkyls, Methylenes, and Aryls from Ketones (Section 72)

SECTION 65: ALKYLS, METHYLENES, AND ARYLS FROM ALKYLS, METHYLENES, AND ARYLS

OTBS, NMe_2 — 1. *t*-BuLi , cyclohexane rt , 72 h / 2. MeI → OTBS, NMe_2 54%

Kraus, G.A.; Kim, J. *J. Org. Chem.* **2002**, *67*, 2358.

O — Bu, Bu , neat , 120°C / 10% $P(O)Bu_3$, 5% $Pd(PPh_3)_4$ → Bu, Bu, O 70%

Nakamura, I.; Siriwardana, A.I.; Saito, S.; Yamamoto, Y. *J. Org. Chem.* **2002**, *67*, 3445.

Br, N_2, CO_2Me / cat $Rh_2(S\text{-DOSP})_4$, 50°C → CO_2Me, Br (80 89% ee + CO_2Me, Br : 20) 62% 85% ee

Davies, H.M.L.; Jin, Q.; Ren, P.; Kovalsky, A.Yu. *J. Org. Chem.* **2002**, *67*, 4165.

Br OTHP / LiCKOR , −78°C → OTHP 62%

Revell, J.D.; Ganesan A. *J. Org. Chem.* **2002**, *67*, 6250.

$SiMe_2(OH)$, N, Boc — I, NO_2 / 2 eq NaO*t*-Bu , CuI , rt cat $Pd_2(dba)_3$•$CHCl_3$ toluene , 6h → N, Boc, NO_2 84%

Denmark, S.E.; Baird, J.D. *Org. Lett.* **2004**, *6*, 3649.

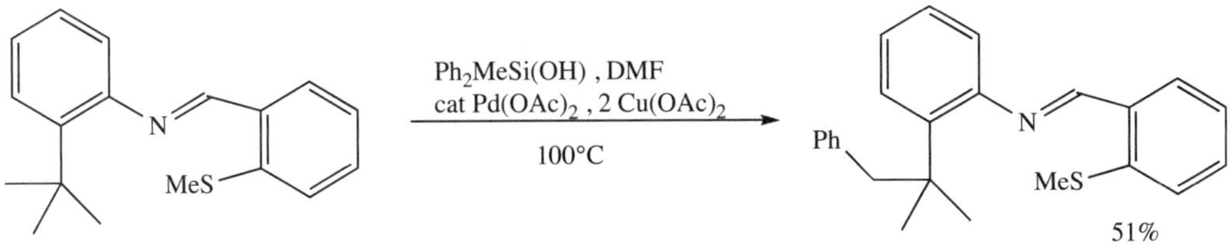

$Ph_2MeSi(OH)$, DMF
cat $Pd(OAc)_2$, 2 $Cu(OAc)_2$
100°C

N MeS Ph N MeS

51%

Sezen, B.; Franz, R.; Sames, D. *J. Am. Chem. Soc.* ***2002***, *124*, 13372.

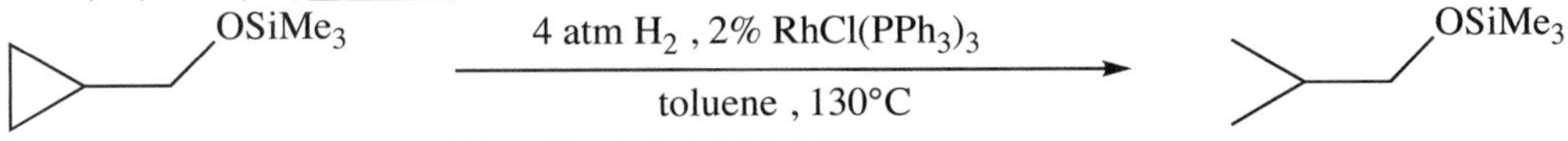

$OSiMe_3$ 4 atm H_2 , 2% $RhCl(PPh_3)_3$ toluene , 130°C $OSiMe_3$

Bart, S.C.; Chirik, P.J.. *J. Am. Chem. Soc.* ***2003***, *125*, 886.

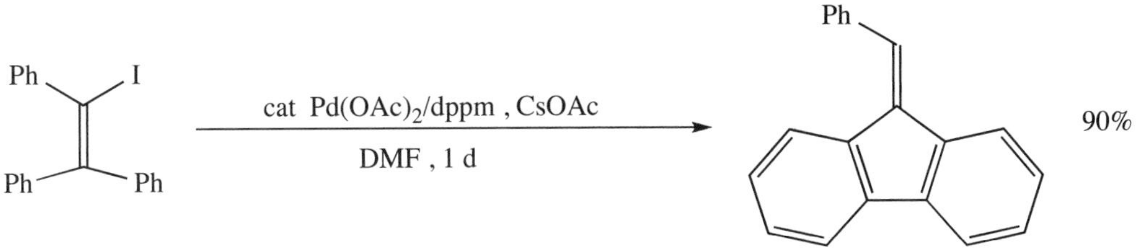

Ph I Ph Ph cat $Pd(OAc)_2$/dppm , CsOAc DMF , 1 d Ph

90%

Huang, Q.; Campo, M.A.; Yao, T.; Tian, Q.; Larock, R.C. *J. Org. Chem.* ***2004***, *69*, 8258.

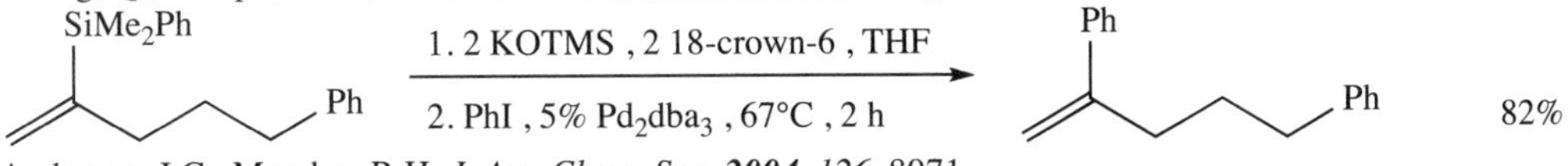

$SiMe_2Ph$ Ph

1. 2 KOTMS , 2 18-crown-6 , THF
2. PhI , 5% Pd_2dba_3 , 67°C , 2 h

Ph Ph

82%

Anderson, J.C.; Munday, R.H. *J. Am. Chem. Soc.* ***2004***, *126*, 8971.

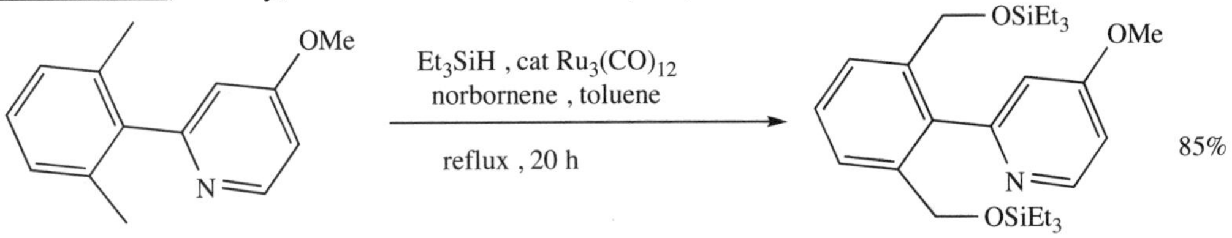

OMe N

Et_3SiH , cat $Ru_3(CO)_{12}$
norbornene , toluene
reflux , 20 h

$OSiEt_3$ OMe N $OSiEt_3$

85%

Kakiuchi, F.; Tsuchiya, K.; Matsumoto, M.; Mizushima, E.; Chatani, N. *J. Am. Chem. Soc.* ***2004***, *126*, 12792.

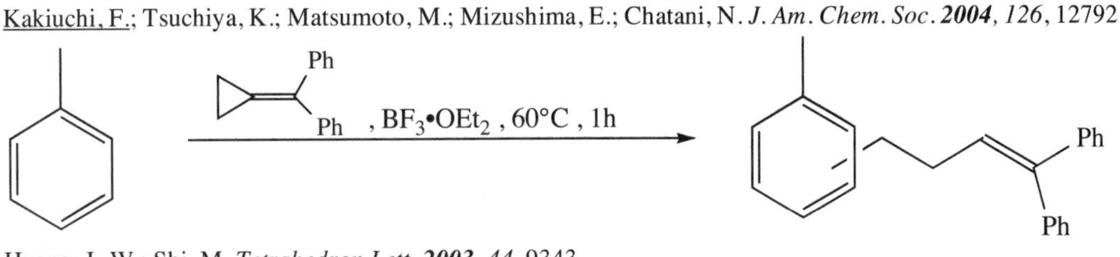

Ph Ph , BF_3•OEt_2 , 60°C , 1h

Ph Ph

Huang, J.-W.; Shi, M. *Tetrahedron Lett.* ***2003***, *44*, 9343.

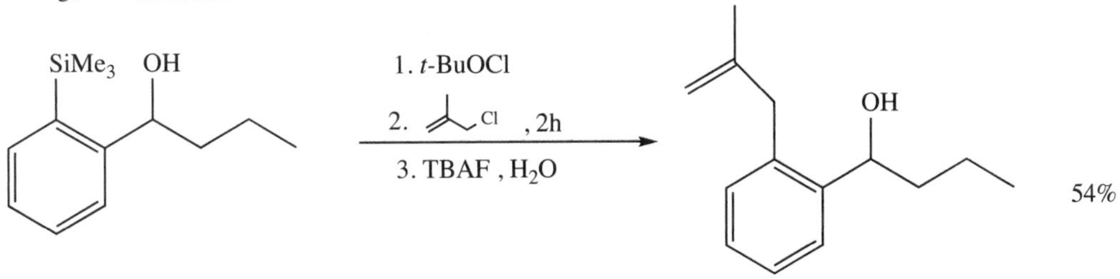

$SiMe_3$ OH

1. *t*-BuOCl
2. Cl , 2h
3. TBAF , H_2O

OH

54%

Taguchi, H.; Takami, K.; Tsubouchi, A.; Takeda, T. *Tetrahedron Lett.* ***2004***, *45*, 429.

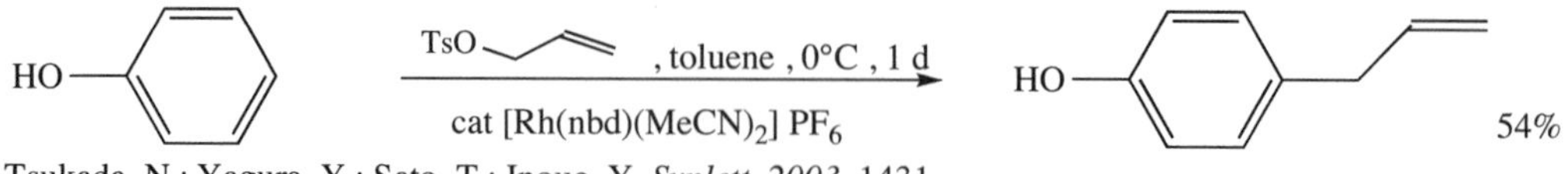

HO–C6H4 (phenol) → TsO-allyl, toluene, 0°C, 1 d; cat $[Rh(nbd)(MeCN)_2]\ PF_6$ → 4-allylphenol (HO) 54%

Tsukada, N.; Yagura, Y.; Sato, T.; Inoue, Y. *Synlett 2003,* 1431.

SECTION 66: ALKYLS, METHYLENES, AND ARYLS FROM AMIDES

NO ADDITIONAL EXAMPLES

SECTION 67: ALKYLS, METHYLENES, AND ARYLS FROM AMINES

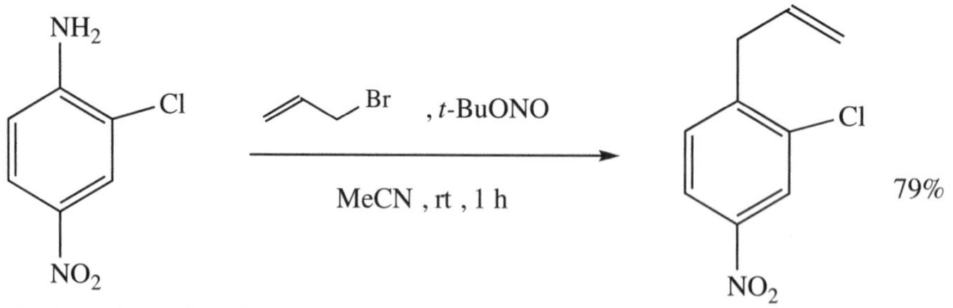

NH_2, Cl, NO_2 → allyl bromide (Br), *t*-BuONO; MeCN, rt, 1 h → Cl, NO_2 79%

Ek, F.; Axelsson, O.; Wistrand, L.G.; Frejd, T. *J. Org. Chem.* **2002**, *67*, 6376.

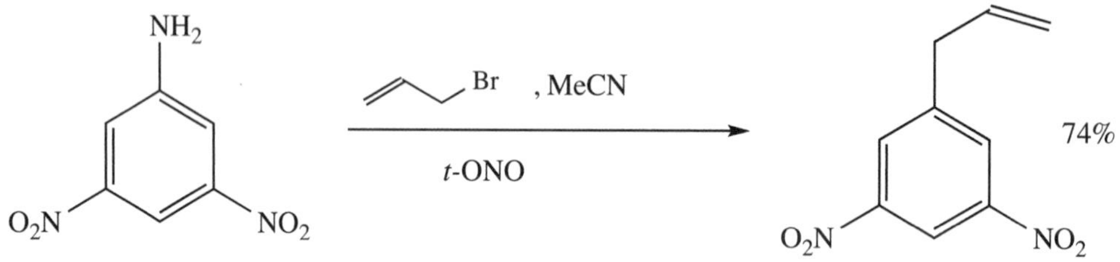

NH_2, O_2N, NO_2 → allyl bromide (Br), MeCN; *t*-ONO → O_2N, NO_2 74%

Ek, F.; Wistrand, L.G.; Frejd, T. *J. Org. Chem.* **2003**, *68*, 1911.

SECTION 68: ALKYLS, METHYLENES, AND ARYLS FROM ESTERS

ASYMMETRIC CONVERSIONS

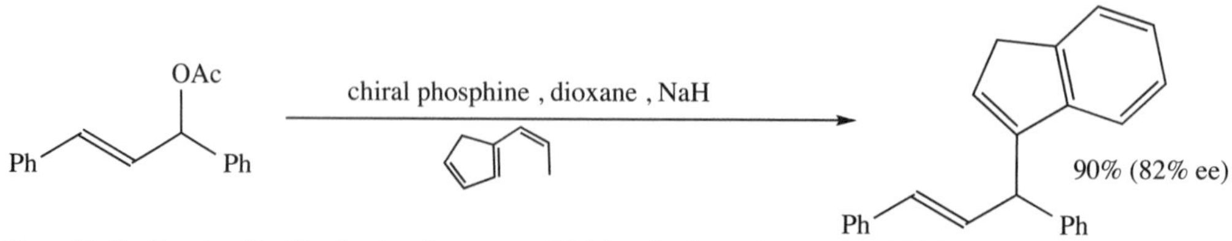

Ph, OAc, Ph → chiral phosphine, dioxane, NaH → Ph, Ph 90% (82% ee)

Hayashi, T.; Suzuka, T.; Okada, A.; Kawatsura, M. *Tetrahedron: Asymmetry* **2004**, *15*, 545.

Ph–CH=CH–CH(OAc)–Ph → (cat $[Pd(\eta^3\text{-}C_3H_5)Cl]_2$ / chiral ferrocene, 3 eq $CH_2(CO_2Me)_2$, 3 eq BSA; $CHCl_3$, rt) → Ph–CH=CH–CH($CH(CO_2Me)_2$)–Ph 92% (93% ee)

Mancheño, O.G.; Preigo, J.; Cabrera, S.; Arrayás, R.G.; Llamas, T.; Cerretero, J.C. *J. Org. Chem.* ***2003***, *68*, 3679.

Ph–CH=CH–CH$_2$–OCO_2Me → (cat $[Ir(cod)Cl]_2$, phosphoramidite ligand; $NaCH(CO_2Me)_2$, LiCl, THF, 25°C) → Ph–CH($CH(CO_2Me)_2$)–CH=CH$_2$ 82% (98% ee)

Alexakis, A.; Polet, D. *Org. Lett.* ***2004***, *6*, 3529.

Ph–CH(OAc)–CH=CH–Ph → ($CH_2(CO_2Me)_2$, BSA, KOAc, DCM; cat Pd-nanoparticles, rt, 1 d) → Ph–CH($CH(CO_2Me)_2$)–CH=CH–Ph 56% (97% ee)

Jansat, S.; Gómez, M.; Philippòl, K.; Muller, G.; Guiu, E.; Cleaver, C.; Castillón, S.; Chaudret, B. *J. Am. Chem. Soc.* ***2004***, *126*, 1592.

Ph–CH=CH–CH(OAc)–Ph → (cat P,N-paracyclophane ligand, cat $[Pd(C_3H_5)Cl]_2$; cat LiOAc, 3 BSA, 3 $CH_2(CO_2Me)_2$, toluene) → Ph–CH=CH–CH($CH(CO_2Me)_2$)–Ph 98% (90% ee)

Wu, X.-W.; Yuan, K.; Sun, W.; Zhang, M.-J.; Hou, X.-L. *Tetrahedron: Asymmetry* ***2003***, *14*, 107.

CH$_2$=CH–C(CH$_3$)(Ph)(OAc) → (2% $[Pd(C_3H_5)Cl]_2$, chiral ferrocenyl ligand; $CH_2(CO_2Me)_2$, BSA, ether, 25°C, 19 h) → CH$_2$=CH–C(CH$_3$)(Ph)–CH(CO_2Me)(CO_2Me) 66% (73% ee)

Hou, X.-L.; Sun, N. *Org. Lett.* ***2004***, *6*, 4399.

Ph–CH=CH–CH(OAc)–Ph → ($CH_2(CO_2Me)_2$, cat $[PdCl(alyl)]_2$, THF; ferrocenyl phosphine-amide ligand, Li_2CO_3) → Ph–CH=CH–CH($CH(CO_2Me)_2$)–Ph 88% (97% ee)

Boaz, N.W.; Ponasik Jr. J.A.; Large, S.E.; Debenham, S.D. *Tetrahedron: Asymmetry* ***2004***, *15*, 2151.

Ph–CH=CH–CH(OAc)–Ph → ($CH_2(CO_2Me)_2$, KOAc, BSA, DCM; cat $[PdCl(\eta^3\text{-}C_3H_5)]_2$, alcohol-thiol ligand) → Ph–CH=CH–CH($CH(CO_2Me)_2$)–Ph quant (94% ee)

Okuyama, Y.; Nakano, H.; Takahashi, K.; Hongo, H.; Kabuto, C. *Chem. Commun.* ***2003***, 524.

Ph–CH=CH–CH(OAc)–Ph → (amino phosphonate ligand, LiOAc, $CH_2(CO_2Me)_2$, BSA, toluene, 20°C; cat $[Pd(allyl)Cl]_2$, 12h) → Ph–CH=CH–CH($CH(CO_2Me)_2$)–Ph 88% (87% ee)

Chen, G.; Li, X.; Zhang, H.; Gong, L.; Mi, A.; Cui, X.; Jiang, Y.; Choi, M.C.K.; Chan, A.S.C. *Tetrahedron: Asymmetry* ***2002***, *13*, 809.

Ph–CH=CH–CH(OAc)–Ph → (cat P,N-ferrocenyl ligand, cat $[PdCl(C_3H_5)]_2$; $CH_2(CO_2Me)_2$, 4.6 KOAc, 2 BSA, DCM) → Ph–CH=CH–CH($CH(CO_2Me)_2$)–Ph 94% (92% ee)

Kloetzing, R.J.; Lotz, M.; Knochel, P. *Tetrahedron: Asymmetry* ***2003***, *14*, 255.

Ph–CH=CH–CH(OAc)–Ph → (xylofuranose-based phosphinooxathiane ligand; cat $[PdCl(\eta^3\text{-}C_3H_5)]_2$, KOAc, BSA, DCM) → Ph–CH=CH–CH($CH(CO_2Me)_2$)–Ph 93% (90% ee)

Nakano, H.; Yokoyama, J.-i.; Okuyama, Y.; Fujita, R.; Hongo, H. *Tetrahedron: Asymmetry* ***2003***, *14*, 2361.

Ph–CH=CH–CH(OAc)–Ph → (1-phosphonobornadiene imine ligand, BSA; $CH_2(CO_2Me)_2$, cat KOAc, DCM) → Ph–CH=CH–CH($CH(CO_2Me)_2$)–Ph quant (93% ee)

Mercier, F.; Brebion, F.; Dupont, R.; Mathey, F. *Tetrahedron: Asymmetry* ***2003***, *14*, 3137.

$CH_2(CO_2Me)_2$, BSA , LiOAc , THF / chiral phosphine , 25°C — 98% (83% ee)

Tollabi, M.; Framery, E.; Goux-Henry, C.; Sinou, D. *Tetrahedron: Asymmetry* **2003**, *14*, 3329.

cat $[Pd(\eta^3\text{-}C_3H_5)Cl]_2$, LiOAc , 1 d / chiral phosphine , base / $CH_2(CO_2Me)_2$, BSA , 0°C , toluene — 94% (94% ee)

Mino, T.; Tanaka, Y.; Akita, K.; Sakamoto, M.; Fuita, T. *Heterocycles* **2003**. *60*, 9.

$CH_2(CO_2Me)_2$, NaH , DMF , rt / cat $Cp^*Ru(MeCN)_3$ PF_6 — 94%ee

Trost, B.M.; Fraisse, P.L.; Ball, Z.T. *Angew. Chem. Int. Ed.* **2002**, *41*, 1059.

2 $NaCH(CO_2Me)_2$, THF , 25°C , 3h / cat $IrCl(cod)_2$/amidite ligand — 91% (69%ee)

Bartels, B.; García-Yebra, C.; Helmchen, G. *Eur. J. Org. Chem.* **2003**, 1097.

NONASYMMETRIC CONVERSIONS

2 eq BuMgBr , ether / 10% BuCN — 72% (86:14 *anti:syn*)

Belelie, J.L.; Chong, J.M. *J. Org. Chem.* **2002**, *67*, 3000.

Ph_3In , 1% Pd_2dba_3 , 4% PPh_3 / THF , reflux — 85%

Baker, L.; Minehan, T. *J. Org. Chem.* **2004**, *69*, 3957.

5% $[PdCl(\pi\text{-allyl})]_2$, dppe / Li_2CO_3, DCM — 72%

Bandini, M.; Melloni, A.; Umani-Ronchi, A. *Org. Lett.* **2004**, *6*, 3199.

, THF , reflux , 2 h / polymer-encarcerated Pd catalyst, cat. PPh_3 — 88%

Akiyama, R.; Kobayashi, S. *J. Am. Chem. Soc.* **2003**, *125*, 3412.

, CsF , dioxane / 10% $Ni(cod)_2$, 10% IMeS-HCl / 80°C , 12 h — 83%

Blakey, S.B.; MacMillan, D.W.C. *J. Am. Chem. Soc.* **2003**, *125*, 6046.

$OCO_2CH(CF_3)_2$ cat $TpRh(C_2H_4)_2$, PhLi / Ph_3ZnLi, LiBr, ether (3 : 1) 88%

Evans, P.A.; Uraguchi, D. *J. Am. Chem. Soc.* **2003**, *125*, 7158.

Ph OCO_2Me — $MeCH(CO_2Et)_2$, BSA, THF / 1% ferrocenyl diphosphine/dppf, cat KOAc, 80°C, 1 d — CO_2Et, CO_2Et, Ph 95%

Kuwano, R.; Kondo, Y.; Matsuyama, Y. *J. Am. Chem. Soc.* **2003**, *125*, 12104.

OAc — $SiMe_3$ / BF_3•OEt_2, DCM — (93:7 *syn:anti*) 83%

Smith, D.M.; Tran, M.B.; Woerpel, K.A. *J. Am. Chem. Soc.* **2003**, *125*, 14149.

OAc, Cl — Me_3Si / BF_3•OEt_2 — Cl 90% (14:86 *cis:trans*)

Ayala, L.; Lucero, C.G.; Romero, J.A.C.; Tabacco, S.A.; Woerpel, K.A. *J. Am. Chem. Soc.* **2003**, *125*, 15521.

OAc — $SiMe_3$, I_2, DCM / rt, 30 min — 82%

Yadav, J.S.; Reddy, B.V.S.; Rao, K.V.; Raj, K.S.; Rao, P.P.; Prasad, A.R.; Gunasekar, D. *Tetrahedron Lett.* **2004**, *45*, 6505.

OAc, Ph, Ph — 3 $CH_2(CO_2Me)_2$, NaOAc, BSA / cat $[Pd(\eta^3(C_3H_5)Cl]_2$ /ferrocenylphosphine-imine — $CH(CO_2Me)_2$, Ph, Ph

Hu, X.; Dai, H.; Hu, X.; Chen, H.; Wang, J.; Bai, C.; Zheng, Z. *Tetrahedron: Asymmetry* **2002**, *13*, 1687.

Ph OCO_2Me — I, In, THF, 65°C / cat $Pd_2(dba)_3$•$CHCl_3$, P(*o*-Tol)$_3$ — Ph 80%

Lee, P.H.; Sung, S.-y.; Lee, K.; Chang, S. *Synlett* **2002**, 146.

OTf — PhTi(O*i*-Pr)$_3$, THF, reflux / cat $[PdCl(\pi\text{-allyl})]_2$ chiral ferrocenyl phosphine anion — Ph 98%

Han, J.W.; Tokunaga, N.; Hayashi, T. *Synlett* **2002**, 871.

OCO_2Et, Ph — Cp*Ru(cod)Cl, $NaOCH(CO_2Me)_2$ chiral bis(imine) ligand, THF / rt, 16 h — $CH(CO_2Me)_2$, Ph + Ph CO_2Me CO_2Me (18 : 1) quant

Renaud, J.-L.; Bruneau, C.; Demersean, B. *Synlett* **2003**, 408.

SECTION 69: ALKYLS, METHYLENES, AND ARYLS FROM ETHERS, EPOXIDES, AND THIOETHERS

The conversion ROR → RR' (R' = alkyl, aryl) is included in this section.

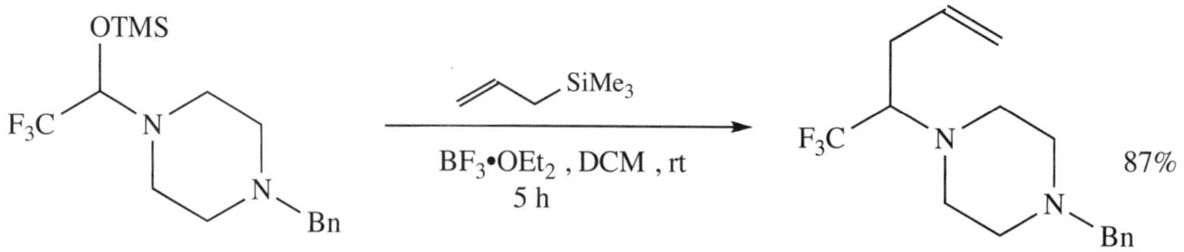

OTMS; F_3C; N; N; Bn → (allyl $SiMe_3$; BF_3•OEt_2, DCM, rt; 5 h) → F_3C; N; N; Bn 87%

Billard, T.; Langlois, B.R. *J. Org. Chem.* ***2002***, *67*, 997.

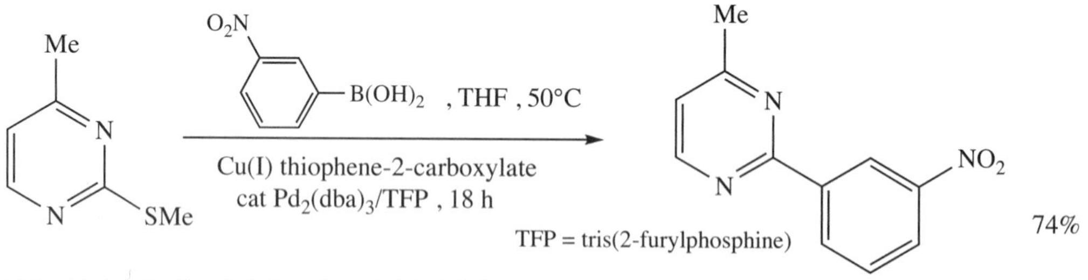

Me; N; N; SMe → (O_2N-phenyl-$B(OH)_2$, THF, 50°C; Cu(I) thiophene-2-carboxylate; cat $Pd_2(dba)_3$/TFP, 18 h) → Me; N; N; NO_2 74%

TFP = tris(2-furylphosphine)

Liebeskind, L.S.; Srogl, J. *Org. Lett.* ***2002***, *4*, 979.

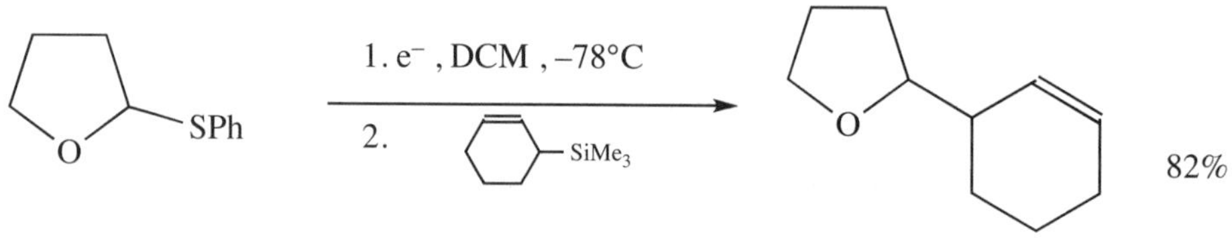

O; SPh → (1. e^-, DCM, –78°C; 2. cyclohexenyl-$SiMe_3$) → O 82%

Suzuki, S.; Matsumoto, K.; Kwamura, K.; Suga, S.; Yoshinda, J.-i. *Org. Lett.* ***2004***, *6*, 3755.

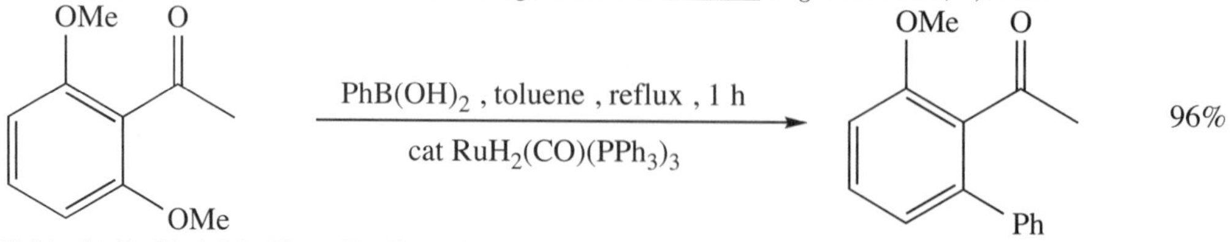

OMe; O; OMe → ($PhB(OH)_2$, toluene, reflux, 1 h; cat $RuH_2(CO)(PPh_3)_3$) → OMe; O; Ph 96%

Kakiuchi, F.; Usai, M.; Ueno, S.; Chatani, N.; Murai, S. *J. Am. Chem. Soc.* ***2004***, *126*, 2706.

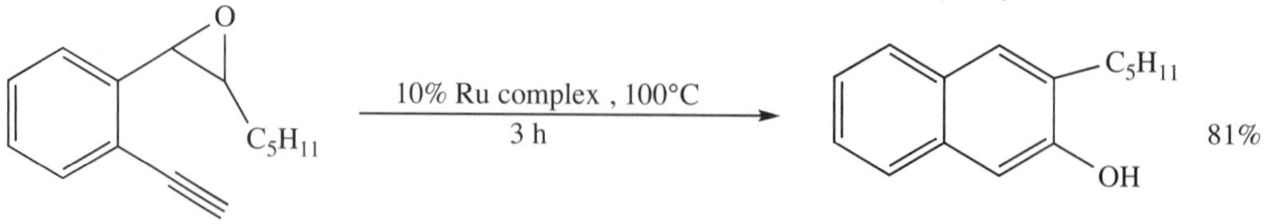

O; C_5H_{11} → (10% Ru complex, 100°C; 3 h) → C_5H_{11}; OH 81%

Madhushaw, R.J.; Lin, M.-Y.; Sohel, SMd.A.; Liu, R.-S. *J. Am. Chem. Soc.* ***2004***, *126*, 6895.

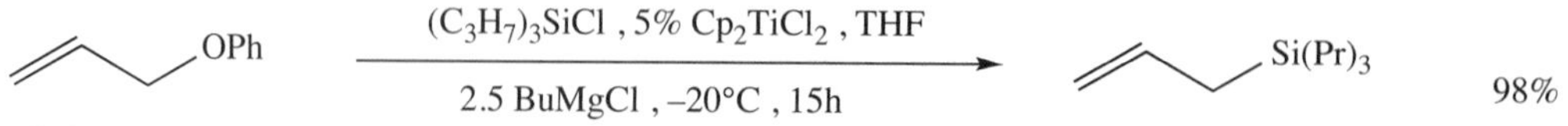

allyl-OPh → ($(C_3H_7)_3SiCl$, 5% Cp_2TiCl_2, THF; 2.5 BuMgCl, –20°C, 15h) → allyl-$Si(Pr)_3$ 98%

Nii, S.; Terao, J.; Kambe, N. *Tetrahedron Lett.* ***2004***, *45*, 1699.

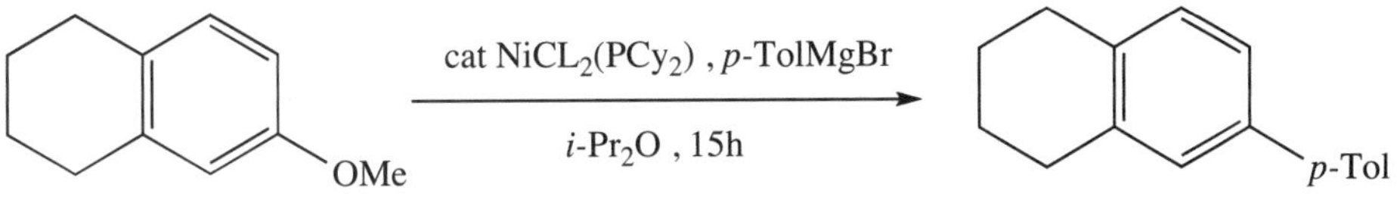

OMe → p-Tol (tetrahydronaphthalene); cat $NiCL_2(PCy_2)$, p-TolMgBr; i-Pr_2O , 15h; 93%

Dankwardt, J.W. *Angew. Chem. Int. Ed.* **2004**, *43*, 2428.

SECTION 70: ALKYLS, METHYLENES, AND ARYLS FROM HALIDES AND SULFONATES

The replacement of halogen by alkyl or aryl groups is included in this section. For the conversion of RX → RH (X = halogen), see Section 160 (Hydrides from Halides and Sulfonates).

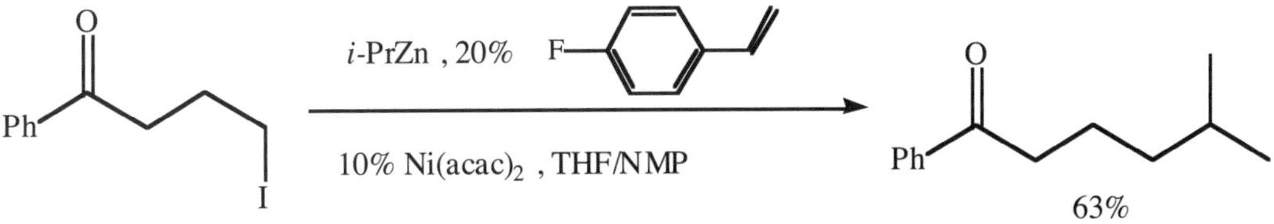

Ph, O, I; i-PrZn , 20% F–(4-vinylphenyl); 10% $Ni(acac)_2$, THF/NMP; Ph, O; 63%

Jensen, A.E.; Knochel, P. *J. Org. Chem.* **2002**, *67*, 79.

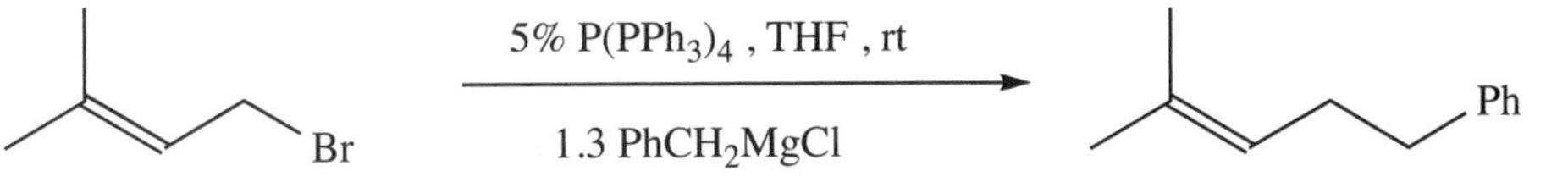

Br; 5% $P(PPh_3)_4$, THF , rt; 1.3 $PhCH_2MgCl$; Ph; 78%

Rosales, V.; Zambrano, J.L.; Demuth, M. *J. Org. Chem.* **2002**, *67*, 1167.

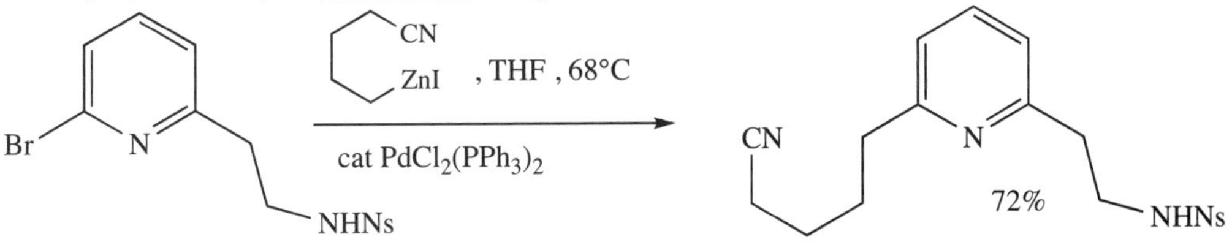

Br, N, NHNs; CN, ZnI , THF , 68°C; cat $PdCl_2(PPh_3)_2$; CN, N, NHNs; 72%

Skerlij, R.T.; Zhou, Y.; Wilson, T.; Bridger, G.J. *J. Org. Chem.* **2002**, *67*, 1407.

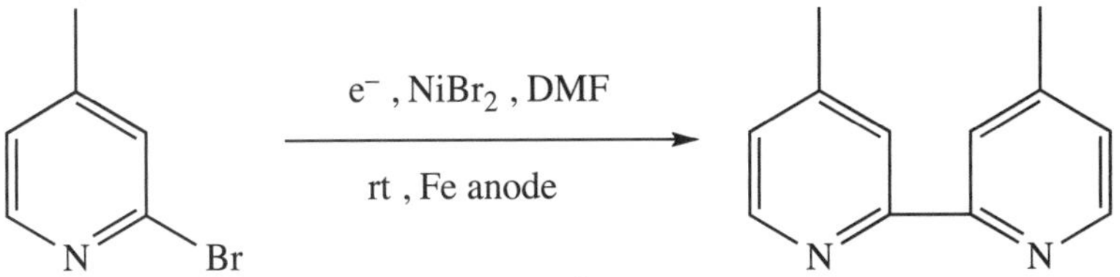

N, Br; e^- , $NiBr_2$, DMF; rt , Fe anode; N, N; 81%

de França, K.W.R.; Navarro, M.; Léonel, É.; Durandetti, M.; Nédélec, J.-Y. *J. Org. Chem.* **2002**, *67*, 1838.

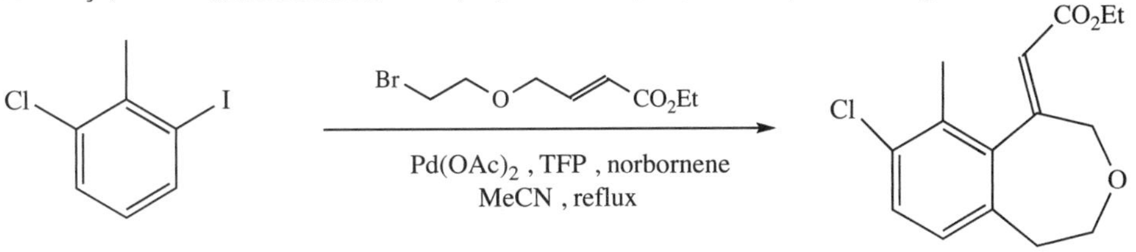

Cl, I; Br–O–CO_2Et; $Pd(OAc)_2$, TFP , norbornene; MeCN , reflux; CO_2Et, Cl, O; 84%

Zhang, J.; Li, C.-J. *J. Org. Chem.* **2002**, *67*, 3972.

$B(OH)_2$, K_2CO_3

Cl

0.5% $[HO\text{–}P(t\text{-}Bu)_2]_2PdCl_2$
THF, reflux

92%

Li, G.Y. *J. Org. Chem.* ***2002****, 67*, 3643.

OTf

I, LiCl, DMF

2% $Pd_2dba_3{\bullet}CHCl_3$, 16% PPh_3
100°C, 1 h

92%

Lee, K.; Lee, J.; Lee, P.H. *J. Org. Chem.* ***2002****, 67*, 8265.

F

1. PhMgCl, 5% $Ni(acac)_2$
5% dppp, THF, rt, 18 h

2. H_2O

N

Ph

82%

N

Mongin, F.; Mojovic, L.; Guillamet, B.; Trécourt, F.; Quéquiner, G. *J. Org. Chem.* ***2002****, 67*, 8991.

Br

$PhB(OH)_2$, H_2O, microwaves

cat $Pd(OAc)_2$

Bu — Ph

79%

Leadbeater, N.E.; Marco, M. *J. Org. Chem.* ***2003****, 68*, 888.

Ph Br

Bu_3In, THF, –30°C → rt

$Cu(OTf)_2\text{–}P(OEt)_3$

Bu

Ph (82 : 18) 85%

+ Ph Bu

Rodríguez, D.; Sestelo, J.P.; Sarandeses, L.A. *J. Org. Chem.* ***2003****, 68*, 2518.

cat $PdCl_2(PhCN)_2$, DMF
2 tetrakis(dimethylamino)ethylene

2 MeO — Br

50°C, 4 h

MeO — OMe

98%

Kuroboshi, M.; Waki, Y.; Tanaka, H. *J. Org. Chem.* ***2003****, 68*, 3938.

MeO — BF_3K

Br — CN

cat $Pd(OAc)_2$, 3 eq K_2CO_3
MeOH, reflux

MeO — CN

99%

Molander, G.A.; Biolatto, B. *J. Org. Chem.* ***2003****, 68*, 4302.

TfO — Ac

Cl BF_3K

9% $PdCl_2(dppf)$, 3 eq Cs_2CO_3
THF-H_2O, reflux, 3 d

Cl — Ac

72%

Molander, G.A.; Yun, C.-S.; Ribagorda, M.; Biolatto, B. *J. Org. Chem.* ***2003****, 68*, 5534.

PhBr

$PhB(OH)_2$, TBAB, Na_2CO_3

H_2O, microwaves

Ph–Ph

90%

Leadbeater, N.E.; Marco, M. *J. Org. Chem.* ***2003****, 68*, 5660.
Leadbeater, N.E.; Marco, M. *Org. Lett.* ***2002****, 4*, 2973.

Cl / Me / N → PhI , H_2O , Cy_2NHMe 140°C / 6% Pd-chiral phosphorous ligand → Ph / Me / N 91%

Wolf, C.; Lerebourse, R. *J. Org. Chem.* **2003**, *68*, 7551.

MeO—OTf → Ph–Si(OCH$_2$CH$_2$)$_3$N / TBAF , 10-20 H_2O , THF / cat. $Pd(dba)_2$, reflux , 12 h → MeO—Ph 90%

Riggleman, S.; DeShong, P. *J. Org. Chem.* **2003**, *68*, 8106.

t-Bu / OH + Br / OMe → cat t-Bu / OP(i-Pr)$_2$ / 5% [$RhCl(PPh_3)_3$] , Cs_2CO_3 → t-Bu / OH / OMe quant

Bedford, R.B.; Limmert, M.E. *J. Org. Chem.* **2003**, *68*, 8669.

—I → $CH_2(CO_2Et)_2$, 1.5 Cs_2CO_3 / 5% CuI , THF , 70°C , 1 d / 10% 2-phenylphenol → CO_2Et / CO_2Et 91%

Hennessy, E.J.; Buchwald, S.L. *Org. Lett.* **2002**, *4*, 269.

N / Cl → EtO / EtO / O / Zn / 2 / cat Pd(dppf)Cl_2 , cat P(t-Bu)$_3$ / THF , 50°C → OHC / O / N 92%

Gauthier Jr. D.R.; Szumigala Jr. R.H.; Dormer, P.G.; Armstrong III, J.D.; Volante, R.P.; Reider, P.J. *Org. Lett.* **2002**, *4*, 375.

Br—CO_2Et → $(EtO)_3SiH$, DMF , 80°C / cat Rh(cod)(MeCN)$_2$ BF_4 / NEt_3 → $(EtO)_3Si$—CO_2Et

Murata, M.; Ishikura, M.; Nagata, M.; Watanabe, S.; Masuda, Y. *Org. Lett.* **2002**, *4*, 1843.

Br → 2 $PhBF_3K$, MeOH / 5% Pd(OAc)$_2$, reflux / 3 K_2CO_3 , 2 h → Ph 88%

Molander, G.A.; Bidatto, B. *Org. Lett.* **2002**, *4*, 1867.

PhI + PhB(OH)$_2$ → 2% resin-supported Pd catalyst / K_2CO_3 , H_2O , 50°C , 10 h → Ph–Ph 97%

Uozumi, Y.; Nakai, Y. *Org. Lett.* **2002**, *4*, 2997.

PhCl + PhCH=CH_2 → (new Pd carbene catalyst; Bu_4NBr, NaOAc, 1 d, 140°C) PhCH=CHPh 72%

Selvakumar, K.; Zapf, A.; Meller, M. *Org. Lett.* ***2002***, *4*, 3031.

O_2N–C6H4–Br → ($PhB(OH)_2$, *i*-Pr_2NH, aq EtOH; $Pd_x([PW_{11}O_{39}])^{-7}{}_y$ nanoparticles) O_2N–C6H4–Ph >99%

Kogan, V.; Aizenshtat, Z.; Popovitz-Bito, R.; Neumann, R. *Org. Lett.* ***2002***, *4*, 3529.

2-bromopyridine + tris(3-pyridyl)boroxine → (10% $Pd(PPh_3)_4$, 2M Na_2CO_3; dioxane, reflux) 2,3'-bipyridine 72%

Cioffi, C.L.; Spencer, W.T.; Richards, J.J.; Herr, R.J. *J. Org. Chem.* ***2004***, *69*, 2210.

NC–C6H4–Br + 2 allyl acetate → (MeCN, Py, $FeBr_2$; 40% $CoBr_2$, 10% Mn, 50°C) NC–C6H4–$CH_2CH=CH_2$ 75%

Gomes, P.; Gosmini, C.; Périchon, J. *Org. Lett.* ***2003***, *5*, 1043.

PhS(O)CH_2Br → ($PhB(OH)_2$, cat $Pd(PPh_3)_4$; aq 2M Na_2CO_3, MeOH, reflux, 16 h) PhS(O)CH_2Ph 80%

Rodríguez, N.; Cuenca, A.; de Arellano, C.R.; Medio-Simón, M.; Asensio, G. *Org. Lett.* ***2003***, *5*, 1705.

1-bromo-2-methylnaphthalene + 2-methyl-1-naphthaleneboronic acid ($B(OH)_2$) → (cat $Pd(OAc)_2$, K_3PO_4, toluene; chiral ferrocenyl monophosphine ligand) 2,2'-dimethyl-1,1'-binaphthyl 65% (54% ee)

Jensen, J.F.; Johannsen, M. *Org. Lett.* ***2003***, *5*, 3025.

methyl thiophene-3-carboxylate (CO_2Me) + 3-bromonitrobenzene (NO_2) → ($Pd(PPh_3)_4$, KOAc; toluene, 110°C) 2-(3-nitrophenyl)-3-(MeO_2C)thiophene 67%

Glover, B.; Harvey, K.A.; Liu, B.; Sharp, M.J.; Tymoschenko, M.F. *Org. Lett.* ***2003***, *5*, 301.

MeO–C6H4–$SiMe_2(OH)$ → (PhI, 5%$[(allyl)PdCl]_2$, toluene; Cs_2CO_3, 10% Ash_3, 90°C) MeO–C6H4–Ph + Ph-Ph 96:4 (87%)

Denmark, S.E.; Ober, M.H. *Org. Lett.* ***2003***, *5*, 1357.

$PhB(OH)_2$, 55°C
2% $PXPd_2$, 3 K_2CO_3

(9 : 1) 61%

Yang, W.; Wang, Y.; Coret, J.R. *Org. Lett.* **2003**, *5*, 3131.

0.28 Me_4InLi , 4% $Pd(PPh_3)_4$
THF , 60°C

94%

Lee, P.H.; Lee, S.W.; Seomoon, D. *Org. Lett.* **2003**, *5*, 4963.

SO_2Cl — $PhB(OH)_2$, cat $Pd(PPh_3)_4$ / 3 K_2CO_3 , THF , reflux — Ph

55%

Dubbaka, S.R.; Vogel, P. *Org. Lett.* **2004**, *6*, 95.

CO_2Et — O_2N–C6H4–Br , AcOH / cat $PdCl_2(PPh_3)_4$, NMP , H_2O / 100°C — CO_2Et, NO_2

91%

Park, C.-H.; Ryabova, V.; Seregin, I.V.; Sromek, A.W.; Gevorgyan, V. *Org. Lett.* **2004**, *6*, 1159.

$PhB(OH)_2$, 3 *t*-BuONa
1% (IPr)Pd(allyl)Cl
dioxane , 60°C , 3h

97%

Navarro, O.; Kaur, H.; Mahjoor, P.; Nolan, S.P. *J. Org. Chem.* **2004**, *69*, 3173.

OTf — BuMgCl , $Fe(acac)_3$ — Bu

80%

Scheiper, B.; Bonnekessel, M.; Krause, H.; Fürstner, A. *J. Org. Chem.* **2004**, *69*, 3943.

$PhB(OH)_2$, EtOH , aerobic
Cs_2CO_3 , rt , 2h
2% chiral amine ligand (DAPCy)

91%

Tao, B.; Boukin, D.W. *J. Org. Chem.* **2004**, *69*, 4330.

MgBr — $C_8H_{17}Br$, 5% $Fe(acac)_3$ / ether , reflux , 0.5h — C_8H_{17}

70%

Nagano, T.; Hayashi, T. *Org. Lett.* **2004**, *6*, 1297.

SMe, Br + IZn (from iodide), N Boc

3% $Cl_2Pd(dppf)$, DMA
6% CuI, 223°C, 30 min

SMe, N Boc 62%

Corely, E.G.; Conrad, K.; Murry, J.A.; Savarin, C.; Holko, J.; Boice, G. *J. Org. Chem.* **2004**, *69*, 5120.

O_2N — Cl

Bu_3SnPh, 1.5% $Pd_2(dba)_3$
6% $P(i\text{-}BuNCH_2CH_2)_3N$
2.2 CsF, dioxane, 110°C, 30h

O_2N — Ph 86%

Su, W.; Urgaonkar, S.; Verkade, J.G. *Org. Lett.* **2004**, *6*, 1421.

$C_8H_{17}O_2SO$, O

1. MeO — B(OH)$_2$
Pd(dppf)Cl_2, K_2CO_3, 50°C
microwaves, 10 min
2. fluorous solid phase extraction

MeO, O 90%

Zhang, W.; Chen, C.H.-T.; Lu, Y.; Nagashima, T. *Org. Lett.* **2004**, *6*, 1473.

Br, Me

$PhB(OH)_2$, 2 K_3PO_4, aq Tol
5% polymer-incarcerated Pd
5% $(o\text{-Tol})_3P$, reflux, 2 h

Ph, Me 88%

Okamoto, K.; Akiyama, R.; Kobayashi, S. *Org. Lett.* **2004**, *6*, 1987.

OMe, Cl, OMe

t-Bu — BF_3K
0.5% $Pd(OAc)_2$/S-PHOS
3 K_2CO_3, MeOH, 12 h

t-Bu, OMe, OMe 93%

Barder, T.E.; Buchwald, S.L. *Org. Lett.* **2004**, *6*, 2649.

MeO, B(OH)$_2$ + CN, Br

3% Fibre catalyst
K_2CO_3, EtOH
microwaves

MeO, CN 98%

Fibre catalyst = polymer-supported Pd reagent

Wang, Y.; Sauer, D.R. *Org. Lett.* **2004**, *6*, 2793.

O_2N, I

$PhB(OH)_2$, 3% $Pd(OAc)_2$
6% DABCO, 3 K_2CO_3
acetone, 110°C

O_2N 95%

Li, J.-H.; Liu, W.-J. *Org. Lett.* **2004**, *6*, 2809.

MeO / Br / F → ($PhB(OH)_2$, EtOH-H_2O , 80°C; cat $PdCl_2(PPh_3)_2$, K_2CO_3) → MeO / Ph / F 93%

Colacot, T.J.; Shea, H.A. *Org. Lett.* ***2004**, 6*, 3731.

N^- Na^+ → (PhBr , $ZnCl_2$, THF , 100°C; cat $Pd(OAc)_2$, $P(t\text{-}Bu)_2(4\text{-}PhC_6H_4)$) → NH / Ph 90%

Rieth, R.D.; Mankand, N.P.; Calimano, E.; Sadighi, J.P. *Org. Lett.* ***2004**, 6*, 3981.

PhBr → (5% $Pd(dba)_2$, TBAF , THF , microwaves; 5% $PCy_2(2\text{-}PhC_6H_4)$, 120°C; Ph, O, Si, $\overset{+}{N}HEt_3$) → Ph–Ph 89%

Seganish, W.M.; DeShong, P. *Org. Lett.* ***2004**, 6*, 4379.

CO_2Et / Cl / CO_2Et → ($B(OH)_2$, Cl, , 3 min; 1% $[(t\text{-}Bu)_2P(OH)]_2PdCl_2$, THF; 3 K_2CO_3 , 100°C , microwaves) → CO_2Et / CO_2Et / Cl 70%

Poondra, R.R.; Fischer, P.M.; Turner, N.J. *J. Org. Chem.* ***2004**, 69*, 6920.

Ac / Cl → ($PhB(OH)_2$, cat $Pd(OAc)_2$, 95°C; ferrocenyl monophosphine ligand; 3 K_3PO_4 , dioxane , 1 d) → Ac / Ph 87%

Baille, C.; Zhang, L.; Xiao, J. *J. Org. Chem.* ***2004**, 69*, 7779.

Br / Me → ($(HO)_2B$, F; 2% Na_2PdCl_4 , 2% *t*-Bu-Amphos; 2 Na_2CO_3 , H_2O , rt) → Me / F 84%

DeVasher, R.B.; Moore, L.R.; Shaughnessy, K.H. *J. Org. Chem.* ***2004**, 69*, 7919.

Ph / Cl → (1.2 Ph_3In , 5% Pd_2dba_3; THF , reflux) → Ph / Ph 85%

Rodríguez, D.; Sestelo, J.P.; Sarandeses, L.A. *J. Org. Chem.* ***2004**, 69*, 8136.

$C_{10}H_{21}Br$ → (BuMgCl , 1% $NiCl_2$, 0°C , 30 min; 1,3-butadiene) → $C_{14}H_{30}$ quant

Terao, J.; Watanabe, H.; Ikumi, A.; Kuniyasu, H.; Kambe, N. *J. Am. Chem. Soc.* ***2002**, 124*, 4222.

Cl → ($PhSnBu_3$, 3% $Pt(Pt\text{-}Bu)_2$, 100°C; 2.2 CsF , dioxane) → Ph 94%

Littke, A.F.; Schwarz, L.; Fu, G.C. *J. Am. Chem. Soc.* ***2002**, 124*, 6343.

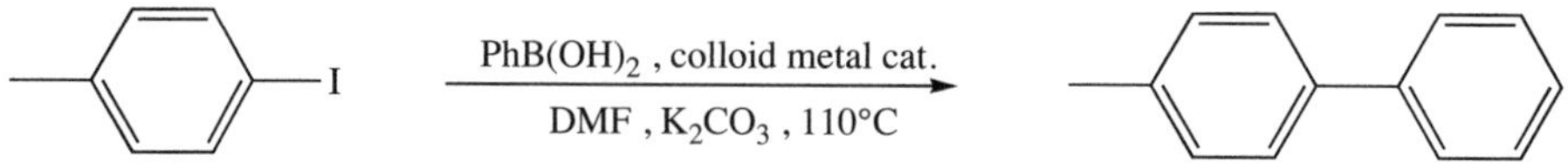

$PhB(OH)_2$, colloid metal cat. / DMF , K_2CO_3 , 110°C — quant

Thathagar, M.B.; Beckers, J.; Rothenberg, G. *J. Am. Chem. Soc.* ***2002****, 124*, 11858.

C_8H_{17}-Br → $PhB(OH)_2$, $Pd(OAc)_2$, 3 KO*t*-Bu / *t*-AmOH , PMe(*t*-Bu)$_2$ → C_8H_{17}-Ph 87%

Kirchhoff, J.H.; Netherton, M.R.; Hills, I.D.; Fu, G.C. *J. Am. Chem. Soc.* ***2002****, 124*, 13662.

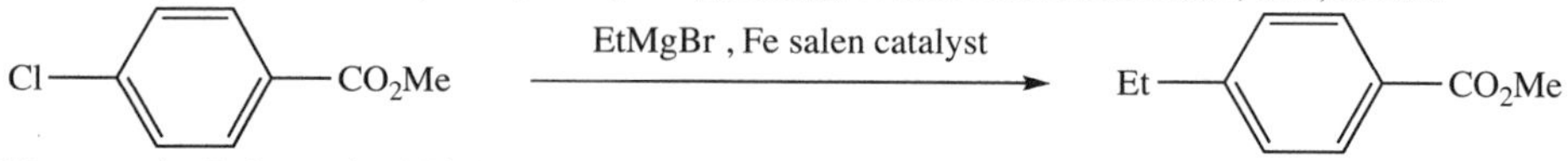

Cl—C6H4—CO_2Me → EtMgBr , Fe salen catalyst → Et—C6H4—CO_2Me >95%

Fürstner, A.; Leitner, A.; Méndez, M.; Krause, H. *J. Am. Chem. Soc.* ***2002****, 124*, 13856.

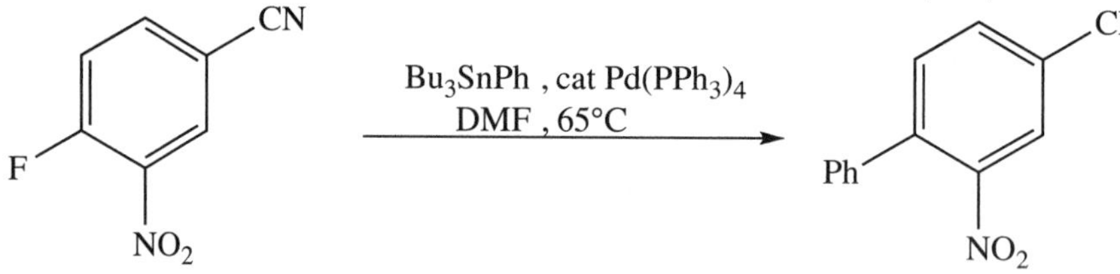

CN, F, NO_2 → Bu_3SnPh , cat $Pd(PPh_3)_4$ / DMF , 65°C → CN, Ph, NO_2 56%

Kim, Y.M.; Yu, S. *J. Am. Chem. Soc.* ***2003****, 125*, 1696.

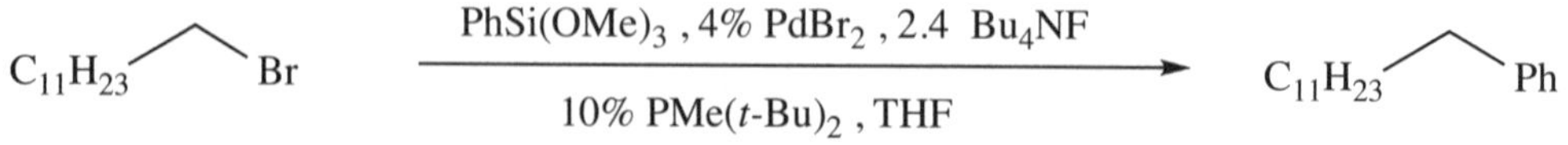

$C_{11}H_{23}$ Br → $PhSi(OMe)_3$, 4% $PdBr_2$, 2.4 Bu_4NF / 10% PMe(*t*-Bu)$_2$, THF → $C_{11}H_{23}$ Ph

Lee, J.-Y.; Fu, G.C. *J. Am. Chem. Soc.* ***2003****, 125*, 5616.

C_3H_7MgBr → $C_8H_{17}F$, 1,3-butadiene , THF / 2% CuCl , 25°C , 6 h → C_3H_7–C_8H_{17} 97%

Terao, J.; Ikumi, A.; Kuniyasu, H.; Kambe, N. *J. Am. Chem. Soc.* ***2003****, 125*, 5646.

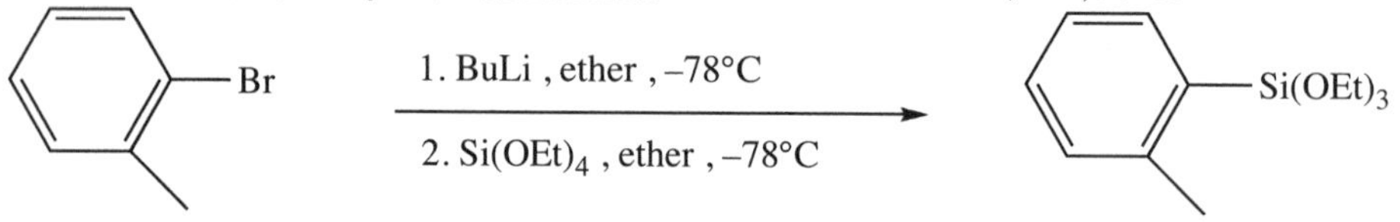

Br → 1. BuLi , ether , –78°C / 2. $Si(OEt)_4$, ether , –78°C → $Si(OEt)_3$ 79%

Manoso, A.S.; Ahn, C.; Soheili, A.; Handy, C.J.; Correia, R.; Seganish, W.M.; DeShong, P. *J. Org. Chem.* ***2004****, 69*, 8305.

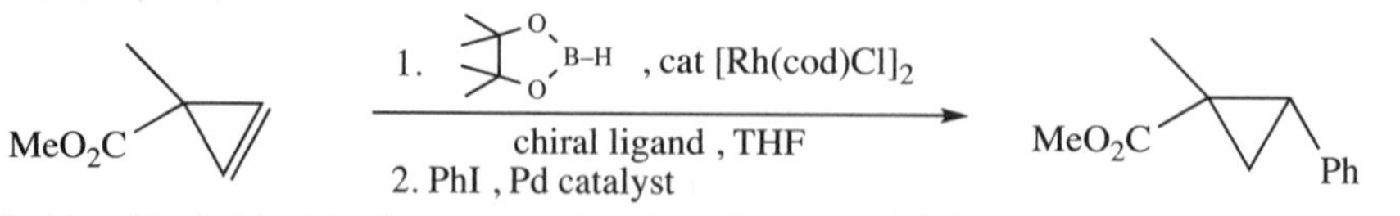

MeO_2C → 1. B–H , cat $[Rh(cod)Cl]_2$ / chiral ligand , THF / 2. PhI , Pd catalyst → MeO_2C Ph 76% (1S,2S)

Rubina, M.; Rubin, M.; Gevorgyan, V. *J. Am. Chem. Soc.* ***2003****, 125*, 7198.

$C_{10}H_{21}$-I → BuZnBr , 2% $Pd_2(dba)_3$/8% PCy_3 , NMI / THF/NMP , 80°C , 14 h → $C_{10}H_{2\text{-}1}$Bu 87%

Zhou, J.; Fu, G.C. *J. Am. Chem. Soc.* ***2003****, 125*, 12527.

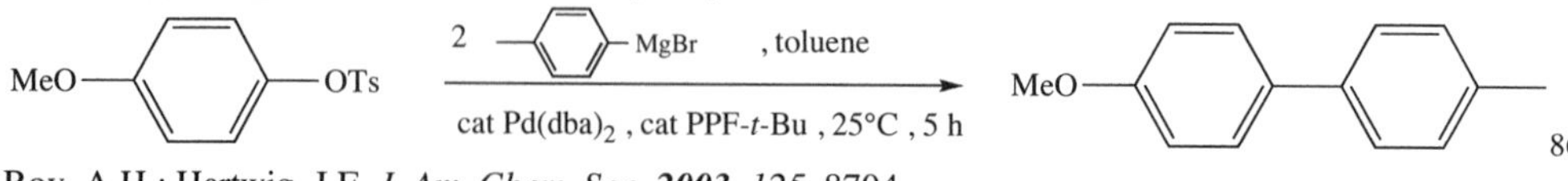

MeO—C6H4—OTs → 2 MgBr , toluene / cat $Pd(dba)_2$, cat PPF-*t*-Bu , 25°C , 5 h → MeO 86%

Roy, A.H.; Hartwig, J.F. *J. Am. Chem. Soc.* ***2003****, 125*, 8704.

I

5% $Pd(OAc)_2$, 5% dppm

2 CsO_2Ct-Bu , DMF , 110°C , 3 d

Ph

40%

Campo, M.A.; Huang, Q.; Yao, T.; Tian, Q.; Larock, R.C. *J. Am. Chem. Soc.* ***2003****, 125*, 11506.

Ts — N — Br

IZn , DMA , rt

4% $Ni(cod)_2$, 8% S-BuPybox

Ts — N

66%

Zhou, J.; Fu, G.C. *J. Am. Chem. Soc.* ***2003****, 125,* 14726.

Cl

$B(OH)_2$, chiral Pd catalyst

0.6 NaO*t*-Bu , *i*-PrOH , rt , 50 min

77%

Navarro, O.; Kelly III, R.A.; Nolan, S.P. *J. Am. Chem. Soc.* ***2003****, 125,* 16194.

I + CO_2Me Br

2-norbornene , CH_2=$CHCO_2Me$
cat $Pd(OAc)_2$, K_2CO_3 , 105°C

DMF , 1 d

CO_2Me

80%

Faccini, F.; Motti, E.; Catellani, M. *J. Am. Chem. Soc.* ***2004****, 126,* 78.

Br

$PhB(OH)_2$, 4% $Ni(cod)_2$, 60°C

8% bathophenathridine , 1.6 *t*-BuOK
sec-BuOH , 5 h

Ph

91%

Zhou, J.; Fu, G.C. *J. Am. Chem. Soc.* ***2004****, 126*, 1340.

Br

$PhSiF_3$, cat $NiBr_2$•diglyme

cat bathophenantroine , DMSO
3.8 CsF , 60°C

Ph

80%

Powell, D.A.; Fu, G.C. *J. Am. Chem. Soc.* ***2004****, 126*, 7788.

Cl

O

EtMnCl , THF , –50°C , 1 h

Et

O

97%

Cahiez, G.; Luart, D.; Lecomte, F. *Org. Lett.* ***2004****, 6*, 4395.

MeO — OTs

$PhB(OH)_2$, K_3PO_4 , THF , rt , 8 h

3% $Ni(cod)_2$, 12% PCy_3

MeO — Ph

86%

Tang, Z.-Y.; Hu, Q.-S. *J. Am. Chem. Soc.* ***2004****, 126*, 3058.

Me$_2$N–C$_6$H$_4$–Br → 1. H–B(pinacol), 5% Pd(OAc)$_2$, 100°C, dioxane, 2 h; 2. Br–C$_6$H$_4$–OMe, 5% Pd(OAc)$_2$, 8 CsF, dioxane → Me$_2$N–C$_6$H$_4$–C$_6$H$_4$–OMe 62%

Broutin, P.-E; Cerna, I.; Camaniello, M.; Leroux, F.; Colobert, F. *Org. Lett.* ***2004**, 6*, 4419.

O$_2$N–C$_6$H$_4$–Cl → PhB(OH)$_2$, 3% imidazolium salt, 40°C, MeOH, toluene; Pd catalyst, 10% Bu$_4$NBr, 3 KOMe → O$_2$N–C$_6$H$_4$–Ph 80%

Fairlamb, I.J.S.; Kapdi, A.R.; Lee, A.F. *Org. Lett.* ***2004**, 6*, 4435.

C$_6$H$_{13}$, SiMe$_3$, I, I, C$_6$H$_{13}$ → MeMgBr, toluene; cat Pd(Ph$_3$)$_4$ → C$_6$H$_{13}$, SiMe$_3$, Me, I, C$_6$H$_{13}$ 60%

Uemura, M.; Takayama, Y.; Sato, F. *Org. Lett.* ***2004**, 6*, 5001.

OMe, Br, O → cat Pd(OAc)$_2$, 2 K$_2$CO$_3$; bis(phosphazine) ligand, 125°C, DMF → MeO, O 90% (95% ee) 91%

Campeau, L.-C.; Parisien, M.; Leblanc, M.; Fagnou, K. *J. Am. Chem. Soc.* ***2004**, 126*, 9186.

1-iodonaphthalene → Pd/C, KOH-H$_2$O, reflux; PhMe, SiCl$_2$ → 1-phenylnaphthalene 72%

Huang, T.; Li, C-J. *Tetrahedron Lett.* ***2002**, 43*, 403.

2 PhI → NEt$_3$, DMF, cat palldocycle, 130°C, 1 d → Ph–Ph quant

Silveira, P.B.; Lando, V.R.; Dupont, J.; Monteiro, A.L. *Tetrahedron Lett.* ***2002**, 43*, 2327.

F$_3$C–C$_6$H$_4$–Br → CH$_2$(SO$_2$Ph)$_2$, dioxane, 70°C, 1.5 NaH; 2% Pd$_2$dba$_3$•CHCl$_3$, 12% PPh$_3$, 3 h → F$_3$C–C$_6$H$_4$–CH(SO$_2$Ph)$_2$ 85%

Kashin, A.N.; Mitin, A.V.; Beletskaya, I.P.; Wife, R. *Tetrahedron Lett.* ***2002**, 43*, 2539.

2-chloroquinoline (N, Cl) → PhMgBr, THF, –30°C, 1h; 10% Fe(acac)$_3$ → 2-phenylquinoline (N, Ph) 65%

Quintin, J.; Franck, X.; Hocquemiller, R.; Figadère, B. *Tetrahedron Lett.* ***2002**, 43*, 3547.

MeO–C$_6$H$_4$–Br → PhB(OH)$_2$, 0.5% NiCl$_2$•6 H$_2$O; 2 K$_3$PO$_4$, dioxane, 60h → 74%

Zim, D.; Monteiro, A.L. *Tetrahedron Lett.* ***2002**, 43*, 4009.

4-Me-C6H4-I $\xrightarrow[\text{2\% Pt(PPh}_3)_4\text{ , 120°C}]{\text{PhB(OH)}_2\text{ , 2 Cs}_2\text{CO}_3\text{ , DMF , 12h}}$ 4-Me-C6H4-Ph 66%

Oh, C.H.; Lim, Y.M.; You, C.H. *Tetrahedron Lett.* ***2002***, *43*, 4645.

Br-C6H4-OMe $\xrightarrow[\text{2\% (oxazoline ligand) , dioxane , 80°C}]{\text{PhBOH)}_2\text{ , 2 Cs}_2\text{CO}_3}$ Ph-C6H4-OMe 80%

Tao, B.; Boykin, D.W. *Tetrahedron Lett.* ***2002***, *43*, 4955.

PhI $\xrightarrow[\text{K}_2\text{CO}_3\text{ , DMF , 1d}]{\text{PhB(OH)}_2\text{ , cat Pd}^{+2}\text{-Sepiolite , 100°C}}$ Ph-Ph 80%

Shimizu, K.-i.; Kan-no, T.; Kodama, T.; Hagiwara, H.; Kitayama, Y. *Tetrahedron Lett.* ***2002***, *43*, 5653.

t-Bu-C6H4-I $\xrightarrow[\text{microwaves , 195°C , 2h}]{\text{PhI , 10\% CuI , 2 Cs}_2\text{CO}_3\text{ , NMP}}$ *t*-Bu-C6H4-Ph 90%

He, H.; Wu, Y.-J. *Tetrahedron Lett.* ***2003***, *44*, 3445.

Br-C6H4-CH(Me)NH2 + imidazole $\xrightarrow{\text{cat CuI , K}_2\text{CO}_3\text{ , 2h, microwavaes}}$ (imidazol-1-yl)-C6H4-CH(Me)NH2 76%

Wu, Y.-J.; He, H.; L'ttereux, A. *Tetrahedron Lett.* ***2003***, *44*, 4217.

C_5H_{11}-C≡C-CH=CH-Cl $\xrightarrow[\text{cat PdCl}_2\text{(dppf) , ZnCl}_2\text{ , THF , reflux}]{\text{MeO-C}_6\text{H}_4\text{-MgBr}}$ C_5H_{11}-C≡C-CH=CH-C6H4-OMe 95%

Peyrat, J.-F.; Thomas, E.; L'Hermite, N.; Alami, M.; Brion, J.-D. *Tetrahedron Lett.* ***2003***, *44*, 6703.

PhBr $\xrightarrow{\text{Et}_3\text{Al , CoCl}_3\text{ , cat PdCl}_2\text{(PPh}_3)_2\text{ , THF , 22°C}}$ PhEt 71%

Shenglof, M.; Gelman, D.; Molander, G.A.; Blum, J. *Tetrahedron Lett.* ***2003***, *44*, 8593.

Br-C6H4-CH_2Br $\xrightarrow[\text{cat Pd(PPh}_3)_4\text{ , aq Na}_2\text{CO}_3]{\text{PhB(OH)}_2\text{ , EtOH , toluene}}$ Br-C6H4-CH_2Ph 84%

Langle, S.; Abarbri, M.; Duchêne, A. *Tetrahedron Lett.* ***2003***, *44*, 9255.

PhCH2Br $\xrightarrow[\text{cat Pd complex}]{\text{CH}_2\text{=CHCH}_2\text{SnBu}_3\text{ , toluene , 60°C , 2d}}$ PhCH2CH2CH=CH2 31% (better yield with Sn compunds)

Crawforth, C.M.; Fairlamb, I.J.S.; Taylor, R.J.K. *Tetrahedron Lett.* ***2004***, *45*, 461.

Br-C6H4-CO_2Et $\xrightarrow[\text{dioxane , reflux , 2h}]{\text{Me}_2\text{Zn , cat d(dppe)Cl}_2}$ Me-C6H4-CO_2Et 95%

Herbert, J.M. *Tetrahedron Lett.* ***2004***, *45*, 817.

2 PhI $\xrightarrow{\text{Zn , NH}_4\text{OCHO , NaOH , MeOH , reflux}}$ Ph–Ph 96%

Abiraj, K.; Srinivasa, G.R.; Gowda, D.C. *Tetrahedron Lett.* ***2004***, *45*, 2081.

4-Me-C6H4-Cl $\xrightarrow[\text{1,3-bis(\textit{tert}-butyl)imidazolium chloride}]{\text{PhB(OH)}_2\text{ , 3 KOMe , 3\% Pd(dba)}_2\text{ , toluene , MeOH , 10\% TBAB , 40°C , 1d}}$ 4-Me-C6H4-Ph 75%

Arentsen, K.; Caddick, S.; Cloke, G.N.; Herring, A.P.; Hitchcock, P.B. *Tetrahedron Lett.* ***2004***, *45*, 3511.

MeO / Br (3-bromoanisole) $\xrightarrow[\text{aq DMF , rt , 1h}]{PhB(OH)_2 \text{ , Pd(0)-NaY}}$ MeO / Ph 97%

Artok, L.; Bulut, H. *Tetrahedron Lett.* ***2004**, 45*, 3881.

I—C6H4—CO_2Et $\xrightarrow[\text{2. Br—C6H4—CHO cat } PdCl_2(PPh_3)_2 \text{ , 100°C}]{\text{1, Zn/Cu , THF}}$ EtO_2C—C6H4—C6H4—CHO 87x85%

Mutule, I.; Suna,, E. *Tetrahedron Lett.* ***2004**, 45*, 3909.

2 PhI $\xrightarrow{2\ NEt_3 \text{ , [bmim] } PF_6 \text{ , 120°C , 1d}}$ Ph-Ph 71%

Park, S.B.; Alper, H. *Tetrahedron Lett.* ***2004**, 45*, 5515.

$PhCH_2Br$ $\xrightarrow[\text{2.5h , rt}]{PhB(OH)_2 \text{ , cat } PdCl_2 \text{ , aq acetone}}$ $PhCH_2Ph$ 95%

Bandgar, B.P.; Bettigeri, S.V.; Phopase, J. *Tetrahedron Lett.* ***2004**, 45*, 6959.

Ph/CH=CH/CH2Cl $\xrightarrow[\text{DCM , –78°C , 4h}]{\text{MeMgBr , 3\% CuBr , 3.3\% L}}$ (chiral branched product) (89 : 11) quant + Ph/CH=CH/Et

L = chiral phosphoramidite

Tissot-Croset, K.; Alexakis, A. *Tetrahedron Lett.* ***2004**, 45*, 7375.

Br—C6H4—CHO $\xrightarrow[\text{0.5\% Pd resin , 90°C}]{1.5\ PhB(OH)_2 \text{ , } 3\ K_3PO_4 \text{ , DMF , 1d}}$ Ph—C6H4—CHO quant

Phan, N.T.S.; Brown, D.H.; Styring, P. *Tetrahedron Lett.* ***2004**, 45*, 7915.

$PhCH_2Br$ $\xrightarrow[\text{cat } Pd(OAc)_2\text{, } PCy_3]{PhB(OH)_2 \text{ , } K_3PO_4 \text{ , toluene , 80°C}}$ $PhCH_2Ph$ 94%

Nobre, S.M.; Monteiro, A.L. *Tetrahedron Lett.* ***2004**, 45*, 8225.

MeO—C6H4—I $\xrightarrow[\text{dioxane , 100°C}]{PhB(OH)_2 \text{ , cat Pd° , } Cs_2CO_3}$ MeO—C6H4—Ph 81%

Pd° = macrocyclic triolefinic Pd° complex anchored to silica

Blanco, B.; Mehdi, A.; Moreno-Mañas, M.; Pleixats, R.; Reyé, C. *Tetrahedron Lett.* ***2004**, 45*, 8789.

(2-OMe, 6-Me-phenyl)Li + (2-Cl-anisole, OMe / Cl) $\xrightarrow[\text{12h}]{\text{65°C , THF}}$ biaryl (OMe, Me) (82 : 18) 80% + biaryl (OMe, Me, OMe)

Becht, J.-M.; Gissot, A.; Wagner, A.; Mioskowski, C. *Tetrahedron Lett.* ***2004**, 45*, 9331.

Ph/CH2CH2/$B(OH)_2$ $\xrightarrow[3\ K_2CO_3 \text{ , } THF/H_2O \text{ , reflux}]{\text{PhBr , cat } PdCl_2(dppf)\text{•}CH_2Cl_2}$ Ph/CH2CH2/Ph 73%

Molander, G.A.; Yun, C.-S. *Tetrahedron* ***2002**, 58*, 1465.

Br—C6H4—C(O)Me $\xrightarrow[\text{130°C , 20h , xylene , } K_2CO_3]{BuB(OH)_2 \text{ , cat } [Pd(C_3H_5)Cl]_2\text{/Tedicup}}$ Bu—C6H4—C(O)Me 92%

Kondolff, I.; Doucet, H.; Santelli, M. *Tetrahedron* ***2004**, 60*, 3813.

2-Bromopyridine $\xrightarrow{\text{2 (TMS)}_3\text{SiH , PhH , 2 AIBN}}$ 2-phenylpyridine 41%

Núñez, A.; Sánchez, A.; Burgos, C.; Alvarez-Builla, J. *Tetrahedron* **2004**, *60*, 6217.

MeO–C_6H_4–Br $\xrightarrow[\text{EtOH , rt , 4h}]{\text{PhB(OH)}_2\text{ , cat NHC-Pd complex}}$ MeO–C_6H_4–Ph 98%

NHC = tetradentate N-heterocyclic carbene

Zhao, Y.; Zhou, Y.; Ma, D.; Liu, J.; Li, L.; Zhang, T.Y.; Zhang, H. *Org. Biomol. Chem.* **2003**, *1*, 1643.

2-Iodophenyl allyl ether $\xrightarrow{\text{SmI}_2\text{ , H}_2\text{O , NEt}_3\text{ , 20°C}}$ 3-methyl-2,3-dihydrobenzofuran >99%

Dahlén, A.; Petersson, A.; Hilmersson, G. *Org. Biomol. Chem.* **2003**, *1*, 2423.

MeO–C_6H_4–I $\xrightarrow[\text{(bmim) BF}_4\text{ , ultrasound}]{\text{PhB(OH)}_2\text{ , MeOH , 20 min}}$ MeO–C_6H_4–Ph 92%

Rajagopal, R.; Jarikote, D.V.; Srinivasan, K.V. *Chem. Commun.* **2002**, 616.

PhI $\xrightarrow[\text{C}_{10}\text{H}_{23}\text{(C}_6\text{H}_{13}\text{)}_3\text{PCl (ionic liquid) , 1h}]{\text{PhB(OH)}_2\text{ , cat Pd}_2\text{dba}_3\text{•CHCl}_3\text{ , K}_3\text{PO}_4\text{ , 50°C}}$ Ph-Ph 95%

McNulty, J.; Capretta, A.; Wilson, J.; Dyck, J.; Adjabeng, G.; Robertson, A. *Chem. Commun.* **2002**, 1986.

2,4-Dinitrofluorobenzene (O_2N, F, NO_2) $\xrightarrow[\text{reflux}]{\text{PhB(OH)}_2\text{ , 5\% Pd}_2\text{dba}_3\text{ , 16h ; 20\% PMe}_3\text{ , 2.2 Cs}_2\text{CO}_3\text{ , DME}}$ 2,4-dinitrobiphenyl (O_2N, Ph, NO_2) 85%

Widdowson, D.A.; Wilhelm, T. *Chem. Commun.* **2003**, 578.

2-Chlorobenzonitrile $\xrightarrow[\text{3\% NiCL}_2\text{(dppf) , microwaves}]{\text{ClZn–C}_6\text{H}_4\text{–OMe , 140°C , THF}}$ 2-cyano-4'-methoxybiphenyl (CN, OMe) 81%

Walla, P.; Kappe, C.O. *Chem. Commun.* **2004**, 564.

MeO_2C–C_6H_4–Br $\xrightarrow[\text{PhH , 30h}]{\text{cat [Cp*IrHCl]}_2\text{ , KO}t\text{-Bu , 80°C}}$ MeO_2C–C_6H_4–NMe_2 66%

Fujita, K.-i.; Nonogawa, M.; Yamaguchi, R. *Chem. Commun.* **2004**, 1926.

Cyclohexyl chloride $\xrightarrow[\text{Fe salen catalyst}]{\text{4-Me–C}_6\text{H}_4\text{–MgBr , ether , 45°C , 30 min}}$ 4-cyclohexyltoluene 90%

Bedford, R.B.; Bruce, D.W.; Frost, R.M.; Goodby, J.W.; Hird, M. *Chem. Commun.* **2004**, 2822.

Ph–CH=CH–CH₂–Br $\xrightarrow[\text{cat Cu(OTf)}_2\text{C}_6\text{H}_6\text{ , spirophosphoramidite ligand}]{\text{Et}_2\text{Zn , digylme , 30°C , 72h}}$ Ph–CH(Et)–CH=CH₂ + Ph–CH=CH–CH₂–Et (88 : 12) 62%

71% ee

Shi, W.-J.; Wang, L.-X.; Fu, Y.; Zhu, S.-F.; Zhou, Q.-L. *Tetrahedron: Asymmetry* **2003**, *14*, 3867.

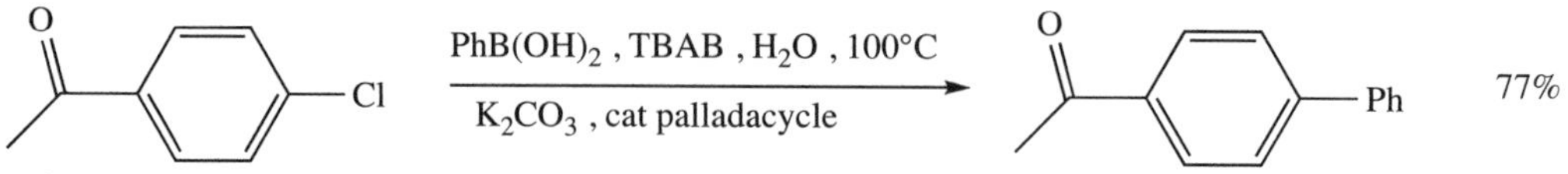

$PhB(OH)_2$, TBAB , H_2O , 100°C
K_2CO_3 , cat palladacycle
Cl
Ph
77%

Boltella, L.; Nájera, C. *Angew. Chem. Int. Ed.* **2002**, *41*, 179.

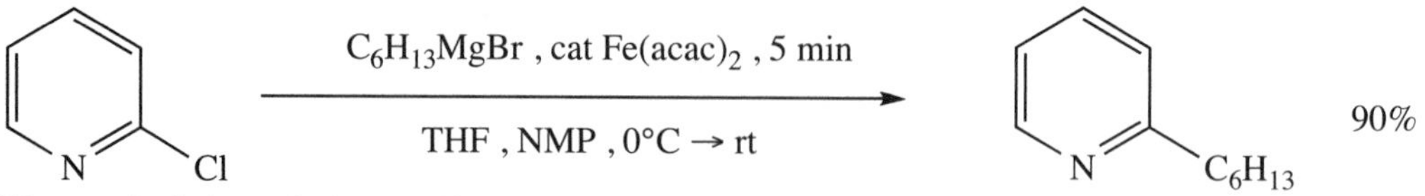

$C_6H_{13}MgBr$, cat $Fe(acac)_2$, 5 min
THF , NMP , 0°C → rt
N Cl
N C_6H_{13}
90%

Fürstner, A.; Leitner, A. *Angew. Chem. Int. Ed.* **2002**, *41*, 609.

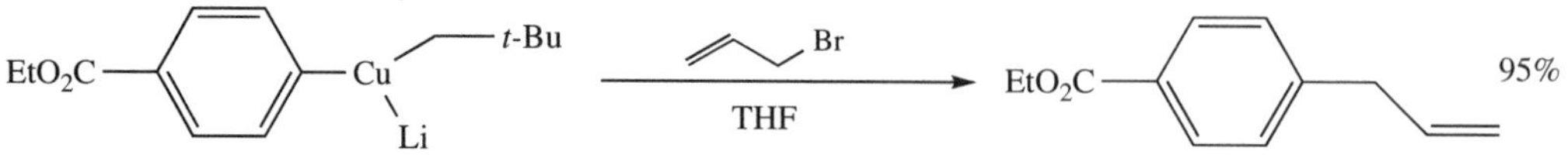

EtO_2C Cu *t*-Bu Li
Br
THF
EtO_2C
95%

Piazza, C.; Knochel, P. *Angew. Chem. Int. Ed.* **2002**, *41*, 3263.

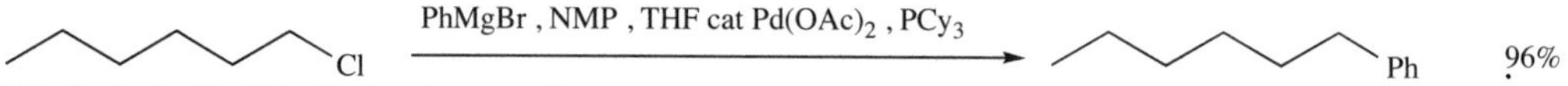

PhMgBr , NMP , THF cat $Pd(OAc)_2$, PCy_3
Cl
Ph
96%

Frisch, A.C.; Shaikh, N.; Zapf, A.; Beller, M. *Angew. Chem. Int. Ed.* **2002**, *41*, 4056.

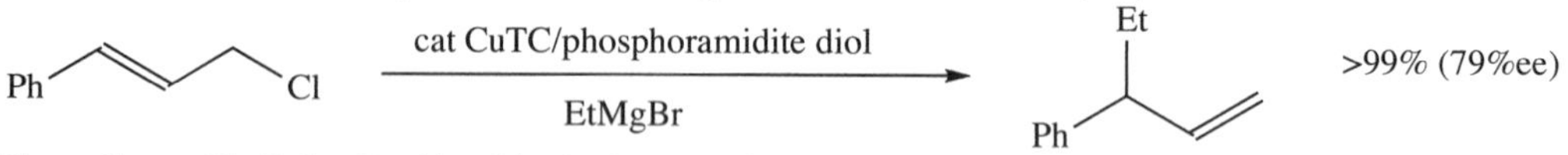

cat CuTC/phosphoramidite diol
EtMgBr
Ph Cl
Et Ph
>99% (79%ee)

Tissot-Croset, K.; Polet, D.; Alexakis, A. *Angew. Chem. Int. Ed.* **2004**, *43*, 2426.

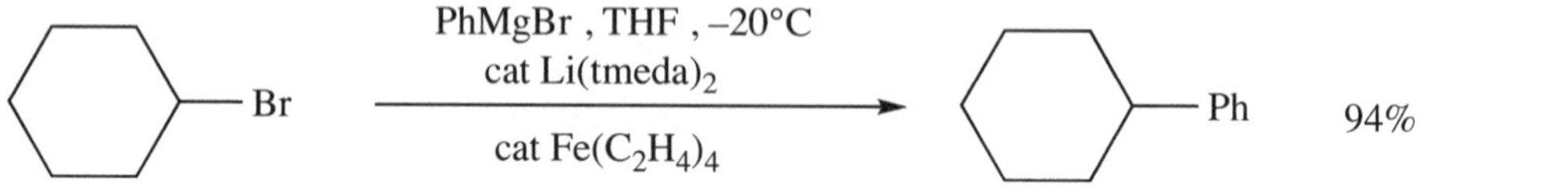

PhMgBr , THF , –20°C
cat $Li(tmeda)_2$
cat $Fe(C_2H_4)_4$
Br
Ph
94%

Martin, R.; Fürstner, A. *Angew. Chem. Int. Ed.* **2004**, *43*, 3955.

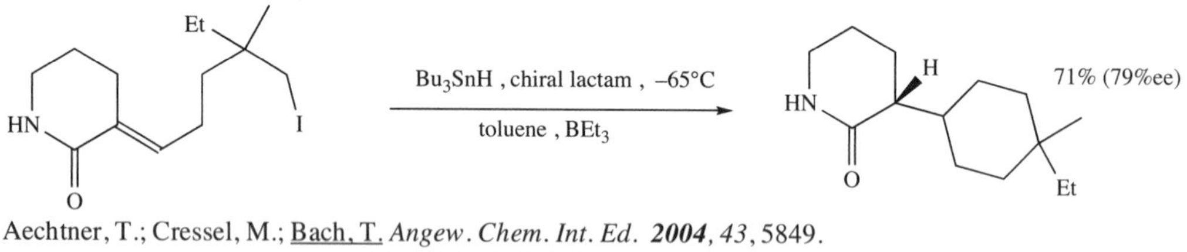

Et HN O I
Bu_3SnH , chiral lactam , –65°C
toluene , BEt_3
HN O H Et
71% (79%ee)

Aechtner, T.; Cressel, M.; Bach, T. *Angew. Chem. Int. Ed.* **2004**, *43*, 5849.

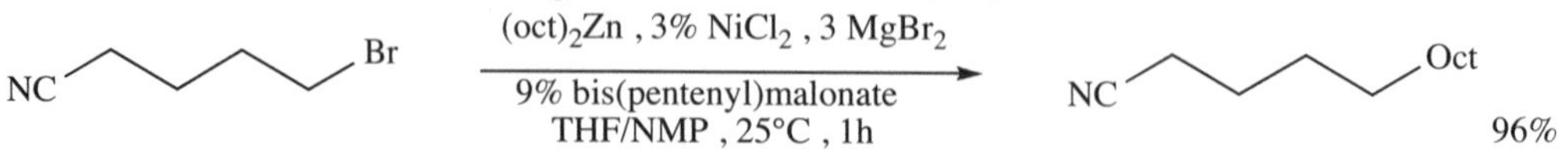

NC Br
$(oct)_2Zn$, 3% $NiCl_2$, 3 $MgBr_2$
9% bis(pentenyl)malonate
THF/NMP , 25°C , 1h
NC Oct
96%

Terao, J.; Todo, H.; Watanabe, H.; Ikumi, A.; Kambe, N. *Angew. Chem. Int. Ed.* **2004**, *43*, 6180.

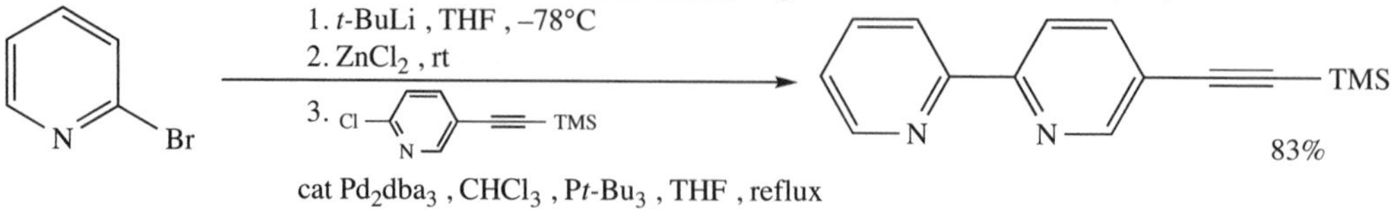

N Br
1. *t*-BuLi , THF , –78°C
2. $ZnCl_2$, rt
3. Cl N TMS
cat Pd_2dba_3 , $CHCl_3$, Pt-Bu_3 , THF , reflux
N N TMS
83%

Lützen, A.; Hapke, M. *Eur. J. Org. Chem.* **2002**, 2292.

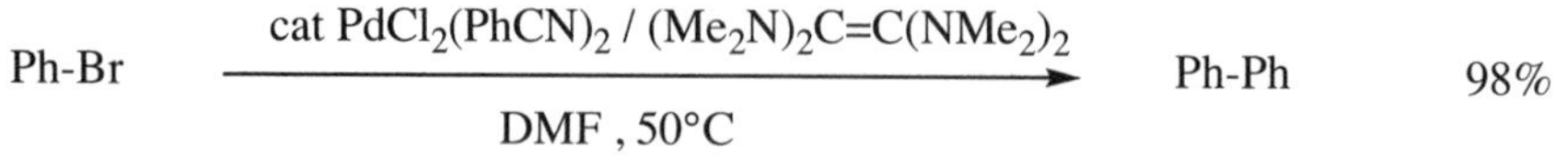

Ph-Br
cat $PdCl_2(PhCN)_2$ / $(Me_2N)_2C{=}C(NMe_2)_2$
DMF , 50°C
Ph-Ph
98%

Kuroboshi, M.; Waki, Y.; Tanaka, H. *Synlett* **2002**, 637.

I—C6H4—OMe → Ph—C6H4—OMe 96%

$PhB(OH)_2$, H_2O , CTAB / cat $Pd(PPh_3)_4$, BuOH / K_2CO_3 , 3 h

Arcadi, A.; Cerichelli, G.; Chiarini, M.; Correa, M.; Zorzan, D. *Eur. J. Org. Chem.* ***2003***, 4080.

t-Bu—C6H4—Br → *t*-Bu—C6H4—CH=CH2 87%

CH_2=CHB$(OH)_2$, K_2CO_3 , xylene / cat $[Pd(C_3H_5)Cl]_2$/Tedicyp

Peyroux, E.; Berthiol, F.; Doucet, H.; Santelli, M. *Eur. J. Org. Chem.* ***2004***, 1075.

Me—C6H4—Br → Me—C6H4—Ph 99%

$PhB(OH)_2$, 2% $Pd(OAc)_2$, DMF / 3% bis(hydrazone) , Cs_2CO_3 , H_2O

Mino, T.; Shirae, Y.; Sakamoto, M.; Fujita, T. *Synlett* ***2003***, 882.

2-Bromoanisole (OMe, Br) → 2-phenylanisole (OMe, Ph) 98%

Ph_2BF_2K , 3% $Pd(OAc)_2$/4 PPh_3 , 6h / $MeOCH_2CH_2OH$, 3 Na_2CO_3

Ito, T.; Iwai, T.; Mizuno, T.; Ishino, Y. *Synlett 2003,* 1435.

2-Bromoanisole (Br, OMe) → 2-phenylanisole (Ph, OMe)

$PhB(OH)_2$, 2M K_2CO_3 , EtOH / cat (bead)-Pd , 65°C

Shieh, W.-C.; Shekhar, R.; Blacklock, T.; Tedesco, A. *Synth. Commun.* ***2002***, *32,* 1059.

Cl—C6H4—Me → Ph—C6H4—Me 92%

–(Si(Me)Ph-O–$)_n$, toluene , 120°C , 2d / 5% $PdCl_2(PCy_3)$- , 3 K_2CO_3 , H_2O

Koike, T.; Mori, A. *Synlett* ***2003,*** 1850.

2-Chloroquinoline (N, Cl) → 2-phenylquinoline (N, Ph) 84%

PhMgCl , 5% $CoCl_2$, ether / –40°C

Korn, T.J.; Cahiez, G.; Knochel, P. *Synlett* ***2003***, 1892.

Br—CH=CH—C6H4—Me → Ph—CH(Me)—CH=CH—C6H4—Me 50% (78% ee)

PhCH(Me)MgCl , $PhCF_3$ / Pd/chiral phosphine / –10°C , 3 h

Horibe, H.; Kazuta, K.; Kotoku, M.; Kondo, K.; Okuno, H.; Murakami, Y.; Aoyama, T. *Synlett* ***2003***, 2047.

$PhCH_2Br$ → $PhCH_2Ph$ 99%

$PhB(OH)_2$, cat $[Pd(C_3H_5)Cl]_2$, xylene / cat tetraphosphine, K_2CO_3 , 130°C

Chahen, L.; Doucet, H.; Santelli, M. *Synlett* ***2003***, 1668.

PhBr → Ph—CH(CN)2 90%

$CH_2(CN)_2$, 1% $PdCl_2$/dppb , NaH / Py , 85°C , 12 h

Gao, C.; Tao, X.; Qian, Y.; Huang, J. *Synlett* ***2003***, 1716.

N-Bn-2-piperidone → 3-Ph-N-Bn-2-piperidone: 2.5 LiHMDS , PhBr , 80°C, toluene , phosphine ligand, 3% $Pd(OAc)_2$ — 87%

Cossy, J.; de Filippis, A.; Pardo, D.G. *Synlett* ***2003***, 2171.

CH_2=CHPh — PhI , cat $Pd(PPh_3)_2Cl_2$, K_2CO_3 , TBAB; H_2O , microwaves → PhCH=CHPh 91%

Wang, J.-X.; Liu, Z.; Hu, Y.; Wei, B.; Bai, L. *Synth. Commun.* ***2002***, *32*, 1607.

4-Me-C₆H₄-Br — $PhB(OH)_2$, 1% $PdCl_2$, K_2CO_3; Py , reflux , 3 h; no ligand → 4-Me-C₆H₄-Ph quant

Tao, X.; Zhao, Y.; Shen, D. *Synlett* ***2004***, 359.

Ph-I — HCO_2 $NHEt_3$, NaOH , MeOH , reflux; Zn , 2 h → Ph-Ph 95%

Abiraj, K.; Srinivasa, G.R.; Gowda, D.C. *Synlett* ***2004***, 877.

Cl-C₆H₄-CN — Bi complex , 2 CsF , PPh_3 , NMP; 10% $Pd(OAc)_2/PPh_3$, 100°C , 12 h → Ph-C₆H₄-CN 93%

Yamazaki, O.; Tanaka, T.; Shimada, S.; Suzuki, Y.; Tanaka, M. *Synlett* ***2004***, 1921.

4-Me-C₆H₄-Cl — $(HO)_2B$-C₆H₄-OMe; 5% $Pd(OAc)_2$, 2 *t*-BuOK , DMF; 5% tetrazole ligand , 0.1 TBAB → 4-Me-C₆H₄-C₆H₄-OMe 88%

Gupta, A.K.; Rim, C.Y.; Oh, C.H. *Synlett* ***2004***, 2227.

Ph-I — 5% Pd on KF/Al_2O_3 , 100°C, 4 h; 4-Me-C₆H₄-$B(OH)_2$, solvent free → 4-Me-C₆H₄-Ph 99%

Kabalka, G.W.; Wang, L.; Pagni, R.M.; Hair, C.M.; Namboodiri, V. *Synthesis* ***2003***, 217.

1-Cl-naphthalene — ethylated indium complex, Pd cat; 78°C , 8 h → 1-Et-naphthalene 99%

Shenglof, M.; Gelman, D.; Heymer, B.; Schumann, H.; Molander, G.A.; Blum, J. *Synthesis* ***2003***, 302.

$(PhCH_2)_2C(CH_2Cl)_2$ — $Cp_2Ti[P(OEt)_3]_2$; rt , 2 h → 1,1-dibenzylcyclopropane 93%

Takeda, T.; Shimane, K.; Fujiwara, T.; Tsubouchi, A. *Chem. Lett.* ***2002***, *31*, 290.

MeO-C₆H₄-Br — $PhB(OH)_2$, $PdCl_2$, 5% Bu_4NBr , EtOH; K_3PO_4 , rt → MeO-C₆H₄-Ph 84%

Deng, Y.; Gong, L.; Mi, A.; Li, H.; Jiang, Y. *Synthesis* ***2003***, 337.

Ph, Br, O; Bu_3Sn; $BF_3{\bullet}OEt_2$, DCM; Ph, O — 79%

Miyake, H.; Hirai, R.; Nakajima, Y.; Sasaki, M. *Chem. Lett.* ***2003****, 32,* 164.

I, O, N, Me; 5-10% $Pd(OAc)_2$, P ligand; Bu_4NCl , AcONa , DMF; 100°C , 30 min; microwaves; O, N, Me — 99%

Sørensen, U.S.; Pombo, E. *Helv. Chim. Acta* ***2004****, 87,* 82.

Br, N, Me, O; $Pd(OAc)_2$, dppp , Bu_3P , DMF; Ag_2CO_3 , reflux; N, Me, O — 95%

Harayama, T.; Toko, H.; Nishioka, H.; Abe, H.; Takeuchi, Y. *Heterocycles* ***2003****, 59,* 541.

Cl, N, Cl, N; cat $Pd(PPh_3)_2Cl_2$; Ph, N, Ph, N — 91%

Schultheiss, N.; Bosch, E. *Heterocycles* ***2003****, 60,* 1891.

I, N, NHMe; 2 OAc , 5% $Pd(OAc)_2$, LiCl; K_2CO_3 , DMF , 120°C; N, N, Me, Me — 65%

Sung, C.; Seo, J.Y.; Yum, E.E.; Sung, N.-D. *Heterocycles* ***2004****, 63,* 631.

$PhCH_2OCOCF_3$; CH_2=$CHCO_2Et$, DMF , 100°C; 5% $Pd(OAc)_2$, 20% PPh_3 , 39 h; Ph, CO_2Et — 75%

Narahashi, H.; Yamamoto, A.; Shimizu, I. *Chem. Lett.* ***2004****, 33,* 348.

$PhSi(allyl)_3$; 1. 5 TBAF , THF-H_2O , rt , 1 h; 2. 2.5% PhCl , $[\eta^3\text{-}C_3H_5)PdCl]_2$, 80°C; 10% chiral biaryl phosphine; Ph–Ph — 95%

Sahoo, A.K.; Nakao, Y.; Hiyama, T. *Chem. Lett.* ***2004****, 33,* 632.

n-$C_7H_{15}OTs$ + PhMgBr; cat $NiCl_2$, 1,3-butadiene , THF; 25°C , 3 h; n-C_7H_{15}–Ph — 85%

Terao, J.; Naitoh, Y.; Kuniyasu, H.; Kambe, N. *Chem. Lett.* ***2003****, 32,* 890.

REVIEWS:

"Palladium-Catalyzed Reactions of Aryl Halides with Soft, Non-Organometallic Nucleophiles"
Prim, D.; Campagne, J.-M.; Joseph, D.; Andrioleth, B. *Tetrahedron* **2002**, *58*, 2041.

"Palladium-Catalyzed Coupling Reactions of Aryl Chlorides"
LIttke, A.F.; Fu, G.C. *Angew. Chem. Int. Ed.* **2002**, *41*, 4176.

"New Catalytic Approaches in the Stereoselective Friedel-Crafts Alkylation Reaction"
Bandini, M.; Melloni, A.; Umani-Ronchi, A. *Angew. Chem. Int. Ed.* **2004**, *43*, 550.

"Ruthenium, Rhodium, and Palladium-Catalyzed Carbon-Carbon Bond Formation Involving C-H Activation and Addition on Unsaturated Substrates: Reactions and Mechanistic Aspects"
Fokin, A.A.; Schrener, P.R. *Chem. Rev.* **2002**, *102*, 1731.

"Nucleophilic Substitution Reactions by Electron Transfer"
Rossi, R.A.; Pierini, A.B.; Peñéñory, A.B. *Chem. Rev.* **2003**, *103*, 71.

"The Asymmetric Intramolecular Heck Reaction in Natural Product Total Synthesis"
Dounay, A.B.; Overman, L.E. *Chem. Rev.* **2003**, *103*, 2945.

"π-Nucleophilicity in Carbon-Carbon Bond-Forming Reactions"
Mayr, H.; Kempf, B.; Ofial, A.R. *Acc. Chem. Res.* **2003**, *36*, 66.

"Palladium-Catalyzed Arylation of Carbonyl Compounds and Nitriles"
Culkin, D.A.; Hartwig, J.F. *Acc. Chem. Res.* **2003**, *36*, 234.

"Palladium Catalysts for Suzuki Cross-Coupling"
Bellina, F.; Carpita, A.; Rossi, R. *Synthesis* **2004**, 2419.

SECTION 71: ALKYLS, METHYLENES, AND ARYLS FROM HYDRIDES

This section lists examples of the reaction of RH → RR' (R,R' = alkyl or aryl). For the reaction C=CH → C=C-R (R = alkyl or aryl), see Section 209 (Alkenes from Alkenes). For alkylations of ketones and esters, see Section 177 (Ketones from Ketones) and Section 113 (Esters from Esters).

Ph–I —[5% Pd/C, Zn, CO_2 / H_2O, 18 h]→ Ph–Ph >99%

Li, J.-H.; Xie, Y.-X.; Yin, D.-L. *J. Org. Chem.* **2003**, *68*, 9867.

NO_2 —[BuLi, THF, 0°C, e^- / DMF, Et_4NBF_4]→ Bu —— NO_2 41%

Gallardo, I.; Guirado, G.; Marquet, J. *J. Org. Chem.* **2003**, *68*, 631.

NO_2 —[1. Cl, NaH; 2. MeI]→ EtO_2C —— NO_2 60%

Lawrence, N.J.; Liddle, J.; Bushell, S.M. *J. Org. Chem.* **2002**, *67*, 457.

CO_2Et / CO_2Et —— PhBr , NaH , THF , 70°C / 2% $Pd(dba)_2$, 4% $P(t\text{-}Bu)_3$ ——→ Ph—CH(CO_2Et)$_2$

Beare, N.A.; Hartwig, J.F. *J. Org. Chem.* ***2002**, 67*, 541.

1. 1.4 eq *t*-BuLi , ether
2. Me_2SO_4

N, Cl → N, Cl 45%

Fort, Y.; Rodriguez, A.L. *J. Org. Chem.* ***2003**, 68*, 4918.

N, S —— $PhNH_2$, Cs_2CO_3 , 150°C / 5% $Co(OAc)_2$, 10% IMeS , 9h / 0.2 CuI , dioxane ——→ N, S, Ph 90%

Sezen, B.; Sames, D. *Org. Lett.* ***2003**, 5*, 3607.

$B(OH)_2$ + Cl—OMe —— K_3PO_4 , toluene / 80°C , 14 h ——→ OMe 79%

Percec, V.; Golding, G.M.; Smidrkal, J.; Weichold, O. *J. Org. Chem.* ***2004**, 69*, 3447.

N, N, Me —— 0.5% $Pd(OAc)_2$, 2% PPh_3 / PhI , DMA , 2 eq CsOAc / 125°C ,1 d ——→ N, N, Me, Ph 85%

Lane, B.S.; Sames, D. *Org. Lett.* ***2004**, 6*, 2897.

OMe —— 1. BuLi / 2. $(EO)_3SiOTf$ ——→ OMe, $Si(OEt)_3$ 88%

Seganish, W.M.; DeShong, P. *J. Org. Chem.* ***2004**, 69*, 6790.

O —— Ph—B(O)(O) , cat $RuH_2(cod)(PPh_3)_3$ / toluene , reflux ——→ Ph, O, Ph 60%

Kakiuchi, F.; Kan, S.; Igi, K.; Chatani, N.; Murai, S. *J. Am. Chem. Soc.* ***2003**, 125*, 1698.

—— Ru catalyst , CH_2=CH_2 , 90°C ——→ Et no yield

Lail, M.; Arrowood, B.N.; Gunnoe, T.B. *J. Am. Chem. Soc.* ***2003**, 125*, 7506.

N H —— 1. O, P, OMe, OMe, O , 10% Sc catalyst / 2. DBU , MeOH ——→ N H, CO_2Me 83% (83% ee)

Evans, D.A.; Scheidt, K.A.; Fandrick, K.R.; Lam, H.W.; Wu, J. *J. Am. Chem. Soc.* ***2003**, 125*, 10780.

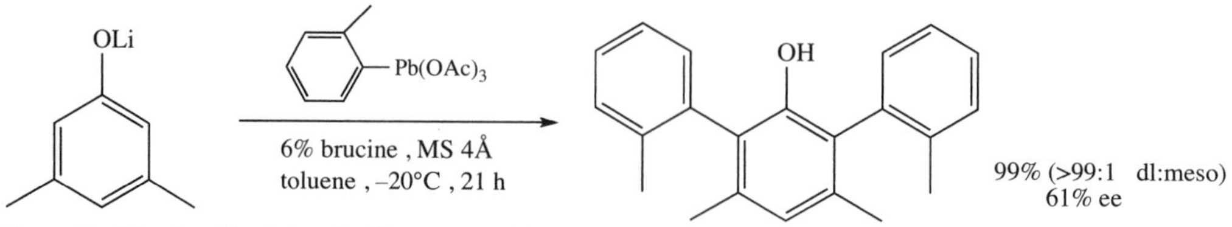

OLi; $Pb(OAc)_3$; 6% brucine, MS 4Å; toluene, –20°C, 21 h; OH; 99% (>99:1 dl:meso) 61% ee

Kano, T.; Ohyabu, Y.; Saito, S.; Yamamoto, H. *J. Am. Chem. Soc.* ***2002**, 124*, 5365.

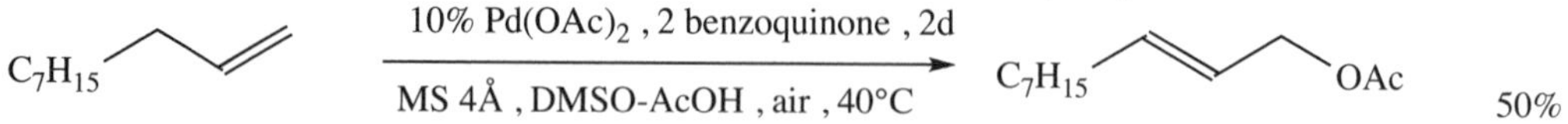

C_7H_{15}; 10% $Pd(OAc)_2$, 2 benzoquinone, 2d; MS 4Å, DMSO-AcOH, air, 40°C; C_7H_{15}; OAc; 50%

Chen, M.S.; White, M.C. *J. Am. Chem. Soc.* ***2004**, 126*, 1346.

$CH_2(CO_2Et)_2$; PhBr, DMAC, Na_2PdCl_4, $Ba(OH)_2$; $PhCH(CO_2Et)_2$; 99%

Aramendía, M.A.; Borau, V.; Jimenez, C.; Marinas, J.M.; Ruiz, J.B.; Urbano, F.J. *Tetrahedron Lett.* ***2002**, 43*, 2847.

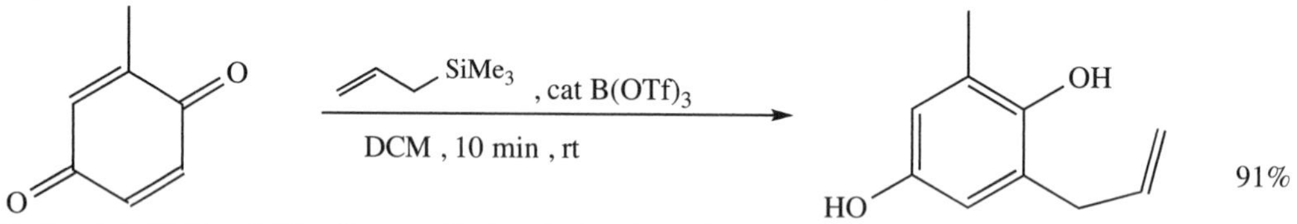

O; O; $SiMe_3$, cat $B(OTf)_3$; DCM, 10 min, rt; OH; HO; 91%

Yadav, J.S.; Reddy, B.V.S.; Swamy, T. *Tetrahedron Lett.* ***2003**, 44*, 4861.

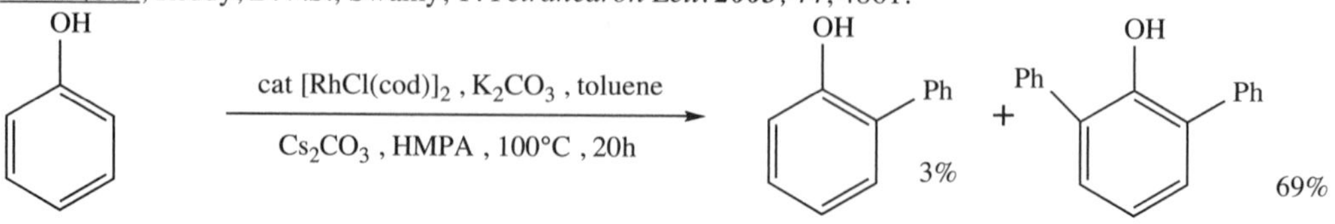

OH; cat $[RhCl(cod)]_2$, K_2CO_3, toluene; Cs_2CO_3, HMPA, 100°C, 20h; OH; Ph; 3%; + Ph; OH; Ph; 69%

Oi, S.; Watanabe, S.; Fukita, S.; Inoue, Y. *Tetrahedron Lett.* ***2003**, 44*, 8665.

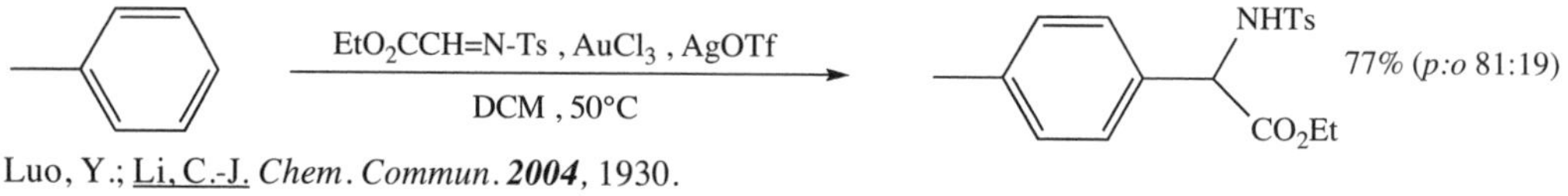

$EtO_2CCH{=}N\text{-}Ts$, $AuCl_3$, AgOTf; DCM, 50°C; NHTs; CO_2Et; 77% (*p*:*o* 81:19)

Luo, Y.; Li, C.-J. *Chem. Commun.* ***2004**,* 1930.

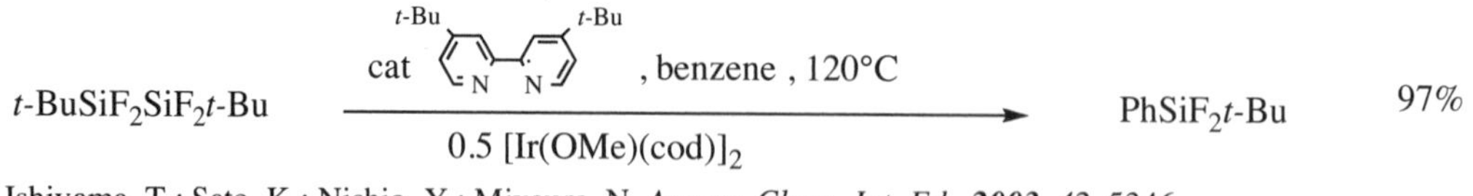

t-$BuSiF_2SiF_2$*t*-Bu; cat (*t*-Bu, *t*-Bu, N, N bipyridine), benzene, 120°C; 0.5 $[Ir(OMe)(cod)]_2$; $PhSiF_2$*t*-Bu; 97%

Ishiyama, T.; Sato, K.; Nishio, Y.; Miyaura, N. *Angew. Chem. Int. Ed.* ***2003**, 42*, 5346.

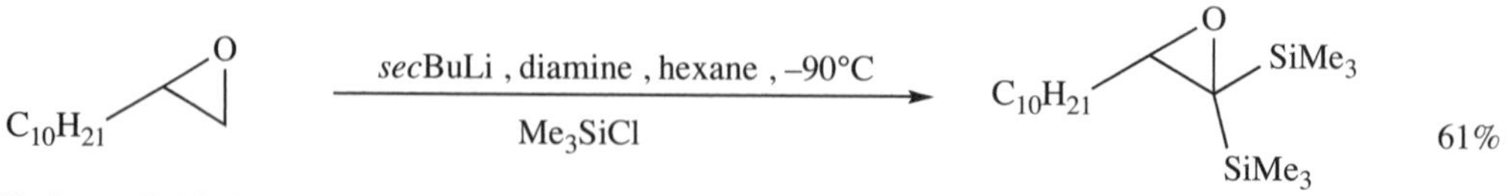

$C_{10}H_{21}$; O; *sec*BuLi, diamine, hexane, –90°C; Me_3SiCl; $C_{10}H_{21}$; O; $SiMe_3$; $SiMe_3$; 61%

Hodgson, D.M.; Krton, E.H.M. *Synlett* ***2004**,* 1610.

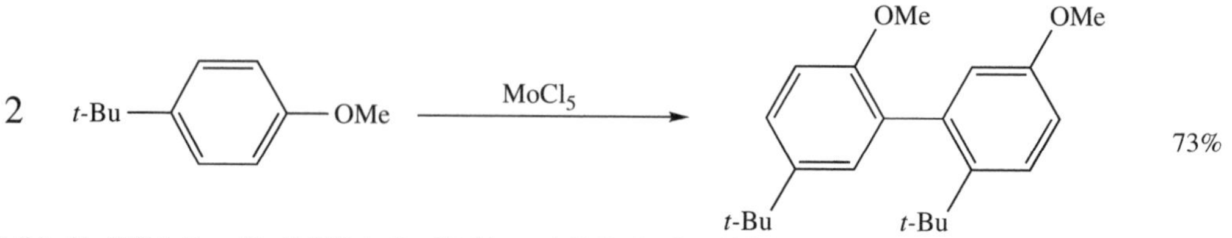

2 *t*-Bu—C6H4—OMe; $MoCl_5$; OMe; OMe; *t*-Bu; *t*-Bu; 73%

Mirk, D.; Wibbeling, B.; Fröhlich, R.; Waldvogel, S.R. *Synlett* ***2004**,* 1970.

SECTION 72: ALKYLS, METHYLENES, AND ARYLS FROM KETONES

The conversions $R_2C=O \rightarrow R\text{-}R$, R_2CH_2, R_2CHR', etc. are listed in this section.

$PhSeSiMe_3$, Bu_3SnH, PhH; cat AIBN, 80°C, 12 h — 73%

Nishiyama, Y.; Kajimoto, H.; Kotani, K.; Nishida, T.; Sonoda, N. *J. Org. Chem.* ***2002***, *67*, 5696.

1. allyl bromide, InCl•3 TMSCl, $MeNO_2$, rt, 2h; 2. TBAF — 64%

Onishi, Y.; Ito, T.; Yasuda, M.; Baba, A. *Tetrahedron* ***2002***, *58*, 8227.

TFA, rt — 77%

Watanuki, S.; Sakamoto, S.; Harada, H.; Kikuchi, K.; Kuramochi, T.; Kawaguchi, K.; Okazaki, T.; Tsukamoto, S. *Heterocycles* ***2004***. *62*, 127.

acetophenone, bmim BF_4; 5%$FeCl_3$•6 H_2O, 100°C, 10 h — 78%

Wang, J.; Fan, X.; Zhang, X.; Han, L. *Can. J. Chem.* ***2004***, *82*, 1192.

SECTION 73 ALKYLS, METHYLENES, AND ARYLS FROM NITRILES

1. LiBDD, THF, −78°C; 2. cyclohexenone; 3. HCl — 60% (9:1 dr)

LiDBB = lithium di-*tert*-butylphenylide

Wolckenhauer, S.A.; Rychnovsky, S.D. *Org. Lett.* ***2004***, *6*, 2745.

MeMgO*t*-Bu, THF, 60°C; cat $NiCl_2(PMe_3)_2$ — 66%

Miller, J.A.; Dankwardt, J.W. *Tetrahedron Lett.* ***2003***, *44*, 1907.

1. 2 eq allyl bromide; 2. 2 eq Zn; 3. 0.1 AcOH, THF — 65%

Bernardi, L.; Bonini, B.F.; Capito, E.; Dessole, G.; Fochi, M.; Comes-Franchini, M.; Ricci, A. *Synlett* **2003**, 1778.

MeO–C$_6$H$_4$–CN → MeO–C$_6$H$_4$–Ph: PhMgO*t*-Bu, cat $Cl_2Ni(PMe_3)_2$; THF, 60°C, 2 h — 91%

Miller, J.A.; Dankwardt, J.W.; Penney, J.M. *Synthesis* **2003**, 1643.

PhCOCN → PhCOCH$_2$CH=CH$_2$: allyl bromide, Yb, THF, 8 h — 80%

Baikuntha, M.G.; Gogoi, J.; Prajapati, D.; Sandhu, J.S. *New J. Chem.* **2003**, *27*, 1038.

SECTION 74: ALKYLS, METHYLENES AND ARYLS FROM ALKENES

The following reaction types are included in this section:

A. Hydrogenation of Alkenes (and Aryls)
B. Formation of Aryls and Heteroaryls
C. Alkylations and Arylations of Alkenes
D. Conjugate Reduction of α,β-Unsaturated Carbonyl Compounds and Nitriles
E. Conjugate Alkylations
F. Cyclopropanations, including Halocyclopropanations
G. Cyclobutanations, including Halocyclobutanations

0.1 M SmI_2; THF-HMPA-acetone — 60%

Rivkin, A.; Nagashima, T.; Curran, D.P. *Org. Lett.* **2003**, *5*, 419.

$LiAlH_4$, ether, 40°C — 85% (80:20 dr)

Zohar, E.; Marek, I. *Org. Lett.* **2004**, *6*, 341.

$(EtO)_2MeSiH$, sealed tube; PtO_2, 85°C, 20 h — >95%

Sabourault, N.; Mignani, G.; Wagner, A.; Mioskowski, C. *Org. Lett.* **2002**, *4*, 2117.

1. 9-BBN, $PhCH(Br)CO_2Me$
2. $PdCl_2(BINAP)$, K_3PO_4 decyl chloride — 42%

Fujioka, T.; Nakamura, T.; Yorimitsu, H.; Oshima, K. *Org. Lett.* **2002**, *4*, 2285.

50% $RuCl_3$•x H_2O, 60°C; 10% AgOTf, DCE — (9 : 1) 63%

Young, S.W.; Pastine, S.J.; Sames, D. *Org. Lett.* **2004**, *6*, 581.

(bmim) Cl, $AlCl_3$, 8h — 82%

Qiao, K.; Deng, Y. *Tetrahedron Lett.* **2003**, *44*, 2191.

SECTION 74A: HYDROGENATION OF ALKENES (AND ARYLS)

Reduction of aryls to dienes is listed in Section 377 (Alkene-Alkene).

ASYMMETRIC HYDROGENATIONS

NHAc, CO_2Me → NHAc, CO_2Me

H_2, $Rh(cod)_2PF_6$, 1 d; phosphoramidite ligand — quant (99% ee)

Peña, D.; Minnaard, A.J.; de Vries, J.G.; Feringa, B.L. *J. Am. Chem. Soc.* **2002**, *124*, 14552.

AcHN, Me, CO_2Me → AcHN, Me, CO_2Me

$[Rh(cod)_2]BF_4$, monophosphite ligand; H_2, DCM — 98% (96.7% ee)

Huan, H.; Zhang, Z.; Luo, H.; Bai, C.; Hu, X.; Chen, H. *J. Org. Chem.* **2004**, *69*, 2355.

NHAc, OMOM, OMe → OMe, OMOM, NHAc

$Rh(cod)_2OTf$, MeOH, H_2; chiral bis(phosphine) — 95% ee

Le, J.C.-D.; Pagenkopf, B.L. *J. Org. Chem.* **2004**, *69*, 4175.

H_2 , cat Ir (chiral ligand)

quant (81% ee)

Yue, T.-Y.; Nugent, W.A. *J. Am. Chem. Soc.* **2002**, *124*, 13692.

0.6% Ir-imidazolylidene-oxazoline complex

H_2 (50 bar) , DCM , 25°C , 2 h

MeO MeO 99% (79% ee)

Perry, M.C.; Cu, X.; Powell, M.T.; Hou, D.-R.; Reibenspies, J.H.; Burgess, K. *J. Am. Chem. Soc.* **2003**, *125*, 113.

AcHN CO_2Et AcHN CO_2Et

50 atm H_2 , chiral Ru catalyst

EtOH , rt , 18 h

99% ee

Tang, W.; Wu, S.; Zhang, X. *J. Am. Chem. Soc.* **2003**, *125*, 9570.

NHAc Ph CO_2Me

cat Th/chiral O/S ligand , H_2

THF , rt

NHAc Ph CO_2Me 95% ee

Evans, D.A.; Michael, F.E.; Tedrow, J.S.; Campos, K.R. *J. Am. Chem. Soc.* **2003**, *125*, 3534.

Ph NO_2

1.5 $PhSiH_3$, 0.1 PMHS , 1% CuF_2 , rt

1.1% JOSIPHOS , toluene , 16 h

Ph NO_2 76% (96:4 er)

Czekelius, C.; Carreira, E.M. *Org. Lett.* **2004**, *6*, 4575.

AcHN CO_2H

H_2 , MeOH , rt

chiral Rh catalyst

AcHN CO_2H >99% ee

Hoge, G.; Wu, H.-P.; Kissel, W.S.; Pflum, D.A.; Greene, D.J.; Bao, J. *J. Am. Chem. Soc.* **2004**, *126*, 5966.

NHAc

H_2 , cat $Rh(L)_2(cod)\ BF_4$, 30°C , 20h

chiral monophosphite ligand , DCM

Cl Cl NHAc quant (96% ee)

Reetz, M.T.; Mehler, G.; Meiswinkel, A.; Sell, T. *Tetrahedron Lett.* **2002**, *43*, 7941.

NHAc NHAc

[Rh(MonoPhos)(cod)] BF_4 , –20°C

H_2 , 8h

>99% (95% ee)

Jia, X.; Guo, R.; Li, X.; Yao, X.; Chan, A.S.C. *Tetrahedron Lett.* **2002**, *43*, 5541.

Ph Ph

H_2 , cat Ir-proline derivative complex

DCM ,rt

Ph Ph 99% (93:7 *R:S*)

Xu, G.; Gilbertson, S.R. *Tetrahedron Lett* **2003**, *44*, 953.

OMEM OMEM

H_2, Pd catalyst

O O O O >99%

Brandänge, S.; Färnbäck, M.; Leijonmarck, H.; Sundin, A. *J. Am. Chem. Soc.* ***2003***, *125*, 11942.

$CONH_2$ $CONH_2$

Ph NH H_2, PtO_2, THF Ph NH

Et_3N, 3 h 96% (99% ee)

CO_2Me CO_2Me

Ikemoto, N.; Tellers, D.M.; Dreher, S.D.; Liu, J.; Huang, A.; Rivera, N.R.; Njolito, E.; Hsiao, Y.; McWilliams, J.C.; Williams, J.M.; Armstrong III, J.D. *J. Am. Chem. Soc.* ***2004***, *126*, 3048.

phospholane-oxazoine ligand, Ir catalyst

Ph Ph H_2, DCM, rt Ph Ph 91%ee

Tang, W.; Wang, W.; Zhang, X. *Angew. Chem. Int. Ed.* ***2003***, *42*, 943.

H_2 (50 bar), Ir HePHOX complex

Ph Ph Ph >99% (99% ee)

Ph DCM, rt, 2 h Ph

Cozzi, P.G.; Menges, F.; Kaiser, S. *Synlett* ***2003***, 833.

NONASYMMETRIC HYDROGENATIONS

H_2, Ru(η^6-p-cymene)(η^2-TRIPHOS)Cl PF_6

(bmim) BF_4

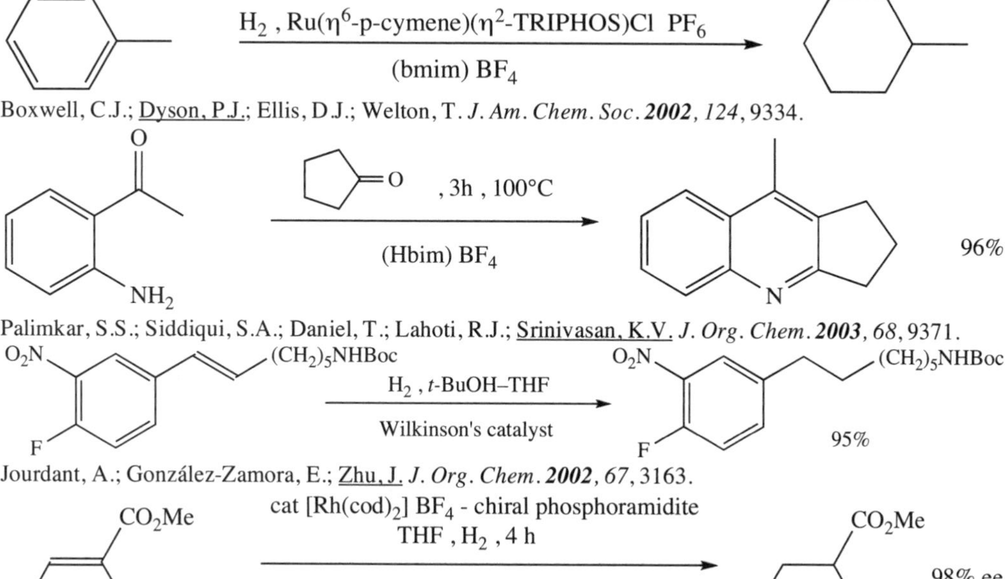

Boxwell, C.J.; Dyson, P.J.; Ellis, D.J.; Welton, T. *J. Am. Chem. Soc.* ***2002***, *124*, 9334.

O

=O , 3h, 100°C

(Hbim) BF_4 96%

NH_2 N

Palimkar, S.S.; Siddiqui, S.A.; Daniel, T.; Lahoti, R.J.; Srinivasan, K.V. *J. Org. Chem.* ***2003***, *68*, 9371.

O_2N $(CH_2)_5NHBoc$ O_2N $(CH_2)_5NHBoc$

H_2, *t*-BuOH–THF

Wilkinson's catalyst

F F 95%

Jourdant, A.; González-Zamora, E.; Zhu, J. *J. Org. Chem.* ***2002***, *67*, 3163.

cat $[Rh(cod)_2]$ BF_4 - chiral phosphoramidite

CO_2Me THF, H_2, 4 h CO_2Me

98% ee

Ph NHAc Ph NHAc

Jia, X.; Li, X.; Xu, L.; Shi, Q.; Yao, X.; Chan, A.S.C. *J. Org. Chem.* ***2003***, *68*, 4539.

OAc / Ph → H_2 , Rh[(TangPhos)(nbd)] SbF_6 → OAc / Ph — 96%

Tang, W.; Liu, D.; Zhang, X. *Org. Lett.* ***2003**, 5*, 205.

MeO_2C / CO_2Me → H_2 (100 psi) , $Rh(cod)_2$ BF_4 , 23°C; monophosphite ligand , DCM → MeO_2C / CO_2Me — quant (96.5% ee)

Hua, Z.; Vassar, V.C.; Ojima, I. *Org. Lett.* ***2003**, 5*, 3831.

Ph / Ph → 5% polymer incarcerated Pd; THF , H_2 , rt → Ph / Ph — quant

Okamoto, K.; Akiyama, R.; Kobayashi, S. *J. Org. Chem.* ***2004**, 69*, 2871.

NHAc / Ph / CO_2Me → $Rh(cod)_2BF_4$, 50 atm H_2 , rt , 2d , DCM; chiral spirophosphoramidite ligand → NHAc / Ph / CO_2Me — 90% (90% ee)

Fu, Y.; Guo, X.-X.; Zhu, S.-F.; Hu, A.-G.; Xie, J.-H.; Zhou, Q.-L. *J. Org. Chem.* ***2004**, 69*, 4648.

CO_2H / Ph / NHAc → H_2 , dendritic pyrophos , MeOH; $Rh(cod)_2BF_4$, toluene , 2 h → CO_2H / Ph / NHAc — quant (98% ee)

Yi, B.; Fan, Q.-H.; Deng, G.-J.; Li, Y.-M.; Qiu, L.-Q.; Chan, A.S.C. *Org. Lett.* ***2004**, 6*, 1361.

CO_2Me / Ph / NHAc → $Rh(cod)_2BF_4$, H_2 , DCM , rt; phosphoramidite ligand → CO_2Me / Ph / NHAc — quant (92% ee)

Hoen, R.; van den Berg, M.; Bernsmann, H.; Minnaard, A.J.; de Vries, J.G. *Org. Lett.* ***2004**, 6*, 1433.

O / Ph → BnO—/=\—OBn; cat Grubbs II , DCM , 25°C → O / Ph / OBn — 34% (1:7 *E:Z*)

Smidt, S.P.; Menges, F.; Pfaltz, A. *Org. Lett.* ***2004**, 6*, 2023.

Ph → SmI_2 , H_2O , 5 min; N,N,N',N",N"'-pentamethylenetriamine → Ph — >99%

Dahlén, A.; Hilmersson, G. *Tetrahedron Lett.* ***2003**, 44*, 2661.

N / Ts → H_2 , 1% $[Rh(ndb)_2]$; 1% (*S,S*)-(*R,R*)-PhTRAP; 10% Cs_2CO_3 , *i*-PrOH; 80°C → N / Ts — 94% (97% ee)

Kuwano, R.; Kaneda, K.; Ito, T.; Sato, K.; Kurokawa, T.; Ito, Y. *Org. Lett.* ***2004**, 6*, 2213.

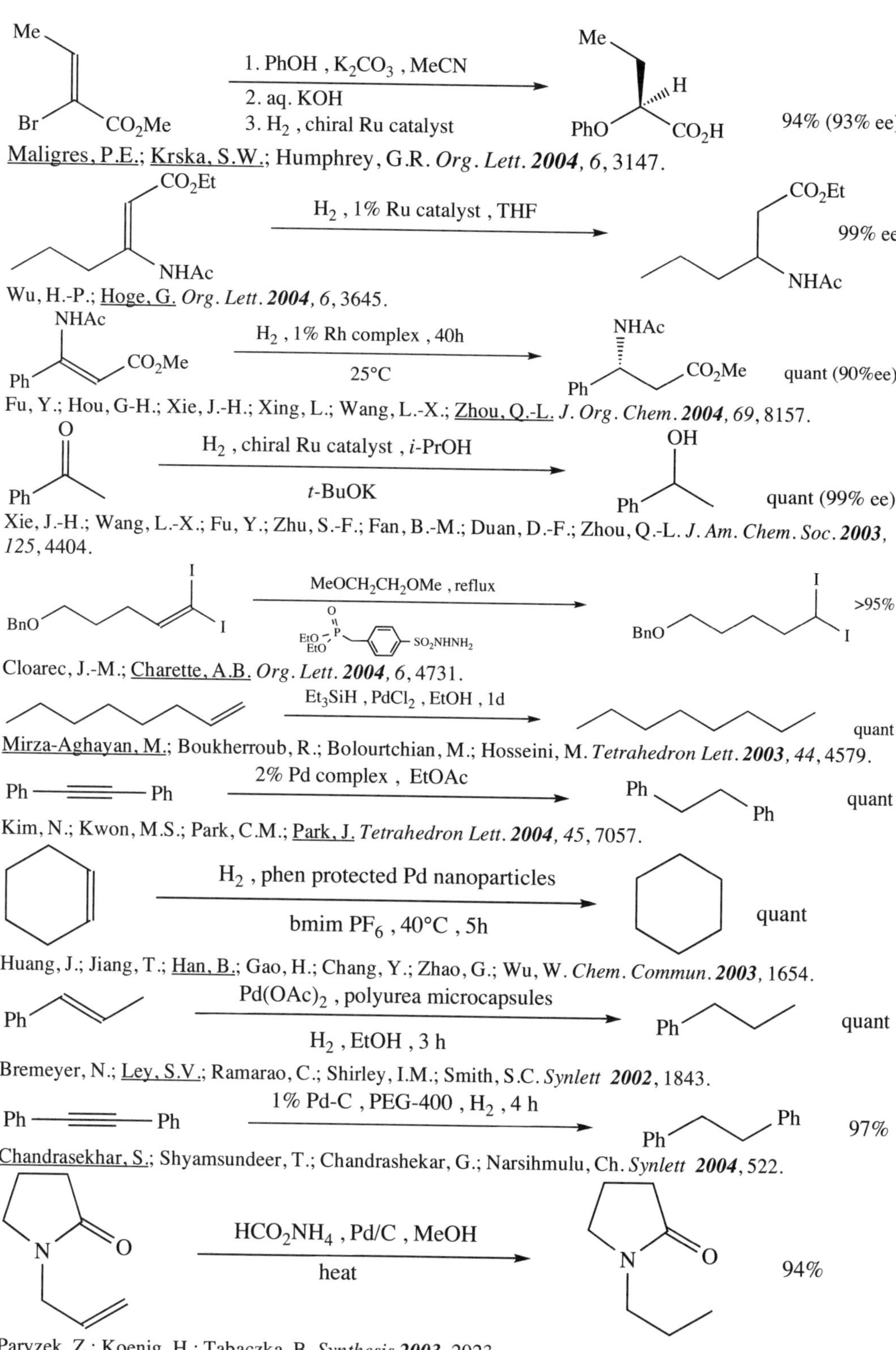
Me
Br
CO_2Me
1. PhOH , K_2CO_3 , MeCN
2. aq. KOH
3. H_2 , chiral Ru catalyst
Me
H
PhO
CO_2H
94% (93% ee)
Maligres, P.E.; Krska, S.W.; Humphrey, G.R. *Org. Lett.* ***2004**, 6*, 3147.
CO_2Et
NHAc
H_2 , 1% Ru catalyst , THF
CO_2Et
99% ee
NHAc
Wu, H.-P.; Hoge, G. *Org. Lett.* ***2004**, 6*, 3645.
NHAc
Ph
CO_2Me
H_2 , 1% Rh complex , 40h
25°C
NHAc
Ph
CO_2Me
quant (90%ee)
Fu, Y.; Hou, G-H.; Xie, J.-H.; Xing, L.; Wang, L.-X.; Zhou, Q.-L. *J. Org. Chem.* ***2004**, 69*, 8157.
O
Ph
H_2 , chiral Ru catalyst , *i*-PrOH
t-BuOK
OH
Ph
quant (99% ee)
Xie, J.-H.; Wang, L.-X.; Fu, Y.; Zhu, S.-F.; Fan, B.-M.; Duan, D.-F.; Zhou, Q.-L. *J. Am. Chem. Soc.* ***2003**, 125*, 4404.
I
BnO
I
$MeOCH_2CH_2OMe$, reflux
O
EtO
EtO
P
SO_2NHNH_2
I
>95%
BnO
I
Cloarec, J.-M.; Charette, A.B. *Org. Lett.* ***2004**, 6*, 4731.
Et_3SiH , $PdCl_2$, EtOH , 1d
quant
Mirza-Aghayan, M.; Boukherroub, R.; Bolourtchian, M.; Hosseini, M. *Tetrahedron Lett.* ***2003**, 44*, 4579.
Ph
Ph
2% Pd complex , EtOAc
Ph
Ph
quant
Kim, N.; Kwon, M.S.; Park, C.M.; Park, J. *Tetrahedron Lett.* ***2004**, 45*, 7057.
H_2 , phen protected Pd nanoparticles
bmim PF_6 , 40°C , 5h
quant
Huang, J.; Jiang, T.; Han, B.; Gao, H.; Chang, Y.; Zhao, G.; Wu, W. *Chem. Commun.* ***2003**,* 1654.
Ph
$Pd(OAc)_2$, polyurea microcapsules
H_2 , EtOH , 3 h
Ph
quant
Bremeyer, N.; Ley, S.V.; Ramarao, C.; Shirley, I.M.; Smith, S.C. *Synlett* ***2002***, 1843.
Ph
Ph
1% Pd-C , PEG-400 , H_2 , 4 h
Ph
Ph
97%
Chandrasekhar, S.; Shyamsundeer, T.; Chandrashekar, G.; Narsihmulu, Ch. *Synlett* ***2004***, 522.
N
O
HCO_2NH_4 , Pd/C , MeOH
heat
N
O
94%
Paryzek, Z.; Koenig, H.; Tabaczka, B. *Synthesis* ***2003***, 2023.

REVIEWS:

"Chiral Monodentate Phosphorous Ligands for Rhodium-Catalyzed Asymmetric Hydrogenation"
Jerphagnon, T.; Renaud, J.-L.; Bruneau, C. *Tetrahedron: Asymmetry* **2004**, *15*, 2101.

"Asymmetric Hydrogenations"
Knowles, W.S. *Angew. Chem. Int. Ed.* **2002**, *41*, 1999.

"On the Mechanism of Stereoselection in Rh-Catalyzed Asymmetric Hydrogenation: A General Approach for Predicting the Sense of Enantioselectivity"
Gridnev, I.D.; Imamoto, T. *Acc. Chem. Res.* **2004**, *37*, 633.

SECTION 74B: FORMATION OF ARYLS AND HETEROARYLS

NHSiMe$_2$*t*-Bu → 1. BuLi, THF, PhH, 60°C; 2. PhCO$_2$Et → SiMe$_2$*t*-Bu, N, Ph 71%

Jacobson, M.A.; Williard, P.G. *J. Org. Chem.* **2002**, *67*, 32.

SMe, I → 1. PhC≡CH, cat Pd/C; 2. Br$_2$ → Br, Ph, S 92%

Yue, D.; Larock, R.C. *J. Org. Chem.* **2002**, *67*, 1905.

O, O → 40% Pd(PPh$_3$)$_4$, 80% dppf; toluene, 2.4 mM, 80°C, 30 min → O, O 86% 65%

Kawasaki, T.; Saito, S.; Yamamoto, Y. *J. Org. Chem.* **2002**, *67*, 2653.

O, C$_7$H$_{15}$ → 1% TpRu(phth)$_3$(MeCN)Cl; 0.5 NEt$_3$, DCM, 12 h → O, C$_7$H$_{15}$ 84%

Tp = trispyrazolylborate

Lo, C.-Y.; Guo, H.; Lian, J.-J.; Shen, F.-M.; Liu, R.S. *J. Org. Chem.* **2002**, *67*, 3930.

N*t*-Bu, Ph → 6 I$_2$, MeCH, rt, 30 min; 3 eq NaOCO$_2$Me → N, Ph, I 68%

Huang, Q.; Hunter, J.A.; Larock, R.C. *J. Org. Chem.* **2002**, *67*, 3437.

CO_2Me CO_2Me → (Bu allene, cat $Ni(dppe)Br_2$, Zn; MeCN, 80°C, 8 h) CO_2Me C_5H_{11} CO_2Me 88%

Shanmugasundaram, M.; Wu, M.-S.; Jeanmohan, M.; Huang, C.-W.; Cheng, C.-H. *J. Org. Chem.* ***2002***, *67*, 7724.

N—*t*-Bu Ph → (PhI, K_2CO_3, 100°C; cat $Pd(PPh_3)_4$, DMF; 12 h) N Ph Ph 49%

Dai, G.; Larock, R.C. *J. Org. Chem.* ***2003***, *68*, 920.

N → (10% $Co_2(CO)_8$, CO, THF; 105°C, 2 d) N 59%

Jacob, J.; Jones, W.D. *J. Org. Chem.* ***2003***, *68*, 3563.

CHO Br N Ph → (1. *t*-$BuNH_2$, 100°C, 8 h; 2. DMF, cat $Pd(OAc)_2/PPh_3$, 100°C, Na_2CO_3, 10 h) N Ph N 93%

Zhang, H.; Larock, R.C. *J. Org. Chem.* ***2003***, *68*, 5132.

O Ph Ph O → (HCOOH, 5% Pd/C, cat H_2SO_4; PEG-200, microwaves, 1 min) Ph O Ph 95%

Rao, H.S.P.; Jothilingam, S. *J. Org. Chem.* ***2003***, *68*, 5392.

O Ph → (H—≡—CH_2NH_2; $NaAuCl{\bullet}2\ H_2O$, EtOH; 78°C, 5 h) Ph N 74%

Abbiati, G.; Arcadi, A.; Bianchi, G.; DeGiuseppe, S.; Marinelli, F.; Rossi, E. *J. Org. Chem.* ***2003***, *68*, 6959.

CO_2Et I → (Ph—≡—Ph; cat. $Pd(OAc)_2/PPh_3$; DMF, 80°C, NEt_3, 12 h) CO_2Et Ph Ph 86%

Huang, Q.; Larock, R.C. *J. Org. Chem.* ***2003***, *68*, 7342.

cat Pde_2dba_3•$CHCl_3$, DCE
cat $P(OPh)_3$, 80°C, 12 h

Et Et Et Et 63%

Tsukada, N.; Sugawara, S.; Nakaoka, K.; Inoue, Y. *J. Org. Chem.* ***2003***, *68*, 5961.

OAc CO_2Me NO_2 dioxane, 6 atm CO cat $[Cp^*Fe(CO)_2]_2$, 150°C 41 h CO_2Me 65%

O'Dell, D.K.; Nicholas, K.M. *J. Org. Chem.* ***2003***, *68*, 6427.

EtO_2C—≡—CO_2Et PhI, cat. $Pd(OAc)_2$, DMF Ag_2CO_3, 120°C, 8 h CO_2Et CO_2Et CO_2Et CO_2Et 90%

Kawasaki, S.; Satoh, T.; Miura, M.; Nomura, M. *J. Org. Chem.* ***2003***, *68*, 6836.

NHBn Bu PdI_2, 2 KI, DMA 25°C, 15 h Bu N Bn 80%

Gabriele, B.; Salerno, G.; Fazio, A. *J. Org. Chem.* ***2003***, *68*, 7853.

$PhCH_2CHO$, $TiCl_4$, DCM rt, 3 h Ph 88%

Kabalka, G.W.; Ju, Y.; Wu, Z. *J. Org. Chem.* ***2003***, *68*, 7915.

O_2, 50% activated C xylene, 120°C, 15 h 86%

Nakamichi, N.; Kawabata, H.; Hiyashi, M. *J. Org. Chem.* ***2003***, *68*, 8272.

MeO_2C MeO_2C NC CN, 10% Et_4NCl, DMF, rt cat $[Cp^*Ru(MeCN)_3]PF_6$, 0.16h MeO_2C MeO_2C N CN 84%

Varela, J.A.; Castedo, L.; Saá, C. *J. Org. Chem.* ***2003***, *68*, 8595.

\+ C_6H_{13}—≡—SPh 1. $CoBr_2(dppe)$, ZnI, Zn 2. DDQ SPh C_6H_{13} 86%

Hilt, G.; Lüers, S.; Harms, K. *J. Org. Chem.* ***2004***, *69*, 624.

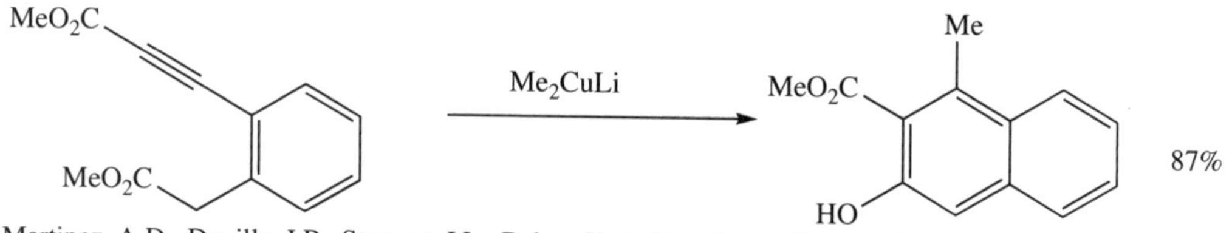

MeO_2C … MeO_2C — Me_2CuLi → Me, MeO_2C, HO — 87%

Martinez, A.D.; Deville, J.P.; Stevens, J.L.; Behar, V. *J. Org. Chem.* ***2004***, *69*, 991.

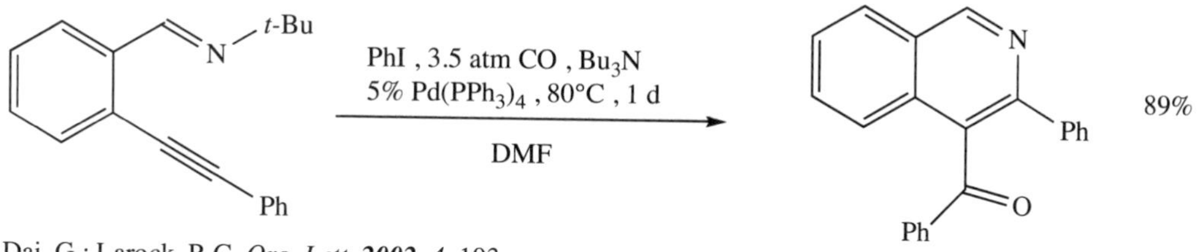

t-Bu, N, Ph — PhI, 3.5 atm CO, Bu_3N; 5% $Pd(PPh_3)_4$, 80°C, 1 d; DMF → N, Ph, O, Ph — 89%

Dai, G.; Larock, R.C. *Org. Lett.* ***2002***, *4*, 193.

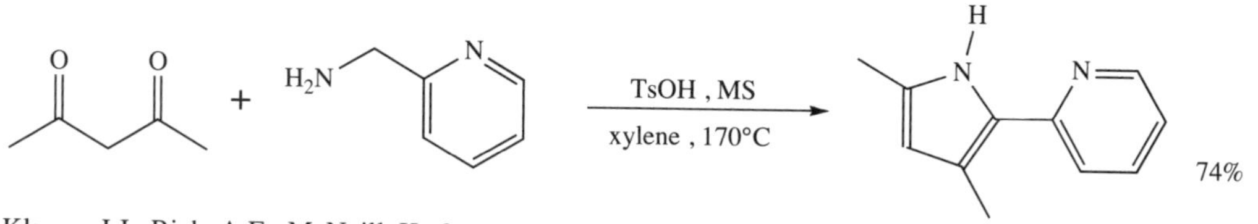

O O + H_2N, N — TsOH, MS; xylene, 170°C → H, N, N — 74%

Klappa, J.J.; Rich, A.E.; McNeill, K. *Org. Lett.* ***2002***, *4*, 435.

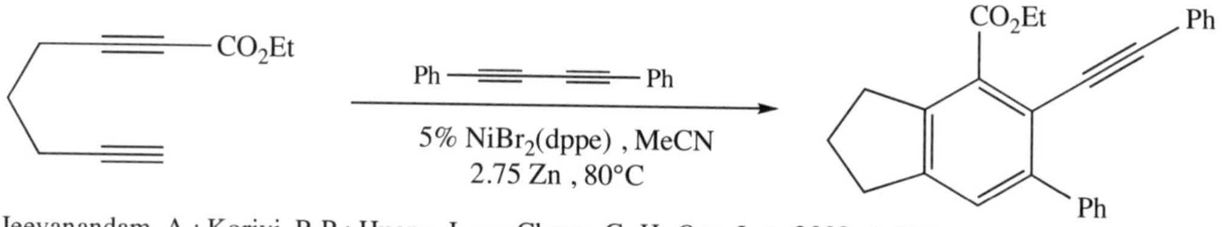

CO_2Et — Ph—≡—≡—Ph; 5% $NiBr_2$(dppe), MeCN; 2.75 Zn, 80°C → CO_2Et, Ph, Ph

Jeevanandam, A.; Korivi, R.P.; Huang, I.-w.; Cheng, C.-H. *Org. Lett.* ***2002***, *4*, 807.

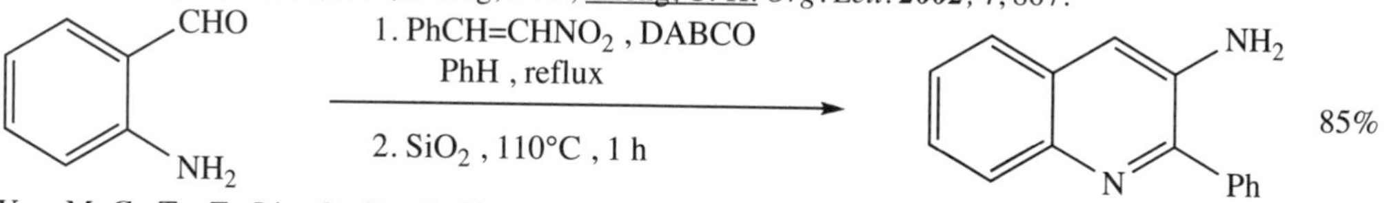

CHO, NH_2 — 1. $PhCH{=}CHNO_2$, DABCO; PhH, reflux; 2. SiO_2, 110°C, 1 h → NH_2, N, Ph — 85%

Yan, M.-C.; Tu, Z.; Lin, C.; Ko, S.; Hsu, J.; Yao, C.-F. *Org. Lett.* ***2002***, *4*, 1565.

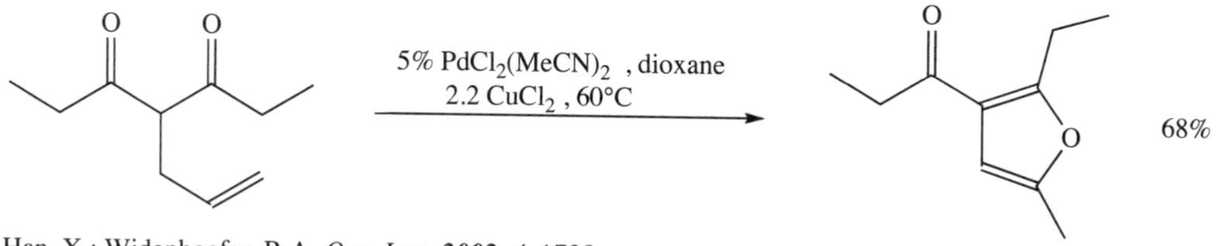

O O — 5% $PdCl_2(MeCN)_2$, dioxane; 2.2 $CuCl_2$, 60°C → O, O — 68%

Han, X.; Widenhoefer, R.A. *Org. Lett.* ***2002***, *4*, 1738.

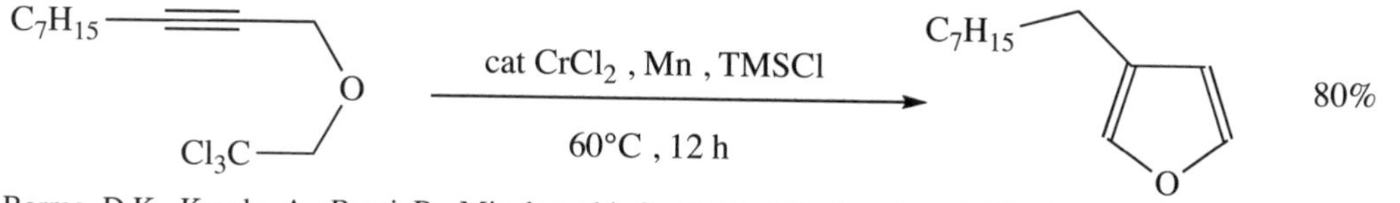

C_7H_{15}, O, Cl_3C — cat $CrCl_2$, Mn, TMSCl; 60°C, 12 h → C_7H_{15}, O — 80%

Barma, D.K.; Kundu, A.; Baati, R.; Mioskowski, C.; Falck, J.R. *Org. Lett.* ***2002***, *4*, 1387.

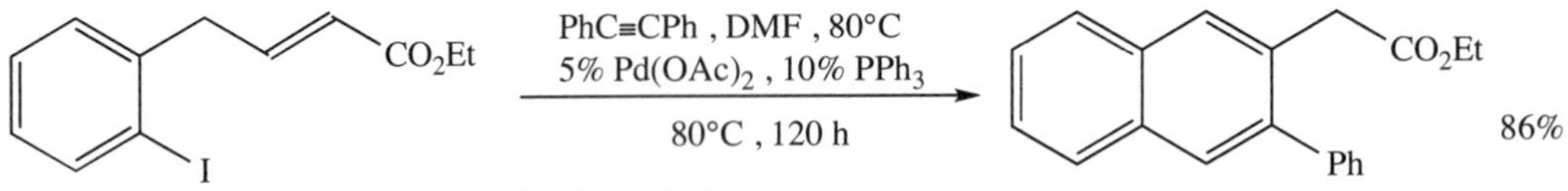

CO_2Et I — PhC≡CPh , DMF , 80°C; 5% $Pd(OAc)_2$, 10% PPh_3; 80°C , 120 h → CO_2Et Ph 86%

Huang, Q.; Larock, R.C. *Org. Lett.* ***2002***, *4*, 2505.

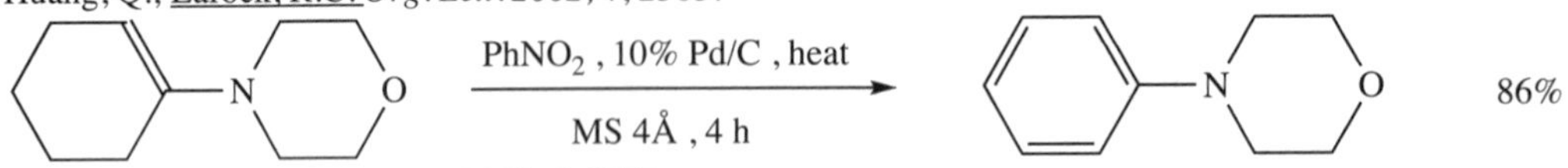

N O — $PhNO_2$, 10% Pd/C , heat; MS 4Å , 4 h → N O 86%

Cossy, J.; Belotti, D. *Org. Lett.* ***2002***, *4*, 2557.

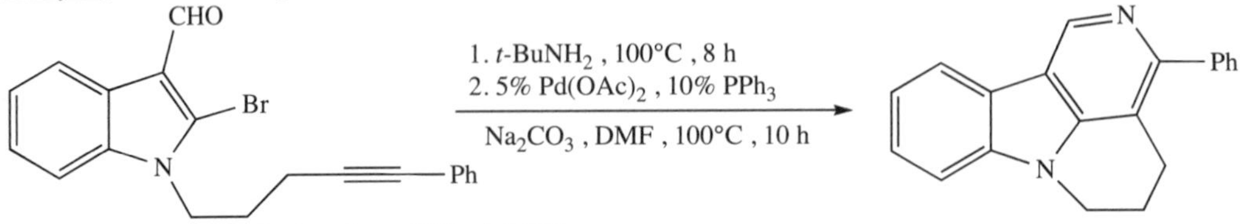

CHO Br N Ph — 1. *t*-$BuNH_2$, 100°C , 8 h; 2. 5% $Pd(OAc)_2$, 10% PPh_3; Na_2CO_3 , DMF , 100°C , 10 h → N Ph N

Zhang, H.; Larock, R.C. *Org. Lett.* ***2002***, *4*, 3035.

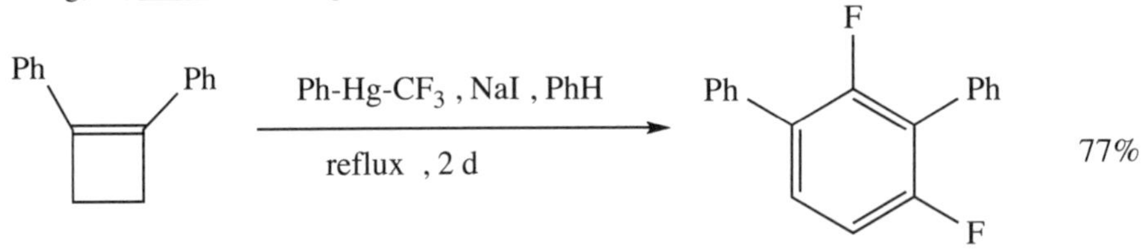

Ph Ph — Ph-Hg-CF_3 , NaI , PhH; reflux , 2 d → F Ph Ph F 77%

Morrison, H.M.; Rainbolt, J.E.; Lewis, S.B. *Org. Lett.* ***2002***, *4*, 3871.

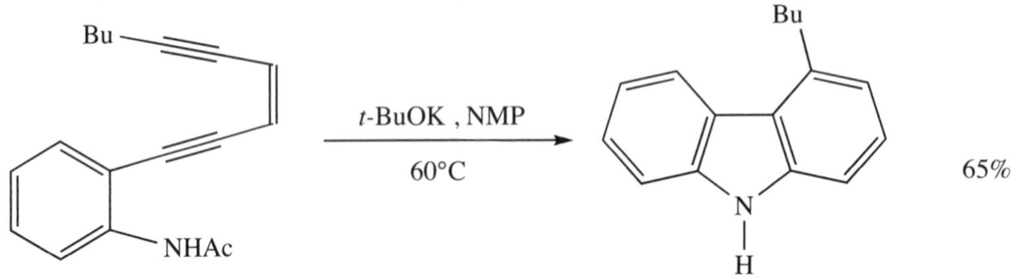

Bu NHAc — *t*-BuOK , NMP; 60°C → Bu N H 65%

Lee, C.-Y.; Lin, C.-F.; Lee, J.-L.; Chium C.-C.; Lu, W.-D.; Wu, M.-J. *J. Org. Chem.* ***2004***, *69*, 2106.

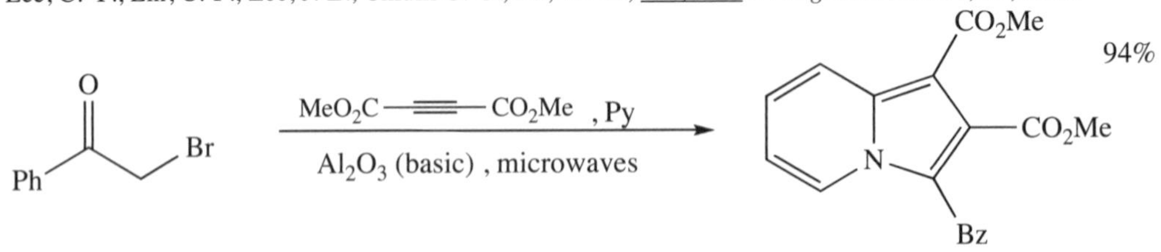

O Ph Br — MeO_2C—≡—CO_2Me , Py; Al_2O_3 (basic) , microwaves → CO_2Me CO_2Me N Bz 94%

Bora, U.; Saikia, A.; Boruah, R.C. *Org. Lett.* ***2003***, *5*, 435.

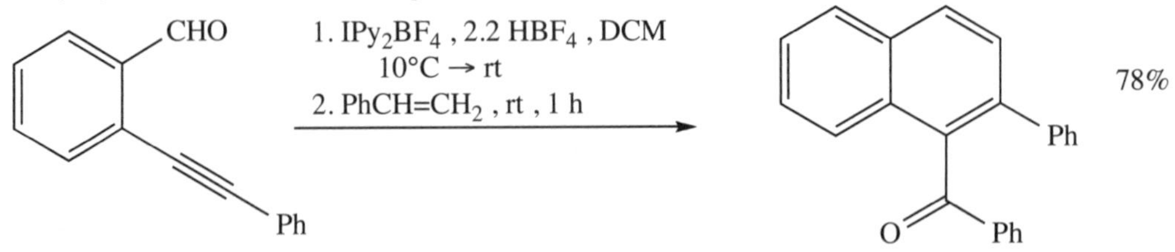

CHO Ph — 1. IPy_2BF_4 , 2.2 HBF_4 , DCM 10°C → rt; 2. PhCH=CH_2 , rt , 1 h → Ph O Ph 78%

Barluenga, J.; Vásquez-Villa, H.; Ballesteros, A.; González, J.M. *Org. Lett.* ***2003***, *5*, 4121.

2 $C_{10}H_{21}$—≡ → EtO_2C—≡—CO_2Et , DCM , rt , 1h / cat $[Rh(cod)_2]BF_4$, Hf-BINAP → 92x88% + 2 other isomers

Tanaka, K.; Shirasaka, K. *Org. Lett.* **2003**, *5*, 4697.

BnHN—…≡—CO_2Et → $C_6H_{13}CHO$, toluene, reflux , NH_4Cl , 15h → 82%

Fayol, A.; Zhu, J. *Org. Lett.* **2004**, *6*, 115.

→ EtO_2C—≡—CO_2Et / tetralin , heat , 5h → 67%

Kranjc, K.; Stefane, B.; Polanc, S.; Kocevar, M. *J. Org. Chem.* **2004**, *69*, 3190.

$ZrCp_2$ + → DMPU , THF , 50°C, CuCl , 3h → 62%

Zhou, X.; Li, Z.; Wang, H.; Kitamura, M.; Kanno, K.-i.; Nakajima, K.; Takahashi, T. *J. Org. Chem.* **2004**, *69*, 4559.

→ 1. $W(CO)_5(thf)$, THF, reflux / 2. 3 NMO , DCM , rt , 1h → 70%

Sangu, K.; Fuchibe, K.; Akiyama, T. *Org. Lett.* **2004**, *6*, 353.

→ 5% $CpCo(CO)_2$, hν / xylene , heat → 70%

Chouraqui, G.; Petit, M.; Aubert, C.; Malacria, M. *Org. Lett.* **2004**, *6*, 1579.

BnNH$_2$, AcOH , microwaves

70%

Minetto, G.; Raveglia, L.F.; Taddei, M. *Org. Lett.* **2004**, *6*, 389.

EtCCEt . cat. Pd(OAc)$_2$, refllux

2.5 agOAc , toluene , 2d

92%

Huang, W.; Zhou, X.; Kanno, K.-i.; Takahashi, T. *Org. Lett.* **2004**, *6*, 2429.

, TsOH , MS 4Å

20%

DBU , *i*-PrOH , THF

57%

Bharadwaj, A.R.; Scheidt, K.A. *Org. Lett.* **2004**, *6*, 2465.

30% Pd(PPh$_3$)$_4$, 120°C , 3 d

neat , pressure vessel

69%

Siriwardana, A.I.; Kathriarachchi, K.K.A.D.S.; Nakamura, I.; Gridnev, I.D.; Yamamoto, Y. *J. Am. Chem. Soc.* **2004**, *126*, 13898.

NEt$_3$, TiCl$_4$, DCM

0°C → rt

75%

Srinivas, G.; Periasamy, M. *Tetrahedron Lett.* **2002**, *43*, 2785.

5% Hg(OTf)$_2$, PhH

98%

Imagawa, H.; Kurisaki, T.; Nishizawa, M. *Org. Lett.* **2004**, *6*, 3679.

1. CH$_2$I$_2$, Et$_2$Zn
2. CO , –23°C , 3 h
3. H$^+$

63%

Takahashi,k T.; Ishikawa, M.; Huo, S. *J. Am. Chem. Soc.* **2002**, *124*, 388.

CHO

$C_3H_7C\equiv CH$, cat $AuCl_3$
DCE , 1.5 h , 80°C

Ph (92 CHO : CHO 8) 91%

C_3H_7 + C_3H_7

Asao, N.; Nogami, T.; Lee, S.; Yamamoto, Y. *J. Am. Chem. Soc.* ***2003***, *125*, 10921.

O Ph

cat $AuCl_3$, MeOH , DC

O Ph

88%

OMe

Yao, T.; Zhang, X.; Larock, R.C. *J. Am. Chem. Soc.* ***2004***, *126*, 11164.

H_2N

O EtO_2C , DMSO , 170°C

pressure tube , microwaves , 20 min

Ph

EtO_2C

87%

N Ph

Bagley, M.C.; Lunn, R.; Xiong, X. *Tetrahedron Lett.* ***2002***, *43*, 8331.

BzO

Cl OMe

Py , xylene , 140°C
20h

Ph

Ph Cl

OMe 81%

Hamura, TA.; Morita, M.; Matsumoto, T.; Suzuki, K. *Tetrahedron Lett* ***2003***, *44*, 167.

Ph Ph

Ph

W catalyst , aq EtOH , 25°C

Ph + Ph

Ph (69 Ph Ph 31) 84%

Yong, L.; Butenschön, H. *Chem. Commun.* ***2002***, 2852.

C_3H_7 C_3H_7

C_3H_7 C_3H_7

cat $Co_3(CO)_3$ (μ^3-CH)

toluene , reflux , 1h

C_3H_7 C_3H_7

92%

C_3H_7 C_3H_7

Sugihara, T.; Wakabayashi, A.; Nagai, Y.; Takao, H.; Imayawa, H.; Nishizawa, M. *Chem. Commun.* ***2002***, 576.

CHO

1. i. LDA ii. benzphenone

2. TsOH

65%

PPh_3 Br

Ph

Kraus, G.A.; Choudhury, P.K. *Synlett* ***2004***, 97.

Et—≡—Et , 5% Ni(acac)$_2$, THF

10% biaryl bis(phosphine) , 40% $AlMe_3$
PhOH , rt , 2h

77%

Ikeda, S.; Kondo, H.; Arii, T.; Odashima, K. *Chem. Commun.* ***2002***, 2422.

N-CN , cat CoCp(CO)$_2$, 1d

dioxane , reflux

MeO_2C MeO_2C N 88%

Boñaga, L.V.R.; Zhang, H.-C.; Maryannoff, B.E. *Chem. Commun.* ***2004***, 2394.

Ph

$NHCO_2t$-Bu

IPy_2BF_4 , HBF_4 , DCM

–60°C , 15h

I Ph N CO_2t-Bu 68%

Barluenga, J.; Trincado, M.; Rubio, E.; González, J.M. *Angew. Chem. Int. Ed.* ***2003***, *42*, 2406.

O Ph NH_2

O O OEt , cat $NaAuCl_4$

EtOH , 40°C , 6 h

Ph CO_2Et N 93%

Arcadi, A.; Chiarini, M.; Di Giuseppe, S.; Marinelli, F. *Synlett* ***2003***, 203.

cat Ru(tmp)(O)$_2$, 31 h

10 H_2SO_4 , rt

97%

Tanaka, H.; Ikeno, T.; Yamada, T. *Synlett* ***2003***, 576.

O O

HC≡CH , cat RhCl(PPh$_3$)$_3$

toluene , 20°C

O O 68%

Witulski, B.; Zimmermann, A. *Synlett* ***2002***, 1855.

Ph O NH_2

2-butanone , cat Bi(OTf)$_3$

EtOH , rt , 16 h

Ph N 75%

Yadav, J.S.; Reddy, B.V.S.; Premalatha, K. *Synlett* ***2004***, 963.

Ph—≡—Ph $\xrightarrow[\text{rt , 1 d}]{PdCl_2\text{ , }CuCl_2\text{ , }CO_2\text{ , }H_2O}$ hexaphenylbenzene 90%

Li, J.-H.; Xie, Y.-X. *Synth. Commun.* **2004**, *34*, 1737.

$N{=}PPh_3$ … CO_2Me $\xrightarrow[110°C\text{ , 2 d}]{PhCOCH_2Br\text{ , }NEt_3\text{ , toluene}}$ Ph … N H … CO_2Me 57%

Palacios, F.; Herrán, E.; Rubiales, G. *Heterocycles* **2002**, *58*, 89.

REVIEWS:

"Palladium-Catalyzed Intramolecular Arylation Reaction: Mechanism and Application for the Synthesis of Polyarenes"
Echavarren, A.M.; Gómez-Lor, B.; González, J.J.; de Frutos, Ó. *Synlett* **2003**, 585.

"New Advances in Selected Transition Metal-Catalyzed Annulation"
Rubin, M.; Sromek, A.W.; Gevorgyan, V. *Synlett* **2003**, 2265.

SECTION 74C: ALKYLATIONS AND ARYLATIONS OF ALKENES

$\xrightarrow{PhSnBu_3\text{ , }Pd(OAc)_2\text{ ; }DMF\text{ , }O_2\text{ , }NaOAc}$ Ph 82%

Parrish, J.P.; Jung, Y.C.; Shin, S.I.; Jung, K.W. *J. Org. Chem.* **2002**, *67*, 7127.

MeO_2C MeO_2C $\xrightarrow[1\%\ Co(OAc)_2\text{ , AcOH , 90°C ; }N_2/O_2\text{ , 5 h}]{=C(CH_3)C_5H_{11}\text{ , 2\% }Mn(OAc)_2}$ MeO_2C MeO_2C C_5H_{11} 34%

Hirase, K.; Iwahama, T.; Sakaguchi, S.; Ishii, Y. *J. Org. Chem.* **2002**, *67*, 970.

I … O O $\xrightarrow{h\nu\text{ , 10\% }(Bu_3Sn)_2\text{ , PhH ; 3 eq }BF_3{\bullet}OEt_2\text{ , 20°C}}$ I … O O 80%

Wang, J.; Li, C. *J. Org. Chem.* **2002**, *67*, 1271.

Et_3SiH, DCM
10% $B(C_6F_5)_3$
$SiEt_3$
94%

Rubin, M.; Schwier, T.; Cevorgyan, V. *J. Org. Chem.* **2002**, *67*, 1936.

Bn N SeMe O NC
TTMSS/AIBN
TTMSS = $HSi(SiMe_3)_3$
H Bn N NC H 27% + H Bn N NC H 42%

Quirante, J.; Vila, X.; Escalano, C.; Bonjoch, J. *J. Org. Chem.* **2002**, *67*, 2323.

Me Si Me H N
C_6H_{13}
5% $RhCl(PPh_3)_3$, MeCN
Me Me Si N C_6H_{13} 86%

Itami, K.; Mitsudo, K.; Nishino, A.; Yoshida, J.-i. *J. Org. Chem.* **2002**, *67*, 2645.

Ph O H O O H OTf
2 eq IZn CO_2Me
5% $Pd(PPh_3)_4$, PhH, rt
Ph O H O O H CO_2Me

Kadota, I.; Takamura, H.; Sato, K.; Yamamoto, Y. *J. Org. Chem.* **2002**, *67*, 3494.

MeO Br
CO_2Bu, $PdCl_2$, $P(o\text{-}Tol)_3$
microwaves, NEt_3
bmim PF_6, 220°C, 20 min
MeO CO_2Bu 61%

Vallin, K.S.A.; Gmilsson, P.; Larhed, M.; Hallberg, A. *J. Org. Chem.* **2002**, *67*, 6243.

Br CO_2Me
O
DABCO, rt, 15 min
O CO_2Me 82%

Basavaiah, D.; Sharada, D.S.; Kumaragurubaran, N.; Reddy, R.M. *J. Org. Chem.* **2002**, *67*, 7135.

Br CN
C_8H_{17} BF_3K
$PdCl_2(dppf){\bullet}CH_2Cl_2$, reflux
i-PrOH, H_2O, *t*-$BuNH_2$, 6 h
C_8H_{17} CN 87%

Molander, G.A.; Bernardi, C.R. *J. Org. Chem.* **2002**, *67*, 8424.

Ts O O
$PhB(OH)_2$, cat $PdCl_2(PPh_3)_2$
THF, KF, H_2O, 60°C
Ph O O 63%

Wu, J.; Zhu, Q.; Wang, L.; Fathi, R.; Yang, Z. *J. Org. Chem.* **2003**, *68*, 670.

SiMe$_3$ OH

Ph

1. i-BuOCu
2. PhI , cat Pd(PPh$_3$)$_4$
3. aq TBAF

Ph OH

Ph

78%

Taguchi, H.; Ghoroku, K.; Tadaki, M.; Isubouchi, A.; Takeda, T. *J. Org. Chem.* ***2002**, 67*, 8450.

O

CO$_2$Et

Fe

CF$_3$SO$_3$H , rt , 5 h

Fe

CO$_2$Et

83%

Plazuk, D.; Zakrzewski, J. *J. Org. Chem.* ***2002**, 67*, 8672.

O

CO$_2$Et

PPh$_3$, toluene , rt
72 h

O

CO$_2$Et +

O

CO$_2$Et

(80 : 20) 78%

Du, Y.; Lu, X.; Yu, Y. *J. Org. Chem.* ***2002**, 67*, 670.

Br

1. 2.2 eq t-BuLi , hexane-Bu$_2$O , –78°C
2. TMEDA
3. 45°C 4. MeOH

H

90% (*cis/trans* = 1.2)

Bailey, W.F.; Daskapan, T.; Rampalli, S. *J. Org. Chem.* ***2003**, 68*, 1334.

Ph

PhBr , 0.1% Pd(OAc)$_2$, DMA

no ligands K$_3$PO$_4$, 140°C

Ph Ph

95%

Yao, Q.; Kinney, E.P.; Yang, Z. *J. Org. Chem.* ***2003**, 68*, 7518.

I

MeO

O

cat Cp$_2$TiCl$_2$, Mn , THF
TMSCl , 1d , 65°C

O

MeO

83% (80:20 *trans:cis*)

Zhu, .L; Hirao, T. *J. Org. Chem.* ***2003**, 68*, 1633.

C$_6$H$_{13}$

C$_6$H$_{13}$

HC(CN)MeCO$_2$Et , EtOH
cat Pd(PPh$_3$)$_4$/dppf
100°C

C$_6$H$_{13}$

C$_6$H$_{13}$

CO$_2$Et

CN

82%

Nakamura, I.; Bajracharya, G.B.; Yamamoto, Y. *J. Org. Chem.* ***2003**, 68*, 2297.

Ph

CO$_2$Et

Br , 130°C , Bu$_4$NBr

Pd-nanoparticles

CO$_2$Et

Ph

90% (98:2 *E:Z*)

Caló, V.; Nacci, A.; Monopoli, A.; Lawra, S.; Cioffi, N. *J. Org. Chem.* ***2003**, 68*, 2929.

$SiPh_2Me$, F; I–C_6H_4–NO_2, 5% CuI, DMF; 2.3 eq CsF, 5% $Pd(PPh_3)_4$, 2.5 h; F, NO_2 86%

Hanamoto, T.; Kobayashi, T. *J. Org. Chem.* ***2003***, *68*, 6354.

CH_2=CH_2, DCM, NaBARF, 2h; 55°C, $[(allyl)NiBr]_2$/L; Ph; Ph 85% (27%ee)

Rajanbabu, T.V.; Nomura, N.; Jin, J.; Nandi, M.; Park, H.; Sun, X. *J. Org. Chem.* ***2003***, *68*, 8413.

$SiEt_3$; BuMgCl, cat Cp_2TiCl_2; Cl, 0°C, 2 h; H_2O; $SiEt_3$ 71%

Nii, S.; Terao, J.; Kambe, N. *J. Org. Chem.* ***2004***, *69*, 573.

CO_2Et, O_2, acac, 90°C; cat $Pd(OAc)_2$, $HPMo_{11}V$, 2.5 h; NaOAc, EtCOOH; CO_2Et 70% (15:41:44 *o:m:p*)

Tani, M.; Sakaguchi, S.; Ishii, Y. *J. Org. Chem.* ***2004***, *69*, 1221.

Ac, Br; Ph, BF_3K, 2% $PdCl_2(dppf)$•DCM; NEt_3, *n*-PrOH, reflux, 8 h; Ac, Ph

Molander, G.A.; Rivero, M.R. *Org. Lett.* ***2002***, *4*, 107.

OBn, O; I–C_6H_4–OMe; *t*-BuLi, $ZnCl_2$, ether; –78°C → 25°C; OBn, O, OMe 81%

Steinhuebel, D.P.; Fleming, J.J.; DuBois, J. *Org. Lett.* ***2002***, *4*, 293.

Bu, OTs; 5% Cp_2ZrCl_2, THF; 5 BuMgCl, 55°C, 3.5 h; Bu 68% (>25:1 *anti:syn*)

Cesati III, R.R.; de Armas, J.; Hoveyda, A.H. *Org. Lett.* ***2002***, *4*, 395.

$PhN_2^+ BF_4^-$; PhCH=CH_2, 2% $Pd(OAc)_2$•imid; THF, rt, 5h; Ph, Ph 77%

Andrus, M.B.; Song, C.; Zhang, J. *Org. Lett*, ***2002***, *4*, 2079.

Ph, NO_2; $BuZnCH_2SiMe_3$, NMP, THF; 2.5 TMSBr, cat. $Cu(OTf)_2$; –30°C; Ph, NO_2, Bu 72%

Rimkus, A.; Sewald, N. *Org. Lett.* ***2002***, *4*, 3289.

PhN_2BF_4 —[$H_2C{=}CHCO_2Et$, EtOH, 1.25 h; 5% Pd-triolefin macrocyclic cat.]→ $PhCH{=}CHCO_2Et$ (97% recovery of catalyst) 95%

Masllorens, J.; Moren-Mañas, M.; Pla-Quintana, A.; Roglans, A. *Org. Lett.* **2003**, *5*, 1559.

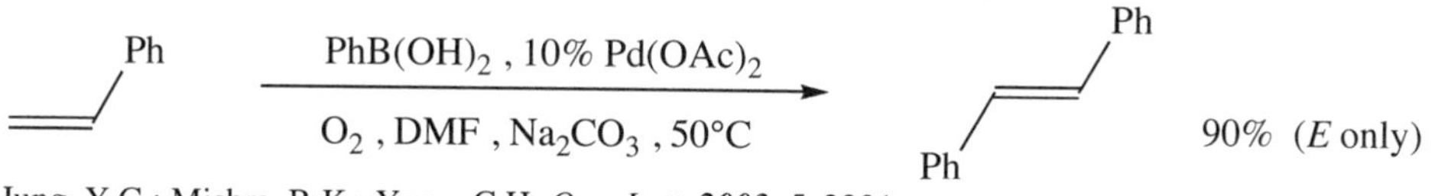

Ph —[$PhB(OH)_2$, 10% $Pd(OAc)_2$; O_2, DMF, Na_2CO_3, 50°C]→ Ph, Ph 90% (*E* only)

Jung, Y.C.; Mishra, R.K.; Yoon, C.H. *Org. Lett.* **2003**, *5*, 2231.

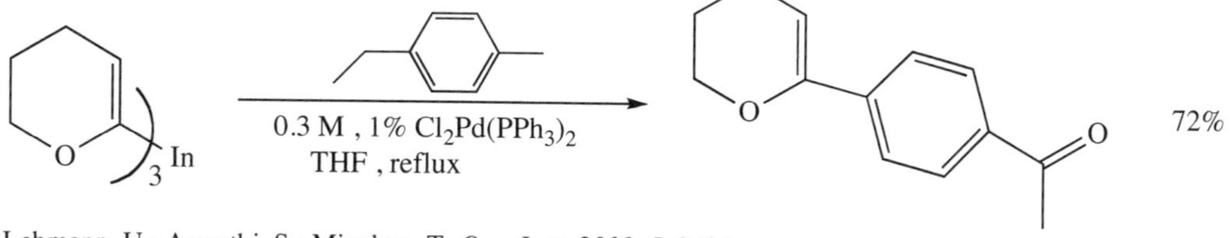

O, In, 3 —[0.3 M, 1% $Cl_2Pd(PPh_3)_2$; THF, reflux]→ O, O 72%

Lehmann, U.; Awasthi, S.; Minehan, T. *Org. Lett.* **2003**, *5*, 2405.

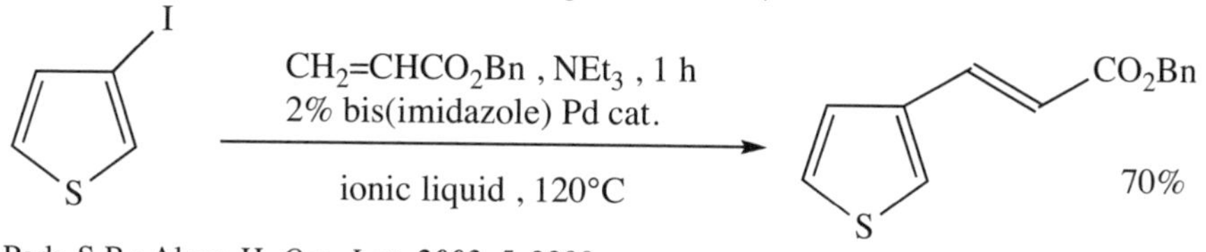

I, S —[$CH_2{=}CHCO_2Bn$, NEt_3, 1 h; 2% bis(imidazole) Pd cat.; ionic liquid, 120°C]→ CO_2Bn, S 70%

Park, S.B.; Alper, H. *Org. Lett.* **2003**, *5*, 3209.

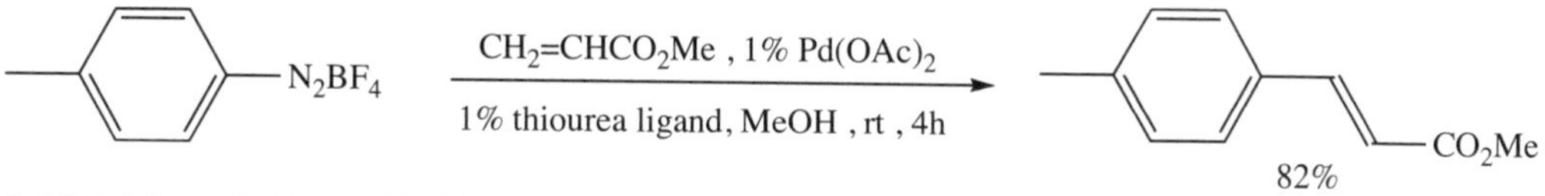

N_2BF_4 —[$CH_2{=}CHCO_2Me$, 1% $Pd(OAc)_2$; 1% thiourea ligand, MeOH, rt, 4h]→ CO_2Me 82%

Dai, M.; Liang, B.; Wang, C.; Chen, J.; Yang, Z. *Org. Lett.* **2004**, *6*, 221.

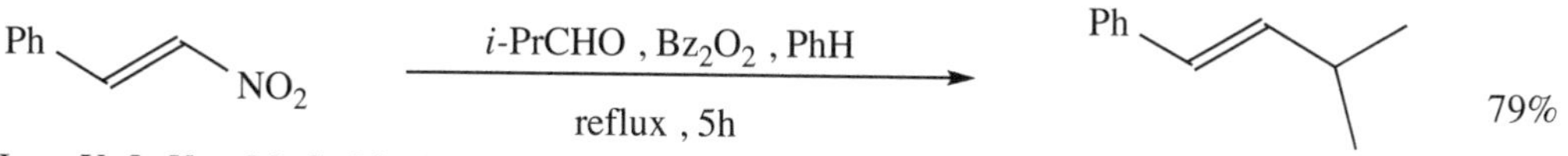

Ph, NO_2 —[*i*-PrCHO, Bz_2O_2, PhH; reflux, 5h]→ Ph 79%

Jang, Y.-J.; Yan, M.-C.; Lin, Y.-F.; Yao, C.-F. *J. Org. Chem.* **2004**, *69*, 3961.

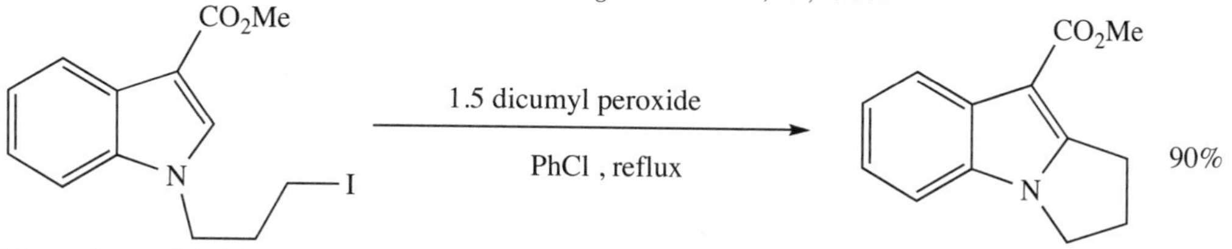

CO_2Me, N, I —[1.5 dicumyl peroxide; PhCl, reflux]→ CO_2Me, N 90%

Menes-Arzate, M.; Martínez, R.; Cruz-Almanza, R.; Muchowski, J.M.; Osornio, Y.M.; Miranda, L.D. *J. Org. Chem.* **2004**, *69*, 4001.

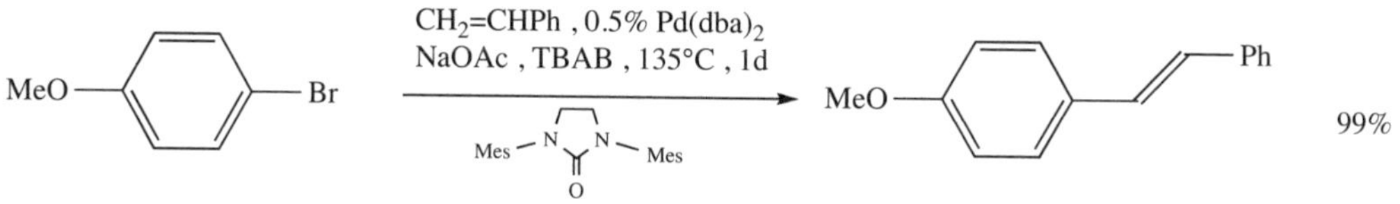

MeO, Br —[$CH_2{=}CHPh$, 0.5% $Pd(dba)_2$; NaOAc, TBAB, 135°C, 1d; Mes–N, N–Mes, O]→ MeO, Ph 99%

Yang, D.; Chen, Y.-C.; Zhu, N.-Y. *Org. Lett.* **2004**, *6*, 1577.

$PhB(OH)_2$ → (CH_2=CHOBu , dioxane , O_2 , NMM; cat $Pd(OAc)_2$, dmphen) → BuO(Ph)C=CH_2 51%

Andappan, M.M.S.; Nilsson, P.; von Schenck, H.; Larhed, M. *J. Org. Chem.* **2004**, *69*, 5212.

2-(CO_2Et)cyclopentanone → ((*p*-tolyl)$_3$Bi-CH=CHPh , TMG; toluene , –50°C → rt) → 2-CO_2Et-2-(CH=CHPh)cyclopentanone 90%

Matano, Y.; Imahori, H. *J. Org. Chem.* **2004**, *69*, 5505.

PhCH=CH–Si$(SiMe_3)_3$ → (1. H_2O_2, NaOH , aq THF; 2. cat $Pd(PPh_3)_4$, PhI , 40°C) → PhCH=CHPh 75%

Wnuk, S.F.; Garcia Jr. P.I.; Wang, Z. *Org. Lett.* **2004**, *6*, 2047.

4-MeC$_6$H$_4$CH=CHNO_2 → (Et_2Zn , toluene , –65°C; $Cu(OTf)_2$-phosphoramidite ligand) → 4-MeC$_6$H$_4$CH(Et)CH$_2NO_2$ quant (98% ee)

Choi, H.; Hua, Z.; Ojima, I. *Org. Lett.* **2004**, *6*, 2689.

PhCH=CHNO_2 → (3 Et_2Zn , 1% $(CuOTf)_2$–PhH; toluene , 22°C; 2% chiral dipeptide phosphine) → PhCH(Et)CH$_2NO_2$ 79% (95% ee)

Mampreian, D.M.; Hoveyda, A.H. *Org. Lett.* **2004**, *6*, 2829.

PhBr → (CH_2=CHCO_2Bn , Na_2CO_3 , 140°C; Pd catalyst , DMF , 70 h) → PhCH=CHCO_2Bn 89%

Yao, Q.; Kinney, E.P.; Zheng, C. *Org. Lett.* **2004**, *6*, 2997.

PhBr → (CH_2=CHCO_2Bu , DMA , NEy_3; palladacycle catalyst , 120°C , 8 h) → PhCH=CHCO_2Bu 91%

Xiong, Z.; Wang, N.; Dai, M.; Li, A.; Chen, J.; Yang, Z. *Org. Lett.* **2004**, *6*, 3337.

1,5-cyclooctadiene → (PhI , cat $Pd_2(dba)_3$; $CH_2(CO_2Me)_2$, $NaHCO_3$, 80°C; Bu_4NCl , DMSO) → bicyclo[3.3.0]octane bearing Ph and $CH(CO_2Me)_2$ (H, H) 63%

Hulin, B.; Newton, L.S.; Cabral, S.; Walker, A.J.; Bordner, J. *Org. Lett.* **2004**, *6*, 4343.

N-methacryloyl-oxazolidin-2-one → ($MgBr_2$, bis(oxazolidine) ligand; *i*-PrI , Bu_3SnH , BEt_3 , O_2; DCM , –78°C) → N-(2,4-dimethylpentanoyl)oxazolidin-2-one 76% (42% ee)

Sibi, M.P.; Sausker, J.B. *J. Am. Chem. Soc.* **2002**, *124*, 984.

$C_{12}H_{25}Br$ → CH_2=CHPh , 1.6 Me_3SiCH_2MgCl / cat $CoCL_2$(dpph) , ether , 20°C , 8h → $C_{12}H_{25}$CH=CHPh 76%

Ikeda, Y.; Makamura, T.; Yorimitsu, H.; Oshima, K. *J. Am. Chem. Soc.* **2002**, *124*, 6514.

Et(Br)C=C(Br)Et → 2 (2,6-dimethylphenyl)MgBr , THF , heat / cat $PdCl_2(PPh_3)_2$, 8 h → 96%

Et Et

Rathore, R.; Deselnicu, M.I.; Brns, C.L. *J. Am. Chem. Soc.* **2002**, *124*, 14832.

CH_2=CHCO$_2$Me → PhH , O_2 , cat $Pd(OAc)_2$, NaOAc , $EtCO_2H$ / cat $H_7PMo_8V_4O_{40}$, 2,4-pentanedione , 90°C , 2 h → PhCH=CHCO$_2$Me 65%

Yokota, T.; Tani, M.; Sakaguchi, S.; Ishii, Y. *J. Am. Chem. Soc.* **2003**, *125*, 1476.

PhP(O)(OH)OH → CH_2=CHPh , THF-dioxane , 100°C , 1 d / cat $Pd(OAc)_2$, Me_3NO•2 H_2O , TBAF → PhCH=CHPh 95%

Inoue, A.; Shinokubo, H.; Oshima, K. *J. Am. Chem. Soc.* **2003**, *125*, 1484.

1-hexene → $PhSiH_3$ / Cp*(i-Pr_3P)Ru(H) SiH_2Ph/Li[B(C_6F_5)$_4$] → SiH_2Ph 77%

Glaser, P.B.; Tilley, T.D. *J. Am. Chem. Soc.* **2003**, *125*, 13640.

O I Ph → 5% $Pd(OAc)_2$, 5% dppm / 2 CsO_2t-Bu , DMF , 100°C → O 88%

Huang, Q.; Fazio, A.; Dai, G.; Campo, M.A.; Larock, R.C. *J. Am. Chem. Soc.* **2004**, *126*, 7460.

O O → PhCH=CH_2 , DCM / cat $AuCl_3$/AgOTf → O O Ph 74%

Yao, X.; Li, C.-J. *J. Am. Chem. Soc.* **2004**, *126*, 6884.

Bn-N → 5% $[RhCl(cod)_2]_2$, toluene , 50°C / 15% chiral aminophosphine , 9 h → Bn-N 94% (95% ee)

Thalji, R.K.; Ellman, J.A.; Bergman, R.G. *J. Am. Chem. Soc.* **2004**, *126*, 7192.

1. CuCl , (S)-tol-BINAL , NaO*t*-Bu , Ph_2SiH_2, THF , 0°C
2. 5% $Pd(OAc)_2$, 10% Ph-Ph(*t*-Bu_2P=O), CsF , 1.4 4-Br-*t*-butylbenzene , THF , rt

74%

Chae, J.; Yun, J.; Buchwald, S.L. *Org. Lett.* **2004**, *6*, 4809.

5% $PdCl_2(MeCN)_2$, 3 $CuCl_2$, MeOH

MeO_2C

83%

Liu, C.; Widenhofer, R.A. *J. Am. Chem. Soc.* **2004**, *126*, 10250.

i-PrOCOCl , DCM

$Et_3Al_2Cl_3$, –15°C → rt

67%

Biermann, U.; Metzger, J.O. *J. Am. Chem. Soc.* **2004**, *126*, 10319.

PhI , 3% $Pd(OAc)_2$, NEt_3

Ph + Ph (80 : 20) 63%

Nifant'ev, I.E.; Sitnikov, A.A.; Andriukhova, N.V.; Laishevtsev, I.P.; Luziikov, Y.N. *Tetrahedron Lett.* **2002**, *43*, 3213.

Ph — PhI , 130°C , DMF , 2 K_2CO_3 , 20h / cat Tedicyp-Pd complex → Ph — C_8H_{17} 88%

Berthiol, F.; Doucet, H.; Santelli, M. *Tetrahedron Lett.* **2003**, *44*, 1221.

CH_2=$CHCO_2Bn$ — 2.5% Pd-fibroin complex / H_2 , THF , 7d → $CH_3CH_2CO_2Bn$ 91%

Sajiki, H.; Ikawa, T.; Hirota, K. *Tetrahedron Lett.* **2003**, *44*, 8437.

CO_2Me — PhI , DMF , K_2CO_3 / cat $[Pd(C_3H_5)Cl]_2$/tedicyp → Ph, CO_2Me 88%

Kondolff, I.; Doucet, H.; Santelli, M. *Tetrahedron Lett.* **2003**, *44*, 8487.

$B(OH)_2$ — CH_2=$CHSO_2Ph$, 60°C , 15h , DMF / 2 Na_2CO_3 , 10% $Pd(OAc)_2$, O_2 → SO_2Ph 84%

Kabalka, G.W.; Guchhait, S.K. *Tetrahedron Lett.* **2004**, *45*, 4021.

MeO — Br — OAc , cat $CoBr_2$, e^- , bpy / Bu_4NBF_4 , rt , MeCN/Py → OMe 61%

Gomes, P.; Gosmini, C.; Périchon, J. *Tetrahedron* **2003**, *59*, 2999.

PhI — CH_2=$CHCO_2Et$, Alquot-336 , 100°C , 4h / 5% Pd/C , NEt_3 , neat → Ph — CO_2Et quant

Perosa, A.; Tundo, P.; Selva, M.; Zinovyev, S.; Testa, A. *Org. Biomol. Chem.* **2004**, *2*, 2249.

OTs — CH_2=$CHCO_2Bn$, cat Pd_2dba_3•$CHCl_3$/PCy_3 ; 80°C , 12h → CO_2Bn 89%

Tsukada, N.; Sato, T.; Inoue, Y. *Chem. Commun.* ***2003***, 2404.

O, OBu — N, O , $PhB(OH)_2$, 2.5 $Cu(OAc)_2$; cat Ru arene complex , DCM , rt → Ph, Ph 66%

Farrington, E.J.; Brown, J.M.; Barnard, C.F.J.; Rowsell, E. *Angew. Chem. Int. Ed.* ***2002***, *41*, 169.

CO_2Bu — $PhSi(OMe)_3$/TBAF , toluene , H_2O , 120°C ; 5% $[IrCl(cod)]_2$, 1d → Ph, CO_2Bu 71%

Koike, T.; Du, X.; Sanada, J.; Danda, Y.; Mori, A. *Angew. Chem. Int. Ed.* ***2003***, *42*, 89.

O, Ph, Cl — CH_2=CHPh , cat $[RhCl(C_2H_4)]_2$; *o*-xylene , reflux , 6h → Ph, Ph 86%

Sugihara, T.; Satoh, T.; Miura, M.; Nomura, M. *Angew. Chem. Int. Ed.* ***2003***, *42*, 4672.

O, Br — $PhCH$=CH_2 , Pd/C , NaOAc ; NMP , 140°C , 40h → O, Ph 90%

Heidenreich, R.G.; Köhler, K.; Kraufer, J.G.E.; Pietsch, J. *Synlett* ***2002***, 1118.

O — PhCHO , Al , TMSCl , 5% $InCl_3$; THF , 3 h → Ph, O 56% + 18% diol

Ohe, T.; Ohse, T.; Mori, K.; Ohtaka, S.; Uemura, S. *Bull. Chem. Soc. Jpn.* ***2003***, *76*, 1823.

Ph, CHO — $NiCl_2$•6 H_2O , $NaBH_4$ → Ph, CHO 70%

Khurana, J.M.; Sharma, P. *Bull. Chem. Soc. Jpn.* ***2004***, *77*, 549.

X, OMe — CH_2=$CHGaCl_2$, DMSO , THF ; cat $Pd_2(dba)_3$•$CHCl_3$ → OMe 97%

Mikami, S.; Yorimitsu, H.; Oshima, K. *Synlett* ***2002***, 1137.

MeO_2C, CO_2Me, O — EtI , BEt_3 , O_2 , –78°C ; $Yb(OTf)_3$, Bu_3SnH → MeO_2C, MeO_2C, Et, O 55% (>50:1 *trans:cis*)

Sibi, M.P.; Patil, K.; Rheault, T.R. *Eur. J. Org. Chem.* ***2004***, 372.

OH — 0.5 $Ti(O\textit{i}\text{-}Pr)_4$, *i*-PrMgBr → OH, OH 68%

Isakov, V.E.; Kulinkovich, O.G. *Synlett* ***2003***, 967.

(EtO)Me$_2$SiH , toluene , 50°C

chiral Rh catalyst , 2% AgBF$_4$, 1 d

MeO (EtO)Me$_2$Si MeO (38 : MeO 62) 92% SiMe$_2$(OEt)

Tsuchiya, Y.; Uchimura, H.; Kobayshi, K.; Nishiyama, H. *Synlett* ***2004***, 2099.

MeO—I ; CN , Pd(OAc)$_2$, KOAc , 80°C ; Bu$_4$NBr , DMF , 2 d

Ph CN MeO 84% (100% *E*)

Masllorens, J.; Moreno-Mañas, M.; Pla-Quintana, A.; Pleixats, R.; Roglans, A. *Synthesis* ***2002***, 1903.

SECTION 74D: CONJUGATE REDUCTION OF α,β-UNSATURATED CARBONYL COMPOUNDS AND NITRILES

ASYMMETRIC REDUCTIONS

CHO ; , cat chiral pyrrolidines ; ether

O O H >90% (70% ee)

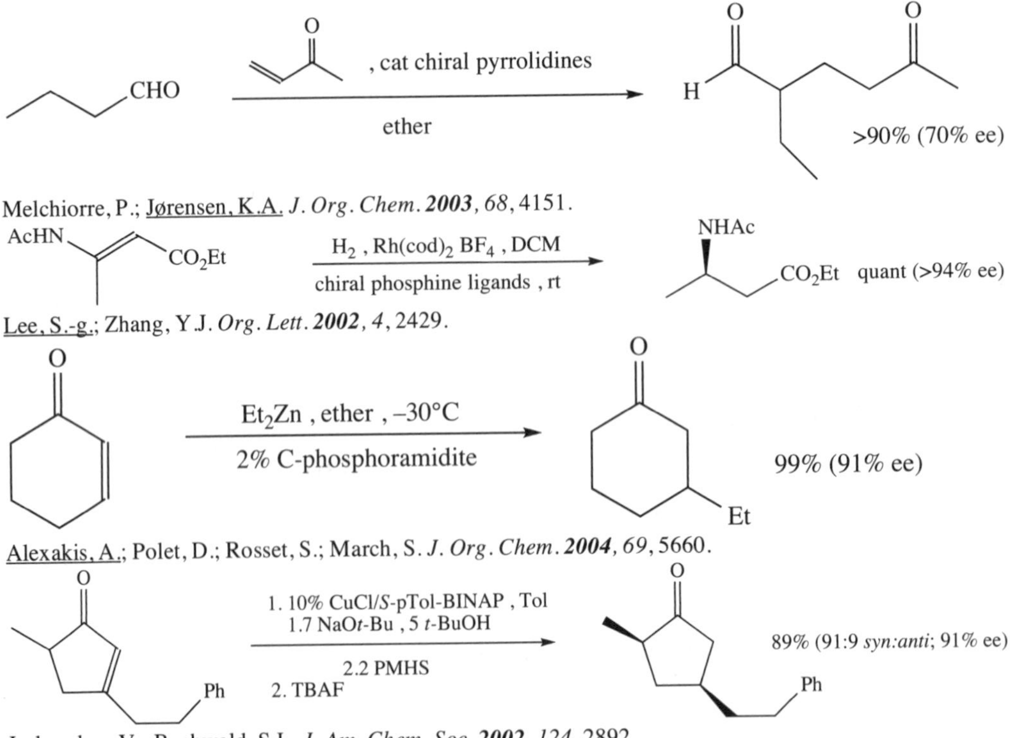

Melchiorre, P.; Jørensen, K.A. *J. Org. Chem.* ***2003***, *68*, 4151.

AcHN CO$_2$Et ; H$_2$, Rh(cod)$_2$ BF$_4$, DCM ; chiral phosphine ligands , rt

NHAc CO$_2$Et quant (>94% ee)

Lee, S.-g.; Zhang, Y.J. *Org. Lett.* ***2002***, *4*, 2429.

O ; Et$_2$Zn , ether , –30°C ; 2% C-phosphoramidite

O Et 99% (91% ee)

Alexakis, A.; Polet, D.; Rosset, S.; March, S. *J. Org. Chem.* ***2004***, *69*, 5660.

O Ph ; 1. 10% CuCl/*S*-pTol-BINAP , Tol 1.7 NaO*t*-Bu , 5 *t*-BuOH ; 2.2 PMHS 2. TBAF

O Ph 89% (91:9 *syn:anti*; 91% ee)

Jurkauskas, V.; Buchwald, S.L. *J. Am. Chem. Soc.* ***2002***, *124*, 2892.

O

O

4 Ph_3SiH , 5% CuCl , cat *t*-BuONa

5% S-p-Tol-BINAP , toluene , 23°C , 2 d

O

50% (80% ee)

Bn Bn

Hughes, G.; Kimura, M.; Buchwald, S.L. *J. Am. Chem. Soc.* ***2003***, *125*, 11253.

Ph CO_2Et cat CuH-L* , PHMS , *ti*-BuOH Ph CO_2Et

L* = chiral bis(phsophino) aryl

96% (90% ee)

Lipshutz, B.H.; Servesko, J.M.; Taft, B.R. *J. Am. Chem. Soc.* ***2004***, *126*, 8352.

NH_2 H_2 , chiral Rh catalyst NH_2

Ph CO_2Me TFMe , 50°C Ph CO_2Me 98% (96% ee)

Hsiao, Y.; Rivera, N.R.; Rosner, T.; Krska, S.W.; Niolito, E.; Wang, F.; Sun, Y.; Armstrong III, J.D.; Grabowski, E.J.J.; Tillyer, R.D.; Spindler, F.; Malan, C. *J. Am. Chem. Soc.* ***2004***, *126*, 9918.

CO_2Me polymer supported Rh catalyst CO_2Me

H_2 (40 atm) , toluene

96% ee

NHAc NHAc

Wang, X.; Ding, K. *J. Am. Chem. Soc.* ***2004***, *126*, 10524.

H_2 , cat $[RhL_2BF_4]$

MeO_2C NHAc L = chiral monodentate P ligand MeO_2C NHAc

Reetz, M.T.; Mehler, G. *Tetrahedron Lett.* ***2003***, *44*, 4593.

cat $Rh(cod)_2BF_4$, H_2 , DCM , 12h

AcHN CO_2Me BINOL-phosphite AcHN CO_2Me 89% ee

Huang, H.; Liu, X.; Chen, S.; Chen, H.; Zheng, Z. *Tetrahedron: Asymmetry* ***2004***, *15*, 2011.

NC CN NC CN

L* , $[RuCL_2(cymene)]_2$, TEAF

THF , 30°C , 4h

99% (85% ee)

Chen, Y.-C.; Xue, D.; Deng, J.-G.; Cui, X.; Zhu, J.; Jiang, Y.-Z. *Tetrahedron Lett.* ***2004***, *45*, 1555.

CO_2Me H_8-monoPhos , $Rh(cod)_2BF_4$ COMe

NHAc H_2 , toluene , rt , 2h NHAc 99% (95% ee)

MeO MeO

Zeng, Q.; Liu, H.; Cui, X.; Mi, A.; Jiang, Y.; Li, X.; Choi, M.C.K.; Chan, A.S.C. *Tetrahedron: Asymmetry* ***2002***, *13*, 115.

CO_2Me cat Rh(cod) BF_4 , chiral ligand CO_2Me

Ph NHAc H_2 , MeOH , rt Ph NHAc quant (57% ee)

Komarov, I.V.; Monsees, A.; Kadyrov, R.; Fischer, C.; Schmidt, U.; Börner, A. *Tetrahedron: Asymmetry* ***2002***, *13*, 1615.

Ph–CH=C(NHAc)CO_2Me — H_2 , MeOH , Ru catalyst , rt / cat chiral bipyridyl diphosphine → PhCH_2CH(NHAc)CO_2Me quant (90% ee)

Wu, J.; Pai, C.C.; Kwok, W.H.; Guo, R.W.; Au-Yueng, T.T.L.; Yeung, C.H.; Chan, A.S.C. *Tetrahedron: Asymmetry* ***2003**, 14*, 987.

reductase from *Nicotiana tabacum* / 1d — 86% (>99% ee)

Shimoda, K.; Kubota, N.; Itamada, H. *Tetrahedron: Asymmetry* ***2004**, 15*, 2443.

Ph–CH=C(NHAc)CO_2Me — H_2 , cat $[Rh(cod)_2]$ BF_4 , toluene / chiral biaryl phosphine → PhCH_2CH(NHAc)CO_2Me >99% (95% ee)

Junge, K.; Hagemann, B.; Enthaler, S.; Spannenberg, A.; Michalik, M.; Oehme, G.; Monsees, A.; Riermeier, T.; Beller, M. *Tetrahedron: Asymmetry* ***2004**, 15*, 2621.

$(CH_2)_4OTBS$ — 1% ferrocenyl N-P ligand , 2% CuCl / PHMS , 2% NaO*t*-Bu , toluene , –78°C → $(CH_2)_4OTBS$ 96% (98% ee)

Lipshutz, B.H.; Servesko, J.M. *Angew. Chem. Int. Ed.* ***2003**, 42*, 4789.

Ph–CH=C(NHAc)CO_2Me — H_2 , cat $[Rh(cod)_2]$ OTf , MeOH , rt / cat chiral phosphine phosphinate , 18h → PhCH_2CH(NHAc)CO_2Me quant (81% ee)

Monsees, A.; Laschat, S. *Synlett* **2002**, 1011.

NONASYMMETRIC REDUCTIONS

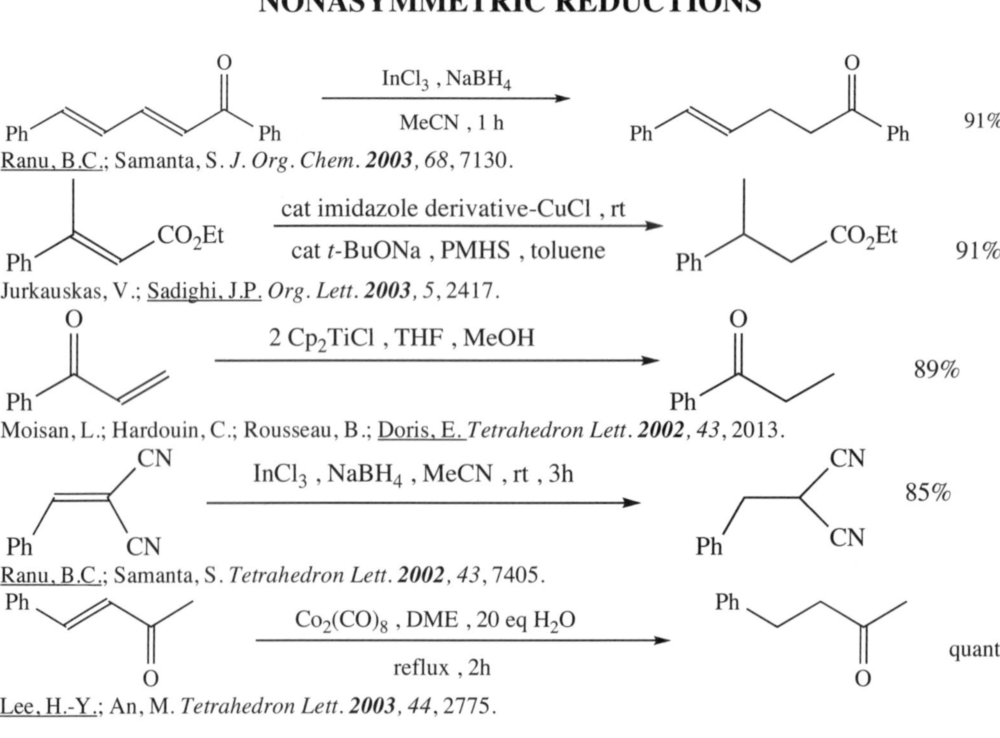

Ranu, B.C.; Samanta, S. *J. Org. Chem.* ***2003**, 68*, 7130.

Jurkauskas, V.; Sadighi, J.P. *Org. Lett.* ***2003**, 5*, 2417.

Moisan, L.; Hardouin, C.; Rousseau, B.; Doris, E. *Tetrahedron Lett.* ***2002**, 43*, 2013.

Ranu, B.C.; Samanta, S. *Tetrahedron Lett.* ***2002**, 43*, 7405.

Lee, H.-Y.; An, M. *Tetrahedron Lett.* ***2003**, 44*, 2775.

polymer supported formate, 70°C, 10h — 82%

Basu, B.; Bhuiyan, Md.M.H.; Das, P.; Hossain, I. *Tetrahedron Lett.* **2003**, *44*, 8931.

Bu_3SnH, $MgBr_2 \bullet OEt_2$, DCM, rt, 5h — 66%

Hirasawa, S.; Nagano, H.; Kameda, Y. *Tetrahedron Lett.* **2004**, *45*, 2207.

CO, H_2O, 20% Se, DMF, 90°C, 2 h — 99%

Tian, F.; Lu, S. *Synlett* **2004**, 1953.

1. $InCl_3$, NaH_4, MeCN 2. MeOH — 94%

Ranu, B.C.; Samanta, S. *Tetrahedron* **2003**, *59*, 7901.

CH_2=CHPh, $NaBO_3$, K_2CO_3, 10h; imidazoidine carbene ligated Pd cat. — 85%

Liu, J.; Zhao, Y.; Zhou, Y.; Li, L.; Zhang, T.Y.; Zhang, H. *Org. Biomol. Chem.* **2003**, *1*, 3227.

H_2, Pd/C, $MgBr_2$, THF — 78% (1:1 *syn:anti*)

Nagano, H.; Yokota, M.; Iwazaki, Y. *Tetrahedron Lett.* **2004**, *45*, 3035.

$NaBH_4$, aq THF, DABCO, 15 min; $NaBH_4$, MeOH — 91%

Ar = 3-nitrophenyl

Patra, A.; Batra, S.; Bhaduri, A.P. *Synlett* **2003**, 1611.

Ph_4Sn, cat $PdCl_2$/ 2 LiCl, AcOH, 25°C, 20 h — 63% + Ph-Ph 24%

Ohe, T.; Uemura, S. *Bull. Chem. Soc. Jpn.* **2003**, *76*, 1423.

SECTION 74E: CONJUGATE ALKYLATIONS

ASYMMETRIC ALKYLATIONS

cat $[RhCl(C_2H_4)_2]_2$, 10% KOH; cat bis(phosphine) ligand, 50°C; $PhB(OH)_2$, aq dioxane — 980% (90% ee)

Shintani, R.; Ueyama, K.; Yamada, I.; Hayashi, T. *Org. Lett.* **2004**, *6*, 3425.

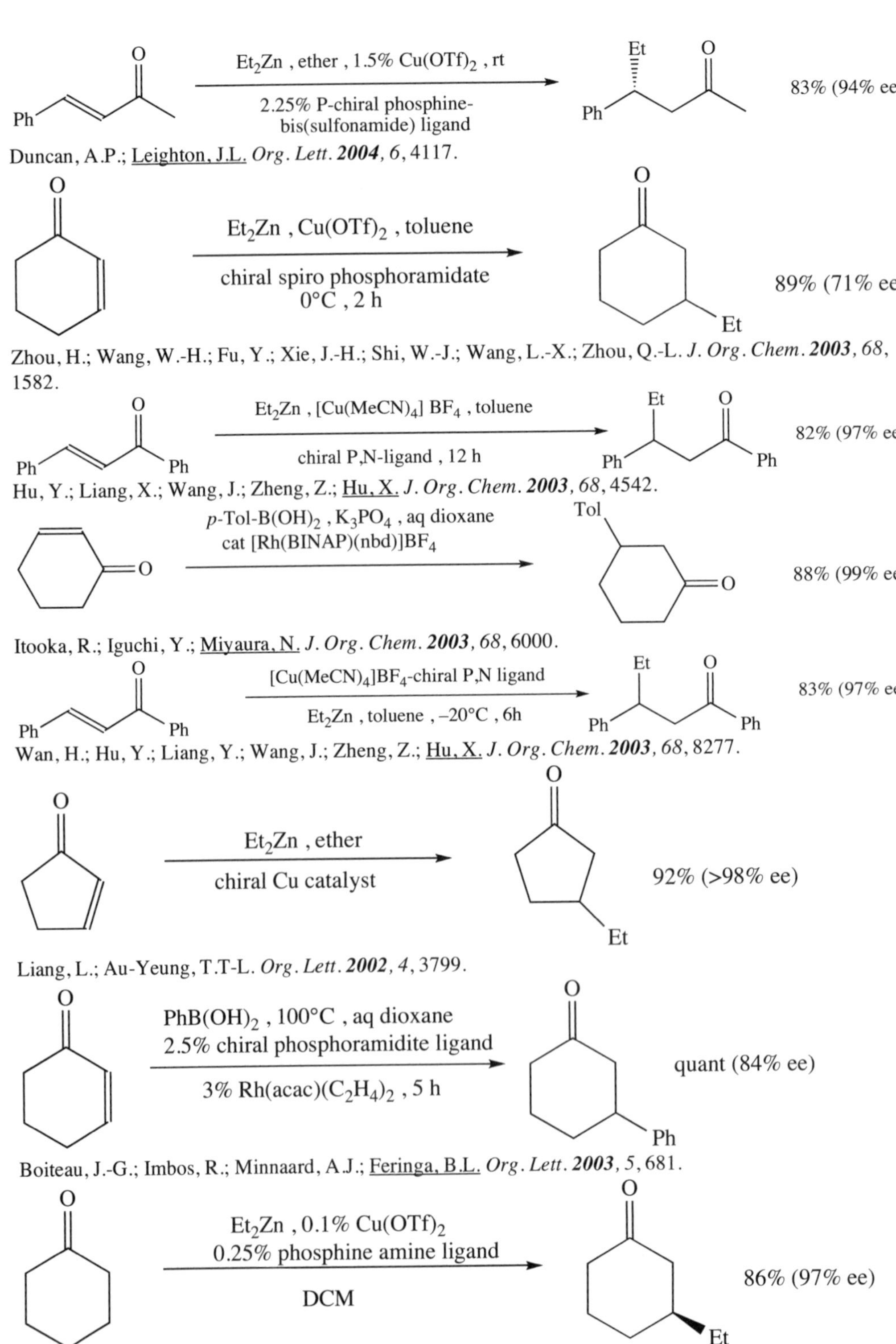

Duncan, A.P.; Leighton, J.L. *Org. Lett.* ***2004***, *6*, 4117.

Zhou, H.; Wang, W.-H.; Fu, Y.; Xie, J.-H.; Shi, W.-J.; Wang, L.-X.; Zhou, Q.-L. *J. Org. Chem.* ***2003***, *68*, 1582.

Hu, Y.; Liang, X.; Wang, J.; Zheng, Z.; Hu, X. *J. Org. Chem.* ***2003***, *68*, 4542.

Itooka, R.; Iguchi, Y.; Miyaura, N. *J. Org. Chem.* ***2003***, *68*, 6000.

Wan, H.; Hu, Y.; Liang, Y.; Wang, J.; Zheng, Z.; Hu, X. *J. Org. Chem.* ***2003***, *68*, 8277.

Liang, L.; Au-Yeung, T.T-L. *Org. Lett.* ***2002***, *4*, 3799.

Boiteau, J.-G.; Imbos, R.; Minnaard, A.J.; Feringa, B.L. *Org. Lett.* ***2003***, *5*, 681.

Krauss, I.J.; Leighton, J.L. *Org. Lett.* ***2003***, *5*, 3201.

Et_2Zn , ether , –20°C , 4 h

copper thiolate catalyst

>99% (83% ee)

Et

Arink, A.M.; Braam, T.W.; Keeris, R.; Jastrzebski, J.T.B.H.; Benhaim, C.; Rosset, S.; Alexakis, A.; van Koten, G. *Org. Lett.* ***2004***, *6*, 1959.

SO_2(2-Py)

$PhB(OH)_2$, aq. dioxane

cat Rh(acac((C_3H_4)

Chiraphos

Ph

SO_2(2-Py)

97% (81% ee)

Mauleón, P.; Carretero, J.C. *Org. Lett.* ***2004***, *6*, 3195.

3% Rh(acac)(C_2H_4)$_2$, 3 $PhB(OH)_2$

7.5% phosphoramidite , 100°C

aq dioxane

>98% ee

Ph

Boiteau, J.-G.; Minnaard, A.J.; Feringa, B.L. *J. Org. Chem.* ***2003***, *68*, 9481.

$PhSi(OMe)_3$, aq dioxane

90°C , 20 h

cat [Rh(cod)(MeCN)$_2$] BF_4

Ph

76% (98% ee)

Oi, S.; Taira, A.; Honma, Y.; Inoue, Y. *Org. Lett.* ***2003***, *5*, 97.

CH_2=$CHBF_3K$, 4% Rh(acac)(eth)$_2$

10% phosphoramidite ligand

EtOH , reflux

99% (88% ee)

Duursma, A.; Boiteau, J.-G.; Lefort, L.; Boogers, J.A.F.; de Vries, A.H.M.; de Vries, J.G.; Minnaard, A.J.; Feringa, B.L. *J. Org. Chem.* ***2004***, *69*, 8045.

(CH_2)$_5$

$PhB(OH)_2$, cat Rh(acac)(C_2H_4)$_2$

phosphine , aq dioxane , 100°C , 1 h

(CH_2)$_5$

Ph

99% (97% ee)

Kuriyama, M.; Nagai, K.; Yamada, K.-i.; Mmiwas, Y.; Taga, T.; Tomioka, K. *J. Am. Chem. Soc.* ***2002***, *124*, 8932.

1. PhTi(O*i*-Pr)$_3$, THF , 20°C

cat [Rh(OH)(S-BINAP)]$_2$

2. MeOH

99% ee

Ph

Hayashi, T.; Tokunaga, N.; Han, J.W. *J. Am. Chem. Soc.* ***2002***, *124*, 12102.

MeO

O

HO

2% Et_2Zn, 1% BINOL derivative
THF, –38°C, 20 h

93% (78:22 *syn:anti*)
(95% ee; 93% ee)

Harada, S.; Kumagai, N.; Kinoshita, T.; Matsunaga, S.; Shibasaki, M. *J. Am. Chem. Soc.* ***2003***, *125*, 2582.

2% phosporamidite ligand, –55°C

78% (96% ee)

Duursma, A.; Minnaard, A.J.; Feringa, B.L. *J. Am. Chem. Soc.* ***2003***, *125*, 3700.

3 $CH_2(CO_2Me)_2$, THF, –20°C, 36 h
Chinchona alkaloid catalyst

97% (96% ee)

Li, H.; Wang, Y.; Tang, L.; Deng, L. *J. Am. Chem. Soc.* ***2004***, *126*, 9906.

$CH_2(CO_2Me)_2$, chiral Ru catalyst
t-BuOH, 40°C, 1 d

87% (82% ee)

Watanabe, M.; Murata, K.; Ikariya, T. *J. Am. Chem. Soc.* ***2003***, *125*, 7508.

$CH_2(CO_2Me)_2$, chiral Ru catalyst
toluene, –20°C, 2 d

97% (95% ee)

Watanabe, M.; Ikagawa, A.; Wang, H.; Murata, K.; Ikariya, T. *J. Am. Chem. Soc.* ***2004***, *126*, 11148.

$PhB(OH)_2$, chiral Rh catalyst, 1 h
0.5 KOH, aq dioxane, 30°C

94% (96% ee)

Hayashi, T.; Ueyama, K.; Tokunaga, N.; Yoshida, K. *J. Am. Chem. Soc.* ***2003***, *125*, 11508.

PhZnCl, cat $[RhCl(BINAP)]_2$
THF, 20°C

95% (>99.5% ee)

Shitani, R.; Tokunaga, N.; Doi, H.; Hayashi, T. *J. Am. Chem. Soc.* ***2004***, *126*, 6240.

$C_3H_7NO_2$, toluene, Cs_2CO_3
1% chiral ammonium salt

99% (1:19 *anti:syn*)
96% ee:76% ee

Ooi, T.; Fujioka, S.; Maruoka, K. *J. Am. Chem. Soc.* ***2004***, *126*, 11790.

C_3H_7MgBr, $CuBr{\bullet}SMe_2$, –75°C
t-BuOMe, 6% JOSIPHOS

84% (90% ee)

López, F.; Harutyayan, S.R.; Minnaard, A.J.; Feringa, B.L. *J. Am. Chem. Soc.* ***2004***, *126*, 12784.

Et_2Zn , 1% $Cu(OTf)_2$, THF , –30°C
2% chiral ferrocenyl ligand , 20 h
quant (87% ee)

Reetz, M.T.; Gosberg, A.; Moulin, D. *Tetrahedron Lett.* **2002**, *43*, 1189.

$PhB(OH)_2$, 3% $Rh(acac)(C_2H_4)_2$
cat H_2O , 100°C , dioxane , 3% (*S*)-P-Phos
99% (99% ee)

Shi, Q.; Xu, L.; Li, X.; Wang, R.; Au-Yeung, T.T.-L.; Chan, A.S.C.; Hayashi, T.; Cao, R.; Hong, M. *Tetrahedron Lett.* **2003**, *44*, 6505.

$ZnPh_2$, cat $Cu(OTf)_2$, –60°C , 16h
cat phosphoramidite ligand , toluene
quant (94% ee)

Peña, D.; López, F.; Harutyuyan, S.R.; Minnaard, A.J.; Feringa, B.L. *Chem. Commun.* **2004**, 1836.

Ph_3Bi , aq MeOH , –5°C
cat dicationic PdH complex
85% (95% ee)

Nishikata, T.; Yamamoto, Y.; Miyaura, N. *Chem. Commun.* **2004**, 1822.

Me_2Zn , cat $Cu(OTf)_2$/ligand , –40°C
toluene , 8h
85% (93% ee)

Choi, Y.H.; CHoi, J.Y.; Yang, H.Y.; Kim, Y.H. *Tetrahedron: Asymmetry* **2002**, *13*, 801.

cat $Cu(OT)_2$/biaryl ligand
Et_2Zn , ether , 0°C , 2h
95% (69% ee)

Kang, J.; Lee, J.H.; Lim, D.S. *Tetrahedron: Asymmetry* **2003**, *14*, 305.

cat Ph-BINAP(cod) PF_6 , aq PhH
$PhBF_3K$, 110°C
91% (68% ee)

Moss, R.J.; Wadsworth, K.J.; Chapman, C.J.; Frost, C.G. *Chem. Commun.* **2004**, 1984.

Et_2Zn , toluene , –50°C → 0°C , 2h
cat $Cu(OTf)_2$/phosphoramidite

Bn N O → Bn N O Et 65% (95% ee)

Pineschi, M.; Del Moro, F.; Gini, F.; Minnaard, A.J.; Feringa, B.L. *Chem. Commun.* ***2004***, 1244.

$AlEt_3$, cat $Cu(OTf)_2$, ether 0°C
BINOL derivative

O → O Et 92% (94% ee)

Su, L.; Li, X.; Chan, W.L.; Jia, X.; Chan, A.S.C. *Tetrahedron: Asymmetry* ***2003***, *14*, 1865.

Et_2Zn , CuBr•SMe_2 , ether , –10°C
2h

O → O Et 98% (>97% ee)

Breit, B.; Laungani, A.Ch. *Tetrahedron: Asymmetry* ***2003***, *14*, 3823.

Et_2Zn , $Cu(OTf)_2$, phosphite ligand
toluene , 40°C

Ph O Ph → Et O Ph Ph 66% (48% ee)

Scafato, P.; Labano, S.; Cunsolo, G.; Rosini, C. *Tetrahedron: Asymmetry* ***2003***, *14*, 3873.

i-$PrNO_2$, DMSO , rt , 5 d
15% dipeptide piperazine additive

O → O NO_2 53% (29% ee)

Tsogoeva, S.B.; Jagtap, S.B. *Synlett* ***2004***, 2624.

Et_2Zn , ether , –70°C , 1h
cat Cu O O , phosphoramidite ligand

O → O Et quant (99.5% ee)

Alexakis, A.; Polet, D.; Benhaim, C.; Rosset, S. *Tetrahedron: Asymmetry* ***2004***, *15*, 2199.

Et_2Zn , cat $Cu(OTf)_2$, ether , –30°C
chiral Schiff base ligand

O → O Et 93% (79% ee)

Liang, L.; Yan, M.; Li, Y.-M.; Chan, A.S.C. *Tetrahedron: Asymmetry* ***2004***, *15*, 2575.

O
Ph Ph
Et_2Zn , $[Cu(MeCN)_4]BF_4$, 12h
toluene , phosphite-pyyridine ligand
Et O
Ph Ph
72% (94% ee)

Hu, Y.; Liang, X.; Wang, J.; Zheng, Z.; Hu, X. *Tetrahedron: Asymmetry* ***2003***, *14*, 3907.

OHC O Ph
THF , rt , 1d
cat (Me, O, N, Bn, NH_2)
OHC O Ph
99% (97%ee) (24:1 *anti:syn*)

Fonseca, M.T.H.; List, B. *Angew. Chem. Int. Ed.* ***2004***, *43*, 3958.

O
2 $PhBF_3K$, Rh catalyst/chiral ligand
$Rh(cod)_2\ PF_6$, toluene/H_2O , 10°C
O Ph
99% (95%ee)

Pucheault, M.; Darses, S.; Genêt, J.-P. *Eur. J. Org. Chem.* ***2002***, 3552.

O
$PhB(OH)_2$, Na_2CO_3 , ethylene glycol
cat $Rh(acac)(C_2H_4)_2$, cat Digm-BINAP
120°C
O Ph
quant (98% ee)

Amengual, R.; Michelet, V.; Genêt, J.-P. *Synlett* ***2002***, 1791.

O
$PhB(OH)_2$, KOH , aq. dioxane
cat $Rh(acac)C_2H_4)_2$. 50°C , 6 h
phosphoramidite
Ph O
95% (98% ee)

Iguchi, Y.; Itooka, R.; Miyaura, N. *Synlett* ***2003***, 1040.

O
Ph NO_2 , MeOH , rt , 4d
cat L-proline
O Ph NO_2
70% (76% ee)

Enders, D.; Seki, A. *Synlett* ***2002***, 26.

O
C_5H_{11} $B(OH)_2$, dioxane/H_2O
3% $Rh(acac)(C_2H_4)_2$, 3% S-BINAP
100°C
O C_5H_{11}
94% ee

Hayashi, T. *Pure Appl. Chem.* ***2004***. *76*, 465.

O
Ph
$CH_2(CO_2Bn)_2$, 10% chiral amino acid
165h , rt
$(BnO_2C)_2HC$ O
Ph
86% (99%ee)

Halland, N.; Aburel, P.S.; Jørgensen, K.A. *Angew. Chem. Int. Ed.* ***2003***, *42*, 661.

Et_2Zn , cat $Cu(OTf)_2$, PhH , 0°C
triamide phosphane , toluene

95% (95%ee)

Hird, A.W.; Hoveyda, A.H. *Angew. Chem. Int. Ed.* **2003**, *42*, 1276.

cat. chiral Cu complex , Et_2Zn
toluene

90% (93% ee)

Mizutani, H.; Degrado, S.J; Hoveyda, A.H. *J. Am. Chem. Soc.* **2002**, *124*, 779.

NONASYMMETRIC ALKYLATIONS

1. *t*-BuMgCl
2. BuMgCl

80%

Fleming, F.F.; Wang, Q.; Zhang, Z.; Steward, O.W. *J. Org. Chem.* **2002**, *67*, 5953.

, cat $B(OTf)_3$
MeCN , rt , 1h

95%

Reddy, A.V.; Ravinder, K.; Goud, T.V.; Krishnaiah, P.; Raju, T.V.; Venkajeswarlu, Y. *Tetrahedron Lett.* **2003**, *44*, 6257.

10% DMAP , 10% $LiClO_4$
3 BnOH , 85 h , –20°C
frozen water in autoclave at 200 MPa

97%

Hayashi, Y.; Nishimura, K. *Chem. Lett.* **2002**, *31*, 296.

, 10% $NiBr_2$•3 H_2O
e^- , DMF , Py , 70°C

74%

Condon, S.; Dupré, D.; Falgayrac, G.; Nédélec, J.-Y. *Eur. J. Org. Chem.* **2002**, 105.

$InBr_3$, 3 eq *i*-PrOH , DCM , rt

85%

Bandini, M.; Cozzi, P.G.; Giacomini, M.; Melchiorre, P.; Selva, S.; Umani-Ronchi, A. *J. Org. Chem.* **2002**, *67*, 3700.

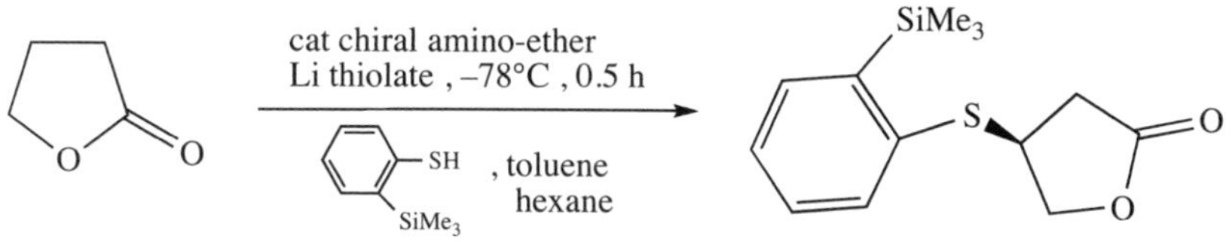

cat chiral amino-ether, Li thiolate, –78°C, 0.5 h; 2-($SiMe_3$)C$_6$H$_4$SH, toluene, hexane — 99% (7% ee)

Nishimura, K.; Tomioka, K. *J. Org. Chem.* **2002**, *67*, 435.

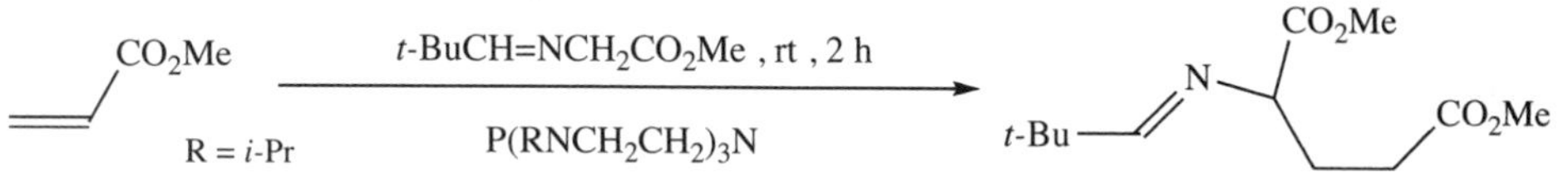

CO_2Me — *t*-BuCH=NCH$_2$CO$_2$Me, rt, 2 h; P(RNCH$_2$CH$_2$)$_3$N, R = *i*-Pr — *t*-Bu, N, CO_2Me, CO_2Me — 72%

Kisanga, P.B.; Ilankumaran, P.; Fetlerly, B.M.; Verkade, J.G. *J. Org. Chem.* **2002**, *67*, 3555.

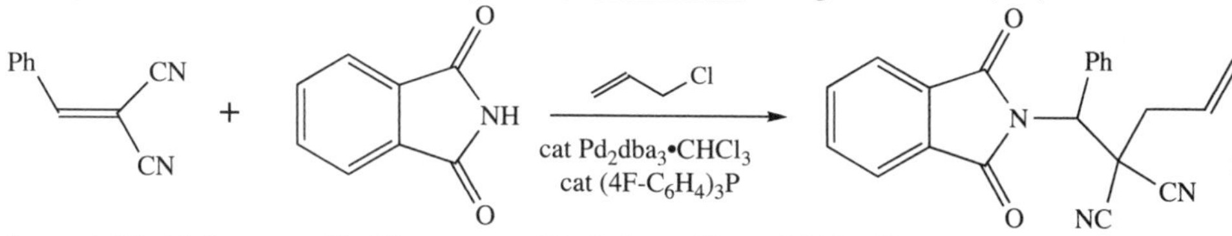

Ph, CN, CN + phthalimide (O, NH, O) — allyl chloride (Cl); cat Pd_2dba_3•$CHCl_3$, cat (4F-C_6H_4)$_3$P — O, N, O, Ph, NC, CN — >99%

Aoyagi, K.; Nakamura, H.; Yamamoto, Y. *J. Org. Chem.* **2002**, *67*, 5977.

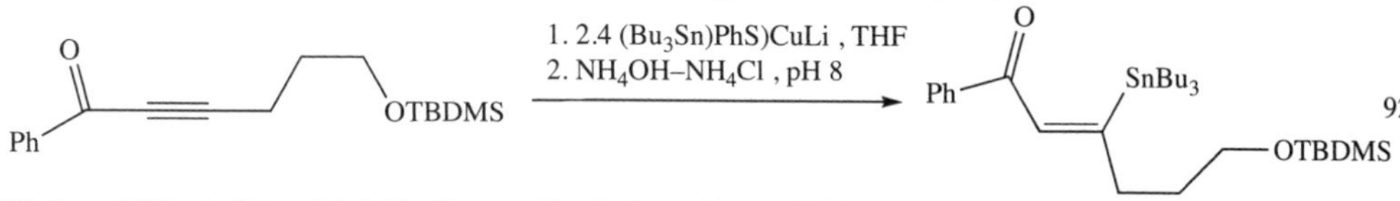

O, Ph, OTBDMS — 1. 2.4 (Bu_3Sn)PhS)CuLi, THF; 2. NH_4OH–NH_4Cl, pH 8 — O, Ph, $SnBu_3$, OTBDMS — 92%

Nielsen, T.E.; de Dios, M.A.C.; Tanner, D. *J. Org. Chem.* **2002**, *67*, 7309.

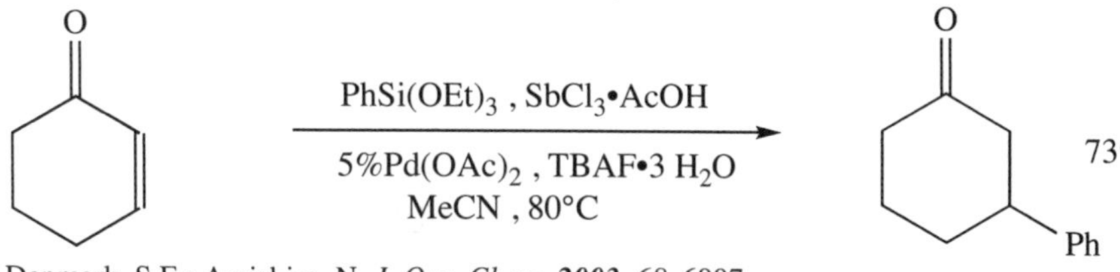

O — PhSi(OEt)$_3$, $SbCl_3$•AcOH; 5%Pd(OAc)$_2$, TBAF•3 H_2O, MeCN, 80°C — O, Ph — 73%

Denmark, S.E.; Amishiro, N. *J. Org. Chem.* **2003**, *68*, 6997.

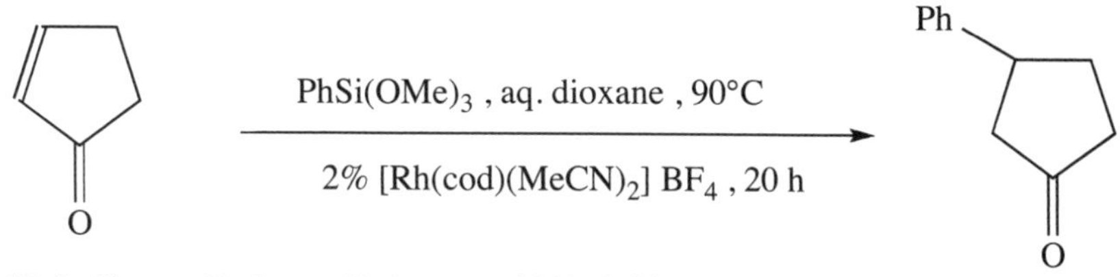

O — PhSi(OMe)$_3$, aq. dioxane, 90°C; 2% [Rh(cod)(MeCN)$_2$] BF_4, 20 h — Ph, O — 87%

Oi, S.; Honma, Y.; Inoue, Y. *Org. Lett*, **2002**, *4*, 667.

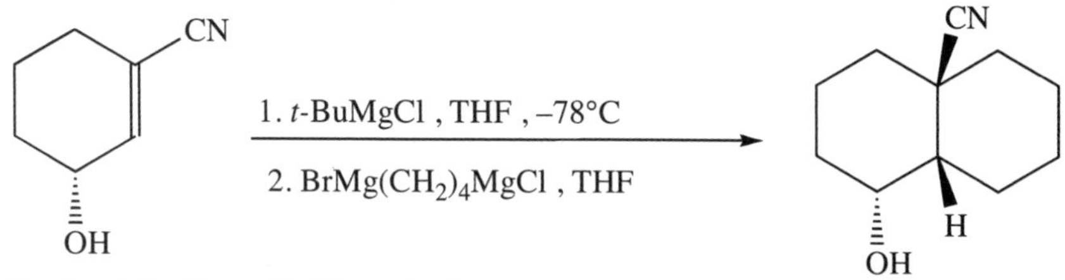

CN, OH — 1. *t*-BuMgCl, THF, –78°C; 2. BrMg(CH$_2$)$_4$MgCl, THF — CN, H, OH — 64%

Fleming, F.F.; Zhang, Z.; Wang, Q.; Steward, O.W. *Org. Lett.* **2002**, *4*, 2493.

Ph CN 1. Cp_2Zr (zirconacyclopentane), 0°C, 1 h; 2. 2.5 I_2, 20°C, 3 h → Ph, Bu, CN, CN 73%

Liu, Y.; Shen, B.; Kotora, M.; Nakajima, K.; Takahashi, T. *J. Org. Chem.* **2002**, *67*, 7010.

CO_2t-Bu [Cp_2ZrCl_2/CH_2=$CHCH_2MgCl$] −78°C → 25°C, cyclohexyl–I → CO_2t-Bu 84%

Hirano, K.; Fujita, K.; Shinokubo, H.; Oshima, K. *Org. Lett.* **2004**, *6*, 593.

BEt_3, 10 eq *i*-Pr–I, THF; also with Et_3Al → 91%

Liu, J.-Y.; Jang, Y.-T.; Lin, W.-W.; Liu, J.-T.; Yao, C.-F. *J. Org. Chem.* **2003**, *68*, 4030.

NH indole; methyl vinyl ketone, NaI, SiO_2; $CeCl_3$•7 H_2O → 96%

Bartoli, G.; Bartolacci, M.; Bosco, M.; Foglia, G.; Giuliani, A.; Marcantoni, E.; Sambri, L.; Torregiani, E. *J. Org. Chem.* **2003**, *68*, 4594.

N, O + O, Ph; Al_2O_3, microwaves, 5 min → Ph, O, O 83%

Sharma, U.; Bora, Y.; Boruah, R.C.; Sandhu, J.S. *Tetrahedron Lett.* **2002**, *43*, 143.

O, MeO, O, I; Bu_3SnH, AIBN, toluene, 85°C → O, MeO, O 87%; TsOH toluene → HO, O 98%

Clive, D.L.J.; Fletcher, S.P.; Liu, D. *J. Org. Chem.* **2004**, *69*, 3282.

O, O, N, O; Bu_3SnH, $Yb(OTf)_3$, O_2, BEt_3, THF, −78°C, 2-iodobutane → O, O, N, O 80% (1.8:1 *syn:anti*)

Sibi, M.P.; Rheault, T.R.; Chandramouli, S.V.; Jasperse, C.P. *J. Am. Chem. Soc.* **2002**, *124*, 2924.

copper(I) thiophene-2-carboxylate

Et_2Zn , ether

Et

>99% (99% ee)

Alexakis, A.; Benhaim, C.; Rosset, S.; Humam, M. *J. Am. Chem. Soc.* ***2002***, *124*, 5262.

CO_2Me Ph CO_2Me

1. $SIPhMe_2)_2$, 10% $(CuOTf)_2$-PhH , DME 20% PBu_3

2. aq TsOH

$SiMe_2Ph$ Ph CO_2Me CO_2Me

56%

Clark, C.T.; Lake, J.F.; Schedit, K.A. *J. Am. Chem. Soc.* ***2004***, *126*, 84.

N-Bn Ph

1. CO_2Me , cat $RhCl(PPh_3)_3$

toluene , 150°C , 2 h

2. H_3O^+

O CO_2Me

94%

Lim, S.-G.; Ahn, J.-A.; Jun, C.-H. *Org. Lett.* ***2004***, *6*, 4687.

2 , chiral W catalyst

rt , 10 h

94%

Wang, H.-S.; Yu, S.J. *Tetrahedron Lett.* ***2002***, *43*, 1051.

Ph NO_2

Ph_4Sn , AcOH , cat $PdCl_2$/LiCl

cat $Bi(NO_3)_3$•5 H_2O , 25°C

Ph Ph NO_2

81%

Ohe, T.; Uemura, S. *Tetrahedron Lett.* ***2002***, *43*, 1269.

C_6H_{13}

Me_2CuLi , TMSCl , DCM

6h

C_6H_{13}

quant

Asao, N.; Lee, S.; Yamamoto, Y. *Tetrahedron Lett.* ***2003***, *44*, 4265.

NH_2

$AlCL_3$, PhH , 25°C , 2h

Ph NH_2

99%

Koltunov, K.Yu.; Walspurger, S.; Sommer, J. *Tetrahedron Lett.* ***2004***, *45*, 3547.

Ph Ph

$PhSnMe_3$, H_2O

2% $[Rh(cod)(MeCN)_2]$ BF_4

Ph Ph Ph

98%

Oi, S.; Moro, M.; Ito, H.; Honma, Y.; Miyano, S.; Inoue, Y. *Tetrahedron* ***2002***, *58*, 91.

NO_2

Et_2Zn , 2% $Cu(OTf)_2$, toluene

cat chiral phosporamidite −45°C , 29 min

NO_2

quant

Duursma, A.; Minnaard, A.J.; Feringa, B.L. *Tetrahedron* ***2002***, *58*, 5773.

Ph, CN, CO_2Me — $MeNO_2$, Al_2O_3 , microwaves → Ph, CN, O_2N, CO_2Me 70%

Michaud, D.; Hamelin, J.; Texier-Boullet, T. *Tetrahedron* ***2003***, *59*, 3323.

OMe, Ph — 1. $Cp_2Zr(H)Cl$, DCM , rt; 2. $BF_3•OEt_2$, 0°C → rt → Ph 80%

Gandon, V.; Szymoniak, J. *Chem. Commun.* ***2002***, 1308.

O, Ph, Ph — $ZrCp_2Cl$, dioxane , rt; 2% $[RhCl(cod)]_2$ → Ph, O, Ph 93%

Kakuuchi, A.; Taguchi, T.; Hanzawa, Y. *Tetrahedron* ***2004***, *60*, 1293.

O, O, O, N, Ph — 2 $SnBu_3$, cat $Sc(OTf)_2$; DCM , –78°C → –20°C → O, O, O, N, Ph 85% (10:1 dr)

Williams, D.R.; Mullins, R.J.; Miller, N.A. *Chem. Commun.* ***2003***, 2220.

O, Ph, Ph — $SnBu_3$, $TaCl_5$, MeCN; –40°C , 2h → O, Ph, Ph 70%

Shibata, I.; Kano, T.; Kanazawa, N.; Fukuoka, S.; Baba, A. *Angew. Chem. Int. Ed.* ***2002***, *41*, 1389.

O + Ph_2, P, Pd – $(PhCN)_2(PF_3)_2$, P, Ph_2 — $PhB(OH)_2$, THF; 20°C, 23h → O, Ph + O, Ph (>99 : 1) 63%

Nishikata, T.; Yamamoto, Y.; Miyaura, N. *Angew. Chem. Int. Ed.* ***2003***, *42*, 2768.

OH — PhBr , TBAB , 2% $NaCD_2H$; Pd-benzothiazole catalyst; $NaHCO_3$, 130°C , 4h → CHO, Ph + Ph, CHO (70 : 30) 63%

Caló, V.; Nacci, A.; Monopoli, A.; Spinelli, M. *Eur. J. Org. Chem.* ***2003***, 1382.

Ph, CO_2Et — $–(Si(Me)(Ph)\text{-}O\text{-})_n–$; cat $Rh(OH)(cod)_2$, aq K_2CO_3; toluene , 120°C → Ph, Ph, CO_2Et 78%

Koike, T.; Du, X.; Mori, A.; Osakada, K. *Synlett* ***2002***, 301.

MeO_2C, CO_2Me — $PhB(OH)_2$, 2% $[Rh(cod)Cl]_2$; H_2O , Na_2CO_3 , SDS , 80°C → Ph, MeO_2C, CO_2Me 85%

Wadsworth, K.J.; Wood, F.K.; Chapman, C.J.; Frost, C.G. *Synlett* ***2004***, 2022.

CH_2=$CHCO_2$ Me , cat B(OTf)$_3$
MeCN , 30 min
CO_2Me
96%

Varala, R.; Alam, M.M.; Adapa, S.R. *Synlett* ***2003***, 720.

I
CH_2=$CHSO_2Ph$, AIBN , H_2O
N-ethylpiperidine hypophosphite
100°C
SO_2Ph
83%

Jang, D.O.; Cho, D.H. *Synlett* ***2002***, 1523.

O
CN
2 BrMg , THF , 0°C
O
CN
79%

Kung, L.-R.; Tu, C.-H.; Shia, K.-S.; Liu, H.-J. *Chem. Commun.* ***2003***, 2490.

O
Ph Ph
H N , MeOH , rt , 3 h
ultrasound , 10% CAN
H N
O
Ph Ph
81%

Ji, S.-J.; Wang, S.-Y. *Synlett* ***2003***, 2074.

O
NO_2
CH_2=CHCHO , rt , 8 h
Yb(OTf)$_3$, H_2O
O
NO_2
CHO
85%

Miranda, S.; López-Alvarado, P.; Giorgi, G.; Rodriguez, J.; Avendaño, C.; Menéndez, J.C. *Synlett* ***2003***, 2159.

O
4 NaN_3 , 4 AcOH , H_2O
20% PBu_3 , rt
O
N_3
90%

Xu, L.-W.; Xia, C.-G.; Li, J.-W.; Zhou, S.-L. *Synlett* ***2003***, 2246.

O
Ph Ph
H N , 10% I_2
rt , 4 h
H N
O
Ph Ph
86%

Wang, S.-Y.; Ji, S.-J.; Loh, T.-P. *Synlett* ***2003***, 2377.

$CH_2(CO_2Et)_2$, 10% K_2CO_3
55 min , 6 h
high speed vibration milling
98%

Zhang, Z.; Dong, Y.-W.; Wang, G.-W.; Komatsu, K. *Synlett* ***2004***, 61.

, $NaAuCl_4$•2 H_2O
ETOH , 1.5 h
88%

Arcadi, A.; Bianchi, G.; Chiarini, M.; D'Anniballe, G.; Marinelli, F. *Synlett* ***2004***, 944.

$(PhMe_2Si)_2Zn$, cat. CuCN
THF , –78°C → 0°C
$SiMe_2Ph$
90%

Oestreich, M.; Weiner, B. *Synlett* ***2004***, 2139.

, no solvent
microwaves , thiourea catalyst
96%

Dessole, G.; Herrera, R.P.; Ricci, A. *Synlett* ***2004***, 2374.

10 , 10% $NiBr_2$•3 H_2O , e^-
DMF-Py , 75°C , 7 h
83%

Condon, S.; Dupré, D.; Lachaise, I.; Nédélec, J.-Y. *Synthesis* ***2002***, 1752.

$PhB(OH)_2$
1. $ZnEt_2$, toluene , 60°C
2. , toluene , 0°C
92%

Dong, L.; Xu, Y.-J.; Gong, L.-Z.; Mi, A.-Q.; Jiang, Y.-Z. *Synthesis* ***2004***, 1057.

p-TsOH , toluene
reflux , 5 h
74%

Horiguchi, Y.; Saitoh, T.; Kashiwagi, T.; Katura, L.; Itagaki, M.; Oda, J.; Sano, T. *Heterocycles* ***2002***, *57*, 1063.

, 2% $ZrCl_4$, rt

95%

Smitha, G.; Patnaik, S.; Reddy, Ch.S. *Synthesis* **2004**, 711.

$PhSi(OMe)_3$, aq 1,4-dioxane
cat $Pd(dba)_2$/dppe/$Cu(BF_4)_2$
75°C

(43 : 57) 60%

Nishikata, T.; Yamamoto, Y.; Miyaura, N. *Chem Lett.* **2003**, *32*, 752.

REVIEWS:

"Unsaturated Nitriles: Conjugate Additions of Carbon Nucleophiles to a Recalcitrant Class of Acceptors"
Fleming, F.F.; Qang, Q. *Chem. Rev.* **2003**, *103*, 2035.

"Rhodium-Catalyzed Asymmetric 1,4-Addition and Its Related Asymmetric Reactions"
Hayashi, T.; Yamasaki, K. *Chem. Rev.* **2003**, *103*, 2829.

SECTION 74F: CYCLOPROPANATIONS, INCLUDING HALOCYCLOPROPANATIONS

CF_3CO_2H , $Zn(CH_2I)_2$
DCM , rt , 1 h

quant

Lorenz, J.C.; Long, J.; Yang, Z.; Xue, S.; Xie, Y.; Shi, Y. *J. Org. Chem.* **2004**, *69*, 327.

1. MeLi , THF , 0°C
2. , rt , 3 h

81%

Huang, Z.-Z.; Ye, S.; Xia, W.; Yu, Y.-H.; Tang, Y. *J. Org. Chem.* **2002**, *67*, 3096.

$Ti(Oi\text{-}Pr)_4$, EtMgBr
ether , 2 H_2O

74%

Bertus, P.; Szymoniak, J. *J. Org. Chem.* **2002**, *67*, 3965.

, 5% $Cr(CO)_5$(thf)
THF , rt

82%

Miki, K.; Yokoi, T.; Nishino, F.; Kato, Y.; Washitake, Y.; Ohe, K.; Uemura, S. *Org. Lett.* **2002**, *4*, 1557.

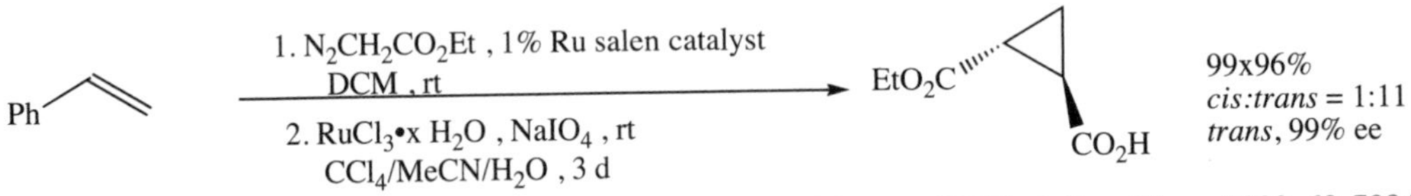

Ph
1. $N_2CH_2CO_2Et$, 1% Ru salen catalyst DCM , rt
2. $RuCl_3$•x H_2O , $NaIO_4$, rt CCl_4/MeCN/H_2O , 3 d
EtO_2C CO_2H
99x96%
cis:trans = 1:11
trans, 99% ee

Miller, J.A.; Hennessy, E.J.; Marshall, W.J.; Scialdone, M.A.; Nguyen, S.B.T. *J. Org. Chem.* ***2003***, *68*, 7884.

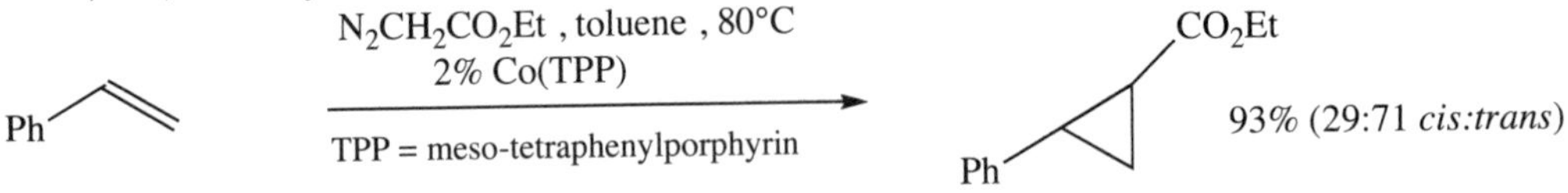

Ph
$N_2CH_2CO_2Et$, toluene , 80°C
2% Co(TPP)
TPP = meso-tetraphenylporphyrin
CO_2Et Ph
93% (29:71 *cis:trans*)

Huang, L.; Chen, Y.; Gao, G.-Y.; Zhang, X.P. *J. Org. Chem.* ***2003***, *68*, 8179.

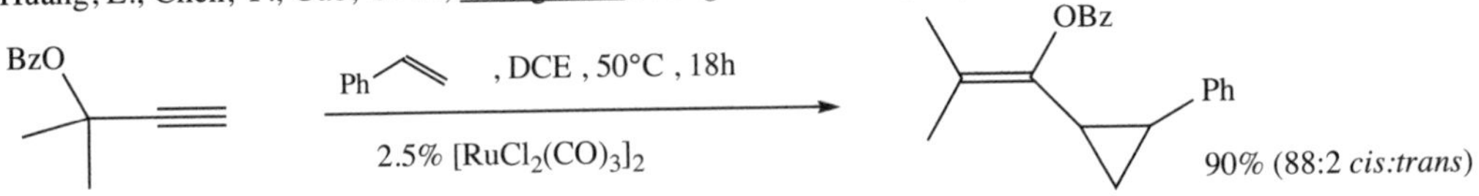

BzO
Ph , DCE , 50°C , 18h
2.5% $[RuCl_2(CO)_3]_2$
OBz Ph
90% (88:2 *cis:trans*)

Miki, K.; Ohe, K.; Uemura, S. *J. Org. Chem.* ***2003***, *68*, 8505.

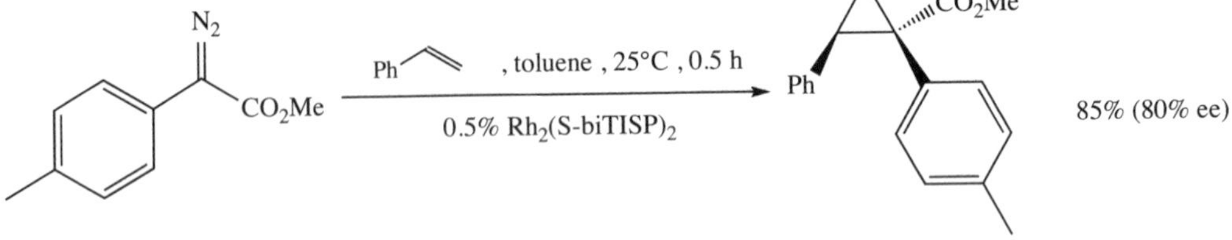

N_2 CO_2Me
Ph , toluene , 25°C , 0.5 h
0.5% $Rh_2(S\text{-biTISP})_2$
CO_2Me Ph
85% (80% ee)

Nagashima, T.; Davies, H.M.L. *Org. Lett*, ***2002***, *4*, 1989.

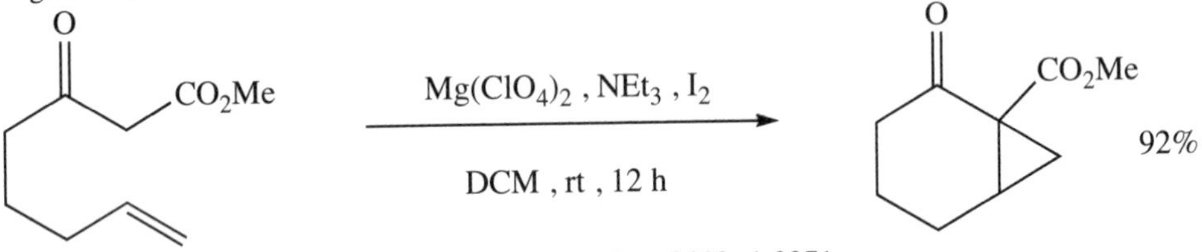

O CO_2Me
$Mg(ClO_4)_2$, NEt_3 , I_2
DCM , rt , 12 h
O CO_2Me
92%

Yang, D.; Gao, Q.; Lee, C.-S.; Cheung, K.-K. *Org. Lett.* ***2002***, *4*, 3271.

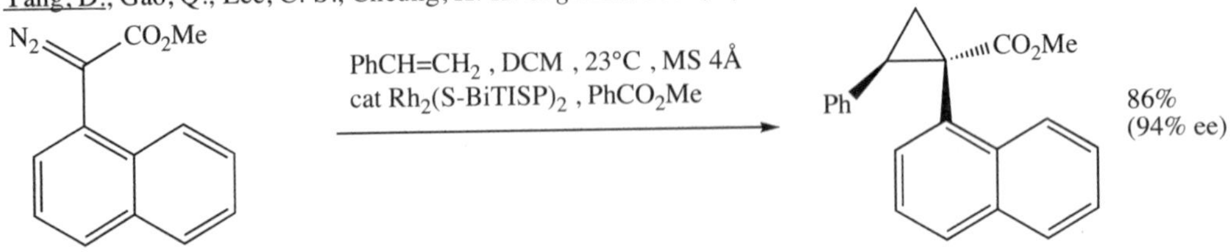

N_2 CO_2Me
$PhCH{=}CH_2$, DCM , 23°C , MS 4Å
cat $Rh_2(S\text{-BiTISP})_2$, $PhCO_2Me$
CO_2Me Ph
86%
(94% ee)

Davies, H.M.L.; Venkataraman, C. *Org. Lett.* ***2003***, *5*, 1403.

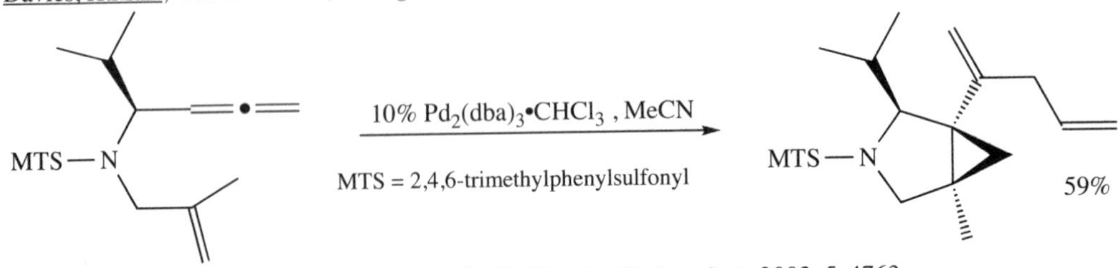

MTS—N
10% $Pd_2(dba)_3$•$CHCl_3$, MeCN
MTS = 2,4,6-trimethylphenylsulfonyl
MTS—N
59%

Ohno, H.; Takeoka, Y.; Miyamura, K.; Kadoh, Y.; Tanaka, T. *Org. Lett.* ***2003***, *5*, 4763.

$(OC)_5Cr$ carbene (NMe_2, CO_2Me); toluene, reflux, 1 d

Ph NMe$_2$ CO$_2$Me 47% (>95% de)

Barluenga, J.; Aznar, F.; Gutiérrez, I.; García-Granda, S.; Llorca-Baragaño, M.A. *Org. Lett.* ***2002***, *4*, 4273.

N_2 Ph CO_2Me — Ph—≡ , hexane, 23°C; 1% Rh(S-DOSP)$_4$ — CO_2Me Ph Ph 62% (90% ee)

Davies, H.M.L.; Lee, G.H. *Org. Lett.* ***2004***, *6*, 1233.

PO(OMe)$_2$ N_2 OMe — Ph⌒ , EtCHMe$_3$, reflux; Rh$_2$(S-BiTISP)$_2$ — Ph PO(OMe)$_2$ OMe 84% (92% ee)

Davies, H.M.L.; Lee, G.H. *Org. Lett.* ***2004***, *6*, 2117.

OMe O — 1. BuMgBr, ether, Ti(O*i*-Pr)$_4$; 2. aq. H$^+$ — O Et 46%

Masalov, N.; Feng, W.; Cha, J.K. *Org. Lett.* ***2004***, *6*, 2365.

Ph⌒ + N_2=CH–PO(O*i*-Pr)$_2$ — Ru porphyrin catalyst; DCM, 2 h — PO(O*i*-Pr)$_2$ Ph 96% (*trans* : *cis* 95:5) 87% ee 4% ee

Ferrand, Y.; LeMaux, P.; Simonneaux, G. *Org. Lett.* ***2004***, *6*, 3211.

C_6H_{13} CO_2Me CO_2Me — I–C$_6$H$_4$–CO$_2$Me; 5% Pd(PPh$_3$)$_4$, 4 eq K$_2$CO$_3$ 9 h, 10% Bu$_4$NBr, MeCN reflux — CO_2Me C_6H_{13} Me MeO$_2$C CO$_2$Me 77% (6:94 *cis:trans*)

Ma, S.; Jiao, N.; Yang, Q.; Zheng, Z. *J. Org. Chem.* ***2004***, *69*, 6463.

Ph O — CH$_2$=CHO*t*-Bu, THF, rt; 5% Cr(CO)$_5$(thf), 2h — Ph O O*t*-Bu 90% (60:40 *cis:trans*)

Miki, K.; Nishino, F.; Ohe, K.; Uemura, S. *J. Am. Chem. Soc.* ***2002***, *124*, 5260.

OH — $Me_2S{=}CHCO_2Et$, MnO_2, DCM / reflux, MS 4Å → EtO_2C … CHO 77% (*trans:cis* 3.1:1)

Oswald, M.F.; Raw, S.A.; Taylor, R.J.K. *Org. Lett.* ***2004***, *6*, 3997.

N_2 CO_2Me CO_2Me — $PhCH{=}CH_2$, PhI=O, MgO / cat $Rh_2L^*_4$, MS → Ph CO_2Me CO_2Me 75% (82% ee)

L* = chiral amino acid derivatives

Müller, P.; Ghanem, A. *Org. Lett.* ***2004***, *6*, 4347.

Ph — N_2CHCO_2Et / cat. Cu(I)-homoscorpionate → CO_2Et Ph >98% (98:2 *cis:trans*), 96% de

Díaz-Requejo, M.M.; Caballero, A.; Belderraín, T.R.; Nicasio, M.C.; Trofimenko, S.; Pérez, P.J. *J. Am. Chem. Soc.* ***2002***, *124*, 978.

OTIPS — $t\text{-}Bu_2SiCl_2$, Li → OTIPS $Si(t\text{-}Bu)_2$ 61% (97:3 ds)

Driver, T.G.; Franz, A.K.; Woerpel, K.A. *J. Am. Chem. Soc.* ***2002***, *124*, 6524.

Me_2N — MeO_2C CHO, $CHCl_3$, 20°C / 10% imidazolidinone derivative → Me_2N CO_2Me CHO 96% (95% ee)

Paras, N.A.; MacMillan, D.W.C. *J. Am. Chem. Soc.* ***2002***, *124*, 7894.

Et—≡—Et — 1. 3 Cp_2ZrHCl, DCM, rt 2. 3 Me_2Zn, toluene, –78°C 3. $PhCH{=}N\text{-}P({=}O)Ph_2$, toluene, rt → $NHP({=}O)Ph_2$ Ph Et Et 46% (96:4 dr)

Wipf, P.; Kendall, C.; Stephensen, C.R.J. *J. Am. Chem. Soc.* ***2003***, *125*, 761.

N_2 SO_2Ph O — 10% CuOTf, toluene, rt, 2 h / 15% bis(oxazoline) ligand → SO_2Ph O 91% (65% ee)

Honma, M.; Sawada, T.; Fujisawa, Y.; Utsugi, M.; Watanabe, H.; Umino, A.; Matsumura, T.; Hagihara, T.; Takano, M.; Nakada, M. *J. Am. Chem. Soc.* ***2003***, *125*, 2860.

O NHBz C_3H_7 — Ph NC CO_2Et, salen-Al complex / cyclohexane, *t*-BuOH, 23°C, 19 h → O NHBz C_3H_7 Ph CO_2Et CN 98% (97% ee) (14:1 dr)

Taylor, M.S.; Jacobsen, E.N. *J. Am. Chem. Soc.* ***2003***, *125*, 11204.

C_9H_{19} — CHI_3, $CrCl_2$, TEDDA / THF, 25°C → I C_9H_{19} 93% (95:5 *trans:cis*)

Takai, K.; Toshikawa, S.; Inoue, A.; Kokumai, R. *J. Am. Chem. Soc.* ***2003***, *125*, 12990.

1. LDA, LiBr
2. Ph–CH=CH–CO_2Me

CO_2Me CO_2Me + Ph TMS Ph TMS

(4 : 96) 98%
81% ee

Liao, W.-W.; Li, K.; Tang, Y. *J. Am. Chem. Soc.* ***2003***, *125*, 13030.

Et_2Zn, CH_2I_2, DCM

BocHN O CO_2Me N

Ph Ph 83% (75% ee)

Long, J.; Yuan, Y.; Shi, Y. *J. Am. Chem. Soc.* ***2003***, *125*, 13632.

TIPSO

1. Cp_2ZrHCl
2. Me_2Zn
3. Ph–C≡C–CH=CH–P(O)Ph Ph

OTIPS Ph Ph P Ph O 66%

Wipf, P.; Stephenson, C.R.J.; Okumura, K. *J. Am. Chem. Soc.* ***2003***, *125*, 14694.

$Hg(CF_3)_2$/NaI, THF, reflux

O N F F 93%

Nowak, I.; Cannon, J.F.; Robins, M.J. *Org. Lett.* ***2004***, *6*, 4767.

LTMP, *t*-BuOH, 10 h

HO 81%

Hodgson, D.M.; Chung, Y.K.; Paris, M.-M. *J. Am. Chem. Soc.* ***2004***, *126*, 8664.

Ph–CH=CH–NO_2, $CHCl_3$, 1 d
5% 2-nitrobenzenesulfonic acid
10% chiral diamine

O Ph NO_2 95% (98:2 dr (99% ee for *syn*)

Ishii, T.; Fujioka, S.; Skiguchi, Y.; Kotsuki, H. *J. Am. Chem. Soc.* ***2004***, *126*, 9558.

1% $(PPh_3P)Au\ SbF_6$, DCM, rt

Ph Ph H Ph Ph 94%

Luzung, M.R.; Marcham, J.P.; Toste, F.D. *J. Am. Chem. Soc.* ***2004***, *126*, 10858.

C_5H_{11} — ≡ —[N_2CHCO_2Et , DCM , 27°C; chiral Ru catalyst]→ C_5H_{11} (cyclopropene) CO_2Et 90% (95% ee)

Lou, Y.; Horikawa, M.; Kloster, R.A.; Hawryluk, N.A.; Corey, E.J. *J. Am. Chem. Soc.* ***2004***, *126*, 8916.

Ph —[N_2CHCO_2Et; ciral Ru half-sandwich complex]→ Ph (cyclopropane) CO_2Et 85% (1.85 *cis:trans*)

Tatusaus, O.; Delfosse, S.; Demonceau, A.; Noels, A.F.; Núñez, R.; Viñas, C.; Teixidor, F. *Tetrahedron Lett.* ***2002***, *43*, 983.

Ph —[N_2CHCO_2Et , 1% chiral Ru catalyst , DCM]→ Ph (cyclopropane) CO_2Et 88% (90 : 10 *cis:trans*) 83% ee 96% ee

Tang, W.; Hu, X.; Zhang, X. *Tetrahedron Lett.* ***2002***, *43*, 3075.

O —[DMSO , KOH , (bmim) PF_6 , rt]→ O 96%

Chandrasekhar, S.; Jagadeshwar, N.V. *Tetrahedron Lett.* ***2003***, *44*, 3629.

BzO —[CH_2=CHPh , 5% $[RuCl_2(CO)_3]_2$; toluene , 60°C , 18h]→ OBz Ph 81% (88:12 dr)

Miki, K.; Ohe, K.; Uemura, S. *Tetrahedron Lett.* ***2003***, *44*, 2019.

OMe Ph Ph —[2 Bui , Cp_2ZrCl_2 , THF , 20°C]→ Ph Ph 80%

Gandon, V.; Laroche, C.; Szymoniak, J. *Tetrahedron Lett.* ***2003***, *44*, 4827.

Ph —[$EtO_2CCH=N_2$, DCM , rt; cat $Cu(OTf)_2$/ligand]→ Ph CO_2Et 85% (75:25 *cis:trans*) 24%ee 24%ee

Wong, W.-L.; Lee, W.-S.; Kwong, H.-L. *Tetrahedron: Asymmetry* ***2002***, *13*, 1485.

Ph O NEt_2 —[Sm , CH_2I_2]→ Ph O NEt_2 79%

Concellón, J.M.; Rodríguez-Solla, H.; Gómez, C. *Angew. Chem. Int. Ed.* ***2002***, *41*, 1917.

O Cl O Ph —[20% DABCO , 2 Na_2CO_3; MeCN , 50°C , 5h]→ O H O Ph H 84% (>95:5 dr)

Bremeyer, N.; Smith, S.C.; Ley, S.V.; Gaunt, M.J. *Angew. Chem. Int. Ed.* ***2004***, *43*, 2681.

OH —[Et_2Zn , I_2CMe_2 , DCM , 10 h; −10°C → rt]→ OH 71%

Charette, A.B.; Wilb, N. *Synlett* ***2002***, 176.

Ph_2CCl_2 , DMF , Mg , –5°C

via Barbier, not carbene

64%

Oudeyer, S.; Auziz, A.; Léonel, E.; Paugam, J.P.; Nédélec, J.-Y. *Synlett* ***2003***, 485.

1. DABCO , rt , 30 min

2. CH_2=$CHCO_2$*t*-Bu , NAOH , MeCN , 80°C

79% (>95:5 *trans:cis*)

Papageorgiou, C.D.; Ley, S.V.; Gaunt, M.J. *Angew. Chem. Int. Ed.* ***2003***, *42*, 828.

Me_2S(=O)-I , NaH

microwaves , solid state

87%

Xu, Q.-h.; Chen, B.-h.; Ma, Y.-x.; Liu, W.-Y. *Org. Prep. Proceed. Int.* ***2002***, *34*, 194.

Me_3SiCHI_2 , $CrCl_2$, THF

TMEDA , 25°C , 6 h

89% (69:31 *t:c*)

Takai, K.; Hirano, M.; Toshikawa, S. *Synlett* ***2004***, 1304.

3 Zn , *t*-BuOH-H_2O , reflux

99%

Sakuma, D.; Togo, H. *Synlett* ***2004***, 2501.

5% Ru(salen) catalyst

hv , THF , rt , 16 h

62% (79% ee)

Saha, B.; Uchida, T.; Katsuki, T. *Chem Lett.* ***2002***, *31*, 846.

PhI , 3% $Pd(Ph_3)_4$, DMF

1.5 K_2CO_3 , heat , 100°C , 8 h

89%

Oh, C.H.; Rhim, C.Y.; Song, C.H.; Ryu, J.H. *Chem. Lett.* ***2002***, *31*, 1140.

REVIEW:

"Stereoselective Cyclopropanation Reactions"

Lebel, H.; Marcoux, J.-F.; Molinaro, C.; Charette, A.B. *Chem. Rev.* ***2003***, *103*, 977.

SECTION 74G: CYCLOBUTANATIONS, INCLUDING HALOCYCLOBUTANATIONS

1. $Ph_3P{=}CH_2$, BuLi
2. PhCHO, toluene, –40°C

62%

Okuma, K.; Kamahri, Y.; Tsubakihara, K.; Yoshihara, K.; Tanaka, Y.; Shioji, K. *J. Org. Chem.* ***2002****, 67*, 7355.

CO_2HFIP, 20% $EtAlCl_2$

DCM, –78°C

HFIP = 1,1,1,3,3,3-hexafluoroisopropyl

74% (60:40 *trans:cis*)

Takasu, K.; Ueno, M.; Inanaga, K.; Ihara, M. *J. Org. Chem.* ***2004****, 69*, 517.

$PtCl_2$, toluene, 80°C

71%

Marion, F.; Coulomb, J.; Courillong, C.; Fensterbank, L.; Malacria, M. *Org. Lett.* ***2004****, 6*, 1509.

, DCM

5% $AgNTf_2$

77%

Sweis, R.F.; Schramm, M.P.; Kozmin, S.A. *J. Am. Chem. Soc.* ***2004****, 126*, 7442.

Lawessons' reagent, DMF

62%

Hewton, C.E.; Kimber, M.C.; Taylor, D.K. *Tetrahedron Lett.* ***2002****, 43*, 3199.

REVIEW:

"Enantiomerically Pure Cyclobutane Derivatives and Their Use in Organic Synthesis"
Lee-Ruff, E.; Mladenova, G. *Chem. Rev.* ***2003****, 103*, 1449.

SECTION 75: ALKYLS, METHYLENES, AND ARYLS FROM MISCELLANEOUS COMPOUNDS

$PhB(OH)_2$ —[$Mn(OAc)_3$, benzene]→ Ph–Ph 95%

Demir, A.A.; Reis, Ö.; Emrullahoglu, M. *J. Org. Chem.* ***2003****, 68*, 578.

$O_2N-C_6H_4-NO_2$ → (t-BuOK , t-BuOH / BEt_3) → $Et-C_6H_4-NO_2$ 85%

Palani, N.; Jayaprakash, K.; Hoz, S. *J. Org. Chem.* **2003**, *68*, 4388.

$Ge-C_6H_4-OMe$ → (PhI , 4 eq TBAF , $CHCl_3$, 5 h / cat. $Pd_2(dba)_3$•$CHCl_3$, NMP / cat. $P(2\text{-furyl})_3$, 100°C) → $Ph-C_6H_4-OMe$ 90%

Nakamura, T.; Kinashita, H.; Shinokubo, H.; Oshima, K. *Org. Lett.* **2002**, *4*, 3165.

Mesityl-SO_2Ph → (t-Bu-C_6H_4-MgBr / $NiCl_2$–dppf , THF reflux) → t-Bu-C_6H_4-mesityl 56%

Cho, C.-H.; Yun, H.-S.; Park, K. *J. Org. Chem.* **2003**, *68*, 3017.

2-MeO-C_6H_4-N=N–N(pyrrolidine) → (MeO-C_6H_4-$B(OH)_2$ / BF_3•OEt_2 , $P(t\text{-Bu})_3$, DME / cat $Pd_2(dba)_3$) → 2-MeO-C_6H_4-C_6H_4-OMe 81%

Saeki, T.; Son, E.-C.; Tamao, K. *Org. Lett.* **2004**, *6*, 617.

Ph-CH(NHCHO)-SO_2Tol → (Et_2Zn , cat [2.3]-paracyclophane , 10°C / hexane , no ligand , 36 h) → Ph-CH(NHCHO)-Et >99% (95% ee)

Dahmen, S.; Bräse, S. *J. Am. Chem. Soc.* **2002**, *124*, 5940.

1-Cyclohexenyl-SO_2Ph → ($PhTi(Oi\text{-}Pr)_3$, THF , 40°C , 12 h / cat $[Rh(OH)(BINAP)]_2$) → 3-phenylcyclohexene 94% (>99% ee)

Yoshida, K.; Hayashi, T. *J. Am. Chem. Soc.* **2003**, *125*, 2872.

$C_{12}H_{25}$-$SiMe_2OH$ → (supercritical H_2O , 390°C , 3 h) → $C_{12}H_{25}$-H 68%

Itami, K.; Terakawa, K.; Yoshida, J.-i.; Kajimoto, O. *J. Am. Chem. Soc.* **2003**, *125*, 6058.

1-Naphthyl-SO_2Cl → (F-C_6H_4-$SnBu_3$, toluene , 6 h / 1.5% Pd_2dba_3 , 5% TFP , reflux / 10% CuBr-SMe_2) → 1-(4-fluorophenyl)naphthalene 90%

Dubbaka, S.R.; Vogel, P. *J. Am. Chem. Soc.* **2003**, *125*, 15292.

PhCH=CHCH$_2$O-P(O)(OEt)$_2$ → (3 Et_2Zn , 5% $(CuOTf)_2$•PhH , THF , –15°C / 10% chiral amino acid , 1 d) → PhCH(Et)CH=CH$_2$ 61% (95% ee)

Kacprzynski, M.A.; Hoveyda, A.H. *J. Am. Chem. Soc.* **2004**, *126*, 10676.

3,5-dimethylphenyl–$B(OH)_2$ $\xrightarrow[BrCH_2CO_2Et\text{ , 3 KF , 100°C}]{\text{aq dioxane , 2.5\% }PdCl_2(BINAP)}$ 3,3',5,5'-tetramethylbiphenyl 95%

Lei, A.; Zhang, X. *Tetrahedron Lett.* **2002**, *43*, 2525.

$PhB(OH)_2$ $\xrightarrow[\text{23°C , 3h}]{\text{cat }Pd(OAc)_2\text{ , }O_2\text{ , NaOAc , }H_2O}$ Ph–Ph 95%

Parrish, J.P.; Jung, Y.C.; Floyd, R.J.; Jung, K.W. *Tetrahedron Lett.* **2002**, *43*, 7899.

Ph$CH_2CH_2CH_2B(OH)_2$ $\xrightarrow{Ag_2O\text{ , cat }CrCl_2\text{ , THF}}$ Ph$(CH_2)_6$Ph 98%

Falck, J.R.; Mohapatra, S.; Bondlela, M.; Venkataraman, S.K. *Tetrahedron Lett.* **2002**, *43*, 8149.

2 $PhB(OH)_2$ $\xrightarrow[\text{80°C , 18h}]{\text{DMSO , }O_2\text{ , cat }Pd(OAc)_2\text{•dppp}}$ Ph–Ph 91%

Yoshida, H.; Yamaryo, Y.; Ohshita, J.; Kunai, A. *Tetrahedron Lett.* **2003**, *44*, 1541.

PhO$(CH_2)_5S(O_2)CH_2CH{=}CH_2$ $\xrightarrow[\text{PhCl , lauryl peroxide}]{PhCH{=}CHSO_2Ph\text{ , 110°C}}$ PhO$(CH_2)_6CH{=}CH$Ph 72%

Kim, S.; Lim, C.J. *Angew. Chem. Int. Ed.* **2002**, *41*, 3265.

4-$MeC_6H_4CH(NHCHO)SO_2Tol$ $\xrightarrow[\text{paracyclophane N,O-ligand}]{Et_2Zn\text{ , }ZnPh_2}$ 4-$MeC_6H_4CH(NHCHO)Ph$ 98% (97%ee)

Hermanns, N.; Dahmen, S.; Bolm, C.; Bräse, S. *Angew. Chem. Int. Ed.* **2002**, *41*, 3692.

$PhB(OH)_2$ $\xrightarrow[H_2O\text{ , air , rt}]{\text{10\% }Pd(PPh_3)_2Cl_2\text{ , }Bu_4NF\text{ •}H_2O\text{ , THF}}$ Ph–Ph 77%

Punna, S.; Díaz, D.D.; Finn, M.G. *Synlett* **2004**, 2351.

CHAPTER 6

PREPARATION OF AMIDES

SECTION 76: AMIDES FROM ALKYNES

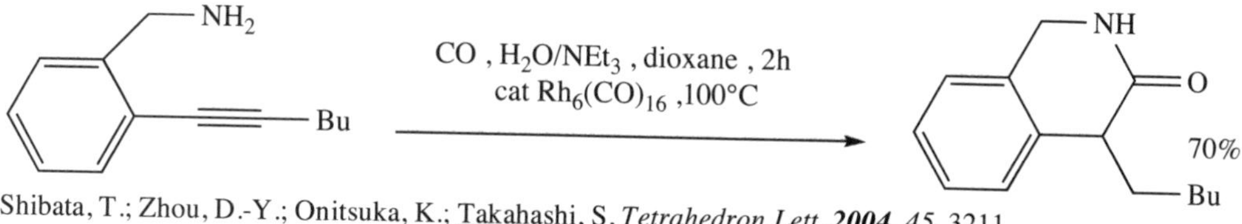

Shibata, T.; Zhou, D.-Y.; Onitsuka, K.; Takahashi, S. *Tetrahedron Lett.* ***2004**, 45*, 3211.

SECTION 77: AMIDES FROM ACID DERIVATIVES

O
Ph OH
1. CDI
2. H_2N Ph
O
Ph N Ph
H

Vaidyanathan, R.; Kalthod, V.G.; Ngo, D.P.; Manley, J.M.; Lapekas, S.P. *J. Org. Chem.* ***2004**, 69*, 2565.

O
C_5H_{11} OH
1. DCM , $F_3SN(CH_2CH_2OMe)_2$
iPr_2NEt
2. $NHEt_2$, DCM
O
C_5H_{11} NEt_2
98%

White, J.M.; Tunoori, A.R.; Turunen, B.J.; Georg, G.I. *J. Org. Chem.* ***2004**, 69*, 2573.

O
BocHN OH
Me
O N - Me , aq $MeNH_2$, 1 h
O Cl
O
, THF , 20°C
O
BocHN NHMe
Me
>98%

Shendage, D.M.; Frölich, R.; Haufe, G. *Org. Lett.* ***2004**, 6*, 3675.

O
N Cl
O
2 BuMgBr , CuCN , ether
−30°C
O
N Bu
O
79%

Lemoucheux, L.; Seitz, T.; Rouden, J.; Lasne, M.-C. *Org. Lett.* ***2004**, 6*, 3703.

Ph, Cl, O + N—Ts, EtO$_2$C, H → chiral ammonium catalyst, toluene, –78°C → Ts, N, O, EtO$_2$C, Ph 95% ee

Taggi, A.E.; Hafez, A.M.; Wack, H.; Young, B.; Perraris, D.; Lectka, T. *J. Am. Chem. Soc.* **2002**, *124*, 6626.

Cl, N, OTf$^-$; PhCMe$_2$NH$_2$, 3 NEt$_3$, DCM; Ph, O, OH → Ph, O, N H, Ph 69%

Crosignani, S.; Gonzalez, J.; Swinnen, D. *Org. Lett.* **2004**, *6*, 4579.

Ph, O, OH + Ph, *, n, O, + NH, O, O, BF$_4^-$ → O, Ph, NH$_2$ 78%

Chinchilla, R.; Dodsworth, D.J.; Nájera, C.; Soriano, J.M. *Tetrahedron Lett.* **2003**, *44*, 463.

O, OH → imidazole, 100 sec, microwaves → O, NH$_2$ 85%

Khalafi-Nezhad, A.; Mokhtari, B.; Rad, M.N.S. *Tetrahedron Lett.* **2003**, *44*, 7325.

O, N, N, +, N—Me, I$^-$, O, O → PhCH$_2$CO$_2$H, NEt$_3$, MeCN, rt, 16h → O, N, Ph, O, O 92%

Grzyb, J.A.; Batey, R.A. *Tetrahedron Lett.* **2003**, *44*, 7485.

MeO, O, Cl → aminocyclohexane, In, MeOH, rt, 4h → MeO, O, HN 82%

Cho, D.H.; Jang, D.O. *Tetrahedron Lett.* **2004**, *45*, 2285.

HO, Ph, O → NH$_4$HCO$_3$, HCONH$_2$, microwaves, 10 min → H$_2$N, Ph, O 75%

Peng, Y.; Song, G. *Org. Prep. Proceed. Int.* **2002**, *34*, 95.

Ph, O, Cl + Ts, N, CO$_2$Et → 10% benzoyl quinine, –10°C, NaHCO$_3$, 15-crown-5, toluene → Ts, N, O, EtO$_2$C, Ph 58% (92% ee, 12:1 dr)

Shah, M.H.; France, S.; Lectka, T. *Synlett* **2003**, 1937.

PhCOOH → 1. ClCO$_2$CCl$_3$, NEt$_3$, DCM, 0°C; 2. MeONHMe, HCl, 25°C, 1 h → O, Ph, N(Me)OMe 97%

Kim, M.; Lee, H.; Han, K.-J.; Kay, K.-Y. *Synth. Commun.* **2003**, *33*, 4013.

NH_2 ... CO_2H → 1. NsTf , DCM , $CuSO_4$, H_2O / MeOH , K_2CO_3 2. $HSCH_2PPh_2$•BH_3 , DCC , DCM , 0°C 3. DABCO , aq. THF , 70°C → caprolactam (HN, O) 95x100x84%

David, O.; Meester, W.J.N.; Bieräugel, H.; Schoemaker, H.E.; Hiemstra, H.; van Maarsevoen, J.H. *Angew. Chem. Int. Ed.* **2003**, *42*, 4373.

cyclohexyl–CO_2H → H_2NC(O)NH_2 , toluene / 10% (F_3C)$_2C_6H_3$–$B(OH)_2$, aceotropic reflux → cyclohexyl–C(O)–NH–C(O)–NH–3,5-(CF_3)$_2C_6H_3$ 82%

Maki, T.; Ishihara, K.; Yamamoto, H. *Synlett* **2004**, 1355.

phthalic anhydride → HC=O)NH_2 , 3 min / microwaves → phthalimide (NH) 90%

Kacprzak, K. *Synth. Commun.* **2003**, *33*, 1499.

succinic anhydride → bmim PF_6 , $PhNH_2$, 20 min → N–Ph succinimide 96%

Le, Z.-G.; Chen, Z.-C.; Hu, Y.; Zheng, Q.-G. *Synthesis* **2004**, 995.

SECTION 78: AMIDES FROM ALCOHOLS AND THIOLS

Ph–CH_2CH_2–OH → $HN(SO_2Me)_2$, CHMDS , rt / THF → Ph–CH_2CH_2–$N(SO_2Me)_2$ 84%

Dastrup, D.M.; van Brunt, M.P.; Weinreb, S.M. *J. Org. Chem.* **2003**, *68*, 4112.

HO–CH_2CH_2–C_6H_4–Br → BnO_2CNSO_2 NEt_3 , PhH , 36h / $NHEt_3Cl$, 85°C → CbzHN–CH_2CH_2–C_6H_4–Br 71%

Wood, M.R.; Kim, J.Y.; Books, K.M. *Tetrahedron Lett.* **2002**, *43*, 3887.

SECTION 79: AMIDES FROM ALDEHYDES

PhCHO → (1. I_2, aq. NH_3, rt; 2. aq. H_2O_2, rt) → PhC(O)NH_2 98%

Shie, J.-J.; Fang, J.-M. *J. Org. Chem.* **2003**, *68*, 1158.

$Cl(CH_2)_3C(CH_3)_2CHO$ → ($H_2NCH_2SnBu_3$, 110°C, 6 h; N-phenylmaleimide, toluene) → 78%

Pearson, W.H.; Stoy, P.; Mi, Y. *J. Org. Chem.* **2004**, *69*, 1919.

PhCHO → (allyl-$SiMe_3$, $CbzNH_2$, 5% I_2; MeCN, 25°C, 25 min) → NHCbz 82%

Phukan, P. *J. Org. Chem.* **2004**, *69*, 4005.

PhCHO → ($LiNHC_6H_{13}$, 10% LDA, THF) → PhC(O)NHC_6H_{13} 71%

Ishihara, K.; Yano, T. *Org. Lett.* **2004**, *6*, 1983.

PhCHO → (1. NH_3, EtOH, –10°C, 2 h; 2. allyl-B(pin), –10°C → rt) → PhCH(NH_2)$CH_2CH=CH_2$ 84%

Sugiura, M.; Hirano, K.; Kobayashi, S. *J. Am. Chem. Soc.* **2004**, *126*, 7182.

PhCOOH → (1. MsCl, 3 NEt_3; 2. 1.5 MeONHMe, THF, 0°C) → PhC(O)N(Me)OMe 61%

Woo, J.C.S.; Fenster, E.; Dake, G.R. *J. Am. Chem. Soc.* **2004**, *126*, 8984.

PhCHO (also with ketones) + Hantzsch ester (CO_2Et) → (MeO–C_6H_4–NH_2; 2% $Sc(OTf)_3$, THF, rt, 4h) → MeO–C_6H_4–NHBn 98%

Itoh, T.; Nagata, K.; Kurihara, A.; Miyazaki, M.; Ohsawa, A. *Tetrahedron Lett.* **2002**, *43*, 3105.

Cyclohexyl-CHO → (2-pyridylmethylamine (NH_2), THF; polymer-supported $C_6H_4CH_2\overset{+}{N}Et_3$ $\overset{-}{B}H(OAc)_3$) → 2-pyridyl-CH_2-NH-CH_2-cyclohexyl 90%

Bhattacharyya, S.; Rana, S.; Gooding, O.W.; Labadie, J. *Tetrahedron Lett.* **2003**, *44*, 4957.

Cyclohexyl-CHO → (allyl-$SiMe_3$, $CbzNH_2$, 25°C, 5h; MeCN, 1% $Bi(OTf)_3$•n H_2O) → cyclohexyl-CH(NHCbz)$CH_2CH=CH_2$ 86%

Ollevier, T.; Ba, T. *Tetrahedron Lett.* **2003**, *44*, 9003.

PhCHO $\xrightarrow[\text{cat } RuCl_2(PPh_3)_3]{Ph_3P{=}NTs\text{ , 6h , rt}}$ PhCH=N-Ts 75%

Jain, S.L.; Sharma, V.B.; Sain, B. *Tetrahedron Lett.* ***2004**, 45*, 4341.

PhCHO $\xrightarrow[\text{SDS-}H_2O]{\text{allyl-}SnBu_3\text{ , }PhNH_2\text{ , 0.1 sn}Cl_2{\bullet}2\,H_2O}$ Ph–CH(NHPh)–CH₂CH=CH₂ 80%

Akiyama, T.; Onuma, Y. *J. Chem. Soc. Perkin Trans. 1* ***2002**,* 1157.

SECTION 80: AMIDES FROM ALKYLS, METHYLENES, AND ARYLS

1,2-Diphenylcyclopropane $\xrightarrow[\text{mDCB , }Et_4NBF_4\text{ , aq. acetone}]{NH_3\text{ , PhH , h}\nu}$ Ph–CH₂CH₂CH(NH₂)–Ph 71%

Yasuda, M.; Kojima, R.; Tsutsui, H.; Utsunomiya, D.; Ishii, K.; Jinnouchi, K.; Shiragami, T.; Yamashita, T. *J. Org. Chem.* ***2003**, 68*, 7618.

SECTION 81: AMIDES FROM AMIDES

Conjugate reductions of unsaturated amides are listed in Section 74D (Conjugate Reduction of α,β-Unsaturated Carbonyl Compounds and Nitriles).

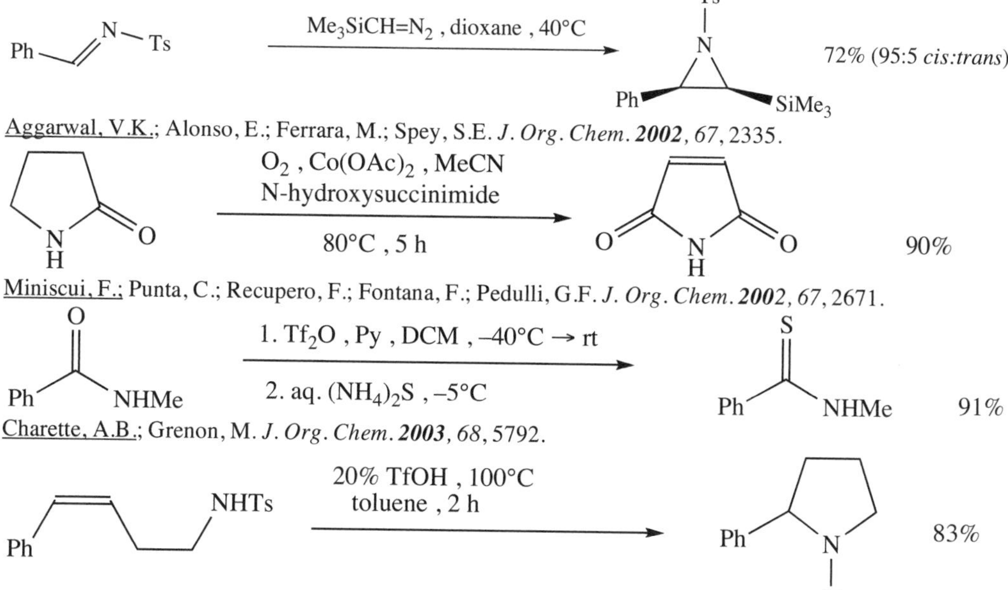

Aggarwal, V.K.; Alonso, E.; Ferrara, M.; Spey, S.E. *J. Org. Chem.* ***2002**, 67*, 2335.

Miniscui, F.; Punta, C.; Recupero, F.; Fontana, F.; Pedulli, G.F. *J. Org. Chem.* ***2002**, 67*, 2671.

Charette, A.B.; Grenon, M. *J. Org. Chem.* ***2003**, 68*, 5792.

Schlummer, B.; Hartwig, J.F. *Org. Lett.* ***2002**, 4*, 1471.

F—C6H4—CH=NTs → (1. 3% $Cu(MeCN)_4BF_4$, toluene, chiral binaphthylthiophospholamide; 2. Et_2Zn, toluene) → F—C6H4—CH(NHTs)Et 96% (90% ee)

Wang, C.-J.; Shi, M. *J. Org. Chem.* ***2003***, *68*, 6229.

PhCH=NSO_2Tol → (3 Et_2Zn, toluene, 0°C, 4h; cat amidophosphine/$Cu(OTf)_2$) → PhCH(Et)NHSO_2Tol 79% (63% ee) + 20% $BnNHSO_2Tol$

Soeta, T.; Nagai, K.; Fujihara, H.; Kuriyama, M.; Tomioka, K. *J. Org. Chem.* ***2003***, *68*, 9723.

Me_3Si N-Me pyrrolidinone → (1. MeO—C6H4—NO_2, TASF/THF; 2. DDQ) → OMe, O_2N aryl N-Me pyrrolidinone 85%

Rege, P.D.; Johnson, F. *J. Org. Chem.* ***2003***, *68*, 6133.

NBoc, Boc, N, SMe, O, $SiMe_3$ → (MeO-C6H4-$B(OH)_2$; Cu(I) thiophene-2-carboxylate, cat Pd, THF, reflux) → NBoc, Boc, N, OMe, O, Me_3Si 81%

Kusturin, C.L.; Liebeskind, L.S.; Neumann, W.L. *Org. Lett.* ***2002***, *4*, 983.

Ph-C(=Se)-NH-CH2Ph → (1. 2 BuLi, THF, 0°C; 2. bromocyclohexane, 0°C) → Ph-C(=Se)-NH-CH(Ph)(cyclohexyl) 79%

Murai, T.; Aso, H.; Kato, S. *Org. Lett.* ***2002***, *4*, 1407.

N-acyl oxazolidinone (Bn), Ph → ($BnONH_2$, 0.1M, THF, 0.3 $Sm(OTf)_2$) → BnOHN-C(=O)-CH2CH2-Ph 70%

Sibi, M.P.; Hasegawa, H.; Ghorpade, S.R. *Org. Lett.* ***2002***, *4*, 3343.

CH2=CHCH2C(=O)NHTs → (2 CH3CH=CHBr, 5% $Pd(OAc)_2$; 11% $Pd(OTol)_3$, 2 Na_2CO_3, 2 Bu_4Cl, MeCN, 90°C) → N-Ts 5-propenyl pyrrolidin-2-one 82%

Pinho, P.; Minnaard, A.J.; Feringa, B.L. *Org. Lett.* ***2003***, *5*, 259.

$Et_2P(=O)H$, H_2O

V-501 (water sol. initiator)

97%

Khan, T.A.; Tripoli, R.; Crawford, J.T.; Martin, C.G.; Murphy, J.A. *Org. Lett.* ***2003***, *5*, 2971.

2 LiHMDS , 2.2 $ZnCl_2$, –20°C
2 PhBr , 5% $Pd(dab)_2$, THF

7.5% biaryl phosphine , 65°C

98%

Cossy, J.; de Filippis, A.; Pardo, D.G. *Org. Lett.* ***2003***, *5*, 3037.

cat $Pd(PCy_3)_2$, toluene , 0.25 M

70°C , 4h

74%

Wolfe, J.P.; Ney, J.E. *Org. Lett.* ***2003***, *5*, 4607.

1. 3 Et_3Al

2. H_2O

67%

El Kaim, L.; Grimaud, L.; Lee, A.; Perroux, Y.; Tiria, C. *Org. Lett.* ***2004***, *6*, 381.

e⁻ , $Et_3N{\bullet}3\ HF$, MeCN

Bu_4NBF_4

41% + 42%

Shen, Y.; Atobe, M.; Fuchigami, T. *Org. Lett.* ***2004***, *6*, 2441.

, dioxane , 1% CuI , 110°C

2 K_3PO_4 , 1,2-diamine ligand , 23 h

97%

Klapars, A.; Huang, X.; Buchwald, S.L. *J. Am. Chem. Soc.* ***2002***, *124*, 7421.

, $Ti(NMe_2)_4$, 2 h

toluene , 90°C

reversible

Eldred, S.E.; Stone, D.A.; Gellman, S.H.; Stahl, S.S. *J. Am. Chem. Soc.* ***2003***, *125*, 3422.

cat $Pd(OAc)_2$/phosphine , NEt_3
toluene , 80°C
O
Cl
N
Me
O
N
Me
94%

Hennessy, E.J.; Buchwald, S.L. *J. Am. Chem. Soc.* ***2003***, *125*, 12084.

I + Br NH2 O
cat $Pd(OAc)_2$, TFP , DMF
105°C , 2 K_2CO_3 , 1 d
2-norbornene
H N O
86%

Ferraccioli, R.; Carenzi, D.; Rombolà, O.; Catellani, M. *Org. Lett.* ***2004***, *6*, 4759.

N—Ts
5 $PhB(OH)_2$, 3% $Rh(acac)(C_2H_4)_2$
cat chiral amido phosphorane
dioxane , 100°C
Ph
NHTs
95% (76:24 er)

Kuriyama, M.; Soeta, T.; Hao, X.; Chen, Q.; Tomioka, K. *J. Am. Chem. Soc.* ***2004***, *126*, 8128.

MeO O
PhCH=N-Boc , –35°C , DCE
cat chiral phosphonic acid
NHBoc
MeO O
Ph
87% (97% ee)

Uraguchi, D.; Sorimachi, K.; Terada, M. *J. Am. Chem. Soc.* ***2004***, *126*, 11804.

N—Ts
OH , KSF-Montmorillonite clay
DCM , rt , 5h
NHTs
89%

Yadav, J.S.; Reddy, B.V.S.; Balanarsaiah, E.; Raghavendra, S. *Tetrahedron Lett.* ***2002***, *43*, 5105.

$C_6H_{13}NHCO_2Me$ —[$C_8H_{17}OH$, 160°C , 2h / 1% $Bi(O_3SCF_3)_3$]→ $C_6H_{13}NHCO_2C_8H_{17}$ 92%

Jousseaume, B.; Laporte, C.; Toupance, T.; Bernard, J.-M. *Tetrahedron Lett.* ***2002***, *43*, 6305.

S
NH2
$Bi(NO_3)_3$•5 H_2O , MeCN , reflux
10 min
O
NH2
97%

Mohammadpoor-Baltork, I.; Khodaci, M.M.; Nikoofar, K. *Tetrahedron Lett.* ***2003***, *44*, 591.

O
Ph—B
O
2% $Pd(dba)_2$, 6% PCy_3 , 3.4 eq K_3PO_4
Br NMe2 O , THF , 70°C
Ph NMe2 O
89%

Lu, T.-Y.; Xue, C.; Luo, F.-T. *Tetrahedron Lett.* ***2003***, *44*, 1587.

B(OH)2
O
Br NBu2 , cat $Pd(PPh_3)_4$, toluene
K_3PO_4•3 H_2O , 6% Cu_2O , 6% PPh_3
90°C , 1d
O
NBu2
79%

Duan, Y.-Z.; Deng, M.-Z. *Tetrahedron Lett.* ***2003***, *44*, 3423.

$PhSO_2NH_2$ —PhI, 10% CuI, microwaves, 2 K_2CO_3, 3h→ $PhSO_2NHPh$ 90%

He, H.; Wu, Y.-J. *Tetrahedron Lett.* **2003**, *44*, 3385.

$PhSO_2NH_2$ —Ac_2O, MeCN, 3% H_2SO_4, 60°C→ $PhSO_2NHAc$ 95%

Martin, M.T.; Roschangar, F.; Eaddy, J.F. *Tetrahedron Lett.* **2003**, *44*, 5461.

N—Ts —NaN_3, $MeCN/H_2O$, 80°C; also with NaCN→ NHTs, N_3 99%

Bisai, A.; Pandey, G.; Pandey, M.K.; Singh, V.K. *Tetrahedron Lett.* **2003**, *44*, 5839.

$SO_2NHt\text{-}Bu$ —$BiCl_3$, DCM, rt, 30 min→ SO_2NH_2 88%

Wan, Y.; Wu, X.; Kannan, M.A.; Alterman, M. *Tetrahedron Lett.* **2003**, *44*, 4523.

N, O —1. Grubbs' I; 2. $RuCl_3$, $NaIO_4$→ N—H, O 78%

Alcaide, B.; Almendros, P.; Alonso, J.M. *Tetrahedron Lett.* **2003**, *44*, 8693.

O, Ph, NMe_2 —1. $(COCl)_2$, DCM, 6h; 2. $(Et_4N)_2WSe_4$→ Se, Ph, NMe_2 65%

Saravanan, V.; Mukherjee, C.; Das, S.; Chandrasekaran, S. *Tetrahedron Lett.* **2004**, *45*, 681.

O, N—H —PhI, K_3PO_4, dioxane, 1d; CuI, 100°C, glycine→ O, N—Ph 97%

Deng, W.; Wang, Y.-F.; Zou, Y.; Liu, L.; Guo, Q.-X. *Tetrahedron Lett.* **2004**, *45*, 2311.

Ph, N, Ts —$PhSnMe_3$, H_2O, air, 35°C, sonication; cat $Rh_2(cod)_2Cl_2$→ NHTs, Ph, Ph 81%

Ding, R.; Zhao, C.H.; Chen, Y.J.; Lu, L.; Wang, D.; Li, C.J. *Tetrahedron Lett.* **2004**, *45*, 2995.

CHO —1. wet Al_2O_3, $NH_2OH•HCl$, 45 min; 2. $MeSO_2Cl$, 100°C→ O, NH_2 90%

Sharghi, H.; Sarvari, M.H. *Tetrahedron* **2002**, *58*, 10323.

NC—Br + TBSO, HN, O —cat $Pd(OAc)_2$, dppf, 3.5h; NaO*t*-Bu, toluene→ TBSO, NC, N, O 95%

Browning, R.G.; Badarinarayana, V.; Mahmud, H.; Lovely, C.J. *Tetrahedron* **2004**, *60*, 359.

$PhSO_2NH_2$ → (Ac_2O, cat K10-FeO, MeCN; 15 min) → $PhSO_2NHAc$ 92%

Singh, D.U.; Singh, P.R.; Samant, S.D. *Tetrahedron Lett.* **2004**, *45*, 4805.

Ts, N, Ph, Ph → (BF_3, $CHCl_3$, –15°C, 4d) → N–Ts, Ph, Ph 96%

Sugihara, Y.; Iimura, S.; Nakayama, J. *Chem. Commun.* **2002**, 134.

OTMS, Ph, N, Cbz → (1. BF_3•OEt_2, DCM, –78°→–20°C; 2. $SiMe_3$) → Ph, N, Cbz 83%

Suh, Y.-G.; Shin, D-Y.; Jung, J.-K.; Kim, S.-H. *Chem. Commun.* **2002**, 1064.

NHTs, CO_2Me → (0.4 TfOH, $CHCl_3$, 0°C, 15 min) → CO_2Me, N, Ts 95%

Haskins, C.M.; Knight, D.W. *Chem. Commun.* **2002**, 2724.

PhCH=N-Ts → (Et_2Zn, toluene, MS 4Å, –20°C; 10% $Cu(OT)_2$-oxazoline ligand) → NHTs, Ph, Et 58% (77% ee)

Li, X.; Cun, L.-F.; Gong, L.-Z.; Mi, A.-Q.; Jiang, Y.-Z. *Tetrahedron: Asymmetry* **2003**, *14*, 3819.

O, P—xylyl, N, xylyl, Ph → (6% CuCl, 6% NaOMe, 3 TMDS; 6% R-(–)-DTBM-SEG-PHDS; 3.3 *t*-BuOH, toluene, rt, 17h) → O, P—xylyl, HN, xylyl, Ph 99% (96%ee)

Lipshutz, B.H.; Shimizu, H. *Angew. Chem. Int. Ed.* **2004**, *43*, 2228.

S, I → (O, N H, O; K_3PO_4, dioxane, 110°C; cat CuI, EDTA, 1d) → S, N, O 95%

Kang, S.-K.; Kim, D.-H.; Park, J.-N. *Synlett* **2002**, 427.

Ts, N, Ph → ($TMSN_3$, 10% $InCl_3$, MeCN; rt, 6 h) → NHTs, Ph, N_3 + N_3, Ph, NHTs (18 : 82) 92%

Yadav, J.S.; Reddy, B.V.S.; Kumar, G.M.; Murthy, Ch.V.S.R. *Synth. Commun.* **2002**, *32*, 1797.

PhI → (benzamide, 10% CuI, KF/Al_2O_3; 10% 10-phenantholine) → PhNHBz 99%

Hosseinzadeh, R.; Tajbakhsh, M.; Mohadjerani, M.; Mehdinejad, H. *Synlett* **2004**, 1517.

Me_3Si–allyl, DCM; 0.6 $NbCl_5$ — 84%

Kleber, C.; Andrade, Z.; Matos, R.A.F. *Synlett* **2003**, 1189.

NaN_3, aq. MeCN, 2.5 h; 1% phosphomolybdic acid/SiO_2 — 95%

Kumar, G.D.K.; Baskaran, S. *Synlett* **2004**, 1719.

MeO–C_6H_4–SO_2NH*i*-Pr → 6 H_5IO_6, MeCN, rt; 5% $Cr_3(OAc)_7(OH)_2$ → MeO–C_6H_4–SO_2NH_2 — 93%

Xu, L.; Zhang, S.; Trudell, M.L. *Synlett* **2004**, 1901.

t-BuOC(O)NHBn → 1. 4 acetyl bromide, 2 MeOH, DCM; 2. K_2CO_3 → H_3CC(O)NHBn — quant

Nazih, A.; Heissler, D. *Synthesis* **2002**, 203.

NaN_3, $LiClO_4$, MeCN; heat, 6 h — 90%

Yadav, J.S.; Reddy, B.V.S.; Parimala, G.; Reddy, P.V. *Synthesis* **2002**, 2383.

MeC(S)NH_2 → Oxone, solid phase, 5 min → MeC(O)NH_2 — 95%

Mohammadpoor-Baltork, I.; Sadeghi, M.M.; Esmayilpour, K. *Synth. Commun.* **2003**, *33*, 953.

MeC(S)NH_2 → Oxone, solid phase, 5 min → MeC(O)NH_2 — 95%

Mohammadpoor-Baltork, I.; Sadeghi, M.M.; Esmayilpour, K. *Synth. Commun.* **2004**, *34*, 953.

SO_2NH_2 → $PhI(OAc)_2$, MgO, DCM; chiral Rh catalyst → SO_2–N–H + SO_2–N–H (99 : 1) 95% (66% ee)

Fruit, C.; Müller, P. *Helv. Chim. Acta* **2004**, *87*, 1607.

PhS–CH=CH–C(O)N(Me)CH(Me)Ph → LDA, –20°C → rt; THF, 1 h → Me, Ph, N–Me, O — 80%

Naitoh, R.; Nakamura, Y.; Katano, E.; Nakamura, Y.; Okada, E.; Asaoka, M. *Heterocycles* **2004**, *63*, 1009.

REVIEW:

"Metalated Sulfonamides and Their Synthetic Applications"
Familoni, O.B. *Synlett* ***2002***, 1181.

SECTION 82: AMIDES FROM AMINES

Bn
H_2N CO_2H
CO_2Me O O—N O O
Na_2CO_3, aq. MeCN, rt
Bn N CO_2H
quant

Casimir, J.R.; Guichard, G.; Briand, J.-P. *J. Org. Chem.* ***2002***, *67*, 3764.

NH_2
1. MeCN, Et_4NClO_4, e^-
2. CO_2
3. EtI
$NHCO_2Et$ 73%

Feroci, M.; Casadei, M.A.; Orsini, M.; Palombi, L.; Inesi, A. *J. Org. Chem.* ***2003***, *68*, 1548.

PhI
HN, $Mo(CO)_6$, 3 eq DBU
cat $Pd(OAc)_2$, 100°C, microwaves
O Ph N 81%

Wannberg, J.; Larhed, M. *J. Org. Chem.* ***2003***, *68*, 5750.

Ph N Ph
$CH_2=CHCO_2C_6H_5$, Et_2MeSiH
2.5% $[(cod)IrCl]_2$, 10% $P(OPh)_3$
60°C, 6 h
Ph N O Ph Me
68% (>20:1 *trans:cis*)

Townes, J.A.; Evans, M.A.; Queffelec, J.; Taylor, S.J.; Morken, J.P. *Org. Lett.* ***2002***, *4*, 2537.

Ph N *t*-Bu
CO, dendrimer catalyst
400 psi, PhH
Ph O N *t*-Bu
quant

Lu, S.-M.; Alper, H. *J. Org. Chem.* ***2004***, *69*, 3558.

N Ph
(salen)Cr(III)/DMAP cat.
CO_2, 100°C, 14 h
O N O Ph
90%

Miller, A.W.; Nguyen, S.T. *Org. Lett.* ***2004***, *6*, 2301.

cat $[Cp^*RhCl_2]_2$, K_2CO_3
100°C , acetone , 20 h
OH
NH_2
N
H
O
80%

Fujita, K.-i.; Takahashi, Y.; Owaki, M; Yamamoto, K.; Yamaguchi, R. *Org. Lett.* ***2004***, *6*, 2785.

CN
N—H
Ph, Et C=C=O , toluene , rt
2% chiral ferrocenyl compound
CN
N
O
Et
Ph
91% (81% ee)

Hodous, B.L.; Fu, G.C. *J. Am. Chem. Soc.* ***2002***, *124*, 10006.

$PhOCH_2CO_2H$
PhCH=N-Bn , NEt_3 , DCM
N^+ Cl Cl^-
PhO Ph
N Bn
O
65%

Donati, D.; Morelli, C.; Procheddu, A.; Taddei, M. *J. Org. Chem.* ***2004***, *69*, 9317.

NH_2
N
H
1. CHO , –30°C , MS 3Å
2. AcCl , 2,6-lutidine , ether
chiral thiourea catalyst
N — Ac
N
H
65% (93% ee)

Taylor, M.S.; Jacobsen, E.N. *J. Am. Chem. Soc.* ***2004***, *126*, 10558.

$C_{10}H_{21}NH_2$ $\xrightarrow[130°C\text{ , }17\text{ h}]{(MeO)_2C=O\text{ , supercritical }CO_2}$ $C_{10}H_{21}NHCO_2Me$ 83%

Selva, M.; Tundo, P.; Perosa, A. *Tetrahedron Lett.* ***2002***, *43*, 1217.

$BuNH_2$ $\xrightarrow[80°C\text{ , }8h]{(MeO)_2C=O\text{ , cat }Yb(OTf)_3}$ $BuNHCO_2Me$ 93%

Curini, M.; Epifano, F.; Maltese, F.; Rosati, O. *Tetrahedron Lett.* ***2002***, *43*, 4895.

Bn
N
Bu Bu
1. LiI
2. CO_2
O
Bn N O
Bu Bu
>80%

Hancock, M.T.; Pinhas, A.R. *Tetrahedron Lett.* ***2003***, *44*, 5457.

H
N
N
O
$PhCH_2NH_2$, AcCl
$Ti(Oi\text{-}Pr)_4\text{-}NaBH_4$
$PhCH_2NHAc$ 98%

Bhattacharyya, S.; Gooding, O.W.; Labadie, J. *Tetrahedron Lett.* ***2003***, *44*, 6099.

$PhNH_2$ $\xrightarrow{CO_2\text{ , }PPh_3/DEAD\text{ , }BuOH}$ $PhNHCO_2Bu$ 90%

Chaturvedi, D.; Kumar, A.; Ray, S. *Tetrahedron Lett.* ***2003***, *44*, 7637.

Ph–CH=N–Ph —[$BF_3•OEt_2$, mcpba]→ PhC(O)NPh_2 82%

An, G.-i.; Kim, M.; Kim, J.Y.; Rhee, H. *Tetrahedron Lett.* **2003**, *44*, 2183.

$RuCl_3$, $NaIO_4$, aq. EtOAc 45%

CO_2Me MeO_2SO NHBoc

Patel, S.; Mishra, B.K. *Tetrahedron Lett.* **2004**, *45*, 1371.

$PhNH_2$ —[1. HCl, EtOAc, dioxane; 2. 130°C, $(COCl)_2$, 2h; 3. MeOH]→ PhHNC(O)OMe 85%

Oh, L.M.; Spoors, P.G.; Goodman, R.M. *Tetrahedron Lett.* **2004**, *45*, 4769.

$BuNH_2$ —[$AcNHSO_2Me$, heat]→ AcNHBu 90%

Coniglio, S.; Aramini, A.; Cesta, M.C.; Colagioia, S.; Curti, R.; D'Alessandro, F.; D'Anniballe, G.; D'Elia, V.; Nano, G.; Orlando, V.; Allegretti, M. *Tetrahedron Lett.* **2004**, *45*, 5375.

NH_2 NO_2 —[BiOCl, AcCl, DCM, 5 min]→ NHAc NO_2 90%

Ghosh, R.; Maiti, S.; Chakraborty, A. *Tetrahedron Lett.* **2004**, *45*, 6775.

OH —[1. 1.3 BTC, 7.5 Py, DCM; 2. 5 Me_2CHCH_2OH, DCM 7.5 Py, rt; 3. 5 $PhCH_2NH_2$, DCM, rt]→ Ph, N H, O 88%

Mormeneo, D.; Llebaria, A.; Delgado, A. *Tetrahedron Lett.* **2004**, *45*, 6831.

$PhNH_2$ —[$PhCH_2CO_2H$, microwaves, 30 min]→ PhHNC(O)CH_2Ph 52%

Perreux, L.; Loupy, A.; Volatron, F. *Tetrahedron* **2002**, *58*, 2155.

H_2N —[BuLi, THF]→ HN + H_2N (75 : 25) 81%

Ates, A.; Quinet, C. *Eur. J. Org. Chem.* **2003**, 1623.

NMe_2 O —[1. KOH, air, 40 h; 2. H_3O^+]→ NMe_2 O 79% + OH O

García-Valverde, M.; Pedrosa, R.; Vicente, M. *Synlett* **2002**, 2092.

Ac_2O , H_2O / sodium dodecyl sulfate → NHAc 71%

Naik, S.; Bhattacharjya, G.; Talukdar, B.; Patel, B.K. *Eur. J. Org. Chem.* **2004**, 1254.

mcpba , BF_3•OEt_2 / rt , 2 h → Ph(C=O)NHBn 80%

An, G.-i.; Rhee, H. *Synlett* **2003**, 876.

$PhNH_2$ + (Cl, Cl, O, N, N, Ac) DCM, 17°C , 5 min → PhNHAc 98%

Kang, Y.-J.; Chung, H.-A.; Kim, J.-J.; Yoon, Y.-J. *Synthesis* **2002**, 733.

PhNHMe KF-Al_2O_3 , $CHCl_3$, rt , 18 h → Ph—N(CHO)Me 93%

Mihara, M.; Ishino, Y.; Minakata, S.; Komatsu, M. *Synthesis* **2003**, 2317.

SECTION 83: AMIDES FROM ESTERS

1. $C_5H_9NH_2$, 11 eq EDC•HCl EtOH , DCM , rt , 18 h / 2. DMF , microwaves , 170°C → N—C_5H_9 80%

Martin, B.; Sekljic, H.; Chassaing, C. *Org. Lett.* **2003**, *5*, 1857.

EtO, N_3, Br — 1. In , EtOH / 2. NEt_3 , reflux → 82%

Panchaud, P.; Ollivier, C.; Renaud, P.; Zigmantas, S. *J. Org. Chem.* **2004**, *69*, 2755.

$PhCH_2NH_2$ MeO_2COMe , cat Sc(OTf)$_3$, 20°C → $PhCH_2NHCO_2Me$ 75%

Distaso, M; Quaranta, E. *Tetrahedron* **2004**, *60*, 1531.

N—H + (Cl, N, N, N, Cl, Cl) HCOOH , NMM , cat DMAP / DCM , microwaves , 3 min → N—CHO 95%

DeLuca, L.; Giacomelli, G.; Porcheddu, A.; Salaries, M. *Synlett* **2004**, 2570.

Ph(CH$_2$)C(O)OMe $BnNH_2$, InI_3 , 7 h → Ph(CH$_2$)C(O)NHBn 92%

Ranu, B.C.; Dutta, P. *Synth. Commun.* **2003**, *33*, 297.

$BuNH_2$ → (ethyl acetate, KO*t*-Bu, 95°C; microwaves) → AcNHBu 70%

Zradni, F.-Z.; Hamelin, J.; Derdour, A. *Synth. Commun.* **2002**, *32*, 3525.

NO_2 → ($PhCO_2Et$, SmI_2, THF, rt) → NHBz 92%

Wang, X.; Guo, H.; Xie, G.; Zhang, Y. *Synth. Commun.* **2004**, *34*, 3001.

O O O → (NH_2, toluene; Pd_2dba_3, rt, 1 d) → N O O 70%

Shi, M.; Shen, Y.-M.; Chen, Y.-J. *Heterocycles* **2002**, *57*, 245.

SECTION 84: AMIDES FROM ETHERS, EPOXIDES, AND THIOETHERS

OMe Ph OTIPS → (1. CSI, Na_2CO_3, DCM; 2. Na_2SO_3) → $NHCO_2Me$ Ph OTIPS 83% (4.5:1 *anti:syn*)

Kim, J.D.; Zee, O.P.; Jung, Y.H. *J. Org. Chem.* **2003**, *68*, 3721.

O Ph O*t*-Bu → (1. $BnMe_2NOH$, MeOH, THF; 2. TsOH; 3. $TolCH_2NH_2$, HOBt; 4. TsOH) → O Ph NHTol 89%

Kawahata, N.H.; Brokes, J.; Makara, G.M. *Tetrahedron Lett.* **2002**, *43*, 7221.

SECTION 85: AMIDES FROM HALIDES AND SULFONATES

MeO Br → ($HCONH_2$, 5 bar CO; 120°C, excess PPh_3; cat $PdCl_2(PPh_3)_2$, 8 h; dioxane) → MeO O NHCHO 58%

Schnyder, A.; Indolese, A.F. *J. Org. Chem.* **2002**, *67*, 594.

Me Br → ($BnNH_2$, imidazole, KO*t*-Bu; cat $Pd(OAc)_2$/dppf, DMF; 180°C, microwave) → Me NHBn O 63%

Wan, Y.; Alterman, M.; Larhed, M.; Hallberg, A. *J. Org. Chem.* **2002**, *67*, 6232.

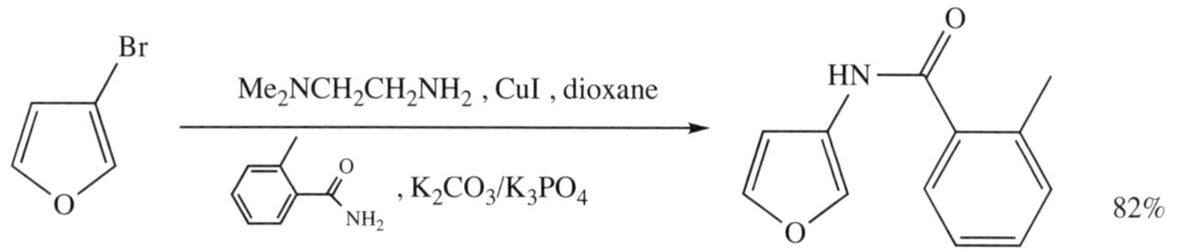

$Me_2NCH_2CH_2NH_2$, CuI , dioxane
, K_2CO_3/K_3PO_4
NH$_2$
82%

Padwa, A.; Crawford, K.R.; Rashatasakhon, P.; Rose, M. *J. Org. Chem.* ***2003***, *68*, 2609.

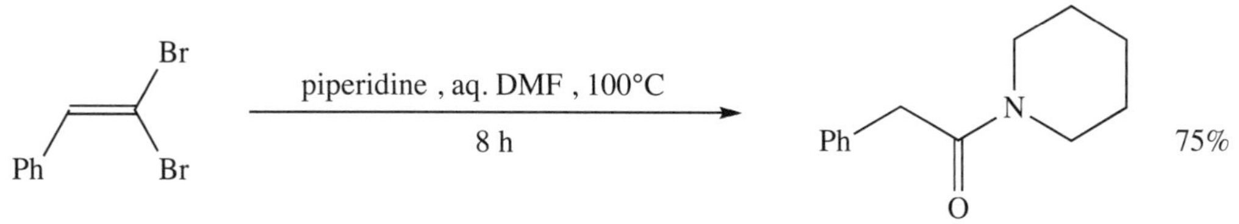

Br Br Ph
piperidine , aq. DMF , 100°C
8 h
Ph N O
75%

Shen, W.; Kunzer, A. *Org. Lett.* ***2002***, *4*, 1315.

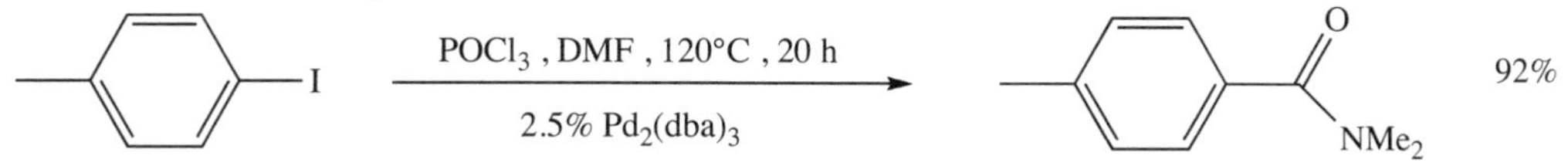

I
$POCl_3$, DMF , 120°C , 20 h
2.5% $Pd_2(dba)_3$
O NMe_2
92%

Hosoi, K.; Nozaki, K.; Hiyama, T. *Org. Lett.* ***2002***, *4*, 2849.

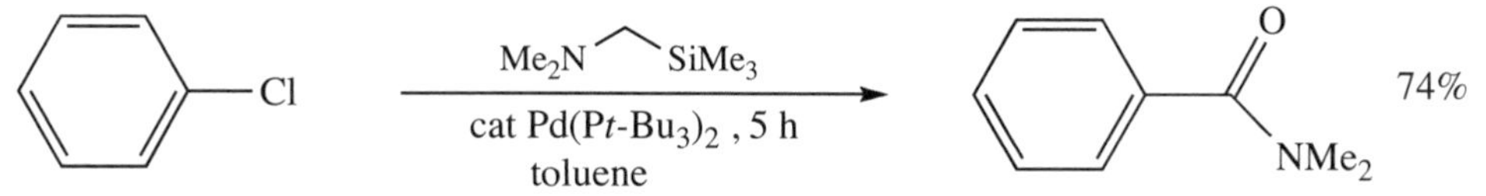

Cl
Me_2N $SiMe_3$
cat $Pd(Pt\text{-}Bu_3)_2$, 5 h
toluene
O NMe_2
74%

Cunico, R.F.; Maity, B.C. *J. Org. Chem.* ***2002***, *67*, 4357.

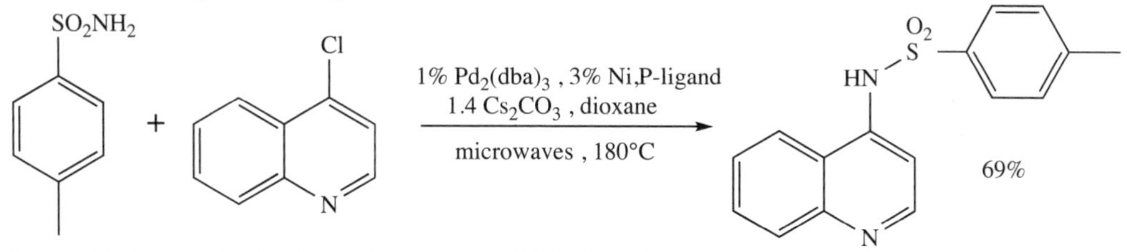

SO_2NH_2 + Cl N
1% $Pd_2(dba)_3$, 3% Ni,P-ligand
1.4 Cs_2CO_3 , dioxane
microwaves , 180°C
O_2 S HN N
69%

Burton, G.; Cao, P.; Li, G.; Rivero, R. *Org. Lett.* ***2003***, *5*, 4373.

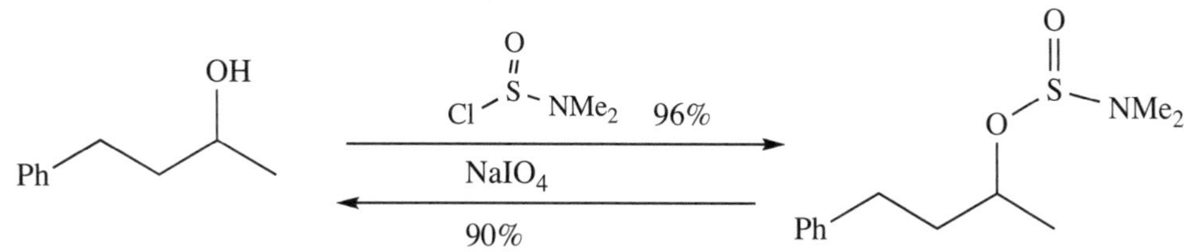

OH Ph
O Cl S NMe_2 96%
$NaIO_4$
90%
O S NMe_2 O Ph

Barma, D.K.; Bandyopadhyay, A.; Capdevila, J.H.; Falck, J.R. *Org. Lett.* ***2003***, *5*, 4755.

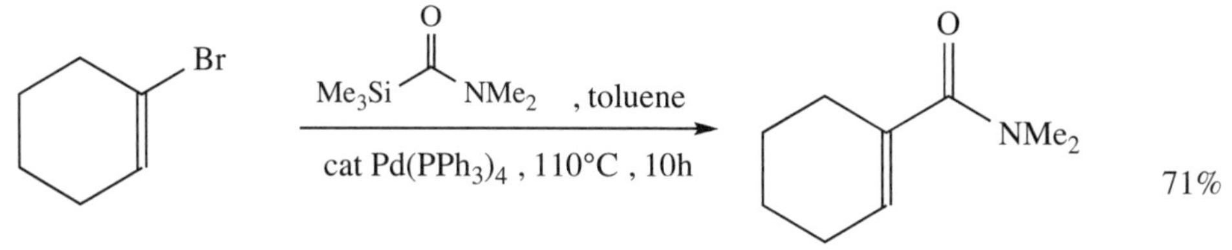

Br
O Me_3Si NMe_2 , toluene
cat $Pd(PPh_3)_4$, 110°C , 10h
O NMe_2
71%

Cunico, R.F.; Maity, B.C. *Org. Lett.* ***2003***, *5*, 4947.

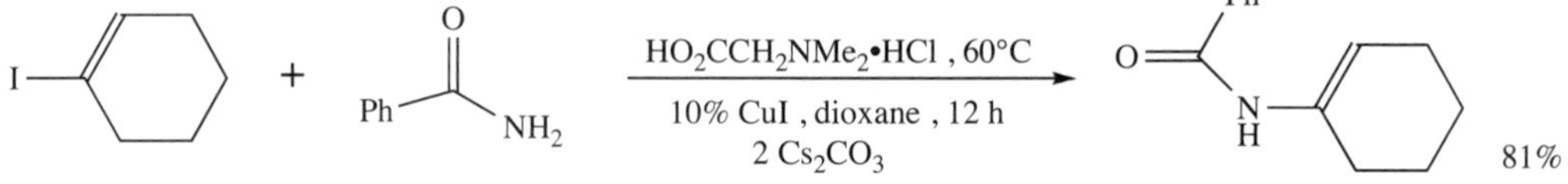

$HO_2CCH_2NMe_2$•HCl , 60°C
10% CuI , dioxane , 12 h
2 Cs_2CO_3

81%

Pan, X.; Cai, Q.; Ma, D. *Org. Lett.* ***2004***, *6*, 1809.

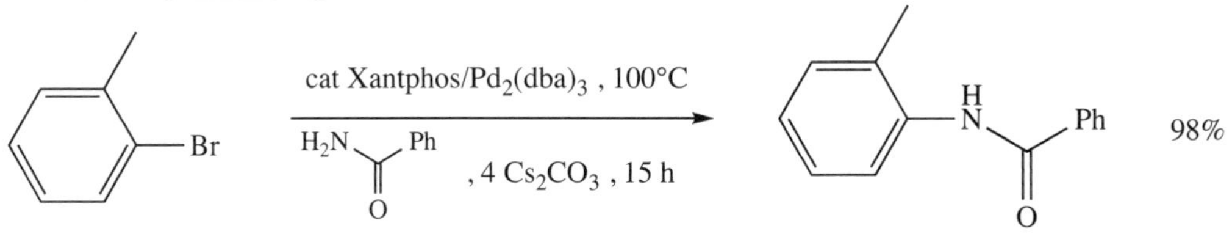

cat Xantphos/$Pd_2(dba)_3$, 100°C
, 4 Cs_2CO_3 , 15 h

98%

Yin, J.; Buchwald, S.L. *J. Am. Chem. Soc.* ***2002***, *124*, 6043.

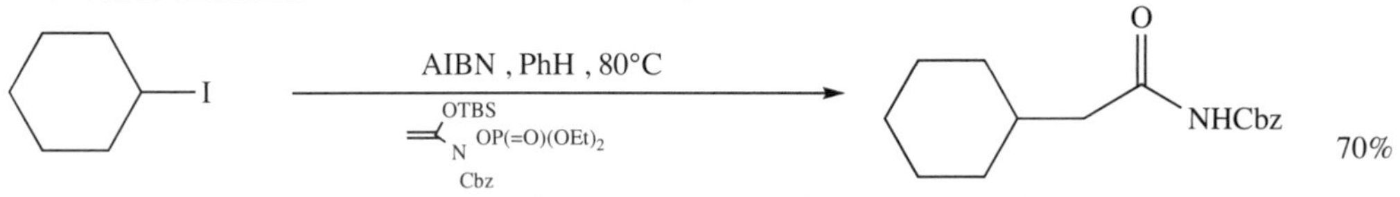

AIBN , PhH , 80°C

NHCbz

70%

Kim, S.; Lim, C.J.; Song, C.; Chung, W.-j. *J. Am. Chem. Soc.* ***2002***, *124*, 14306.

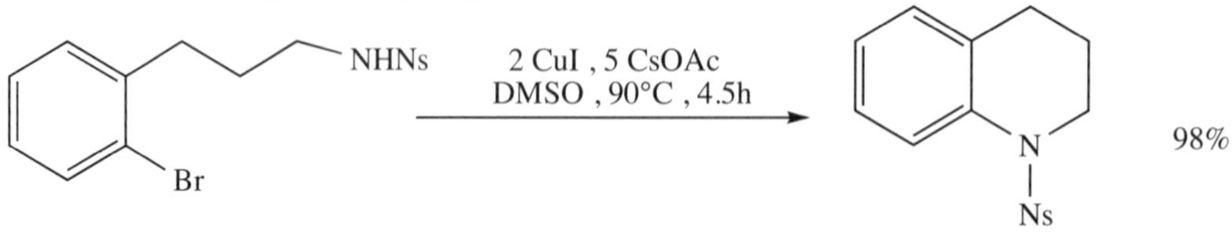

NHNs
2 CuI , 5 CsOAc
DMSO , 90°C , 4.5h

98%

Yamada, K.; Kubo, T.; Tokuyama, H.; Fukuyama, T. *Synlett* ***2002***, 231.

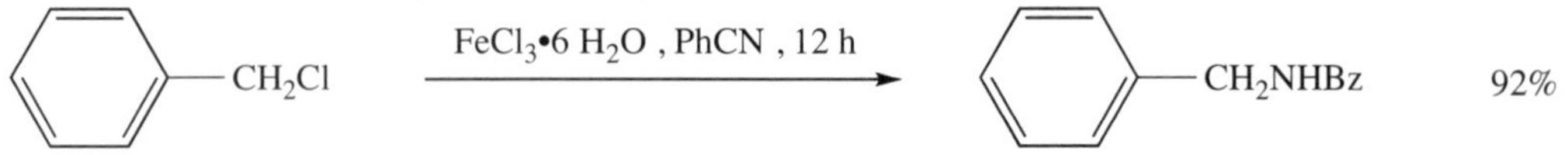

CH_2Cl → $FeCl_3$•6 H_2O , PhCN , 12 h → CH_2NHBz 92%

Karabulut, H.R.F.; Kacan, M. *Synth. Commun.* ***2002***, *32*, 2345.

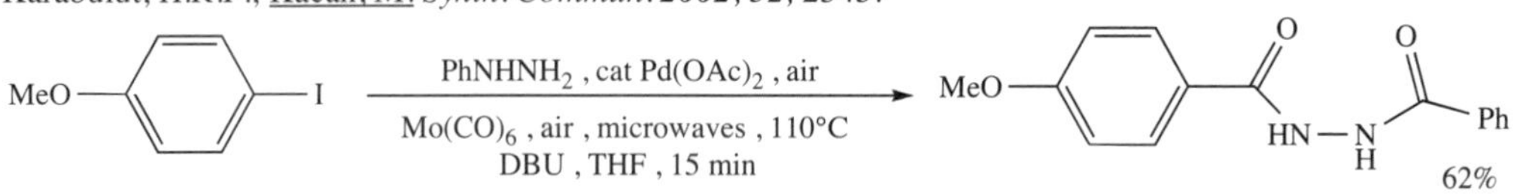

$PhNHNH_2$, cat $Pd(OAc)_2$, air
$Mo(CO)_6$, air , microwaves , 110°C
DBU , THF , 15 min

62%

Herrero, M.A.; Wannberg, J.; Larhed, M. *Synlett* ***2004***, 2335.

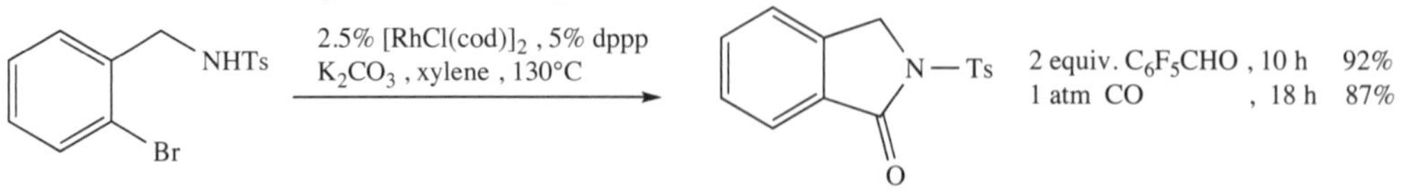

2.5% $[RhCl(cod)]_2$, 5% dppp
K_2CO_3 , xylene , 130°C

2 equiv. C_6F_5CHO , 10 h 92%
1 atm CO , 18 h 87%

Morimoto, T.; Fujioka, M.; Fuji, K.; Tsutsumi, K.; Kakiuchi, K. *Chem. Lett.* ***2003***, *32*, 154

SECTION 86: AMIDES FROM HYDRIDES

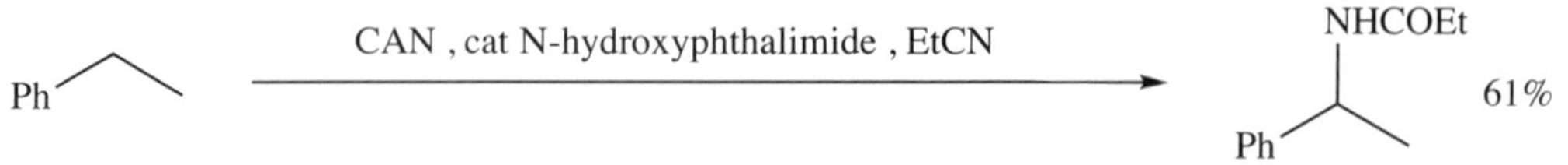

CAN , cat N-hydroxyphthalimide , EtCN

NHCOEt

61%

Sakaguchi, S.; Hirabayashi, T.; Ishii, Y. *Chem. Commun.* ***2002***, 516.

MeO-C6H4 → (ClSO$_2$0NMe$_2$, DCE , 1d; cat In(OTf)$_3$, 100°C) → MeO-C6H4-SO$_2$NMe$_2$ 99%

Frost, C.G.; Hartley, J.P.; Griffin, D. *Synlett* **2002**, 1928.

cat Sc(OTf)$_3$, DEAD , 1.5h

2.5 h with In(OTf)$_3$ gives 85% yield

MeO, CO$_2$Et, N, NHCO$_2$Et, MeO 90%

Yadav, S.; Reddy, B.V.S.; Veerendhar, G.; Rao, R.S.; Nagaiah, K. *Chem. Lett.* **2002**, *31*, 318.

OH → (PhN=C=S , bmim Cl / AlCl$_3$, rt , 8 h) → S, PhHN, OH 80%

Naik, P.U.; Nara, S.J.; Harjani, K.R.; Salunkhe, M.M. *Can. J. Chem.* **2003**, *81*, 1057.

SECTION 87: AMIDES FROM KETONES

O, N, Boc, OBoc → (1. TFA; 2. hν) → O, N 50%

Zeng, Y.,; Smith, B.T.; Hershberger, J.; Aubé, J. *J. Org. Chem.* **2003**, *68*, 8065.

O → (EtO$_2$C-N=N-CO$_2$Et , rt , 10h; 10% L-proline) → O, NHCO$_2$Et, N, CO$_2$Et + O, NHCO$_2$Et, N, CO$_2$Et (91 : 9) 80% 93% ee

Kumaragurubaran, N.; Juhl, K.; Zhuang, W.; Bøgevig, Z.; Jørgensen, K.A. *J. Am. Chem. Soc.* **2002**, *124*, 6254.

O → (1. N$_3$, Ph, OH; BF$_3$•OEt$_2$; 2. aq KOH) → Ph, O, HO, N + Ph, OH, O, N (78 : 22) 97%

Sahasrabudhe, K.; Gracias, V.; Rurness, K.; Smith, B.T.; Katz, C.E.; Reddy, D.S.; Aubé, J. *J. Am. Chem. Soc.* **2003**, *125*, 7914.

O, Ph, Ph → (NH$_2$OH•HCl , (CO$_2$H)$_2$, 100°C , 4h) → O, Ph, NHPh 91%

Chandrasekhar, S.; Gopalaiah, K. *Tetrahedron Lett.* **2003**, *44*, 7437.

i-PrNH$_2$, neat , rt , 15h — quant

Ballini, R.; Bosica, G.; Fiorini, D. *Tetrahedron* ***2003***, *59*, 1143.

ZnO , $NH_2OH{\bullet}HCl$, 140°C , 1h; solvent free — 95%

Sharghi, H.; Hosseini, M. *Synthesis* ***2002***, 1057.

P_2O_5/SiO_2 , microwaves; solid phase — 88%

Eshghi, H.; Gordi, Z. *Synth. Commun.* ***2003***, *33*, 2971.

$NH_2OH{\bullet}HCl$, silica chloride; microwaves , 3 min — 90%

Srinivas, K.V.N.S.; Mahender, I.; Das, B. *Chem. Lett.* ***2003***, *32*, 738.

SECTION 88: AMIDES FROM NITRILES

1. 8 MeOH , 8 H_2SO_4 , 100°C , 1h; 2. H_2O — 73%

Lebedev, M.Y.; Erman, M.B. *Tetrahedron Lett.* ***2002***, *43*, 1397.

H_2SO_4 , t-BuOAc; 42°C , 2h — 95%

Reddy, K.L. *Tetrahedron Lett.* ***2003***, *44*, 1453.

Ph-CN → 0.1 $NiCl_2{\bullet}6H_2O$, 7 $NaBH_4$, 2 Boc_2O , MeOH; 15h , 0°C → rt → $PhCH_2NHBoc$ 80%

Caddick, S.; Judd, D.B.; Lewis, A.K.de K.; Reich, M.T.; Williams, M.R.V. *Tetrahedron* ***2003***, *59*, 5417.

PhCN → cat $Ru(OH)_x/Al_2O_3$, H_2O , 413 K — 99%

Yamaguchi, K.; Matsushita, M.; Mizuno, N. *Angew. Chem. Int. Ed.* ***2004***, *43*, 1576.

EtO_2CCH_2CN → Ph_2CHOH , HCOOH , heat → $AcNHCHPh_2$ 80%

Gullickson, G.C.; Lewis, D.E. *Synthesis* ***2003***, 681.

Ph–C≡N → (MnO_2-SiO_2, microwaves, 14 min; solvent free) → $PhC(O)NH_2$ 65%

Khadilkar, B.M.; Madyar, V.R. *Synth. Commun.* ***2002***, *32*, 1731.

2-cyanopyridine → ($(NH_4)_2S$, MeOH, microwaves) → pyridine-2-carbothioamide ($C(S)NH_2$) 98%

Bagley, M.C.; Chapaneri, K.; Glover, C.; Merritt, E.A. *Synlett* ***2004***, 2615.

$PhCH_2CN$ → (HS, Dowex-1x8 SH^-; MeOH, H_2O, 3 h) → $PhCH_2C(S)NH_2$ 69%

Liboska, R.; Zyka, D.; Bobek, M. *Synthesis* ***2002***, 1649.

SECTION 89: AMIDES FROM ALKENES

SES = $-SO_2CH_2CH_2SiMe_3$

α-methylstyrene → (Py-N oxide, DCM, 0°C; Mn salen complex, 9 h; SES-Cl, $AgClO_4$) → N-SES aziridine (Ph, Me) 70% (83% ee)

Nishimura, M.; Minakata, S.; Talahashi, T.; Oderaotoshi, Y.; Komatsu, M. *J. Org. Chem.* ***2002***, *67*, 2107.

6-acetyl-2,2-dimethylchromene → (5% Cu^+/chiral 2,2'-diamino-6,6'-dimethylbiphenyl; PhINTs, –40°C) → N-Ts aziridine 87% (99% ee)

Gillesie, K.M.; Sander, C.J.; O'Shaughnessy, P.; Wetmoreland, I.; Thickett, C.P.; Scott, P. *J. Org. Chem.* ***2002***, *67*, 3450.

$Bu–N^+(=CH_2)–CO_2Me$ → (1. CH_2=$CHC_{10}H_{21}$; 2. NEt_3) → 3-butyl-6-decyl-1,3-oxazinan-2-one ($C_{10}H_{21}$) 72%

Suga, S.; Nagaki, A.; Tsutsui, Y.; Yoshida, J.-i. *Org. Lett.* ***2003***, *5*, 945.

cyclopentene → (PhInTs, (L)$CuCl_n$, DCM; $NaBPh_4$) → N–Ts bicyclic aziridine 66%

L = a pyridinophane

Vedernikov, A.N.; Caluton, K.G. *Org. Lett.* ***2003***, *5*, 2591.

styrene → ($TolSO_2N(Br)Na$, MS 5Å; cat Fe(TPP)Cl, MeCN) → 2-phenyl-N-Ts aziridine 68%

Vyas, R.; Gao, G.-Y.; Hardin, J.D.; Zhang, X.P. *Org. Lett.* ***2004***, *6*, 1907.

Ph–CH=CH$_2$ + 6-methylpyridine-3-SO$_2$NH$_2$ → PhI(OAc)$_2$, MeCN; Cu(MeCN)$_2$ClO$_4$, MS, 25°C, 12h → 84%

Han, H.; Bae, I.; Yoo, E.J.; Lee, J.; Do, Y.; Chang, S. *Org. Lett.* ***2004***, *6*, 4109.

$H_2NSO_3CH_2CCl_3$, PhI(OAc)$_2$•MgO; cat. Rh$_2$(tfacam)$_4$, PhH → N-$SO_2CH_2Cl_3$ 85%

Guthikonda, K.; DuBois, J. *J. Am. Chem. Soc.* ***2002***, *124*, 13672.

PhI=NTs, MeCN, 0°C; 2% AgNO$_3$, *t*-BuS*t*-Py → N-Ts 66%

Cui, Y.; He, C. *J. Am. Chem. Soc.* ***2003***, *125*, 16202.

Ph; PhSiH$_3$ EtOH, chiral Co catalyst; Boc-N=N-Boc → Ph, N-Boc, NHBoc 86%

Waser J.; Carreira, E.M. *J. Am. Chem. Soc.* ***2004***, *126*, 5676.

O, S, N–Ts, NH$_2$; cyclohexene, PhI=O, MeCN, –20°C; 10% Cu(MeCN)$_4$ PF$_6$ → O, S, N–Ts, N 81% (50% de)

Dichenna, P.H.; Robert-Peillard, F.; Dauban, P.; Dodd, R.H. *Org. Lett.* ***2004***, *6*, 4503.

TsNIK, MeCN, MS 5Å; 5% CuCl, 25°C → N-Ts 75%

Jain, S.L.; Sain, B. *Tetrahedron Lett.* ***2003***, *44*, 575.

Ph; 20% NBS, Chloramine-T; MeCN, 25°C → N-Ts, Ph 65%

Thakur, V.V.; Sudalai, A. *Tetrahedron Lett.* ***2003***, *44*, 989.

PhCN; NaN$_3$, fluoroapatite, 3h → O, Ph, NH$_2$ 95%

Solhy, A.; Smahi, A.; El Badaour, H.; Elaabar, B.; Amoukal, A.; Tikad, A.; Sebti, S.; Macquarrie, D.J. *Tetrahedron Lett.* ***2003***, *44*, 4031.

5 PhCH=CH$_2$; H$_2$NO$_2$S–C$_6$H$_4$–NO$_2$; 5% [Cu(MeCN)$_4$]ClO$_4$, PhI(OAc)$_2$; PhH, 6% Phbox derivative → Ph, N–S(O)(O)–C$_6$H$_4$–NO$_2$ 94% (75% ee)

Kwong, H.-L.; Liu, D.; Chan, K.-Y.; Lee, C.-S.; Huang, K.-H.; Che, C.-M. *Tetrahedron Lett.* ***2004***, *45*, 3965.

pyrrolidine, H_2O, rt
cat. $PdCl_2(PPh_3)_2/PPh_3$
30 min

99%

Huh, D.H.; Jeong, J.S.; Lee, H.B.; Ryu, H.; Kim, Y.G. *Tetrahedron* ***2002***, *58*, 9925.

1. t-BuO_2CNBr_2, DCM
2. 12 Na_2SO_3, 10°C

87%

Sliwinska, A.; Zwierzak, A. *Tetrahedron* ***2003***, *59*, 5927.

PhI=NTs, bmim BF_4, 25 min
cat $Cu(acac)_2$

88%

Kantam, M.L.; Neeraja, V.; Kavita, B.; Haritha, Y. *Synlett* ***2004***, 525.

$TolSO_2N_3$, Ru(salen)(CO) catalyst
DCM, MS 4Å, rt, 1 d

71% (87% ee)

Omura, K.; Murakami, M.; Uchida, T.; Irie, R.; Katsuki, T. *Chem. Lett.* ***2003***, *32*, 354.

$BnNH_2$, 3 t-$BuBH_2$
DMSO, rt, 1h

96%

Matsuo, J.; Yamanaka, H.; Kawana, A.; Mukaiyama, T. *Chem. Lett.* ***2003***, *32*, 392.

$PhTe^+Me_2\ I^-$, AgOAc, MeCN
10% $Pd(OAc)_2$, 50°C, 6h

99%

Hirabayashi, K.; Nara, Y.; Shimizu, T.; Kamigata, N. *Chem. Lett.* ***2004***, *33*, 1280.

2 Cl(Na)N–Ts, $AgNO_3$, PhH, Ar
60°C, 3h

92% (6:94 *cis:trans*)

Minakata, S.; Kano, D.; Fukuoka, R.; Oderaotoshi, Y.; Komatsu, M. *Heterocycles* ***2003***. *60*, 289.

5% $[PdCl_2(MeCN)_2]$, 1.5 $CuCl_2$
CO/O_2, PhOH, THF

88%

Mizutani, T.; Ukaji, Y.; Inomata, K. *Bull. Chem. Soc. Jpn.* ***2003***, *76*, 1251.

SECTION 90: AMIDES FROM MISCELLANEOUS COMPOUNDS

Ph-N=C=O → 1. $LiAlHSeH$, rt , 1h; 2. MeI , rt , 2h → PhHN–C(=O)–SeMe 70%

Koketsu, M.; Ishida, M.; Takakura, N. *J. Org. Chem.* ***2002***, *67*, 486.

PhCH=N-OH → Chloramine-T → Ph–C(=O)–N(Tos)(Tol) 20%

Padmavathi, V.; Reddy, K.V.; Padmaja, A.; Venugopalan, P. *J. Org. Chem.* ***2003***, *68*, 1567.

TFA , 70-110°C; – CH_2=CH_2 63%

Cordero, F.M.; Pisaneschi, F.; Salvati, M.; Paschetta, V.; Ollivier, J.; Salaün, J.; Brandi, A. *J. Org. Chem.* ***2003***, *68*, 3271.

CO_1 , DBU , MeCN; PBu_3 , DBAD

DBAD = di-*tert*-butyl azodicarboxylate 92%

Dinsmore, C.J.; Mercer, S.P. *Org. Lett.* ***2004***, *6*, 2885.

PhCH=N-Ts toluene , rt; ferrocenyl catalyst 88% (98% ee)

Hodous, B.L.; Fu, G.C. *J. Am. Chem. Soc.* ***2002***, *124*, 1578.

PhN_3 → Pd/C , HCO_2NH_4 , MeCN , reflux → PhNHCHO 79%

Reddy, P.G.; Baskaran, S. *Tetrahedron Lett.* ***2002***, *43*, 1919.

Ph–C(=N-OH)–Ph → $(HBO_2)_n$, 145°C , 17h → Ph–C(=O)–Ph 92%

Chandrasekhar, S.; Gopalaiah, K. *Tetrahedron Lett.* ***2002***, *43*, 2455.

$PhCH_2N_3$ → HSBz , MeCN PPh_3 → $BzNHCH_2Ph$

Park, S.-D.; Oh, J.-H.; Lim, D. *Tetrahedron Lett.* ***2002***, *43*, 6309.

$PhNO_2$ → Ac_2O , AcOH , In , MeOH , 2h → PhNHAc 92%

Kim, B.H.; Han, R.; Piao, F.; Jun, Y.M.; Baik, W.; Lee, B.M. *Tetrahedron Lett.* ***2003***, *44*, 77.

Ph-C=C=O → (0.1 NHC carbene [N,N'-bis(2,6-di-i-Pr-phenyl)imidazolin-2-ylidene], THF, rt 1h) → 1,3,5-triphenyl-1,3,5-triazinane-2,4,6-trione 99%

i-Pr i-Pr N N i-Pr i-Pr

Duong, H.A.; Cross, M.J.; Louie, J. *Org. Lett.* ***2004**, 6*, 4679.

N-OH Ph → $Cl_3C\text{-}CH(OH)_2$, 125°C , neat , 3h → O Ph 95%

Chandrasekhar, S.; Gopalaiah, K. *Tetrahedron Lett.* ***2003**, 44*, 755.

Br CO_2Et → PhCH=N-Ph Zn , cat. Cp_2TiCl_2 / THF , rt , 6 min → Ph Ph O (80% (91:9 *cis:trans*)) + NHPh Ph CO_2Et 12%

Chen, L.; Zhao, G.; Ding, Y. *Tetrahedron Lett.* ***2003**, 44*, 2611.

O O O N Me Ph → TFA , DCM , 35°C , 16h → PhHN Me O 75%

His, S.; Meyer, C.; Cossy, J.; Emeric, G.; Greiner, A. *Tetrahedron Lett.* ***2003**, 44*, 8581.

OH N Ph → 10% NH_2SO_3H , MeCN , 90°C , 6h → PhNHAc 96%

Wang, B.; Gu, Y.; Luo, C.; Yang, T.; Yang, L.; Suo, J. *Tetrahedron Lett.* ***2004**, 45*, 3369.

OH N Ph → MoO_3 , SiO_2 → O Ph

Dongare, M.K.; Bhagwat, V.V.; Ramana, C.V.; Gurgar, M.K. *Tetrahedron Lett.* ***2004**, 45*, 4759.

$PhNO_2$ → Sn/NH_4Cl , Boc_2O , MeOH / ultrasound → PhNHBoc 92%

Chandrasekhar, S.; Barishmulu, Ch.; Jagadeshwar, V. *Synlett* **2002**, 771.

N-OH Ph Ph → cat. $RuCl_3$, MeCN , 2.5h → O Ph NHPh 91%

De, S.K. *Synth. Commun.* ***2004**, 34*, 3431.

PhN_3 → AlI_3 , Ac_2O , MeCN , reflux → PhNHAc 85%

Bez, G. *Synth. Commun.* ***2002**, 32*, 3625.

OH NO$_2$ — Se , CO , NMP , THF / 150°C , 5 h → O N H O 91%

Nishiyama, Y.; Naitoh, Y.; Sonoda, N. *Synlett* ***2004***, 886.

N-OH Ph — $Y(OTf)_3$, MeCN , reflux , 2.5 h → O Ph NHMe 90%

De, K. *Org. Prep. Proceed. Int.* ***2004***, *36*, 383

CN Ph N O — mcpba , Me_2S , aq KOH / aq MeCN , 1.5 h → O Ph N O 73%

Yokoshima, S.; Kubo, T.; Tokuyama, H.; Fukuyama, T. *Chem. Lett.* ***2002***, 122.

Me O N — I_2 , $CuCl_2$, rt / 3 d , MeCN → Me O N I Cl 40% + Me O N I OH 35%

Terao, K.; Takechi, Y.; Kunishima, M.; Tani, S.; Ito, A.; Yamasaki, C.; Fukuzawa, S. *Chem. Lett.* ***2002***, *31*, 522.

SECTION 90A: PROTECTION OF AMIDES

O HN O — $(t\text{-BuCl}_2)_2O$, 2 $MgBr_2{\bullet}OEt_2$, NEt_3 / DCM , 0°C , 4 h → O O *t*-Bu N O 80%

Yamada, S.; Yaguchi, S.; Matsuda, K. *Tetrahedron Lett.* ***2002***, *43*, 647.

O N H OMe OMe — 4 TsOH , toluene → O NH$_2$ quant

Chern, C.-Y.; Huang, Y.-P.; Kan, W.M. *Tetrahedron Lett* ***2003***, *44*, 1039.

CHAPTER 7

PREPARATION OF AMINES

SECTION 91: AMINES FROM ALKYNES

Ph—≡—Me → 1. $Cp*_2tiMe_2$, $BnNH_2$ toluene , 114°C , 1 d; 2. $NaBH_3CN$, $ZnCl_2$ MeOH , 25°C , 20 h → NHBn Ph Me 70% + BnHN Ph Me 24%

Jeutling, A.; Doye, S. *J. Org. Chem.* ***2002***, *67*, 1961.

Ph—≡ → PhNO , autoclave , PhH , 170°C; CO , cat $[Cp*Ru(CO)_2]_2$, 1 d → Ph N H 53%

Penoni, A.; Volkmann, J.; Nicholas, K.M. *Org. Lett.* ***2002***, *4*, 695.

Ph—≡ → $PhNH_2$, 1% $H_3PW_{12}O_{40}$, 70°C , 2 h; 0.2% $(Ph_3P)AuMe$ → Ph N Ph 98%

Mizushima, E.; Hayashi, T.; Tanaka, M. *Org. Lett.* ***2003***, *5*, 3349.

Bu—≡ → 1. t-$BuNH_2$, C_6D_6 , 65°C 5% bis(amidate) Ti; 2. $LiAlH_4$, ether → t-BuHN Bu (>49 + : NHt-Bu Bu 1) 82%

Zhang, Z.; Schaafer, L.L. *Org. Lett.* ***2003***, *5*, 4733.

$PhNH_2$ → 1-octyne , 5% Ti catalyst , 1d; 100°C , toluene → NHPh 92%

Khedkar, V.; Tillack, A.; Beller, M. *Org. Lett.* ***2003***, *5*, 4767.

Ph—≡—Ph → $PhNH_2$, C_5D_5Cl; 5% $(Bn)_3Ta{=}NCMe_3$ 135°C , 30 h → Ph Ph Ph (3 + : NHPh Ph Ph 1) >95%

Anderson, L.L.; Arnold, J.; Bergman, R.G. *Org. Lett.* ***2004***, *6*, 2519.

Bu—≡—CH$_2$—≡ → PhNH$_2$, 100°C , 1 d / 10% Ti(NMe$_2$)$_2$(dpma) → 2-Me-5-Bu-N-Ph pyrrole 56%

Ramanathan, B.; Keith, A.J.; Armstrong, D.; Odom, A.L. *Org. Lett.* **2004**, *6*, 2957.

C_4H_9—≡ → 1. *t*-BuNH$_2$, toluene , 85°C , 2h; $Cp_2Ti(\eta^2Me_3Si$-C≡C-$SiMe_3)$ / 2. BuLi , –70°C→ rt , 13h → C_4H_9CH$_2$CH(NH*t*-Bu)C_4H_9 63%

Castro, I.G.; Tillack, A.; Hartung, C.G.; Beller, M. *Tetrahedron Lett.* **2003**, *44*, 3217.

Ph—≡— → PhNH$_2$, chiral Ti catalyst / toluene → Ph—N=C(Me)CH$_2$Ph 41%

Shi, Y.; Hall, C.; Ciszewski, J.T.; Cao, C.; Odom, A.L. *Chem. Commun.* **2003**, 586.

Bu—≡ → cat [$Cp_2Ti(\eta^2$-ME_3Si-C≡C-$SiMe_3$] , *t*-BuNH$_2$ / 85°C , 2h → BuCH$_2$CH=N–*t*-Bu 90%

Tillack, A.; Castro, I.G.; Hartung, C.G.; Beller, M. *Angew. Chem. Int. Ed.* **2002**, *41*, 2541.

2-Cl-C$_6$H$_4$—≡—C_3H_7 → 1. PhCH(Me)NH$_2$, 110°C , 1d; cat Cp_2TiMe_2 / 2. cat Pd_2dba_3 , *t*-BuOK , dioxane; 110°C , 10h → 2-C_3H_7-N-(CH(Me)Ph)indole 65%

Siebeneicher, H.; Bytschkov, I.; Doye, S. *Angew. Chem. Int. Ed.* **2003**, *42*, 3042.

Ph—≡—Ph → 1. *t*-BuNH$_2$, cat Cp_2TiMe_2 , 105°C , 1d; 2. NaBH$_3$CN , ZnCl$_2$, MeOH , 25°C , 20h → PhCH$_2$CH(Ph)NH*t*-Bu 81%

Pohlki, F.; Heutling, A.; Bytschkov, I.; Hotopp, T.; Doye, S. *Synlett* **2002**, 799.

REVIEW:

"Development of the Ti-Catalyzed Intermolecular Hydroamination of Alkynes"
Doye, S. *Synlett* **2004**, 1653.

SECTION 92: AMINES FROM ACID DERIVATIVES

NO ADDITIONAL EXAMPLES

SECTION 93: AMINES FROM ALCOHOLS AND THIOLS

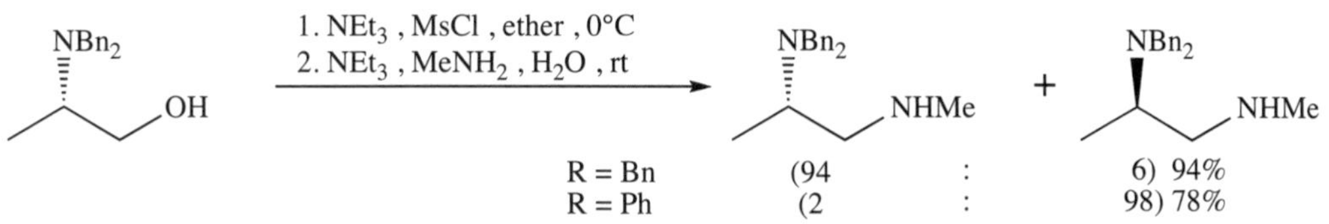
NBn2
OH
1. NEt3 , MsCl , ether , 0°C
2. NEt3 , MeNH2 , H2O , rt
NBn2
NHMe
+
NBn2
NHMe
R = Bn (94 : 6) 94%
R = Ph (2 : 98) 78%

Shi, M.; Xu, B. *J. Org. Chem.* **2002**, *67*, 294.

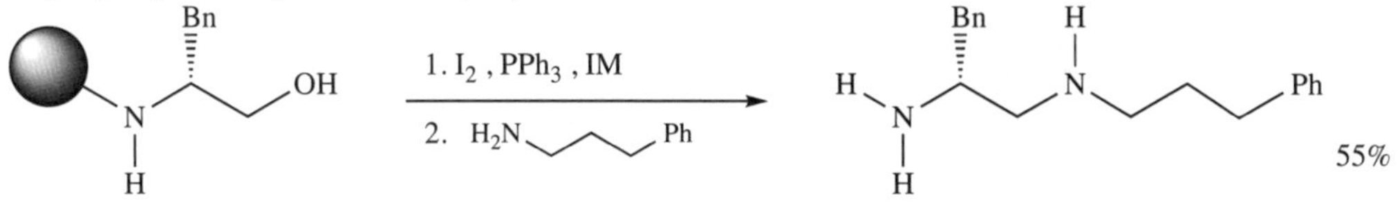
Bn
OH
N
H
1. I2 , PPh3 , IM
2. H2N Ph
Bn
H
H
N
N
Ph
H
55%

Olsen, C.A.; Witt, M.; Jaroszewski, J.W.; Franzyk, H. *Org. Lett.* **2004**, *6*, 1937.

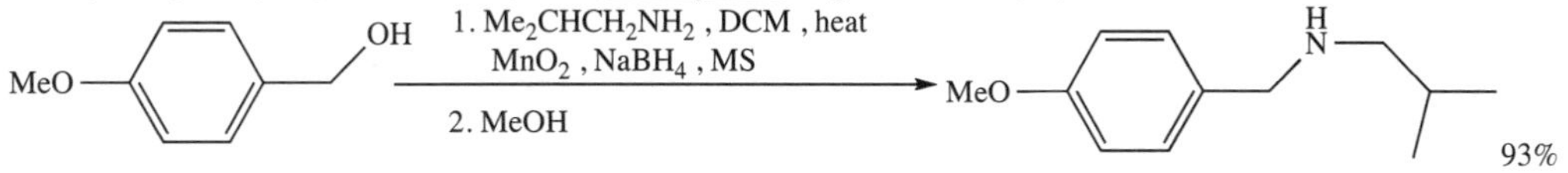
MeO
OH
1. Me2CHCH2NH2 , DCM , heat
MnO2 , NaBH4 , MS
2. MeOH
MeO
H
N
93%

Kanno, H.; Taylor, R.J.K. *Tetrahedron Lett.* **2002**, *43*, 7337.

$$PhNH_2 \xrightarrow[110°C\text{ , }17h]{PhCH_2OH\text{ , cat }[Cp^*IrCl_2]_2\text{ , }5\%\ K_2CO_3} PhNHCH_2Ph \quad 88\%$$

Fujita, K.-i.; Li, Z.; Ozeki, N.; Yamaguchi, R. *Tetrahedron Lett.* **2003**, *44*, 2687.

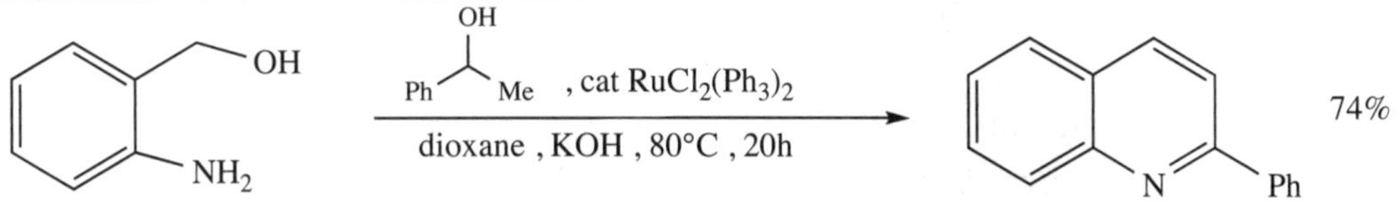
OH
NH2
OH
Ph Me
, cat RuCl2(Ph3)2
dioxane , KOH , 80°C , 20h
N
Ph
74%

Cho, C.S.; Kim, B.T.; Choi, H.-J.; Kim, T.-J.; Shim, S.C. *Tetrahedron* **2003**, *59*, 7997.

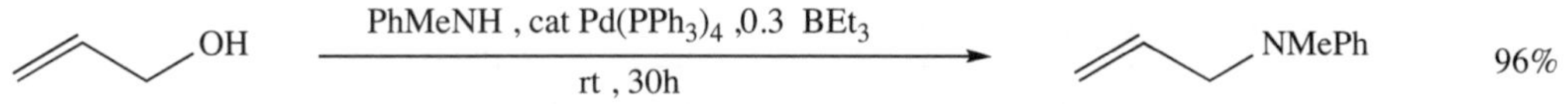
OH
PhMeNH , cat Pd(PPh3)4 ,0.3 BEt3
rt , 30h
NMePh
96%

Kimura, M.; Futamata, M.; Shibata, K.; Tamaru, Y. *Chem. Commun.* **2003**, 234.

$$PhCH_2OH \xrightarrow[\text{cat }[IrCl(cod)]_2/dppf]{Ph_3P{=}NPh\text{ , }110°C\text{ , 1d , cat }K_2CO_3} PhCH_2NHPh \quad 91\%$$

Cami-Kobeci, G.; Williams, J.M.J. *Chem. Commun.* **2004**, 1072.

$$PhCH_2OH \xrightarrow[\text{overnight , reflux}]{MeONH_2\text{-}Cl\text{ , MS 4Å , }MnO_2\text{ , DCM}} PhCH{=}NH\text{-}OMe \quad 62\%\ (12:1\ E:Z)$$

Kanno, H.; Taylor, R.J.K. *Synlett* **2002**, 1287.

SECTION 94: AMINES FROM ALDEHYDES

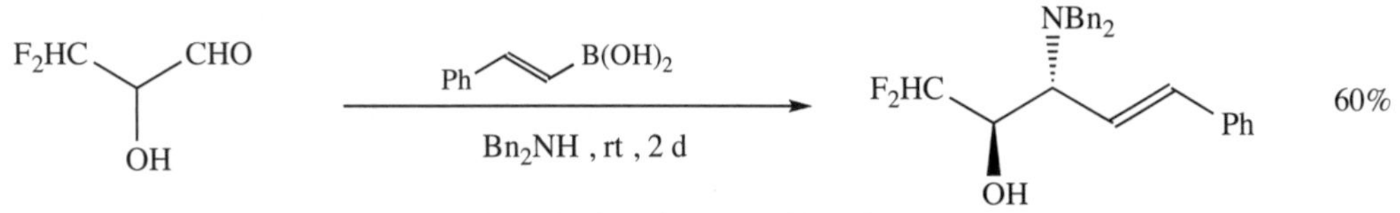
F2HC
CHO
OH
Ph
B(OH)2
Bn2NH , rt , 2 d
NBn2
F2HC
Ph
OH
60%

Prakash, G.K.S.; Mandal, M.; Schweizer, S.; Petasis, N.A.; Olah, G.A. *J. Org. Chem.* **2002**, *67*, 3718.

$$PhCHO \xrightarrow{[Rh(cod)Cl]_2\text{ , TPPTS , }NH_3\text{ , }H_2} PhCH_2NH_2 \quad 97\%$$

Gross, T.; Seayad, A.M.; Ahmad, M.; Beller, M. *Org. Lett.* **2002**, *4*, 2055.

PhCHO → 1. $C_3H_7NH_2$, MS 4Å , rt , 15 h; 2. CH_2=$CHCH_2SnBu_3$, $TaCl_5$; MgO , 3 h → Ph, NH_2 71%

Shibata, I.; Nose, K.; Sakamoto, K.; Yasuda, M.; Baba, A. *J. Org. Chem.* ***2004***, *69*, 2185.

PhCHO → H_2N, SnBu$_3$; 2% $La(OTf)_3$, $PhCO_2H$, MeCN → HN, Ph 67%

Aspinall, H.C.; Bissett, J.S.; Greeves, N.; Levin, D. *Tetrahedron Lett.* ***2002***, *43*, 323.

CHO → *p*-anisidine , THP , 50°C , 10 h; , cat $Ni(acac)_2$ → NH*p*-anisidine 89% (8:1 *syn:anti*)

Kimura, M.; Miyachi, A.; Kojima, K.; Tanaka, S.; Tamaru, Y. *J. Am. Chem. Soc.* ***2004***, *126*, 14360.

PhCHO → $SnBu_3$, $PhNH_2$ $LiClO_4$, rt; MeCN , 3h → NHPh, Ph 90%

Yadav, J.S.; Reddy, B.V.S.; Reddy, P.S.R.; Rao, M.S. *Tetrahedron Lett.* ***2002***, *43*, 6245.

PhCHO → $TsNH_2$, TFAA , DCM , reflux , 12h → PhCH=N-Ts 88%

Lee, K.Y.; Lee, C.G.; Kim, J.N. *Tetrahedron Lett.* ***2003***, *44*, 1231.

PhCHO → $PhNH_2$, pic-NH_3 , MeOH , AcOH → $PhCH_2NHPh$ 95%

Sato, S.; Sakamoto, T.; Miyazawa, E.; Kitugawa, Y. *Tetrahedron* ***2004***, *60*, 7899.

PhCHO → Bu_3Sn , 2% $Bi(OTf)_3$, MeCN; $PhCO_2H$, rt , 10 min → OH, Ph 90%

Choudary, B.M.; Chidara, S.; Sekhar, Ch.V.R. *Synlett* ***2002***, 1694.

O_2N—C$_6$H$_4$—CHO → $SnBu_3$, 20% $Sc(OTf)_3$; O_2N—C$_6$H$_4$—NH_2 , $PhCO_2H$, PEG, rt , 4 h → NO_2, O_2N, N—H 90%

Choudary, B.M.; Jyothi, K.; Madhi, S.; Kantam, M.L. *Synlett* ***2004***, 231.

PhCHO → $PhNH_2$, bmim PF_6; 15 min → PhCH=NHPh 95%

Andrade, C.K.Z.; Takada, S.C.S.; Alves, L.M.; Rodrigues, J.P.; Suarez, P.A.Z.; Brandão, R.F; Soares, V.C.D. *Synlett* ***2004***, 2135.

PhCHO → (allyl-$SnBu_3$, $PhNH_2$, bmim BF_4; rt, 4.5 h) → NHPh, Ph 92%

Yadav, J.S.; Reddy, B.V.S.; Raju, A.K. *Synthesis* **2003**, 883.

CHO → (NH_2, N; $Ti(Oi\text{-}Pr)_4$, THF, 8 h, rt) → N H, N 78%

Kumpaty, H.J.; Bhattacharyya, S.; Rehr, E.W.; Gonzalez, A.M. *Synthesis* **2003**, 2206.

N CHO + NH_2 → (MeOH, reflux, 4 h) → N, N 60%

Bianchini, C.; Lee, H.M.; Mantovani, G.; Meli, A.; Oberhauser, W. *New J. Chem.* **2002**, *26*, 387.

Related Method: Section 102 (Amines from Ketones)

SECTION 95: AMINES FROM ALKYLS, METHYLENES, AND ARYLS

NO ADDITIONAL EXAMPLES

SECTION 96: AMINES FROM AMIDES

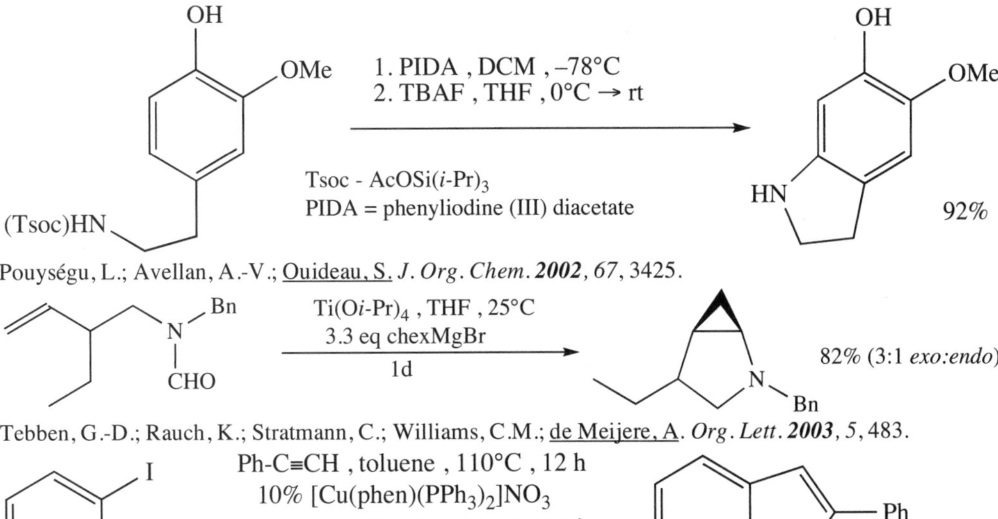

Pouységu, L.; Avellan, A.-V.; Quideau, S. *J. Org. Chem.* **2002**, *67*, 3425.

Tebben, G.-D.; Rauch, K.; Stratmann, C.; Williams, C.M.; de Meijere, A. *Org. Lett.* **2003**, *5*, 483.

I, NHAc → (Ph-C≡CH, toluene, 110°C, 12 h; 10% $[Cu(phen)(PPh_3)_2]NO_3$; 2 K_3PO_4) → N H, Ph 78%

Cacchi, S.; Fabrizi, G.; Parisi, L.M. *Org. Lett.* **2003**, *5*, 3843.

MeO, MeO, NHAc; $POCl_3$, bmim PF_6 , 10°C , 1h; MeO, MeO, N; 87%

Judeh, Z.M.A.; Ching, C.B.; Bu, J.; McCluskey, A. *Tetrahedron Lett.* ***2002***, *43*, 5089.

CO_2Me, Me, N, N, H, $COCF_3$; toluene , 120°C, 38 h; CO_2Me, Me, N, H; 60% + 21% of isomer

Miyata, O.; Takeda, N.; Naito, T. *Heterocycles* ***2002***, *57*, 1101.

Related Method: Section 105A (Protection of Amines)

SECTION 97: AMINES FROM AMINES

O, N—H; $(HO)_2B$—, OMe; 1.5 $Cu(OAc)_2$, 2 Py , DCM, air , rt , 3 d; O, N, OMe; 92%

Yu, S.; Saenz, J.; Srirangam, J.K. *J. Org. Chem.* ***2002***, *67*, 1699.

H, N, Ph; PhBr , Pd cat , 110°C , NaO*t*-Bu, toluene , 1 h; Ph, N, Ph; 93%

Prashad, M.; Mak, X.Y.; Liu, Y.; Repic, O. *J. Org. Chem.* ***2003***, *68*, 1163.

N—H; $PhB(OH)_2$, cat $Cu(OAc)_2$, 20% myristic acid , 2,6-lutidine, toluene , rt; N—Ph; 87%

Sasaki, M.; Dalili, S.; Yudin, A.K. *J. Org. Chem.* ***2003***, *68*, 2045.

N—Bn; $BnNH_2$, 10% $B(C_6F_5)_3$, MeCN , 65°C; NHBn, NHBn; quant

Watson, I.D.G.; Yudin, A.K. *J. Org. Chem.* ***2003***, *68*, 5160.

N, Cl; HN, O, , MeCN, Na_2CO_3 , 70°C; N, N, O; 90%

Faul, M.M.; Kobierski, M.E.; Kopach, M.E. *J. Org. Chem.* ***2003***, *68*, 5739.

Ph–CH2CH2–NH2 → (BuBr, CsOH, DMF; MS 4Å, 23°C, 21 h) Ph–CH2CH2–NHBu 89% + Ph–CH2CH2–NBu$_2$ 10%

Salvtore, R.N.; Nagle, A.S.; Jung, K.W. *J. Org. Chem.* ***2002***, *67*, 674.

NH_2 / OH → (NaY-Faujasite, methyl acetate, 90°C, 7 h) NHMe / OH 89%

Selva, M.; Tundo, P.; Perosa, A. *J. Org. Chem.* ***2003***, *68*, 7374.

N / H •HCl → (1. NaOH, C_6D_6; 2. 10% $Cp^*_2NdCH(TMS)_2$; 3. HCl) N •HCl 85% (>50:1 dr)

Molander, G.A.; Pack, S.K. *J. Org. Chem.* ***2003***, *68*, 9214.

C / NH_2 / Ph → (1. cat $Ti(NMe_2)_4$, PhH, 22 h, 75°C; 2. NaOH, MeOH) Ph / N 95%

Ackermann, L.; Bergman, R.G. *Org. Lett.* ***2002***, *4*, 1475.

Bu / Ph / N-NMe$_2$ → (1. $Cr(CO)_5$ Me OMe, THF, 65°C; 2. HCl) Bu / O / Ph / N—NMe$_2$ 62%

Zhang, Y.; Herdon, J.W. *Org. Lett.* ***2003***, *5*, 2043.

NH_2 / Cl → (2% Ni(0), 2% SIPr, 100°C; dioxane, 1.5 NaO*t*-Bu) N / H 98%

Omar-Amrani, R.; Thomas, A.; Brenner, E.; Schneider, R.; Fort, Y. *Org. Lett.* ***2003***, *5*, 2311.

N-H → (AcO, cat $[Pd(\eta^3\text{-}C_3H_7)Cl]_2$/BINAP; THF, rt) N 45%

Watson, I.D.G.; Styler, S.A.; Yudin, A.K. *J. Am. Chem. Soc.* ***2004***, *126*, 5086.

N_2 / CO_2Et / CO_2Et + Me / N / Me → (5% $Cu(acac)_2$; toluene) Me / CO_2Et / N / CO_2Et / Me 98%

Roberts, E.; Sançon, J.P.; Sweeney, J.B.; Workman, J.A. *Org. Lett.* ***2003***, *5*, 4775.

Morpholine N–OBz → Ph_2Zn , 1.25% $[Cu(OTf)]_2$•PhH / THF, 25°C , 1 h → N-phenylmorpholine 91%

Berman, A.M.; Johnson, J.S. *J. Am. Chem. Soc.* ***2004***, *126*, 5680.

Ph_2CH–N(OH)Me → In , reflux , EtOH/aq. NH_4Cl / 4h → Ph_2CH–NHMe quant

Cicchi, S.; Bonanni, M.; Cardona, F.; Revuelta, J.; Goti, A. *Org. Lett.* ***2003***, *5*, 1773.

(4-amino-2-chlorophenyl)(2-methylphenyl)methanone → 2-bromotoluene , NaO*t*-Bu , 3 min / 2% $Pd(OAc)_2$/phoshine ligand microwaves , *t*-BuOH/toluene 150°C → N-(2-methylphenyl) amine 78%

Jensen, T.A.; Liang, X.; Tanner, D.; Skjaerbaek, N. *J. Org. Chem.* ***2004***, *69*, 4936.

4-methyl-NMe_2-benzene + 4-Br-C_6H_4–$C(N_2)CO_2Me$ → Ru(II) catalyst / DCM , rt → Me–N(4-tolyl)–CH_2–CH(4-BrC$_6$H$_4$)–CO_2Me 67% (53% ee)

Davies, H.M.L.; Jin, Q. *Org. Lett.* ***2004***, *6*, 1769.

Cyclohexyl–NH_2 → 4-Me-C_6H_4–$B(OH)_2$ rt , 1 d / polymer bound Cu-catalyst DCM , MS 4Å , 2.6 NEt_3 → cyclohexyl–NH–(4-tolyl) 54%

Chiang, G.C.H.; Olsson, T. *Org. Lett.* ***2004***, *6*, 3079.

$PhCH_2NH_2$ → HO(CH$_2$)$_4$OH / cat $[Cp^*IrCl_2]_2$, $NaHCO_3$, 110°C toluene , 17 h → N-benzylpyrrolidine 72%

Fujita, K.-i.; Fujii, T.; Yamaguchi, R. *Org. Lett.* ***2004***, *6*, 3525.

Indole N–H → 2-bromotoluene , 5% CuI , 110°C / 20% diamine , 2.5 K_3PO_4 , 1 d toluene → N-(2-methylphenyl)indole 99%

Antilla, J.C.; Klapars, A.; Buchwald, S.L. *J. Am. Chem. Soc.* ***2002***, *124*, 11684.

2-Methylquinoline → [Ir(cod)Cl]/R-MeO-Biphep / H_2 , I_2 , toluene , rt → 2-methyl-1,2,3,4-tetrahydroquinoline (N–H) 94% (94% ee)

Wang, W.-B.; Lu, S.-M.; Yang, P.-Y.; Han, X.-W.; Zhou, Y.-G. *J. Am. Chem. Soc.* ***2003***, *125*, 10536.

I, NH_2 (2-iodoaniline) → 1. (2-(trimethylsilyl)phenyl triflate: OTf, $SiMe_3$), 3 CsF, MeCN; 2. 5% $Pd(OAc)_2$, 10% PCy_3, 100°C, 1 d → N–H (carbazole) 77%

Liu, Z.; Larock, R.C. *Org. Lett.* **2004**, *6*, 3739.

C, NH_2, Ph → 5% Ti complex, 75°C; C_6D_6, 1 h → Ph, N 84%

Ackermann, L.; Bergman, R.G.; Loy, R.N. *J. Am. Chem. Soc.* **2003**, *125*, 11956.

NH_2 → chiral La catalyst, C_6D_6, –30°C → NH 74% ee

Hong, S.; Tian, S.; Metz, M.V.; Marks, T.J. *J. Am. Chem. Soc.* **2003**, *125*, 14768.

F_3C–C_6H_4–C≡C–CO_2Et → Tol–CH=N–C_6H_4–OMe, 65°C; TolCOCl, *i*-Pr_2NEt, 4 atm CO, 16 h; 6% Pd catalyst, 15% $P(oTol)_3$ MeCN/THF → MeO, CO_2Et, Tol, N, Tol, An 81%

Dhawan, R.; Arndtsen, B.A. *J. Am. Chem. Soc.* **2004**, *126*, 468.

Ph, NH_2 → H_2, 10% PD/C, 5 MeCN; MeOH, 1 d → Ph, NHEt 95%

Sajiki, H.;; Ikawa, T.; Hirota, K. *Org. Lett.* **2004**, *6*, 4977.

PhBr → $PhNH_2$, cat Pd(π-allyl)(diphosphinidine cyclobutene); toluene, rt, 12 h → PhNHPh 98%

Gajare, A.S.; Toyota, K.; Yoshifuji, M.; Ozawa, F. *J. Org. Chem.* **2004**, *69*, 6504.

NH_2 → 1. PhBr, 0.5% Pd_2dba_3, toluene, 1% *t*-Bu_2P(*o*-biphenyl), 80°C; 2 NaO*t*-Bu; 2. 2% dpe-PHOS, toluene, 105°C, (2-bromonaphthalene) Br → N, Ph 88%

Lira, R.; Wolfe, J.P. *J. Am. Chem. Soc.* **2004**, *126*, 13906.

Ph_3PH → CsOH, EtBr, DMF, MS 4Å; 23°C, 1d → Ph_2PEt 83%

Honakeer, M.T.; Sandefur, B.J.; Hargett, J.L.; McDaniel, A.L.; Salvatore, R.N. *Tetrahedron Lett.* **2003**, *44*, 8373.

PhI , 5% $RhCl(CO)[P(fur)_3]_2$, 2 TBE

Cs_2CO_3 , dioxane

N H

N Ph

78%

Sezen, B.; Sames, D. *J. Am. Chem. Soc.* ***2004**, 126*, 13244.

Br

N H

BnOH , $Me_2P=CHCN$, toluene

110°C , 15 h

Br

N CH_2Ph

80%

Bombrun, A.; Casi, G. *Tetrahedron Lett.* ***2002**, 43*, 2187.

N H

Br , Zn , 4.5 h

THF

N H 70% + N H 10%

Yadav, J.S.; Reddy, B.V.S.; Reddy, P.M. *Tetrahedron Lett.* ***2002**, 43*, 5185.

O

Ph N H O

1. I_2 , MeOH

2. Zn , MeOH

$PhCH_2NH_2$ 8%

Vatèle, J.-M. *Tetrahedron Lett.* ***2003**, 44*, 9127.

$PhNH_2$

$PhB(OH)_2$, cat $Cu(OAc)_2$

MeOH , air , reflux

PhNHPh 78%

Lan, J.-B.; Zhang, G.-L.; Yu, X.-Q.; You, J.-S.; Chen, L.; Yan, M.; Xie, R.-G. *Synlett* **2004**, 1095.

N

cat $[Cp^*IrCl_2]_2$, *i*-PrOH , reflux

$HClO_3$, 17h

N H

93%

Fujita, K.; Kitatsuji, C.; Furukawa, S.; Yamaguchi, R. *Tetrahedron Lett.* ***2004**, 45*, 3215.

Ph

H_2N OMe

1. PhCHO

2. $Me_2S=CH_2$, THF-HMPA

Ph

N OMe

Ph

89% (95:5 *S*:*R*)

Higashiyama, K.; Matsumura, M.; Shiogama, A.; Yamauchi, T.; Ohmiya, S. *Heterocycles* ***2002**, 58*, 85.

N — Bn

1. O Cl O , $CHCl_3$

2. $(BnNEt_3)_2MsSO_4$, MeCN , rt

N — H

90x97%

Bhat, R.G.; Ghosh, Y.; Chandrasekaran, S. *Tetrahedron Lett.* ***2004**, 45*, 7983.

$PhNH_2$

BuI , microwaves , 10°C , 0.1 KI

MeCN

PhNHBu + $PhNBu_2$

(88 : 12) 84%

Romera, J.L.; Cid, J.M.; Trabanco, A.A. *Tetrahedron Lett.* ***2004**, 45*, 8797.

Ph—N(Ph)—NH_2 → Ph-C≡C-Ph , cat $TiCl_4$/t-$BuNH_2$; toluene , 105°C , 2d → 1,2,3-triphenylindole 55%

Ackermann, L.; Born, R. *Tetrahedron Lett.* **2004**, *45*, 9541.

O_2N—C_6H_4—NH_2 → 1. acetone , $B_{10}H_{14}$, Pd/C; 2. $B_{10}H_{14}$ 3. HCHO , $B_{10}H_{14}$ → O_2N—C_6H_4—N($CHMe_2$)Me 93%

Jung, Y.J.; Bae, J.W.; Park, E.S.; Chang, Y.M.; Yoon, C.M. *Tetrahedron* **2003**, *59*, 10331.

NH_2 → cat [Y{N($SiMe_2H$)$_2$}$_3$ (thf)$_2$] , d_8-toluene; chiral biaryl diamine , 60°C , 14d → NH 50% ee

O'Shaughnessy, P.N.; Scott, P. *Tetrahedron: Asymmetry* **2003**, *14*, 1979.

Ph$CH_2CH_2NH_2$ → ClCH_2C(O)NEt_2 , CsOH , pH 12; DMF → PhCH_2CH_2N(H)CH_2C(O)NEt_2

Loeser, E.; Prasad, K.; Repic, O. *Synth. Commun.* **2002**, *32*, 403.

O N—H (morpholine) → 10 $(COCl)_2$, $(HCHO)_n$; 100°C (1 h) , 120°C (10 min) → O N—Me quant

Rosenau, T.; Potthast, A.; Röhrling, J.; Hofinger, A.; Sixxa, H.; Kosma, P. *Synth. Commun.* **2002**, *32*, 457.

indole (N—H) → cat $Cu(OAc)_2$, Ph_2I^+ BF_4^- , DMF; K_2CO_3 , 150°C , 7 h → N—Ph indole 52%

Zhou, T.; Chen, Z.-C. *Synth. Commun.* **2002**, *32*, 903.

indole (N—H) → Ph-epoxide , DMPU , Cs_2CO_3 → N-(1-phenylpropyl)indole 82%

DMPU = 1,3-dimethyl-3,4,5,6-tetrahydro-2(1H)pyrimimidone

Fink, D.M. *Synlett* **2004**, 2394.

Bn—N N—H → $BnCH_2CH_2CH_2Cl$, Zn , THF; rt → Bn—N N—$CH_2CH_2CH_2$Cl 92%

Murty, M.S.R.; Jyothirmai, B.; Krishna, P.R.; Yadav, J.S. *Synth. Commun.* **2003**, *33*, 2483.

isoindoline N—Bn → H_3PO_2 , AcOH , I_2; reflux , 12 h → isoindoline N—H 75%

Meng, G.; He, Y.-P.; Chen, F.-E. *Synth. Commun.* **2003**, *33*, 2593.

$O_2N-C_6H_4-NH_2$ —[$PhCHOMe)_2$, $B_{10}H_{14}$]→ $BnHN-C_6H_4-NO_2$ 96%

Park, E.S.; Lee, J.H.; Kim, S.J.; Yoon, C.M. *Synth. Commun.* **2003**, *33*, 3387.

Tr / N / C_8H_{17} C_8H_{17} —[1. Li , 10% C_8H_8 , THF; 2. H_2O]→ H / N / C_8H_{17} C_8H_{17} 61%

Behloul, C.; Guijarro, D.; Yus, M. *Synthesis* **2004**, 1274.

N-H pyrrole —[MeI , KOH , bmim PF_6; 40°C , 2 h]→ N-CH_3 pyrrole 82%

Le, Z.-G.; Chen, Z.-C.; Hu, Y.; Zheng, Q.-G. *Synthesis* **2004**, 1951.

NHPMP, HO, Ph, OTBS —[1. 2 $Im_2C{=}O$, MeCN , reflux; 2. 150°C , 0.05 mbar]→ TBSO, N, OMe, Ph 74%

Münch, A.; Wendt, B.; Christmann, M. *Synlett* **2004**, 2751.

Br, N —[5% Grubbs I , toluene; reflux]→ Br, N, H 81%

Alcaide, B.; Almendros, P.; Alonso, J.M.; Luna, A. *Synthesis* **2004**, 668.

PhCHO + $PhNH_2$ —[Bi_3SnH , SiO_2; rt , 1 h]→ $PhCH_2NHPh$ quant

Hiroi, R.; Miyoshi, N.; Wada, M. *Chem. Lett.* **2002**, *31*, 274.

ALKYLATION OF IMINES AND RELATED COMPOUNDS

Ph–N=CH–Ph —[$CH_2{=}CHOEt$, CF_3CH_2OH , 1h]→ H N, Ph, OEt 92% (9:1 *cis:trans*)

Spanedda, M.V.; Hoang, V.D.; Crousse, B.; Bonnet-Delpon, D.; Bégué, J.-P. *Tetrahedron Lett.* **2003**, *44*, 217.

Ph–CH=N–Ph —[N_2, H, CO_2Et , THF , rt; 5% Rh-chiral bis(oxazoline)]→ Ph, N, EtO_2C, Ph 73% (75:25 *cis:trans*)

Krumper, J.R.; Gerisch, M.; Suh, J.M.; Bergman, R.G.; Tilley, T.D. *J. Org. Chem.* **2003**, *68*, 9705.

Ph–N=CH–Ph → H_2 , DCM , rt / chiarl Ir catalyst, with a P-oxazoline ligand → HN(Ph)–CH(CH$_3$)Ph 98% (90% ee)

Trifonova, A.; Diesen, J.S.; Chapman, C.J.; Andersson, P.G. *Org. Lett.* ***2004**, 6*, 3825.

PhCH=N-OMe → allyl bromide , In / H_2O , rt → NHOMe 70%

Bernardi, L.; Cerè, V.; Femoni, C.; Pollicino, S.; Ricci, A. *J. Org. Chem.* ***2003**, 68*, 3348.

F_3C–CH=N–Bn → allyl bromide , DMF / Zn , cat TMSCl → F_3C, NHBn 95%

Legros, J.; Meyer, F.; Colifoeuf, M.; Crousse, B.; Bonnet-Delpon, D.; Bégué, J.-P. *J. Org. Chem.* ***2003**, 68*, 6444.

Ph–CH=N–Bn → $(CH_2{=}CHCH_2)_4Si$, 4 h / chiral bis(π-allyl)Pd cat TBAF–MeOH → NHBn, Ph 86% (91% ee)

Fernandes, R.A.; Yamamoto, Y. *J. Org. Chem.* ***2004**, 69*, 735.

Ph–CH=N–Ph → 1. $AlEt_3$, cat. $Eu(dpm)_3$ / 2. MeOH , NaOH → Et, Ph, N(H)Ph 73%

Tsvelikhovsky, D.; Gelman, D.; Molander, G.A.; Blum, J. *Org. Lett.* ***2004**, 6*, 1995.

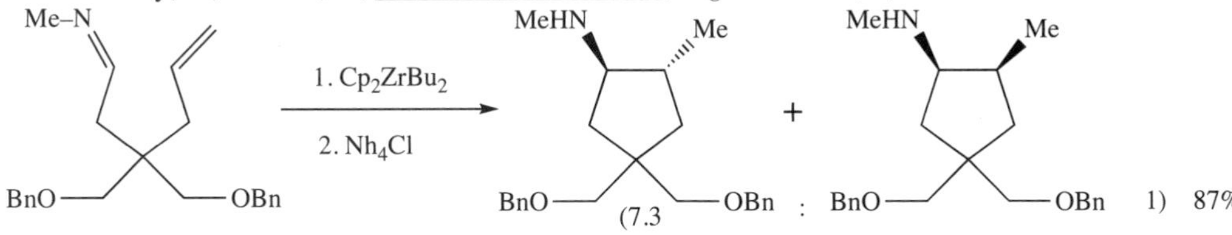

Makabe, M.; Sato, Y.; Mori, M. *J. Org. Chem.* ***2004**, 69*, 6238.

C_7H_{15}–CH=N—NHBz → allyl-$SiCl_3$, DCM , –78°C / (R)-Me-Tol-sulfone , 1 h → NHNHBz, C_7H_{15} 81% (88% ee)

Kobayashi, S.; Ogawa, C.; Konishi, H.; Sugiura, M. *J. Am. Chem. Soc.* ***2003**, 125*, 6610.

Ph–CH=N-Bn → Bu_3Sn-allyl , H_2O , 0°C / chiral Pd catalyst , 2 d → NHBn 86% (85% ee)

Fernandes, R.A.; Stimac, A.; Yamamoto, Y. *J. Am. Chem. Soc.* ***2003**, 125*, 14133.

Ph–CH=N–Ph → allyl bromide , [Al , e^-] / $InCl_3$, Bu_4NBr , THF → NHPh, Ph 86%

Hilt, G.; Smolko, K.I.; Waloch, C. *Tetrahedron Lett.* ***2002**, 43*, 1437.

PhCH=N-Ph → (allyl-$SnBu_3$, $NbCl_5$, DCM) → PhCH(NHPh)CH$_2$CH=CH$_2$ 82%

Andrade, C.K.Z.; Oliveira, G.R. *Tetrahedron Lett.* **2002**, *43*, 1935.

PhCH=N-OBn → (Bu_3SnH, BF_3•OEt_2, DCM, 20°C) → $PhCH_2NHOBn$ 95%

Ueda, M.; Miyabe, H.; Namba, M.; Nakabayashi, T.; Naito, T. *Tetrahedron Lett.* **2002**, *43*, 4369.

PhCH=N-Ph → (1. allyl-Br, Ga, ultrasound; 2. H_2O) → PhCH(NHPh)CH$_2$CH=CH$_2$ 94%

Andrews, P.C.; Peatt, A.E.; Raston, C.L. *Tetrahedron Lett.* **2004**, *45*, 243.

PhCH=N-PMP → (1. TMSCl; 2. Zn, DMI, $C_7H_{15}I$, AcOEt, rt, 18h) → PhCH(NHPMP)C_6H_{13} 70%

Iwai, T.; Ito, T.; Mizuno, T.; Ishino, Y. *Tetrahedron Lett.* **2004**, *45*, 1083.

2-Ph-2H-azirine → (amino alcohol ligand, cat [RuCl(*p*-cymene)]$_2$; *i*-PrOH, *i*-PrOK, 2h) → 2-Ph-aziridine 80% (70%ee)

Roth, P.; Andersson, P.G.; Somfai, P. *Chem. Commun.* **2002**, 1752.

PhC(Me)=N-OBn → (BH_3•oxazaborolidine, THF; 10°C, 1d) → PhCH(Me)NHOBn + PhCH(Me)NH_2 (77 : 23) 88%; 94%ee

Krezeminshi, M.P.; Zaidlewicz, M. *Tetrahedron: Asymmetry* **2003**, *14*, 1463.

PhC(Me)=N-Ph → (bis(Rh catalyst), *i*-PrOH, 70°C; PhH, 90 min) → PhCH(Me)NHPh 97%

Samec, J.S.M.; Bäckvall, J.-E. *Chem. Eur. J.* **2002**, *8*, 2955

Indol-3-yl-CH$_2$CH$_2$CH=N-Ph → (Yb(OTf)$_3$, TMSCl; DCM-THF, 5 h, rt) → 1-Ph-tetrahydro-β-carboline >99%

trace product without both reagents
also works with nitrone substrates

Tsuji, R.; Yamanaka, M.; Nishida, A.; Nakagawa, M. *Chem Lett.* **2002**, *31*, 428

cyclic imine → (H_2, S-BINAP, Ir*I)-F_4-phthalimide; mesitylene-DCM-MeOH) → NH amine 50% conversion (82% ee)

Mortimoto, T.; Suzuki, N.; Achiwa, K. *Heterocycles* **2004**, *63*, 2097.

REVIEWS:

"Nucleophilic Ring Opening of Aziridines"
Hu, X.E. *Tetrahedron* ***2004***, *60*, 2701.

"Chiral Heterocycles by Iminium Ion Cyclization"
Royer, J.; Bonin, M.; Mocouin, L. *Chem. Rev.* ***2004***, *104*, 2311.

SECTION 98: AMINES FROM ESTERS

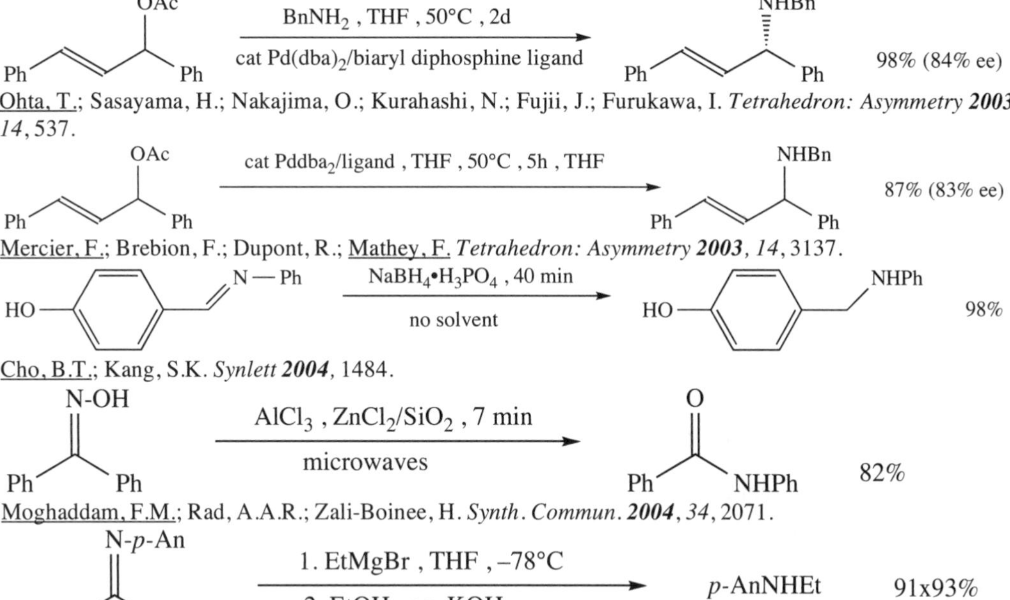

Ohta, T.; Sasayama, H.; Nakajima, O.; Kurahashi, N.; Fujii, J.; Furukawa, I. *Tetrahedron: Asymmetry* ***2003***, *14*, 537.

Mercier, F.; Brebion, F.; Dupont, R.; Mathey, F. *Tetrahedron: Asymmetry* ***2003***, *14*, 3137.

Cho, B.T.; Kang, S.K. *Synlett* ***2004***, 1484.

Moghaddam, F.M.; Rad, A.A.R.; Zali-Boinee, H. *Synth. Commun.* ***2004***, *34*, 2071.

Niwa, Y.; Takayama, K.; Shimizu, M. *Bull. Chem. Soc. Jpn.* ***2002***, *75*, 1819.

SECTION 99: AMINES FROM ETHERS, EPOXIDES, AND THIOETHERS

N Me N N Me — O Ph , $TsOH \cdot H_2O$, 3.5 d; toluene , reflux — N Ph 60%

Tsuchiya, Y.; Kumamoto, T.; Ishikawa, T. *J. Org. Chem.* ***2004***, *69*, 8504.

OTMS OEt — $PhNH_2$, AcOH; MeOH — NHPh OEt — BH_3/THF — NHPh 75%

Yoshida, Y.; Umezu, K.; Hamada, Y.; Atsumi, N.; Tabuchi, F. *Synlett* ***2003***, 2139.

LTMP , THF

75%

Hodgson, D.M.; Bray, C.D.; Kindon, N.D. *J. Am. Chem. Soc.* ***2004***, *126*, 6870.

SECTION 100: AMINES FROM HALIDES AND SULFONATES

PhBr → PhNBu$_2$ 92%

Bu_2NH , aq. KOH , cat cetylMe_3NBr

cat Pd[P(*t*-Bu)$_3$]$_2$, 20 h

Kuwano, R.; Utsunomiya, M.; Hartwig, J.F. *J. Org. Chem.* ***2002***, *67*, 6479.

NC–C$_6$H$_4$–Br → NC–C$_6$H$_4$–N(pyrrolidine) 96%

N–H , toluene , 80°C

NaO*t*-Bu , cat Pd(OAc)$_2$

cat bicylic triaminophosphine

Urgaonkar, S.; Nagarajan, M.; Verkade, J.G. *J. Org. Chem.* ***2003***, *68*, 452.

NC–C$_6$H$_4$–Cl → NC–C$_6$H$_4$–N(morpholine) O 83%

HN O , Ni/C , BuLi

dppf , 1 h

Tasler, S.; Lipshutz, B.H. *J. Org. Chem.* ***2003***, *68*, 1190.

PhI → Ph_3P 83%

Ph_2PH , cat CuI , K_2CO_3

toluene , reflux

Van Allen, D.; Venkataraman, D. *J. Org. Chem.* ***2003***, *68*, 4590.

NC–C$_6$H$_4$–Cl → NC–C$_6$H$_4$–N(morpholine) O 91%

H·N O , 1.5 *t*-BuONa

110°C , 1 d

2% $Pd_2(dba)_3$, 8% triaminophospine ligand (bicyclic)

Urgaonkar, S.; Xu, J.-H.; Verkade, J.G. *J. Org. Chem.* ***2003***, *68*, 8416.

Ph–I → Ph–NHBu 91%

$BnNH_2$, 5% CuI , 2 K_3PO_4 , *i*-PrOH

2 $HOCH_2CH_2OH$, 80°C , 8 h

Kwong, F.Y.; Klapars, A.; Buchwald, S.L. *Org. Lett.* ***2002***, *4*, 581.

Cl → NHPh quant

$PhNH_2$, 0.5% [Pd(IPr)Cl_2]$_2$

DME , KO*i*-Am , 80°C , 0.6 h

Viviu, M.S.; Kissling, R.M.; Stevens, E.D.; Nolan, S.P. *Org. Lett.* ***2002***, *4*, 2229.

I → PPh$_2$ 86%

Ph_2PH , 5% Pd/C , 180°C , 5 min

Stadler, A.; Kappe, C.O. *Org. Lett.* ***2002***, *4*, 3541.

Bu_2NH, $LiN(TMS)_2$, THF
65°C, cat $Pd_2(dba)_3$
cat biarylamino-phosphine
HO Cl → HO NBu_2 64%

Harris, M.C.; Huang, X.; Buchwald, S.L. *Org. Lett.* ***2002***, *4*, 2885.

RINK NH_2
1. Br–C₆H₄–NO_2, dioxane/*t*-BuOH
10% Pd_2DBA_2, 3% BINAP
10 NaO*t*-Bu, 80°C, 2 h
2. 5% TFA, DCM
→ H_2N–C₆H₄–NO_2 57%

Weigand, K.; Pelka, S. *J. Org. Chem.* ***2002***, *67*, 4689.

Br
$H_2NC_6H_{13}$, 5% CuI, 20% L
2 K_3PO_4, DMF, Ar, 90°C
L = diethyl salicylamide
→ NHC_6H_{13} 91%

Kwong, F.Y.; Buchwald, S.L. *Org. Lett.* ***2003***, *5*, 793.

Cl
$H_2NC_6H_{11}$, toluene
1.5 NaO*t*-Bu, 5% $Pd(OAc)_2$
10% $P[N(i\text{-}Bu)CCH_2CH_3]_3N$
→ NHC_6H_{11} 91%

Urgaonkar, S.; Nagarajan, M.; Verkade, J.G. *Org. Lett.* ***2003***, *5*, 815.

Br
$BnNH_2$, microwaves, toluene
cat $Pd_2(dba)_3$, PPFA, NaO*t*-Bu
→ NHBn 80%

Wang, T.; Magnin, D.R.; Hamann, L.G. *Org. Lett.* ***2003***, *5*, 897.

Cl
HN O, chiral Pd catalyst
toluene, 80°C, 2 h
→ N O 97%

Zim, D.; Buchwald, S.L. *Org. Lett.* ***2003***, *5*, 2413.

PhI
$C_6H_{13}NH_2$, 10% CuI, K_2CO_3, 14 h
20% L-proline, DMSO, 60°C
→ Ph-NHC_6H_{13} 87%

Ma, D.; Cai, Q.; Zhang, H. *Org. Lett.* ***2003***, *5*, 2453.

Ph-Br
$PhNH_2$, KO*t*-Bu, DMSO, 10 min
microwaves
→ Ph-NHPh 61%

Shi, L.; Wang, M.; Fan, C.-A.; Zhang, F.-M.; Tu, Y.-Q. *Org. Lett.* ***2003***, *5*, 3515.

N F
N - BH_3Li
THF, 65°C, 2h
→ N N 81%

Thomas, S.; Roberts, S.; Pasumaansky, L.; Gamsey, S.; Singaram, B. *Org. Lett.* ***2003***, *5*, 3867.

Ph-I → (2 BNH_2 , CuI , 2.5 CsOAc; DMF , 90°C) → Ph-NHBu 93%

Okano, K.; Tokuyama, H.; Fukuyama, T. *Org. Lett.* **2003**, *5*, 4987.

Piperidine N—H → (PhOTf , NMP , reflux; microwaves , 204°C , 45 min) → N-phenylpiperidine 83%

Xu, G.; Wang, Y.-G. *Org. Lett.* **2004**, *6*, 985.

EtO_2C–C_6H_4–MgCl → (1. Br–C_6H_4–NO_2 , THF , –20°C; 2. $FeCl_3$, $NaBH_4$, –20°C → rt) → EtO_2C–C_6H_4–NH–C_6H_4–Br 73%

Sapountzis, I.; Knochel, P. *J. Am. Chem. Soc.* **2002**, *124*, 9390.

3-O_2N–C_6H_4–Cl → ($PhNH_2$, KOH/H_2O , 110°C; cat $Pd_2(dba)_3$, 2% phosphine ligand) → 3-O_2N–C_6H_4–NHPh 93%

Huang, X.; Anderson, K.W.; Zim, D.; Jiang, L.; Klepars, A.; Buchwald, S.L. *J. Am. Chem. Soc.* **2003**, *125*, 6653.

4-Acetyl–C_6H_4–Br → (2.3 Ph_3 , 10% $Pd(OAc)_2$, 115°C; neat , 1.5d) → 4-Acetyl–C_6H_4–PPh_2 44%

Kwong, F.Y.; Lai, C.W.; Chan, K.S. *Tetrahedron Lett.* **2002**, *43*, 3537.

PhBr → (1. Me_2C=C=$NOSO_2Mes$, Mg , toluene ether , 75°C, 23h; 2. 6% HCl , 1d) → $PhNH_2$ 59%

Erdik, E.; Daskapan, T. *Tetrahedron Lett.* **2002**, *43*, 6237.

PhI → ($PhNH_2$, CuI , toluene , KO*t*-Bu) → Ph_3N 70% + Ph_2NH 7%

Kelkar, AA.; Patil, N.M.; Chaudhari, R.V. *Tetrahedron Lett.* **2002**, *43*, 7143.

2-Bromopyridine → (morpholine , alumina-KF , 100°C; cat Pd-phosphine , toluene) → 2-morpholinopyridine 70%

Basu, B.; Jha, S.; Mridha, N.K.; Bhuiyan, Md.M.H. *Tetrahedron Lett.* **2002**, *43*, 7967.

I–C_6H_4–I → (piperidine , 10% CuI , $Me_2NCH_2CH_2OH$; 2 K_3PO_4•H_2O , 60°C , 40 min) → I–C_6H_4–N(pyrrolidine) 78%

Lu, Z.; Twieg, R.J.; Huang, S.D. *Tetrahedron Lett.* **2003**, *44*, 6289.

O_2N–C_6H_4–Br → (piperidine , (bmim) PF_6; rt , 8h) → O_2N–C_6H_4–N(piperidine) 90%

Yadav, J.S.; Reddy, B.V.S.; Basak, A.K.; Narsaiah, A.V. *Tetrahedron Lett.* **2003**, *44*, 2217.

Ph–CH₂CH₂–I → (7M NH_3, MeOH, 130°C; microwaves) → Ph–CH₂CH₂–NH_2 93%

Saulnier, M.G.; Zimmermann, K.; Struzynski, C.P.; Sang, X.; Velaparthi, U.; Wittman, M.; Frennesson, D.B. *Tetrahedron Lett.* ***2004**, 45*, 397.

MeO–C₆H₄–Br → (piperidine, toluene, NaO*t*-Bu, 1d; cat $Pd(OAc)_2$, 110°C; phosphonyl-imidazole ligand) → MeO–C₆H₄–N(morpholine) 96%

Singer, R.A.; Tom, N.J.; Frost, H.N.; Simon, W.M. *Tetrahedron Lett.* ***2004**, 45*, 4715.

PhCl + HN(piperazine)NH → (cat $Ni(OAc)_2$, bipy, *t*-AmOH; NaH, THF, $PhCH{=}CH_2$, 0.8h) → Ph–N(piperazine)NH 61%

Brenner, E.; Schneider, R.; Fort, Y. *Tetrahedron* ***2002**, 58*, 6913.

Br, Ph, O, N, H, CO_2Bn → (Bn_2NH, NEt_3, TBAI, DCM; rt) → Bn_2N, O, Ph, N, H, CO_2Bn 95% (93:7 *R:S*)

Nam, J.; Chang, J.-y.; Shin, E.-k.; Kim, H.J.; Kim, Y.; Jang, S.; Park, Y.S. *Tetrahedron* ***2004**, 60*, 6311.

PhCl → (morpholine, NaO*t*-Bu, 150°C, toluene, 10 min; cat $Pd(OAc)_2$, chiral amine) → Ph–N(morpholine)O 91%

Maes, B.U.W.; Loones, K.T.J.; Hustyn, S.; Diels, G.; Rombouts, G. *Tetrahedron* ***2004**, 60*, 11559.

MeO_2C–C₆H₄–Br → (Me_3SiNMe_2, Cs_2CO_3, toluene, 60°C; cat $Pd(OAc)_2$, $P(t\text{-}Bu)_2(o\text{-biphen})$) → MeO_2C–C₆H₄–NMe_2 78%

Smith, C.J.; Farly, T.R.; Holmes, A.B.; Shute, R.E. *Chem. Commun.* ***2004**,* 1976.

NC–C₆H₄–I → (H_2N–C₆H₄–EtO_2C, cat $Pd(OAc)_2$/BINAP; Cs_2CO_3, toluene, 1h) → NC–C₆H₄–NH–C₆H₄–EtO_2C 95%

Maes, B.U.W.; Loones, K.T.J.; Jonckers, T.H.M.; Lemière, G.L.F.; Dommisse, R.A.; Haemers, A. *Synlett* ***2002**,* 1995.

3-chloropyridine (N, Cl) → (PhNHMe, NaO*t*-Bu, toluene, 10 min; 2% $Pd(OAc)_2/PhC_6H_4PCy_2$, 200°C; microwaves) → 3-pyridyl–N(Me)Ph 84%

Maes, B.U.W.; Loones, K.T.J.; Lemière, G.L.F.; Dommisse, R.A. *Synlett* ***2003**,* 1822.

Cl–C₆H₃(Cl)–NO_2 → ($BnNH_2$, DMSO, NEt_3; microwaves, 8 min) → Cl–C₆H₃(NHBn)–NO_2 quant

Li, W.; Yun, L.; Wang, H. *Synth. Commun.* ***2002**, 32*, 2657.

H–N(morpholine)O → (PhI, $CsOH{\bullet}H_2O$, DMSO, 120°C) → Ph–N(morpholine)O 95%

Varala, R.; Ramu, E.; Alam, M.M.; Adapa, S.R. *Synlett* ***2004**,* 1747.

BnNH$_2$, BINAP , DMF , 4 min

microwaves , 130°C , t-BuOK

Br

NHBn

80%

Wan, Y.; Alterman, M.; Hallberg, A. *Synthesis* **2002**, 1597.

S

NH_2

Br

LDA , THF , –78°C → rt

5 h

S

N
H

89%

Mukherjee, C.; Biehl, E. *Heterocycles* **2004**, *63*, 2309.

SECTION 101: AMINES FROM HYDRIDES

OMe

O_2N NO_2

1. tetramethylhydrazonium iodide
2. NH_3

OMe

NH_2

O_2N NO_2

+

OMe

O_2N NO_2

NH_2

(90 : 10) 90%

Rozhkov, V.V.; Shevelev, S.A.; Chervin, I.I.; Mitchell, A.R.; Schmidt, R.D. *J. Org. Chem.* **2003**, *68*, 2498.

PhI=NTs , Tp^{Br3}Cu(MeCN) , rt

NHTs

65%

Díaz-Requejo, M.M.; Belderraín, T.R.; Nicasio, M.C.; Trofimenko, S.; Pérez, P.J. *J. Am. Chem. Soc.* **2003**, *125*, 12078.

Ph

PhNO$_2$, [Cp$_2$Fe(CO)$_2$]$_2$, CO

Ph

PhHN

42%

Srivstava, R.S.; Kolel-Veetil, M.; Nicholas, K.M. *Tetrahedron Lett.* **2002**, *43*, 931.

SECTION 102: AMINES FROM KETONES

O

Ph

HCO$_2$NH$_4$, MeOH , 2 h

cat [Cp*RhCl$_2$]$_2$

NH_2

Ph

97%

Kitamura, M.; Lee, D.; Hayashi, S.; Tanaka, S.; Yoshimura, M. *J. Org. Chem.* **2002**, *67*, 8685.

O

Ph

1.2 eq p-anisidine , 1000 psi H_2
Ir-chiral ligand , cat I_2

1.5 eq Ti(Oi-Pr)$_4$

OMe

HN

Ph

99% (98% ee)

Chi, Y.; Zhou, Y.-G.; Zhang, X. *J. Org. Chem.* **2003**, *68*, 4120.

$PhNH_2$, Montmorillonite KSF; 10 h — 96%

O O N Ph

Banik, B.K.; Samajdar, S.; Banik, I. *J. Org. Chem.* **2004**, *69*, 213.

O NHBoc

1. HCO_2H, 16 h
2. NEt_3
3. 0.25% [(*p*-cymene)$RuCl]_2$
0.5% (*R,R*)-TsDPED, 20h
HCO_2H, MeCN

HN — 98%

Williams, G.D.; Pike, R.A.; Wade, C.E.; Wills, M. *Org. Lett.* **2003**, *5*, 4227.

EtO_2C Cl NO_2

1. acetophenone, K_3PO_4, 50°C
cat $Pd_2(dba)_3$, 20% 4-methoxyphenol
2. NaH, MeI, THF

Ph EtO_2C N H — 61%

Rutherford, J.; Rainka, M.P.; Buchwald, S.L. *J. Am. Chem. Soc.* **2002**, *124*, 15168.

NH_2 Br

1. acetophenone
2. Bu_3SnH, AIBN, PhH, 80°C

N Ph — 86%

Viswanathan, R.; Prabhakaran, E.N.; Plotkin, M.A.; Johnston, J.N. *J. Am. Chem. Soc.* **2003**, *125*, 163.

O Ph N_3

2.5 eq SmI_2, THF, rt, 5 min

Ph Ph N H — 72%

Fan, X.; Zhang, Y. *Tetrahedron Lett.* **2002**, *43*, 1863.

NH_2

acetone, NaOH, $CHCl_3$, 15°C
PTC, overnight

N — 77%

Lai, J.T. *Tetrahedron Lett.* **2002**, *43*, 1965.

O O

$C_3H_7NH_2$, (bmim) I, 30 min

N C_3H_7 — 96%

Wang, B.; Gu, Y.; Luo, C.; Yang, T.; Yang, L.; Suo, J. *Tetrahedron Lett.* **2004**, *45*, 3417.

1. NH_3 , Ti(O*i*-Pr)$_4$, EtOH , 25°C , 6h
2. $NaBH_4$, 25°C , 3h

88%

Miriyala, B.; Bhattacharyya, S.; Williamson, J.S. *Tetrahedron* **2004**, *60*, 1463.

, PhH , rt
cat Sc(OTf)$_3$, MS 5Å , 1d

75%

Itoh, T.; Nagata, K.; Miyazaki, M.; Ishikawa, H.; Kurihara, A.; Ohsawa, A. *Tetrahedron* **2004**, *60*, 6649.

, AIBN , reflux
heptane/PhCl , 15 min

54%

Quiclet-Sire, B.; Wendeborn, F.; Zard, S.Z. *Chem. Commun.* **2002**, 2214.

1. $BnNH_2$, $BnNH_2$•BH_3 , THF MS 4Å , rt
2. NaOMe , MeOH , reflux

80%

Peterson, M.A.; Bowman, A.; Morgan, S. *Synth. Commun.* **2002**, *32*, 443.

$TsNH_2$, 0.5 Zr(OTf)$_4$, DCE
60°C , 2 d

80%

Shi, M.; Yang, Y.-H.; Xu, B. *Synlett* **2004**, 1622.

$TiCl_4$-Zn , THF

72%

Shi, D.; Shi, C.; Wang, X.; Zhuang, Q.; Tu, S.; Hu, H. *Synlett* **2004**, 2239.

MeI , KOH , NH_2OH•HCl

85%

Li, C.; Zhang, H.; Cui, Y.; Zhang, S.; Zhao, Z.; Choi, M.C.K.; Chan, A.S.C. *Synth. Commun.* **2003**, *33*, 543.

REVIEW:

"Homogeneous Rhodium(I)-Catalyzed Reductive Aminations"
Riermeier, T.H.; Dingerdissen, U.; Börner, A. *Org. Prep. Proceed. Int.* **2004**, *36*, 99.

Related Method: Section 94 (Amines from Aldehydes)

SECTION 103: AMINES FROM NITRILES

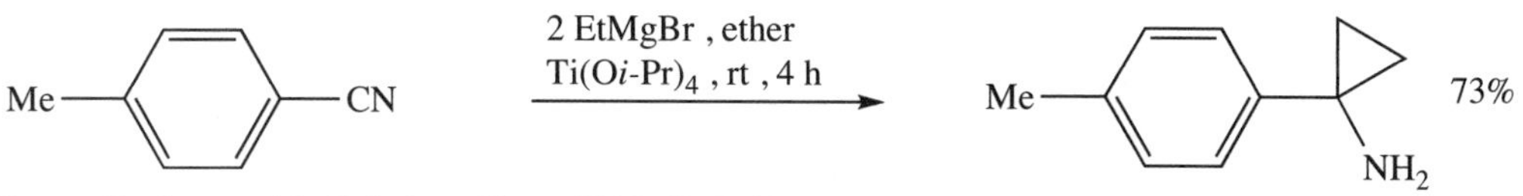

Me—C6H4—CN → (2 EtMgBr, ether; Ti(O*i*-Pr)$_4$, rt, 4 h) → Me—C6H4—C(cyclopropyl)NH$_2$ 73%

Bertus, P.; Szymoniak, J. *J. Org. Chem.* **2003**, *68*, 7133.

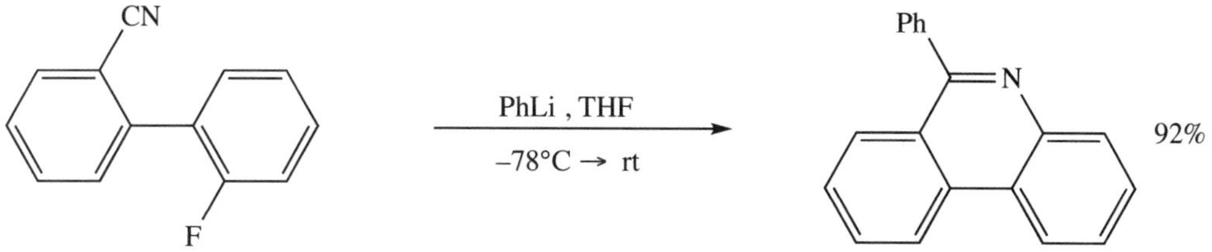

CN, F, PhLi, THF, –78°C → rt, Ph, N, 92%

Lysén, M.; Kristensen, J.L.; Vedsø, P.; Begtrup, M. *Org. Lett.* **2002**, *4*, 257.

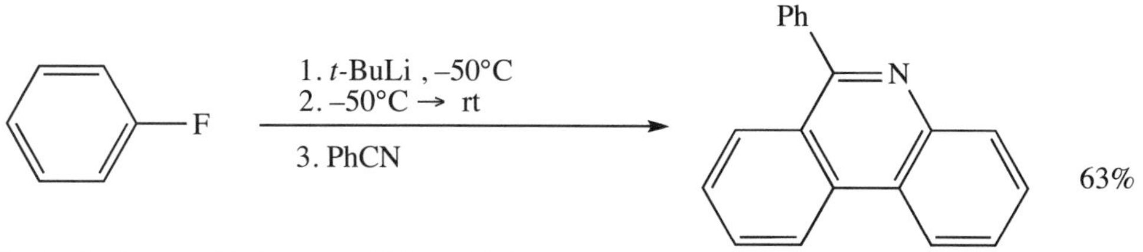

F, 1. *t*-BuLi, –50°C; 2. –50°C → rt; 3. PhCN, Ph, N, 63%

Pawla, J.; Begtrup, M. *Org. Lett.* **2002**, *4*, 2687.

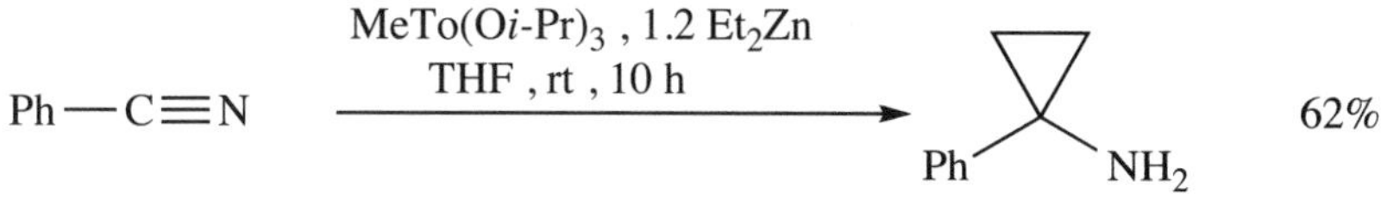

Ph—C≡N → (MeTo(O*i*-Pr)$_3$, 1.2 Et$_2$Zn; THF, rt, 10 h) → Ph, NH$_2$ 62%

Wiedemann, S.; Frank, D.; Winsel, H.; de Meijere, A. *Org. Lett.* **2003**, *5*, 753.

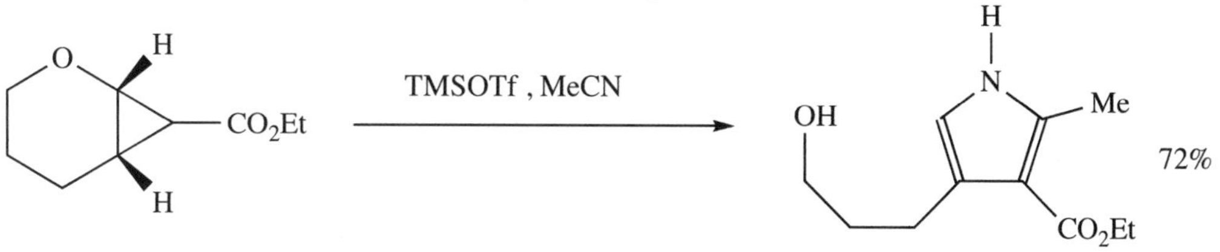

O, H, CO$_2$Et, H → (TMSOTf, MeCN) → OH, H, N, Me, CO$_2$Et 72%

Yu, M.; Pagenkopf, B.L. *Org. Lett.* **2003**, *5*, 5099.

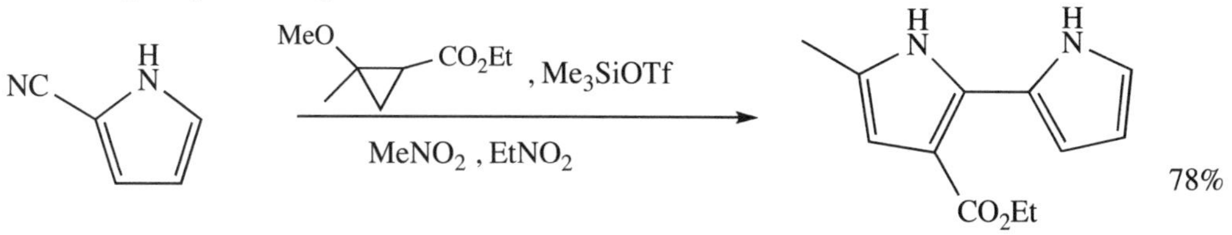

NC, H, N + MeO, CO$_2$Et, Me$_3$SiOTf; MeNO$_2$, EtNO$_2$ → H, N, H, N, CO$_2$Et 78%

Yu, M.; Pantos, G.D.; Sessler, J.L.; Pagenkopf, B.L. *Org. Lett.* **2004**, *6*, 1057.

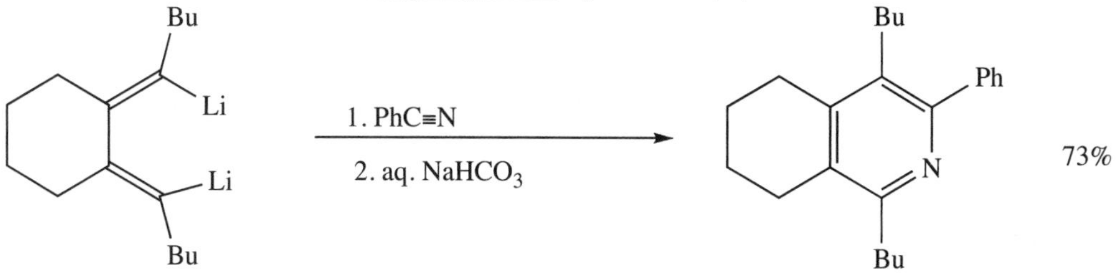

Bu, Li, Li, Bu → (1. PhC≡N; 2. aq. NaHCO$_3$) → Bu, Ph, N, Bu 73%

Chen, J.; Song, Q.; Wang, C.; Xi, Z. *J. Am. Chem. Soc.* **2002**, *124*, 6228.

NHBoc

1. t-BuLi , ether , –78°C
2. PhCN , –25°C
3. 12M HCl , EtOAc

H N Ph t-Bu 65%

Coleman, C.M.; O'Shea, D.F. *J. Am. Chem. Soc.* **2003**, *125*, 4054.

PhCN —$NiCl_2$, $NaBH_4$, EtOH , rt→ $PhCH_2NH_2$ 82%

Khurana, J.M.; Kukreja, G. *Synth. Commun.* **2002**, *32*, 1265.

BuCN —Ni(R) , i-PrOH / 20% KOH , reflux→ $BuCH_2N{=}CMe_2$ —1. 10% HCl 2. KOH / 3. HCl→ $BuCH_2NH_3^+ Cl^-$ 93%

Mebane, R.C.; Jensen, D.R.; Rickerd, K.R.; Gross, B.H. *Synth. Commun.* **2003**, *33*, 3373.

Ph N Ph CN —SmI_2 , THF , 18 h→ N Ph Ph Ph N Ph 80%

Thakur, A.J.; Prajapati, D.; Sandhu, J.S. *Chem. Lett.* **2004**, *33*, 102.

SECTION 104: AMINES FROM ALKENES

O CO_2R^* —$NsONHCO_2Et$, CuO / DCM→ O CO_2R^* N CO_2Et 91% (99% de)

R* = (–)-8-phenylmenthol

Fioravanti, S.; Morreale, A.; Pellacani, L.; Tardella, P.A. *J. Org. Chem.* **2002**, *67*, 4972.

Ph —5% $Ni[P(OEt)_3]_4$, 0.5 Ph_2PH , NEt_3 / 130°C , 20 h→ Ph PPh_2 >99%

Shulyupin, M.O.; Kazankova, M.A.; Beletskaya, I.P. *Org. Lett.* **2002**, *4*, 761.

Ph —TosMIC , NaOt-Bu , DMSO / 50°C , 18 h→ Ph N—H 47%

Smith, N.D.; Huang, D.; Cosford, B.D.P. *Org. Lett.* **2002**, *4*, 3537.

—$PhNH_2$, $TiCl_4$ / toluene , 169°C→ MHPh + H_2N (65 : 35) 48%

Ackermann, L.; Kaspar, L.T.; Gschrei, C.J. *Org. Lett.* **2004**, *6*, 2575.

e⁻, MeCN, $NHEt_3OAc$. rt; N-aminophthalimide — 85%

Siu, T.; Yudin, A.K. *J. Am. Chem. Soc.* ***2002***, *124*, 530.

morpholine, 5% [Rh(cod)(DPEphos)] BF_4; toluene, 70°C, 2 d — 71% (75:25 amine:enamine)

Utsunomiya, M.; Kuwano, R.; Kawatsura, M.; Hartwig, J.F. *J. Am. Chem. Soc.* ***2003***, *125*, 5608.

piperidine, MeOH/toluene, 7 bar CO, 5 h; 33 bar H_2, cat $Rh(cod)_2\ BF_4$/Xantphos, 125°C — >99%

Ahmed, M.; Seayad, A.M.; Jackstell, R.; Beller, M. *J. Am. Chem. Soc.* ***2003***, *125*, 10311.

morphiline, 20% CF_3SO_3H, dioxane; 5% $Pd(O_2CCF_3)_2$, 10-% dpf, 120°C, 1 d — 75%

Utsunomiya, M.; Hartwig, J.F. *J. Am. Chem. Soc.* ***2003***, *125*, 14286.

H-N(morpholine), dioxane, 100°C, 1 d; $Ru(cod)(Meallyl)_2/dppp_{ent}$; 10% TOH — 96%

Utsonomiya, M.; Hartwig, J.F. *J. Am. Chem. Soc.* ***2004***, *126*, 2702.

PhI, Et_2NH, 10% dppe, THF; 5% $Pd(OAc)_2$, 40°C, 3.5h — 42%

Onitsuka, K.; Suzuki, S.; Takahashi, S. *Tetrahedron Lett.* ***2002***, *43*, 6197.

piperidine, THF, CO, H_2, 12h; cat Rh(IMes)(cod)Cl, 95°C — (62 : 38) 99%

Seayad, A.M.; Selvakumar, K.; Ahmed, M.; Beller, M. *Tetrahedron Lett.* ***2003***, *44*, 1679.

1. 2.2 eq. EtMgCl, $Ti(Oi\text{-}Pr)_4$, ether; 2. $BF_3{\bullet}OEt_2$ — 69%

Laroche, C.; Bertus, P.; Szymoniak, J. *Tetrahedron Lett.* ***2003***, *44*, 2485.

SECTION 105: AMINES FROM MISCELLANEOUS COMPOUNDS

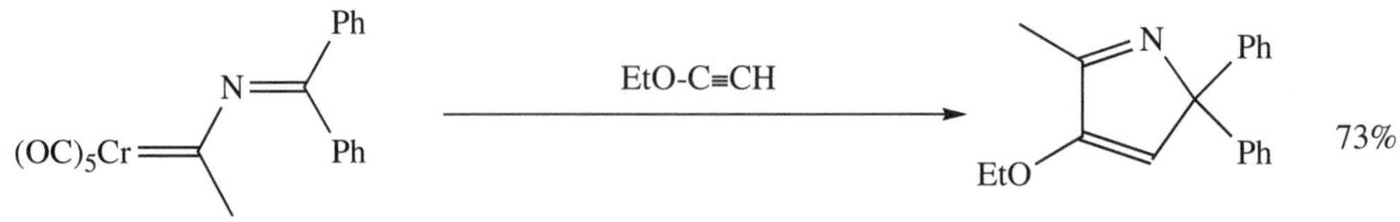
Ph
N
$(OC)_5Cr$
Ph
EtO-C≡CH
N
Ph
Ph
EtO
73%

Campos, P.J.; Sampedro, D.; Rodríguez, M.A. *J. Org. Chem.* ***2003***, *68*, 4674.

$PhBF_3^{-}\,^{+}K$ —(1. $Cu(OAc)_2$, 10% H_2O , DCM , MS 4Å; 2. $BuNH_2$, O_2 , 1d , heat)→ PhNHBu 89% (92% from $PhB(OH)_2$)

Quach, T.D.; Batey, R.A. *Org. Lett.* ***2003***, *5*, 4397.

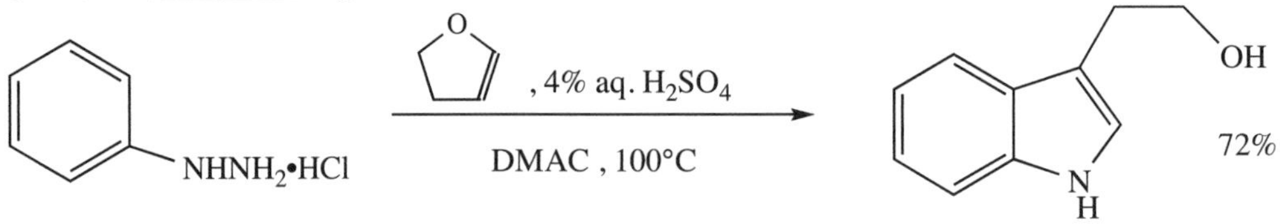
O
, 4% aq. H_2SO_4
$NHNH_2$•HCl
DMAC , 100°C
OH
N
H
72%

Campos, K.R.; Woo, J.C.S.; Lee, S.; Tillyer, R.D. *Org. Lett.* ***2004***, *6*, 79.

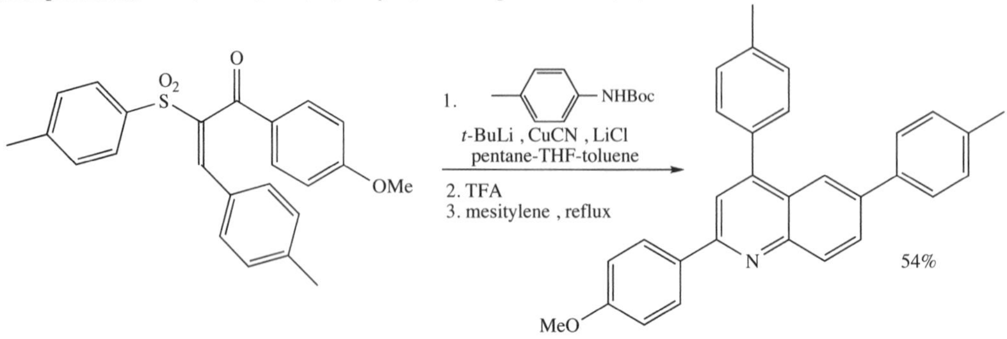
O
O_2
S
OMe
1.
NHBoc
t-BuLi , CuCN , LiCl
pentane-THF-toluene
2. TFA
3. mesitylene , reflux
N
MeO
54%

Swenson R.E.; Sowin, T.J.; Zhang, H.Q. *J. Org. Chem.* ***2002***, *67*, 9182.

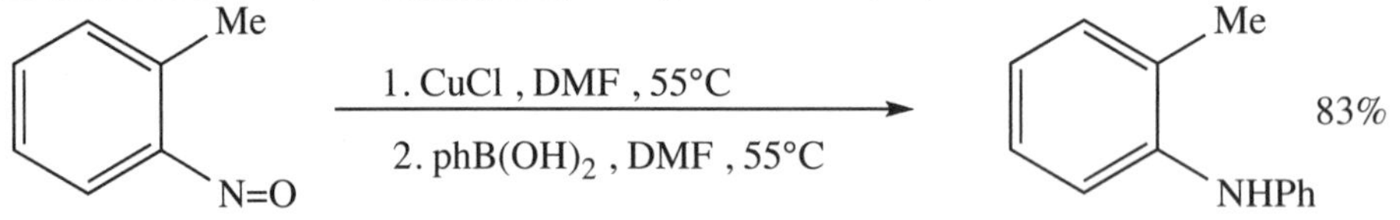
Me
N=O
1. CuCl , DMF , 55°C
2. $phB(OH)_2$, DMF , 55°C
Me
NHPh
83%

Yu, Y.; Srogl, J.; Liebeskind, L.S *Org. Lett.* ***2004***, *6*, 2631.

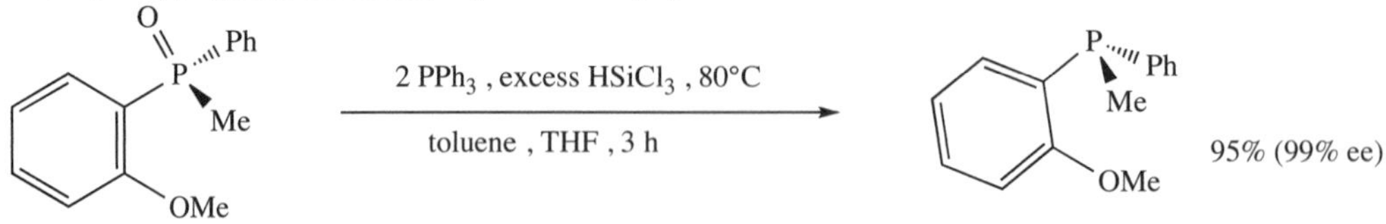
O
P
Ph
Me
OMe
2 PPh_3 , excess $HSiCl_3$, 80°C
toluene , THF , 3 h
P
Ph
Me
OMe
95% (99% ee)

Wu, H.-C.; Yu, J.-Q.; Spencer, J.B. *Org. Lett.* ***2004***, *6*, 4675.

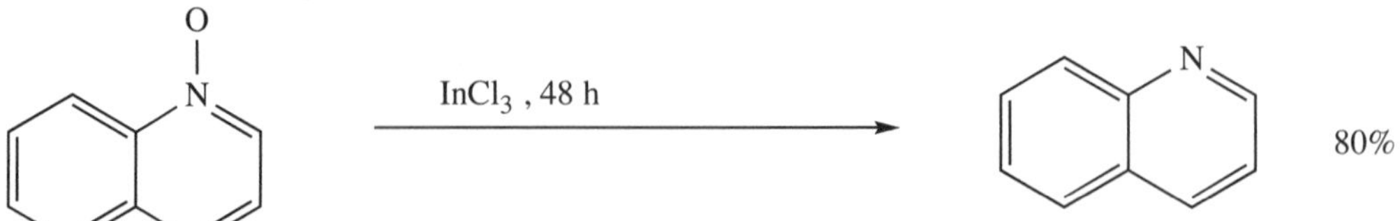
O
N
$InCl_3$, 48 h
N
80%

Ilias, Md.; Barman, D.C.; Prajapati, D.; Sandhu, J.S. *Tetrahedron Lett.* ***2002***, *43*, 1877.

Cl–C$_6$H$_4$–N_3 → ($FeCl_3$, NaI , MeCN , rt , 10 min) → Cl–C$_6$H$_4$–NH_2 95%

Kamal, A.; Ramana, K.V.; Ankati, H.B.; Ramana, Q.V. *Tetrahedron Lett.* ***2002**, 43*, 6861.

Ph-N=O → (1. PhMgCl , THF , –20°C; 2. $NaBH_4$, $FeCl_2$) → PhNHPh 68%

Kopp, F.; Sapountzis, I.; Knochel, P. *Synlett* ***2003***, 885.

PhCH=CH$_2$ → (TBAF , $Ph_2P(Me_2t$-Bu) , DMF , rt, 15 min) → PhCH$_2$CH$_2$Ph 89%

Hayashi, M.; Matsuura, Y.; Watanabe, Y. *Tetrahedron Lett.* ***2004**, 45*, 9167.

Cl–C$_6$H$_4$–N_3 → (Sm , $NiCl_2$•6 H_2O , THF, 40°C , 2.5 h) → Cl–C$_6$H$_4$–NH_2 85%

Wu, H.; Chen, R.; Zhang, Y. *Synth. Commun.* ***2002**, 32*, 189.

4-*t*-Bu-pyridine N-oxide → (HCO_2^- NH_4^+ , Ni(R), 50°C , 4 h) → 4-*t*-Bu-pyridine 82%

Balicki, R.; Maciejewski, G. *Synth. Commun.* ***2002**, 32*, 1681.

Pyridine N-oxide → (PMHS - Pd/C , EtOH , rt; PMHS = polymethylhydrosiloxane) → pyridine 90%

Chandrasekhar, S.; Reddy, Ch.R.; Rao, R.J.; Rao, J.M. *Synlett* ***2002***, 349.

n-C$_8$H$_{17}$–N_3 → ($BHCl_2$•SMe_2 , DCM) → n-C$_8$H$_{17}$–NH_2 95%

Salunkhe, A.M.; Ramachandran, P.V.; Brown, H.C. *Tetrahedron* ***2002**, 58*, 10059.

HC≡C–C(=O)Ph → (microwaves , PhC(=NH)NH_2, Na_2CO_3 , 400 min) → 2,4-diphenylpyrimidine 98%

Bagley, M.C.; Hughes, D.D.; Taylor, P.H. *Synlett* ***2003***, 259.

PhCH$_2$CH(NH$_2$)CO$_2$H → (1. NBS , pH 5 , DMF; 2. $NiCl_2$•6 H_2O , $NaBH_4$, rt) → PhCH$_2$CH$_2$NH$_2$ 76%

Laval, G.; Golding, B.T. *Synlett* ***2003**,* 542.

4-*t*-Bu-pyridine N-oxide → (Zn/HCO_2NH_4, MeOH , 4 h) → 4-*t*-Bu-pyridine 81%

Balicki, R.; Cybulski, M.; Maciejewski, G. *Synth. Commun.* ***2003**, 33*, 4137.

4-*t*-Bu-pyridine N-oxide → ($TiCl_4$, In , THF , rt) → 4-*t*-Bu-pyridine 95%

Yoo, B.W.; Choi, K.H.; Choi, K.I.; Kim, J.H. *Synth. Commun.* ***2003**, 33*, 4105.

Ph-N=N-Ph —— Zn , HCO_2NH_4 , MeOH / rt , 3 min ——> 2 $PhNH_2$ 95%

Gowda, S.; Abiraj, K.; Gowda, D.C. *Tetrahedron Lett.* **2002**, *43*, 1329.

$PhCH_2N_3$ —— Zn , NH_4Cl , EtOH , H_2O / reflux , 10 min ——> $PhCH_2NH_2$ 87%

Lin, W.; Zhang, X.; He, Z.; Jin, Y.; Gong, L.; Mi, A. *Synth. Commun.* **2002**, *32*, 3279.

Ph-N=N-Ph —— $NH_2NH_2{\bullet}H_2O$, EtOH ——> Ph-NH-NH-Ph 92%

Zhang, C.-R.; Wang, Y.-L. *Synth. Commun.* **2003**, *33*, 4205.

PhCH=N-OH —— Mg/HCO_2NH_4 , MeOH , rt ——> $PhCH_2NH_2$ 91%

Abiraj, K.; Gowda, D.C. *Synth. Commun.* **2004**, *34*, 599.

Quinoline N-oxide —— Ga , H_2O , reflux , 4 h ——> quinoline 83%

Han, J.H.; Choi, K.I.; Kim, J.H.; Yoo, B.W. *Synth. Commun.* **2004**, *34*, 3197.

Ph_2C=N-OH —— NH_2NH_2 , H_2O , EtOH , reflux ——> Ph_2C=N-NH_2 95%

Pasha, M.A.; Nanjundaswamy, H.M. *Synth. Commun.* **2004**, *34*, 3827.

N-OAc oxime ester (Ph…CN) —— cat 1,5-naphthalenediol , AcOH / 1,4-cyclohexadiene , 1,4-dioxane reflux ——> pyrroline (Ph, CN) 69%

Yoshida, M.; Kitamura, M.; Narasaka, K. *Chem. Lett.* **2002**, 144.

1,3-dimethylimidazolidin-2-ylidene =N-Ph —— CsOH , H_2O , 150°C , 1 h / ethylene glycol , HCl/ether ——> $PhNH_2$ 95%

Kitamura, M.; Chiba, S.; Narasaka, K. *Bull. Chem. Soc. Jpn.* **2003**, *76*, 1063.

AMINES FROM NITRO COMPOUNDS

$PhNO_2$ —— $MeCO_2$ NH_4 , Pb , MeOH , rt ——> Ph-N=N-Ph 91%

Srinivasa, G.R.; Abiraj, K.; Gowda, D.C. *Synth. Commun.* **2003**, *33*, 4221.

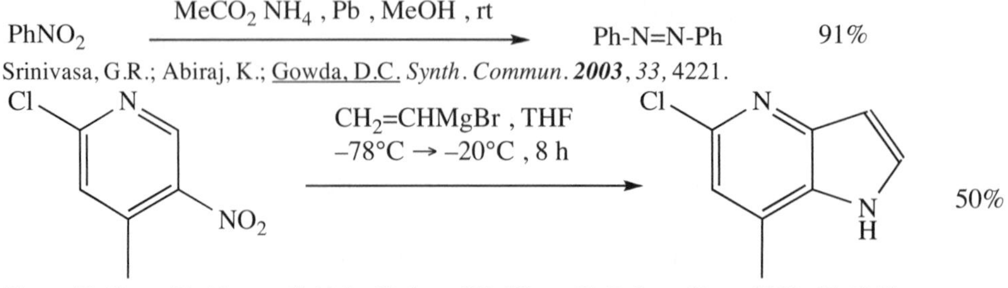

Zhang, Z.; Yang, Z.; Meanwell, N.A.; Kadow, J.F.; Wang, T. *J. Org. Chem.* **2002**, *67*, 2345.

BnHN–CH_2CH_2–CH(OP(=O)(OPh)$_2$)–C(CH_3)$_2$$NO_2$ —[Bu_3SnH , AIBN , PhH , 80°C]→ N-Bn 2-isopropylpyrrolidine 43% (61% ee)

Crich, D.; Shirai, M.; Rumthao, S. *Org. Lett.* ***2003***, *5*, 3767.

AcHN (Ph, NO_2 arene) —[2% $Pd(OAc)_2$, 4% phen , 140°C; 70 psi CO , DMF , 16h]→ AcHN carbazole 96%

Smitrovich, J.H.; Davies, I.W. *Org. Lett.* ***2004***, *6*, 533.

$PhNO_2$ —[H_2 , cat $FeSO_4$•7 H_2O , H_2O; 150°C , 2.4 h]→ $PhNH_2$ 99%

Deshpande, R.M.; Mahajan, A.N.; Diwakaar, M.M.; Ozarde, S.; Chaudhari, R.V. *J. Org. Chem.* ***2004***, *69*, 4835.

$Ph_2P(=O)O$–CH_2–C(CH_3)(NO_2)–$CH_2CH_2CH_2$NHBn —[Ph_3SnH , AIBN , PhH , reflux]→ N-Bn 3-methylpiperidine 90%

Crich, D.; Ranganathan, K.; Neelamkavil, S.; Huang, X. *J. Am. Chem. Soc.* ***2003***, *125*, 7942.

Resin–O–CH_2–C_6H_4–NO_2 —[Al/$NiCl_2$•6 H_2O; 80 min]→ HO–CH_2–C_6H_4–NH_2 85%

Kamal, A.; Reddy, K.L.; Deviah, V.; Reddy, G.S.K. *Tetrahedron Lett.* ***2003***, *44*, 4741.

3-nitrophenol (OH, NO_2) —[Ni(R)/N_2H_4 , HCOOH , MeOH; rt , 2 min]→ 3-aminophenol (OH, NH_2) 94%

Gowda, S.; Gowda, D.C. *Tetrahedron* ***2002***, *58*, 2211.

$PhNO_2$ —[PhC≡CH , CO , 170°C , 2d; cat $[Cp^*Ru(CO)_2]_2$]→ 3-phenylindene (Ph) 39%

Penoni, A.; Nicholas, K.M. *Chem. Commun.* ***2002***, 484.

1-bromo-2-nitrobenzene (NO_2, Br) —[1.PhLi , –78°C; 2. Br–CH_2CH=C(CH_3)$_2$; 3. CH_2=CHMgBr , DME , –40°C]→ 7-prenylindole (N-H) 50%

Pirrung, M.C.; Wedel, M.; Zhao, Y. *Synlett* ***2002***, 143.

3 Li–C_6H_4–NMe_2 —($PhNO_2$, THF, –78°C)→ PhHN–C_6H_4–NMe_2 52%

Yang, T.; Cho, B.P. *Tetrahedron Lett.* **2003**, *44*, 7549.

$PhNO_2$ —(Al_2O_3, NaHS, microwaves, H_2O)→ $PhNH_2$ 95%

Kanth, S.R.; Reddy, G.V.; Rao, V.V.V.N.S.R.; Maitraie, P.; Narsaiah, B.; Rao, P.S. *Synth. Commun.* **2002**, *32*, 2849.

O_2N–C_6H_4–NO_2 —(1. PhMgCl, THF, –20°C; 2. $FeCl_2$, $NaBH_4$, –20°C → rt)→ O_2N–C_6H_4–NHPh 67%

Sapountzis, I.; Knochel, P. *Synlett* **2004**, 955.

O_2N–C_6H_4–CH_3 —(TBAB, 7.5 min, 90°C; 3 $SnCl_2$•2 H_2O)→ H_2N–C_6H_4–CH_3 81%

De, P. *Synlett* **2004**, 1835.

2-iodo-4-nitrotoluene —(H_2, 5% Pt/C, EtOAc, 4h; 0.4 ZnI_2, 22°C)→ 3-iodo-4-methylaniline 97%

Wu, G.; Huang, M.; Richards, M.; Poirer, M.; Wen, X.; Draper, R.W. *Synthesis* **2003**, 1657.

$PhNO_2$ —(Fe nanoparticles, H_2O, 210°C)→ $PhNH_2$ 95%

Wang, L.; Li, P.; Wu, Z.; Yan, J.; Wang, M.; Ding, Y. *Synthesis* **2003**, 2001.

$PhNO_2$ —($FeCl_3$•6 H_2O/In, ultrasound, 1.5 h; H_2O/MeOH)→ $PhNH_2$ 94%

Yoo, B.W.; Choi, J.W.; Hwang, S.K.; Kim, D.Y.; Baek, H.; Choi, K.I.; Kim, J.H. *Synth. Commun.* **2003**, *33*, 2985.

REVIEWS:

"*N-tert*-Butanesulfinyl Imines: Versatile Intermediates for the Asymmetric Synthesis of Amines"
Ellman, J.A.; Owens, T.D.; Tang, T.P. *Acc. Chem. Res.* **2002**, *35*, 984.

"Preparation, Properties and Synthetic Applications of 2H-Azirines"
de Retana, A.M.O.; de Marigorta, E.M.; de los Santos, J.M. *Org. Prep. Proceed. Int.* **2002**, *34*, 219.

SECTION 105A: PROTECTION OF AMINES

MeO_2C–CH_2–CH(NHCbz)–CO_2Me —(1. NaHMDS, THF-HMA; 2. Cnz-Cl)→ MeO_2C–CH_2–CH($N(Cbz)_2$)–CO_2Me 80%

Hernández, J.N.; Martín, V.S *J. Org. Chem.* **2004**, *69*, 3590.
Hernández, J.N.; Ramírez, M.A.; Martín, V.S. *J. Org. Chem.* **2003**, *68*, 743.

N-Bn indole → indole: O_2 , DMSO , KO*t*-Bu; 25°C , 15 min — 87%

Haddach, A.A.; Kelleman, A.; Deaton-Rewoliwski, M.V. *Tetrahedron Lett.* **2002**, *43*, 399.

$(C_8H_{17})_2NAc$ → $(C_8H_{17})_2NH$: $LiBHEt_3$, THF , 0°C → rt — 76%

Tanaka, H.; Ogasawara, K. *Tetrahedron Lett.* **2002**, *43*, 4417.

N-Boc indole → indole (N–H): 5 TBAF , THF , 8 h , reflux — 91%

Routier, S.; Saugé, L.; Ayerbe, N.; Coudert, G.; Mérour, J.-Y. *Tetrahedron Lett.* **2002**, *43*, 589.

PhCH₂OC(O)NHCH₂CO₂H → H_2NCH₂CO_2H: cat hydroxyapatite bound Pd , MeOH; H_2 , 40°C , 2h — 94%

Murata, M.; Hara, T.; Mori, K.; Ooe, M.; Mizugaki, T.; Ebitani, K.; Kaneda, K. *Tetrahedron Lett.* **2003**, *44*, 4981.

HO_2CCH₂CH(NHBoc)CO_2Bn → HO_2CCH₂CH(NH_2)CO_2Bn: aq. 85% H_3PO_4 — 94%

Li, B.; Bemish, R.; Buzon, R.A.; Chiu, C.K.-F.; Colgan, S.T.; Kissel, W.; Le, T.; Leeman, K.R.; Newell, L.; Roth, J. *Tetrahedron Lett.* **2003**, *44*, 8113.

5-(NHBoc)indane → 5-(NH_2)indane: NaO*t*-Bu , H_2O , reflux; 2-methyltetrahydrofuran — quant

Tom, N.J.; Simon, W.M.; Frost, H.N.; Ewing, M. *Tetrahedron Lett.* **2004**, *45*, 905.

Ph(CO₂Me)CH–N(Boc)(Bz) → Ph(CO₂Me)CH–NH(Bz): $CeCl_3$•7 H_2O , NaI , MeCN , 5h , rt — 87%

Yadav, J.S.; Reddy, B.V.S.; Reddy, K.S. *Synlett* **2002**, 468.

N-Cbz proline methyl ester (CO_2Me) → proline methyl ester (N–H): 1.5 Et_2AlCl , DCM , 0°C , 1h — 81%

Tsujimoto, T.; Murai, A. *Synlett* **2002**, 1283.

N-Boc indole → indole (N–H): $Sn(OTf)_2$, DCM — 89%

Bose, D.S.; Kumar, K.C.; Reddy, A.V.N. *Synth. Commun.* **2003**, *33*, 445.

CHAPTER 8

PREPARATION OF ESTERS

SECTION 106: ESTERS FROM ALKYNES

$(CO)_3(Cp)W$ — alkynyl ether epoxide → 25% $BF_3•OEt_2$, DCM, –40°C → bicyclic lactone 82%

Madhushaw, R.J.; Li, C.-L.; Su, H.-L.; Hu, C.-C.; Lush, S.-F.; Liu, R.-S. *J. Org. Chem.* **2003**, *68*, 1872.

PhSe—≡—Ph → TsOH, DCM, reflux → PhSe(C=O)CH$_2$Ph 81%

Sheng, S.; Liu, X. *Org. Prep. Proceed. Int.* **2002**, *34*, 499.

SECTION 107: ESTERS FROM ACID DERIVATIVES

The following types of reactions are found in this section:

1. Esters from the reaction of alcohols with carboxylic acids, acid halides, and anhydrides.
2. Lactones from hydroxy acids.
3. Esters from carboxylic acids and halides, sulfoxides, and miscellaneous compounds

Ph–CH=CH–CO_2H → (resin)–CH$_2$–N=C(NH*c*-Hex)(OBn), MeCN, microwaves, 125°C, 3 min → Ph–CH=CH–CO_2Bn 96%

Crosignani, S.; White, P.D.; Steinauer, R.; Linclau, B. *Org. Lett.* **2003**, *5*, 853.
Crosignani, S.; White, P.D.; Linclau, B. *Org. Lett.* **2002**, *4*, 2961.

Ph–CH$_2$–CO_2H + (resin)–CH$_2$–CH=C(NH*c*-hex)(OMe) → microwaves, MeCN, 120°C, 5 min → Ph–CH$_2$–CO_2Me 90%

Crosignani, S.; White, P.D.; Linclau, B. *J. Org. Chem.* **2004**, *69*, 5897.

$(MeO)_2C{=}O$, DBU , reflux , 1 d — 99%

Shieh, W.C.; Dell, S.; Repic, O. *J. Org. Chem.* **2002**, *67*, 2188.

$(CH_2)_7$ CO_2H O OH

1. 1.5 EtOC≡CH , toluene 2% $[RuCl_2(p\text{-cymeme})]_2$
2. 10% CSA , 0.005 M

69%

Trost, B.M.; Chisholm, J.D. *Org. Lett.* **2002**, *4*, 3747.

Ph CHO , TMSQ — 84% (91% de) >99% ee

Zhu, C.; Shen, X.; Nelson, S.G. *J. Am. Chem. Soc.* **2004**, *126*, 5352.

$CH_2{=}CHCO_2H$ — MeOH , reflux , cat. I_2 , 12 h → $CH_2{=}CHCO_2Me$ 95%

Ramalijga, K.; Vijayalakshmi, P.; Kaimal, T.N.B. *Tetrahedron Lett.* **2002**, *43*, 879.

$PhCO_2H$ — $(MeO)_2C{=}O$, DBU , 4h → $PhCO_2Me$ 99%

(+ microwaves = 12min, 99%)

Shieh, W.-C.; Dell, S.; Repic, O. *Tetrahedron Lett.* **2002**, *43*, 5607.

$PhCH_2CO_2H$ — CAN , MeOH , rt , 12h → $PhCH_2CO_2Me$ 76%

Pan, W.-B.; Chang, F.-R.; Wei, L.-M.; Wu, M.J.; Wu, Y.-C. *Tetrahedron Lett.* **2003**, *44*, 331.

CO_2H — *i*-PrOH , C_6F_{14} , 2% DPAT, 80°C , 1d → $CO_2i\text{-Pr}$ 65%

DAPT = diphenylammonium triflate

Gacem, B.; Jenner, G. *Tetrahedron Lett.* **2003**, *44*, 1391.

CH_3CO_2H — aa imidazolium mesylate (MMIM) , 2h, $PhCH_2Cl$, 90°C → $CH_3CO_2CH_2Ph$ >95%

Brinchi, L.; Germani, R.; Savelli, G. *Tetrahedron Lett.* **2003**, *44*, 6583.

OH — cat $InCl_3$, Ac_2O , neat , rt , 2 min → OAc 95%

Charkraborti, A.K.; Gulhane, R. *Tetrahedron Lett.* **2003**, *44*, 6749.

$C_{11}H_{23}O_2H$ — $PhCH_2CH_2CH_2OH$, neat, Ti(III)- Montmorillonite → $C_{11}H_{23}O_2CH_2CH_2CH_2Ph$ 96%

Kawabata T.; Mizugaki, T.; Ebitani, K.; Kaneda, K. *Tetrahedron Lett.* **2003**, *44*, 9205.

10% PPh_3 , BnOH , toluene; reflux , 1d — OBn 79%

Adair, G.R.A.; Edwards, M.G.; Williams, J.M.J. *Tetrahedron Lett.* ***2003***, *44*, 5523.

O_2N–C6H4–CH2OH; O_2N–C6H4–CO_2H , DCM; polymer–$PPh_2OSO_2CF_3$ — 95% (NO_2)

Elson, K.E.; Jenkins, I.D.; Loughlin, W.A. *Tetrahedron Lett.* ***2004***, *45*, 2491.

$C_9H_{19}CO_2H$ —hν , MeOH , CCl_4 , 12h→ $C_9H_{19}CO_2Me$ 99%

Hwu, J.R.; Hsu, C.-Y.; Jain, M.L. *Tetrahedron Lett.* ***2004***, *45*, 5151.

Ph CO_2H —BnOH , toluene , 18h; 0.1 $HfCl_4(thf)_2$→ Ph OBn >99%

Ishihara, K.; Nakayama, M.; Ohara, S.; Yamamoto, H. *Tetrahedron* ***2002***, *58*, 8179.

$C_{11}H_{23}CO_2H$ —$C_{12}H_{21}SH$, 10% TOH , toluene; reflux , 6h→ $C_{11}H_{23}$C(O)$SC_{12}H_{25}$ 97%

Iimura, S.; Manabe, K.; Kobayashi, S. *Chem. Commun.* ***2002***, 94.

CH3C(O)Cl —MeOH , DABCO , 2 min; solvent free→ CH3C(O)OMe 75%

Hajipour, A.R.; Mazloumi, Gh. *Synth. Commun.* ***2002***, *32*, 23.

Ph CO_2H —BnOH , Boc_2O , DMAP , 50°C; $MeNO_2$, 16h→ Ph C(O)O Ph 99%

Gooßen, L.J.; Döhring, A. *Synlett* ***2004***, 263.

$PhCH_2COOH$ —silica chloride, MeOH , heat→ $PhCH_2CO_2Me$ 96%

Srinivas, K.V.N.S.; Mahender, I.; Das, B. *Synthesis* ***2003***, 2479.

PhCOOH —BuOH , $AlCl_3$/NaI , MeCN→ PhC(O)OBu 71%

Karade, N.N.; Shirodkaar, S.G.; Potrekar, R.A.; Karade, H.N. *Synth. Commun.* ***2004***, *34*, 391.

Ph COOH —BnOH , cat DMAP; DCM , NEt_3; (Me, O, NO_2 anhydride)$_2$O→ Ph CO_2Bn 95% + benzyl 2-methyl-6-nitrobenzoate

Kubota, M. *Chem. Lett.* ***2002***, *31*, 286.

PhCOOH —Ph_2POBn , DCM , rt , 0.5 n; 2,6-dimetyl-1,4-benzoquinone (DMBQ)→ PhCOOMe 98%

Mukaiyama, T.; Shintou, T.; Kikuchi, W. *Chem. Lett.* ***2002***, *31*, 1126.

Ph_2PNMe_2 —[1. 1.1 *i*-PrOH , DCE , 100°C , 7 h; 2. PhCOOH , DMBQ , DCE , rt , 1 h]→ PhCOO*i*-Pr 94%

DMBQ = 2,6-dimethyl-1,4-benzoquinone

Mukaiyama, T.; Kikuchi, W.; Shintou, T. *Chem. Lett.* ***2003***, *32,* 300

$Ph(CH_2)_2CO_2H$ —[DMAP , DCM , rt , 6 h; di-2-thienyl carbonate]→ $Ph(CH_2)_2CO_2CH_2CH_2CH_2Ph$ 94%

Mukaiyama, T.; Oohashi, Y.; Fukumto, K. *Chem. Lett.* ***2004***, *33,* 552.

BuCOOH —[1. di-2-thienyl carbonate , 5% DMAP, MeCN , rt; 2. 1.2 $PhCH_2CH_2CH_2OH$, 1.05 I_2]→ $BuCO_2CH_2CH_2CH_2Ph$ 86%

Oohashi, Y.; Fukumoto, K.; Mukaiyama, T. *Chem. Lett.* ***2004***, *33,* 968.

PhCOOH —[Ph_2POBn , 2,6-dimethylbenzoquinone; DCM ,rt , 3 h]→ $PhCO_2Bn$ 90%

Shintou, T.; Kikuchi, W.; Mukaiyama, T. *Bull. Chem. Soc. Jpn.* ***2003***, *76,* 1645.

Further examples of the reaction RCO_2H + R'OH → RCO_2R' are included in Section 108 (Esters from Alcohols and Thiols) and in Section 30A (Protection of Carboxylic Acid Derivatives).

SECTION 108: ESTERS FROM ALCOHOLS AND THIOLS

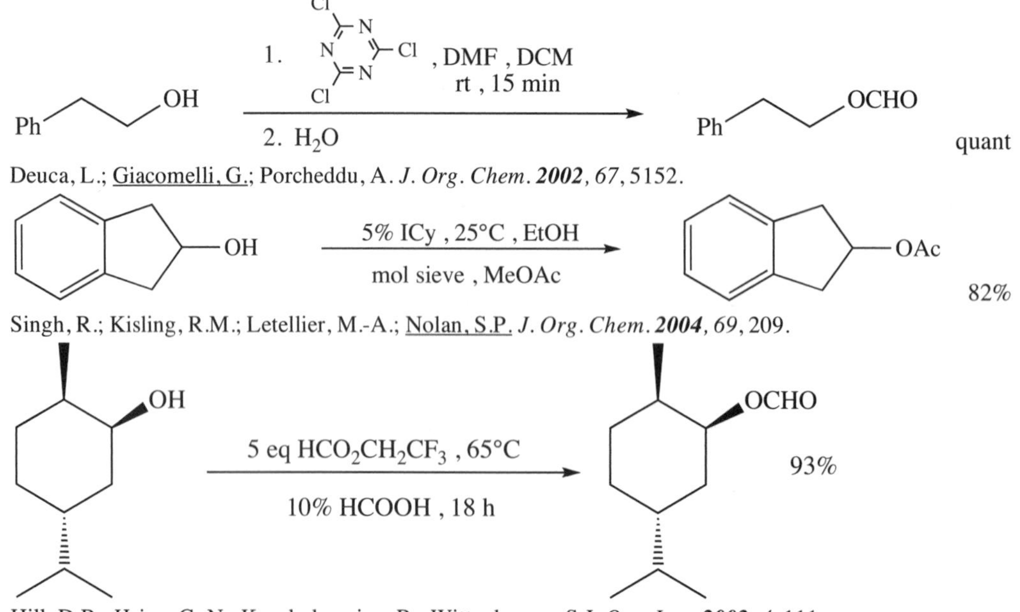

Deuca, L.; Giacomelli, G.; Porcheddu, A. *J. Org. Chem.* ***2002***, *67*, 5152.

Singh, R.; Kisling, R.M.; Letellier, M.-A.; Nolan, S.P. *J. Org. Chem.* ***2004***, *69*, 209.

Hill, D.R.; Hsiao, C.-N.; Kurukulasuriya, R.; Wittenberger, S.J. *Org. Lett.* ***2002***, *4*, 111.

Ir catalyst , acetone , rt
amino alcohol ligand
OH OH O O >99%

Suzuki, T.; Morita, K.; Tsuchida, M.; Hiroi, K. *Org. Lett.* ***2002***, *4*, 2361.

cat. OsO_4 , Oxone , DMF
OH O O 73%

Schomaker, J.M.; Travis, B.R.; Borhan, B. *Org. Lett.* ***2003***, *5*, 3089.

1% $Pd(OAc)_2$, (+)-DIOP
CO/H_2 , 500 psi/100 psi , DCM
OH O O 75% (81% ee)

Dong, C.; Alper, H. *J. Org. Chem.* ***2004***, *69*, 5011.

O=N NHAc BF_4
Py , mol. sieves , DCM
Ph O OH Ph O O O OPh 68%

Merbouh, N.; Bobbitt, J.M.; Brückner, C. *J. Org. Chem.* ***2004***, *69*, 5116.

1. 1.5 BCl_3 , 4% $AlCl_3$
70°C , 6 h
2. $Pd(OAc)_2$, CO , 16 h
DMSO , MeOH , rt
HO O O 99x90%

Zhou, Q.J.; Worm, K.; Dole, R.E. *J. Org. Chem.* ***2004***, *69*, 5147.

TEMPO/bis(acetoxy)iodobenzene
DCM , 20h , rt
TBDPSO OH OH TBDPSO O O 84%

Hansen, T.M.; Florence, G.J.; Lugo-Mas, P.; Chen, J.; Abrams, J.N.; Forsyth, C.J. *Tetrahedron Lett.* ***2003***, *44*, 57.

1% $Ce(OTf)_3$, 1h
Ac_2O
OH OAc >98%

Dalpozzo, R.; DeNino, A.; Maiuolo, L.; Procopio, A.; Nardi, M.; Bartoi, G.; Romeo, R. *Tetrahedron Lett.* ***2003***, *44*, 5621.

OH , 10% $LiBF_4$, MeCN
rt , 3.5h
O Ph N_2 O Ph O 95%

Yadav, J.S.; Reddy, B.V.S.; Vishbumurthy, P. *Tetrahedron Lett.* ***2003***, *44*, 5691.

0.5% chiral Ir complex , acetone-DCM
30°C , 45h
OH OH O O 99% (80% ee)

Suzuki, T.; Morita, K.; Matsuo, Y.; Hiroi, K. *Tetrahedron Lett.* ***2003***, *44*, 2003.

MeO—C_6H_4—CH_2OH → (5% $CoCL_2$•6 H_2O , *p*-TsOH / DCE , reflux) → MeO—C_6H_4—CH_2OTs 95%

Velusamy, S.; Kumar, J.S.K.; Punniyamurthy, T. *Tetrahedron Lett.* ***2004**, 45*, 203.

OH → (3 Bz_2O , MeCN , 25h / cat $Yb(OTf)_3$) → OBz 99%

Dumeunier, R.; Markó, I.E. *Tetrahedron Lett.* ***2004**, 45*, 825.

$PhCH_2OH$ → (5% $RuCl_3$, Ac_2O , MeCN , rt , 110 min) → $PhCH_2OAc$ 98%

De, K. *Tetrahedron Lett.* ***2004**, 45*, 2919.

$PhCH_2CH_2OH$ → (silica chloride , TsOH , DCM / reflux , 2h) → $PhCH_2CH_2OTs$ 86%

Das, B.; Reddy, V.S.; Reddy, M.R. *Tetrahedron Lett.* ***2004**, 45*, 6717.

OH → (BiOCl , AcCl , neat , 5 min , 2.5h) → OAc 98%

Ghosh, R.; Maiti, S.; Chakraborty, A. *Tetrahedron Lett.* ***2004**, 45*, 6775.

$C_6H_{13}CH_2OH$ → (PCC , Al_2O_3 , neat , 3h) → $C_6H_{13}CO_2C_7H_{15}$ 71%

Bhar, S.; Chaudhuri, S.K. *Tetrahedron* **2003**, *59*, 3493.

OH → (CH_2=CHOAc , $PdCl_2/CuCl_2$ / toluene , rt) → OAc 89%

Bosco, J.W.J.; Saikia, A.K. *Chem. Commun.* ***2004**,* 1116.

$PhCH_2OH$ → (Ph_3P=$CHCO_2Bn$, 5% $[IrCl(cod)]_2$/dppp / 5% Cs_2CO_3 , toluene , 150°C , 72h) → $PhCH_2CH_2CO_2Bn$ 79%

Edwards, M.G.; Williams, J.M.J. *Angew. Chem. Int. Ed.* ***2002**, 41*, 4740.

$PhCH_2OH$ → (benzoyl chloride , THF , cat $LiClO_4$ / 25°C , 7h) → $PhCH_2OBz$ 80%

Bandgar, B.P.; Kamble, V.T.; Sadavarte, V.S.; Uppalla, L.S. *Synlett* **2002**, 735.

Ph OH → (20 MnO_2 , 5 NaCN , rt , 2d / MeOH , AcOH) → Ph CO_2Me 84%

Foot, J.S.; Kanno, H.; Giblin, G.M.P.; Taylor, R.J.K. *Synlett* **2002**, 1293.

Ph OH → (Ac_2O , 1% $Mg(ClO_4)_2$ / 20°C , 10 min) → Ph OAc 99%

Bartoli, G.; Bosco, M.; Dalpozzo, R.; Marcantoni, E.; Massaccesi, M.; Sambri, L. *Eur. J. Org. Chem.* **2003**, 4611.

Bartoli, G.; Bosco, M.; Dalpozzo, R.; Marcantoni, E.; Massaccesi, M.; Rinaldi, S.; Sambri, L. *Synlett* **2003**, 39.

OH → (3.5 PhIO , KBr , MeOH / H_2O , rt , 12 h) → O OMe 75%

Tohma, H.; Maegawa, T.; Kita, Y. *Synlett* **2003**, 723.

OH → OAc (4-hydroxyquinoline → 4-acetoxyquinoline)

$BiOCl_4 \cdot x\ H_2O$, Ac_2O , rt; 30 min , neat — 90%

Chakraborti, A.K.; Gulhane, R.; Shivani *Synlett* ***2003***, 1805.

$PhCH_2OH$ — Ac_2O , Sn(IV)(tpp)$(ClO_4)_2$; rt , 5 min → $PhCH_2OAc$ 99%

Tangestaninejad, S.; Habibi, M.H.; Mirkhani, V.; Moghadam, M. *Synth. Commun.* ***2002***, *32*, 1337.

Ph, OH → Ph, OAc

Ac_2O , WO_3/ZrO_2 , 2 h — 97%

Reddy, B.M.; Sreekanth, P.M. *Synth. Commun.* ***2002***, *32*, 2815.

OH → OAc

Ac_2O , neat , 1% $ZrCl_4$; 15 min — 95%

Chakraborti, A.K.; Gulhane, R. *Synlett* ***2004***, 627.

OH → O, OMe

trichloroisocyanuric acid; MeOH , DCM — 81%

α,ω-diols are converted to lactones

Hiegel, G.A.; Gilley, C.B. *Synth. Commun.* ***2003***, *33*, 2003.

MeO—C₆H₄—CH_2OH — 2 Ac_2O , $NiCl_2$, microwaves , 7 min → MeO—C₆H₄—CH_2OAc 98%

Constantinou-Kokotou, V.; Peristeraki, A. *Synth. Commun.* ***2004***, *34*, 4227.

OH, OH — 2% chiral Ru(salen) catalyst , air , hν; EtOAc , rt → OH, O 77% — PDC , DCM; MS 4Å → O, O 58% ee

Shimizu, H.; Nakata, K.; Katsuki, K. *Chem. Lett.* ***2002***, *31*, 1080

Shimizu, H.; Nakata, K.; Katsuki, K. *Chem. Lett.* ***2003***, *32*, 480.

$PhCH_2OH$ — Ac_2O , $K_5CoW_{12}O_{40} \cdot 3\ H_2O$, 1 min , rt; solvent free → $PhCH_2OAc$ 99%

Habibi, M.H.; Tangestaninejad, S.; Mirkhani, V.; Yadollahi, B. *Synth. Commun.* ***2002***, *32*, 863.

Ph, OH — *p*-TsOH , $ZrCl_4$, DCM; reflux , 6 h → Ph, OTs 95%

Das, B.; Reddy, V.S. *Chem. Lett.* ***2004***, *33*, 1428.

REVIEWS:

"Lewis Acid Catalyzed Arylation Reactions: Scope and Limitations"
Chandra, K.L.; Saravanan, P.; Singh, R.K.; Singh, V.K. *Tetrahedron* ***2002***, *58*, 1369.

"Fluorous Mitsunobu Reagents and Reactions"
Dandapani, S.; Curran, D.P. *Tetrahedron* **2002**, *58*, 3855.

"Recent Advances in the Mitsunobu Reaction: Modified Reagents and the Quest for Chromatography-Free Separation"
Dembinski, R. *Eur. J. Org. Chem.* **2004**, 2163.

Further examples of the reaction ROH → RCO_2R' are included in Section 107 (Esters from Acid Derivatives) and in Section 45A (Protection of Alcohols and Thiols).

SECTION 109: ESTERS FROM ALDEHYDES

PhCHO —(Na perborate, MeOH, 0°C; V_2O_5, 70% $HClO_4$, 0.3 h)→ $PhCO_2Me$ 97%

Gopinath, R.; Barkakaty, B.; Talukdar, B.; Patel, B.K. *J. Org. Chem.* **2003**, *68*, 2944.

Ph–CH₂–CH(Br)–CHO —(20% triazolium salt (N-Ph, Cl⁻), BnOH, NEt_3; toluene, 25°C, 4 h)→ Ph–CH₂–CH₂–CO_2Bn 80%

Reynolds, N.T.; de Alaniz, J.R.; Rovis, T. *J. Am. Chem. Soc.* **2004**, *126*, 9518.

Ph–CH=CH–CHO —(Br–C₆H₄–CHO, THF/*t*-BuOH, 7% DBU; 8% imidazolium salt, 25°C)→ γ-lactone (5-(4-bromophenyl)-4-phenyl-dihydrofuran-2-one) 79% (4:1 dr)

Sohn, S.S.; Rosen, E.L.; Bode, J.W. *J. Am. Chem. Soc.* **2004**, *126*, 14370.

PhCHO —(30% H_2O_2, titanosilicate TS-1; MeOH, reflux, 5h)→ $PhCO_2Me$ 65%

Chavan, S.P.; Dantale, S.W.; Gouande, C.A.; Venkataman, M.S.; Praveen, C. *Synlett* **2002**, 267.

PhCHO —(Oxone, NaBr, 1 d; aq. MeOH, rt)→ $PhCO_2Me$ 71%

Koo, B.-S.; Kim, E.-H.; Lee, K.-J. *Synth. Commun.* **2002**, *32*, 2275.

C_3H_7-$(CH_2)_7$-CHO —(4 PHPB, MeOH, H_2O; 14 h)→ C_3H_7-$(CH_2)_7$-CO_2Me 89%

Sayama, S.; Onami, T. *Synlett* **2004**, 2739.

PhCHO —(MeOH, 4 H_2O_2, rt, 0.5 70% $HClO_4$; sulphated SnO_2 supported on the SBA-A-zeolite)→ $PhCO_2Me$ 98%

Qian, G.; Zhao, R.; Ji, D.; Lu, G.; Qi, Y.; Suo, J. *Chem Lett.* **2004**, *33*, 834.

phthalaldehyde (C₆H₄(CHO)₂) —(1% bowl-shaped $R_3SiOAlMe_2$; toluene, 21°C, 30 min; R = 2,6-diphenylbenzyl)→ phthalide 99%

Shirakawa, S.; Takai, J.; Sasaki, K.; Miura, T.; Maruoka, K. *Heterocycles* **2003**. *59*, 57.

Related Method: Section 117 (Esters from Ketones)

SECTION 110: ESTERS FROM ALKYLS, METHYLENES AND ARYLS

No examples of the reaction R-R → RCO_2R' or $R'CO_2R$ (R,R' = alkyl, aryl, etc.) occur in the literature. For the reaction R-H → RCO_2R' or $R'CO_2R$, see Section 116 (Esters from Hydrides).

CH_3 N — $PhI(OAc)_2$, cat $Pd(OAc)_2$ / AcOH , 100°C → N OAc 88%

Dick, A.R.; Hull, K.L.; Sanford, M.S. *J. Am. Chem. Soc.* ***2004***, *126*, 2300.

SECTION 111: ESTERS FROM AMIDES

O_2N—C6H4—NHAc — 2 $NaNO_2$, Ac_2O/AcOH / 0°C → rt → O_2N—C6H4—OAc 93%

Glatzhofer, D.T.; Roy, R.R.; Cossey, K.N. *Org. Lett.* ***2002***, *4*, 2349.

S O OH N S Ph C_5H_{11} — 2 BnOH , 0.1 DMAP , 5h / DCM , 13 h → O OH BnO C_5H_{11} 93%

Wu, Y.; Sun, Y.-P.; Yang, Y.-Q.; Hu, Q.; Zhang, Q. *J. Org. Chem.* ***2004***, *69*, 6141.

O O N H — 5% $Mg(NTf_2)_2$, bis(oxazoline) ligand / BnNHOH , DCM , –40°C → O O N — Bn 95% (96% ee; 96% de)

Sibi, M.P.; Prabagaran, N.; Ghorpade, S.G.; Jasperse, C.P. *J. Am. Chem. Soc.* ***2003***, *125*, 11796.

SECTION 112: ESTERS FROM AMINES

NO ADDITIONAL EXAMPLES

SECTION 113: ESTERS FROM ESTERS

Conjugate reductions and conjugate alkylations of unsaturated esters are given in Section 74 (Alkyls, Methylenes, and Aryls from Alkenes).

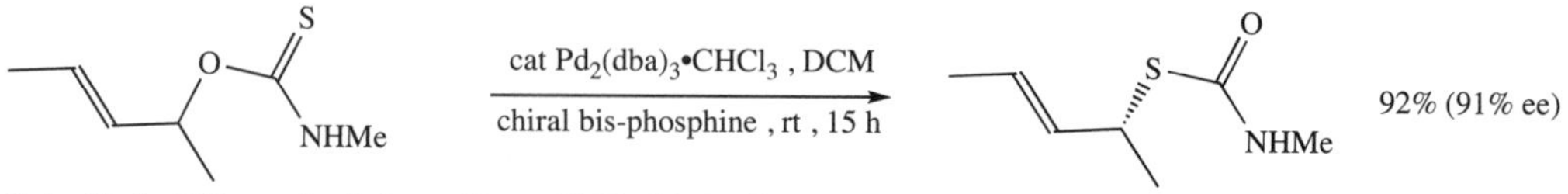

cat $Pd_2(dba)_3$•$CHCl_3$, DCM; chiral bis-phosphine , rt , 15 h — 92% (91% ee)

Gais, H.-J.; Böhme, A. *J. Org. Chem.* **2002**, *67*, 1153.

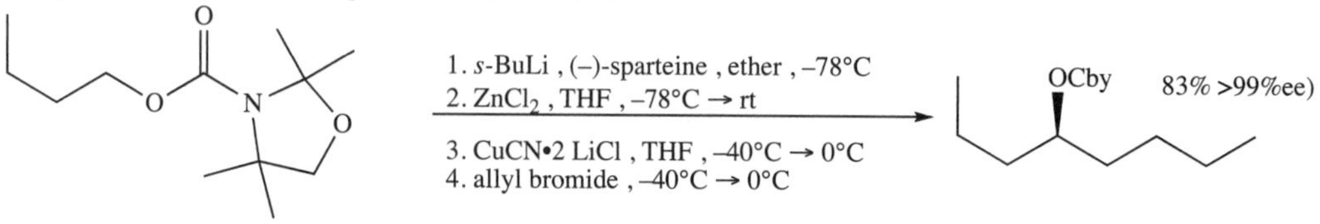

1. *s*-BuLi , (–)-sparteine , ether , –78°C
2. $ZnCl_2$, THF , –78°C → rt
3. CuCN•2 LiCl , THF , –40°C → 0°C
4. allyl bromide , –40°C → 0°C

OCby 83% >99%ee)

Papillon, J.P.N.; Taylor, R.J.K. *Org. Lett.* **2002**, *4*, 119.

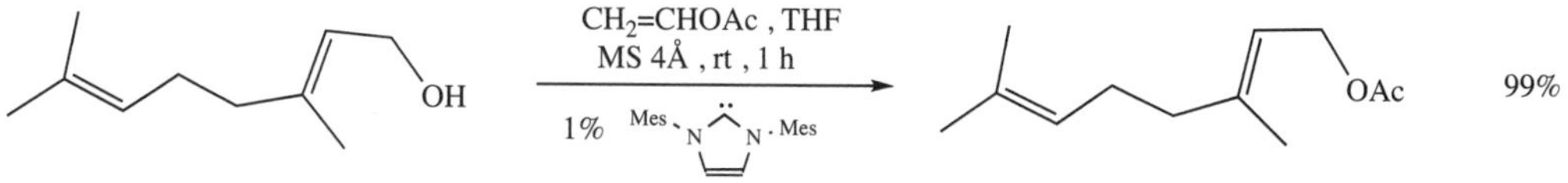

CH_2=CHOAc , THF; MS 4Å , rt , 1 h; 1% Mes-N N-Mes — OAc 99%

Grasa, G.A.; Kissling, R.M.; Nolan, S.P. *Org. Lett.* **2002**, *4*, 3583.

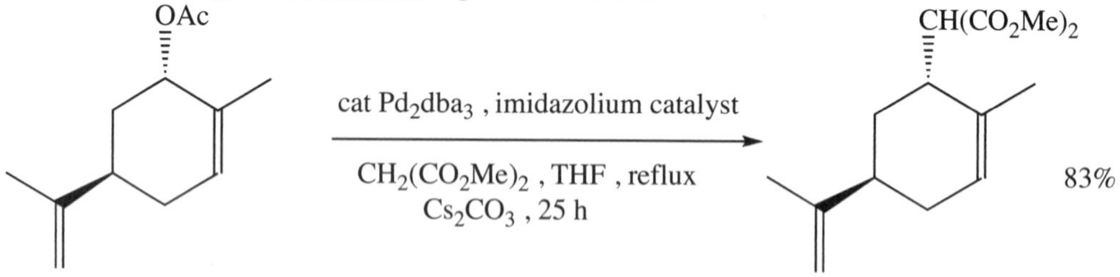

OAc → $CH(CO_2Me)_2$

cat Pd_2dba_3 , imidazolium catalyst; $CH_2(CO_2Me)_2$, THF , reflux; Cs_2CO_3 , 25 h — 83%

Sato, Y.; Yoshino, T.; Mori, M. *Org. Lett.* **2003**, *5*, 31.

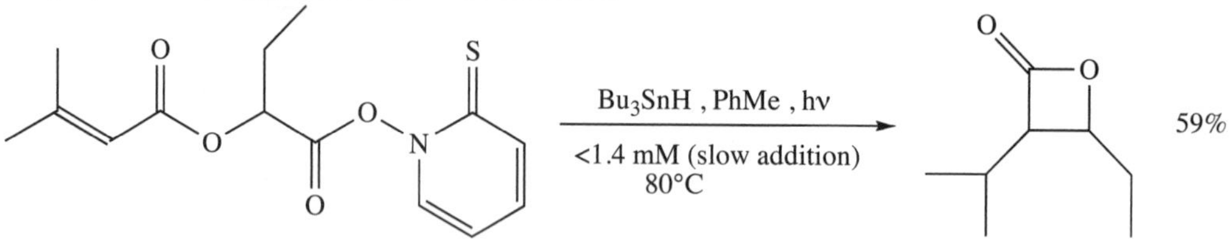

Bu_3SnH , PhMe , hν; <1.4 mM (slow addition); 80°C — 59%

Castle, K.; Hau, C.-S.; Sweeney, J.B.; Tindall, C. *Org. Lett.* **2003**, *5*, 757.

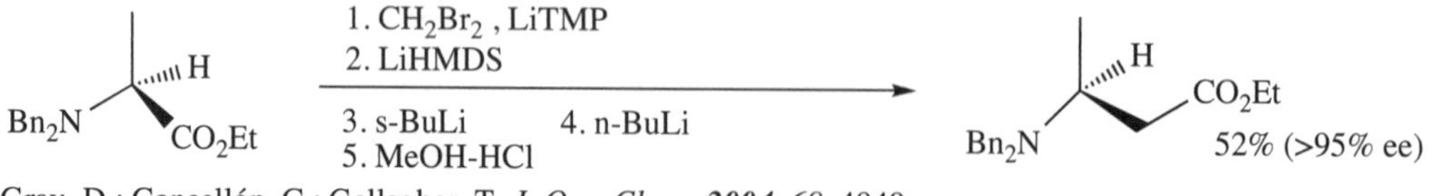

Bn_2N (H) CO_2Et → 1. CH_2Br_2 , LiTMP; 2. LiHMDS; 3. s-BuLi; 4. n-BuLi; 5. MeOH-HCl → Bn_2N (H) CO_2Et 52% (>95% ee)

Gray, D.; Concellón, C.; Gallagher, T. *J. Org. Chem.* **2004**, *69*, 4849.

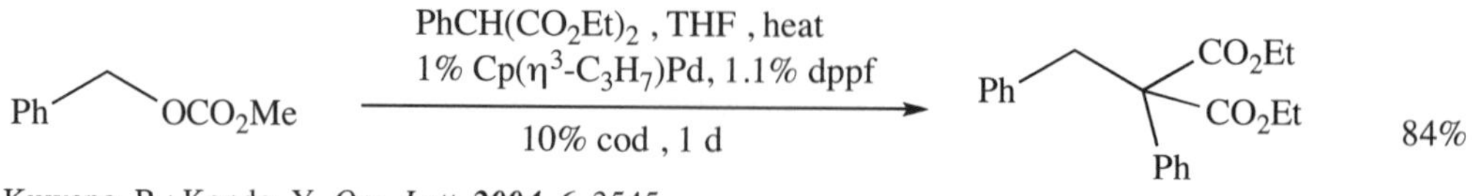

Ph OCO_2Me → $PhCH(CO_2Et)_2$, THF , heat; 1% $Cp(\eta^3\text{-}C_3H_7)Pd$, 1.1% dppf; 10% cod , 1 d → Ph, CO_2Et, CO_2Et, Ph 84%

Kuwano, R.; Kondo, Y. *Org. Lett.* **2004**, *6*, 3545.

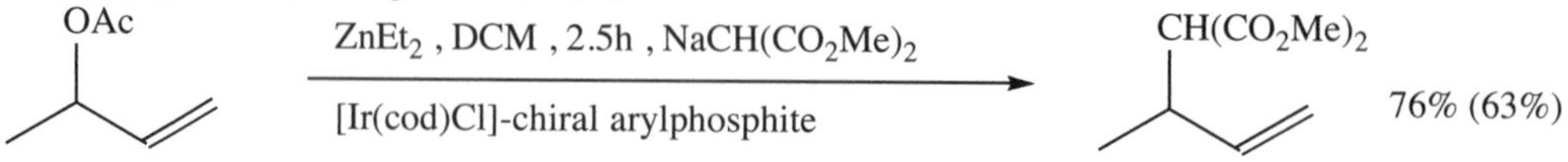

OAc → $ZnEt_2$, DCM , 2.5h , $NaCH(CO_2Me)_2$; [Ir(cod)Cl]-chiral arylphosphite → $CH(CO_2Me)_2$ 76% (63%)

Kinoshita, N.; Marx, K.H.; Tanaka, K.; Tsubaki, K.; Kawabata, T.; Yoshikai, N.; Nakamura, E.; Fuji, K. *J. Org. Chem.* **2004**, *69*, 7960.

Ph–CH(OAc)–CH=CH–Ph → MeCH(CO₂Et)₂ , cat [Pd(allyl)Cl]₂ , THF / 5% chiral bis(oxazoline) ligand , 0°C / BSA , KOAc , MeCN → Ph–CH(CMe(CO₂Me)₂)–CH=CH–Ph >95% (85%ee)

Molander, G.A.; Burke, J.P.; Carroll, P.J. *J. Org. Chem.* ***2004***, *69*, 8062.

t-Bu–C₆H₄–Br → CH_3CO_2*t*-Bu , 2.3 LiHMDS / cat Pd(dba)₂•carbene ligand / toluene , rt , 12 h → *t*-Bu–C₆H₄–CH_2CO_2*t*-Bu 85%

Jørgensen, M.; Lee, S.; Liu, X.; Wolkowski, J.P.; Hartwig, J.F. *J. Am. Chem. Soc.* ***2002***, *124*, 12557.

PhC(=O)OEt → P_4S_{10} , hexamethyldioxolane / xylene → PhC(=S)OEt 73%

Curphey, T.J. *Tetrahedron Lett.* ***2002***, *43*, 371.

γ-valerolactone → Lawessons' reagent , hexamethyl disiloxane / microwaves (3x30 sec) , 120°C , neat → thionolactone 80x95%

Flilppi, J.-J.; Fernandez, X.; Lizzani-Cavelier, L.; Loiseau, A.-M. *Tetrahedron Lett.* ***2003***, *44*, 6647.

Ph–CH(OCO₂Et)–CH=CH–Ph → (η^3-allyl)Pd-BINAP/SbF₆ , 18-crown-6 / 3 NaO₂C*t*-Bu , 0°C , 7h → Ph–CH(OCO₂*t*-Bu)–CH=CH–Ph 90% (89% ee)

Faller, J.W.; Wilt, J.C. *Tetrahedron Lett.* ***2004***, *45*, 7613.

$PhCO_2Me$ → $BrCH_2CO_2Et$, Zn , THF / 20 h , reflux → $PhCO_2Et$ 54%

Coskun, N.; Er, M. *Tetrahedron* ***2003***, *59*, 3481.

$BrCH_2CO_2Et$ → $PhB(OH)_2$, cat Pd)PPh₃)₄ , cat Cu / toluene , 80°C , K_3PO_4 → $PhCH_2CO_2Et$ 85%

Liu, X.-x.; Deng, M.-z. *Chem. Commun.* ***2002***, 622.

N-(2-hydroxyethyl)phthalimide → EtO_2CCH_2Cl , THF / K_2CO_3 , 2 d → N-(2-(chloroacetoxy)ethyl)phthalimide 56%

Janczewski, D.; Synoradzki, L.; Wlostowski, M. *Synlett* ***2003***, 420.

$BzNHCH_2CO_2Et$ → CAN , MeOH , 53 h , rt → $BzNHCH_2CO_2Me$ 85%

Stefane, B.; Kocevar, M.; Polanc, S. *Synth. Commun.* ***2002***, *32*, 1703.

$CH_3COCH_2CO_2Me$ → HO–CH₂CH₂–Ph , 1 d / silica-supported Si(OMe)–CH₂CH₂CH₂–NEt₂ → $CH_3COCH_2CO_2CH_2CH_2Ph$ 56%

Hagiwara, H.; Koseki, A.; Isobe, K.; Shimizu, K.-i.; Hoshi, T.; Suzuki, T. *Synlett* ***2004***, 2188.

$CH_3COCH_2CO_2Me$ → 3% I_2 , BuOH , toluene , reflux → $CH_3COCH_2CO_2Bu$ 81%

Chavan, S.P.; Kale, R.R.; Shivasankar, K.; Chandake, S.I.; Benjamin, S.B. *Synthesis* ***2003***, 2695.

MeHN–C(=S)–OMe → (1. triphosgene, Py, DCM; 2. PhOH) → MeHN–C(=O)–OPh 32%

Joshi, U.M.; Patkar, L.N.; Rajappa, S. *Synth. Commun.* **2004**, *34*, 33.

Ph–C(=Cr(CO)$_5$)–OMe → (MeCHO, DMAP, hν; 30 psi CO, THF) → β-lactone (Me, OMe, Ph) 55% (8:1 *syn:anti*)

Merlic, C.A.; Doroh, B.C. *J. Org. Chem.* **2003**, *68*, 6056.

Methyl 2-bromobenzoate (Br, CO_2Me) → (PhCHO, cat [NiBr$_2$(dppe)]; THF, 2.75 Zn, reflux, 1d) → 3-phenylphthalide 86%

Rayabarapu, D.K.; Chang, H.-T.; Cheng, C.-H. *Chem. Eur. J.* **2004**, *10*, 2991.

1-ethoxy-1-(trimethylsiloxy)cyclopropane (OEt, OTMS) → (1. PhCHO, 10% Ti compund; 2. TsOH, PhH) → 5-phenyl-γ-butyrolactone (Ph) 80% (65% ee)

Burke, E.D.; Lim, N.K.; Gleason, J.L. *Synlett* **2003**, 390.

REVIEW:

"Transesterification/Arylation Reactions Catalyzed by Molecular Catalysts"
Grasa, G.A.; Singh, R.; Nolan, S.P. *Synthesis* **2004**, 971.

SECTION 114: ESTERS FROM ETHERS, EPOXIDES, AND THIOETHERS

$PhCH_2OCPh_3$ → (AcBr, DCE, rt, 1 h) → $PhCH_2OAc$ 96%

Kobayashi, K.; Watahiki, T.; Oriyama, T. *Synthesis* **2003**, 484.

Propylene oxide → (CO, triglyme, 80°C, 1 d; [PPN][Co(CO)$_4$]-BF$_3$•OEt$_2$) → β-butyrolactone 77%

Getzler, Y.D.Y.L.; Mahadevan, V.; Lobkovsky, E.B.; Coates, G.W. *J. Am. Chem. Soc.* **2002**, *124*, 1174.

Alkenyl epoxide → (1. BF$_3$•OEt$_2$, DCM, 0°C; 2. KHF$_2$, MeCN, H$_2$O$_2$) → γ-lactone 89%

Movassaghi, M.; Jacobsen, E.N. *J. Am. Chem. Soc.* **2002**, *124*, 2456.

Ph–(CH$_2$)$_3$–OTBS → (5% $ZrCl_4$, Ac_2O , MeCN; rt , 5 min) → Ph–(CH$_2$)$_3$–OAc 84%

Reddy, Ch.S.; Smitha, G.; Chandrasekhar, S. *Tetrahedron Lett.* **2003**, *44*, 4693.

OTMS, OMe → (1. ethylene oxide , hexane/DCM; –60°C → –20°C; 2. *p*-TsOH , rt) → O, O, Ph 62%

Maslak, V.; Matovic, R.; Saicic, R.N. *Tetrahedron Lett.* **2002**, *43*, 5411.

$OSiMe_3$, OEt → (LPDE , MeOTf; rt , 1 h) → Me_3Si, O, OEt 76%

Heydari, A.; Alijanianzadeh, R. *Chem. Lett.* **2003**, *32*, 226

O → ([(tpp)Ct(thf)$_2$] [Co(CO)$_4$]; 900 psi CO , 60°C , 6h) → O, O >99%

Schmidt, J.A.R.; Mahadevan, V.; Getzler, Y.D.Y.L.; Coates, G.W. *Org. Lett.* **2004**, *6*, 373.

SECTION 115: ESTERS FROM HALIDES AND SULFONATES

PhI → (cat dendrimer-PdCl$_2$, 100 psi CO; 105°C , 4.3 NEt$_3$, 6 h) → $PhCO_2Me$ 90%

Antebi, S.; Arya, P.; Manzer, L.E.; Alper, H. *J. Org. Chem.* **2002**, *67*, 6623.

H_2N–C$_6$H$_4$–Br → (3% (BINAP)PdCl$_2$, MeOH; 1.3 NEt$_3$, 50 psi CO , 2 d; 100°C) → H_2N–C$_6$H$_4$–CO_2Me 50%

Albaneze-Walker, J.; Bazaral, C.; Lavey, T.; Dormer, P.G.; Murry, J.A. *Org. Lett.* **2004**, *6*, 2097.

OTMS, OMe → (5% Pd(dba)$_2$, 10% P(*t*-Bu)$_3$, PhBr; 0.5 ZnF$_2$, DMF , 80°C , 12 h) → Ph, CO_2Me quant

Liu, X.; Hartwig, J.F. *J. Am. Chem. Soc.* **2004**, *126*, 5182.

Ph-I → (PhSeSnBu$_3$, CO , cat Pd(Ph$_3$)$_4$; toluene , 80°C , 5 h) → O, Ph, SePh 89%

Hasegawa, M.; Ishii, H.; Fuchigami, T. *Tetrahedron Lett.* **2002**, *43*, 1503.

N, Br → (BrZrCH$_2$CO$_2$*t*-Bu , microwaves; cat Pd(PPh$_3$)$_4$, THF) → N, CO$_2$*t*-Bu 90%

Bentz, E.; Moloney, M.G.; Westaway, S.M. *Tetrahedron Lett.* **2004**, *45*, 7395.

$Me(CH_2)_7Br$ —— MIM , $MeCO_2^-$,2h / MIM = an imidazolium mesylate ——→ $MeCO_2C_8H_{17}$ >95%

Brinchi, L.; Germani, R.; Savelli, G. *Tetrahedron Lett.* **2003**, *44*, 2027.

Br, Ph, O —— $(Et_4N)_2CO_2$, MeCN / rt ——→ O, Ph, O, O, Ph 54%

Mucciante, V.; Rossi, L.; Feroci, M.; Sotgiu, G. *Synth. Commun.* **2002**, *32*, 1205.

Ph-I —— CO , EtOH , DBU , 120°C , Pd cat ——→ $PhCO_2Et$ 95%

Ramesh, C.; Kubota, Y.; Miwa, M.; Sugi, Y. *Synthesis* **2002**, 2171.

Ph-I —— EtOH , Pd/C , CO , NEt_3 , 140-°C , PhH ——→ $PhCO_2Et$ 80%

Ramesh, C.; Nakamura, R.; Kubota, Y.; Miwa, M.; Sugi, Y. *Synthesis* **2003**, 501.

$PhCH_2Cl$ —— $AcSK/SiO_2$, PhH , 50°C , 3 h ——→ PhSAc 90%

Aoyama, T.; Takido, T.; Kodomari, M. *Synth. Commun.* **2003**, *33*, 3817.

n-$C_{10}H_{21}$–I —— 50 atm CO , 0.3 AIBN , H_2O / 5 H_3PO_2 , 7 $NaHCO_3$, 75°C / 0.2 CTAB , 15 h ——→ n-$C_{10}H_{21}$–CO_2n-$C_{10}H_{12}$ 74%

Sugiura, M.; Hagio, H.; Kobayashi, S. *Chem. Lett.* **2003**, *32*, 898.

Related Method: Section 25 (Acid Derivatives from Halides and Sulfonates).

SECTION 116: ESTERS FROM HYDRIDES

This section contains examples of the reaction R-H → RCO_2R' or $R'CO_2R$ (R = alkyl, aryl, etc.).

—— N_2CHCO_2Et , syringe pump addition / $TpBr_3$-Cu(NCMe) , DCM , EDA ——→ CO_2Et + CO_2Et (80 : 20) 53%

Caballero, A.; Díaz-Requeso, M.M.; Belderraín, T.R.; Nicasio, M.C.; Trofimenko, S.; Pérez, J. *J. Am. Chem. Soc.* **2003**, *125*, 1446.

MeO, N —— $PhI(OAc)_2$, 5% $Pd(OAc)_2$ / $AcOH/Ac_2O$, 100°C ——→ MeO, N, OAc 61%

Desai, L; Hull, K.L.; Sanford, M.S. *J. Am. Chem. Soc.* **2004**, *126*, 9542.

—— $[(HOCH_2CH_2NHCOCH_2NCH_2]_2$, 2 d / $Cu(MeCN)_2$ BF_4 , H_2O , 80°C ——→ OBz 67%

LeBras, J.; Muzart, J. *Tetrahedron Lett.* **2002**, *43*, 431.

$PhNMe_2$ —[$PhCH_2CO_2Et$, $TiCl_4$, DCM]→ Me_2N–C_6H_4–CH(CO_2Et)Ph 89%

Periasamy, M.; KishoreBabu, N.; Jayakumar, K.N. *Tetrahedron Lett.* ***2003***, *44*, 8939.

$PhCO_2H$ —[$PhCH_3$, $NaBrO_3/NaHSO_3$, AcOEt , H_2O / 40h , rt]→ $PhCO_2CH_2Ph$ 95%

Khan, K.M.; Maharvi, G.M.; Hayat, S.; Zia-Ullah; Choudhary, M.I.; Atta-ur-Rahman *Tetrahedron* ***2003***, *59*, 5549.

cyclohexene —[5.2 1,3-propanediol , 0.2 $Cu(OTf)_2$ / 0.45 L-proline , 72h , 1.5 $PhCO_2H$, rt]→ 3-OBz-cyclohexene 70% (34% ee)

LeBras, J.; Muzart, J. *Tetrahedron: Asymmetry* ***2003***, *14*, 1911.

cyclohexene —[Cu_2O , $C_{11}F_{23}CO_2H$, 3 h / *t*-BuO_3CPh , $(CF_3)_2CHOH$]→ cyclohexenyl–O_2CPh 72%

Fache, F.; Piva, O. *Synlett* ***2002***, 2035.

REVIEW:

"Copper Catalyzed Allylic Oxidation with Peresters"
Andrus, M.B.; Lashley, J.C. *Tetrahedron* ***2002***, *58*, 845.

Also via: Section 26 (Acid Derivatives from Hydrides) and Section 41 (Alcohols and Thiols from Hydrides).

SECTION 117: ESTERS FROM KETONES

3-O_2N-C_6H_4-C(O)Et —[DMA , DMF , MeOH / reflux , 16 h]→ 3-O_2N-C_6H_4-C(O)OMe 87%

Zhang, N.; Vozzolo, J. *J. Org. Chem.* ***2002***, *67*, 1703.

Br–C(O)–CH_2CH_3 —[PhCH_2CH_2CHO , 2% chiral Al catalyst / benzotrifluoride , *i*-Pr_2NEt , –25°C]→ β-lactone (PhCH_2CH_2) 71% (95:5 *syn:anti*) (90% ee)

Nelson, S.G.; Zhu, C.; Shen, X. *J. Am. Chem. Soc.* ***2004***, *126*, 14.

C_8H_{17}–cyclobutanone —[UHP , 5% Zr-salen complex , DCM , rt]→ C_8H_{17}–γ-lactone 81% (63% ee)

Watanabe, A.; Uchida, T.; Ito, K.; Katsuki, T. *Tetrahedron Lett.* ***2002***, *43*, 4481.

cyclopentanone —[2% $MeReO_3$, 6 eq 50% aq. H_2O_2 , 60°C / [bmim] BF_4 , 1d]→ δ-valerolactone 70%

Bernini, R.; Coratti, A.; Fabrizi, G.; Goggiamani, A. *Tetrahedron Lett.* ***2003***, *44*, 8991.

O_2 , compressed CO_2
t-BuCHO , rt
74%

Bolm, C.; Palazzi, C.; Francio, G.; Leitner, W. *Chem. Commun.* **2002**, 1588.

MeO
H_2O_2 , cat NaOAc , CF_3CH_2OH
cat chiral bis(flavin) , MeOH , H_2O
MeO
67% (61%ee)

Murahashi, S.-I.; Ono, S.; Imada, Y. *Angew. Chem. Int. Ed.* **2002**, *41*, 2366.

Cl
Acinetobacter BL21(DE3) (pMM4)
Cl
78%(>99% ee)

Mihovilovic, M.D.; Müller, B.; Kayser, M.M.; Stewart, J.D.; Stanetty, P. *Synlett* **2002**, 703.

Cl
20% Me_2AlCl , cumene hydroperoxide
chiral binapthyl diol catalyst
toluene , –30°C
Cl
92% (84% ee)

Bolm, C.; Frison, J.-C.; Zhang, Y.; Wulff, W.D. *Synlett* **2004**, 1619.

cat $Bi(OTf)_3$, mcpba
DCM , rt
90%

Alam, M.M.; Varala, R.; Adapa, S.R. *Synth. Commun.* **2003**, *33*, 3035.

mcpba , [bmim] BF_4
2 h
87%

Yadav, J.S.; Reddy, B.V.S.; Basak, A.K.; Narsaiah, A.V. *Chem. Lett.* **2004**, *33*, 248.

REVIEW:

"The Baeyer-Villiger Reaction: New Developments Toward Greener Procedures"
ten Brink, G.-J.; Arends, I.W.C.E.; Shedon, R.A. *Chem. Rev.* **2004**, *104*, 4105.

Also via: Section 27 (Acid Derivatives from Ketones).

SECTION 118: ESTERS FROM NITRILES

PhCN $\xrightarrow{BF_3 \cdot OEt_2\text{ , MeOH , reflux , 23 h}}$ $PhCO_2Me$ 82%

Jyachitra, G.; Yasmoen, N.; Rao, K.S.; Ralte, S.L.; Srinivasan, R.; Singh, A.K. *Synth. Commun.* **2003**, *33*, 3461.

SECTION 119: ESTERS FROM ALKENES

cat $Pd(OAc)_2$, cat PPMoV
MeOH, NH_4Cl, $MeSO_3H$
5 atm CO, air, 60°C
CO_2Me
Ph
CO_2Me
45%

NPMoV = molybdovanadophosphate

Yokota, T.; Sakaguchi, S.; Ishii, Y. *J. Org. Chem.* **2002**, *67*, 5005.

CO_2H
Ph
$PhI(OAc)_2$, DCM
ca. 4 M
Ph
O
O
AcO
87%

Boye, A.C.; Meyer, D.; Ingison, C.K.; French, A.N.; Wirth, T. *Org. Lett.* **2003**, *5*, 2157.

N
OH, toluene, 20h
cat $Rh_4(CO)_{12}$, 100°C, 3 atm CO
O
O
N
87%

Yokota, K.; Tatamidani, H.; Fukumoto, Y.; Chatani, N. *Org. Lett.* **2003**, *5*, 4329.

ICH_2CO_2Et, 3 $PhSO_2N_3$, 80°C
6% di-*tert*-butyl hyponitrite, 4h
$(Bu_3Sn)_2$
N_3
CO_2Et
79%

Panchaud, P.; Renaud, P. *J. Org. Chem.* **2004**, *69*, 3205.

Bu
DMF, 135°C, 3 h
5% $Ru_3(CO)_{12}$
N
OCHO
Bu
CO_2CH_2(2-Py)
+
CO_2CH_2(2-Py)
Bu
(74 : 26) 98%

Ko, S.; Na, Y.; Chang, S. *J. Am. Chem. Soc.* **2002**, *124*, 750.

H
O
O
N
, 8h
cat $Ru_3(CO)_{12}$, DMF, 135°C
O
O
N
70%

Na, Y.; Ko, S.; Hwang, L.K.; Chang, S. *Tetrahedron Lett.* **2003**, *44*, 4475.

MeO
I
CH_2=$CH(OEt)_2$, DMF, 90°C
2 Bu_3N, Bu_4NCl, 3 h
3% $Pd(OAc)_2$
MeO
CO_2Et
91%

Battistuzzi, G.; Cacchi, S.; Fabrizi, G.; Bernini, R. *Synlett* **2003**, 1133.

Also via: Section 44 (Alcohols and Thiols from Alkenes).

SECTION 120: ESTERS FROM MISCELLANEOUS COMPOUNDS

t-Bu–S–S–t-Bu → (cat $VO(acac)_2$, aq H_2O_2 / cat chiral Schiff base ligand) → t-Bu–S(=O)–S–t-Bu 87% (87% ee)

Blum, S.A.; Berman, R.G.; Ellman, J.A. *J. Org. Chem.* ***2003***, *68*, 150.

$M(CO)_5$ … OMe, $Cr(CO)_3$ → (HO allyl / NaH) → 75%

Kamikawa, K.; Tachibana, A.; Shimizu, Y.; Uchida, K.; Furusho, M.; Uemura, M. *Org. Lett.* ***2004***, *6*, 4307.

$(CO)_5Cr$=C(OMe)Ph → (cat Bu_4NF , air , DME / rt) → O=C(OMe)Ph 84%

Barluenga, J.; Andina, F.; Fernández-Rodríguez, M.A.; García-García, P.; Merino, I. *J. Org. Chem.* ***2004***, *69*, 7352.

OH → (*o*-iodoxybenzoic acid , PhH) → O 80%

Moorthy, J.N.; Singhal, N.; Mal, P. *Tetrahedron Lett.* ***2004***, *45*, 309.

Ph … OTMS, OTMS → (epoxide , DCM-hexane , $TiCl_4$) → O, O, Ph 62%

Maslak, V.; Matovic, R.; Saicic, R.N. *Tetrahedron* ***2004***, *60*, 8957.

Ph–S(=O)–Ph → (1. Tf_2O , 1.2 Proton Sponge 1.2 BnOH , toluene , –78°C / 2. $NaCH(CO_2Me)_2$, –78°C → rt) → Ph–CH_2–$CH(CO_2Me)_2$ 94%

Takuwa, T.; Yoshitaka, J.; Onishi, Y.; Matsuo, J.; Mukaiyama, T. *Chem. Lett.* ***2004***, *33*, 8.

CHAPTER 9

PREPARATION OF ETHERS, EPOXIDES, AND THIOETHERS

SECTION 121: ETHERS, EPOXIDES, AND THIOETHERS FROM ALKYNES

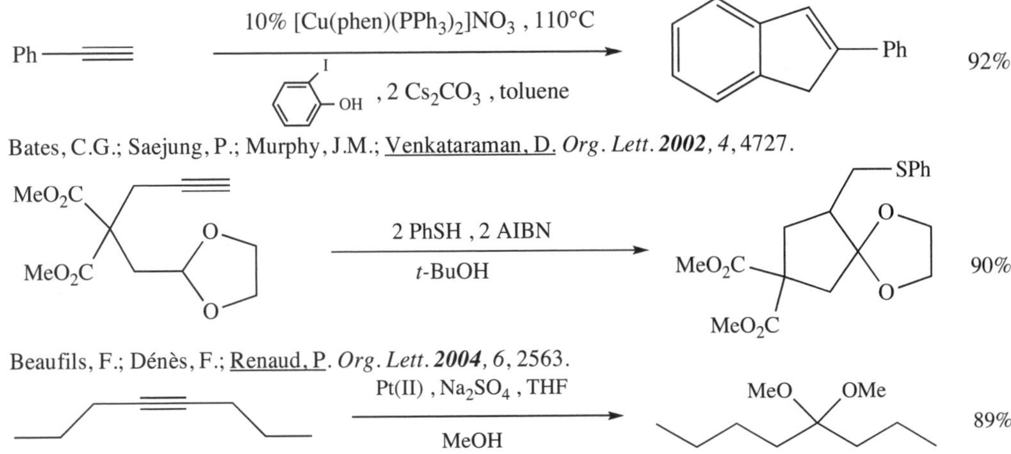

Bates, C.G.; Saejung, P.; Murphy, J.M.; Venkataraman, D. *Org. Lett.* **2002**, *4*, 4727.

Beaufils, F.; Dénès, F.; Renaud, P. *Org. Lett.* **2004**, *6*, 2563.

Hartman, J.W.; Sperry, L. *Tetrahedron Lett.* **2004**, *45*, 3787.

SECTION 122: ETHERS, EPOXIDES, AND THIOETHERS FROM ACID DERIVATIVES

Ph⌒SO$_2$Cl → (1. Zn , AcOH , H_2O; 2. Me_3SiCl , reflux; 3. PhCH(OH)Me) → Ph⌒S⌒CH(Ph)Me 78%

Martin, M.T.; Thomas, A.M.; York, D.G. *Tetrahedron Lett.* **2002**, *43*, 2145.

SECTION 123: ETHERS, EPOXIDES, AND THIOETHERS FROM ALCOHOLS AND THIOLS

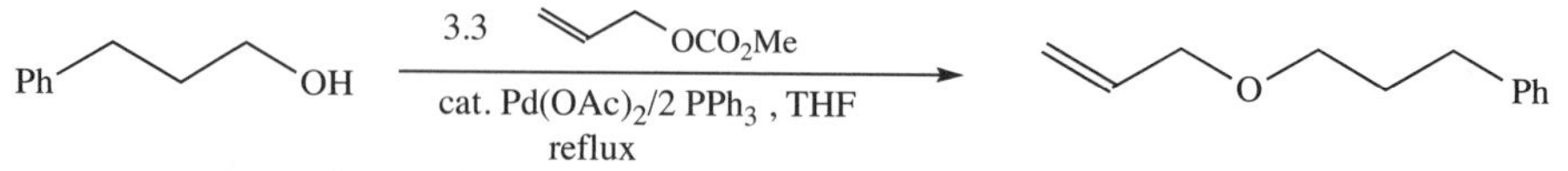

Ph(CH2)3OH → 3.3 CH2=CHCH2OCO2Me, cat. $Pd(OAc)_2$/2 PPh_3, THF, reflux → allyl 3-phenylpropyl ether >95%

Haight, A.R.; Stoner, E.J.; Peterson, M.J.; Grover, V.K. *J. Org. Chem.* **2003**, *68*, 8092.

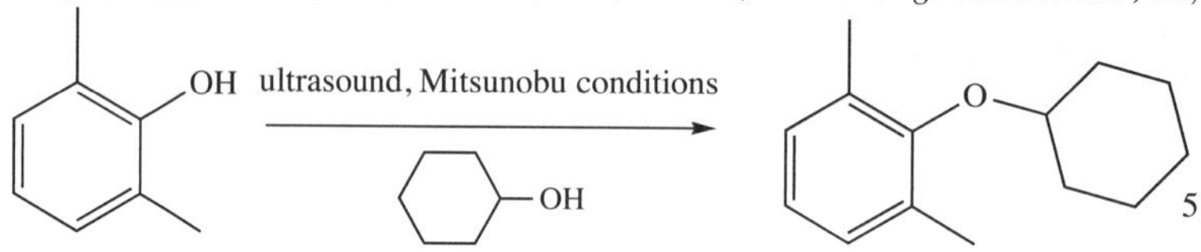

OH ultrasound, Mitsunobu conditions

OH

O

51% (32% without sonication)

Lepore, S.D.; He, Y. *J. Org. Chem.* **2003**, *68*, 8261.

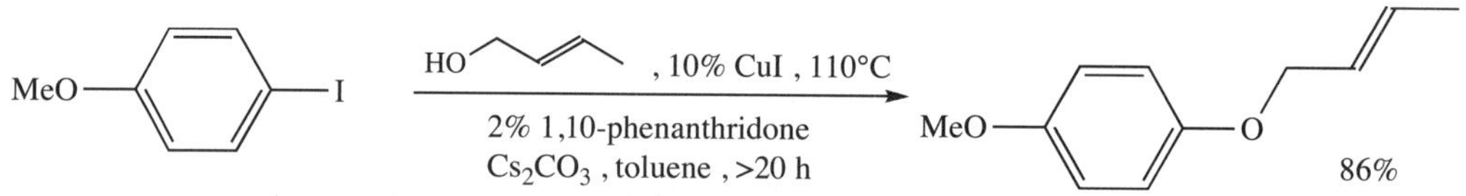

MeO I HO , 10% CuI , 110°C

2% 1,10-phenanthridone

Cs_2CO_3 , toluene , >20 h

MeO O 86%

Wolter, M.; Nordmann, G.; Job, G.E.; Buchwald, S.L. *Org. Lett.* **2002**, *4*, 973.

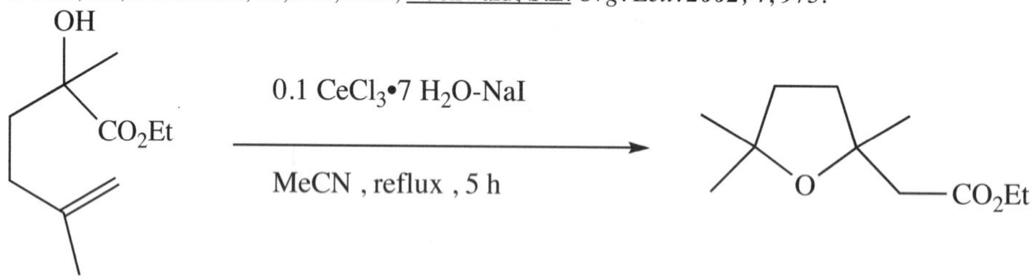

OH

CO_2Et

0.1 $CeCl_3$•7 H_2O-NaI

MeCN , reflux , 5 h

O CO_2Et 91%

Martta, E.; Foresti, E.; Marcelli, T.; Peri, F.; Righi, P.; Scardovi, N.; Rosini, G. *J. Org. Chem.* **2002**, *67*, 4451.

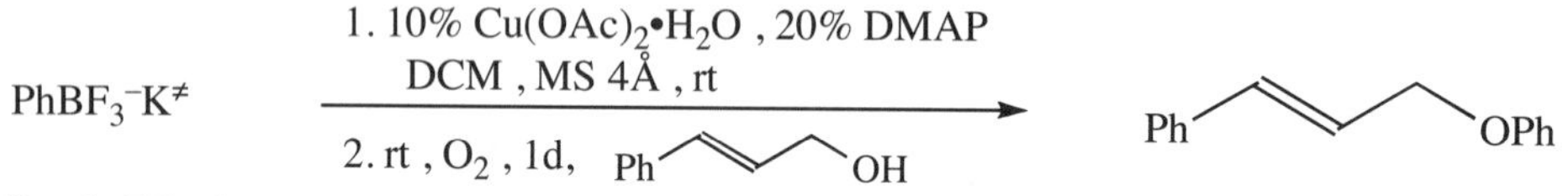

$PhBF_3^-K^{\neq}$

1. 10% $Cu(OAc)_2$•H_2O , 20% DMAP

DCM , MS 4Å , rt

2. rt , O_2 , 1d, Ph OH

Ph OPh 89%

Quach, T.D.; Batey, R.A. *Org. Lett.* **2003**, *5*, 1381.

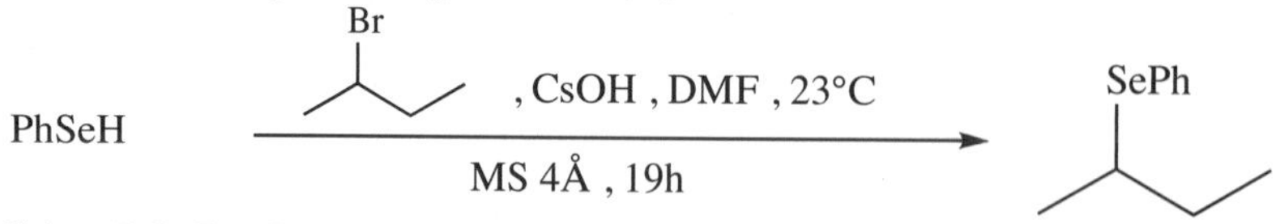

Br

PhSeH , CsOH , DMF , 23°C

MS 4Å , 19h

SePh 88%

Cohen, R.J.; Fox, D.L.; Salvatore, R.N. *J. Org. Chem.* **2004**, *69*, 4265.

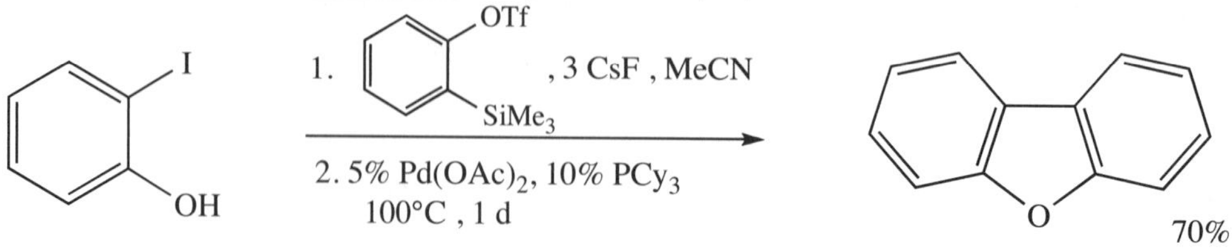

I OTf

1. , 3 CsF , MeCN

SiMe3

OH 2. 5% $Pd(OAc)_2$, 10% PCy_3

100°C , 1 d

O 70%

Liu, Z.; Larock, R.C. *Org. Lett.* **2004**, *6*, 3739.

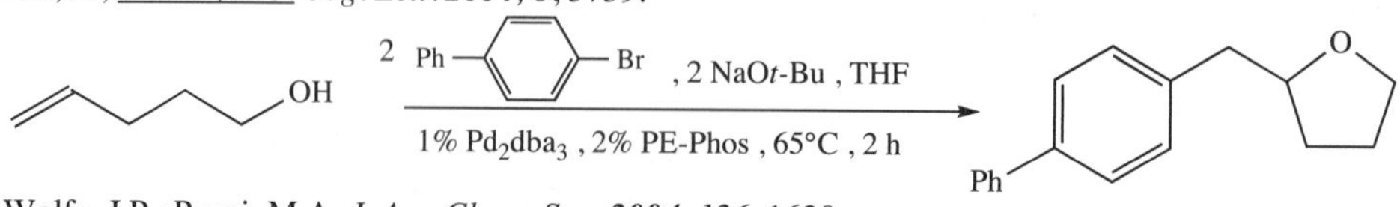

OH 2 Ph Br , 2 NaO*t*-Bu , THF

1% Pd_2dba_3 , 2% PE-Phos , 65°C , 2 h

O

Ph 70%

Wolfe, J.P.; Rossi, M.A. *J. Am. Chem. Soc.* **2004**, *126*, 1620.

PPh$_2$ (polymer-bound) → 1. DEAD , DCM , rt; 2. 2-naphthol (OH); 3. HO–(piperidine)N-Bn , NEt$_3$ → 72%

Lizarzabru M.E.; Shuttleworth, S.J. *Tetrahedron Lett.* ***2002***, *43*, 2157.

PhOH —30 (MeO)$_2$C=O , 1.5 K_2CO_3 , 0.6 Bu_4NBr / 5 h→ PhOMe 99%

Ouk, S.; Thiebaud, S.; Borredon, E.; Legars, P.; Lecomte, L. *Tetrahedron Lett.* ***2002***, *43*, 2661.

Ph_2CHOH —MeOH , $Fe(NO_3)_3$•9 H_2O , 70°C→ Ph_2CHOMe 88%

Namboodiri, V.V.; Varma, R.S. *Tetrahedron Lett.* ***2002***, *43*, 4593.

MeO_2C–CH$_2$CH$_2$–Br —EtSH , CaF-Celite , MeCN , rt→ MeO_2C–CH$_2$CH$_2$–S–Et 60%

Shah, S.T.A.; Khan, K.M.; Heinrich, A.M.; Voelter, W. *Tetrahedron Lett.* ***2002***, *43*, 8281.

MeO-C$_6$H$_4$-CH$_2$O–C(=NH)CCl$_3$ —1.6 HO–CH$_2$CH$_2$C(O)CH$_3$, toluene / cat La(OTf)$_3$→ MeO-C$_6$H$_4$-CH$_2$O–CH$_2$CH$_2$C(O)CH$_3$ 88%

Rai, A.N.; Basu, A. *Tetrahedron Lett.* ***2003***, *44*, 2267.

O_2N-C$_6$H$_4$-F —HO-C$_6$H$_4$-C(O)CH$_3$ / K_2CO_3 , DMF , 120°C→ O_2N-C$_6$H$_4$–O–C$_6$H$_4$-C(O)CH$_3$ 99%

Mitchell, L.H.; Barvian, N.C. *Tetrahedron Lett.* ***2004***, *45*, 5669.

PhOH —Zn/DMF , BuBr , microwaves , 2.5 min→ Ph-OBu 85%

Paul, S.; Gupta, M. *Tetrahedron Lett.* ***2004***, *45*, 8825.

OBn, OH, CO$_2$Et arene + F, NO$_2$, OTBS, OTBS, OH arene —LiOH , DMF / rt , 1d→ OBn, NO$_2$, CO$_2$Et, OTBS, OTBS, OH product 92%

Ankala, S.V.; Fenteany, G. *Synlett* ***2003***, 825.

PhSH —PhI . 5% CuI , 20% N-methylglycine / dioxane , 2.5 KOH , 100°C , 1 d→ Ph-S-Ph 96%

Deng, W.; Zou, Y.; Wang, Y.-F.; Liu, L.; Guo, Q.-X. *Synlett* ***2004***, 1254.

C$_6$H$_5$–OH —1.5 Ph_2POBn , DMBQ / DCM , rt, 0.5 h→ C$_6$H$_5$–OBn 92%

DMBQ = 2,6-dimethyl-1,4-benzoquinone

Shintou, T.; Kikuchi, W.; Mukaiyama, T. *Chem. Lett.* ***2003***, *32*, 22.

$PhCH_2OH$ —$ClPPh_2$ / BuLi→ $PhCH_2OPPh_2$ —ArSSAr / $CHCl_3$, rt→ $PhCH_2S$-Ar 98%

Ar = -Cl-phenyl

Mukaiyama, T.; Ikegai, K. *Chem. Lett.* ***2004***, *33*, 1522.

PhBr , cat $CuCO_3$, $Cu(OH)_2$, H_2O

K_3PO_4 , THF , 115°C

O^-Na^+

Ghosh, R.; Samuelson, A.G. *New J. Chem.* **2004**, *28*, 1390.

SECTION 124: ETHERS, EPOXIDES, AND THIOETHERS FROM ALDEHYDES

Bn, Et_3SiO, CHO, OH, Bn, $SiPhMe_2$

$SiPhMe_2$

2 eq BF_3•OEt_2

62%

Angle, S.R.; Belanger, D.S.; El-Said, N.A. *J. Org. Chem.* **2002**, *67*, 7699.

PhCHO

I , *t*-BuOH , H_2O , 2 d

NaOH , rt , S

O, Ph

85% (2.3:1 *t:c*)

Zanardi, J.; Lamazure, D.; Minière, S.; Reboul, V.; Metzner, P. *J. Org. Chem.* **2002**, *67*, 9083.

Cl, CHO, Cl

$C_8H_{15}-S^+(Me)CO_2^-$

DCE , 60°C

Cl, O, Cl

85%

Forbes, D.C.; Standen, M.C.; Lewis, D.L. *Org. Lett.* **2003**, *5*, 2283.

HO, Ph, $SiPh_2t$-Bu

1. TFA

2. C_3H_7CHO

$SiPh_2t$-Bu, Ph, OH

72%

Yadav, V.K.; Kumar, M.V. *J. Am. Chem. Soc.* **2004**, *126*, 8652.

OH, Ph

1% $PtCl_2(CH_2{=}CH_2)_2$, DCE , 70°C

2% $P(4\text{-}CF_3\text{-}C_6H_4)_3$

O, Ph

66% (8:1)

Qian, H.; Han, X.; Widenhoefer, R.A. *J. Am. Chem. Soc.* **2004**, *126*, 9536.

Ph, O, O, Ph, O, O, S

O_2N–C_6H_4–CHO , BnBr , rt

NaOH , 2d

O_2N, O

71% (84% ee)

Winn, C.L.; Bellenie, B.R.; Gordman, J.M. *Tetrahedron Lett.* **2002**, *43*, 5427.

SiMe$_3$ / OTf aryl + 2,6-dimethoxybenzaldehyde (OMe, CHO, OMe) → KF-18-crown-6 , THF, 10°C , 20 h → MeO, OMe, O 61%

Yoshida, H.; Watanabe, M.; Rukushima, H.; Ohshita, J.; Kunai, A. *Org. Lett.* **2004**, *6*, 4049.

PhCHO → $PhCH_2Br$, NaOH , NaI , 2d / chiral ferrocenyl sulfide → O, Ph, Ph 61% (76:24 *trans:cis*) 90% ee

Minière, S.; Reboul, V.; Metzner, P.; Fochi, M.; Bonini, B.F. *Tetrahedron: Asymmetry* **2004**, *15*, 3275.

PhCHO → $BuOSiHMe_2$, cat TMSI , DCM / 2h , 0°C → rt → $PhCH_2OBu$ quant

Miura, K.; Ootsuka, K.; Suda, S.; Nishikari, H.; Hosomi, A. *Synlett* **2002**, 313.

PhCHO → 1. cat $CeCl_3$, 2 Et_2Zn , THF , rt / 2. aq NH_4Cl → OTMS, Ph, Et 93%

Fischer, S.; Groth, U.; Jeske, M.; Schütz, T. *Synlett* **2002**, 1922.

PhCHO → $BiCl_3$, EtOH , Et_3SiH / DCM , rt , 3 h → Ph, O 74%

Wada, M.; Nagayama, S.; Mizutani, K.; Hiori, R.; Miyoshi, N. *Chem. Lett.* **2002**, 248.

PhCHO + $ClCH_2CO_2Et$ → cat PsTEAC , THF , KOH → O, Ph, CO_2Et 82% (6.5:1 *cis:trans*)

Wang, Z.-T.; Xu, L.-W.; Xia, C.-G.; Wsng, H.-Q. *Helv. Chim. Acta* **2004**, *87*, 1958.

PhCHO → chiral bis(sulfide) , 3 NaOH , 3 BnBr / *t*-BuOH - H_2O , 3 d → O, Ph, Ph 67% (70:30 *cis:trans*)

Ishizaki, M.; Hoshino, O. *Heterocycles* **2002**, *57*, 1399.

SECTION 125: ETHERS, EPOXIDES, AND THIOETHERS FROM ALKYLS, METHYLENES, AND ARYLS

BnO, OBn cyclobutane → CAN , MeOH , rt / 2h → OMe, OBn, OMe, BnO 69%

Nir, V.; Rajan, R.; Mohanan, K.; Sheeba, V. *Tetrahedron Lett.* **2003**, *44*, 4585.

SECTION 126: ETHERS, EPOXIDES, AND THIOETHERS FROM AMIDES

NO ADDITIONAL EXAMPLES

SECTION 127: ETHERS, EPOXIDES, AND THIOETHERS FROM AMINES

NO ADDITIONAL EXAMPLES

SECTION 128: ETHERS, EPOXIDES, AND THIOETHERS FROM ESTERS

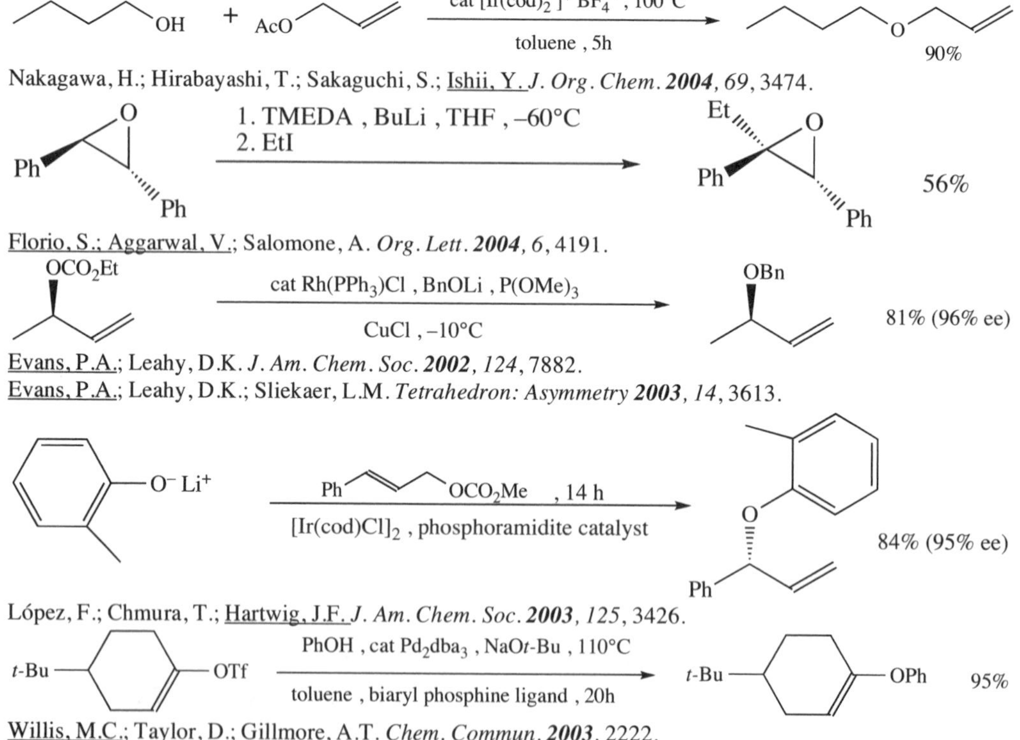

Nakagawa, H.; Hirabayashi, T.; Sakaguchi, S.; Ishii, Y. *J. Org. Chem.* ***2004***, *69*, 3474.

Florio, S.; Aggarwal, V.; Salomone, A. *Org. Lett.* ***2004***, *6*, 4191.

Evans, P.A.; Leahy, D.K. *J. Am. Chem. Soc.* ***2002***, *124*, 7882.
Evans, P.A.; Leahy, D.K.; Sliekaer, L.M. *Tetrahedron: Asymmetry* ***2003***, *14*, 3613.

López, F.; Chmura, T.; Hartwig, J.F. *J. Am. Chem. Soc.* ***2003***, *125*, 3426.

Willis, M.C.; Taylor, D.; Gillmore, A.T. *Chem. Commun.* ***2003***, 2222.

SECTION 129: ETHERS, EPOXIDES, AND THIOETHERS FROM ETHERS, EPOXIDES, AND THIOETHERS

[bmim]PF_6 , H_2O / KSCN , rt — 95%

Yadav, J.S.; Reddy, B.V.S.; Reddy, Ch.S.; Rajasekhar, K. *J. Org. Chem.* **2003**, *68*, 2525.

1.5 Me_3Al , 1.5 NEt_3 / 1.5 TMSSiOTf , –50°C — OTMS — 93%

Shanmugam, P.; Miyashita, M. *Org. Lett.* **2003**, *5*, 3265.

PMBO, OH, O — 10% Me_3S=O I , NaH , DMSO / 80°C , 36 h — PMBO, O, HO — 86%

Schomaker, J.M.; Pulgam, V.R.; Borhan, B. *J. Am. Chem. Soc.* **2004**, *126*, 13600.

$C_7H_{15}CH_2OTMS$ — 10% $Cu(OTf)_2$, PhCHO , Et_3SiH / DCM , 30 min → $C_7H_{15}CH_2OBn$ 85%

Takeda, T.; Shono, T.; Ito, K.; Sasak, H. *Tetrahedron Lett.* **2003**, *44*, 7837.

MeO, OTMS — Nafion-H , hexane , rt , 15 min → (MeO–)$_2$O 90%

Zolfigol, M.A.; Mohammadpor-Baltork, I.; Habibi, D.; Mirjalili, B.F.; Bamoniri, A. *Tetrahedron Lett.* **2003**, *44*, 8165.

Ph, O — $(EtO)_2P(=O)H$, NH_4OAc , S , Al_2O_3 , 30 min → Ph, S 82%

Tamami, B.; Kolahdoozan, M. *Tetrahedron Lett.* **2004**, *45*, 1535.

Ph, OMe, OMe — $Bu_nMe(MgBr)_{n-2}$, 4h / –78°C , BF_3•OEt_2 → Ph, OMe 94%

Hojo, M.; Ushioda, N.; Hosomi, A. *Tetrahedron Lett.* **2004**, *45*, 4499.

Ph, O — 5% $InBr_3$, KSCN , MeCN , 3h → Ph, S 90%

Yadav, J.S.; Reddy, B.V.S.; Baishya, G. *Synlett* **2003**, 396.

Ph, O — thiourea, 0.2 $LiBF_4$ / MeCN , reflux , 20 min → Ph, S 93%

Kazemi, F.; Kiasat, A.R.; Ebrahimi, S. *Synth. Commun.* **2003**, *33*, 595.

SECTION 130: ETHERS, EPOXIDES, AND THIOETHERS FROM HALIDES AND SULFONATES

$C_{11}H_{23}CH_2I$ — PhSeSePh , cat La , THF / cat I_2 → $C_{11}H_{23}CH_2SePh$ 78%

Nishino, T.; Okada, M.; Kuroki, T.; Watanabe, T.; Nishiyama, Y.; Sonoda, N. *J. Org. Chem.* **2002**, *67*, 8696.

Ph–I $\xrightarrow[\text{DMF , 110°C , 30 h}]{\text{0.5 PhSeSePh , cat Cu}_2\text{O-bpy , Mg}}$ Ph–Se–Ph 94%

Taniguchi, N.; Onami, T. *J. Org. Chem.* **2004**, *69*, 915.

3,5-dimethyliodobenzene $\xrightarrow[\textit{i}\text{-PrOH , 20 h}]{\text{PhSH , 5\% CuI , 2 K}_2\text{CO}_3\text{, 2 ethylene glycol , 80°C}}$ 3,5-dimethylphenyl–SPh 92%

Kwong, F.Y.; Buchwald, S.L. *Org. Lett.* **2002**, *4*, 3517.

$PhCH_2Cl$ $\xrightarrow{\text{PhSeSePh , InI , DCM, rt , 12 min}}$ $PhCH_2SePh$ 91%

Ranu, B.C.; Mandal, T.; Samanta, S. *Org. Lett.* **2003**, *5*, 1439.

Cl–C₆H₄–OH $\xrightarrow[\text{K}_2\text{CO}_3\text{ , DMSO , 5 min, microwaves}]{\text{F–C}_6\text{H}_4\text{–CN}}$ Cl–C₆H₄–O–C₆H₄–CN 93%

Li, F.; Wang, Q.; Ding, Z.; Tao, F. *Org. Lett.* **2003**, *5*, 2169.

Ph–C₆H₄–Br $\xrightarrow[\text{30\% \textit{N,N}-dimethylglycine•HCl, dioxane , 50°C , 1d}]{\text{HO–C}_6\text{H}_4\text{–CH}_3\text{ (meta), CuI , Cs}_2\text{CO}_3}$ Ph–C₆H₄–O–C₆H₄–CH₃ (meta) 90%

Ma, D.; Cai, Q. *Org. Lett.* **2003**, *5*, 3799.

2-Br–C₆H₄–CF₃ $\xrightarrow[\text{3. PhCH}_2\text{Br}]{\text{1. BuLi , –78°C; 2. S}}$ 2-SBn–C₆H₄–CF₃ 82%

Ham, J.; Yang, I.; Kang, H. *J. Org. Chem.* **2004**, *69*, 3236.

PhI $\xrightarrow[\text{10\% NiPr-bpy , 2 eq Zn}]{\text{0.5 (BuS)}_2\text{ , DMF , 110°C , 2 d}}$ Ph–S–Bu 81%

Taniguchi, N. *J. Org. Chem.* **2004**, *69*, 6904.

PhBr $\xrightarrow[\text{2 \textit{i}-Pr}_2\text{NEt , dioxane , reflux}]{\text{PhSH , 2.5\% Pd}_2\text{dba}_3\text{ , 5\% XANTPHOS}}$ Ph-S-Ph 85%

Itoh, T.; Mase, T. *Org. Lett.* **2004**, *6*, 4587.

Se $\xrightarrow[\text{3. BuBr , 20°C , 6h}]{\text{1. 2.1 NaH , DMF , 70°C; 2. 2 BuBr , 20°C , 2h}}$ BuSeBu + BuSeSeBu (89 : 11) 67%

Krief, A.; Derock, M. *Tetrahedron Lett.* **2002**, *43*, 3083.

PhI $\xrightarrow[\text{NaO\textit{t}-Bu , 110°C , toluene}]{\text{PhSeOH , 10\% CuI , 10\% neocuproine}}$ PhSePh 90%

Gujadhar, R.K.; Venkataraman, D. *Tetrahedron Lett.* **2003**, *44*, 81.

MeO–C_6H_4–Br → (F–C_6H_4–$SeSnBu_3$, 16h; cat CuI-phen, DMF, 110°C) → MeO–C_6H_4–Se–C_6H_4–F 90%

Beletskaya, I.P.; Sigeev, A.S.; Peregudov, A.S.; Petrovskii, P.V. *Tetrahedron Lett.* ***2003***, *44*, 7039.

O_2N–C_6H_4–Cl → (PhOH, KOH, 150°C, 10 min; microwaves) → O_2N–C_6H_4–OPh 87%

Chaouchi, M.; Loupy, A.; Margue, S.; Petit, A. *Eur. J. Org. Chem.* ***2003***, 1278.

1,2-bis(chloromethyl)benzene (Cl, Cl) → (Al_2O_3, hexane, reflux; 1 h) → 1,3-dihydroisobenzofuran (O) 91%

Mihara, M.; Ishino, Y.; Minakata, S.; Komatsu, M. *Synlett* ***2002***, 1526.

2-iodotoluene (I, Me) → (0.5 $(PhSe)_2$, Mg, DMF, 30 h; cat Cu_2O-bpy, 110°C) → 2-(SePh)toluene (SePh, Me) 92%

Taniguchi, N.; Onami, T. *Synlett* ***2003***, 829.

I–C_6H_4–CN → (BmiI, PhOH, 110°C, 30 h; BmiI = 1-butyl-3-methylimidazolium iodide) → PhO–C_6H_4–CN 99%

Luo, Y.; Wu, J.X.; Ren, R.X. *Synlett* ***2003***, 1734.

t-Bu–C_6H_4–I → (PhSH, 10% CuI, 2 h; microwaves, 2 Cs_2CO_3) → *t*-Bu–C_6H_4–SPh 89%

Wu, Y.-J.; He, H. *Synlett* ***2003***, 1789.

CH_3–C_6H_4–OH → (PhI, CuI, K_2CO_3; Raney Ni alloy, 12 h) → CH_3–C_6H_4–OPh 99%

Xu, L.-W.; Xia, C.-G.; Li, J.-W.; Hu, X.-X. *Synlett* ***2003***, 2071.

PS-SPh → (1. Yb, rt; 2. 3 allyl bromide (Br)) → PhS–$CH_2CH=CH_2$ 85%

Su, W.; Li, Y.; Zhang, Y. *Synth. Commun.* ***2002***, *32*, 2101.

O_2N–C_6H_4–F → (PhOTBS, Cs_2CO_3, 100°C; DMF) → O_2N–C_6H_4–OPh 92%

Cui, S.-L.; Jiang, Z.-Y.; Wang, Y.-G. *Synlett* ***2004***, 1829.

O_2N–C_6H_4–Cl → (PhOH, KOH, microwaves) → O_2N–C_6H_4–OPh 72%

Rebeiro, G.L.; Khadilkar, B.M. *Synth. Commun.* ***2003***, *33*, 1405.

$PhO^- K^+$ + PhI → (CuCl, 100°C, bmim BF_4) → Ph-O-Ph 85%

Chauhan, S.M.S.; Jain, N.; Kumar, A.; Srinivas, K.A. *Synth. Commun.* ***2003***, *33*, 3607.

Cl–C_6H_4–NO_2 —(PhOH , KOH , microwaves)→ Ph-O–C_6H_4–NO_2

Rebeiro, G.L.; Khadilkar, B.M. *Synth. Commun.* **2004**, *34*, 1405.

PhOH —($PhCH_2Cl$, KOH , bmim PF_6)→ $PhOCH_2Ph$ 92%

Xu, Z.Y.; Xu, D.Q.; Liu, B.Y. *Org. Prep. Proceed. Int.* **2004**, *36*, 156.

3-Bromoquinoline (Br, N) —(KOMe , 18-crown-6 / microwaves)→ 3-methoxyquinoline (OMe, N) 58%

Lloung, M.; Loupy, A.; Marque, S.; Petit, A. *Heterocycles* **2004**, *63*, 297.

REVIEW:

"Renaissance of Ullmann and Goldberg Reactions: Progress in Copper Catalyzed C–N, C–O and C–S Coupling"
Kunz, K.; Scholz, U.; Ganzer, D. *Synlett* **2003**, 2428.

Related Method: Section 123 (Ethers, Epoxides and Thioethers from Alcohols and Thiols).

SECTION 131: ETHERS, EPOXIDES, AND THIOETHERS FROM HYDRIDES

Indole (N–H) —(PhSH , cat $VO(acac)_2$, O_2 / toluene , 60°C , 12 h)→ 3-(SPh)indole (N–H) 44%

Maeda, Y.; Koyabu, M.; Nishimura, T.; Uemura, S. *J. Org. Chem.* **2004**, *69*, 7688.

SECTION 132: ETHERS, EPOXIDES, AND THIOETHERS FROM KETONES

NO ADDITIONAL EXAMPLES

Related Method: Section 124 (Epoxide, Ethers and Thioethers from Aldehydes).

SECTION 133: ETHERS, EPOXIDES, AND THIOETHERS FROM NITRILES

NO ADDITIONAL EXAMPLES

SECTION 134: ETHERS, EPOXIDES, AND THIOETHERS FROM ALKENES

ASYMMETRIC COMPOUNDS

Oxone , K_2CO_3 , chiral ketone
pH 8 buffer , −10°C , 3.5 h

Ph Ph O 92% (81% ee)

Tian, H.; She, X.; Yu, H.; Shu, L.; Shi, Y. *J. Org. Chem.* ***2002***, *67*, 2435.

30% chiral ketone , 4 Oxone
aq. MeCN , 12 eq K_2CO_3 , 0°C

Ph Ph O 80% (88% ee)

Denmark, S.E.; Matsuhashi, H. *J. Org. Chem.* ***2002***, *67*, 3479.

HO Bu SiMe$_2$Ph
$TiCl_4$, DCM , $CHCl_3$
rt , 9.5 h

O Bu SiMe$_2$Ph 75%

Miura, K.; Hondo, T.; Okajima, S.; Nakagawa, T.; Takahashi, T.; Hosomi, A. *J. Org. Chem.* ***2002***, *67*, 6082.

Ph Ph
10% (N CO$_2$Et, F, H, O) , MeCN
Oxone , aq. Na_2EDTA , rt

Ph O Ph 88% (76% ee)

Armstrong, A.; Ahmed, G.; Donminguiz-Fernandz, B.; Hayter, B.R.; Wailes, J.S. *J. Org. Chem.* ***2002***, *67*, 8610.

Ph Ph
cyclodextrin O O O
Oxone , rt , aq. Na_2•EDTA , MeCN
$NaHCO_3$

Ph O Ph 99% (31% ee)

Chan, W.-K.; Yu, W.-Y.; Che, C.-M.; Wong, M.-K. *J. Org. Chem.* ***2003***, *68*, 6576.

Ph
Oxone , chiral ketone , K_2CO_3
DME/DMM , AcOH , −10°C

O Ph 71% (83% ee)

Shu, L.; Wang, P.; Gan, Y.; Shi, Y. *Org. Lett.* ***2003***, *5*, 293.

Ph
5% (N, O, Ph, O)
2 TPPP , 40°C , MeCN

Ph O 67% ee

Page, P.C.B.; Barros, D.; Buckley, B.R.; Ardakani, A.; Marples, B.A. *J. Org. Chem.* ***2004***, *69*, 3595.

Ph
5% iminium salt catalyst , 2 Oxone
4 Na_2CO_3 , aq. MeCN , 0°C

Ph O 69% (91% ee)

Page, P.C.B.; Buckley, B.R.; Blacker, A.J. *Org. Lett.* ***2004***, *6*, 1543.

Ph–CH=CH–Ph → (chiral carobhydrate aldehyde , MeCN / Bu_4NHSO_4 , 0°C , 4h , aq. Na_2EDTA 0.05M $N_2B_4O_7$) → Ph-epoxide-Ph 54% (93.5%ee)

Bez, G.; Zhao, C.-G. *Tetrahedron Lett.* ***2003***, *44*, 7403.

Ph–CH=CH–Ph → (5% Ru(*i*-Pr_2-Pybox)(Pydic) , toluene , rt / $PhI(OAc)_2$, *t*-BuOH , H_2O) → Ph-epoxide-Ph 84% (575 ee)

Tse, M.K.; Bhor, S.; Klawonn, M.; Döbler, C.; Beller, M. *Tetrahedron Lett.* ***2003***, *44*, 7479.

1-phenylcyclohexene → (2 Oxone , 10 $NaHCO_3$, chiral amine / 0.5 pyridine , aq. MeCN) → 1-phenylcyclohexene oxide 89% (59% ee)

Aggarwal, V.K.; Lopin, C.; Sandrinelli, F. *J. Am. Chem. Soc.* ***2003***, *125*, 7596.

Ph–CH=CH–Ph + chiral bis(oxazolidinone) ketone → (Oxone , aq. EDTA , 2 h / Bu_4N HSO_5 , MeCN-DMM) → Ph-epoxide-Ph 93% (64% ee)

Matsumoto, K.; Tomioka, K. *Tetrahedron Lett.* ***2002***, *43*, 631.

Ph–CH=CH–Ph → (keto bile acid derivative , Oxone , 20°C / pH 8 ($NaHCO_3$) , aq MeCN , EDTA) → Ph-epoxide-Ph 90% (54% ee)

Bortolini, O.; Fantin, G.; Fogagnolo, M.; Mari, L. *Tetrahedron: Asymmetry* ***2004***, *15*, 3831.

Ph–CH=CH–Ph → (10% cellulose complex , Oxone , pH 7.5 / aq. MeCN) → Ph-epoxide-Ph 97% (76% ee)

Shing, T.K.M.; Leung, G.Y.C.; Yeung, K.W. *Tetrahedron Lett.* ***2003***, *44*, 9225.

α-methylstyrene → (Mn(salen)/mesoporous silica-41 / DCM) → epoxide 99% (70% ee)

Xiang, S.; Zhang, Y.; Xin, Q.; Li, C. *Chem. Commun.* ***2002***, 2696.

α-methylstyrene → (2 Oxone , 4 Na_2CO_3 , aq. MeCN , 0°C / chiral iminium salt , 5 min) → epoxide 64% (20% ee)

Page, P.C.B.; Rassias, G.A.; Barros, D.; Ardakani, A.; Bethell, D.; Merifield, E. *Synlett* ***2002***, 580.

NONASYMMETRIC COMPOUNDS

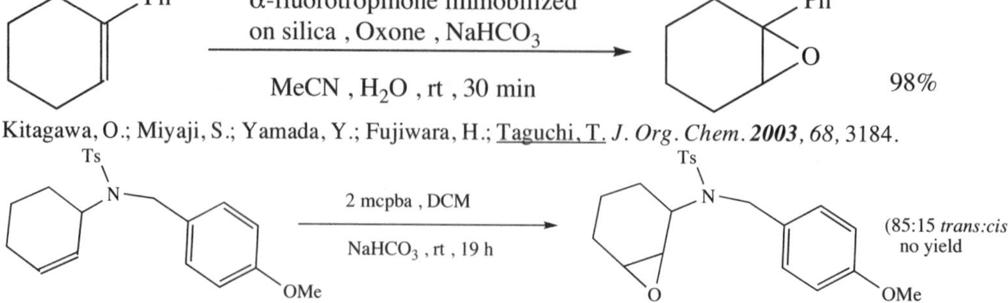

Kitagawa, O.; Miyaji, S.; Yamada, Y.; Fujiwara, H.; Taguchi, T. *J. Org. Chem.* ***2003***, *68*, 3184.

O'Brien, P.; Childs, A.C.; Ensor, G.J.; Hill, C.L.; Kirby, J.P.; Deardon, M.J.; Oxenford, S.J.; Rosser, C.M. *Org. Lett.* ***2003***, *5*, 4955.

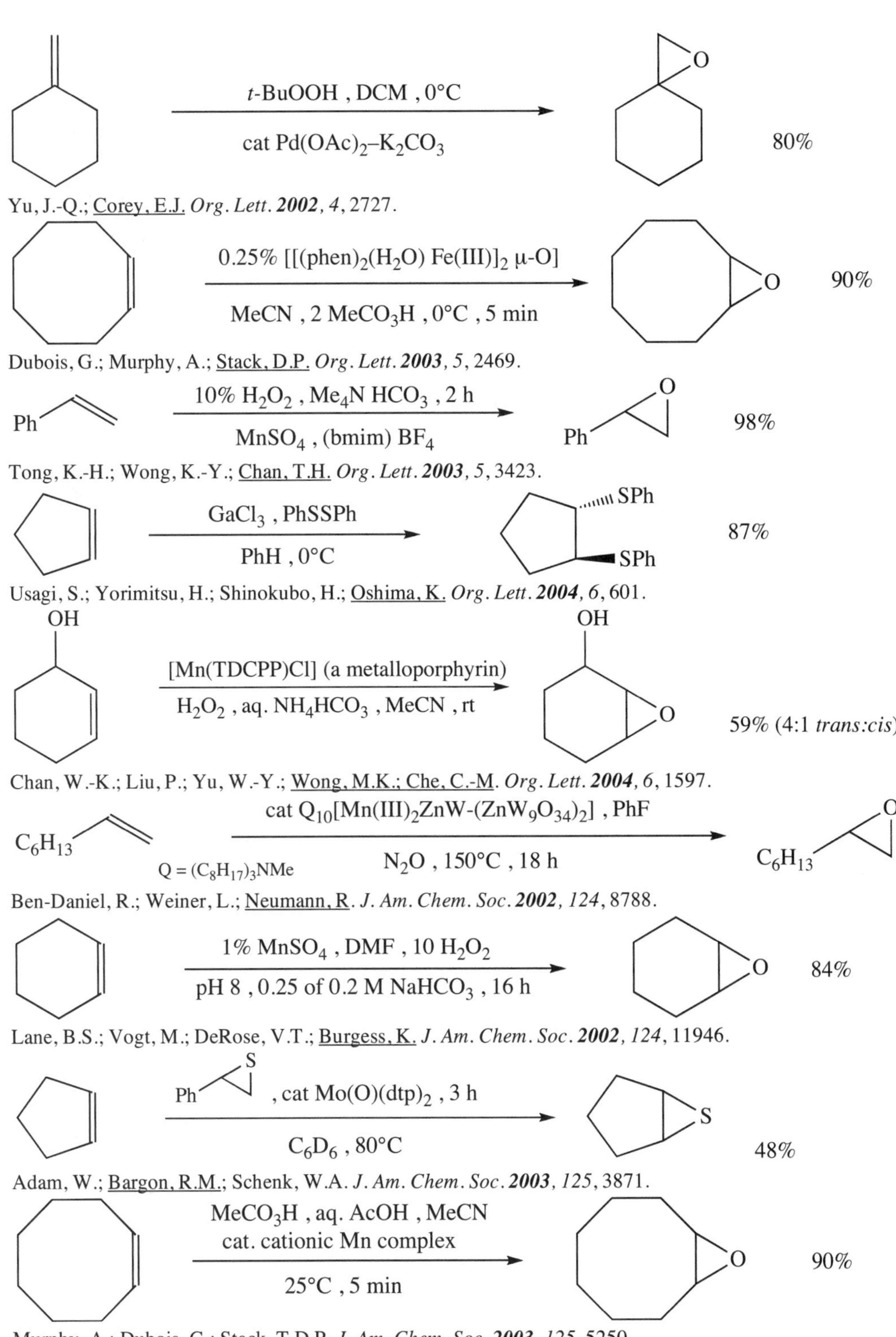
t-BuOOH , DCM , 0°C
cat $Pd(OAc)_2$–K_2CO_3
O
80%
Yu, J.-Q.; Corey, E.J. *Org. Lett.* **2002**, *4*, 2727.
0.25% [[(phen)$_2$(H_2O) Fe(III)]$_2$ μ-O]
MeCN , 2 $MeCO_3H$, 0°C , 5 min
O
90%
Dubois, G.; Murphy, A.; Stack, D.P. *Org. Lett.* **2003**, *5*, 2469.
Ph
10% H_2O_2 , $Me_4N\ HCO_3$, 2 h
$MnSO_4$, (bmim) BF_4
Ph
O
98%
Tong, K.-H.; Wong, K.-Y.; Chan, T.H. *Org. Lett.* **2003**, *5*, 3423.
$GaCl_3$, PhSSPh
PhH , 0°C
SPh
SPh
87%
Usagi, S.; Yorimitsu, H.; Shinokubo, H.; Oshima, K. *Org. Lett.* **2004**, *6*, 601.
OH
[Mn(TDCPP)Cl] (a metalloporphyrin)
H_2O_2 , aq. NH_4HCO_3 , MeCN , rt
OH
O
59% (4:1 *trans:cis*)
Chan, W.-K.; Liu, P.; Yu, W.-Y.; Wong, M.K.; Che, C.-M. *Org. Lett.* **2004**, *6*, 1597.
C_6H_{13}
cat $Q_{10}[Mn(III)_2ZnW\text{-}(ZnW_9O_{34})_2]$, PhF
Q = $(C_8H_{17})_3NMe$
N_2O , 150°C , 18 h
O
C_6H_{13}
Ben-Daniel, R.; Weiner, L.; Neumann, R. *J. Am. Chem. Soc.* **2002**, *124*, 8788.
1% $MnSO_4$, DMF , 10 H_2O_2
pH 8 , 0.25 of 0.2 M $NaHCO_3$, 16 h
O
84%
Lane, B.S.; Vogt, M.; DeRose, V.T.; Burgess, K. *J. Am. Chem. Soc.* **2002**, *124*, 11946.
S
Ph
, cat Mo(O)(dtp)$_2$, 3 h
C_6D_6 , 80°C
S
48%
Adam, W.; Bargon, R.M.; Schenk, W.A. *J. Am. Chem. Soc.* **2003**, *125*, 3871.
$MeCO_3H$, aq. AcOH , MeCN
cat. cationic Mn complex
25°C , 5 min
O
90%
Murphy, A.; Dubois, G.; Stack, T.D.P. *J. Am. Chem. Soc.* **2003**, *125*, 5250.

Ru(II)-TMP(CO), hν; aq. MeCN — 97%

Funyu, S.; Isobe, T.; Takagi, S.; Tryk, D.A.; Inoue, H. *J. Am. Chem. Soc.* **2003**, *125*, 5734.

BF_3K — 1. (dimethyldioxirane); 2. Br–C₆H₄–CN, THF/H_2O, 3 h; cat $PdCl_2/dppf\text{-}CH_2Cl_2$, 3 Cs_2CO_3, reflux — CN, 80%

Molander, G.A.; Ribagorda, M. *J. Am. Chem. Soc.* **2003**, *125*, 11148.

Ph, Ph — $KHSO_5$, $NaHCO_3$, aq. MeCN, 12 h, rt; cat. (N^+ BF_4^- Me) — O, Ph, Ph, 81%

Bohé, L.; Kammoun, M. *Tetrahedron Lett.* **2002**, *43*, 803.

2 eq 30% H_2O_2, 2,2'-bipy, 0°C; $MeReO_3$ — O, 91%

Iskra, J.; Bonnet-Delpon, D.; Bégué, J.-P. *Tetrahedron Lett.* **2002**, *43*, 1001.

H_2O_2, MeCN, hydrotalcite; microwaves — O, 97%

Pillai, U.R.; Sahle-Demessie, E.; Varma, R.S. *Tetrahedron Lett.* **2002**, *43*, 2909.

Ph — (bmim) PF_6-DCM, 1% Mn-porphyrin; $PhI(OAc)_2$, rt, 4h — O, Ph, 89%

Li, Z.; Xia, C.-G. *Tetrahedron Lett.* **2003**, *44*, 2069.

Ph — 10% (O_2N, N, Me, Me, Me, BF_4), Oxone, 6h; aq. MeCN, $NaHCO_3$ — O, Ph, 85%

Bohé, L.; Kammoun, M. *Tetrahedron Lett.* **2004**, *45*, 747.

$SiMe_3$ — *i*-PrCHO, Ru catalyst, O_2; DCM, $NaHCO_3$ — $SiMe_3$, O, 72%

Srikanth, A.; Nagendrappa, G.; Chandrasekaran, S. *Tetrahedron* **2003**, *59*, 7761.

OH — $H_3PW_{12}O_{40}$, Py, 30% H_2O_2; rt — O, OH, 83%

Yamada, Y.M.A.; Tabafa, H.; Ichinohe, M.; Takahashi, H.; Ikegami, S. *Tetrahedron* **2004**, *60*, 4087.

H_2O_2, Fe porphyrin derivative; bmim Br — O, 81%

Srinivas, K.A.; Kumar, A.; Chauhan, S.M.S. *Chem. Commun.* **2002**, 2456.

O_2 , Ru complex , DCE , 3.5 h

i-PrCHO , rt

O 90%

Qi, J.Y.; Qiu, L.Q.; Lam, K.H.; Yip, C.W.; Zhou, Z.Y.; Chan, A.S.C. *Chem. Commun.* ***2003***, 1058.

cat $PPh_4[MoO(O_2)(saloxH]$, MeCN

25% $NaHCO_3$, 6 H_2O_2 , rt

O 92%

Gharah, N.; Chakraborty, S.; Mukherjee, A.K.; Bhattacharyya, R. *Chem. Commun.* ***2004***, 2630.

O O

$[F_2/H_2O/MeCN]$, rt

O 85%

CO_2Et CO_2Et

Rozen, S.; Golan, E. *Eur. J. Org. Chem.* ***2003***, 1915.

i-PrCHO , cat Pt/C , O_2 , rt

DCM , 6 h

O 92%

Chen, W.; Yamada, J.; Matsumoto, K. *Synth. Commun.* ***2002***, *32*, 17.

mcpba , DCM , 8 h

–50°C → –10°C

O 71%

Srinivasan, R.; Chandrasekharan, M.; Vani, P.V.S.N.; Chida, A.S.; Singh, A.K. *Synth. Commun.* ***2002***, *32*, 1853.

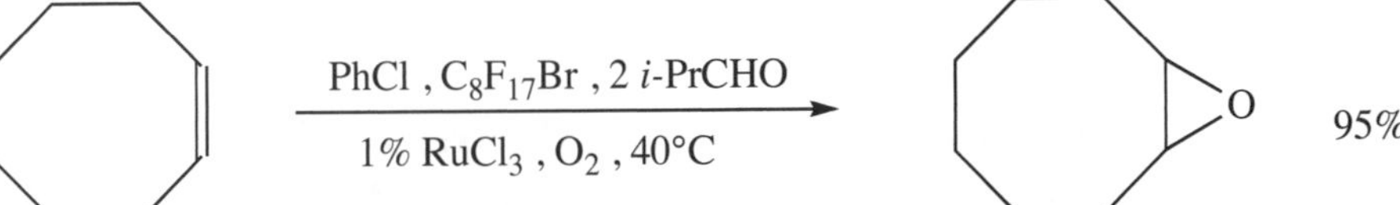

Ragagnin, G.; Knochel, P. *Synlett* ***2004***, 951.

$NaIO_4$, imidazole , 7 h

polymer supported Mn salen catalyst

O 91%

Tangestaninejad, S.; Habibi, M.H. *Synth. Commun.* ***2002***, *32*, 3331.

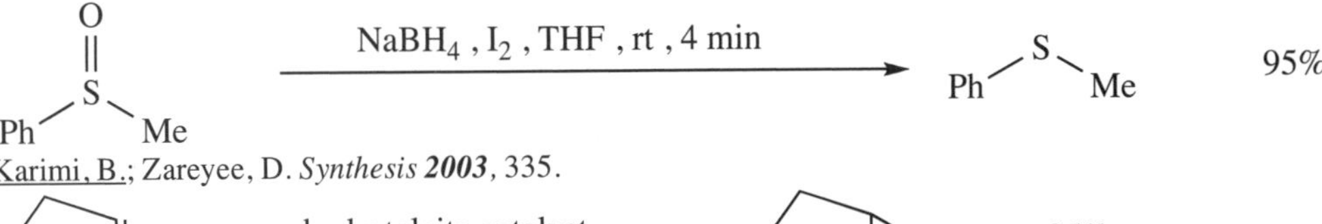

Karimi, B.; Zareyee, D. *Synthesis* ***2003***, 335.

hydrotalcite catalyst

H_2O_2 , MeCN

O >95%

Pillai, U.R.; Sahle-Demessie, E.; Varma, R.S. *Synth. Commun.* ***2003***, *33*, 2017.

Ph–S(=O)–Ph → (20% $MoO_2Cl_2(dmf)_2$, MeCN; $P(OPh)_3$, reflux) → Ph–S–Ph 96%

Sanz, R.; Escribno, J.; Aguado, R.; Pedrosa, M.R.; Arnáiz, F.J. *Synthesis* **2004**, 1629.

MeCH=C(SEt)₂ → (1. $PhCH(OMe)_2$, TMSOTf, −78°C , 0.5 h , toluene; 2. $Me_2Cu(CN)Li_2$, −78°C , 1 h) → PhCH(OMe)CH(Me)C(SEt)₂Me 74% (20:80 *syn:anti*)

also with other nucleophiles

Ichikawa, J.; Saitoh, T.; Tada, T.; Mukaiyama, T. *Chem. Lett.* **2002**, *31*, 996.

REVIEWS:

"A Critical Outlook and Comparison of Enantioselective Oxidation Methodologies of Olefins"
Bonni, C.; Righi, G. *Tetrahedron* **2002**, *58*, 4981.

"Oxidation, Epoxidation and Sulfoxidation Reactions Catalyzed by Haloperoxidases"
Dembitsky, V.M. *Tetrahedron* **2003**, *59*, 4701.

"Organocatalytic Asymmetric Epoxidation of Olefins by Chiral Ketones"
Shi, Y. *Acc. Chem. Res.* **2004**, *37*, 488.

Ketone-Catalyzed Asymmetric Epoxidation Reactions"
Yang, D. *Acc. Chem. Res.* **2004**, *37*, 497.

"Olefin Epoxidation with Inorganic Peroxides: Solutions to Four Long-Standing Controversies on the Mechanism of Oxygen Transfer"
Deubel, D.V.; Frenking, G.; Gisdakis, P.; Herrmann, W.A.; Rösch, N.; Sundemeyer, J. *Acc. Chem. Res.* **2004**, *37*, 645.

SECTION 135: ETHERS, EPOXIDES, AND THIOETHERS FROM MISCELLANEOUS COMPOUNDS

Ph–S⁺(O⁻)–Me → (3 eq $ClP(OEt)_2$, $CHCl_3$; 3 eq NEt_3 , rt , 30 min) → Ph–S–Me 83%

Jie, Z.; Rammoorty, V.; Fischer, B. *J. Org. Chem.* **2002**, *67*, 711.

$PhB(OH)_2$ → ($Cu(OAc)_2$, 5 NEt_3 , DCM , MeCN; 10 eq H_2O , 25°C , 6h) → Ph-O-Ph 95%

Sagar, A.D.; Tale, R.H.; Adude, R.N. *Tetrahedron Lett.* **2003**, *44*, 7061.

Ph–S(=O)–Me → ($CHCl_3$, rt , 15 min; 2,4,4,6-tetrabromocyclohexadienone) → Ph–S–Me 94%

Iranpoor, N.; Firouzabadi, H.; Shaterian, H.R. *J. Org. Chem.* **2002**, *67*, 2826.

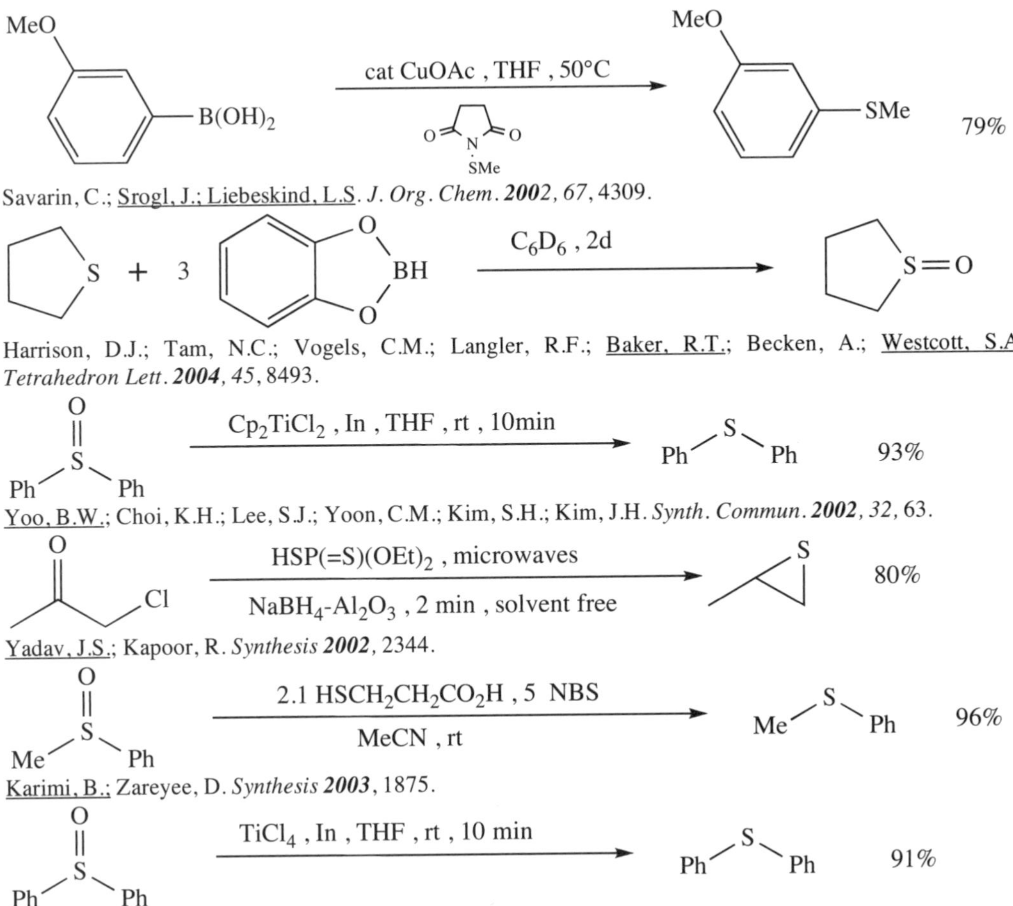

Savarin, C.; Srogl, J.; Liebeskind, L.S. *J. Org. Chem.* ***2002***, *67*, 4309.

Harrison, D.J.; Tam, N.C.; Vogels, C.M.; Langler, R.F.; Baker, R.T.; Becken, A.; Westcott, S.A. *Tetrahedron Lett.* ***2004***, *45*, 8493.

Yoo, B.W.; Choi, K.H.; Lee, S.J.; Yoon, C.M.; Kim, S.H.; Kim, J.H. *Synth. Commun.* ***2002***, *32*, 63.

Yadav, J.S.; Kapoor, R. *Synthesis* ***2002***, 2344.

Karimi, B.; Zareyee, D. *Synthesis* ***2003***, 1875.

You, B.W.; Choi, K.H.; Kim, D.Y.; Choi, K.I.; Kim, J.H. *Synth. Commun.* ***2003***, *33*, 53.

Ph–CH=CH–CO_2t-Bu → 15 $PhCMe_2O_2Li$, chiral ligand / $PhCMe_2O_2H$, toluene , 0°C → Ph-epoxide-CO_2t-Bu 76% (71% ee)

Tanaka, Y.; Nishimura, K.; Tomioka, K. *Heterocycles* ***2002***, *58*, 71.

CHAPTER 10

PREPARATION OF HALIDES AND SULFONATES

SECTION 136: HALIDES AND SULFONATES FROM ALKYNES

NO ADDITIONAL EXAMPLES

SECTION 137: HALIDES AND SULFONATES FROM ACID DERIVATIVES

NO ADDITIONAL EXAMPLES

SECTION 138: HALIDES AND SULFONATES FROM ALCOHOLS AND THIOLS

OH + Cl, N, Cl, N, N, Cl — DMF , DCM , rt → Cl 96%

diols can be converted to chlorohydrins

DeLuca, L.; Giacomelli, G.; Porcheddu, A. *Org. Lett.* **2002**, *4*, 553.

Ph, OH — PBSF-$NEt_3(HF)_3$, NEt_3; 20°C , THF , 26 h → Ph, F 79%

PBSF = perfluoro-1-butanesulfonyl fluoride

Yin, J.; Zarkowsky, D.S.; Thomas, D.W.; Zhao, M.W.; Huffman, M.A. *Org. Lett.* **2004**, *6*, 1465.

C_6H_{13}, OH, $C_{34}H_7$ — 5% $InCl_3$, $HSiMe_2Cl$, benzil; DCM , 8 h → C_6H_{13}, Cl, $C_{34}H_7$ 74%

Yasuda, M.; Yamasaki, S.; Onishi, Y.; Baba, A. *J. Am. Chem. Soc.* **2004**, *126*, 7186.

$C_7H_{15}CH_2OH$ — 1. $[Cl_3CO)_2C=O$, KF , MeCN , 18-crown-6; KF , ether; 2, distill → $C_7H_{15}CH_2F$ 89% overall

Flosser, D.A.; Olofson, R.A. *Tetrahedron Lett.* **2002**, *43*, 4275.

$C_2H_{15}CH_2OH$ —[$P(C_6H_4\text{-}P(OCl)_2C_7H_{15})_3$, CBr_4 , toluene]→ $C_2H_{15}CH_2Br$ 75%

Desmaris, L.; Percina, N.; Cottier, L.; Sinou, D. *Tetrahedron Lett.* ***2003***, *44*, 7589.

Ph(CH₂)₃OH —[*i*-Pr-N=C=N-*i*-Pr , $Cu(OTf)_2$ / THF , 100°C , rt , 1h]→ Ph(CH₂)₃Br 85%

Li, Z.; Crosignani, S.; Linclau, B. *Tetrahedron Lett.* ***2003***, *44*, 8143.

$PhCH_2OH$ —[$ZrCl_4$, NaI , MeCN , 10 min]→ $PhCH_2I$ 95%

Firouzabadi, H.; Iranpoor, N.; Jafarpour, M. *Tetrahedron Lett.* ***2004***, *45*, 7451.

Ph(CH₂)₃OH —[1. *i*-PrN=C=N*i*-Pr , 2% CuCl , THF / microwaves, 100°C , 5 min; 2. AcCl , microwaves, 150°C , 5 min]→ Ph(CH₂)₃Cl quant

Crosignani, S.; Nadal, B.; Li, Z.; Linclau, B. *Chem. Commun.* ***2003***, 260.

$CH_3(CH_2)_8CH_2OH$ + trichloroisocyanuric acid —[PPh_3 , MeCN]→ $CH_3(CH_2)_8CH_2Cl$ 74%

Hiegel, G.A.; Rubino, M. *Synth. Commun.* ***2002***, *32*, 2691.

$PhCH_2OH$ —[NaI , Amberlyst-15, rt , 1h]→ $PhCH_2I$ 95%

Tajbakhsh, M.; Hosseinzadeh, R.; Lasemi, Z. *Synlett* ***2004***, 635.

$PhCH_2OH$ —[CsI/BF_3 , ether , MeCN , 25°C , 5 min]→ $PhCH_2I$ 96%

Hayat, S.; Atta-ur-Rahman; Khan, K.M.; Choudhary, M.I.; Maharvi, G.M.; Zia-Ullah; Bayer, E. *Synth. Commun.* ***2003***, *33*, 2531.

PhCH(OH)CH₂CH₃ —[SiO_2-Cl , $CHCl_3$, 1 d]→ PhCH(Cl)CH₂CH₃ 90%

Firouzabadi, H.; Iranpoor, N.; Karimi, B.; Hazarkhani, H. *Synth. Commun.* ***2003***, *33*, 3671.

PhOH —[1. (5,5-dimethyl-1,3,2-dioxaborinane)$_2$, $PdCl_2/KOAc$, dppf , dioxane , –80°C; 2. NaI , Chloramine-T , 50% aq. THF]→ PhI 90x58%

Thompson , A.L.S.; Kabalka, G.W.; Akula, M.R.; Huffman, J.W. *Synthesis* ***2004***, 547.

$PhCH_2OH$ —[*p*-TsOH , KI , microwaves , 80 sec]→ $PhCH_2I$ 92%

Lee, J.C.; Park, J.Y.; Yoo, E.S. *Synth. Commun.* ***2004***, *34*, 2095.

SECTION 139: HALIDES AND SULFONATES FROM ALDEHYDES

PhCHO —[cat $In(OH)_3$, $HSiMe_2Cl$, $CHCl_3$ / rt , 5.5 h]→ $PhCH_2Cl$ 90%

Onishi, Y.; Ogawa, D.; Yasuda, M.; Baba, A. *J. Am. Chem. Soc.* ***2002***, *124*, 13690.

SECTION 140: HALIDES AND SULFONATES FROM ALKYLS, METHYLENES, AND ARYLS

For the conversion R-H → R-Halogen, see Section 146 (Halides and Sulfonates from Hydrides).

Toluene (CH_3) → AcCl, cat CAN, 7-8 h / MeCN m rt → chlorotoluene (CH_3, Cl) 70% (*para:ortho*, 7:3)

Arcadi, A.; Chiarini, M.; Di Giuseppe, S.; Marinelli, F. *Synlett* **2003**, 221.

Br–C$_6$H$_4$–CH_3 → NBS, cat $(PhCO_2)_2O$ / [bmim] PF_6, 95°C → Br–C$_6$H$_4$–CH_2Br 72%

Togo, H.; Hirai, T. *Synlett* **2003**, 702.

SECTION 141: HALIDES AND SULFONATES FROM AMIDES

NO ADDITIONAL EXAMPLES

SECTION 142: HALIDES AND SULFONATES FROM AMINES

NO ADDITIONAL EXAMPLES

SECTION 143: HALIDES AND SULFONATES FROM ESTERS

Ph–CH=CH–CH$_2$–OAc → AcCl, EtOH, 30 min → Ph–CH=CH–CH$_2$–Cl 96%

Yadav, V.K.; Babu, K.G. *Tetrahedron* **2003**, *59*, 9111.

SECTION 144: HALIDES AND SULFONATES FROM ETHERS, EPOXIDES, AND THIOETHERS

$PhCH_2OTMS$ → SiO_2-Cl, NaI, MeCN, rt, 1h → $PhCH_2I$ 90%

Firouzabadi, H.; Iranpoor, N.; Hazarkhani, H. *Tetrahedron Lett.* **2002**, *43*, 7139.

SECTION 145: HALIDES AND SULFONATES FROM HALIDES AND SULFONATES

C_9H_{19}–CH$_2$–Br → 1. $HC(SMe)_3$, BuLi / 2. 3 BrF_3 → C_9H_{19}–CH(Br)–CF_3 69%

Hageoly, A.; Ben-David, I.; Rozen, S. *J. Org. Chem.* **2002**, *67*, 8430.

Br—C6H4—CH_2CN → (2 NaI , 5% CuI , dioxane; 10% diamine , 110°C , 1 d) → I—C6H4—CH_2CN 97%

Klapars, A.; Buchwald, S.L. *J. Am. Chem. Soc.* ***2002**, 124*, 14844.

$PhCH_2Br$ → (KF , MeCN , Bi_4NHSO_4 , Ph_3SnF) → $PhCH_2F$ 59%

Makosza, M.; Bujok, R. *Tetrahedron Lett.* ***2002**, 43*, 2761.

Me—C6H4—I → (2 $NiCl_2$, DMF; microwaves) → Me—C6H4—Cl 92%

Arvela, R.K.; Leadbeater, N.E. *Synlett* **2003**, 1145.

4-Chloroquinoline (Cl, N) → (1. 4M HCl, THF/dioxane; 2. 10 NaI , MeCN , reflux , 1 d) → 4-Iodoquinoline (I, N) 91%

Wolf, C.; Tumambac, G.E.; Villalobos, C.N. *Synlett* **2003**, 1801.

MeO—C6H4—Cl → (3 *i*-PrMgBr , 10% $FeCl_3$, THF; 50°C , 2 d) → MeO—C6H5 94%

Guo, H.; Kanno, K.; Takahashi, T. *Chem Lett.* ***2004**, 33*, 1356.

SECTION 146: HALIDES AND SULFONATES FROM HYDRIDES

α-Halogenations of aldehydes, ketones and acids are given in Sections 338 (Halide-Aldehyde), 369 (Halide-Ketone), 359 (Halide-Esters) and 319 (Acid-Halide).

C6H5—OMe → ($NMe_4^+ ICl_2^-$, neat; 5 min) → I—C6H4—OMe quant

Hajipour, A.R.; Arbabian, M.; Ruoho, A.E. *J. Org. Chem.* ***2002**, 67*, 8622.

C6H5—NH_2 → (1. BuLi; 2. Me_3SnCl; 3. Br_2 4. aq KF) → Br—C6H4—NH_2 76%

Smith, M.B.; Guo, L.; Okeyo, S.; Stenzel, J.; Yanella, J.; La Chapelle, E. *Org. Lett.* ***2002**, 4*, 2321.

$Me_2C=CH$—Bu → (5% PhSeCl , NCS , DCM; 3h) → $CH_2=C(Me)$—CH(Cl)—Bu 82%

Tunge, J.A.; Mellegaard, S.R. *Org. Lett.* ***2004**, 6*, 1205.

t-BuOI , 40°C — 85%

Montoro, R.; Wirth, T. *Org. Lett.* ***2003***, *5*, 4729.

NCS , BF_3 , H_2O / 100°C — 95%

Prakash, G.K.S.; Mathew, T.; Hoole, D.; Esteves, M.; Wang, Q.; Rasul, G.; Olah, G.A. *J. Am. Chem. Soc.* ***2004***, *126*, 15770.

I_2 , CCl_4 , 2 AlI_3 , CH_2Br_2 , –20°C — 75%

Akhrem, I.; Orlinkov, A.; Vitt, S.; Christyakov, A. *Tetrahedron Lett.* ***2002***, *43*, 1333.

NIS , TFA , MeCN

Castanet, A.-S.; Colobert, F.; Broutin, P.-E. *Tetrahedron Lett.* ***2002***, *43*, 5047.

NBS , [bbim]BF_4 — 98%

bbim = di-n-butylimidazolium tetrafluoroborate

Rajagopal, R.; Jarikote, D.V.; Lahoti, R.J.; Daniel, T.; Srinivasan, K.V. *Tetrahedron Lett.* ***2003***, *44*, 1815.

KI , KIO_3 , dil. HCl , aq MeOH

Adimurthy, S.; Ramachandraiah, G.; Ghosh, K.; Bedekar, A.V. *Tetrahedron Lett.* ***2003***, *44*, 5099.

Py-H CrO_3Br , AcOH , 100°C — 93%

Patwari, S.B.; Baseer, M.A.; Vibhute, Y.B.; Bhusare, S.R. *Tetrahedron Lett.* ***2003***, *44*, 4893.

I_2 , NaI , $Fe(CN)_3$•1.5 N_2O_4/C / DCM , rt, 2h — 96%

Firouzabadi, H.; Iranpoor, N.; Shiri, M. *Tetrahedron Lett.* ***2003***, *44*, 8781.

1. BuLi , THF
2. $B(OMe)_3$
3. Br_2 , –78°C — 91%

Zhao, J.; Jia, X.; Zhai, H. *Tetrahedron Lett.* ***2003***, *44*, 9371.

[hmim] Br_3 (ionic liquid) , 70°C / overnight — 97%

Boñaga, L.V.R.; Zhang, H.-C.; Maryannoff, B.E. *Chem. Commun.* ***2004***, 2394.

MeO— (anisole) —[BR_2 , $(Bu_4N)_2^+$ $OSO_2OOSO_2O^-$ / DCM , 25°C , 1h]→ MeO— —Br 96%

Park, M.Y.; Yang, S.G.; Jadhav, V.; Kim, Y.H. *Tetrahedron Lett.* ***2004***, *45*, 4887.

O, OMe —[I_2 , $PhI(OTf)_2$, MeCN , rt]→ O, OMe, I 86%

Panunzi, B.; Rotiroti, L.; Tingoli, M. *Tetrahedron Lett.* ***2003***, *44*, 8753.

MeO— —[NH_4I , Oxone , 8h]→ MeO— —I 9% (95:5 *o:p*)

Mohan, K.V.V.K.; Narender, N.; Kulkarni, S.J. *Tetrahedron Lett.* ***2004***, *45*, 8015.

—[$PhI(OAc)_2$, I_2 , *t*-BuOH , 0.04M]→ —I 98%

Barluenga, J.; González-Bobes, F.; González, J.M. *Angew. Chem. Int. Ed.* ***2002***, *41*, 2556.

—[Selectfluor , I_2 , MeCN / 61°C , 4 h]→ —I 95%

Stavber, S.; Kralj, P.; Zupan, M. *Synlett* ***2002***, 598.

N, Me —[NBS , cat PBr_3 , THF / −78°C → −10°C]→ Br, N, Me quant

Dvornikova, E.; Kamienska-Trela, K. *Synlett* ***2002***, 1152.

—OMe —[Oxone , KCl , MeCN / 15 h , rt]→ Cl— —OMe 90% (+ 8% *ortho*)

Narender, N.; Srinivasu, P.; Kulkarni, S.J.; Raghavan, K.V. *Synth. Commun.* ***2002***, *32*, 279.

—OMe —[N N⁺−Br , NaI Cl_2 , 10 min / solvent free , rt]→ I— —OMe 98%

Hajipour, A.R.; Ruoho, A.E. *Org. Prep. Proceed. Int.* ***2002***, *34*, 647.

MeO— —[$Bi(NO_3)_3$•5 H_2O , SiO_2 , I_2 / solvent free , rt , 5 min]→ MeO— —I 92%

Alexander, V.M.; Khandekar, A.C.; Samant, S.D. *Synlett* ***2003***, 1895.

MeO—C6H5 → (Oxone , KBr , MeOH; rt , 1 d) → MeO—C6H4—Br 84% (+ 15% ortho)

Narender, N.; Srinivasu, P.; Prasad, M.R.; Kulkarni, S.J.; Raghavan, K.V. *Synth. Commun.* ***2002***, *32*, 2313.

Ph-H → ($PyH^+ BrCl_2^-$, DCM , $FeCl_3$, 1 h) → Ph-Br 86%

Muathen, H.A. *Synthesis* ***2002***, 169.

C6H5—OMe → (I_2 , $BuPPh_3$ S_2O_8 , MeCN; reflux) → I—C6H4—OMe quant

Tajik, H.; Esmaeili, A.A.; Mohammadpoor-Baltork, I.; Ershadi, A.; Tajmehri, H. *Synth. Commun.* ***2003***, *33*, 1319.

C6H5—OMe → (30% H_2O_2 - KI - H_2SO_4; KI , MeOH , rt , 3h) → I—C6H4—OMe 93%

Iskra, J.; Stavber, S.; Zupan, M. *Synthesis* ***2004***, 1869.

C6H5—NH_2 → (bmim Br_3 , –10°C , 2 min) → Br—C6H4—NH_2 90%

Le, Z.-G.; Chen, Z.-C.; Hu, Y.; Zheng, Q.-G. *Synthesis* ***2004***, 2809.

C6H5—OMe → (I_2 , $BuPPh_3S_2O_8$; MeCN , reflux) → I—C6H4—OMe quant

Tajik, H.; Esmaeili, A.A.; Mohammadpoor-Baltork, I.; Ershadi, A.; Tajmehri, A. *Synth. Commun.* ***2004***, *34*, 1319.

C6H5—NH_2 → (NH_4Br , H_2O_2 , AWH) → Br—C6H4—NH_2 61% + 19% *ortho*

Mohan, K.V.V.K.; Narender, N.; Srinivasu, P.; Kulkarni, S.J.; Raghavan, K.V. *Synth. Commun.* ***2004***, *34*, 2143.

HO(Me)C(H)(C_6H_{13}) → (1. TMSCl , NEt_3 , THF , rt , 2 h; 2. $CHCl_3$, 50% aq. NaOH; cat CTAC , 80°C , 18 h; 3. 2% HCl , MeOH , rt , 2 h) → HO(Me)C($CHCl_2$)(C_6H_{13}) 39% + 52% SM

Masaki, Y.; Arasaki, H.; Iwata, I. *Chem. Lett.* ***2003***, *32*, 4.

Me—C6H4—OMe → (NH_4NO_3 , NBS , MeCN; 25°C , 10 min) → Me—C6H3(Br)—OMe 97%

Tanemura, K.; Suzuki, T.; Nishida, Y.; Satsumabayashi, K.; Horaguchi, T. *Chem. Lett.* ***2003***, *32*, 932.

naphthalene → (5 $NaBiO_3$, 10 KCl , 1 h; also with other metal halides) → 1-Cl-naphthalene 82%

Muathen, H.A. *Helv. Chim. Acta* ***2003***, *86*, 168.

SECTION 147: HALIDES AND SULFONATES FROM KETONES

NO ADDITIONAL EXAMPLES

SECTION 148: HALIDES AND SULFONATES FROM NITRILES

NO ADDITIONAL EXAMPLES

SECTION 149: HALIDES AND SULFONATES FROM ALKENES

For halocyclopropanations, see Section 74F (Alkyls from Alkenes: Cylclopropanations).

cat. S-BINAP , H_2O/THF; $CuBr_2$, 0.2 M , 25°C — 95% (95% ee)

El-Qisairi, A.K.; Qaseer, H.A.; Kaatsigras, G.; Lorenzi, P.; Trivedi, U.; Tracz, S.; Hartman, A.; Miller, J.A.; Henry, P.M. *Org. Lett.* ***2003***, *5*, 439.

Oxone , wet Al_2O_3 , KI-NaN_3 , rt — 92%

Curini, M.; Epifano, F.; Marcotullio, M.C.; Rosati, O. *Tetrahedron Lett.* ***2002***, *43*, 1201.

CuO•HBF_4 , I_2 , Et_3SiH , DCM , –30°C , 3h — 65%

Campos, J.; García, B.; Rodríguez, M.A. *Tetrahedron Lett.* ***2002***, *43*, 6111.

H_2O , rt, 40 min — 78%

Mendonca, G.F.; Sanseverino, A.M.; de Mattos, M.C.S. *Synthesis* ***2003***, 45.

SECTION 150: HALIDES AND SULFONATES FROM MISCELLANEOUS COMPOUNDS

Na_2PdCl_4 , 75% MeOH , 40°C — 56%

Willis, D.M.; Strongin, R.M. *Tetrahedron Lett.* ***2000***, *41*, 6271.

MeCN — 95%

Szumigala Jr. D.R.; Devine, P.N.; Gauthier Jr. D.R.; Volante, R.P. *J. Org. Chem.* ***2004***, *69*, 566.

CHAPTER 11

PREPARATION OF HYDRIDES

This chapter lists hydrogenolysis and related reactions by which functional groups are replaced by hydrogen: e.g. $RCH_2X \rightarrow RCH_2\text{-}H$ or R-H.

SECTION 151: HYDRIDES FROM ALKYNES

NO ADDITIONAL EXAMPLES

SECTION 152: HYDRIDES FROM ACID DERIVATIVES

This section lists examples of decarboxylations ($RCO_2H \rightarrow$ R-H) and related reactions.

NO ADDITIONAL EXAMPLES

SECTION 153: HYDRIDES FROM ALCOHOLS AND THIOLS

This section lists examples of the hydrogenolysis of alcohols and phenols (ROH $\rightarrow$ R-H).

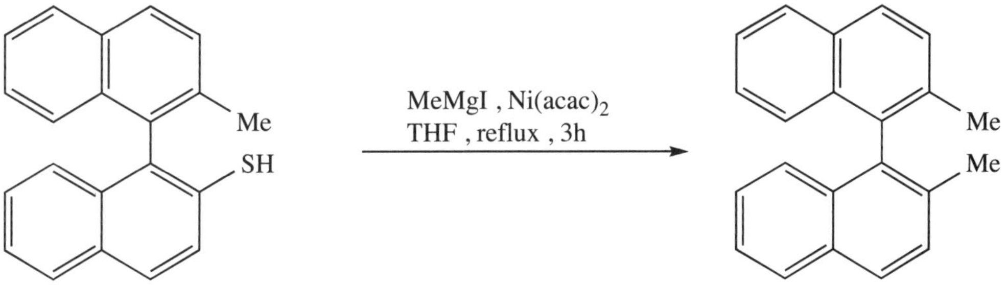

Me SH —[MeMgI , $Ni(acac)_2$ / THF , reflux , 3h]→ Me Me

Cho, Y.-H.; Kina, A.; Shimada, T.; Hayashi, T. *J. Org. Chem.* ***2004***, *69*, 3811.

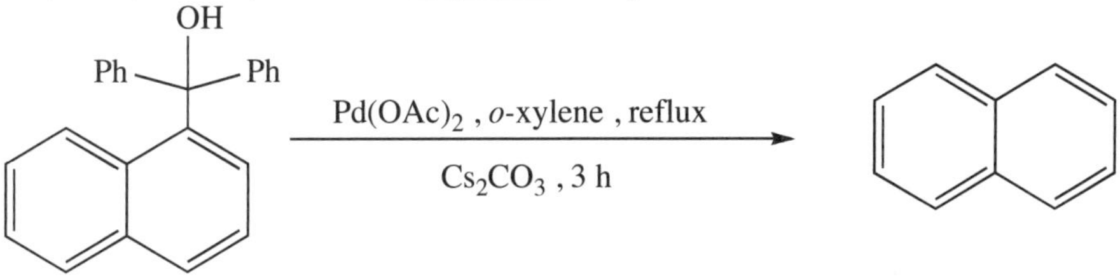

OH Ph Ph —[$Pd(OAc)_2$, *o*-xylene , reflux / Cs_2CO_3 , 3 h]→ 77%

Terao, Y.; Nomoto, M.; Satoh, T.; Miura, M.; Nomura, M. *J. Org. Chem.* ***2004***, *69*, 6942.

Also via: Section 160 (Hydrides from Halides and Sulfonates).

SECTION 154: HYDRIDES FROM ALDEHYDES

For the conversion RCHO → R-Me, etc., see Section 64 (Alkyl, Methylenes, and Aryls from Aldehydes).

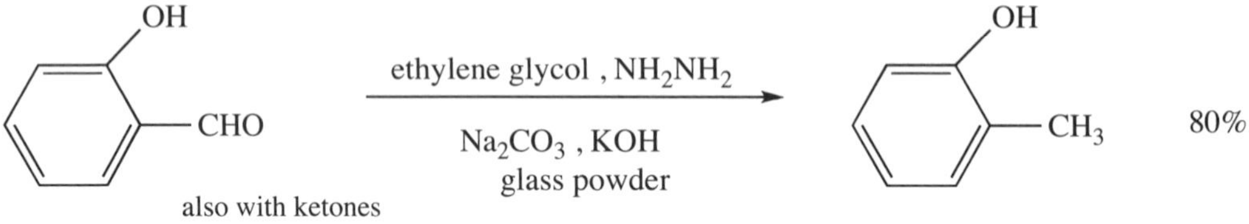

OH, CHO — ethylene glycol , NH_2NH_2 / Na_2CO_3 , KOH / glass powder → OH, CH_3 80%

also with ketones

Jaisankar, P.; Pal, B.; Giri, V.S. *Synth. Commun.* ***2002***, *32*, 2569.

SECTION 155: HYDRIDES FROM ALKYLS, METHYLENES AND ARYLS

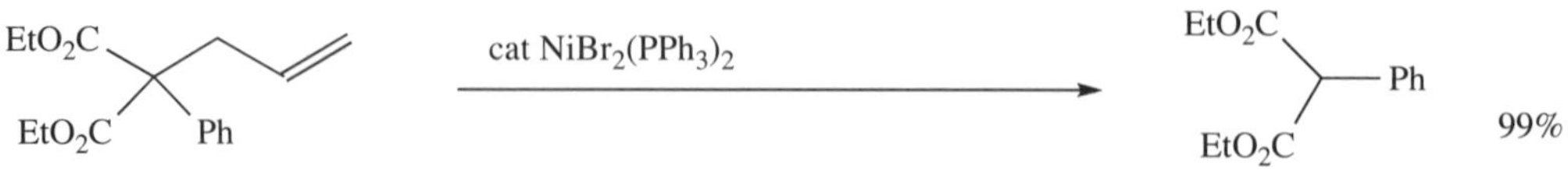

EtO_2C, EtO_2C, Ph — cat $NiBr_2(PPh_3)_2$ → EtO_2C, EtO_2C, Ph 99%

Necas, D.; Tursky, M.; Kotora, M. *J. Am. Chem. Soc.* ***2004***, *126*, 10222.

SECTION 156: HYDRIDES FROM AMIDES

NO ADDITIONAL EXAMPLES

SECTION 157: HYDRIDES FROM AMINES

This section lists examples of the conversion RNH_2 (or R_2NH) → R-H.

NO ADDITIONAL EXAMPLES

SECTION 158: HYDRIDES FROM ESTERS

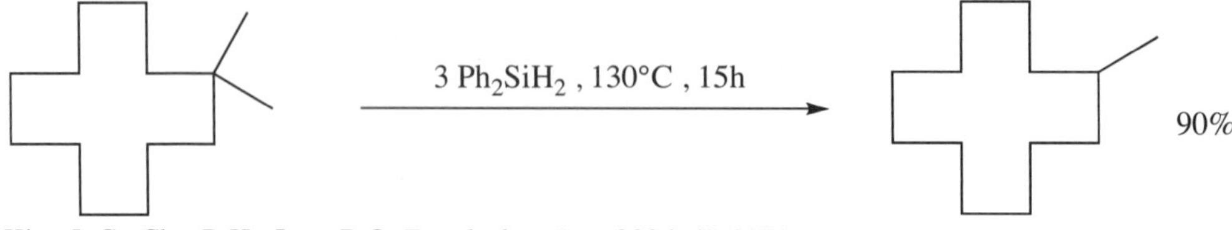

Kim, J.-G.; Cho, D.H.; Jang, D.O. *Tetrahedron Lett.* ***2004***, *45*, 3031.

This section lists examples of the reactions RCO_2R' → R-H and RCO_2R' → R'H.

SECTION 159: HYDRIDES FROM ETHERS, EPOXIDES, AND THIOETHERS

This section lists examples of the reaction R-O-R' → R-H.

NO ADDITIONAL EXAMPLES

SECTION 160: HYDRIDES FROM HALIDES AND SULFONATES

This section lists the reductions of halides and sulfonates, R-X → R-H.

O, I; H_2 , Pd/C , MeOH/NEt_3, 20°C , 2 h; O; 99%

Faucher, N.; Ambroise, Y.; Cintrat, J.-C.; Doris, E.; Pillon, F.; Rousseau, B. *J. Org. Chem.* ***2002***, *67*, 932.

$C_{10}H_{21}$, I; 0.5 La , THF, 67°C , 2 h; $C_{10}H_{21}$, $C_{10}H_{21}$ 64% + $C_{10}H_{21}$ 24%

Nishino, T.; Watanabe, T.; Okada, M.; Nishiyama, Y.; Sonoda, N. *J. Org. Chem.* ***2002***, *67*, 966.

Br, Ph; $InCl_3$, $NaBH_4$, MeCN , rt , 2 h; Ph 90%

Inoue, K.; Sawada, A.; Shibata, I.; Baba, A. *J. Am. Chem. Soc.* ***2002***, *124*, 906.

Br; AIBN , PhH , 14 h; OMe, Me, $SiMe_2t$-Bu, OMe; H 88%

Studer, A.; Amrein, S.; Schleth, F.; Schulte, T.; Walton, J.C. *J. Am. Chem. Soc.* ***2003***, *125*, 5726.

Ph, Br; 0.1 $InCl_3$, Bu_3SnH , THF , 2 h; Ph 82%

Hayashi, N.; Shibata, I.; Baba, A. *Org. Lett.* ***2004***, *6*, 4981.

PhO, I; $SnHMe_2$, AIBN , PhH; PhO 94%

Gastaldi, S.; Stein, D. *Tetrahedron Lett.* ***2002***, *43*, 4309.

PhI; 6 PHMS, 12 KF , THF , rt, 1% $PdCl_2(PPh_3)_2$; PhH 90%

Maleczka Jr. R.E.; Rahaim Jr. R.J.; Teixeira, R.R. *Tetrahedron Lett* ***2002***, *43*, 7087.

Cl, CO_2H; H_2 , NEt_3 , 3% of 10% Pd/C, MeOH , rt , 1h; CO_2H quant

Sajiki, H.; Kume, A.; Hattori, K.; Hirota, K. *Tetrahedron Lett.* ***2002***, *43*, 7247.

Cl—C_6H_4—C(=O)Ph → HCO_2Na , H_2O , cat 10% Pd/C ; rt , 1d → PhC(=O)Ph 63%

Arcadi, A.; Cerichelli, G.; Chiarini, M.; Vico, R.; Zorzan, D. *Eur. J. Org. Chem.* ***2004***, 3404.

Ph-CH=CBr_2 → 1.5 SmI_2 , MeOH , 45°C , 6 h → Ph-CH=CHBr 92% (33:67 *Z*:*E*)

[+ 3 SmI_2, 98% styrene; + 5 SmI_2, 60°C, quant, ethylbenzene]

Wang, L.; Li, P.; Xie, Y.; Ding, Y. *Synlett* ***2003***, 1137.

Ni(R) , *i*-PrOH 51%

Mebane, R.C.; Grimes, K.D.; Jenkins, S.R.; Deardorff, J.D.; Gross, B.H. *Synth. Commun.* ***2002***, *32*, 2049.

Ph—C_6H_4—F → 2 $LiAlH_4$, DME , 5% $NbCl_5$; 85°C → Ph—Ph 91%

Fuchibe, K.; Akiyama, T. *Synlett* ***2004***, 1282.

Br → Bi , NH_4HF_2 , H_2O , rt , 2 h → quant

Lee, Y.J.; Chan, T.H. *Can. J. Chem.* ***2004***, *82*, 71.

SECTION 161: HYDRIDES FROM HYDRIDES

NO ADDITIONAL EXAMPLES

SECTION 162: HYDRIDES FROM KETONES

This section lists examples of the reaction $R_2C\text{-}(C{=}O)R \rightarrow R_2C\text{-}H(R)$.

Ph(C=O)Ph → Polymethylhydrosiloxane ; 5% $B(C_6F_5)_3$, DCM → PhCH_2Ph 90%

Chandrasekhar, S.; Reddy, Ch.R.; Babu, B.N. *J. Org. Chem.* ***2002***, *67*, 9080.

Ph → supercritical *i*-PrOH , 350°C ; 5h → Ph 90%

Hatano, B.; Tagayam H. *Tetrahedron Lett.* ***2003***, *44*, 6331.

Ph(C=O)CH_3 → Ni-Al , H_2O , reflux , 2h → PhCH_2CH_3 99%

Ishimoto, K.; Mitoma, Y.; Negashima, S.; Tashiro, H.; Prakash, G.K.S.; Olah, G.A.; Tashiro, M. *Chem. Commun.* ***2003***, 514.

SECTION 163: HYDRIDES FROM NITRILES

This section lists examples of the reaction, [R-C≡N → R-H (includes reactions of isonitriles (R-N≡C)].

NO ADDITIONAL EXAMPLES

SECTION 164: HYDRIDES FROM ALKENES

NO ADDITIONAL EXAMPLES

SECTION 165: HYDRIDES FROM MISCELLANEOUS COMPOUNDS

NO ADDITIONAL EXAMPLES

CHAPTER 12

PREPARATION OF KETONES

SECTION 166: KETONES FROM ALKYNES

$C_5H_{11}-C\equiv C-C_5H_{11}$ → (2-aminophenol, dioxane, 120°C; cat $Pd(NO_3)_2$) → $C_5H_{11}C(=O)CH_2C_5H_{11}$ 76%

Shimada, T.; Yamamoto, Y. *J. Am. Chem. Soc.* ***2002**, 124*, 12670.

$C_6H_{13}C\equiv CH$ → (cat $(PPh_3PAuMe$, MeOH, 50% H_2SO_4; 70°C, H_2O, 1h) → $C_6H_{13}C(=O)CH_3$ 80%

Mizushima, E.; Sato, K.; Hayashi, T.; Tanaka, M. *Angew. Chem. Int. Ed.* ***2002**, 41*, 4563.

$MeO-C_6H_4-C\equiv CH$ → (H_2O, microwaves, 200°C; 20 min) → $MeO-C_6H_4-C(=O)CH_3$ 94%

Vasudevan, A.; Verzal, M.K. *Synlett* ***2004***, 631.

$PhCH_2C\equiv CH$ → (5% $Hg(OTf)_2$•$(TMU)_2$, 2 H_2O; MeCN/DCM, rt, 12 h) → $PhCH_2C(=O)CH_3$ quant

Nishizawa, M.; Skwarczynski, M.; Imagawa, H.; Sugihara, T. *Chem. Lett.* ***2002***, 12.

SECTION 167: KETONES FROM ACID DERIVATIVES

$PhC(=O)OC(=O)Ph$ → (Ph_2Zn, 5% $Pd(PPh_3)_4$, THF; reflux) → $PhC(=O)Ph$ 85%

Wang, D.; Zhang, Z. *Org. Lett.* ***2003**, 5*, 4645.

C_6H_5OMe → (PhCOCl, ZnO; rt, 10 min) → $PhC(=O)-C_6H_4-OMe$ 95%

Sarvari, M.H.; Sharghi, H. *J. Org. Chem.* ***2004**, 69*, 6953.

AcCl , $AlCl_3$, DCM — 93%

Zhang, Z.; Yang, Z.; Wong, H.; Zhu, J.; Meanwell, N.A.; Kadow, J.F.; Wang, T. *J. Org. Chem.* ***2002***, *67*, 6226.

Ph(C=O)Cl , DCE , 80°C , 3 h; $InCl_3$/Si-MCM-41 catalyst — OMe — 83%

Choudhary, V.R.; Jana, S.K.; Patil, N.S. *Tetrahedron Lett.* ***2002***, *43*, 1105.

1. CuCN , –78°C → 0°C; 2. benzoyl chloride , –78°C — t-Bu, Li, Ph — 58%

Ryu, I.; Ikebe, M.; Sonoda, N.; Yamato, S.-y.; Yamaguchi, G.-h.; Komatsu, M. *Tetrahedron Lett.* ***2002***, *43*, 1257.

t-Bu(C=O)Cl , Zn , toluene; 2.5h , rt — 82%

Yadav, J.S.; Reddy, B.V.S.; Kondaji, G.; Rao, R.S.; Kumar, S.P. *Tetrahedron Lett.* ***2002***, *43*, 8133.

Ph(C=O)Cl — $PhB(OH)_2$, 110°C , K_3PO_4•x H_2O , 4h; cat $PdCl_2(PPh_3)_2$, toluene — Ph(C=O)Ph 91%

Urawa, Y.; Ogura, K. *Tetrahedron Lett.* ***2003***, *44*, 271.

$EtCO_2H$, 15% $Eu[(NTf)_2]_3$; 260°C — 56%

Kawamura, M.; Cui, D.-M.; Hayashi, T.; Shimda, S. *Tetrahedron Lett.* ***2003***, *44*, 7715.

C_5H_{11}(C=O)O(C=O)C_5H_{11} — $PhB(OH)_2$, cat $Pd(OAc)_2$, P(*p*-MeOPh)$_3$; 2.5 H_2O , 16h , 60°C — C_5H_{11}(C=O)Ph 96%

Gooßen, L.J.; Ghosh, K. *Eur. J. Org. Chem.* ***2002***, 3254.

$PhCO_2H$ — 1.2 $PhB(OH)_2$, 1% $[Pd(PPh_3)]_4$, dioxane; 1.2 $(MeOCO)_2O$, 80°C , 6 h — Ph(C=O)Ph 97%

Kakino, R.; Narahashi, H.; Shimizu, I.; Yamamoto, A. *Bull. Chem. Soc. Jpn.* ***2002***, *75*, 1333.

Ph–CH_2CH_2–CO_2H → 10% $Tb(OTf)_3$, PhCl , 250°C , 1h → 1-indanone 95%

Cui, D.-M.; Zhang, C.; Kawamura, M.; Shimada, S. *Tetrahedron Lett.* ***2004**, 45*, 1741.

PhZnBr → AcCl , MeCN , cat $CoBr_2$ → PhC(O)Me 74%

Fillon, H.; Gosmini, C.; Périchon, J. *Tetrahedron* ***2003**, 59*, 8199.

Toluene → $C_7H_{15}CO_2H$, 110°C , 2d → $MeC_6H_4C(O)C_7H_{15}$ 47% (72:22:6 *p:o:m*)

Kaur, J.; Kozhevnikov, I.V. *Chem. Commun.* ***2002***, 2508.

PhC(O)Cl → $NaBPh_4$, cat $[PdCl_2(PPh_3)_2]$, acetone / KF/Al_2O_3 , microwaves → PhC(O)Ph 98%

Wang, J.-X.; Yang, Y.; Wei, B.; Hu, Y.; Fu, Y. *Bull. Chem. Soc. Jpn.* ***2002**, 75*, 1381.

SECTION 168: KETONES FROM ALCOHOLS AND THIOLS

Related Methods: Section 48 (Aldehydes from Alcohols and Thiols).

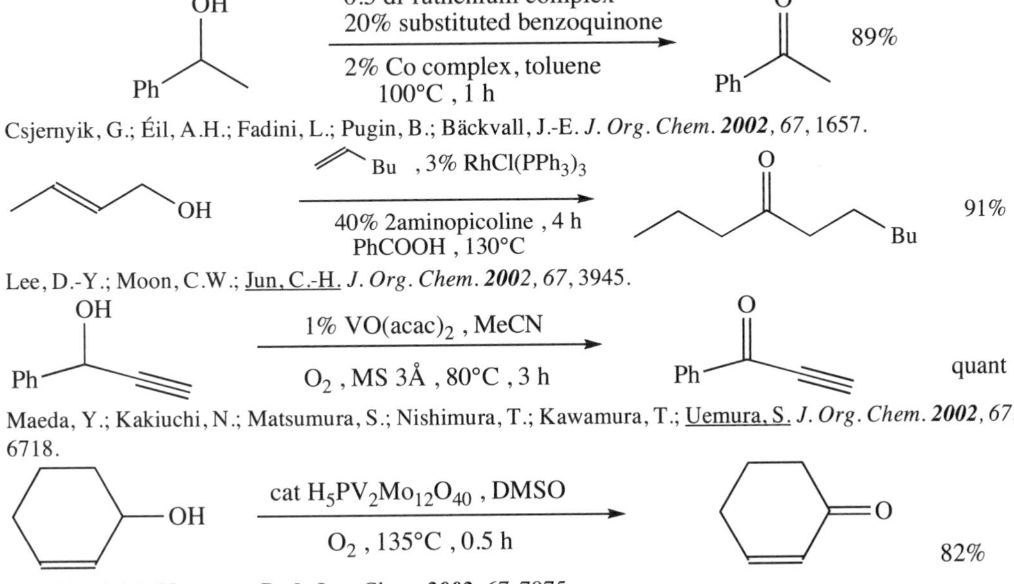

Csjernyik, G.; Éil, A.H.; Fadini, L.; Pugin, B.; Bäckvall, J.-E. *J. Org. Chem.* ***2002**, 67*, 1657.

Lee, D.-Y.; Moon, C.W.; Jun, C.-H. *J. Org. Chem.* ***2002**, 67*, 3945.

Maeda, Y.; Kakiuchi, N.; Matsumura, S.; Nishimura, T.; Kawamura, T.; Uemura, S. *J. Org. Chem.* ***2002**, 67*, 6718.

Khenkin, A.M.; Neumann, R. *J. Org. Chem.* ***2002**, 67*, 7075.

OH / Ph — 1% Cu(II) 2-ethyl hexanoate, 18% allyl ethyl ether, MeCN / O_2, 25°C; also oxidizes primary alcohols → O / Ph 97%

Muldoon, J.; Brown, S.N. *Org. Lett*, ***2002***, *4*, 1043.

OH / Ph — toluene, 105°C, PhCl, K_2CO_3 / $(2\text{-Ph-}C_6H_4)\text{-PCy}_2$, cat $Pd(dba)_2$ → O / Ph 68%

Guram, A.S.; Bei, X.; Turner, H.W. *Org. Lett.* ***2003***, *5*, 2485.

OH — *t*-BuOOH, 1% $Rh_2(OAc)_4$ / DCM, 31h → O 65%

Moody, C.J.; Palmer, F.N. *Tetrahedron Lett.* ***2002***, *43*, 139.

OH — DMSO, $N_2H_4{\bullet}H_2O(H_2O)$, I_2 / MeCN, 80°C → O 95%

Gogoi, P.; Sarmah, G.K.; Konwar, D. *J. Org. Chem.* ***2004***, *69*, 5153.

O / OH — Me, O, F_3C, DCM, $CF_3C(=O)CH_3$, 0°C / 15 min → O / O 94%

D'Accolti, L.; Fusco, C.; Annese, C.; Rella, M.R.; Turteltaub, J.S.; Williard, P.G. *J. Org. Chem.* ***2004***, *69*, 8510.

OH / Ph — toluene, 105°C, K_3PO_4, PhCl / 1% $Pd(dba)_2$, phosphine ligand → O / Ph 68%

Bei, X.; Hagemeyer, A.; Volpe, A.; Sacton, R.; Turner, H.; Guram, A.S. *J. Am. Chem. Soc.* ***2004***, *126*, 8626.

OH / Ph Ph — $(CF_3\text{-}C_6H_4)_3BiCl_2$, DBU, toluene / rt, 5 min → O / Ph Ph 99%

Matano, Y.; Hisanaga, T.; Yamada, H.; Kusakabe, S.; Nomura, H.; Imahori, H. *J. Am. Chem. Soc.* ***2004***, *126*, 8676.

HO — $H_2O_2\text{-}FeBr_3$, MeCN, rt, 1d → O 92%

Martín, S.E.; Garrone, A. *Tetrahedron Lett.* ***2003***, *44*, 549.

OH — cat $Fe(NO_3)_4{\bullet}FeBr_3$, MeCN / air, rt, 1d → O 80%

Martín, S.E.; Suárez, D.F. *Tetrahedron Lett.* ***2002***, *43*, 4475.

PhBr , 5% $Pd(OAc)_2$, Cs_2CO_3 / 5% ferrocenyl-PN ligand , toluene , 50°C — 93% (91% ee)

Matsumura, S.; Maeda, Y.; Mishimura, T.; Uemura, S. *J. Am. Chem. Soc.* ***2003***, *125*, 8862.

Ph OH Ph , 80°C , 18 h / cat $[Ir(cod)(PPh_3)_2]$ PF_6 , H_2 — 39%

Matsuda, I.; Wakamatsu, S.; Komori, K.-i.; Makino, T.; Itoh, K. *Tetrahedron Lett.* ***2002***, *43*, 1043.

cat $Na_{12}[WZn_2(H_2O)_2(ZnW_9O_{34})_2]$ / H_2O , 85°C — 94%

Sloboda-Rozner, D.; Alsters, P.L.; Neumann, R. *J. Am. Chem. Soc.* ***2003***, *125*, 5280.

1. $C_{12}H_{25}SMe$, oxalyl chloride , –60°C / 2. NEt_3 , –40°C — 95%

Nishide, K.; Ohsugi, S.-i.; Fudesaka, M.; Kodama, S.; Node, M. *Tetrahedron Lett* ***2002***, *43*, 5177.

0.03 $PdCl_2$, Bu_4NBr , 120°C , 72h — 91%

Ganchegui, B.; Bouquillon, S.; Hénin, F.; Muzart, J. *Tetrahedron Lett.* ***2002***, *43*, 6641.

Ph_3PBr_2 , DMSO , NEt_3 , DCM / –78°C — 98%

Bisai, A.; Chandrasekhar, M.; Singh, V.K. *Tetrahedron Lett.* ***2002***, *43*, 8355.

Br_2•dioxane , SiO_2 , microwaves , 1 min — 75%

Sharma, V.B.; Jain, S.L.; Sain, B. *Tetrahedron Lett.* ***2003***, *44*, 383.

$NaBrO_3$, Ce(IV)-immobilized on silica / MeCN , H_2O , 2d — 99%

Al-Haq, N.; Sullivan, A.C.; Wilson, J.R.H. *Tetrahedron Lett* ***2003***, *44*, 769.

1.5 PCC , MS 3Å , DCM / rt , 8h — (9 : 1) 70%

Fernandes, R.A.; Kumar, P. *Tetrahedron Lett.* ***2003***, *44*, 1275.

$Ru(O_2CCf_3)_2(CO)(PPh_3)_2$, 130°C / $PhCH_2CH_2OH$, toluene , 5 TFA — 81%

Ligthart, G.B.W.L.; Meijer, R.H.; Donners, M.P.J.; Meuldijk, J.; Vekemans, J.A.J.M.; Hulshof, L.A. *Tetrahedron Lett.* ***2003***, *44*, 1507.

$Mn(Phox)_3$/Oxone , Bu_4NBr / DCM , H_2O , rt , 5 min — 93%

Bagherzadeh, M. *Tetrahedron Lett.* ***2003***, *44*, 8943.

$CuCl_2$, BQC , $CuCl_2$, Na_2CO_3 , Bu_4NCl / TBHP , H_2O

BQC = 2,2'-biquinoline-4,4'-dicarboxylic acid K_2

quant

Ferguson, G.; Ajjou, A.N. *Tetrahedron Lett.* ***2003***, *44*, 9139.

5% $RhCl(CO)(PPh_3)_2$, DMF-H_2O / CH_2=$CHCO_2Me$, microwaves

C_7H_{15} C_4H_9 86%

Takahashi, M.; Oshima, K.; Matsubara, S. *Tetrahedron Lett.* ***2003***, *44*, 9201.

, $(COCl)_2$, DCM / –70°C → –40°C

75%

Choi, M.K.W.C.; Toy, P.H. *Tetrahedron* ***2003***, *59*, 7171.

cat Pd-carbene complex , air , 14h / toluene , MS 3Å , 60°C

97%

Jensen, D.R.; Schultz, M.J.; Mueller, J.A.; Sigman, M.S. *Angew. Chem. Int. Ed.* ***2003***, *42*, 3810.

HNO_3 , DCM , rt , 1h

83%

Strazzolini, P.; Runcio. A. *Eur. J. Org. Chem.* ***2003***, 526.

e^- , –5°C , 20% NaX-sat'd aq. $NaHCO_3$ / PE-N-oxyl

84%

Tanka, H.; Kubota, J.; Itogawa, S.-j.; Ido, T.; Kuroboshi, M.; Shimamura, K.; Uchida, T. *Synlett* ***2003***, 951.

[Cr(II) (salen)Cl] , PhIO , DCM , 30 min / 4-phenylpyridine N-oxide

92%

Kim, S.S.; Kim, D.W. *Synlett* ***2003***, 1391.

PhIO , Mn(III)-salen Cl , MeCN / MS 4Å , rt

89%

Kim, S.S.; Borisova, G. *Synth. Commun.* ***2003***, *33*, 3961.

20% CAN , 10% TEMPO , O_2 / MeCN , reflux , 15 min

86%

aldehydes are also oxidzed

Kim, S.S.; Jung, H.C. *Synthesis* ***2003***, 2135.

1.4 $PhI(OAc)_2$, DCM , rt / 2% Ru(Pybox)(Pydic) , 10 h

96%

Iwasa, S.; Morita, K.; Tajima, K.; Fakhruddin, A.; Nishiyama, H. *Chem. Lett.* ***2002***, *31*, 284.

OH

aluminotungstate catalyst , O_2

isobutyl acetate , 120 h

O

90%

Yamaguchi, K.; Mizuno, N. *New J. Chem.* **2002**, *26*, 972.

OH Ph Ph

O S Cl N-*t*-Bu

2 *i*-Pr_2NEt , DCM , –78°C

O Ph Ph

98%

Matsuo, J.-i.; Kawana, A.; Yamanaka, H.; Kamiyama, H. *Bull. Chem. Soc. Jpn.* **2003**, *76*, 1433.

REVIEWS:

"Palladium Catalyzed Oxidation of Primary and Secondary Alcohols"
Murart, J. *Tetrahedron* **2003**, *59*, 5789.

"Recent Developments in the Aerobic Oxidation of Alcohols"
Zhan, B.-Z.; Thompson, A. *Tetrahedron* **2004**, *60*, 2917.

"Transposition of Allylic Alcohols into Carbonyl Compounds Mediated by Transition Metal Complexes"
Uma, R.; Crévisy, C.; Grée, R. *Chem. Rev.* **2003**, *103*, 27.

"Oxidation of Alcohols with Molecular Oxygen on Solid Catalysts"
Mallat, T.; Baiker, A. *Chem. Rev.* **2004**, *104*, 3037.

"Green, Catalytic Oxidations of Alcohols"
Sheldon, R.A.; Arends, I.W.C.E.; ten Brink, G.-J.; Dijksman, A. *Acc. Chem. Res.* **2002**, *35*, 774.

SECTION 169: KETONES FROM ALDEHYDES

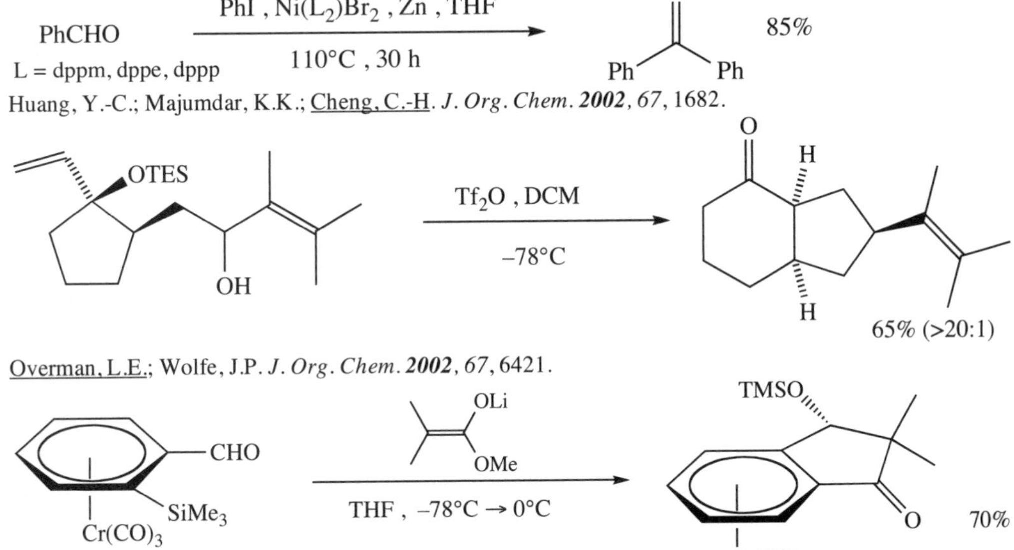

Huang, Y.-C.; Majumdar, K.K.; Cheng, C.-H. *J. Org. Chem.* **2002**, *67*, 1682.

Overman, L.E.; Wolfe, J.P. *J. Org. Chem.* **2002**, *67*, 6421.

Moser, W.H.; Zhang, J.; Lecher, C.S.; Frazier, T.L.; Pink, M. *Org. Lett.* **2002**, *4*, 1981.

CHO

O_2N

$MeO_2C\text{-}C{\equiv}C\text{-}CO_2Me$, 20% Py , THF

60°C , 12 h

O

CO_2Me

O_2N MeO_2C

89%

Li, C.-Q.; Shi, M. *Org. Lett.* ***2003***, *5*, 4273.

PhCHO

N_2CHCO_2Et , $SnCl_2$, EDA

10% $[(\eta^5\text{-}C_5H_5)Fe^+(CO_2)(thf)]BF_4^-$

O

Ph CO_2Et

62%

Dudley, M.E.; Morshed, Md.M.; Brennan, C.L.; Islam, M.S.; Ahmad, M.S.; Atuu, M.-R.; Branstetter, B.; Hossain, M.M. *J. Org. Chem.* ***2004***, *69*, 7599.

PhCHO

cat $Rh(PPh_3)_2(CO)Tol$, 85°C

C_6D_6 , 80 min

O

Ph Tol

68%

Krug, C.; Hartwig, J.F. *J. Am. Chem. Soc.* ***2002***, *124*, 1674.

CHO CO_2Et

O

20% chiral triazolium salt , 25°C

20% KHMDS , xylenes , 1 d

O

CO_2Et

O

94% (94% ee)

Kerr, M.S.; de Alaniz, J.R.; Rovis, T. *J. Am. Chem. Soc.* ***2002***, *124*, 10298.

Ph CHO

Me—≡—Me , cat $RhCl(PPh_3)_3$

cyclohexylamine , 12% $AlCl_3$, 130°C

0.5 2-amino-3-picoline , toluene , 12 h

O

Ph

90%

Lee, D.-Y.; Hong, B.-S.; Cho, E.-G.; Lee, H.; Jun, C.-H. *J. Am. Chem. Soc.* ***2003***, *125*, 6372.

CHO CO_2Me

O

chiral trazolium catalyst , 1 d

2 NEt_3 , toluene , 25°C

O

Et

O

CO_2Me

96% (97% ee)

Kerr, M.S.; Rovis, T. *J. Am. Chem. Soc.* ***2004***, *126*, 8876.

MeO CHO

Me_3SnPh , $P(t\text{-}Bu)_3$, K_2CO_3 , 80°C

cat $[Rh(CH_2{=}CH_2)Cl]_2$, dioxane/acetone

O

MeO Ph

98%

Pucheault, M.; Darses, S.; Genet, J.-P. *J. Am. Chem. Soc.* ***2004***, *126*, 15356.

Me_3Si CHO

$CH_2{=}CHCO_2Me$, DCE

cat Rh(ddpe) ClO_4 , 60°C

Me_3Si CO_2Me

O

71%

Willis, M.C.; McNally, S.J.; Beswick, P.J. *Angew. Chem. Int. Ed.* ***2004***, *43*, 340.

SECTION 170: KETONES FROM ALKYLS, METHYLENES, AND ARYLS

This section lists examples of the reaction, R-CH_2-R' → R(C=O)-R'.

O

5% $Pd(OH)_2$/C , 5 *t*-BuOOH , DCE

0.5 K_2CO_3 , 24°C , 2 d

O O 81%

Yu, J.-Q.; Corey, E.J. *J. Am. Chem. Soc.* ***2003***, *125*, 3232.

Ph O

$NaBrO_3$, Bu_4NHSO_4

aq. MeCN , 1 d

Ph 85%

Shaabani, A.; Bazgir, A.; Abdoli, M. *Synth. Commun.* ***2002***, *32*, 675.

SECTION 171: KETONES FROM AMIDES

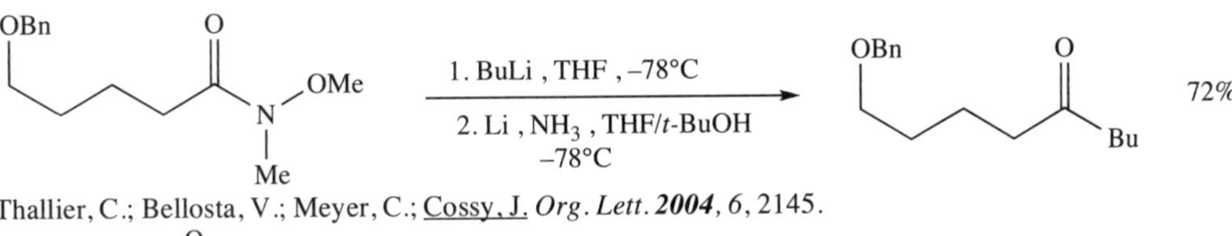

Thallier, C.; Bellosta, V.; Meyer, C.; Cossy, J. *Org. Lett.* ***2004***, *6*, 2145.

O N O

1. $PhMe_2SiLi$, THF , –78°C

2. aq. NH_4Cl

O $SiMe_2Ph$ 72%

Clark, C.T.; Milgram, B.C.; Scheidt, K.A. *Org. Lett.* ***2004***, *6*, 3977.

SECTION 172: KETONES FROM AMINES

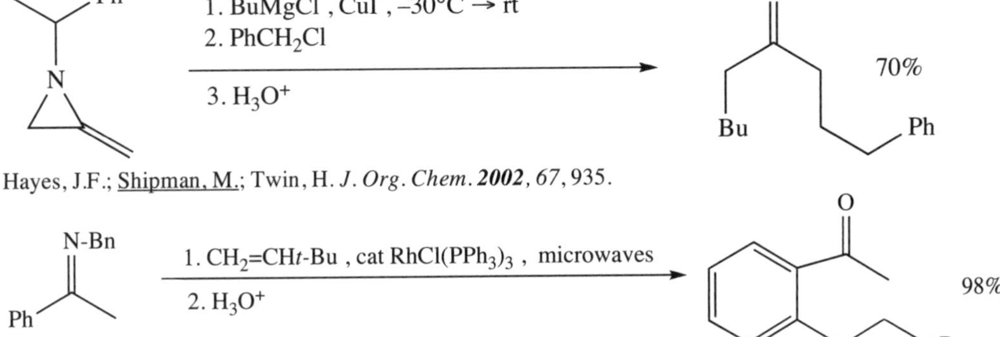

Hayes, J.F.; Shipman, M.; Twin, H. *J. Org. Chem.* ***2002***, *67*, 935.

Vo-Thanh, G.; Lahrache, H.; Loupy, A.; Kim, I.-J.; Chang, D.-H.; Jun, C.-H. *Tetrahedron* ***2004***, *60*, 5539.

HO, NH_2 (3-trimethylsilylcyclopentyl) → $NaNO_2$, AcOH → Me_3Si-cyclohexanone 58% + 3-($SiMe_3$)cyclohexanone 42%

Chow, L.; McClure, M.; White, J. *Org. Biomol. Chem.* ***2004***, *2*, 648.

N-OH, Ph, Ph → $SbCl_5$-DCM, 0°C → rt → O, Ph, Ph 80%

Narsaiah, A.V.; Nagaiah, K. *Synthesis* ***2003***, 1881.

cyclohexanone oxime (=N-OH) + N–Br saccharin → acetone, H_2O / microwaves → cyclohexanone (=O) 97%

Khazaei, A.; Manesh, A.A. *Synthesis* ***2004***, 1739.

SECTION 173: KETONES FROM ESTERS

allyl O_2CCF_3 → PhC(O)$SnMe_3$, THF, rt, 8 h / cat $Pd(O_2CCF_3)_2$ → PhC(O)CH$_2$CH=CH$_2$ 76%

Obora, Y.; Nakanishi, M.; Tokunaga, M.; Tsuji, Y. *J. Org. Chem.* ***2002***, *67*, 5835.

$C_{11}H_{23}$C(O)S-Tol → Bu_3Sn-Tol, CuOP(O)Ph_2 / cat $Pd_2(dba)_2$•TFP, THF, 50°C, 7 h → $C_{11}H_{23}$C(O)Tol 91%

Wittenberg, R.; Srogl, J.; Egi, M.; Liebeskind, L.S. *Org. Lett.* ***2003***, *5*, 3033.

$PhCO_2Me$ → excess CH_2=CHMgBr, THF / 50% $Cu(OAc)_2$, –45°C → PhC(O)CH$_2$CH$_2$CH=CH$_2$ 60%

Hansford, K.A.; Detwiler, J.E.; Lubell, W.D. *Org. Lett.* ***2003***, *5*, 4887.

PhC(O)S-CH$_2$C(O)-N(morpholine) → $C_{11}H_{23}$–B(9-BBN), CuTC, THF / Cs_2CO_3, 5% $Pd(PPh_3)_4$, 45°C, 16h → PhC(O)$C_{11}H_{23}$ 83%

TC = thiophene-2-carboxylate

Yu, Y.; Liebeskind, L.S. *J. Org. Chem.* ***2004***, *69*, 3554.

PhCH$_2$CH$_2$C(O)O-CH$_2$-(2-pyridyl) → Ph–B(O-CH$_2$CH$_2$-O) / cat $Ru_3(CO)_{12}$, 40h, toluene, 120°C → PhCH$_2$CH$_2$C(O)Ph 83%

Tatamidani, H.; Yokota, K.; Kakiuchi, F.; Chatani, N. *J. Org. Chem.* ***2004***, *69*, 5615.

PhB(OH)$_2$, cat Pd(OAc)$_2$
PPh$_3$, dioxane , 50°C , 10h

OPy MeO Ph MeO 97%

Tatamidani, H.; Kakiuchi, F.; Chatani, N. *Org. Lett.* **2004**, *6*, 3597.

cat Pd(PPh$_3$)$_4$, PhH
25°C

Ph Ph Ph Ph 94%

Burger, E.C.; Tunge, J.A. *Org. Lett.* **2004**, *6*, 4113.

OCO$_2$Me
cat RhCl(PPh$_3$)$_3$, cat P(OMe)$_3$
PhC(=O)CH$_2$Li , CuX

Ph Ph Ph 83%

Evans, P.A.; Leahy, D.K. *J. Am. Chem. Soc.* **2003**, *125*, 8976.

OCO$_2$allyl
9h , THF , 25°C , 6.25% *t*-Bu-PHOX
2.5% Pd$_2$(dba)$_3$

90% (89% ee)

Behenna, D.C.; Stolz, B.M. *J. Am. Chem. Soc.* **2004**, *126*, 15044.

IZn(CH$_2$)$_4$CO$_2$Et , 10% Ni(acac)$_2$
THF , toluene , rt , 20 h

SEt (CH$_2$)$_4$CO$_2$Et 73%

Shimizu, T.; Seki, M. *Tetrahedron Lett.* **2002**, *43*, 1039.

Li/NH$_3$
THF

CO$_2$Me + CHO (3 : 1) 84%

Srikrishna, A.; Ramasastry, S.S.V. *Tetrahedron Lett.* **2004**, *45*, 379.

SECTION 174: KETONES FROM ETHERS, EPOXIDES, AND THIOETHERS

10% [TpRu(PPh$_3$)(MeCN)] PF$_6$
100°C , toluene

Ph Bu Ph Bu 94%

Chang, C.-L.; Kumar, M.P.; Liu, R.-S. *J. Org. Chem.* **2004**, *69*, 2793.

OH
NBS , β-cyclodextrin , H$_2$O , rt
aldehydes can also be formed

Ph Ph 98%

Narender, M.; Reddy, M.S.; Rao, K.R. *Synthesis* **2004**, 1741.

OTMS; Ph — Mn(III) salen, PhIO, MeCN; rt, 6 h → Ph(C=O)CH$_3$ 81%

Murahashi, S.-I.; Noji, S.; Hirabayashi, T.; Komiya, N. *Synlett* ***2004***, 1739.

OTMS — FVP; 530°C → 69%

Rüedi, G.; Nagel, M.; Hansen, H.-J. *Org. Lett.* ***2004***, *6*, 2989.

SECTION 175: KETONES FROM HALIDES AND SULFONATES

OH; Br — ZnS, DMSO, 70°C → O 41% + OH; O 39%

Bettadiah, B.K.; Gurudutt, K.N.; Srinivas, P. *J. Org. Chem.* ***2003***, *68*, 2460.

I—C$_6$H$_4$—OMe — Ac$_2$O, *i*-Pr$_2$NEt, DMF; LiCl, 100°C; cat Pd$_2$(dba)$_3$ → O; OMe 70%

Cacchi, S.; Fabrizi, G.; Gavazza, F.; Goggiamani, A. *Org. Lett.* ***2003***, *5*, 289.

Ph-I — Bu$_3$In, 4% Pd(PPh$_3$)$_4$, THF; reflux, CO → O; Ph 87%

Lee, P.H.; Lee, S.W.; Lee, K. *Org. Lett.* ***2003***, *5*, 1103.

Me—C$_6$H$_4$—I — 0.28 Bu$_4$InLi, CO; Pd(PPh$_3$)$_4$, THF; 60°C → Me—C$_6$H$_4$—C(O)Bu + Me—C$_6$H$_4$—Bu (91 : 7) 84%

Lee, S.W.; Lee, K.; Seomoon, D.; Kim, S.; Kim. H.; Kim, H.; Shim, E.; Lee, M.; Le, S.; Kim, M.; Lee, P.H. *J. Org. Chem.* ***2004***, *69*, 4852.

Br — NaIO$_4$, DMF, 150°C, 40h → O 84%

Das, S.; Panigrahi, A.K.; Maikap, G.C. *Tetrahedron Lett.* ***2003***, *44*, 1375.

Ph; F — 2% BF$_3$•OEt$_2$, DCM, –20°C, 18h; OSiMe$_3$; Ph → Ph; O; Ph 89%

Hirabo, K.; Fujita, K.; Yorimitsu, H.; Shinokubo, H.; Oshima, K. *Tetrahedron Lett.* ***2004***, *45*, 2555.

1.2 Bu_3SnOMe , 1.2 CH_2=C(Me)OAc , 100°C

cat $Pd_2(dba)_3$, toluene , chiral phsphine

86%

Liu, P.; Lanza Jr. T.J.; Jewell, J.; Jones, C.P.; Hagmann, W.K.; Lin, L.S. *Tetrahedron Lett.* ***2003**, 44*, 8869.

$Pd(OAc)_2$, imidazolium salt , CO

$PhB(OH)_2$, dioxane , Cs_2CO_3 , 22h

95%

Maerten, E.; Hassouna, F.; Couve-Bonnaire, S.; Mortreux, A.; Carpentier, J.-F.; Castanet, Y. *Synlett* **2003**, 1874.

Ph_3In , CO , Pd catalyst , THF

83%

Pena, M.A.; Sestelo, J.P.; Sarandeses, L.A. *Synthesis* ***2003**,* 780.

Related Methods: Section 177 (Ketones from Ketones)
Section 55 (Aldehydes from Halides and Sulfonates)

SECTION 176: KETONES FROM HYDRIDES

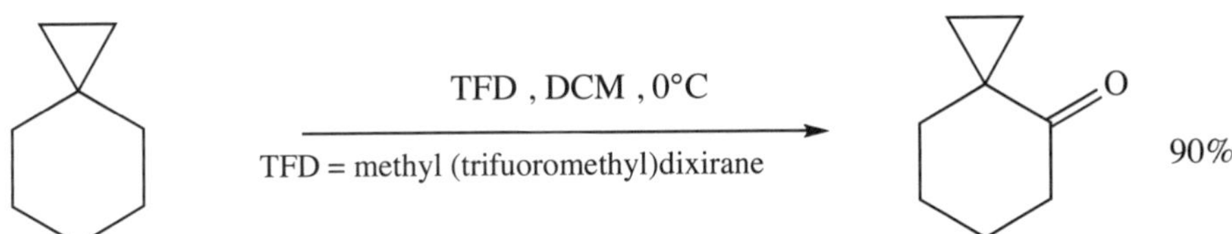

D'Accolti, L.; Dinoi, A.; Fusco, C.; Russo, A.; Curci, R. *J. Org. Chem.* ***2003**, 68*, 7806.

PhCOCl , [bmim] NTf_2

5% Bi_2O_3 , 150°C , 4 h

98% (*o:m:p* 19:2:79)

Gmouth, S.; Yang, H.; Vaultier, M. *Org. Lett.* ***2003**, 5*, 2219.

acetyl chloride , $ZnCl_2$

55°C , DCE , 1.5h

65%

Pal, M.; Dakarapu, R.; Padakanti, S. *J. Org. Chem.* ***2004**, 69*, 2913.

0.1% $Rh_2(cap)_4$, 5 *t*-BuOOH

DCM , 1 h

61%

Catino, A.J.; Forslund, R.E.; Doyle, M.P. *J. Am. Chem. Soc.* ***2004**, 126*, 13622.

Ph⌒Ph → $KMnO_4$/Montmorillonite K10, neat, 20h → PhC(O)Ph 86% [+ ultrasound = 81%; + microwaves = 78%]

Shaabani, A.; Bazgir, A.; Teimouri, F.; Lee, D.G. *Tetrahedron Lett.* **2002**, *43*, 5165.

$Ru(OH)_x$–Al_2O_3, $PhCF_3$; 1 atm O_2, 373 K, 2 h — 92%

Kamata, K.; Kasai, J.; Yamaguchi, K.; Mizuno, N. *Org. Lett.* **2004**, *6*, 3577.

NO_2 — acetyl chloride, EmimCl ($AlCl_3$) (ionic liquid) — NO_2, O, N H — 87%

Yeung, K.-S.; Farkas, M.E.; Qui, Z.; Yang, Z. *Tetrahedron Lett.* **2002**, *43*, 5793.

OMe, OMe — $LiClO_4/Ac_2O$, 60°C, 1.5h — OMe, OMe, O — >99%

Bartoli, G.; Bosco, M.; Marcantoni, E.; Massaccesi, M.; Rinaldi, S.; Sambri, L. *Tetrahedron Lett.* **2002**, *43*, 6331.

benzoyl chloride, 120°C, 5h; Bi(III)-oxychloride — Ph, O — 70% (18:2:80 *o:m:p*)

Répichet, S.; Le Roux, C.; Roques, N.; Dubac, J. *Tetrahedron Lett.* **2003**, *44*, 2037.

Ph⌒Ph — cat Cu complex, 10 30% H_2O_2; MeCN, 80°C, 5h — PhC(O)Ph 87%

Velusamy, S.; Punniyamurthy, T. *Tetrahedron Lett.* **2003**, *44*, 8955.

MnTPFPPC/PhI(OAc)$_2$ (bmim) PF_6, DCM — OH 36% + O 45%

Li, Z.; Xia, C.-G. *Tetrahedron Lett.* **2003**, *44*, 9229.

$SiMe_3$ — $AlCl_3$, BzCl, CS_2, 0°C, 2h — O, $SiMe_3$ — 80%

Georgakilas, V.; Perdikomatis, G.P.; Triantafyllou, A.S.; Siskos, M.G.; Zerkadis, A.K. *Tetrahedron* **2002**, *58*, 2441.

Ph–Et → ($KMnO_4/MnO_2$, DCM, 28h) → PhC(O)Me 89%

Shaabani, A.; Mirzaei, P.; Naderi, S.; Lee, D.G. *Tetrahedron* **2004**, *60*, 11415.

MeO–C$_6$H$_5$ → (cat $Yb[C(SO_2C_4F_4)_3]_3$, 25°C; 2 Ac_2O, $PhCF_3$) → MeO–C$_6$H$_4$–C(O)Me 80% (para only)

Barrett, A.G.M.; Bouloc, N.; Braddock, D.C.; Chadwick, D.; Henderson, D.A. *Synlett* **2002**, 1653.

OH (m-cresol) → (acetyl chloride, neat, $TiCl_4$; 120°C, 1 h) → OH, O (2-hydroxy-4-methylacetophenone) 95%

Bensari, A.; Zaveri, N.T. *Synthesis* **2003**, 267.

Benzene → (AcCl, Zn powder, 60°C; microwaves, 30 sec) → Ph–Ac 95%

Paul, S.; Nanda, P.; Gupta, R.; Loupy, A. *Synthesis* **2003**, 2877.

Ph–OMe → (graphite, $MeSO_3H$, $MeCO_2H$; 80°C, no solvent) → MeC(O)–C$_6$H$_4$–OMe 98%

Sarvari, M.H.; Sharghi, H. *Synthesis* **2004**, 2165.

Benzene → ($Yb(OTf)_3$, rt → 80°C; Cl_3CO–C(O)–C(O)–$OCCl_3$) → PhC(O)Ph 68%

Su, W.; Jin, C. *Synth. Commun.* **2004**, *34*, 4249.

SECTION 177: KETONES FROM KETONES

This section contains alkylations of ketones and protected ketones, ketone transpositions and annulations, ring expansions and ring openings, and dimerizations. Conjugate reductions and Michael alkylations of enone are listed in Section 74 (Alkyls, Methylenes, and Aryls from Alkenes). For the preparation of enamines or imines from ketones, see Section 356 (Amine-Alkene).

3-pyridyl–C(O)–C≡C–$SiMe_3$ → ($CH(CO_2Et)_2$–$(CH_2)_3$–$CH(CN)_2$; KF•2 H_2O, etOH) → 3-pyridyl–C(O)–CH_2–cyclohexane bearing EtO_2C, CO_2Et, CN, CN 71%

Holeman, D.S.; Rasne, R.M.; Grossman, R.B. *J. Org. Chem.* **2002**, *67*, 3149.

$PhMe_2Si$–CH=CH–C(O)Me → 3 eq KF, wet DMSO; 25°C, 18 h → Ph–CH_2CH_2–C(O)Me quant

Jung, M.E.; Piizzi, G. *J. Org. Chem.* **2002**, *67*, 3911.

Ph–C(O)–Ph → P_4S_{10}, HMDO, xylene; reflux, 1.5 h → Ph–C(S)–Ph 93%

HMDO = hexamethyldisioloxane. also for esters, amides, etc.

Curphey, T.J. *J. Org. Chem.* **2002**, *67*, 6461.

5.2 Zn, *t*-amyl-OH/H_2O; 2h — 54%

Br, CO_2Me → CO_2Me

Sugi, M.; Sakuma, D.; Togo, H. *J. Org. Chem.* **2003**, *68*, 7629.

10% $PdCl_2(MeCN)_2$; sealed tube, HCl; $CuCl_2$ — 77%

Wang, X.; Pei, T.; Han, X.; Widenhoefer, R.A. *Org. Lett.* **2003**, *5*, 2699.

CO_2Me → $Yb(OTf)_3$, cat $PdCl_2(MeCN)_2$; dioxane, 50°C, 38 h → CO_2Me 92%

Yang, D.; Li, J.-H.; Gao, Q.; Yan, Y.-L. *Org. Lett.* **2003**, *5*, 2869.

Ph–cyclobutanone → $PhB(OH)_2$, 10% $P(i\text{-}Bu)_3$; 5% $Rh(acac)(C_2H_4)_2$, 100°C; Cs_2CO_3, dioxane → Ph–CH(Me)–CH_2–C(O)–Ph 95%

Matuda, T.; Makino, M.; Murakami, M. *Org. Lett.* **2004**, *6*, 1257.

SMe → 1. BuLi; 2. 2-acetylthiophene; 3. $ZnBr_2$ → SMe 82%

Katritzky, A.R.; Bobrov, S.; Khashab, N. *J. Org. Chem.* **2004**, *69*, 4269.

PhMeN, Me → 4-bromotoluene, toluene; 1% $Pd_2(dba)_3$, NaO*t*-Bu; 2.5% chiral phosphine → PhMeN, Me, Tol 84% (93% ee)

Hamada, T.; Chieffi, A.; Åhman, J.; Buchwald, S.L. *J. Am. Chem. Soc.* **2002**, *124*, 1261.

5 NaN_3 , 0.08M MeOH

55°C , 20h

63%

Snider, B.B.; Duvall, J.R. *Org. Lett.* **2004**, *6*, 1265.

5% [Rh(nbd)dppp] PF_6 , 135°C
10% BHT , m-xylene

89%

Murakami, M.; Itahashi, T.; Ito, Y. *J. Am. Chem. Soc.* **2002**, *124*, 13976.

BuOH , cat $[Ir(cod)Cl]_2$, KOH

PPh_3 , 100°C , 4 h

80%

Taguchi, K.; Nakagawa, H.; Hirabayashi, T.; Sakaguchi, S.; Ishii, Y. *J. Am. Chem. Soc.* **2004**, *126*, 72.

BnOH , 1-dodecene , KOH , 80°C , 20h

cat $RuCl_2(PPh_3)_3$, dioxane

82%

Cho, C.S.; Kim, B.T.; Kim, T.-J.; Shim, S.C. *Tetrahedron Lett.* **2002**, *43*, 7987.

P_4S_{10} , Al_2O_3 , MeCN , reflux , 1.5h

89%

Polshettiwar, V.; Kaushik, M.P. *Tetrahedron Lett.* **2004**, *45*, 6255.

1. FVP , 600°C
2. 1% aq. HCl

73% + 6%

Rüedi, G.; Oberli, M.A.; Nagel, M.; Hansen, H.-J. *Org. Lett.* **2004**, *6*, 3179.

OAc , NaHMDS , DME

cat $[\eta^3\text{-}C_3H_5)PdCl]_2$, chiral diamide

90% (88% ee)

Trost, B.M.; Schroeder, G.M.; Kristensen, J. *Angew. Chem. Int. Ed.* **2002**, *41*, 3492.

1. hν , $W_{10}O_{32}^{-4}$, O_2 , MeCN , 4h

2. PPh_3

with Et_3SiH: 16% alcohol + 84% ketone

quant

Lykakis, I.N.; Orfanopulos, M. *Tetrahedron Lett.* **2004**, *45*, 7645.

Related Method: Section 49 (Aldehydes from Aldehydes)

SECTION 178: KETONES FROM NITRILES

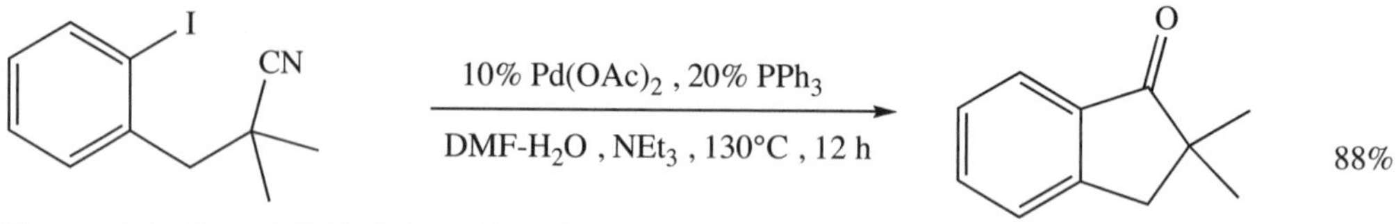

10% $Pd(OAc)_2$, 20% PPh_3; DMF-H_2O , NEt_3 , 130°C , 12 h — 88%

Pletnev, A.A.; Larock, R.C. *J. Org. Chem.* ***2002***, *67*, 9428.

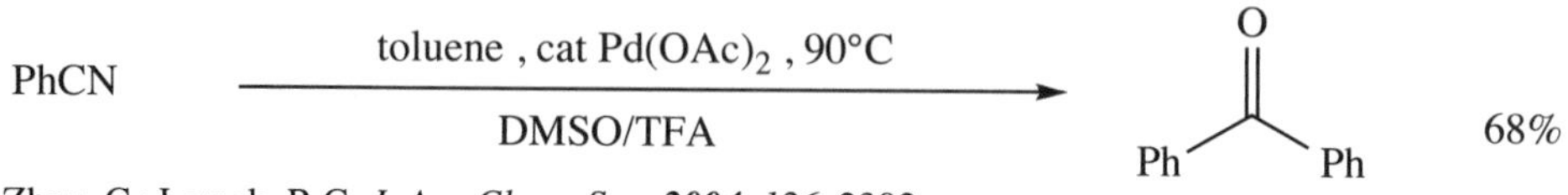

PhCN → toluene , cat $Pd(OAc)_2$, 90°C; DMSO/TFA → Ph(C=O)Ph 68%

Zhou, C.; Larock, R.C. *J. Am. Chem. Soc.* ***2004***, *126*, 2302.

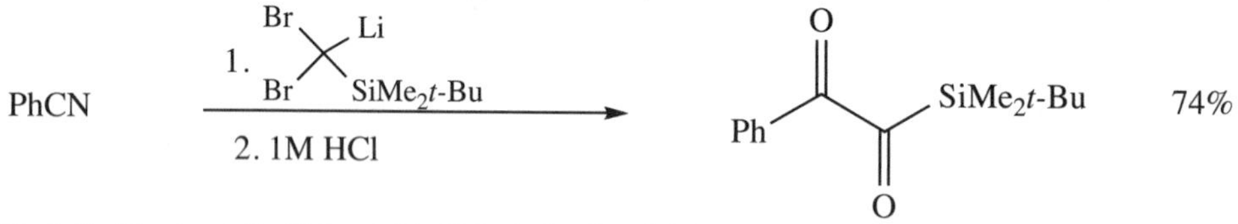

PhCN → 1. $Br_2C(Li)SiMe_2t$-Bu; 2. 1M HCl → PhC(=O)C(=O)$SiMe_2t$-Bu 74%

Yagi, K.; Tsuritani, T.; Takami, K.; Shinokubo, H.; Oshima, K. *J. Am. Chem. Soc.* ***2004***, *126*, 8618.

REVIEW:

"Additions to Metal-Activated Organonitriles"
Kukushkin, V.Yu.; Pombeiro, A.J.L. *Chem. Rev.* ***2002***, *102*, 1771.

SECTION 179: KETONES FROM ALKENES

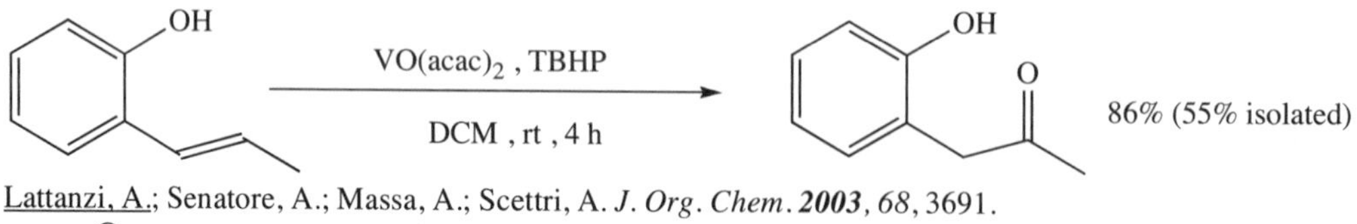

VO(acac)$_2$, TBHP; DCM , rt , 4 h — 86% (55% isolated)

Lattanzi, A.; Senatore, A.; Massa, A.; Scettri, A. *J. Org. Chem.* ***2003***, *68*, 3691.

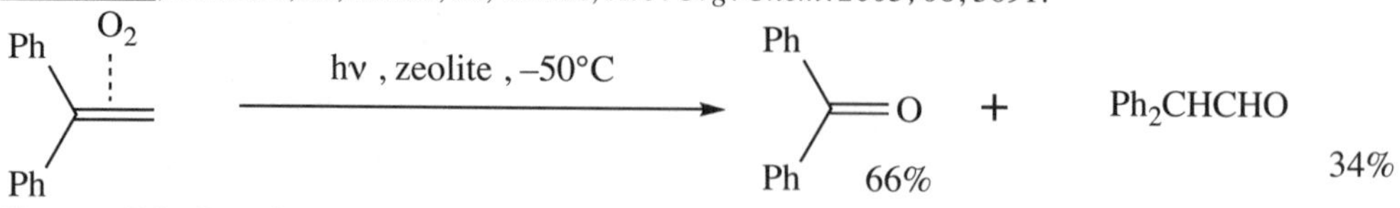

$Ph_2C=CH_2$ / O_2 → hν , zeolite , –50°C → $Ph_2C=O$ 66% + Ph_2CHCHO 34%

Clennan, E.L.; Pan, G.-I. *Org. Lett.* ***2003***, *5*, 4979.

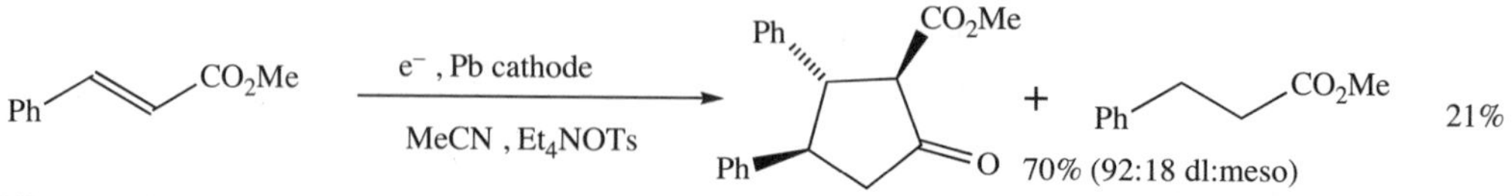

Ph–CH=CH–CO_2Me → e^- , Pb cathode; MeCN , Et_4NOTs → 70% (92:18 dl:meso) + Ph–CH_2CH_2–CO_2Me 21%

Kise, N.; Iitaka, S.; Iwasaki, K.; Ueda, N. *J. Org. Chem.* ***2002***, *67*, 8305.

$CH_2=C(Me)CO_2Bn$ → 1. EtI , Zn , MeCN , rt , 16h; 2. H_3O^+ → Et, Me, CO_2Bn, Me ketone 82%

Yamamoto, Y.; Nakano, S.; Maekawa, H.; Nishiuchi, I. *Org. Lett.* ***2004***, *6*, 799.

OH → $Tl(NO_3)_3$, rt , 15 min; 50% aq. AcOH → OH 71% + OAc 10%

Ferraz, H.M.C.; Longo Jr. L.S.; Zukerman-Schpector, J. *J. Org. Chem.* ***2002***, *67*, 3518.

I → cat $PdCl_2(PPh_3)_2$, 2 pyridine , CO; Bu_4NCl , DMF , 100°C , 8 h → O quant

Gagnier, S.V.; Larock, R.C. *J. Am. Chem. Soc.* ***2003***, *125*, 4804.

Ph, Br, Br → Zn ,H_2O , 275°C , 4h → Ph, O 82%

Wang, L.; Li, P.; Yan, J.; Wu, Z. *Tetrahedron Lett.* ***2003***, *44*, 4685.

Ph, N, N H → , toluene , 170°C; cat $[(C_8H_{14})_2RhCl]_2/PCy_3$ → O + O (87 : 13) quant

Lee, D.-Y.; Lim, I.-J.; Jun, C.-H. *Angew. Chem. Int. Ed.* **2002**, *41*, 3031.

O, CO_2Et → 1. $PhCH=CH_2$, *i*-PrOH , air; $CeCl_3$•7 H_2O , 23°C , 1d; 2. Py-AcCl → O, CO_2Et, Ph, O 71%

Rössle, M.; Werner, T.; Baro, A.; Frey, W.; Christoffers, J. *Angew. Chem. Int. Ed.* **2004**, *43*, 6547.

Ph, Ph → Ac_2O , Mg , TMSCl , DMF → O, Ph, Ph 73%

Nishiguchi, I.; Yammoto, Y.; Sakai, M.; Ohno, T.; Ishino, Y.; Maekawa, H. *Synlett* ***2002***, 759.

→ O_2 , neat , 90°C; 15 h → O 41% + HCHO

Hayashi, Y.; Takeda, M.; Miyamoto, Y.; Shoji, M. *Chem. Lett.* ***2002***, *31*, 414.

See also: Section 134 (Ethers, Epoxides, and Thioethers from Alkenes)
Section 174 (Ketones from Ethers, Epoxides, and Thioethers)

SECTION 180: KETONES FROM MISCELLANEOUS COMPOUNDS

Br, Et, Et, OH → di-*tert*-butylperoxy oxalate, cyclohexane, 10 min, 60°C → reflux → Et, Et, O 93%

Dolenc, D.; Harej, M. *J. Org. Chem.* **2002**, *67*, 312.

N-NHPh, Ph → $KMnO_4$, wet SiO_2, neat, 15 min → O, Ph 90%

Hajipour, A.R.; Adibi, H.; Ruoho, A.E. *J. Org. Chem.* **2003**, *68*, 4553.

Ph, C=O, $(i\text{-Pr})_3Si$ → 1. Me_3SiCHN_2, DCM-hexane, rt; 2. SiO_2, rt → $Si(i\text{-Pr})_3$, O 86%

Dalton, A.M.; Zhang, Y.; Davie, C.P.; Danheiser, R.L. *Org. Lett.* **2002**, *4*, 2465.

NO_2 → 1. 10% chiral phosphine, 3 Me_2Zn, 5% $Cu(OTf)_3$•PhH, toluene, 0°C; 2. 20% H_2SO_4 → O, Me 90% (93% ee)

Luchaco-Cullis, C.A.; Hoveyda, A.H. *J. Am. Chem. Soc.* **2002**, *124*, 8192.

EtO_2C, CO_2Et, I, OEt, O, P, OEt, O → $(Me_2Sn)_2$, PhH, hv, 2 h → EtO_2C, CO_2Et, O 71%

Kim, S.; Cho, C.H.; Lim, C.J. *J. Am. Chem. Soc.* **2003**, *125*, 9574.

NO_2, Ph, C_4H_9 → 2 $NaNO_2$, DMSO-H_2O, 20°C, 148 h → O, Ph, C_4H_9 quant

Gissot, A.; N'Gouela, S.; Matt, C.; Wagner, A.; Mioskowski, C. *J. Org. Chem.* **2004**, *69*, 8997.

NO_2, Ph → DBU, MeCN, 60°C, 4d → O, Ph 60%

Ballini, R.; Bosica, G.; Fiorini, D.; Petrini, M. *Tetrahedron Lett.* **2002**, *43*, 5233.

Cyclohexanone oxime (=N-OH) → $SiBr_4$, wet SiO_2 , 16 min → cyclohexanone (=O) 85%

De, S.K. *Tetrahedron Lett.* **2003**, *44*, 9055.

Ph(C=N–OH)Me → $BiBr_3$-$Bi(OTf)_3$•4 H_2O , MeCN/acetone, H_2O , 2.5h → PhC(=O)Me 68%

Arnold, J.N.; Hayes, P.D.; Kohaus, R.L.; Mohan, R.S. *Tetrahedron Lett.* **2003**, *44*, 9173.

Bicyclic alkylidene sulfoxide (S(=O)Ph) → Bu_3SnH , PhH , reflux , 12h → decalone (=O) 43%

Chuard, R.; Giraud, A.; Renaud, P. *Angew. Chem. Int. Ed.* **2002**, *41*, 4323.

2 $Ph_2I^+ BF_4^-$ → CO , DMF , rt , In, 5% $Pd(OH)_2$ → PhC(=O)Ph + 2 PhI 71%

Zhou, T.; Chen, Z.-C. *Synth. Commun.* **2002**, *32*, 3431.

Ph(C=N-OH)Me → $KMnO_4$, Al_2O_3 , 50°C , 40 min → PhC(=O)Me 78%

Imanzadeh, G.H.; Hajipour, A.R.; Mallakpour, S.E. *Synth. Commun.* **2004**, *34*, 735.

Ph(C=N-OH)Ph → $Hg(NO_3)_2/SiO_2$, THF , 6 h → PhC(=O)Ph 94%

De, S.K. *Synth. Commun.* **2004**, *34*, 2289.

Ph(C=N-OH)Me → bis(tetrabutylammonium) dichromate, microwaves , 1 min → PhC(=O)Me 94%

Murugan, R.; Reddy, B.S.R. *Chem. Lett.* **2004**, *33*, 1038.

Ph(C=N-OH)Ph → WCl_6 , Zn powder , MeCN, 15 min → PhC(=O)Ph 91%

Firouzabadi, H.; Jamalian, A.; Karimi, B. *Bull. Chem. Soc. Jpn.* **2002**, *75*, 1761.

Conjugate reductions and reductive alkylations of enones are listed in Section 74 (Alkyls, Methylenes, and Aryls from Alkenes).

SECTION 180A: PROTECTION OF KETONES

1,1-Dimethoxycyclohexane (OMe, OMe) → 10% I_2 , acetone , 25°C , 5 min; also with acetals → cyclohexanone (=O) 96%

Sun, J.; Dong, Y.; Cao, L.; Wang, X.; Wang, S.; Hu, Y. *J. Am. Chem. Soc.* **2004**, *126*, 8932.

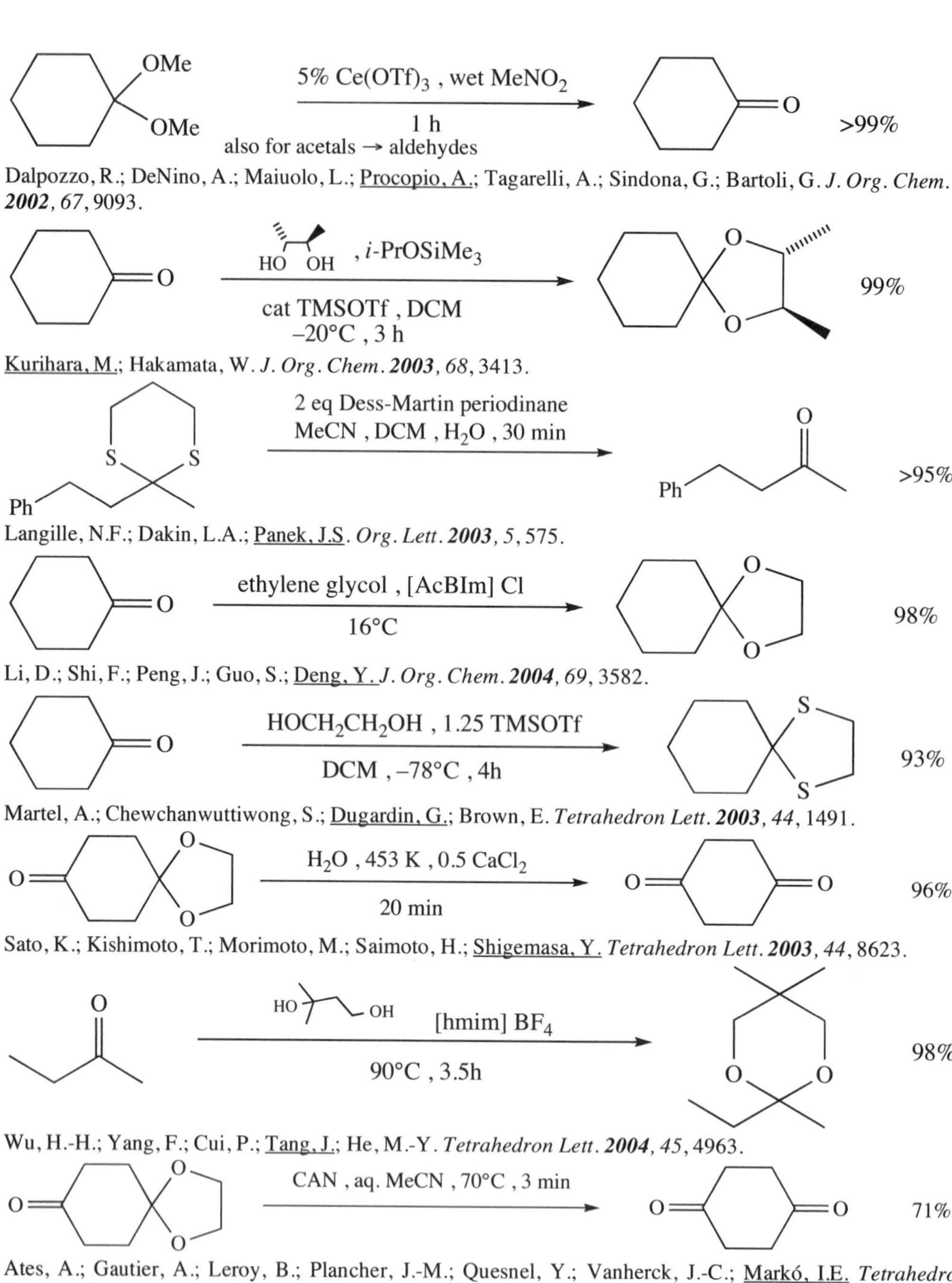

Dalpozzo, R.; DeNino, A.; Maiuolo, L.; Procopio, A.; Tagarelli, A.; Sindona, G.; Bartoli, G. *J. Org. Chem.* ***2002***, *67*, 9093.

Kurihara, M.; Hakamata, W. *J. Org. Chem.* ***2003***, *68*, 3413.

Langille, N.F.; Dakin, L.A.; Panek, J.S. *Org. Lett.* ***2003***, *5*, 575.

Li, D.; Shi, F.; Peng, J.; Guo, S.; Deng, Y. *J. Org. Chem.* ***2004***, *69*, 3582.

Martel, A.; Chewchanwuttiwong, S.; Dugardin, G.; Brown, E. *Tetrahedron Lett.* ***2003***, *44*, 1491.

Sato, K.; Kishimoto, T.; Morimoto, M.; Saimoto, H.; Shigemasa, Y. *Tetrahedron Lett.* ***2003***, *44*, 8623.

Wu, H.-H.; Yang, F.; Cui, P.; Tang, J.; He, M.-Y. *Tetrahedron Lett.* ***2004***, *45*, 4963.

Ates, A.; Gautier, A.; Leroy, B.; Plancher, J.-M.; Quesnel, Y.; Vanherck, J.-C.; Markó, I.E. *Tetrahedron* ***2003***, *59*, 8989.

O

Ph

$HC(OMe)_3$, MeCN

flow through a polymeric Brønsted acid resin

O O

Ph

82%c

Ishinhara, K.; Hasegawa, A.; Yamamoto, H. *Synlett* ***2002***, 1296.

N-OH — HCOOH , SiO_2 , microwaves / solvent free , 2 min → O 86%

Zhou, J.-F.; Tu, S.-J.; Feng, J.-C. *Synth. Commun.* **2002**, *32*, 959.

N-OH — $Bi(NO_3)_3$ - SiO_2 , H_2O/THF / 2 h → O 97%

Samajdar, S.; Basu, M.K.; Becker, F.F.; Banik, B.K. *Synth. Commun.* **2002**, *32*, 1917.

Ph, S, S — ammonium persulfate , 2 min / wet Montmorillonite K10 / microwaves → Ph, O 98%

Ganguly, N.C.; Datta, M. *Synlett* **2004**, 659.

O — 2 $HOCH_2CH_2SH$, DCM , rt , 1 h / Montmorillonite K-10 → S, O 90%

Gogoi, S.; Borah, J.C.; Barua, N.C. *Synlett* **2004**, 1592.

O, O, OH — 10% *p*-TsOH , H_2O , SiO_2 / microwaves , 3 min → O, OH 90%

He, Y.; Johansson, M.; Stermer, O. *Synth. Commun.* **2004**, *34*, 4153.

REVIEW:

"Regeneration of Carbonyl Compounds from Oximes, Hydrazones, Semicarbazones, Acetals, 1,1-Diacetates, 1,3-Dithiolanes, 1,3-Dithianes, and 1,3-Oxathiolanes"
Khoee, S.; Ruoho, A.E. *Org. Prep. Proceed. Int.* **2003**, *35*, 527.

See Section 362 (Ester-Alkene) for the formation of enol esters and Section 367 (Ether-Alkene) for the formation of enol ethers. Many of the methods in Section 60A (Protection of Aldehydes) are also applicable to ketones.

CHAPTER 13

PREPARATION OF NITRILES

SECTION 181: NITRILES FROM ALKYNES

NO ADDITIONAL EXAMPLES

SECTION 182: NITRILES FROM ACID DERIVATIVES

NO ADDITIONAL EXAMPLES

SECTION 183: NITRILES FROM ALCOHOLS AND THIOLS

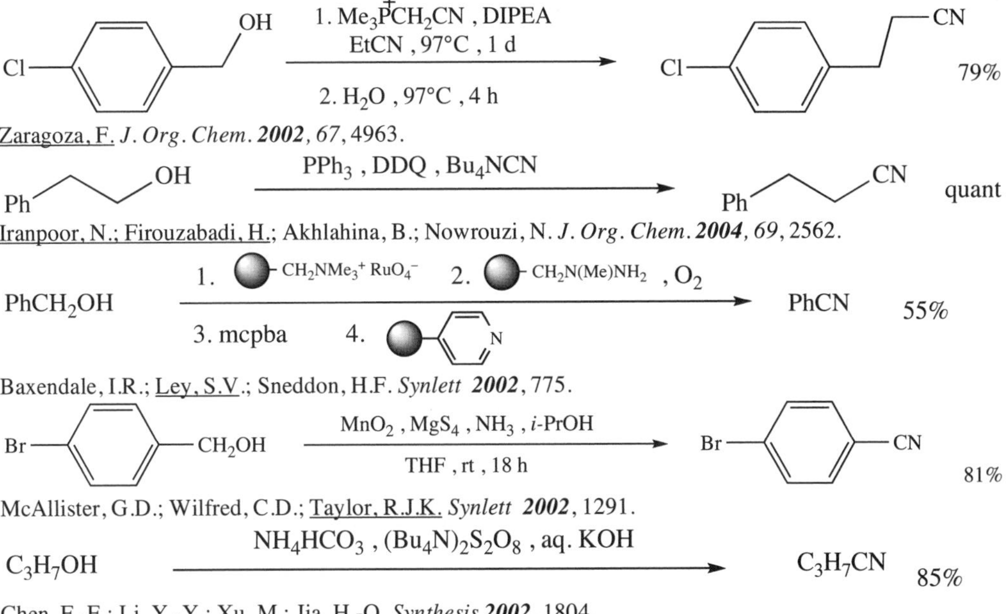

Zaragoza, F. *J. Org. Chem.* ***2002***, *67*, 4963.

Iranpoor, N.; Firouzabadi, H.; Akhlahina, B.; Nowrouzi, N. *J. Org. Chem.* ***2004***, *69*, 2562.

Baxendale, I.R.; Ley, S.V.; Sneddon, H.F. *Synlett* ***2002***, 775.

McAllister, G.D.; Wilfred, C.D.; Taylor, R.J.K. *Synlett* ***2002***, 1291.

Chen, F.-E.; Li, Y.-Y.; Xu, M.; Jia, H.-Q. *Synthesis* ***2002***, 1804.

SECTION 184: NITRILES FROM ALDEHYDES

CH_3CH_2CHO —(1. dry Al_2O_3, $NH_2OH•HCl$, 45 min; 2. $MeSO_2Cl$, 100°C)→ CH_3CH_2CN 95%

Sharghi, H.; Sarvari, M.H. *Tetrahedron* **2002**, *58*, 10323.

PhCHO —(NH_2OH, H-Y-zeolite, microwaves; 2 min)→ PhCN 94%

Srinivas, K.V.N.S.; Reddy, E.B.; Das, B. *Synlett* **2002**, 625.

CH_3CH_2CHO —(CAN, liq. NH_3, H_2O, 0°C)→ CH_3CH_2CN 89%

Bandgar, B.P.; Makone, S.S. *Synlett* **2003**, 262.

$PhCH_2CH_2CHO$ —(NaI, MeCN, $NH_2OH•HCl$; reflux)→ $PhCH_2CH_2CN$ 95%

Ballini, R.; Fiorini, D.; Palmieri, A. *Synlett* **2003**, 1841.

2-Nitrobenzaldehyde (CHO, NO_2) —($NH_2OH•HCl$, NEt_3, $CHCl_3$; $(Cl_3C\text{-}O)_2C=O$, rt)→ 2-nitrobenzonitrile (CN, NO_2) 90%

Bose, D.S.; Goud, P.R. *Synth. Commun.* **2002**, *32*, 3621.

PhCHO —(graphite, $NH_2OH•HCl$, $MeSO_2Cl$, 100°C; solvent free, 90 min)→ PhCN 90%

Sharghi, H.; Sarvari, M.H. *Synthesis* **2003**, 243.

C_3H_7CHO —(TCCA, 1% TEMPO, DCM, 0°C)→ C_3H_7CN 80%

Chen, F.-E.; Kuang, Y.-Y.; Dai, H.-F.; Lu, L.; Huo, M. *Synthesis* **2003**, 2629.

PhCHO —($NH_2OH•HCl$, silica chloride; microwaves, 1 min)→ PhCN 92%

Srinivas, K.V.N.S.; Mahender, I.; Das, B. *Chem. Lett.* **2003**, *32*, 738

SECTION 185: NITRILES FROM ALKYLS, METHYLENES, AND ARYLS

NO ADDITIONAL EXAMPLES

SECTION 186: NITRILES FROM AMIDES

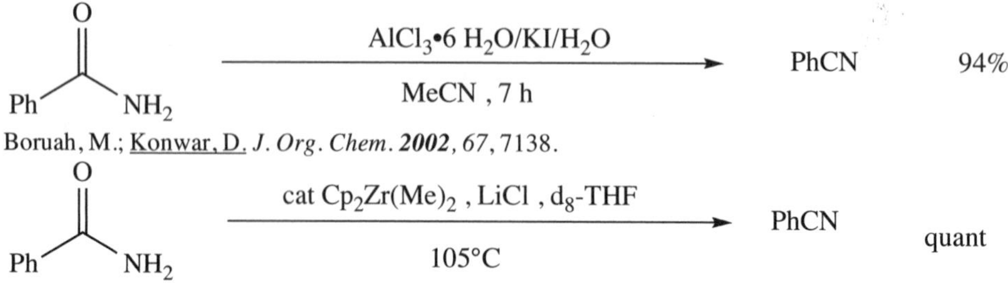

Boruah, M.; Konwar, D. *J. Org. Chem.* **2002**, *67*, 7138.

Ruck, R.T.; Bergman, R.G. *Angew. Chem. Int. Ed.* **2004**, *43*, 5375.

$PhCH_2NH_2$ —[Ru/Al_2O_3, $PhCF_3$, 373 K, O_2, 1h]→ PhCN 82%

Yamaguchi, K.; Mizuno, N. *Angew. Chem. Int. Ed.* ***2003***, *42*, 1479.

$CH_3CH_2C(=O)NH_2$ —[PPh_3, NCS, DCM, rt]→ $CH_3CH_2C{\equiv}N$ 90%

Iranpoor, N.; Firouzabadi, H.; Aghapoor, G. *Synth. Commun.* ***2002***, *32*, 2535.

$(CH_3)_2CHCH_2CH_2NH_2$ —[2% $CuCl_2$, toluene, MS 3Å / 80°C, O_2, 12 h]→ $(CH_3)_2CHCH_2CN$ 75%

Maeda, Y.; Nishimura, T.; Uemura, S. *Bull. Chem. Soc. Jpn.* ***2003***, *76*, 2399.

SECTION 187: NITRILES FROM AMINES

NO ADDITIONAL EXAMPLES

SECTION 188: NITRILES FROM ESTERS

NO ADDITIONAL EXAMPLES

SECTION 189: NITRILES FROM ETHERS, EPOXIDES, AND THIOETHERS

NO ADDITIONAL EXAMPLES

SECTION 190: NITRILES FROM HALIDES AND SULFONATES

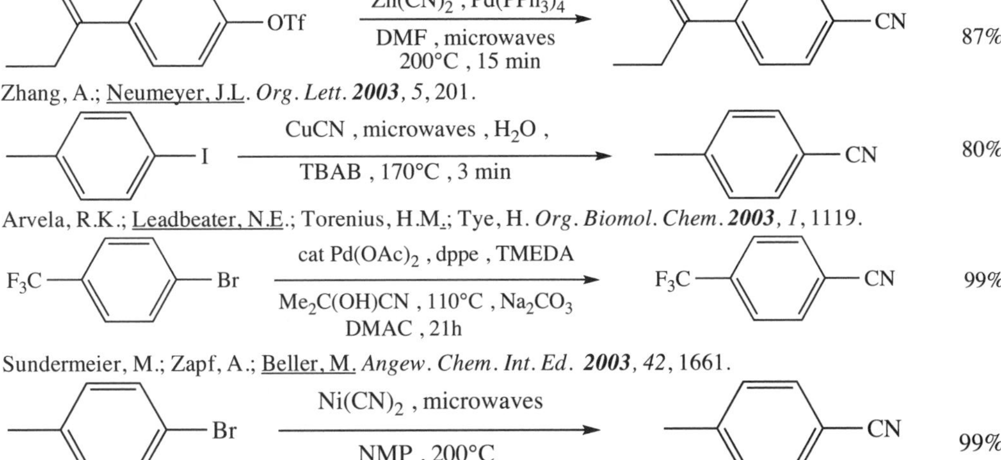

Arvela, R.K.; Leadbeater, N.E. *J. Org. Chem.* ***2003***, *68*, 9122.

PhI → PhCN quant
2 CuCN , bmi I , 90°C , 1 d

Wu, J.X.; Beck, B.; Ren, R.X. *Tetrahedron Lett.* **2002**, *43*, 387.

PhBr → PhCN 89%
cat $Pd_2(dba)_3$/dppf , cat Zn , $Zn(OAc)_2$, 100°C
DMF , H_2O , 6h

Chidambaram, R. *Tetrahedron Lett.* **2004**, *45*, 1441.

Br → CN 97%
KCN , 0.5% $Pd_2(dba)_2$
2.5% $P(t\text{-}Bu)_3$, MeCN
0.14% Bu_3SnCl , 80°C

Yang, C.; Williams, J.M. *Org. Lett.* **2004**, *6*, 2837.

Br, OH, Cl → NC, OH, Cl quant
$Zn(CN)_2$, DMF
cat $Pd(Po\text{-}Tol)_4$
56°C

Marcantonio, K.M.; Frey, L.F.; Liu, Y.; Chen, Y.; Strine, J.; Phenix, B.; Wallace, D.J.; Chen, C.-y. *Org. Lett.* **2004**, *6*, 3723.

Br → CN 90%
10% CuI , 20% KI , 1.2 NaCN , 1 d
$MeNHCH_2CH_2NHMe$, toluene , 110°C

Zanon, J.; Klapars, A.; Buchwald, S.L. *J. Am. Chem. Soc.* **2003**, *125*, 2886.

t-Bu, I → *t*-Bu, CN 94%
SOMe , $TosSO_2Me$
azobis(cyclohexanecarbonitrile)
Ph*t*-Bu , 140°C

Kim, S.; Song, H.-J. *Synlett* **2002**, 2110.

O, Br → O, CN 86%
2.5% $Pd_2(dba)_3$, 1.8 $Zn(CN)_2$
5% P**t*-Bu)$_3$, 0.12 Zn , rt
DMF , 1 h

Ramnauth, J.; Bhardwaj, N.; Renton, P.; Rakhit, S.; Maddaford, S.P. *Synlett* **2003**, 2237.

S, Cl, O → S, CN, O 79%
$Zn(CN)_2$, Zn , DMA , dppf
$Pd_2(dba)_3$, 120°C , 2 d

Ekker, T.; Nemec, S. *Synthesis* **2004**, 23.

SECTION 191: NITRILES FROM HYDRIDES

4-Methoxyquinoline N-oxide → 1. Me_3SCN, THF, rt, 1 h; 2. 10% $Pd(OAc)_2$, DDQ, rt, 1 d → 2-cyano-4-methoxyquinoline N-oxide (OMe, N, O, CN) 66%

Tagawa, Y.; Higuchi, Y.; Yamagata, K.; Shibata, K.; Teshima, D. *Heterocycles* ***2004***, *63*, 2859.

SECTION 192: NITRILES FROM KETONES

NO ADDITIONAL EXAMPLES

SECTION 193: NITRILES FROM NITRILES

Conjugate reductions and Michael alkylations of alkene nitriles are given in Sections 74D and 74E (Conjugate Reductions and Alkylations of α,β-Unsaturated Carbonyl Compounds and Nitriles, respectively).

Ph, Br, CN → Br (allyl bromide), *i*-PrMgBr → Ph, CN (allyl) 72%

Fleming, F.F.; Zhang, Z.; Knochel, P. *Org. Lett.* ***2004***, *6*, 501.

CN (isobutyronitrile) → PhCl, cat $Pd(OAc)_2$/L, $NaN(TMS)_2$; toluene, 90°C, 8h; L = proazaphophatranes → CN, Ph 82%

You, J.; Verkade, J.G. *Angew. Chem. Int. Ed.* ***2003***, *42*, 5051.

SECTION 194: NITRILES FROM ALKENES

NO ADDITIONAL EXAMPLES

SECTION 195: NITRILES FROM MISCELLANEOUS COMPOUNDS

PhCH=N-OH → Al_2O_3, microwaves, 3 h; tetrachloropyridine → PhC≡N 72%

Lingaiah, N.; Narender, R. *Synth. Commun.* ***2002***, *32*, 2391.

Ph, N, O, CN (O-aryl oxime) → 2 $ZrCl_4$, $MeNO_2$; 20 min → Ph, CN 86%

Tsuji, C.; Miyazawa, E.; Sakamoto, T.; Kikugawa, Y. *Synth. Commun.* ***2002***, *32*, 3871.

OMe / CH=N-OEt → ($NiCl_2$, Zn , xylene; reflux , 1 d) → OMe / C≡N 71%

Maeyama, K.; Kobayashi, M.; Kato, H.; Yonezawa, N. *Synth. Commun.* **2002**, *32*, 2519.

MeO—CH=N-OH → (H_2SO_4 , SiO_2 , 3 min; microwaves) → MeO—CN 81%

Sarvari, M.H. *Synthesis* **2004**, 787.

PhCH=N-OH → (NEt_3-DMAD , MeCN , rt , 3 h) → PhCN 92%

Coskun, N. *Synth. Commun.* **2004**, *34*, 1625.

CO_2H / NH_2 → (TCICA , Py , H_2O; TCICA = 1,3,5-trichloro-1,35-triazine[1H,3H,5H]trioxane) → CN 79%

Hiegel, G.A.; Lewis, J.C.; Bae, J.W. *Synth. Commun.* **2004**, *34*, 3449.

PhCH=N-OH → (MeS(C=O)SMe , NEt_3 , dioxane; heat) → Ph-CN 88%

Khan, T.A.; Peruncheralathan, S.; Ila, H.; Junjappa, H. *Synlett* **2004**, 2019.

MeO / MeO / CH=N-OH → (SiO_2 , microwaves , 4 min) → MeO / MeO / CN 95%

Dewan, S.K.; Singh, R.; Kumaar, A. *Synth. Commun.* **2004**, *34*, 2025.

CHAPTER 14

PREPARATION OF ALKENES

SECTION 196: ALKENES FROM ALKYNES

EtO_2C EtO_2C OMe

10% $GaCl_3$, toluene
40°C, 1 h

EtO_2C EtO_2C OMe 80%

Inoue, H.; Chatani, N.; Murai, S. *J. Org. Chem.* ***2002***, *67*, 1414

OH

1. $LiAlH_4$, NaOMe, THF
2. cat $Pd_2(dba)_3$, $AsPh_3$
$(MeO)_2C{=}O$, MeO_2C—C₆H₄—I

MeO_2C OH 78%

Havránek, M.; Dvorák, D. *J. Org. Chem.* ***2002***, *67*, 2125.

CO_2Et EtO_2C Ph

1. $ZnBr_2$, DCM, rt, 13 h
2. NEt_3

CO_2Et Ph Br 64%

Yamazaki, S.; Yamada, K.; Yamabe, S.; Yamamoto, K. *J. Org. Chem.* ***2002***, *67*, 2889.

N Bu

$PhB(OH)_2$, sodium dodecylsulfate
Na_2CO_3, H_2O, 80°C
cat $[Rh(cod)Cl]_2$/phosphine

N Bu Ph 51%

Lautens, M.; Yoshida, M. *J. Org. Chem.* ***2003***, *68*, 762.

2% $[Rh(cod)Cl]_2$, SDS, 80°C
8% triarylphosphine, 3h, H_2O
o-tolyl$B(OH)_2$, Na_2CO_3

TBDPSO OTBDPS 83%

Lautens, M.; Yoshida, M. *Org. Lett.* **2002**, *4*, 123.

C_5H_{11} — $HSiMeCl_2$, EtOH, NEt_3 / 3% $Cp^*RuH_3(PPh_3)$, DCM → Cl_2MeSi (89 : 11) 89% + $SiMeCl_2$ 1:7 *E:Z*

Kawanami, Y.; Sonoda, Y.; Mori, T.; Yamamoto, K. *Org. Lett.* **2002**, *4*, 2825.

CO_2Me — 1. Br, 3% $PdBr_2(PhCN)_2$ THF / 2. 2 MeO-C_6H_4-$B(OH)_2$, rt, 6% $P(t\text{-}Bu)_3$, 2 Cs_2CO_3 → MeO_2C ... OMe 83%

Thadani, A.N.; Rawal, V.H. *J. Org. Chem.* **2002**, *67*, 4317.

OTBS — cat $W(CO)_5(thf)$ / H_2O/THF → O, Me, H 91%

Iwasawa, N.; Miura, T.; Kiyota, K.; Kusama, H.; Lee, K.; Lee, P.H. *J. Org. Chem.* **2002**, *67*, 4463.

MeO_2C, MeO_2C — $HSiMe_2Bn$, DCE, 70°C / cat $[Rh(cod)_2]SbF_6$, R-BIPHEMP → MeO_2C, MeO_2C, $SiMe_2Bn$ 70% (89% ee)

Chakrapani, H.; Liu, C.; Widenhoefer, R.A. *Org. Lett.* **2003**, *5*, 157.

Ph—≡—Ph — $PhB(OH)_2$, PhI, DMF-H_2O, 100°C, 3 h, $KHCO_3$ / cat $PdCl_2(PhCN)_2$ → Ph, Ph, Ph, Ph 92%

Zhou, C.; Emrich, D.E.; Larock, R.C. *Org. Lett.* **2003**, *5*, 1579.

TMS, TMS — OH / $ClTi(Oi\text{-}Pr)_3$, $c\text{-}C_5H_9MgCl$ / THF, −78→0°C→rt → TMS, OH, TMS 85%

Sung, M.J.; Pang, J.-H.; Park, S.-B.; Cha, J.K. *Org. Lett.* **2003**, *5*, 2137.

OTBS

TMS

10% $Co_2(CO)_8$, 10% $P(OMe)_3$

toluene , reflux

OTBS

TMS

61%

Ajamian, A.; Gleason, J.L. *Org. Lett.* ***2003**, 5*, 2409.

OH

3.2 PhMgI , 3 PhI , 14 h
0.05 $Pd(PPh_3)_4$

toluene/THF

Ph

Ph

OH

73%

Tessier, P.E.; Penwell, A.J.; Suza, F.E.S.; Fallis, A.G. *Org. Lett.* ***2003**, 5*, 2989.

N

Ts

CH_2=CHBu , N-phenylmalimide , DCM
10% Grubbs I , reflux

Ph

N

O

O

Bu

Ts

N

86%

Lee, H.-Y.; Kim, H.Y.; Tae, H.; Kim, B.G.; Lee, J. *Org. Lett.* ***2003**, 5*, 3439.

C_3H_7 ——— R

Ph_2SiH_2 , Cp_2TiCl_2/BuLi

THF , 1 h

$SiHPh_2$

C_3H_7 C_3H_7

70%

Takahashi, T.; Bao, F.; Gao, G.; Ogasawara, M. *Org. Lett.* ***2003**, 5*, 3479.

OMe

OMe

2% $Hg(OTf)_2$-$(TMU)_3$

MeCN , rt , 0.8h

TMU = tetramethylurea

OMe

OMe

91%

Nishizawa, M.; Takao, H.; Yadav, V.K.; Imagawa, H.; Sugihara, T. *Org. Lett.* ***2003**, 5*, 4563.

N

Bu

B(pinanyl)

CO_2Me

3% $[Rh(cod)Cl]_2$, 2 Na_2CO_3
6.6% *t*-Bu_3PH BF_4 , aq. dioxane
80°C

Bu

N

CO_2Me

60%

Lautens, M.; Marquardt, T. *J. Org. Chem.* ***2004**, 69*, 4607.

$C_{10}H_{21}$

SDS = sodium dodecyl sulfate

Et_3SiH , H_2O , SDS , rt

0.5% $[Rh(Cl)(nbd)]_2$, dppp

$C_{10}H_{21}$

71% $SiEt_3$

+

$C_{10}H_{21}$

Et_3Si

3%

Sato, A.; Kinoshita, H.; Shinokubo, H.; Oshima, K. *Org. Lett.* ***2004**, 6*, 2217.

OTMS, Ph, Ph, MeO, N, Ts → 20% $BF_3{\bullet}OEt_2$, 0°C; DCM, 10 min → Ph, Ph, C, MeO, N, Ts + Ph, Ph, OMe, C, N, Ts (10 : 1) 99%

Ishikawa, T.; Manabe, S.; Aikawa, T.; Kudo, T.; Saito, S. *Org. Lett.* ***2004***, *6*, 2361.

O, CO_2Et → PhC≡CMe, 10% PhCOOH; toluene, PPh_3; cat $Pd_2(dba)_3{\bullet}CHCl_3$ → O, CO_2Et, Ph 95%

Patil, N.T.; Yamamoto, Y. *J. Org. Chem.* ***2004***, *69*, 6478.

TMS, OH, Et → OEt, $[CpRu(N{\equiv}NMe)_3]PF_6$; DMF → EtO, OH, Et, TMS 73%

López, F.; Cstedo, L.; Mascañas, J.L. *J. Am. Chem. Soc.* ***2002***, *124*, 4218.

$C_{13}H_{27}$, BnO → 8% Ru complex, DCE; 80°C, 12 h → $C_{12}H_{25}$ 75% (*E*/*Z* = 70) + 86% PhCHO

Yeh, K.-L.; Liu, B.; Lo, C.-Y.; Huang, H.-L.; Liu, R.-S. *J. Am. Chem. Soc.* ***2002***, *124*, 6510.

Cl, Bu → 1. $(EtO)_3SiH$, DCM, 0°C → rt; cat $[Cp^*Ru(MeCN)_3]PF_6$; 2. CuI, TBAF, THF → Cl, Bu 59%

Trost, B.M.; Ball, Z.T.; Jöge, T. *J. Am. Chem. Soc.* ***2002***, *124*, 7922.

EtO_2C, EtO_2C → cat $Rh_4(CO)_{12}$, CO, 22°C; Me_2PhSiH, hexane, <1 min → EtO_2C, EtO_2C, $SiMe_2Ph$ 95%

Ojima, I.; Vu, A.T.; Lee, S.-Y.; McCullagh, J.V.; Moralee, A.C.; Fuiwara, M.; Hoang, T.H. *J. Am. Chem. Soc.* ***2002***, *124*, 9164.

MeO_2C, O, Ph → 10% Bu_3P, 0.1 M EtOAc, 110°C; sealed tube → Ph, O, H, MeO_2C, H 76%

Wang, J.-C.; Ng, S.-S.; Krische, M.J. *J. Am. Chem. Soc.* ***2003***, *125*, 3682.

MeO, OH → Ru catalyst, LiOTf, toluene; – CO → MeO 75%

Datta, S.; Chang, C.-L.; Yeh, K.-L.; Kiu, R.-S. *J. Am. Chem. Soc.* ***2003***, *125*, 9294.

Et—≡—Et → (PhH, cat BBu_3, 100°C; chiral Pd catalyst, 4 h) → Ph, Et, Et alkene 86%

Isukada, N.; Mitsuboshi, T.; Setoguchi, H.; Inoue, Y. *J. Am. Chem. Soc.* **2003**, *125*, 12102.

Ph—≡, cat. $In(OTf)_3$, neat; 60°C - 140°C → CO_2Et, Ph product 99%

Nakamura, M.; Endo, K.; Nakamura, E. *J. Am. Chem. Soc.* **2003**, *125*, 13002.

1% $AuCl(PPh_3)_3$, 1% AgOTF; DCM, 15 min, rt → CO_2Me product 94%

Kennedy-Smith, J.J.; Staben, S.T.; Toste, F.D. *J. Am. Chem. Soc.* **2004**, *126*, 4526.

MeO_2C, MeO_2C, Ph, Ph → 3% $Rh(cod)_2OTf$, 3% TsINAP; H_2, DCE, 25°C → MeO_2C, MeO_2C, Ph, Ph 85%

Jang, H.-Y.; Krische, M.J. *J. Am. Chem. Soc.* **2004**, *126*, 7875.

Ph—≡ → 1. 0.1 $InCl_3$, Et_3SnH/BEt_3, MeCN, 0°C; 2. H_3O^+ → Ph 90%

Hayashi, N.; Shibata, I.; Baba, A. *Org. Lett.* **2004**, *6*, 4981.

Ts—N → 10% $PtCl_2$, toluene; 80°C, 2 h → Ts—N 58%

Harrison, T.J.; Dake, G.R. *Org. Lett.* **2004**, *6*, 5023.

C_6H_{13}—≡ → 1. chiral Zn complex THF, rt, 12 h; 2. H_2O → C_6H_{13}, $PhMe_2Si$ 99%

Nakamura, S.; Uchiyama, M.; Ohwada, T. *J. Am. Chem. Soc.* **2004**, *126*, 11146.

OH, O → cat $[Cp^*RuCl(\mu_2\text{-}SPr)_2RuCp^*Cl]$, 60°C; 10% $PtCl_2$, 10% NH_4BF_4, 1 d → H, O 75% (92:98 *syn:anti*)

Nishibayashi, Y.; Yoshikawa, M.; Inada, Y.; Hidai, M.; Uemura, S. *J. Am. Chem. Soc.* **2004**, *126*, 16066.

C_6H_{13} OH — 10% Cp_2TiCl_2, 4 $LiAlH_4$ / THF, reflux → C_6H_{13} OH 75% (10:1 *Z:E*)

Paretny, A.; Campagne, J.-M. *Tetrahedron Lett.* ***2002***, *43*, 1231.

CO_2Me CO_2Me — Cl, 4 LiCl, BSA / 4 LiCl, 0.05 $Pd(OAc)_2$ THF → MeO_2C CO_2Me 87%

Liu, G.; Lu, X. *Tetrahedron Lett.* ***2002***, *43*, 6791.

Ph Ph — $Pd(OAc)_2$, NaOMe, THF, 25°C, 1d → Ph Ph 80%

Wei, L.-L.; Wei, L.-M.; Pan, W.-B.; Leou, S.-P.; Wu, M.-J. *Tetrahedron Lett.* ***2003***, *44*, 1979.

C_3H_7 C_3H_7 — F_3C $B(OH)_2$, H_2O, toluene / cat $[Rh(cod)OH]_2$/MTPPTC, 100°C → F_3C C_3H_7 C_3H_7 99%

Genin, E.; Michelet, V.; Genêt, J.-P. *Tetrahedron Lett.* ***2004***, *45*, 4157.

HO Ph Ph — $B(OH)_2$ / cat $Pd(PPh_3)_4$, dioxane, 100°C → Ph Ph C 67%

Yoshida, M.; Gotou, T.; Ihara, M. *Tetrahedron Lett.* ***2004***, *45*, 5573.

1. $(EtO)_3SiH$, cat $Cp^*Ru(MeCN)_3\ PF_6$ / DCM, rt 2. 2 AgF, aq. THF, MeOH → 84% (90:10 *E:Z*)

Fürstner, A.; Radkowski, K. *Chem. Commun.* ***2002***, 2182.

C_5H_{11} — Et_3SiH, H_2O / Pt(DVDS-P) catalyst → C_5H_{11} $SiEt_3$ 97%

Wu, W.; Li, C.-J. *Chem. Commun.* ***2003***, 1668.

$HO(H_2C)_4$ — $PhB(OH)_2$, cat $Pd(PPh_3)_4$, AcOH / dioxane, 80°C, 10h → $HO(H_2C)_4$ Ph 93%

Oh, C.H.; Jung, H.H.; Kim, K.S.; Kim, N. *Angew. Chem. Int. Ed.* ***2003***, *42*, 805.

C_3H_7 C_3H_7 — MeO $SiEt(OH)_2$ / cat $[Ru(OH)(cod)_2]$, 100°C, 1 d toluene, H_2O → C_3H_7 C_3H_7 MeO 73%

Fujii, T.; Koike, T.; Mori, A.; Osakada, K. *Synlett* ***2002***, 295.

Ph—≡ → (Ph_3SiH, PCy_3, rt; cat $Pd_2(dba)_3$•$CHCl_3$) → Ph–CH=CH–$SiPh_3$ + Ph_3Si–C(Ph)=CH_2 (95 : 5) 55%

Motoda, D.; Shinokubo, H.; Oshima, K. *Synlett* **2002**, 1529.

(Ph–≡–Me, 1.5% $AuCl_3$; 3% $AgSbF_6$, $MeNO_2$, 50°C) Me, Ph 81% (70:30 *Z:E*)

Reetz, M.T.; Sommer, K. *Eur. J. Org. Chem.* **2003**, 3485.

SCN (400°C) SCN, =C= 97%

Banert, K.; Groth, S.; Hückstädt, H.; Lehmann, J.; Schlott, J.; Vrobel, K. *Synthesis* **2002**, 1423.

Me—≡, N—R, NH_2 (MnO_2, DCM; 0°C) Me, =C=, N=N, R + Me, N, N, R

R = Ph 51% 0%

R = Me 34-44% 15-26%

Banert, K.; Haagedorn, M.; Schlott, J. *Chem Lett.* **2003**, *32*, 360.

REVIEW:

"Enyne Methathesis Catalyzed by Ruthenium Carbene Complexes"
Poulsen, C.S.; Madsen, R. *Synthesis* **2003**, 1.

SECTION 197: ALKENES FROM ACID DERIVATIVES

MeO, CO_2H, MeO, OMe ($PhCH{=}CH_2$, 3 Ag_2CO_3, cat $Pd(O_2CCF_3)_2$, 120°C; DMSO-DMF) MeO, Ph, MeO, OMe 91%

Myers, A.G.; Tanaka, D.; Mannion, M.R. *J. Am. Chem. Soc.* **2002**, *124*, 11250.

Ph—≡, $N(i\text{-}Pr)_2$ (cat $Pd_2(dba)_3$•$CHCl_3$, $P(C_6F_5)_3$; dioxane, cat CuI, MeCN, 60°C) Ph, =C= 86% (99% ee)

Nakamura, H.; Kamakura, T.; Ishikura, M.; Biellmann, J.-F. *J. Am. Chem. Soc.* **2004**, *126*, 5958.

Ph, CO_2H ($(t\text{-}BuCO)_2O$, cat $PdCl_2$/DPE-Phos; 110°C, DMPU, 16h) Ph 66%

Gooßen, L.J.; Rodríguez, N. *Chem. Commun.* **2004**, 724.

PhCOOH ($PhCH{=}CH_2$, 0.1 LiCl, cat $PdCl_2$, NMP, 16h; 3 $(t\text{-}BuO_2C)_2O$, 0.1 γ-picoline, 120°C) Ph, Ph 80%

Gooßen, L.J.; Paetzold, J.; Winkel, L. *Synlett* **2002**, 1721.

SECTION 198: ALKENES FROM ALCOHOLS AND THIOLS

OH, C_6H_{13}, CN → MgO, THF, reflux → C_6H_{13}, CN 71% (1:1 *E:Z*)

Fleming, F.F.; Shook, B.C. *J. Org. Chem.* ***2002***, *67*, 3668.

OH, Ph, Ph, Br → In, $InCl_3$, aq. THF, rt; cat $Pd(PPh_3)_4$, 8 h → Ph, Ph 75%

Cho, S.; Kang, S; Keum, G.; Kang, S.B.; Han, S.-Y.; Kim, Y. *J. Org. Chem.* ***2003***, *68*, 180.

OH, O, O → 1. 2.5% Pd(*i*-Pr)$(OAc)_2$, toluene, 60°C; 5% Bu_4NOAc, H_2O, MS 3Å, O_2; 2. 2.5% $RhCl(PPh_3)_3$, PPh_3, *i*-PrOH; $TMSCHN_2$, dioxane, 50°C → O, O 84%

Lebel, H.; Paquet, V. *J. Am. Chem. Soc.* ***2004***, *126*, 11152.

OH → 1. TPAP, NMO, DCM, MS 4Å; 2. $Br^- PPh_3CH_2CH_3$, BuLi, THF → 88% (1:4 *E:Z*)

MacCoss, R.N.; Balskus, E.P.; Ley, S.V. *Tetrahedron Lett.* ***2003***, *44*, 7779.

Bu, OH → PPh_3-I_2, DCM, rt, 2h → Bu 83%

Alvarez-Manzaneda, E.J.; Chahboun, R.; Torres, E.C.; Alvarez, E.; Alvarez-Manzaneda, R.; Haidour, A.; Ramos, J. *Tetrahedron Lett.* ***2004***, *45*, 4453.

O_2N, OH → $(EtO)_2P(=O)CH_2CO_2Et$, THF; 10 MnO_2, 2.2% guanidine derivative; MS 4Å, reflux → O_2N, CO_2Et 75%

Blackburn, L.; Pei, C.; Taylor, R.J.K. *Synlett* ***2002***, 218.

OH → SiO_2-Cl, $CHCl_3$, TMSCl; rt → 95%

Firouzabadi, H.; Iranpoor, N.; Hazarkhani, H.; Karimi, B. *Synth. Commun.* ***2003***, *33*, 3653.

OH / Ph–CH–Ph → ($CH_2=CHCH_2TMS$, 5% $InBr_3$; DCM, rt) → Ph–CH(Ph)–$CH_2CH=CH_2$ 86%

Kim, S.H.; Shin, C.; Pae, A.N.; Koh, H.Y.; Chang, M.H.; Chung, B.Y.; Cho, Y.S. *Synthesis* ***2004***, 1581.

SECTION 199: ALKENES FROM ALDEHYDES

PhCHO → ($BrCH_2CO_2t$-Bu, $NaHSO_3$, K_2CO_3; 2% PEG–Telluride) → Ph–CH=CH–CO_2t-Bu 92% (99:1 *E:Z*)

Huang, Z.-Z.; Ye, S.; Xia, W.; Yu, Y.-H.; Tang, Y. *J. Org. Chem.* ***2002***, *67*, 3096.

Br–CH_2–CO_2Et → (PhCHO, toluene, 36 h; 2 % $BrTeBu_2OTeBu_2Br$, $P(OPh)_3$; K_2CO_3) → Ph–CH=CH–CO_2Et quant (>99:1 *E:Z*)

Huang, Z.-Z.; Tang, Y. *J. Org. Chem.* ***2002***, *67*, 5320.

Ph–CH=CH–CHO → (1. 2.5% $RhCl(PPh_3)_3$, THF, PPh_3, *i*-PrOH; 2. $TMSCHN_2$, THF, 25°C) → Ph–CH=CH–CH=CH_2 >98%

Lebel, H.; Paquet, V. *J. Am. Chem. Soc.* ***2004***, *126*, 329.

PhCHO → ($CH_2(CO_2Et)_2$; [bmim] Cl, $AlCl_3$) → Ph–CH=C(CO_2Et)$_2$ 90%

Harjani, J.R.; Nara, S.J.; Salunkhe, M.M. *Tetrahedron Lett.* ***2002***, *43*, 1127.

Tetrapropyl-aluminacyclopentadiene (C_3H_7 ×4, Al–Cl) → (PhCHO) → 1,2,3,4-tetrapropyl-5-phenylcyclopentadiene 73%

Fang, H.; Zhao, C.; Li, G.; Xi, Z. *Tetrahedron* ***2003***, *59*, 3779.

PhCHO → (cat $NiCl_2(PPh_3)_2$, $PhCH_2ZnCl$, 8h; TMSCl, THF, rt) → Ph–CH=CH–Ph 88%

Wang, J.-X.; Fu, Y.; Hu, Y. *Angew. Chem. Int. Ed.* ***2002***, *41*, 2757.

PhCHO → (5-nonanone, $Yb(OTf)_3$, 60°C, 8h) → Ph–CH=CH–$CH_2CH_2CH_3$ 89%

Curini, M.; Epifano, F.; Maltese, F.; Marcotullio, M.C. *Eur. J. Org. Chem.* ***2003***, 1631.

Ph_3P=C(I^+Ph BF_4^-)(CO_2Et) → (PhCHO, Bu_4NBr, DMF; microwaves, 3 h) → Ph–CH=C(Br)(CO_2Et) 85% (89:11 *Z:E*)

Yu, X.; Huang, X. *Synlett* ***2002***, 1895.

PhCHO → (Ph_3PCH_3 Br/DBU, MeCN, reflux) → PhCH=CH_2 75%

Okuma, K.; Sakai, O.; Shioji, K. *Bull. Chem. Soc. Jpn.* ***2003***, *76*, 1675.

Related Methods: Section 207 (Alkenes from Ketones).

SECTION 200: ALKENES FROM ALKYLS, METHYLENES AND ARYLS

This section contains dehydrogenations to form alkenes and unsaturated ketones, esters, and Amides and also the conversion of aromatic rings to alkenes. Reduction of aryls to dienes is given in Section 377 (Alkene-Alkene). Hydrogenation of aryls to alkanes and dehydrogenations to form aryls are given in Section 74 (Alkyls, Methylenes, and Aryls from Alkenes).

N-Boc pyrrole-2-CO_2i-Pr → 1. Li, di-*tert*-butylbiphenyl, THF; 2. BnBr → N-Boc 2-benzyl-2-CO_2i-Pr-2,5-dihydropyrrole (Ph, CO_2i-Pr, Boc) 80%

Donohoe, T.J.; House, D. *J. Org. Chem.* ***2002***, *67*, 5015.

cycloheptane → Ir catalyst, CH_2=CH*t*-Bu, 200°C → cycloheptene

Göttker-Schnetmann, I.; White, P.; Brookhart, M. *J. Am. Chem. Soc.* ***2004***, *126*, 1804.

MeO–C_6H_4–CHO → C_8H_{17}ZnI, 3% CoCl$(PPh_3)_3$; DMF, –18°C → 0°C → MeO–C_6H_4–CH=CH–C_7H_{15} 60%

Wang, J.-X.; Fu, Y.; Hu, Y.; Wang, K. *Synthesis* ***2003***, 1506.

REVIEWS:

"Allenes from Cyclopropanes and Their Use in Organic Synthesis: Recent Developments"
Sydnes L.K. *Chem. Rev.* ***2003***, *103*, 1133.

"Thermal Rearrangements of Vinylcyclopropanes to Cyclopentenes"
Baldwin, J.E. *Chem. Rev.* ***2003***, *103*, 1197.

SECTION 201: ALKENES FROM AMIDES

Related Methods: Section 65 (Alkyls, Methylenes, and Aryls from Alkyls)
Section 74 (Alkyls, Methylenes, and Aryls from Alkenes)

NO ADDITIONAL EXAMPLES

SECTION 202: ALKENES FROM AMINES

NO ADDITIONAL EXAMPLES

SECTION 203: ALKENES FROM ESTERS

TMS O Ph S Ph — PhH , 180°C , 1.5 h; sealed tube → Ph OTMS Ph (78:22 Z:E) 82% + Ph S Ph TMSO 9%

Choi, J.; Imai, E.; Mihara, M.; Oderaotoshi, Y.; Minakata, S.; Komatsu, M. *J. Org. Chem.* **2003**, *68*, 6164.

OAc Ph CO_2Me — BuI , Zn , aq. NH_4Cl; rt , 10 h → Ph CO_2Me Bu 74% (100% *E*)

Das, B.; Banerjee, J.; Mahender, G.; Majhi, A. *Org. Lett.* **2004**, *6*, 3349.

Ph OAc — 1. cat $PdCl_2$, TFP , MeOH; 2. KF , MeOH , (pinacol o-tolylboronate) → Ph (o-tolyl) 91%

Ortar, G. *Tetrahedron Lett.* **2003**, *44*, 4311.

SECTION 204: ALKENES FROM ETHERS, EPOXIDES AND THIOETHERS

$C_{10}H_{21}$ (epoxide) — PhLi , 2 LTMP , hexane; 0°C → rt , 2 h → $C_{10}H_{21}$ Ph 93% (98:2 *E:Z*)

Hodgson, D.M.; Fleming, M.J.; Stanway, S.J. *J. Am. Chem. Soc.* **2004**, *126*, 12250.

Ph O Ph — PhH $Mo(CO)_6$, reflux , 3h → Ph Ph 84%

Patra, A.; Bandyopadhyay, M.; Mal, D. *Tetrahedron Lett.* **2003**, *44*, 2355.

CH_2=C=CHSnBu_3 — PhI , cat $Pd(PPh_3)_4$, LiCl , DMF; 25°C , 12h → CH_2=C=CHPh 83%

Huang, C.-W.; Shanmugasundaram, M.; Chang, H.-M.; Cheng, C.-H. *Tetrahedron* **2003**, *59*, 3635.

O Ph Ph — 5% [hydrido-tris(3,5-dimethylpyrazolyl)borate] ReO_3; PhH , 75°C , 100 h → Ph Ph .95% conversion

Gable, K.P.; Brown, E.C. *Synlett* **2003**, 2243.

Ph O O NEt_2 — 2.5 SmI_2 → Ph O NEt_2 75%

Concellón, J.M.; Bardales, E. *J. Org. Chem.* **2003**, *68*, 9492.

Bu O O O*i*-Pr Ph — SmI_2 → Bu O O*i*-Pr Ph 68% (93% de)

Concellón, J.M.; Bardales, E. *Org. Lett.* **2002**, *4*, 189.

REVIEW:

"Molybdenum and Tungsten Imide Alkylidene Complexes as Efficient Olefin-Metathesis Catalysts"
Schrock, R.R.; Hoyevda, A.H. *Angew. Chem. Int. Ed.* ***2003***, *42*, 4592.

SECTION 205: ALKENES FROM HALIDES AND SULFONATES

cat $Pd(PPh_3)_4$–K_2CO_3, reflux, DME–H_2O, 2 h; trivinylboroxine •Py — 76%

Kerins, F.; O'Shea, D.F. *J. Org. Chem.* ***2002***, *67*, 4968.

MeO–C_6H_4–$B(OH)_2$ → MeO–C_6H_4–CH=CH_2 ($BrCH_2CH_2Br$, $Pd(OAc)_2$, phosphine, 4 eq KOH; THF, 100°C, 1 h) 82%

Lando, V.R.; Moneiro, A.L. *Org. Lett.* ***2003***, *5*, 2891.

$C_{10}H_{21}Br$ → $C_{10}H_{21}CH=CH_2$ ($Bu_3SnCH=CH_2$, 2.5% [(π-allyl)PdCl]$_2$, MS 3Å; 15% $P(t\text{-}Bu)_2Me$, THF, 1 d, rt) 86%

Menzel, K.; Fu, G.C. *J. Am. Chem. Soc.* ***2003***, *125*, 3718.

EtO_2C~~~Br → EtO_2C~~~CH=CH–Ph (Cp_2ClZr–CH=CH–Ph, NMP:THF; 2.5% $Pd(acac)_2$, 2 LIBr, 55°C, 1 d) 99%

Wiskur, S.L.; Korte, A.; Fu, G.C. *J. Am. Chem. Soc.* ***2004***, *126*, 82.

Ph-Br + (vinyl aluminum complex, NMe_2)$_2$ → Ph-CH=CH_2 (5% $PdCL_2(PPh_3)_2$, THF; 60°C, 12 h) 97%

Schumann, H.; Kaufmann, J.; Schmalz, H.-G.; Böttcher, A.; Gotov, B. *Synlett* ***2003***, 1783.

PhCHBr–CHBrPh → PhCH=CHPh ($InCl_3$, $NaBH_4$, MeCN, –10°C) 85%

Ranu, B.C.; Das, A.; Hajra, A. *Synthesis* ***2003***, 1012.

SECTION 206: ALKENES FROM HYDRIDES

For conversions of methylenes to alkenes (RCH_2R' → $RR'C=CH_2$), see Section 200 (Alkenes, Methylenes, and Aryls from Alkyls).

NO ADDITIONAL EXAMPLES

SECTION 207: ALKENES FROM KETONES

t-BuOK , THF , rt; 5 h — 60%

Fleming, F.F.; Funk, L.A.; Altundas, R.; Sharief, V. *J. Org. Chem.* **2002**, *67*, 9414.

1. $C_3H_7C{\equiv}CC_3H_7/Cp_2ZrEt_2$, rt , 1 h; 2. 20% HCl , 0°C — C_3H_7, C_3H_7 — 90%

Xi, Z.; Guo, R.; Mito, S.; Yan, H.; Kanno, K.; Nakajima, K.; Takahashi, T. *J. Org. Chem.* **2003**, *68*, 1252.

$CH_2{=}CHCO_2Me$, $Co(dppe)Cl_2$/dppe; Zn , MeCN , 80°C — CO_2Me — 66%

Chang, K.-J.; Rayabarapu, D.K.; Cheng, C.-H. *J. Org. Chem.* **2004**, *69*, 4781.

$RhCl(PPh_3)$, PPh_3 , 60°C; TMSCHN$_2$, *i*-PrOH; 1,4-dioxane — Ph — 88%

Lebel, H.; Guay, D.; Paquet, V.; Huard, K. *Org. Lett.* **2004**, *6*, 3047.

t-Bu — $TiCl_4$/Mg , DCM , 0°C , THF — *t*-Bu — 81%

Yan, T.-H.; Tsai, C.-C.; Chien, C.-T.; Cho, C.-C.; Huang, P.-C. *Org. Lett.* **2004**, *6*, 4961.

1. MesMgBr , ClPO(OPh)$_2$; 2. PhMgCl , cat $dCl_2(PPh_3)_2$ — Ph — 85%

Miller, J.A. *Tetrahedron Lett.* **2002**, *43*, 7111.

Ph, $SiMe_3$ — LHMDS , THF , –78°C; $C_{11}H_{23}SO_2Pt$ — $CHC_{11}H_{23}$, Ph, $SiMe_3$ — 79% (65:35 *E:Z*)

Jankowski, P.; Plesniak, K.; Wicha, J. *Org. Lett.* **2003**, *5*, 2789.

Ph, Ph — I , THF; cat $Cp_2Ti[P(OEt)_3]_2$ — Ph, Ph — 64%

Takeda, T.; Shimane, K.; Ito, K.; Saeki, N.; Tsubouchi, A. *Chem. Commun.* **2002**, 1974.

$CH_2(ZnI)_2$, THF-hexane , 60°C
bmim PF_6 , 6 h
84%

Yoshino, H.; Kobata, M.; Yamamoto, Y.; Oshima, K.; Matsubara, S. *Chem. Lett.* **2004**, *33*, 1224.

REVIEW:

"Asymmetric Wittig-Type Reactions"
Rein, T.; Pedersen, T.M. *Synthesis* **2002**, 579.

Related Method: Section 199 (Alkenes from Aldehydes)

SECTION 208: ALKENES FROM NITRILES

NO ADDITIONAL EXAMPLES

SECTION 209: ALKENES FROM ALKENES

, NaH , DMSO
5% $Pd(PPh_3)_4$, 85°C
79%

Ma, S.; Jiao, N.; Zhao, S.; Hou, H. *J. Org. Chem.* **2002**, *67*, 2837.

10% $PdCl_2(MeCN)_2$, DCM
rt , 14 h
90%

Yu, J.; Gaunt, M.J.; Spencer, J.B. *J. Org. Chem.* **2002**, *67*, 4627.

$PhB(OH)_2$, 3% $Pd(PPh_3)_4$
2 eq Cs_2CO_3 , EtOH
60°C , 2 h
85%

Oh, C.H.; Sung, H.R.; Park, S.J.; Ahn, K.N. *J. Org. Chem.* **2002**, *67*, 7155.

3% Grubbs' II catalyst
microwaves
0.02-0.04 M
quant (13% thermally)

Mayo, K.G.; Nearhoof, E.H.; Kiddle, J.J. *Org. Lett.* **2002**, *4*, 1567.

$Me_2C{=}CHMe$, 12 h; 1% Grubbs II , 23°C — OAc — 97%

Chatterjee, A.K.; Sanders, D.P.; Grubbs, R.H. *Org. Lett*, **2002**, *4*, 1939.

BzS — 9 eq $CH_2{=}CHOEt$, 5% Grubbs I; $CH_2{=}CH_2$, PhH , rt — OEt — BzS — 91% (2.7:1 *E:Z*)

Giessert, A.J.; Brazis, N.J.; Diver, S.T. *Org. Lett.* **2003**, *5*, 3819.

Br — Me_3SiCH_2MgCl , $CoCl_2$, dpph — $SiMe_3$ — 99% (>99:1 *E:Z*)

Mizutani, K.; Shinokubo, H.; Oshima, K. *Org. Lett.* **2003**, *5*, 3959.

EtO_2C, EtO_2C, Ph — cat. $[Pd(PPh_3)_3(MeCN)]_2BF_4$; 0.2M , 5% MeOH/THF 65°C , 1.5h — EtO_2C, EtO_2C, H, H, Ph — 91%

Takacs, J.M.; Leonov, A.P. *Org. Lett.* **2003**, *5*, 4317.

$C_{10}H_{21}$, C, $C_{10}H_{21}$ — 1. $Mn(CH_2CH{=}CH_2)_4(MgCl)_2$, rt THF , 3 h; 2. H_3O^+ — $C_{10}H_{21}$, $C_{10}H_{21}$ — 87%

Nishikawa, T.; Shinokubo, H.; Oshima, K. *Org. Lett.* **2003**, *5*, 4623.

0.5 Ph_3SiCl , PhMgBr; cat $Pd(acac)_2$ — $SiPh_3$, $SiPh_3$ — 93% (100% *E*)

Terao, J.; Oda, A.; Kambe, N. *Org. Lett.* **2004**, *6*, 3341.

Ph — $[Ni(allyl)Cl]_2$ - phosphoramidite ligand , DCM; NaBARF , C_2H_4 — Ph, CH_3 — 85% (95% ee)

BARF = tetrakis[3,5-bis(trifluoromethyl)phenyl]borate

Francìo, G.; Faraone, F.; Leitner, W. *J. Am. Chem. Soc.* **2002**, *124*, 736.

Ph, OMOM, $SiMe_3$, MeO, OC_8H_{17} — 2.5 CAN , MeCN , DCE; $NaHCO_3$, MS 4Å , 45°C — OMOM, $C_8H_{17}O$ — 71%

Seiders II, J.R.; Wang, L.; Floreancig, P.E. *J. Am. Chem. Soc.* **2003**, *125*, 2406.

$SiMe_3$, EtO_2C, CO_2Et — 3% Grubbs II , DCM; reflux , 8 h — EtO_2C, EtO_2C, $SiMe_3$ — 98%

also with N or O analogs

Schuman, M.; Gouverneur, V. *Tetrahedron Lett.* **2002**, *43*, 3513.

5% $[RuCl_2(cod)]_n$, EtOH , 80°C
12h
75% (67:33 dr)

Michaut, M.; Santelli, M.; Parrain, J.-L. *Tetrahedron Lett.* ***2003***, *44*, 2157.

cat $CoCl_2(PPh_3)_2/PPh_3/Zn$, MeCN
EtOH , 80°C , 18h
88%

Wang, C.-C.; Lin, R.S.; Cheng, C.-H. *Tetrahedron Lett.* ***2004***, *45*, 6203.

$PhB(OH)_2$, cat $Pd(PPh_3)_4$
10% AcOH , 60°C , 3h
65%

Oh, C.H.; Ahn, T.W.; Reddy, R. *Chem. Commun.* ***2003***, 2622.

polymer-supported Ir catalyst
THF , rt

Baxendale, I.R.; Lee, A.-L.; Ley, S.V. *Synlett* ***2002***, 516.

$CrCl_2$, CCl_4 , THF
0°C , 1 d
60%

Takai, K.; Kokumai, R.; Toshikawa, S. *Synlett* ***2002***, 1164.

PhI , cat heterocyclic carbene palladacycle
NMP , 130-150°C , 1 d
70%

Iyer, S.; Jayanthi, A. *Synlett* ***2003***, 1125.

REVIEWS:

"Olefin Metathesis"
Grubbs, R.H. *Tetrahedron* ***2004***, *60*, 7117.

"Catalytic Enantioselective Diels-Alder Reactions: Methods, Mechanism, Fundamental, Pathways, and Applications"
Corey, E.J. *Angew. Chem. Int. Ed.* ***2002***, *41*, 1651.

"Diels-Alder Reaction in Synthesis"
Nicolaou, K.C.; Snyder, S.A.; Montagnon, T.; Vassilikogiannakis, G. *Angew. Chem. Int. Ed.* ***2002***, *41*, 1669.

"Thermal [1,3]-Carbon Sigmatropic Rearrangements of Vinylcyclobutanes"
Perrin, C.L. *Acc. Chem. Res.* ***2002***, *35*, 279.

"Catalytic C-H/Olefin Coupling"
Kakiuchi, F.; Murai, S. *Acc. Chem. Res.* ***2002***, *35*, 826.

SECTION 210: ALKENES FROM MISCELLANEOUS COMPOUNDS

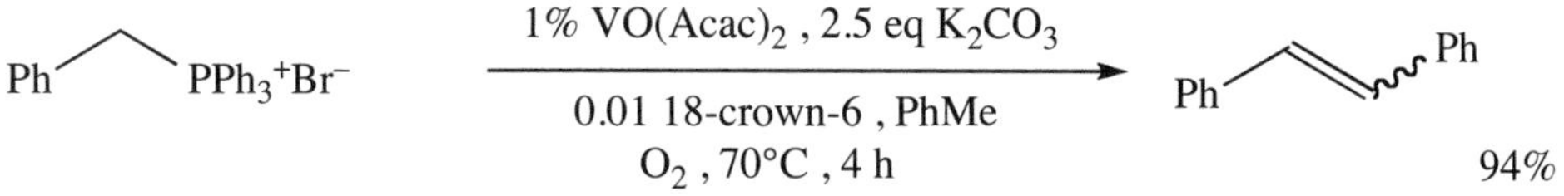

Ph–CH2–$PPh_3^+Br^-$ → 1% $VO(Acac)_2$, 2.5 eq K_2CO_3 / 0.01 18-crown-6 , PhMe / O_2 , 70°C , 4 h → Ph–CH=CH–Ph 94%

O'Brien, P.; Towers, T.D. *J. Org. Chem.* **2002**, *67*, 304.

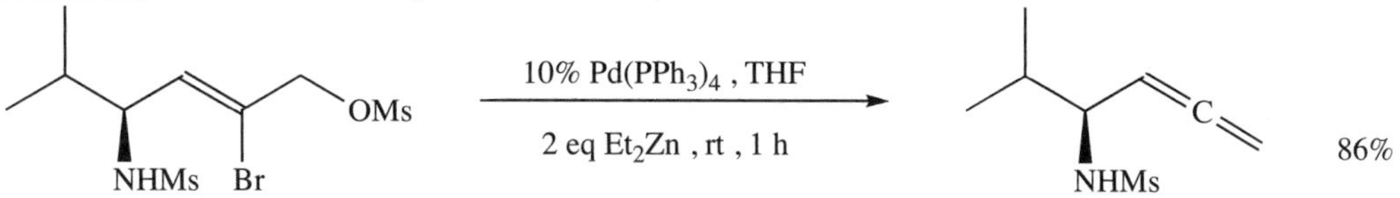

OMs, NHMs, Br → 10% $Pd(PPh_3)_4$, THF / 2 eq Et_2Zn , rt , 1 h → C, NHMs 86%

Ohno, H.; Miyamaura, K.; Tanaka, T.; Oishi, S.; Toda, A.; Takemoto, Y.; Fujii, N.; Kbuka, T. *J. Org. Chem.* **2002**, *67*, 1359.

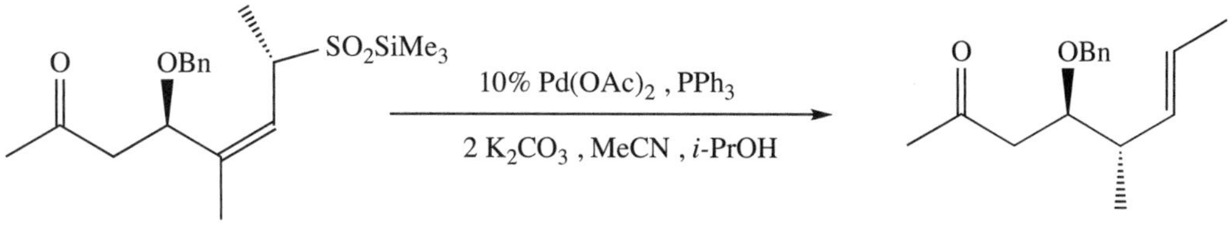

O, OBn, SO_2SiMe_3 → 10% $Pd(OAc)_2$, PPh_3 / 2 K_2CO_3 , MeCN , *i*-PrOH → O, OBn

Huang, X.; Craita, C.; Vogel, P. *J. Org. Chem.* **2004**, *69*, 4272.

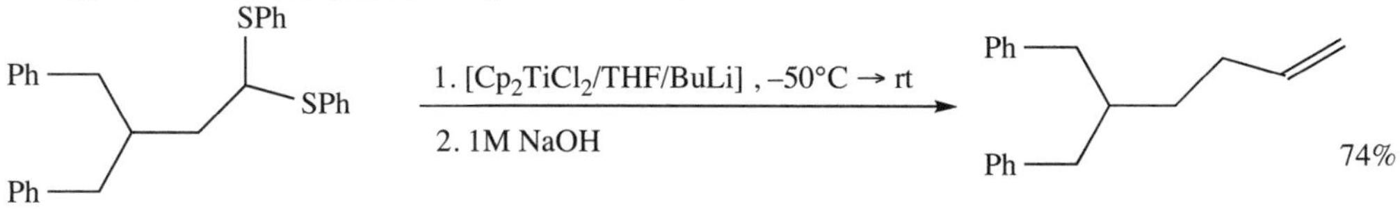

Ph, Ph, SPh, SPh → 1. $[Cp_2TiCl_2/THF/BuLi]$, –50°C → rt / 2. 1M NaOH → Ph, Ph 74%

Tsubouchi, A.; Nishio, E.; Kato, Y.; Fujiwara, T.; Takeda, T. *Tetrahedron Lett.* **2002**, *43*, 5755.

CHAPTER 15

PREPARATION OF OXIDES

This chapter contains reactions that prepare the oxides of nitrogen, sulfur, and selenium. Included are *N*-oxides, nitroso and nitro compounds, nitrile oxides, sulfoxides, selenoxides, and sulfones. Oximes are considered to be amines and appear in those sections. Preparation of sulfonic acid derivatives is described in Chapter 2 and the preparation of sulfonate esters in Chapter 10.

SECTION 211: OXIDES FROM ALKYNES

NO ADDITIONAL EXAMPLES

SECTION 212: OXIDES FROM ACID DERIVATIVES

MeO, CO_2H — HNO_3 , cat. AIBN; 50°C , 4 h → MeO, NO_2 75%

Das, J.P.; Sinha, P.; Roy, S. *Org. Lett.* **2002**, *4*, 3055.

F, SO_2Cl — PhOMe , $InCl_3$/TfOH , 1h → F, O, S, O, OMe 1:1 *o:p*

Garzya, V.; Forbes, I.T.; Lauru, S.; Maragni, P. *Tetrahedron Lett.* **2004**, *45*, 1499.

SECTION 213: OXIDES FROM ALCOHOLS AND THIOLS

O, Cl, P—OPh, OPh — *i*-PrOH , 2% $TiCl_4$, THF; 0.2 M, 1.5 NEt_3 , 1 h → O, *i*-PrO, P—OPh, OPh 90%

Jones, S.; Selitsianos, D. *Org. Lett*, **2002**, *4*, 3671.

OH — O, *t*-Bu, S, Cl , THF , −78°C; 0.5% peptide , Proton Sponge → O, S, *t*-Bu, O 61% (>99% ee)

Evans, J.W.; Fierman, M.B.; Miller, S.J.; Ellman, J.A. *J. Am. Chem. Soc.* **2004**, *126*, 8134.

SECTION 214: OXIDES FROM ALDEHYDES

NO ADDITIONAL EXAMPLES

SECTION 215: OXIDES FROM ALKYLS, METHYLENES, AND ARYLS

Ph-H → (1. [bmim]Cl•ACl_3 , $SOCl_2$, rt , 5 min; 2. 6M HCl) → Ph–S(=O)–Ph 85%

Mohile, S.S.; Potdar, M.K.; Salunkhe, M.M. *Tetrahedron Lett.* **2003**, *44*, 1255.

SECTION 216: OXIDES FROM AMIDES

Tol–S(=O)–NH_2 → (PhI=O , 3 MeOH , MeCN; rt) → Tol–S(=O)(OMe)–NH_2 85%

Leca, D.; Fensterbank, L.; Lacôte, E.; Malacria, M. *Org. Lett.* **2002**, *4*, 4093.

SECTION 217: OXIDES FROM AMINES

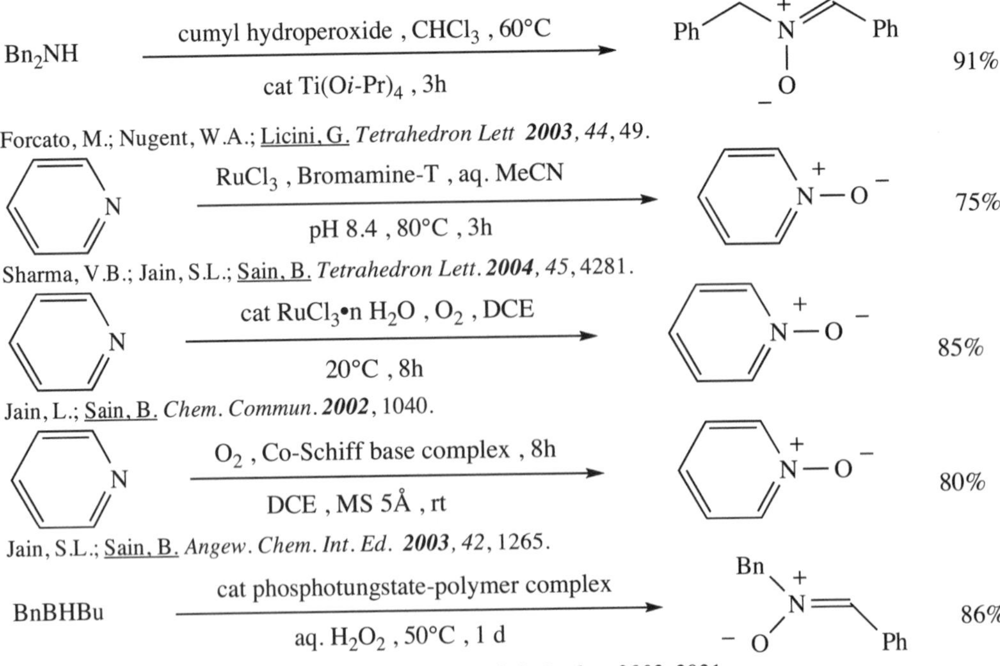

Forcato, M.; Nugent, W.A.; Licini, G. *Tetrahedron Lett* **2003**, *44*, 49.

Sharma, V.B.; Jain, S.L.; Sain, B. *Tetrahedron Lett.* **2004**, *45*, 4281.

Jain, L.; Sain, B. *Chem. Commun.* **2002**, 1040.

Jain, S.L.; Sain, B. *Angew. Chem. Int. Ed.* **2003**, *42*, 1265.

Yamada, Y.M.A.; Tabata, H.; Takahashi, H.; Ikegami, S. *Synlett* **2002**, 2031.

Br—C_6H_4—NH_2 —(AcOH , Pr_4NBrO_3 / 6N H_2SO_4)→ Br—C_6H_4—NO_2 84%

Das, S.S.; Nath, U.; Deb, D.; Das, P.J. *Synth. Commun.* **2004**, *34*, 2359.

SECTION 218: OXIDES FROM ESTERS

OAc

Ph Ph —(cat Pd/phosphilooxazoilne , 2d , rt / $NaSO_2Tol$, THF)→ SO_2Tol Ph Ph 57% (92% ee)

Bondarev, O.G.; Lyubimov, S.E.; Shiryaev, A.A.; Kadilnikov, N.E.; Davankov, V.A.; Gaurilov, K.N. *Tetrahedron: Asymmetry* **2002**, *13*, 1587.

SECTION 219: OXIDES FROM ETHERS, EPOXIDES, AND THIOETHERS

Ph–S– —(OOH, N, Br, S, O_2 / DCM , 20°C , 1 h)→ Ph–S(=O)– 95%

Gelalcha, F.G.; Schulze, B. *J. Org. Chem.* **2002**, *67*, 8400.

Ph–S– —($BnPh_3PHSO_5$, MeCN / reflux , 2 h)→ Ph–S(=O)– 911%

Hajipour, A.R.; Mallakpour, S.E.; Adibi, J. *J. Org. Chem.* **2002**, *67*, 8666.

Ph–S–Me —(hypervalent iodine compound / cat TEAB , $CHCl_3$–H_2O , rt)→ Ph–S(=O)–Me 98%

Shukla, V.G.; Salgaonkar, P.D.; Akamanchi, K.G. *J. Org. Chem.* **2003**, *68*, 5422.

S–Ph —(Flavin catalyst – H_2O_2 / MeOH , rt)→ S(=O)–Ph 92%

Linden, A.A.; Krüger, L.; Bäckvall, J.-E. *J. Org. Chem.* **2003**, *68*, 5890.

Ph–S— —(H_2O_2 , DCM , 10% EtOH , 5 h / 20% $Sc(OTf)_3$)→ Ph–S(=O)— 95%

Matteucci, M.; Bhalay, G.; Bradley, M. *Org. Lett.* **2003**, *5*, 235.

Ph–S–Me —([bis(μ-oxo)Cu(III)]$_2$)→ Ph–S(=O)–Me

Taki, M.; Itoh, S.; Fukuzami, S. *J. Am. Chem. Soc.* **2002**, *124*, 998.

PhS(Me) → chiral diamine/cat $VO(acac)_2$, H_2O_2 / $CHCl_3$, 0°C , 16 h → PhS(O)Me 81% (95% ee)

Sun, J.; Zhu, C.; Dai, Z.; Yang, M.; Pan, Y.; Hu, H. *J. Org. Chem.* ***2004***, *69*, 8500.

PhSeCH_2Ph → O_2 , [$^1O_2$] , hν , rose bengal → PhSe(O)CH_2Ph 88%

Kreif, A.; Lonez, F. *Tetrahedron Lett.* ***2002***, *43*, 6255.

PhS(Me) → cetylMe_3NBr_3 , aq. MeCN , 12h / MoO_4^{-2} , H_2O_2 → PhS(O)Me 93%

Kar, G.; Saika, A.K.; Bora, U.; Dehury, S.K.; Chaudhuri, M.K. *Tetrahedron Lett.* ***2003***, *44*, 4503.

PhS(Me) → pyridium-2,6-dicarboxylic acid chlorochromate / MeCN , rt → PhS(O)Me 95%

Tajbakhsh, M.; Hosseinzadeh, R. *Tetrahedron Lett.* ***2004***, *45*, 1889.

Ph—S—Me → Mn(tpp)OAc , Im , <1 min / Bu_4N peroxymonosulfate → Ph—S(O_2)—Me 96%

Iranpoor, N.; Mohajer, D.; Rezaeifard, A.-R. *Tetrahedron Lett.* ***2004***, *45*, 3811.

Ph—S—Et → H_2O_2 , cat Mn-Me_3TaCN , H_2O_2 / oxalic acid-oxalate , MeCN , H_2O → Ph—S(O_2)—Et quant

Barker, J.E.; Ren, T. *Tetrahedron Lett.* ***2004***, *45*, 4681.

TolS(Me) → Ti(O*i*-Pr)$_4$, DET , DCM , –20°C / 2-methyl-4-C_6H_{13}-furan-5-yl-CH_2OOH → TolS(O)Me 61% (98% ee)

Massa, A.; Siniscalchi, F.R.; Bugatti, V.; Lattanzi, A.; Scettri, A. *Tetrahedron: Asymmetry* ***2002***, *13*, 1277.

PhS(Me) → (DHQD)$_2$-PYR , 49h , 0°C / WO_3-30% H_2O_2 → PhS(O)Me 88% (59% ee)

Thakur, V.V.; Sudalai, A. *Tetrahedron: Asymmetry* ***2003***, *14*, 407.

PhS(Me) → H_2O_2 , DCM/MOH , 0°C , 16h / polymer supported chiral Schiff base-$VO(acac)_2$ complex → PhS(O)Me 75% (56% ee)

Barbarini, A.; Maggi, R.; Muratori, M.; Sartori, G.; Sartorio, R. *Tetrahedron: Asymmetry* ***2004***, *15*, 2467.

PhS(Me) → cat $Fe(acac)_3$(chiral Schiff base) / aq. H_2O_2 , DCM → PhS(O)Me 36% (59%ee)

Legros, J.; Bolm, C. *Angew. Chem. Int. Ed.* ***2003***, *42*, 5487.

Ph–S–Me → (cat $VO(acac)_2$, chiral amino alcohol / aq. H_2O_2 , DCM) → Ph–S(=O)–Me 77% (88% ee)

Ohta, C.; Shimizu, H.; Kondo, A.; Katsuki, T. *Synlett* **2002**, 161.

Ph–S–Me → (cat chiral amino alcohol , $VO(acac)_2$, DCM , rt / 7% H_2O_2 , 1,2,3-trimethoxybenzene , 16h) → Ph–S(=O)–Me 6% (50% ee)

Pelotier, B.; Anson, M.S.; Campbell, I.B.; Macdonald, S.J.F.; Priem, G.; Jackson, R.F.W. *Synlett* **2002**, 1051.

Ph–S–Me → (5% $NbCl_3$(dme).salen , DCM , MS 4Å / UHP) → Ph–S(=O)–Me 94% (86% ee)

Miyazaki, T.; Katsuki, T. *Synlett* **2003**, 1046.

PhCH₂–S–Ph → (Caro's acid , SiO_2 , MeCN) → PhCH₂–S(=O)(=O)–Ph 97%

Lakouraj, M.M.; Movassagh, B.; Ghodrati, K. *Synth. Commun.* **2002**, *32*, 847.

Et–S–Et → (N_2O_4/charcoal , DCM , rt / 20 min) → Et–S(=O)–Et quant

Iranpoor, N.; Firouzabadi, H.; Pourali, A.-R. *Synlett* **2004**, 347.

Ph–S–Me → (H_5IO_6 , $FeCl_3$, MeCN , rt , 2 min) → Ph–S(=O)–Me 99%

Kim, S.S.; Nehru, K.; Kim, S.S.; Kim, D.W.; Jung, H.C. *Synthesis* **2002**, 2484.

Ph–S–Me → (PhIO , Cr(III)-salen , DCM / rt , 2 h) → Ph–S(=O)–Me 99%

Kim, S.S.; Rajaopal, G. *Synthesis* **2003**, 2461.

SECTION 220: OXIDES FROM HALIDES AND SULFONATES

2 $CH_3CH_2NO_2$ → (PhBr , Cs_2CO_3 , DMF, 50°C , 3% $Pd_2(dba)_3$ / 6% 2-di-*t*-butylphosphino-2'-methylbiphenyl) → PhCH(CH₃)NO₂ 90%

Vogl, E.M.; Buchwald, S.L. *J. Org. Chem.* **2002**, *67*, 99.

1-iodonaphthalene → ($MeSO_2Na$, 5% $(CuOTf)_2$•PhH / 10% $MeHNCH_2CH_2NHMe$, DMSO , 110°C , 20 h) → 1-(SO_2Me)naphthalene 96%

Baskin, J.M.; Wang, Z. *J. Org. Chem.* **2002**, *67*, 4423.

MeO—C_6H_4—I → MeO—C_6H_4—SO_2Tol 90%
(p-$CH_3C_6H_4SO_2Na$; cat $Pd_2(dba)_3$, Xantphos , 80°C; Cs_2CO_3 , Bu_4NCl , toluene)

Cacchi, S.; Fabrizi, G.; Goggiamani, A.; Parisi, L.M. *Org. Lett.* ***2002**, 4*, 4719.

$HP(O)(OBu)_2$, 5% CuI , toluene
2 Cs_2CO_3 , 110°C
2% MeNHCH$_2$CH$_2$NJHMe

p-$CH_3C_6H_4$—I → p-$CH_3C_6H_4$—P(O)(OBu)$_2$ 88%

Gelman, D.; Jiang, L.; Buchwald, S.L. *Org. Lett.* ***2003**, 5*, 2315.

p-$CH_3C_6H_4$—SO_2Na → p-$CH_3C_6H_4$—SO_2—C_6H_4—Br 96%
(I—C_6H_4—Br; cat $Pd(dba)_3$, Xantphos , 6h; Cs_2CO_3 , Bu_4NCl , toluene , 80°C)

Cacchi, S.; Fabrizi, G.; Goggiamani, A.; Parisi, L.M.; Bernini, R. *J. Org. Chem.* ***2004**, 69*, 5608.

$PhCH_2Br$ → ($AgNO_2$, H_2O , rt , 2 h) → $PhCH_2NO_2$ 55%

Ballini, R.; Barboni, L.; Giarloi, G. *J. Org. Chem.* ***2004**, 69*, 6907.

Ph—CH$_2$CH$_2$—C(CH$_3$)$_2$—F → Ph—CH$_2$CH$_2$—C(CH$_3$)$_2$—P(=S)Ph$_2$ 73%
(1. $HPPh_2$, BF_3•OEt_2 , DCM , –20°C; 2. S_8)

Hirano, K.; Yorimitsu, H.; Oshima, K. *Org. Lett.* ***2004**, 6*, 4873.

MeO—C_6H_4—OTf → MeO—C_6H_4—SO_2—C_6H_4—CH_3 89%
(p-$CH_3C_6H_4SO_2Na$, 1.5 Cs_2CO_3; cat Pd_2dba_3/Xantphos , Bu_4NCl; toluene , 120°C)

Cacchi, S.; Fabrizi, G.; Goggiamani, A.; Parisi, L.M. *Synlett* ***2003**,* 361.

$PhSO_2Na$ → (BuCl , 70°C , 1 d , bmim BF_4 , H_2O) → $PhSO_2Bu$ 53%

Hu, Y.; Chen, Z.-C.; Le, Z.-G.; Zheng, Q.G. *Synth. Commun.* ***2004**, 34*, 4031.

SECTION 221: OXIDES FROM HYDRIDES

butane → 2-nitrobutane 65%
(3.2 NO_2 , 2 atm air , $PhCF_3$; 70°C , 14 h , glass autoclave; 0.6 N-hydroxyphthalimide)

Nishiwaki, Y.; Sakaguchi, S.; Ishii, Y. *J. Org. Chem.* ***2002**, 67*, 5663.

PhH → $PhNO_2$ 81%
(10% $(4\text{-}NO_2C_6H_4\text{-}SO_3^-)_3Ln^{+3}$; 67% HNO_3 , DCE , reflux , 16h)

Parac-Vogt, T.N.; Binnemans, K. *Tetrahedron Lett.* ***2004**, 45*, 3137.

PhH —[$(H_2N)_2C{=}NH{\bullet}HNO_3$, 0°C, 3h; H_2SO_4]→ $PhNO_2$ 89%

Ramana, M.M.V.; Malik, S.S.; Parihar, J.A. *Tetrahedron Lett.* **2004**, *45*, 8681.

—[bmpy $(NTf)_2$ (ionic liquid, 25°C, 1h; HNO_3-Ac_2O]→ NO_2 63%

Lancaster, N.L.; Llopis-Mestre, V. *Chem. Commun.* **2003**, 2812.

MeO —[$SOCl_2$, $Sc(OTf)_3$; DCM, rt]→ MeO—S(=O)—OMe 90%

Yadav, J.S.; Reddy, B.V.S.; Rao, R.S.; Kumar, S.P.; Nagaiah, K. *Synlett* **2002**, 784.

MeO —[$SOCl_2$, 10% H_2O; rt]→ MeO—S(=O)—OMe 93%

Bandgar, B.P.; Kinkar, S.N.; Kamble, V.T.; Bettigeri, S.V. *Synlett* **2003**, 2029.

OH, F —[Cl, N, O, N - Cl, Cl; $NaNO_2$, wet SiO_2]→ OH, O_2N, NO_2, F 99%

Zolfigol, M.A.; Madrakian, E.; Ghaemi, E. *Synlett* **2003**, 222.

—[40% HNO_3, Bentonite clay; 5 h]→ NO_2 81%

Babulayan, D.; Narayan, G.; Sreekumar, V. *Synth. Commun.* **2002**, *32*, 3565.

F—OH —[silica acetate, N_2O_4; DCM, rt, 5 min]→ F—OH, NO_2 91%

Iranpoor, N.; Firouzabadi, H.; Heydari, R. *Synth. Commun.* **2003**, *33*, 703.

OH —[$Fe(NO_3)_3{\bullet}9\ H_2O$, bbim BF_4, 1.5 h; 1,3-dibutylimidazolium tetrafluoroborate]→ O_2N—OH 80%

Rajogopal, R.; Srinivasan, K.V. *Synth. Commun.* **2004**, *34*, 961.

—[*p*-TsOH, $NaIO_4$, reflux, 6 h]→ —S(=O)(=O)—

Bandgar, B.P.; Kamble, V.T.; Fulse, D.B.; Deshmukh, M.V. *New J. Chem.* **2002**, *26*, 1105.

PhH + Tol-SO_2Cl —($NaIO_4$, heat, 7 h)→ Ph-SO_2-Tol 75%

Bandgar, B.P.; Kambi, V.T. *Chem. Lett.* **2002**, *31*, 1066.

benzene —(HNO_3, imidizolium ionic liquid)→ nitrobenzene (NO_2) 70%

Qiao, K.; Yokoyama, C. *Chem. Lett.* **2004**, *33*, 808.

SECTION 222: OXIDES FROM KETONES

NO ADDITIONAL EXAMPLES

SECTION 223: OXIDES FROM NITRILES

NO ADDITIONAL EXAMPLES

SECTION 224: OXIDES FROM ALKENES

Me–CH=CH–C_5H_{11} —(cat $Mn(OAc)_2$, 110°C, air, 2h; $(EtO)_2P(=O)H$)→ $(EtO)_2P(=O)$–CH(Me)–CH_2–C_5H_{11} 49% + $(EtO)_2P(=O)$–CH(Et)–CH_2–C_4H_9 35%

Tayama, O.; Nakano, A.; Iwahama, T.; Sakaguchi, S.; Ishii, Y. *J. Org. Chem.* **2004**, *69*, 5494.

C_6H_{13}–CH=CH_2 —(50% aq. H_2PO_2H, MeCN, reflux, polymer-supported Pd catalyst)→ $C_6H_{13}CH_2CH_2$–P(=O)(H)OH 70%

Deprèle, S.; Montchamp, J.-L. *Org. Lett.* **2004**, *6*, 3805.

CH_2=CH–C_6H_{13} —(BuO–P(=O)(H)H, XANTPHOS; MeCN, reflux)→ BuO–P(=O)(H)–CH_2CH_2–C_6H_{13} 76%

Deprèle, S.; Montchamp, J.-L. *J. Am. Chem. Soc.* **2002**, *124*, 9386.

1,6-heptadiene —($(EtO)_2P(=O)H$, cyclohexane; 80°C, 6h)→ EtO–P(=O)(OEt)–CH_2–(2-methylcyclopentyl) 87% (3:1)

Jessop, C.M.; Parsons, A.F.; Routledge, A.; Irvine, D. *Tetrahedron Lett.* **2003**, *44*, 479.

Ph–P(=O)(Ph)–H —($C_6H_{13}CH=CH_2$, MeOH, cat BEt_3, 20°C)→ Ph–P(=O)(Ph)–CH_2CH_2–C_6H_{13} 96%

Rey, P.; Rey, P.; Taillades, J.; Rossi, J.C. *Tetrahedron Lett.* **2003**, *44*, 6169.

PhO–P(=O)–H / PhO

20% t-BuOK , DMSO
70°C , 16 h

85%

Bunaksananusorn, T.; Knochel, P. *J. Org. Chem.* ***2004***, *69*, 4595.

SECTION 225: OXIDES FROM MISCELLANEOUS COMPOUNDS

Et_2Zn , toluene , rt
chiral oxazoline

77% (91% ee)

Zhang, X.-M.; Zhang, H.-L.; Lin, W.-Q.; Gong, L.-Z.; Mi, A.-Q.; Cui, X.; Jiang, Y.-Z.; Yu, K.B. *J. Org. Chem.* ***2003***, *68*, 4322.

$PhB(OH)_2$ —[$PhSO_2Cl$, $PdCl_2$, K_2CO_3 / aq. acetone , N_2 , 35 min]→ $PhSO_2Ph$ 94%

Bandgar, B.P.; Bettigeri, S.V.; Phopase, J. *Org. Lett.* ***2004***, *6*, 2105.

$PhB(OH)_2$ —[$AgNO_3$, 30 h , TMSCl / sealed tube , 50°C]→ $PhNO_2$ 98%

Prakash, G.K.S.; Panja, C.; Maathew, T.; Surampudi, V.; Petasis, N.A.; Olah, G.A. *Org. Lett.* ***2004***, *6*, 2205.

cat $Cu(OTf)_2$/Me-DuPhos , 0°C
2 Et_2Zn , toluene , 1 d

96% (93% ee)

Boezio, A.A.; Charette, A.B. *J. Am. Chem. Soc.* ***2003***, *125*, 1692.

$C_{10}H_{21}Br$ —[1. NaN_3 / 2. F_2-H_2O-MeCN]→ $C_{10}H_{21}NO_2$ 98%

Rozen, S.; Carmeli, M. *J. Am. Chem. Soc.* ***2003***, *125*, 8118.

Et_2Zn , toluene , rt
chiral oxazolidine

77% (91% ee)

Zhang, X.; Lin, W.; Gong, L.; Mi, A.; Cui, X.; Jiang, Y.; Choi, M.C.K.; Chan, A.S.C. *Tetrahedron Lett.* ***2002***, *43*, 1535.

Ph B$(OH)_2$ → $MeSO_2^{-}Na^{+}$, $Cu(OAc)_2$ / 2 K_2CO_3 , DMSO , rt / MS 4Å → Ph SO_2Me 97%

Beaulieu, C.; Guay, D.; Wang, Z.; Evans, D.A. *Tetrahedron Lett.* **2004**, *45*, 3233.

Tol N^{+} O^{-} Ph + CO_2Et CO_2Et → 5% $Yb(OTf)_3$ / DCM , rt , 18h → Tol N O Ph EtO_2C CO_2Et 77%

Young, I.S.; Kerr, M.A. *Angew. Chem. Int. Ed.* **2003**, *42*, 3023.

Ph N OH → 4% $MeReO_3$, 3 urea•H_2O_2 , MeOH / rt , 8 d → Ph NO_2 44%

Cardona, F.; Soldaini, G.; Goti, A. *Synlett* **2004**, 1553.

CHAPTER 16

PREPARATION OF DIFUNCTIONAL COMPOUNDS

SECTION 300: ALKYNE - ALKYNE

Ph—≡ → ($BrCH_2CO_2Et$, i-Pr_2NEt, rt; 2% $PdCl_2(PPh_3)_2$, 2% CuI, THF) → Ph—≡—≡—Ph 99%

Lei, A.; Srivastava, M.; Zhang, X. *J. Org. Chem.* ***2002**, 67*, 1969.

Et_3Si—≡ → (Br—≡—CH_2OH; cat cuCl, NH_2OH•HCl) → Et_3Si—≡—≡—CH_2OH 95%

Marino, J.P.; Nguyen, H.N. *J. Org. Chem.* ***2002**, 67*, 6841.

C_5H_{11}—≡—I → (4% $Pd(PPh_3)_4$, DMF; rt, 4 h) → C_5H_{11}—≡—≡—C_5H_{11} 84%

Damle, S.V.; Seomoon, D.; Lee, P.H. *J. Org. Chem.* ***2003**, 68*, 7085.

Ph—≡—(1-OAc-cyclohexyl) → ($Ti(Oi\text{-}Pr)_2Cl_2$, Mg, ether, 10°C) → 1,1'-bis(phenylethynyl)bicyclohexyl 89%

Yang, F.; Zhao, G.; Ding, Y.; Zhao, Z.; Zheng, Y. *Tetrahedron Lett.* ***2002**, 43*, 1289.

Ph—≡ → (CuCl, TMEDA, O_2, [bmim]PF_6, 4.8h) → Ph—≡—≡—Ph 95%

Yadav, J.S.; Reddy, B.V.S.; Reddy, K.B.; Gayathri, K.U.; Prasad, A.R. *Tetrahedron Lett.* ***2003**, 44*, 6493.

Ph—≡—≡—$SiMe_3$ → (1. K_2CO_3, MOH; 2. MeLI, LiBr, $C_5H_{11}I$) → Ph—≡—≡—C_5H_{11} 65%

Fiandanese, V.; Bottalico, D.; Marchese, G.; Punzi, A. *Tetrahedron Lett.* ***2003**, 44*, 9087.

O_2N—C_6H_4—≡ → (cat $Pd(OAc)_2$, 7h; DPEPhos, cat CuI; THF, 60°C) → O_2N—C_6H_4—≡—≡—C_6H_4—NO_2 72%

Oh, C.H.; Reddy, V.R. *Tetrahedron Lett.* ***2004**, 45*, 522.

C_5H_{11}—≡ $\xrightarrow[\text{cat PdCl}_2(\text{PPh}_3)_2\text{ , 2h}]{\text{NEt}_3\text{ , MeCN , cat CuI , 80°C}}$ C_5H_{11}—≡—≡—C_5H_{11} 90%

Fairlamb, I.J.S.; Bäuerlein, P.S.; Marrison, L.R.; Dickinson, J.M. *Chem. Commun.* ***2003***, 632.

SECTION 301: ALKYNE - ACID DERIVATIVES

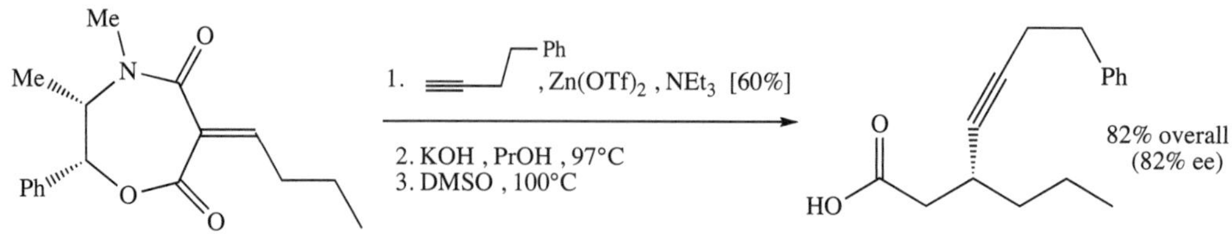

Knöpfel, T.F.; Boyall, D.; Carreira, E.M. *Org. Lett.* ***2004***, *6*, 2281.

SECTION 302: ALKYNE - ALCOHOL, THIOL

Ph—≡ 1. Et_2Zn; 2. PhCHO , *S*-BINOL, $Ti(Oi\text{-}Pr)_4$, rt → Ph—≡—CH(OH)Ph 77% (96% ee)

Moore, D.; Pu, L. *Org. Lett.* ***2002***, *4*, 1855.

PhC(=O)CO_2Me PhCCH , 0.22 chiral amino-alcohol, 0.2 Zn $(OTf)_2$, 0.3 NEt_3 , 70°C; toluene → Ph—≡—C(Ph)(OH)CO_2Me 91% (89% ee)

Jiang, B.; Chen, Z.; Tang, X. *Org. Lett.* ***2002***, *4*, 3451.

PhCHO PhC≡CH , Et_2Zn , $Ti(Oi\text{-}Pr)_4$; S-BINOL → HO, Ph—CH—≡—Ph 77% (96% ee)

Gao, G.; Moore, D.; Xie, R.-G.; Pu, L. *Org. Lett.* ***2002***, *4*, 4143.

2 Ph—≡ 1. Me_2Zn , BINOL ligand, toluene , rt; 2. PhCHO → Ph—≡—CH(Ph)OH 84% (92% ee)

Li, Z.-B.; Pu, L. *Org. Lett.* ***2004***, *6*, 1065.

Ph—≡ PhCHO , $Ti(Oi\text{-}Pr)_4$, Et_2Zn , DCM , rt; camphorsulfonamide ligand → Ph—≡—CH(OH)Ph 93% (97% ee)

Xu, Z.; Chen, C.; Xu, J.; Miao, M.; Yan, W.; Wang, R. *Org. Lett.* ***2004***, *6*, 1193.

PhCHO Ph-C≡CH , Et_2Zn , 10°C; ([2.2]-paracyclophane ligand), hexane/toluene , aq. MeOPEG → Ph—CH(OH)—≡—Ph 51% (91% ee)

Dahmen, S. *Org. Lett.* ***2004***, *6*, 2113.

O

Ph , DMSO , rt , 2 h

10% $BuNMe_3OH$

OH

Ph

95%

Ishikawa, T.; Mizuta, T.; Hagiwara, K.; Aikawa, T.; Kudo, T.; Saito, S. *J. Org. Chem.* ***2003***, *68*, 3702.

PhCHO

1. Br , 2 eq Mn , 30% DIPEA

10% Cr-carbazole ligand , THF , rt

2 eq. TMSCl 160 h

2. TBAF

OH

Ph

75% (71% ee)

Inoue, M.; Nakada, M. *Org. Lett.* ***2004***, *6*, 2977.

Ph

acetophenone , Et_2Zn , toluene , rt

S-BINOL-Ti(O*i*-Pr)$_4$

OH

Ph Ph

Me

67% (85% ee)

Zhou, Y.; Wang, R.; Xu, Z.; Yan, W.; Liu, L.; Kang, Y.; Han, Z. *Org. Lett.* ***2004***, *6*, 4147.

OH

Cl $SiEt_3$, cat $GaCl_3$, 3 h

PhCl , 120°C

OH

$SiEt_3$

90%

Kobayashi, K.; Arisawa, M.; Yamaguchi, M. *J. Am. Chem. Soc.* ***2002***, *124*, 8528.

PhCHO

PhC≡CH , THF , $ZnMe_2$, Ti(O*i*-Pr)$_4$

BINOL , chiral sulfonamide

OH

Ph

Ph

83% (96% ee)

Li, X.; Lu, G.; Kwork, W.H.; Chan, A.S.C. *J. Am. Chem. Soc.* ***2002***, *124*, 12636.

Et O H

Et N

O S O

O O

TBDPSO ZnCl

$BF_3•OEt_2$, THF , 40°C

OTBDPS H O

N S O

O

85% (20:1)

Fleming, J.J.; Fiori, K.W.; DuBois, J. *J. Am. Chem. Soc.* ***2003***, *125*, 2028.

Br_2In $InBr_2$

$C_5H_{11}CHO$, THF

ZnF_2

OH

C_5H_{11}

C_5H_{11}

HO

80%

Miao, W.; Lu, W.; Chan, T.H. *J. Am. Chem. Soc.* ***2003***, *125*, 2412.

Ph I

PhCHO , 20 In , DCM , 1d

HO

Ph

Ph

83%

Augé, J.; Lubin-Germain, N.; Seghrouchni, L. *Tetrahedron Lett.* ***2002***, *43*, 5255.

CHO

PhC≡CH , $ZnCl_2$, NEt_3

toluene

OH

Ph

79%

Jiang, B.; Si, Y.-G. *Tetrahedron Lett.* ***2002***, *43*, 8323.

Br (propargyl bromide) → PhCHO, In, THF-hexane; (–)-cinchonidine → Ph–CH(OH)–CH$_2$–C≡CH 71% (72% ee)

Loh, T.-P.; Lin, M.-J.; Tan, K.-L. *Tetrahedron Lett.* **2003**, *44*, 507.

Ph–C≡CH → 1. Et$_2$Zn, 24°C; 2. PhCHO, 2 Ti(O*i*-Pr)$_4$, 4% cinchonidine, ether → Ph–C≡C–CH$_2$OH 69% (77% ee)

Kamble, R.M.; Singh, V.K. *Tetrahedron Lett.* **2003**, *44*, 5347.

$C_5H_{11}CHO$ → 1. HC≡CCH$_2$Br, THF, ultrasound; 2. 2M HCl → C_5H_{11}–CH(OH)–CH$_2$–C≡CH 96%

Lee, A.S.-Y.; Chu, S.-F.; Chang, Y.-T. *Tetrahedron Lett.* **2004**, *45*, 1551.

C_8H_{17}–CH(Cl)–CHO → (HO)Me$_2$C–C≡C–CH=CH–Ph, Al-BINOL complex, DCM, rt → C_8H_{17}–CH(Cl)–CH(OH)–C≡C–CH=CH–Ph 80%

Ooi, T.; Miura, T.; Ohmatsu, K.; Saito, A.; Maruoka, K. *Org. Biomol. Chem.* **2004**, *2*, 3312.

PhCHO → PhC≡CH, BINOL, THF, 0°C, 18h; Ti(O*i*-Pr)$_4$-Me$_2$Zn → H$^+$ → Ph–CH(OH)–C≡C–Ph 85% (92% ee)

Lu, G.; Li, X.; Chan, W.L.; Chan, A.S.C. *Chem. Commun.* **2002**, 172.

Ph–C≡CH + (CH$_3$)$_2$CH–CHO → Zn(OTf)$_2$, NEt$_3$, toluene, 2h; chiral aminoalcohol, 25°C → (CH$_3$)$_2$CH–CH(OH)–C≡C–Ph 99% (98% ee)

Jiang, B.; Chen, Z.; Xiong, W. *Chem. Commun.* **2002**, 1524.

Me$_2$CHCHO → PhC≡CH, Zn(ODf)$_2$, NEt$_3$, toluene, 25°C; chiral aminoalcohol; Df = diflate → HO–CH(C≡C–Ph)(alkyl) 99% (97% ee)

Chen, Z.; Xiong, W.; Jiang, B. *Chem. Commun.* **2002**, 2098.

(epoxy geranyl benzyl ether, OBn) → PhC≡C-AlMe$_3$Li, 2 BF$_3$•OEt$_2$ → (quaternary C bearing C≡C–Ph and Me)–CH(OH)–CH$_2$OBn 73%

Zhao, H.; Pagenkopf, B.L. *Chem. Commun.* **2003**, 2592.

Cyclohexyl–CH(OH)–C(TMS)=C=CH$_2$ → cyclohexane-CHO, cat In(OTf)$_3$, 1h; DCM, (0.03M), rt → Cyclohexyl–CH(OH)–CH$_2$–C≡C–TMS 98%

Lee, K.-C.; Lin, M.-J.; Loh, T.-P. *Chem. Commun.* **2004**, 2456.

PhCHO → PhC≡CH, BINOL, Ti(O*i*-Pr)$_4$, Me$_2$Zn; THF → Ph–CH(OH)–C≡C–Ph 84% (95% ee)

Lu, G.; Li, X.; Chen, G.; Chan, W.L.; Chan, A.S.C. *Tetrahedron: Asymmetry* **2003**, *14*, 449.

Cl–C6H4–CHO → (Ph–C≡CH, ether, rt, 5h / $InBr_3$, NEt_3) → Ph–C≡C–CH(OH)–C6H4–Cl 82% (49:51 *cis:trans*) 90%

Sakai, N.; Hirasawa, M.; Konakahara, T. *Tetrahedron Lett.* ***2003***, *44*, 4171.

Br–C6H4–CHO → (PhC≡CH, DCM, 0°C / ferrocenyl oxazoline ligand) → Br–C6H4–CH(OH)–C≡C–Ph 86% (86% ee)

Li, M.; Zhu, X.-Z.; Yuan, K.; Cao, B.-X.; Hou, X.-L. *Tetrahedron: Asymmetry* ***2004***, *15*, 219.

PhCHO → (PhC≡CH, Et_2Zn, THF, 0°C / bifunctional LA/LB catalyst) → Ph–CH(HO)–C≡C–Ph 95% (94% ee)

Kang, Y.-F.; Liiu, L.; Wang, R.; Yan, W.-J.; Zhou, Y.-F. *Tetrahedron: Asymmetry* ***2004***, *15*, 3155.

PhC(O)CH3 → (PhC≡CH, Me_2Zn, DCM, 0°C, 2d / 10% $Cu(OTf)_2$, chiral amino alcohol) → Ph(HO)C(CH3)–C≡C–Ph 90% (82%ee)

Lu, G.; Li, X.; Jia, X.; Chan, W.L.; Chan, A.S.C. *Angew. Chem. Int. Ed.* ***2003***, *42*, 5057.

PhCHO → (PhC≡CH, Et_2Zn, Ti(O*i*-Pr)$_4$, toluene / rt, 12h) → Ph–C≡C–CH(OH)–Ph 92% (95%ee)

Xu, Z.; Wang, R.; Xu, J.; Da, C.-s.; Yan, W.-j.; Chen, C. *Angew. Chem. Int. Ed.* ***2003***, *42*, 5747.

SECTION 303: ALKYNE - ALDEHYDE

NO ADDITIONAL EXAMPLES

SECTION 304: ALKYNE - AMIDE

PhCH2CH2NHCO$_2$Me → (1. KHMDS, CuI, Py, rt, 2h / 2. 2 Br-C≡C-Ph, rt, 20h) → PhCH2CH2N(CO$_2$Me)–C≡C–Ph 76%

Dunetz, J.R.; Danheiser, R.L. *Org. Lett.* ***2003***, *5*, 4011.

Ts(Ph)N–C≡CH → (10% CuI, 20% TMEDA, acetone / O_2, rt) → Ts(Ph)N–C≡C–C≡C–N(Ts)Ph 91%

Rodríguez, D.; Castedo, L.; Saá, C. *Synlett* ***2004***, 377.

MeOC(O)NHBn → (Br–C≡C–Ph, 2 K_3PO_4, 65°C / 0.2 $CuSO_4$, 0.4 1,10-phen, toluene) → MeO$_2$C(Bn)N–C≡C–Ph 73%

Zhang, Y.; Hsung, R.P.; Tracey, M.R.; Kurtz, K.C.M.; Vera, E.L. *Org. Lett.* ***2004***, *6*, 1157.

BnO–C(O)–Cl , ≡—CO_2Me

10% CuI , i-Pr_2NEt , MeCN , rt

BnO_2C, N, Et, CO_2Me 68%

Black, D.A.; Arndtsen, B.A. *Org. Lett.* ***2004***, *6*, 1107.

O, NH, Ph

Br—≡—C_6H_{13} , toluene , 110°C

5% CuCN , $MeNHCH_2CH_2NHMe$

2 K_3PO_4

O, N—≡—C_6H_{13}, Ph 74%

Frederick, M.O.; Mulder, J.A.; Tracey, M.R.; Hsung, R.P.; Huang, J.; Jurtz, K.C.M.; Shen, L.; Douglas, C.J. *J. Am. Chem. Soc.* ***2003***, *125*, 2368.

I

1. ≡—Ph , Et_2NH , CuI

cat $Pd(PPh_3)_2Cl_2/PPh_3$

DMF , 120°C , microwaves

2. TFA

Ph, O, NH_2

Erdélyi, M.; Gogoli, A. *J. Org. Chem.* ***2003***, *68*, 6431.

N, OMe, CO_2Me

Ph-C≡CH , CuBr , H_2O , 50°C

N, Ph, CO_2Me 58%

Zhang, J.; Wei, C.; Li, C.-J. *Tetrahedron Lett.* ***2002***, *43*, 5731.

Ts, N, Ph, Cl, Cl

1. BuLi , –78°C

2. $ZnBr_2$, rt

3. Pd catalyst , PhI

Ts, N—≡—Ph, Ph

Rodríguez, D.; Castedo, L.; Saá, C. *Synlett* ***2004***, 783.

N—Ts

Ph-C≡CH , cat Cu(OTf)

ether , 45 h

NHTs, Ph 96%

Ding, C.-H.; Daqi, L.-X.; Hou, X.-L. *Synlett* ***2004***, 1691.

Ts, N, EtO_2C

=C=/ $SnBu_3$, ether , –30°C

1% $[Cu(MeCN)_4]ClO_4$/(R)-tol BINAP

NHTs, EtO_2C + NHTs, EtO_2C, H

(97 : 3) 96%

86% ee

Kagoshima, H.; Uzawa, T.; Akiyama, T. *Chem. Lett.* ***2002***, *31*, 298.

REVIEW:

"In Search of an Atom-Economical Synthesis of Chiral Ynamides"
Mulder, J.A.; Kurtz, K.C.M.; Hsung, R.P. *Synlett 2003,* 1379.

SECTION 305: ALKYNE - AMINE

Bu—≡ → (Ph–CH=N⁺(O⁻)Bn; 0.2 Et_2Zn , toluene , 20°C , 6 h) → Bu–C≡C–CH(Ph)–N(OH)Bn 92%

Pinet, S.; Pandya, S.U.; Chavant, P.Y.; Ayling, A.; Vallee, Y. *Org. Lett.* **2002**, *4*, 1463.

Ph—≡ → (*t*-Bu – N ≡ C , 100°C , 2 d; $(Et_2N)_3U\ BF_4$) → Ph–C≡C–CH=N—*t*-Bu 81%

Barnea, E.; Andrea, T.; Kapon, M.; Berthet, J.-C.; Ephritikhine, M.; Wisen, M.S. *J. Am. Chem. Soc.* **2004**, *126*, 10860.

≡—Ph → ($PhNMe_2$, 5% CuBr , *t*-BuOOH; 100°C , 3 h) → Ph–N(Me)–CH₂–C≡C–Ph 77%

Li, Z.; Li, C.-J. *J. Am. Chem. Soc.* **2004**, *126*, 11810.

Ph—≡ → (Ph_2PCl , 3% $Ni(acac)_2$, NEt_3; toluene , 80°C) → Ph—≡—PPh_2 95%

Beletskaya, I.P.; Afanasiev, V.V.; Kazankova, M.A.; Efimova, I.V. *Org. Lett.* **2003**, *5*, 4309.

Ph—≡ → (1. 1.2 $Zn(OTf)_2$, 1.32 NEt_3 , 1.32 TMPDA, toluene , rt , 2h; 2. PhCH=N–CH₂CH=CH₂ , BzCl , toluene , –78°C; 3. rt , 1h) → PhC(O)–N(allyl)–CH(Ph)–C≡C–Ph 83%

Fischer, C.; Carreira, E.M. *Org. Lett.* **2004**, *6*, 1497.

3-bromoquinolizinium Br^- → (1. TIPS-C≡CH , 10% CuI, 5% $Pd_2Cl_2(PPh_3)_2$, DMF , 60°C , NEt_3 , 14 h; 2. NH_4PF_6) → 3-(TIPS-ethynyl)quinolizinium PF_6^- 80%

García, D.; Cuadro, A.M.; Alvarez-Builla, J.; Vaquero, J.J. *Org. Lett.* **2004**, *6*, 4175.

PhCH=N–Ph → (Ph—≡ , 10% Cu(OTf) , toluene; 10% copper-pybox ligand , 22°C , 4h) → Ph–CH(NHPh)–C≡C–Ph 78% (96% ee)

Wei, C.; Li, C.-J. *J. Am. Chem. Soc.* **2002**, *124*, 5638.

10% peptide chiral ligand, 72h
11% $Zr(Oi\text{-}Pr)_4$, toluene, 22°C
0.6 $Zn(\text{-}C{\equiv}C\text{-}SiMe_3)_2$
2 eq $Zn(CH_2SiMe_3)_2$

90% (81% ee)

Traverse, J.F.; Hoveyda, A.H.; Snapper, M.L. *Org. Lett.* ***2003***, *5*, 3273.

CHO

, Ph-C≡CH
1.5% AgI, 100°C, H_2O

96%

Wei, C.; Li, Z.; Li, C.-J. *Org. Lett.* ***2003***, *5*, 4473.

PhCHO

, Ph-C≡CH, 15% CuI
H_2O, microwaves

90%

Shi, L.; Tu, Y.-Q.; Wang, M.; Zhang, F.-M.; Fan, C.-A. *Org. Lett.* ***2004***, *6*, 1001.

, 2% $PdCl_2(PPh_3)_2$
2% CuI, NEt_3, 60°C
BuMgBr

80%

Kikiya, H.; Yagi, K.; Shinokubo, H.; Oshima, K. *J. Am. Chem. Soc.* ***2002***, *124*, 9032.

PhCHO

piperidine, Ph-C≡CH, cat Au, H_2O

>99%

Wei, C.; Li, C.-J. *J. Am. Chem. Soc.* ***2003***, *125*, 9584.

1. MeOTf, ether, rt
2. PhC≡CLi, ether, 0°C → rt
3. HC≡CMgBr, ether

91%

Murai, T.; Mutoh, Y.; Ohta, Y.; Murakami, M. *J. Am. Chem. Soc.* ***2004***, *126*, 5968.

Ph-C≡CH , CuOTf , bis(oxazolone) ligand
t-BuOOH , 50°C , 2 d

67% (63% ee)

Li, Z.; Li, C.-J. *Org. Lett.* ***2004***, *6*, 4997.

MeO N NH Br , In , aq THF
Bu_4NBr , 7h

80%

Prajapati, D.; Laskar, D.D.; Gogoi, B.J.; Devi, G. *Tetrahedron Lett.* ***2003***, *44*, 6755.

Ph N * Ph — Ph , $ZnCl_2$, NEt_3 , TMSCl
toluene , 60°C

HN * Ph Ph Ph 82% (50%ee)

Jiang, B.; Si, Y.-G. *Tetrahedron Lett.* ***2003***, *44*, 6767.

CHO OH Bu — BF_3K , Bn_2NH , $PhCO_2H$
(bmim) BF_4 , 80°C , 20h

NBn_2 OH Bu 81%

Kabalka, G.W.; Venkataiah, B.; Dong, G. *Tetrahedron Lett.* ***2004***, *45*, 729.

PhCHO PhC≡CH , piperidine , 3% AgI , 100°C
(bmim) BF_4

N Ph Ph 86%

Li, Z.; Wei, C.; Chen, L.; Varma, R.S.; Li, C.-J. *Tetrahedron Lett.* ***2004***, *45*, 2443.

I — OMe HC≡CCH_2Br , pyrrolidine , 0°C
5% $PdCl_2(PPh_3)_2$, 10% CuI

N OMe 80%

Olivi, N.; Spruyt, P.; Peyrat, J.-F.; Alami, M.; Brion, J.-D. *Tetrahedron Lett.* ***2004***, *45*, 2607.

Ph — CH_2O , Me_2NH , aq. DMSO
cat CuI , 10h

Ph — NMe_2 95%

Bieber, L.W.; da Silva, M.F. *Tetrahedron Lett.* ***2004***, *45*, 8281.

Ph–CH=N–Ph + Ph–C≡CH —[chiral bis(imine), CuOTf / toluene]→ PhCH(NHPh)–C≡C–Ph 98% (77% ee)

Benaglia, M.; Negri, D.; Dell'Anna, G. *Tetrahedron Lett.* **2004**, *45*, 8705.

PhCHO —[PhC≡CH, $PhNH_2$, H_2O]→ PhCH(NHPh)–C≡C–Ph 91%

Li, C.-J.; Wei, C. *Chem. Commun.* **2002**, 268.

TMS–C≡CH —[$(C_3H_7)_2NH$, cat $[IrCl(cod)]_2$, dioxane / *i*-PrCHO, 75°C, 15h]→ Bu_2N, C_3H_7 ... C≡C–TMS 97%

Sakaguchi, S.; Mizuta, T.; Furuwan, M.; Kubo, T.; Ishii, Y. *Chem. Commun.* **2004**, 1638.

C_5H_{11}–C≡C–CHO + 2-methoxyaniline (OMe, NH_2) —[Et_2Zn, 10% $Zr(Oi\text{-}Pr)_4$ / cat dipeptide ligand, toluene, *i*-PrOH]→ C_5H_{11}–C≡C–CH(Et)–NH–(2-MeO-phenyl) 68%

Akullian, L.C.; Snapper, M.L.; Hoveyda, A.H. *Angew. Chem. Int. Ed.* **2003**, *42*, 4244.

Ph–C(=N–OMe)–CH_2Cl —[LDA, $MgBr_2$, THF]→ Ph–C(=N–OMe)–C≡C–Ph 76% (>95:5 *Z*:*E*)

Tsuritani, T.; Yagi, K.; Shinokubo, H.; Oshima, K. *Angew. Chem. Int. Ed.* **2003**, *42*, 5613.

Ph, OH, HN ... C≡C–C_5H_{11} —[potassium 3-aminoproyl amide / 0°C, 1,3-diaminopropane, 30 min]→ Ph, OH, HN ... terminal alkyne 77%

Blanchet, J.; Bonin, M.; Micouin, L.; Husson, H.-P. *Eur. J. Org. Chem.* **2002**, 2598.

enamine N(allyl)$_2$ —[Ph–C≡CH, cat CuBr, rt, 16 h / decane/toluene]→ propargylamine N(allyl)$_2$ 98%

Koradin, C.; Gommermann, N.; Polborn, K.; Knochel, P. *Chem. Eur. J.* **2003**, *9*, 2797.

Ph–CH=N–$CHPh_2$ —[TMS–C≡CH, $[IrCl(cod)]_2$, THF / MgI_2, rt]→ Ph_2HCHN–CH(Ph)–C≡C–TMS 92%

Fischer, C.; Carreira, E.M. *Synthesis* **2004**, 1497.

PhCHO + $PhNH_2$ + C_3H_7C≡CH —[bmim PF_6, CuBr / 60°C, 4 h]→ PhHN–CH(Ph)–C≡C–C_3H_7

Yadav, J.S.; Reddy, B.V.S.; Naveenkumar, V.; Rao, R.S.; Nagaiah, K. *New J. Chem.* **2004**, *28*, 335.

SECTION 306: ALKYNE - ESTER

Ph—≡ → (cat $Pd(OAc)_2(PPh_3)_2$, CO (1 atom); e^-, MeCN/MeOH, 2 NaOAc, 8-10 h, rt) → Ph—≡—CO_2Me 56%

Chiarotto, I.; Carelli, I. *Synth. Commun.* **2002**, *32*, 881.

Ph—≡ → (CO/O_2, 50 MeOH, DMF, rt; 0.3 NaOAc, 0.2 PPh_3, 0.1 $PdCl_2$) → Ph—≡—CO_2Me 82%

Izawa, Y.; Shimizu, I.; Yamamoto, A. *Bull. Chem. Soc. Jpn.* **2004**, *77*, 2033.

SECTION 307: ALKYNE - ETHER, EPOXIDE, THIOETHER

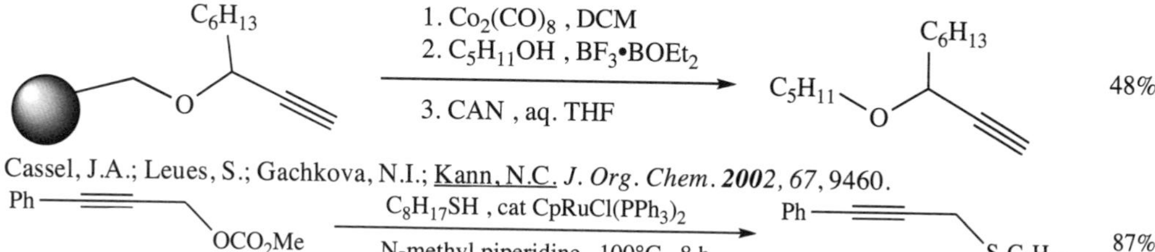

Cassel, J.A.; Leues, S.; Gachkova, N.I.; Kann, N.C. *J. Org. Chem.* **2002**, *67*, 9460.

Kondo, T.; Kanda, Y.; Baba, A.; Fukuda, K.; Nakamura, A.; Wada, K.; Morisaki, Y.; Mitsudo, T.-a. *J. Am. Chem. Soc.* **2002**, *124*, 12960.

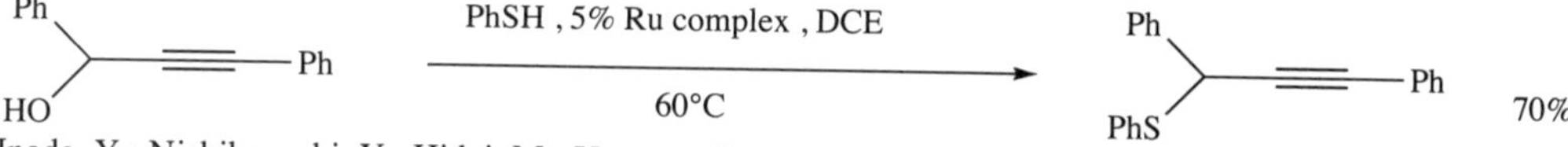

Inada, Y.; Nishibayashi, Y.; Hidai, M.; Uemura, S. *J. Am. Chem. Soc.* **2002**, *124*, 15172.

Bu—≡ → (0.5 PhSSPh, CuI, 72h, DMSO; K_2CO_3, 70°C) → Bu—≡—SPh 96%

Bieber, L.W.; da Silva, M.F.; Menezes, P.H. *Tetrahedron Lett.* **2004**, *45*, 2735.

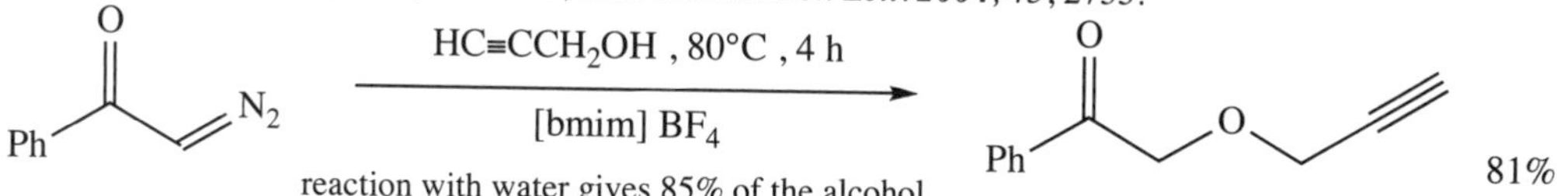

Yadav, J.S.; Reddy, B.V.S.; Srinivas, M. *Chem Lett.* **2003**, *32*, 1060.

SECTION 308: ALKYNE - HALIDE

NO ADDITIONAL EXAMPLES

SECTION 309: ALKYNE - KETONE

1. Cl-C≡C-$SiMe_3$, $GaCl_3$, 5 min; 2. MeOH; H^+ — 73%

Arisawa, M.; Amemiya, R.; Yamaguchi, M. *Org. Lett.* **2002**, *4*, 2209.

Ph—≡ + ClCOCO(O*i*-Pr), 5% CuI, NEt_3, THF, rt, 12 h → Ph—≡—COCO(O*i*-Pr) 70%

Guo, M.; Li, D.; Zhang, Z. *J. Org. Chem.* ***2003***, *68*, 10172.

PhCOCH$_3$ + Ph—≡, 110°C, TEA, 1 h, cat $Pd(OAc)_2$, toluene → PhCO—≡—Ph 75%

Alonso, D.A.; Nájera, C.; Pacheco, M[a].C. *J. Org. Chem.* ***2002***, *67*, 5588.

C_6H_{13}—≡ + MeO-C$_6$H$_4$-I, CO, 2% CuI, 1 d, 5% $PdCl_2(PPh_3)_2$, aq NH_3 (0.5 M) → C_6H_{13}—≡—CO-C$_6$H$_4$-OMe 74%

Ahmed, M.S.M.; Meri, A. *Org. Lett.* ***2003***, *5*, 3057.

MeO-dienyl acid chloride + ≡—$SiMe_3$, 4% CuI, 1h, 2% $Pd(PPh_3)_2Cl_2$, NEt_3, THF → MeO-C$_6$H$_4$-CO—≡—$SiMe_3$ 82%

Karpov, A.S.; Müller, T.J.J. *Org. Lett.* ***2003***, *5*, 3451.

PhCOCl + PhC≡CH, cat $Pd(PPh_3)_2Cl_2$/CuI, H_2O, 65°C, K_2CO_3, 7% Na lauryl sulfate → PhCO—≡—Ph 98%

Chen, L.; Li, C.-J. *Org. Lett.* ***2004***, *6*, 3151.

Ph—≡—C(Ph)(Me)OAc + CH$_2$=C(Me)OTMS, DCM, 25°C, 12 h, cat Ir(cod)[P(OPh)$_3$]OTf → Ph—≡—C(Ph)(Me)CH$_2$COMe 82%

Matsuda, I.; Komori, K.-i.; Itoh, K. *J. Am. Chem. Soc.* ***2002***, *124*, 9072.

C_9H_{18}COCl + I—≡—Ph, 1.2 In, DCE, rt, 1h → C_9H_{18}CO—≡—Ph 80%

Augé, J.; Lubin-Germain, N.; Sehrouchni, L. *Tetrahedron Lett* ***2003***, *44*, 819.

C_6H_{13}—≡ + CH$_2$=CHCOMe, 5% [Rh(acac)CO]$_2$, cat P(2-OMeC$_6$H$_4$)$_3$, PhH, reflux, 1d → C_6H_{13}—≡—CH$_2$CH$_2$COMe 63%

Lerum, R.V.; Chisholm, J.D. *Tetrahedron Lett.* ***2004***, *45*, 6591.

Fe-phthalocyanine catalyst , 40°C; t-BuOOK , 1d — 80%

Pérollier, C.; Sorokin, A.B. *Chem. Commun.* **2002**, 1548.

$PhC\equiv CH$, H_2O , 40h; cat $Pd(OAc)_2/PMe_3$ — 91%

Chen, L.; Li, C.-J. *Chem. Commun.* **2004**, 2362.

$NaIO_4$, 1% $Na_2Cr_2O_7$, 50 h; aq. MeCN , 100°C , 20 HNO_3 — 86%

van de Vondervoot, L.S.; Bouttemy, S; Pardón, J.M.; LeBras, J.; Muzart, J.; Alsters, P.L. *Synlett* **2002**, 243.

1. BuLi , THF , –78°C; 2. $BF_3\bullet OEt_2$, 28°C; 3. (γ-butyrolactone) , –78°C → rt — 92%

Doubsky, J.; Streinz, L.; Leseticky, L; Koutek, B. *Synlett* **2003**, 937.

2 Bu—≡ , 2.5 CuI , DMF; 0.1 $PdCl_2(dppf)$, 0.25 $P(2\text{-furyl})_3$, NEt_3 , 50°C — 94%

Tokuyama, H.; Miyazaki, T.; Yokoshima, S.; Fukuyama, T. *Synlett* 2003, 1512.

Ph—≡—$SiMe_3$, 5% $InBr_3$, DCM; 0°C → rt — 92%

Yadav, J.S.; Reddy, B.V.S.; Reddy, M.; Parimala, G. *Synthesis* **2003**, 2390.

1. $Me_3SiC\equiv C\text{-}Cl$, $GaCl_3$; 2. H^+ — n = 1 , 45% + n = 2 , 14%

Amemiya, R.; Fujii, A.; Arisawa, M.; Yamaguchi, M. *Chem. Lett.* **2003**, *32*, 298

SECTION 310: ALKYNE - NITRILE

NO ADDITIONAL EXAMPLES

SECTION 311: ALKYNE - ALKENE

Ph—≡ → 1% $Pd(OAc)_2$, 2 eq K_2CO_3, 2% imidazolium ligand / DMAc, 80°C, 2 h → (69 : 5 : 26) quant

Yang, C.; Nolan, S.P. *J. Org. Chem.* ***2002**, 67*, 591.

Ph—≡—Ph → 1. Cp_2ZrEt_2 2. Cl—≡—Ph 3. H_3O^+ → 75%

Liu, Y.; Zhong, Z.; Nakajima, K.; Takahashi, T. *J. Org. Chem.* ***2002**, 67*, 7451.

→ 4 eq $C_5H_{11}C{\equiv}CH$, cat $Ni(dppe)Cl_2$ / cat CuI, pyrrolidine, 20 h → 2 ... C_5H_{11} 78%

Silveira, C.C.; Braga, A.L.; Vieira, A.S.; Zeni, G. *J. Org. Chem.* ***2003**, 68*, 662.

→ C_5H_9—≡, cat $PdCl_2(PPh_3)_2$ / K_2CO_3, MeCN, 12 h → 92%

Huang, X.; Zhou, H.; Chen, W. *J. Org. Chem.* ***2004**, 69*, 839.

→ BrZn—≡—CO_2Et / cat $Pd(PPh_3)_4$, 23°C, THF → 84% (>98% *E*)

Negishi, E.; Qian, M.; Zeng, F.; Anastasia, L.; Babinski, D. *Org. Lett.* ***2003**, 5*, 1597.

→ $CH_2{=}CHCO_2Me$, O_2, Py, 2 d, 10% $Pd(acac)_2$, MS 3Å, toluene / 80°C →

Nishimura, T.; Araki, H.; Maeda, Y.; Uemura, S. *Org. Lett.* ***2003**, 5*, 2997.

Ph—≡ → I—CH=CH—CO_2Et, 2 K_2CO_3, toluene / 10% [Cu(bipy)PPh_3Br], 110°C, 8 h → 81%

Bates, C.G.; Saejueng, P.; Venkataraman, D. *Org. Lett.* ***2004**, 6*, 1441.

→ allyl-$SiMe_3$ / cat $B(C_6F_5)_3$, DCM, rt → 95%

Schwier, T.; Rubin, M.; Gevorgyan, V. *Org. Lett.* ***2004**, 6*, 1999.

OTf, SiMe$_3$ → $\diagup\!\!\diagdown$Cl , cat Pd(dba)$_2$–dppe; Bu_3Sn—≡—Bu , CsF , MeCN; 40°C , 10 h → Bu 92%

Jeganmohan, M.; Cheng, C.-H. *Org. Lett.* **2004**, *6*, 2821.

Ph, S, S, SiMe$_3$ → BuCH=CH$_2$; Cp$_2$Ti[P(OEt)$_3$]$_2$ → Ph, Me$_3$Si, Bu 67% (57:43)

Takeda, T.; Kuroi, S.; Ozaki, M.; Tsubouchi, A. *Org. Lett.* **2004**, *6*, 3207.

Ph—≡ → 1. BCl$_3$, DCM; 2. OLi, Ph, Cl → Cl, Cl, Ph, Ph 90% (98:2 *Z*:*E*)

Kabalka, G.W.; Wu, Z.; Ju, Y. *Org. Lett.* **2004**, *6*, 3929.

Ph—≡ → chiral Lu catalyst , THF; PhH , 80°C → Ph, Ph 99%

Nishiura, M.; Hou, Z.; Wakatsuki, Y.; Yamaki, T.; Miyamoto, T. *J. Am. Chem. Soc.* **2003**, *125*, 1184.

OH, O → 5% bis(Ru compound) , 10% NH$_4$BF$_4$; DCE , rt , 20 h → O 74%

Nishibayashi, Y.; Inada, Y.; Hidai, M.; Uemura, S. *J. Am. Chem. Soc.* **2003**, *125*, 6060.

O, OTs → PhC≡CH , cat. Pd(OH)$_2$ /PPh$_3$; DMA/DMF/TEA , rt , 3.5h → O, Ph 91%

Fu, X.; Zhang, S.; Yin, J.; Schumacher, D.P. *Tetrahedron Lett.* **2002**, *43*, 6773.

CO$_2$Me, NH$_2$ → 2 B(OH)$_2$, 2 TEA , MS 4Å; Cu(OAc)$_2$, DCM , rt → CO$_2$Me, HN 67%

Lam, P.Y.S.; Bonne, D.; Vincent, G.; Clark, C.G.; Combs, A.P. *Tetrahedron Lett.* **2003**, *44*, 1691.

Ph–CH=CH–$TeCl_2Bu$ → ($HOCH_2C{\equiv}CH$, cat dCl_2, CuI; NEt_3, rt, 6h) 82%

Braga, A.L.; Lüdtke, D.S.; Vargas, F.; Donato, R.K.; Silveira, C.C.; Stefani, H.A.; Zeni, G. *Tetrahedron Lett.* ***2003***, *44*, 1779.

$CH_2{=}C{=}CH_2$, toluene, 5% TFP; 5% $Pd(OAc)_2$, 5% CuI — 76%

Bruyere, D.; Grigg, R.; Hinsley, J.; Hussain, R.K.; Korn, S.; DeLacierva, C.O.; Sridharan, V.; Wang, J. *Tetrahedron Lett.* ***2003***, *44*, 8669.

cat Grubbs' II, $CH_2{=}CH_2$; toluene, 80°C — 88%

van Otterlo, W.A.L.; Ngidi, E.L.; de Koning, C.B.; Fernandes, M.A. *Tetrahedron Lett.* ***2004***, *45*, 659.

Ph–C≡CH, water, rt, 2d; cat CuBr, $PdCl_2(PPh_3)_2$, toluene — 63%

Chen, L.; Li, C.-J. *Tetrahedron Lett.* ***2004***, *45*, 2771.

Ph–CH=CH–TeBu → (2.5 C_5H_{11}–C≡C–Li, THF, reflux, 3h; 5% $Ni(PPh_3)_2Cl_2$) quant

Raminelli, C.; Gargalak Jr. J.; Silveira, C.C.; Comasseto, J.V. *Tetrahedron Lett.* ***2004***, *45*, 4927.

Me_3Si–C≡CH → (cat $Ni(cod)_2Pt$-Bu_3, CD_3CN; rt, 2h) Me_3Si–CH=CH–C≡C–$SiMe_3$ 97%

Ogoshi, S.; Ueta, M.; Oka, M.-a.; Jurosawa, H. *Chem. Commun.* ***2004***, 2732.

C_5H_{11}–C≡CH → (Ph–CH=C(TeBu)Br; cat $PdCl_2$, MeOH, NEt_3, rt) 75%

Zeni, G.; Perin, G.; Cella, R.; Jacob, R.G.; Braga, A.L.; Silveira, C.C.; Stefani, H.A. *Synlett* ***2002***, 975.

→ (C_5H_{11}(BuTe)C=CH–SMe, $PdCl_2$, CuI; NEt_3, MeOH, 6 h) 85%

Zeni, G.; Nogueira, C.W.; Pena, J.M.; Pilassão, C.; Menezes, P.H.; Braga, A.L.; Rocha, J.B.T. *Synlett* ***2003***, 579.

Ph—≡—H → ($InCl_3$-$NaBH_4$-MeCN, −15°C → rt) Ph–CH=CH–C≡C–Ph 92x86% + 2%, 6% other isomers

Wang, C.-Y.; Su, H.; Yang, D.-Y. *Synlett* ***2004***, 561.

REVIEW:

"Enyne Metathesis (Enyne Bond Reorganization)"
Diver, S.T.; Giessert, A.J. *Chem. Rev.* ***2004***, *104*, 1317.

SECTION 312: CARBOXYLIC ACID - CARBOXYLIC ACID

NO ADDITIONAL EXAMPLES

SECTION 313: CARBOXYLIC ACID - ALCOHOL, THIOL

CO_2SiMe_3 → (1. BuLi, −70°C; 2. PhCHO, −70°C; 3. aq. H^+; aq. OH^-) CO_2H, Ph, OH 95% (55:45 *syn:anti*)

Bellassoued, M.; Grugier, J.; Lensen, N.; Catheline, A. *J. Org. Chem.* ***2002***, *67*, 5440.

O, HO_2C → (allyl Br, In, aq. THF) HO_2C, HO 66%

Kumar, S.; Kaur, P.; Chimni, S.S. *Synlett* ***2002***, 573.

SECTION 314: CARBOXYLIC ACID - ALDEHYDE

NO ADDITIONAL EXAMPLES

SECTION 315: CARBOXYLIC ACID - AMIDE

N—O, Et, OH → (1. $LiAlH_4$; 2. Boc_2O; 3. $NaIO_4$, $RuCl_3$) NHBoc, Et, CO_2H 76% (7:1 dr)

Minter, A.R.; Fuller, A.A.; Mapp, A.K. *J. Am. Chem. Soc.* ***2003***, *125*, 6847.

Cyclohexyl-CHO → (H_2NCOMe, 60 atom CO, NMP, 120°C, 15 h, 10% HCl; 5% K_2PtCl_4, 10% PPh_3) COOH, NHCOMe 53%

Sagae, T.; Sugiura, M.; Hagio, H.; Kobayashi, S. *Chem. Lett.* ***2003***, *32*, 160.

SECTION 316: CARBOXYLIC ACID - AMINE

Ph–CH(NH_2)–CN → (*Rhodococcus* sp. AJ270, pH 7.62; phosphate buffer, 30°C, 4 h) → Ph–CH(NH_2)–$CONH_2$ 48% (>99% ee) + Ph–CH(NH_2)–CO_2H 50% (95% ee)

Wang, M.-X.; Lin, S.-J. *J. Org. Chem.* **2002**, *67*, 6542.

Ph–CH_2–C(=O)–CO_2H → ($BnNH_2$, H_2; cat Rh(P–P*)) → Ph–CH_2–CH(NHBn)–CO_2H 99% (98% ee)

Kadyrov, R.; Riermeier, T.H.; Dingerdissen, U.; Tararov, V.; Börner, A. *J. Org. Chem.* **2003**, *68*, 4067.

$BocNHCH_2CO_2t$-Bu → (5 $ZnBr_2$, DCM, rt, 1d) → $^{+}NH_3CH_2CO_2^{-}$ 80%

Kaul, R.; Brouillette, Y.; Sajjadi, Z.; Hansford, K.A.; Lubell, W.D. *J. Org. Chem.* **2004**, *69*, 6131.

quinoline-2-CO_2H → (1. 3 LTMP, THF, –78°C; 2. D_2O 3. H^+) → 3-D-quinoline-2-CO_2H 57%

Rebstock, A.-s.; Mongin, F.; Trécourt, F.; Quéguiner, G. *Tetrahedron Lett.* **2002**, *43*, 767.

3-Me_2N-toluene → (MeO–C_6H_4–$B(OH)_2$; HOOCCHO, dioxane, reflux, 1d) → (Me_2N, Me-aryl)–CH(CO_2H)–C_6H_4–OMe 50%

Naskar, D.; Roy, A.; Seibel, W.L.; Portlock, D.E. *Tetrahedron Lett.* **2003**, *44*, 5819.

Related Methods: Section 315 (Carboxylic Acid - Amide)
Section 344 (Amide - Ester)
Section 351 (Amine - Ester)

SECTION 317: CARBOXYLIC ACID - ESTER

NO ADDITIONAL EXAMPLES

SECTION 318: CARBOXYLIC ACID - ETHER, EPOXIDE, THIOETHER

NO ADDITIONAL EXAMPLES

SECTION 319: CARBOXYLIC ACID - HALIDE, SULFONATE

NO ADDITIONAL EXAMPLES

SECTION 320: CARBOXYLIC ACID - KETONE

t-BuO(C=O)CO_2Me; NaOMe, THF-DME, rt, 10 min → 95%

Jiang, X.-H.; Song, L.-D.; Long, Y.-Q. *J. Org. Chem.* ***2003***, *68*, 7555.

5% $Pd(OAc)_2$, 6% JOSIPHOS, Ph_2Zn, THF, 23 °C → 84% (95% ee)

Bercot, E.A.; Rovis, T. *J. Am. Chem. Soc.* ***2004***, *126*, 10248.

PhMgCl, toluene, (–)-sparteine; 9h, –78°C → 63% (88% ee)

Shintani, R.; Fu, G.C. *Angew. Chem. Int. Ed.* ***2002***, *41*, 1057.

Also via: Section 360 (Ketone - Ester).

SECTION 321: CARBOXYLIC ACID - NITRILE

NO ADDITIONAL EXAMPLES

Also via: Section 361 (Nitrile - Ester).

SECTION 322: CARBOXYLIC ACID - ALKENE

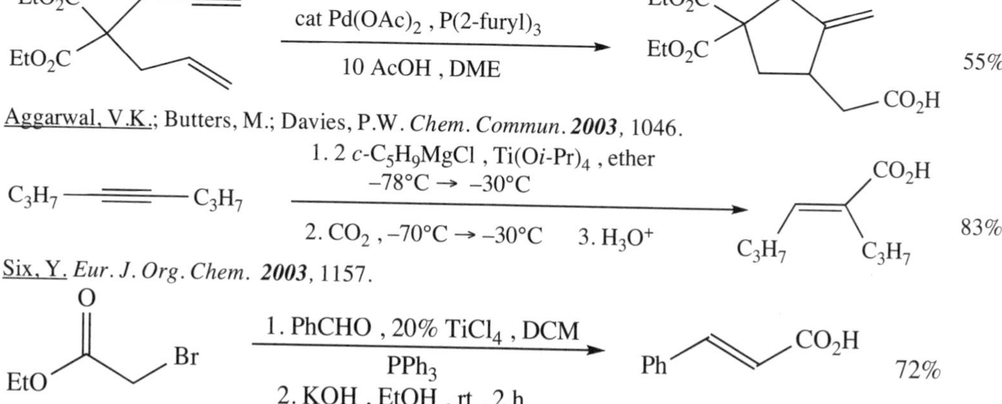

Aggarwal, V.K.; Butters, M.; Davies, P.W. *Chem. Commun.* ***2003***, 1046.

Six, Y. *Eur. J. Org. Chem.* ***2003***, 1157.

Basavaiah, D.; Rao, A.J. *Synth. Commun.* ***2002***, *32*, 195.

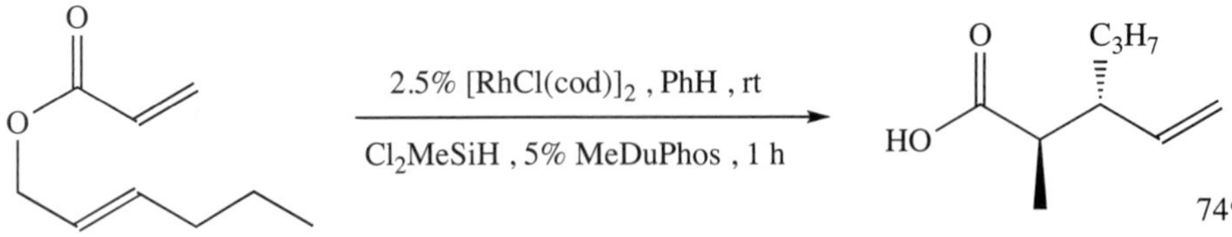

2.5% $[RhCl(cod)]_2$, PhH , rt

Cl_2MeSiH , 5% MeDuPhos , 1 h

74% (11:1 dr)

Miller, S.P.; Morken, J.P. *Org. Lett.* ***2002**, 4*, 2743.

REVIEW:

"New Aspects of the Ireland and Related Claisen Rearrangements"
Chai, Y.; Hong, S.-p.; Lindsay, H.A.; McFarland, C.; McIntosh, M.C. *Tetrahedron* ***2002**, 58*, 2905.

Also via: Section 313 (Alcohol - Carboxylic Acid)
Section 349 (Amide - Alkene)
Section 362 (Ester - Alkene)
Section 376 (Nitrile - Alkene)

SECTION 323: ALCOHOL, THIOL - ALCOHOL, THIOL

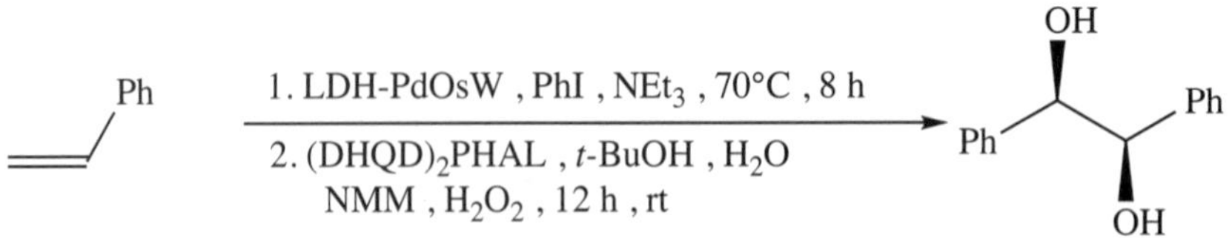

1. LDH-PdOsW , PhI , NEt_3 , 70°C , 8 h
2. $(DHQD)_2PHAL$, *t*-BuOH , H_2O
NMM , H_2O_2 , 12 h , rt

85% (99% ee)

Choudary, B.M.; Chowdari, N.S.; Madhi, S.; Kantam, M.L. *J. Org. Chem.* ***2003**, 68*, 1736.

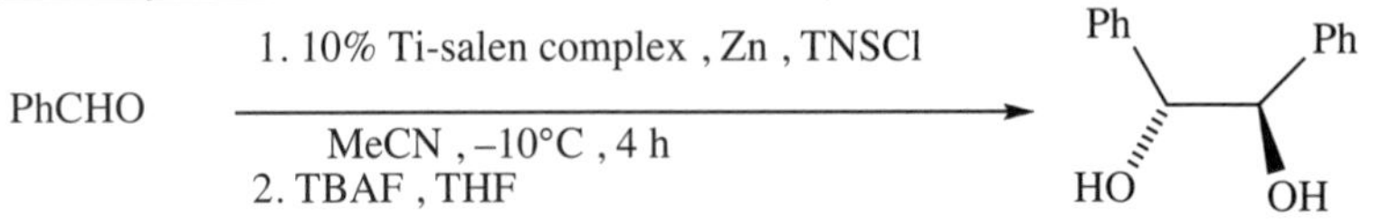

PhCHO

1. 10% Ti-salen complex , Zn , TNSCl
MeCN , –10°C , 4 h
2. TBAF , THF

94% (98:2 *dl* : *meso*)
95% ee

Chatterjee, A.; Bennur, T.H.; Joshi, N.N. *J. Org. Chem.* ***2003**, 68*, 5668.

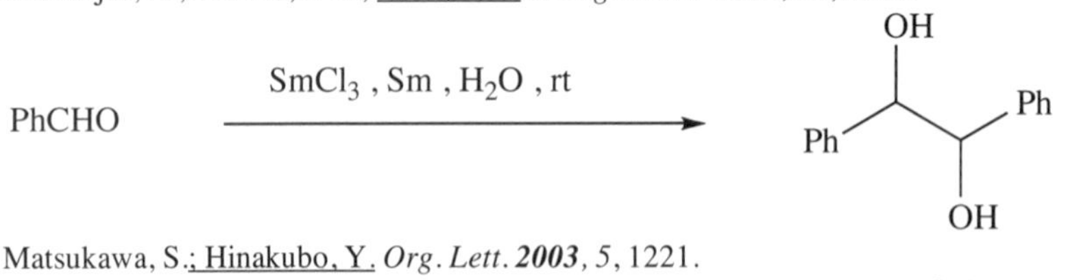

PhCHO

$SmCl_3$, Sm , H_2O , rt

81% (52:48 *dl:meso*)

Matsukawa, S.; Hinakubo, Y. *Org. Lett.* ***2003**, 5*, 1221.

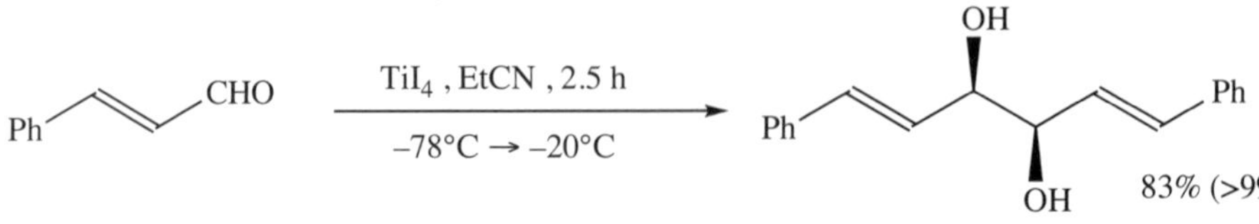

TiI_4 , EtCN , 2.5 h
–78°C → –20°C

83% (>99:1 *dl:meso*)

Shimizu, M.; Goto, H.; Hayakawa, R. *Org. Lett.* ***2002**, 4*, 4097.

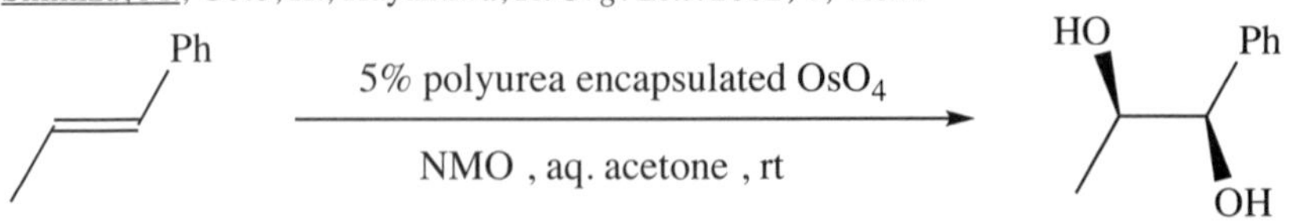

5% polyurea encapsulated OsO_4
NMO , aq. acetone , rt

80%

Ley, S.V.; Ramarao, C.; Lee, A.-L.; Østergaard, N.; Smith, S.C.; Shirley, I.M. *Org. Lett.* ***2003**, 5*, 185.

Ph–CH=CH–CO_2Me → (1.5 $NaIO_4$, 20% H_2SO_4, 0°C; EtOAc/MeCN/H_2O, 2 min) → PhCH(OH)CH(OH)CO_2Me 84%

Plietker, B.; Niggemann, M. *Org. Lett.* ***2003****, 5*, 3353.

Ph–CH=CH_2 → (NaCl, $(DHQD)_2PHAL$, aq. *t*-BuOH; $K_2OsO_4 \cdot 2\ H_2O$, pH 10.9, 1M $NaCoO_2$) → PhCH(OH)CH_2OH 73% (96% ee, *R*)

Junttila, M.H.; Hormi, O.E.O. *J. Org. Chem.* ***2004****, 69*, 4816.

t-BuC(O)CH_2CH(OH)CH_3 → (SmI_2, DME, MeOH) → *t*-BuCH(OH)CH_2CH(OH)CH_3 (15 : 1 *syn:anti*)

Chopade, P.R.; Davis, T.A.; Prasad, E.; Flowers II, R.A. *Org. Lett.* ***2004****, 6*, 2685.

Ph–C≡CH → (1. cat $[PtCl_2(C_2H_4)]_2$, $HSiCl_3$; 2. chiral $[PdCl(\pi\text{-}C_3H_5)]_2$, $HSiCl_3$, chiral phosphine; 3. H_2O_2, KF, $KHCO_3$) → PhCH(OH)CH_2OH 87% (95% ee)

Shimada, T.; Mukaide, K.; Shinohara, A.; Han, J.W.; Hayashi, T. *J. Am. Chem. Soc.* ***2002****, 124*, 1584.

Ph–CH=CH–CH_3 → (chiral Ru catalyst, THF; $(catecholato)B)_2$) → PhCH(OH)CH(OH)CH_3 71% (93% ee)

Morgan, J.B.; Miller, S.P.; Morken, J.P. *J. Am. Chem. Soc.* ***2003****, 125*, 8702.

cyclopentene oxide → (epoxide hydrolase) → *trans*-cyclopentane-1,2-diol 90% ee

Zhao, L.; Han, B.; Huang, Z.; Miller, M.; Huang, H.; Malashock, D.S.; Zhu, Z.; Milan, A.; Robertson, D.E.; Weiner, D.P.; Burk, M.J. *J. Am. Chem. Soc.* ***2004****, 126*, 11156.

2 PhCHO → (1. 3% TBO_xCrCl, Mn, TESCl, MeCN, rt; 2. H^+) → PhCH(OH)CH(OH)Ph 94% (98:2 dl:meso) 97% ee

Takenaka, N.; Xia, G.; Yamamoto, H. *J. Am. Chem. Soc.* ***2004****, 126*, 13198.

OHC–CHO → (2 allyl bromide, $SnCl_2$, KI, H_2O) → CH_2=CHCH_2CH(OH)CH(OH)CH_2CH=CH_2 85%

Samoshin, V.V.; Gremyachinskiy, D.E.; Smith, L.I.; Bliznets, I.V.; Gross, P.H. *Tetrahedron Lett* ***2002****, 43*, 6329.

Ph–CH=CH_2 → (OsO_4, [emim] BF_4, NMO•H_2O, rt, 18h) → PhCH(OH)CH_2OH 91%

Yanada, R.; Takemoto, Y. *Tetrahedron Lett.* ***2002****, 43*, 6849.

NHCOCCl$_3$ (cyclopentene) — 1. OsO_4 , TMEDA , DCM , –78°C; 2. HCl , MeOH → NHCOCCl$_3$, OH, OH 80% (>25:1)

Donohoe, T.J.; Blades, K.; Moore, P.R.; Waring, M.J.; Winter, J.J.G.; Helliwell, M.; Newcombe, N.J.; Stemp, G. *J. Org. Chem.* ***2002***, *67*, 7946.

O — 1. [2.2 Sm/$Br_2CHCHBr_2$, THF , rt , 4h] THF , 20°C; 2. H_3O^+ → OH, HO 74%

Hélion, F.; Lannou, M.-I.; Namy, J.-L. *Tetrahedron Lett.* ***2003***, *44*, 5507.

Ph — fluorous OsO_4 , NMO , Fe-77 , rt; *t*-BuOH/acetone/H_2O → OH, Ph, OH 97%

Huang, Y; Meng, W.-D.; Qing, F.L. *Tetrahedron Lett.* ***2004***, *45*, 1965.

O, Ph — 1. acetone , Mg , TMSCl; 2. 5% H_2SO_4 → HO, Ph, OH 70%

Maekawa, H.; Yamamoto, Y.; Shimada, H.; Yonemura, K.; Nishiguchi, I. *Tetrahedron Lett.* ***2004***, *45*, 3869.

2 PhCHO — cat $NiCl_2$, Mg , TMSCl , THF , rt → OH, Ph, Ph, OH 70% (78:22 *dl:meso*)

Shi, L.; Fan, C.-A.; Tu, Y.-Q.; Wang, M.; Zhang, F.-M. *Tetrahedron* ***2004***, *60*, 2851.

O — 10% Bu_4NHSO_4 , H_2O , 45°C → OH, OH 97%

Fan, R.-H.; Hou, X.-L. *Org. Biomol. Chem.* ***2003***, *1*, 1565.

O, O, Ph — $NaBH_4$, albumin → OH, OH, Ph 93 : OH, OH, Ph 7

Benedetti, F.; Berti, F.; Donati, I.; Fregonese, M. *Chem. Commun.* ***2002***, 828.

Ph — bmim PF_6 , *t*-BuOH-H_2O , 3 $K_3Fe(CN)_6$, rt; cat $K_2OsO_2(OH)_4$, $(DHQD)_2PHAL$ 3 K_2CO_3 , 1d → OH, Ph, OH 88% (97% ee)

Branco, L.C.; Afonso, C.A.M. *Chem. Commun.* ***2002***, 3036.

Ph — NMO , H_2O , toluene , 60°C , 12h; cat Os/Cu-Al-hydrotalcite → OH, Ph, OH quant

Friedrich, H.B.; Govender, M.; Makhoba, X.; Ngcoo, T.D.; Onani, M.O. *Chem. Commun.* ***2003***, 2922.

Ts N O O

2.5 SmI_2 2 *t*-BuOH , THF

−78°C → rt

Ts — N OH OH 62%

Handa, S.; Kachala, M.S.; Lowe, S.R. *Tetrahedron Lett.* ***2004***, *45*, 253.

Ph

bmim PF_6 , acetone , cat flavin , TEAA

cat K_2OsO_4 , cat NMM DMAP

OH Ph OH 85%

Peña, D.; López, F.; Harutyuyan, S.R.; Minnaard, A.J.; Feringa, B.L. *Chem. Commun.* ***2004***, 1494.

2 PhCHO

$TiCl_4(thf)_2$, chiral Schiff base

TMSCl , Mn

OH Ph Ph OH 95% (93:7 *dl:meso*) 88%ee

Li, Y.-G.; Tian, Q.-S.; Zhao, J.; Feng, Y.; Li, M.-J.; You, T.-P. *Tetrahedron: Asymmetry* ***2004***, *15*, 1707.

30% aq. H_2O_2 , 70°C , 20h

Nafion resin supported sulfonic acid catalyst

OH OH 67%

Usui, Y.; Sato, K.; Tanaka, M. *Angew. Chem. Int. Ed.* ***2003***, *42*, 5623.

t-Bu CHO

1. *t*-BuCHO , 2 TMSCl , 2 Mn
cat $CrCl_2$, DMF
2. 2 TBAF , THF

OH *t*-Bu *t*-Bu OH 61% (>98:2 *syn:anti*)

Jung, M.; Groth, U. *Synlett* ***2002***, 2015.

O OH

$AlEt_3$, THF

OH OH 58%

Bin, X.; Wu, B.; Zhao, X.Z.; Jia, Y.X.; Tu, Y.Q.; Li, D.R. *Synlett* ***2003***, 623.

PhCHO

Mg , 0.1 M aq. NH_4Cl , rt

ultrasound

OH Ph Ph OH 75% (55:45 dl:meso)

Li, J.-T.; Bian, Y.-J.; Zang, H.-J.; Li, T.-s. *Synth. Commun.* ***2002***, *32*, 547.

PhCHO

Al-NaOH , MeOH , rt , 20 min

ultrasound

OH Ph Ph OH 91% (51:49 *dl:meso*)

Bian, Y.-J.; Liu, S.-M.; Li, J.-T.; Li, T.-S. *Synth. Commun.* ***2002***, *32*, 1169.

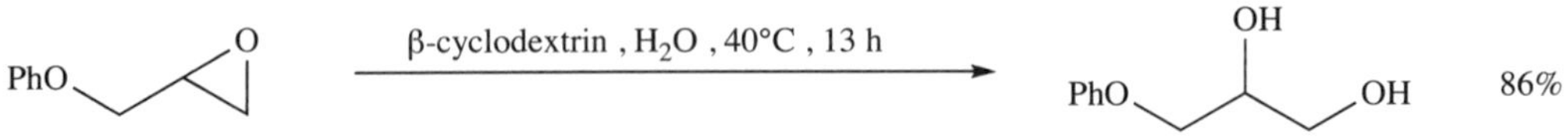

PhO–epoxide → β-cyclodextrin , H_2O , 40°C , 13 h → PhO–CH₂CH(OH)CH₂OH 86%

Reddy, M.A.; Reddy, L.R.; Bhanumathi, N.; Rao, K.R. *Org. Prep. Proceed. Int.* **2002**, *34*, 537.

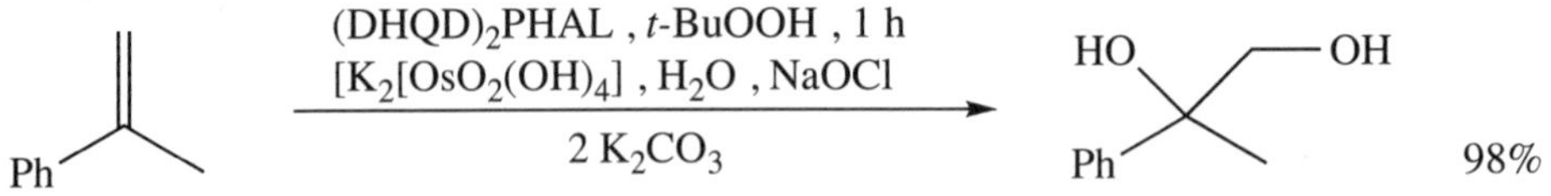

(DHQD)$_2$PHAL , *t*-BuOOH , 1 h
[$K_2[OsO_2(OH)_4]$] , H_2O , NaOCl
2 K_2CO_3
98%

Mehltretter, G.M.; Bhor, S.; Klawonn, M.; Döbler, C.; Sundermeier, U.; Eckert, M.; Militzer, H.-C.; Beller, M. *Synthesis* **2003**, 295.

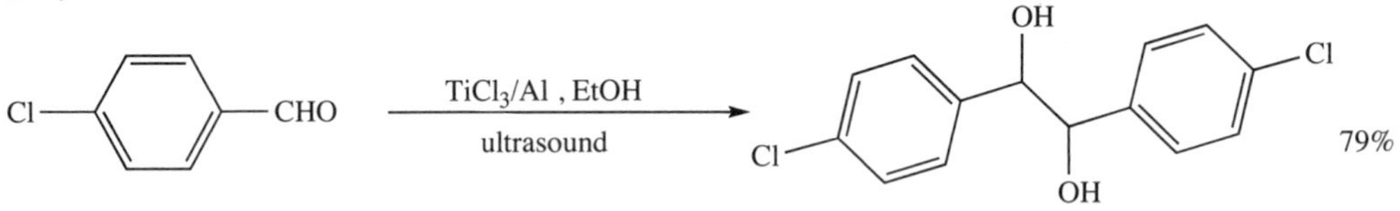

$TiCl_3$/Al , EtOH
ultrasound
79%

Li, J.-T.; Lin, Z.-P.; Qi, N.; Li, T.-S. *Synth. Commun.* **2004**, *34*, 4339.

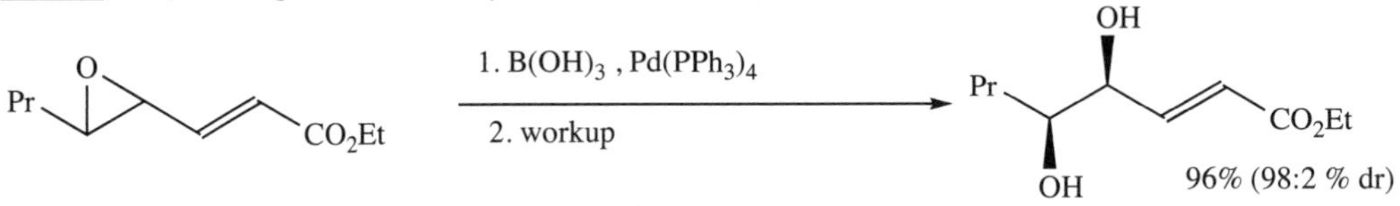

1. $B(OH)_3$, $Pd(PPh_3)_4$
2. workup
96% (98:2 % dr)

Yu, X.-Q.; Hirai, A.; Miyashita, M. *Chem Lett.* **2004**, *33*, 764.

Also via: Section 327 (Alcohol - Ester)
Section 357 (Ester - Ester)

SECTION 324: ALCOHOL, THIOL - ALDEHYDE

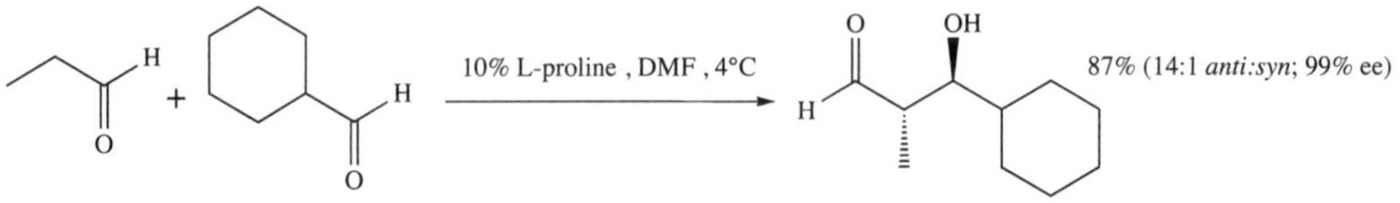

10% L-proline , DMF , 4°C
87% (14:1 *anti:syn*; 99% ee)

Northrup, A.B.; MacMillan, D.W.C. *J. Am. Chem. Soc.* **2002**, *124*, 6798.

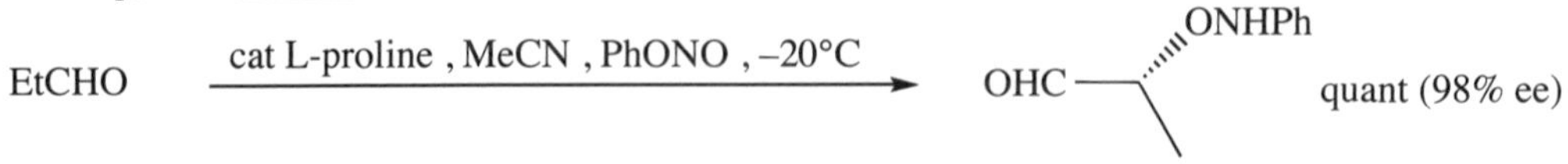

EtCHO → cat L-proline , MeCN , PhONO , –20°C → OHC–CH(ONHPh)CH₃ quant (98% ee)

Hayashi, Y.; Yamaguchi, J.; Hibino, K.; Shoji, M. *Tetrahedron Lett.* **2003**, *44*, 8293.

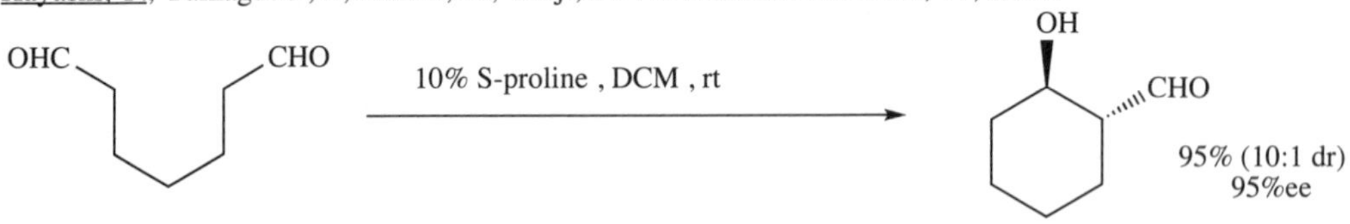

10% S-proline , DCM , rt
95% (10:1 dr)
95%ee

Pidathala, C.; Hoang, L.; Vignola, N.; List, B. *Angew. Chem. Int. Ed.* **2003**, *42*, 2785.

Related Method: Section 330 (Alcohol - Ketone).

SECTION 325: ALCOHOL, THIOL - AMIDE

PhCHO , SmI_2 , THF

NHTs Ph OH 89% (78:22 *syn:anti*)

Tanaka, Y.; Tankiguchi, N.; Uemura, M. *Org. Lett.* ***2002***, *4*, 835.

10% $MgBr_2$•OEt_2 , PhCHO 2 NEt_3 , 1.5. TMSCl 0.4 M , EtOAc , 1 d

85% (19:1 dr)

Evans, D.A.; Downey, C.W.; Shaw, J.T.; Tedrow, J.S. *Org. Lett.* ***2002***, *4*, 1127.

Me_2Zn , cat Ni(cod)$_2$PBu$_3$

70% (>97:3 dr)

Montgomery, J.; Song, M. *Org. Lett.* ***2002***, *4*, 4005.

$PdCl_2(PPh_3)_2$, NEt_3 DMF-H_2O , 80°C

HO–C6H4–CHO 75%

Liu, L.; Wang, X.; Li, C. *Org. Lett.* ***2003***, *5*, 361.

1. $TiCl_4$, (–)-sparteine 2. $TiCl_4$, butanal

74% (97:3 R:S)

Crimmins, M.T.; McDougall, P.J. *Org. Lett.* ***2003***, *5*, 591.

acetone , CbzN=NCbz 20% L-proline , MeCN (0.33M) , rt , 96 h

(45 : 55) 83%

Chowdari, N.S.; Ramachary, D.B.; Barbas III, C.F. *Org. Lett.* ***2003***, *5*, 1685.

4 SmI_2 MeOH

C_7H_{15} ... NMe_2 OH 71%

Concellón, J.M.; Bardales, E. *Org. Lett.* ***2003***, *5*, 4783.

1. 2.5 *i*-PrMgBr, 5% *i*-Pr_2NH
THF, 50°C, 16h
2. PhCHO

SO_2Ph OH Ph 57%

Dinsmore, A.; Billing, D.G.; Mandy, K.; Michael, J.P.; Mogano, D.; Patil, S. *Org. Lett.* ***2004**, 6*, 293.

F_3C OMe F Br Me

PhCHO, Et_2AlCl
DCM, PPh_3
rt, 1d

OH O Ph OMe F CF_3 Me (67 : 33) 84%

Sato, K.; Sekiguchi, T.; Ishihara, T.; Konno, T.; Yamanaka, H. *J. Org. Chem.* ***2004**, 69*, 5041.

2.2 kinetic resolution

$BocNH_2$, air, Co(III) salen catalyst
4-nitrobenzoic acid, TBME, rt, 1 d

OH NHBoc 99% based on amine (>99% ee)

Bartoli, G.; Bosco, M.; Carlone, A.; Locatelli, M.; Melchiorre, P.; Sambri, L. *Org. Lett.* ***2004**, 6*, 3973.

1. MeO–C6H4–CHO
TMSCl, $MgCl_2$
NEt_3, EtOAc
2. TFA, MeOH

Bn OH OMe 91% (32:1 dr)

Evans, D.A.; Tedrow, J.S.; Shaw, J.T.; Downey, C.W. *J. Am. Chem. Soc.* ***2002**, 124*, 392.

1. CbzN=NCbz, 10% S-proline, MeCN
2. $NaBH_4$

HO Cbz N NHCbz 99% (96% ee)

List, B. *J. Am. Chem. Soc.* ***2002**, 124*, 5656.

Ph P Ph O

O OMe OH, cat Et_2Zn
BINOL derivative, MS 3Å
THF, –20°C, 9 h

$Ph_2(O)PHN$ O OMe OH 98% (96:4 *anti:syn*; 98% ee)

Matsunaga, S.; Kumagai, N.; Harada, S.; Shibasaki, M. *J. Am. Chem. Soc.* ***2003**, 125*, 4712.

CHO

1. *t*-BuN≡C, $SiCl_4$, –74°C
chiral bis(phosphoramide)
2. sat'd aq. $NaHCO_3$

OH NH*t*-Bu O 91% (99.9:0.1 er)

Denmark, S.E.; Fan, Y. *J. Am. Chem. Soc.* ***2003**, 125*, 7825.

OTMS F_3C N OMe TMSO

PhCHO, toluene, *t*-BuOMe, 12 h
$Zr(OTf)_4$/(R)-BINOL, PrOH, H_2O
–20°C

OH Ph CO_2Me $NHCOCF_3$ (43 : 57) 46% ee 71%

Kobayashi, J.; Nakamura, M.; Mori, Y.; Yamashita, Y.; Kobayashi, S. *J. Am. Chem. Soc.* ***2004**, 126*, 9182.

1. (salicylaldoxime), cyclohexane, 23°C; Al salen catalyst, 2 d
2. $H_2/Pd(OH)_2/C$, AcOH, EtOH, 23°C

NHBz → OH O NHBz 90% (97% ee)

Möllerstedt, H.; Piqueras, M.C.; Crespo, R.; Ottosson, H. *J. Am. Chem. Soc.* ***2004***, *126*, 13938.

N — Ts → 10% CAN, aq. MeCN → NHTs, OH 95%

Chandrasekhar, S.; Natrsihmulu, Ch.; Sultana, S.S. *Tetrahedron Lett.* ***2002***, *43*, 7361.

EtO CHO; NHCbz, Ph, 10% $Cu(OTf)_2$/chiral diamine; DCM, 0°C → EtO, OH, Ph, N–Cbz 93%

Matsubara, R.; Vital, P.; Nakamura, Y.; Kiyohara, H.; Kobayashi, S. *Tetrahedron* ***2004***, *60*, 9769.

Ph, N*i*-Pr_2; SmI_2, THF, rt, 2h → OH, N*i*-Pr_2, Ph 91%

Concellón, J.M.; Bernad, P.L.; Bardales, E. *Chem. Eur. J.* ***2004***, *10*, 2445.

Me, N, O, Me_2HC; BuLi, TFH, –78°C, 2 h → Bu, N, Me, $CHMe_2$, OH 93%

Jones, S.; Norton, H.C. *Synlett* ***2004***, 338.

Boc, N; 2.2 SmI_2, HMPA, THF; 2 *t*-BuOH, rt, overnight → HO, N, Boc 66%

Hölemann, A.; Reissig, H.-U. *Synlett* ***2004***, 2732.

N, OPMB; 1. LiHMDS, THF, –78°C; 2. PhCHO → OH, Ph, N, OPMB 93% (4:1 *anti:syn*)

Williams, D.R.; Donnell, A.F.; Kammler, D.C. *Heterocycles* ***2004***, *62*, 167.

SECTION 326: ALCOHOL, THIOL - AMINE

$PhNH_2$, MeCN , 82°C , 12h; 5% $ZnCl_2$ — 93%

Pachón, L.D.; Gamez, P.; van Brussel, J.J.M.; Reedijk, J. *Tetrahedron Lett.* **2003**, *44*, 6025.

$PhNH_2$ 5% $ZrCl_4$, neat , rt — quant

Chakraborti, A.K.; Kondaskar, A. *Tetrahedron Lett.* **2003**, *44*, 8315.

aq. NH_4OH , pressure tube; *i*-PrOH , 85°C , 15h — 95%

Pastó, M.; Rodriguez, B.; Riera, A.; Pericàs, M.A. *Tetrahedron Lett.* **2003**, *44*, 8369.

1. LDA , THF , –80°C; 2. (aldehyde) CHO; 3. NH_4Cl , NH_4OH –80°C — 60% (69:31 dr)

Leclerc, E.; Vrancken, E.; Mangeney, P. *J. Org. Chem.* **2002**, *67*, 8928.

polymethylhydrosiloxane THF , TBAF — 80% (*syn* only)

Nadkarni, D.; Hallissey, J.; Mojica, C. *J. Org. Chem.* **2003**, *68*, 594.

(3-iodonitrobenzene) , 2.5% CuI , 16 h; ethylene glycol , 2 K_3PO_4 *i*-PrOH , 75°C — 73%

Job, G.E.; Buchwald, S.L. *Org. Lett.* **2002**, *4*, 3703.

$PhNH_2$, DCM , cat $BiCl_3$ — 78%

Ollevier, T.; Lavie-Compin, G. *Tetrahedron Lett.* **2002**, *43*, 7891.

$BnNH_2$, 0.1 $LiNTf_2$, DCM; 20h , rt — 95%

Cossy, J.; Bellosta, V.; Hamoir, C.; Desmurs, J.-R. *Tetrahedron Lett.* **2002**, *43*, 7083.

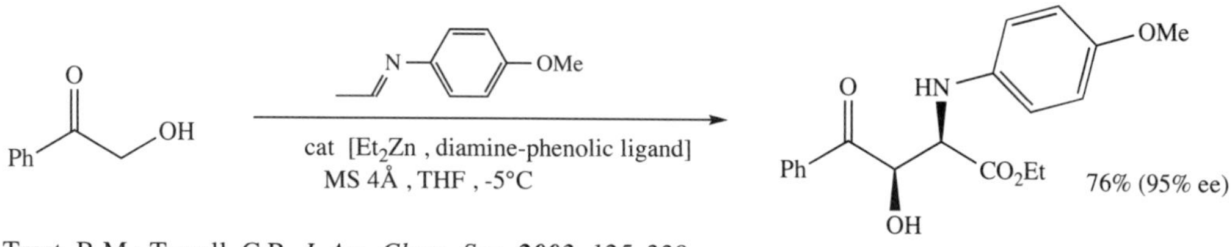
O
OH
Ph
N
OMe
cat [Et2Zn , diamine-phenolic ligand]
MS 4Å , THF , -5°C
O
HN
OMe
Ph
CO2Et
OH
76% (95% ee)

Trost, B.M.; Terrell, C.R. *J. Am. Chem. Soc.* ***2003***, *125*, 338.

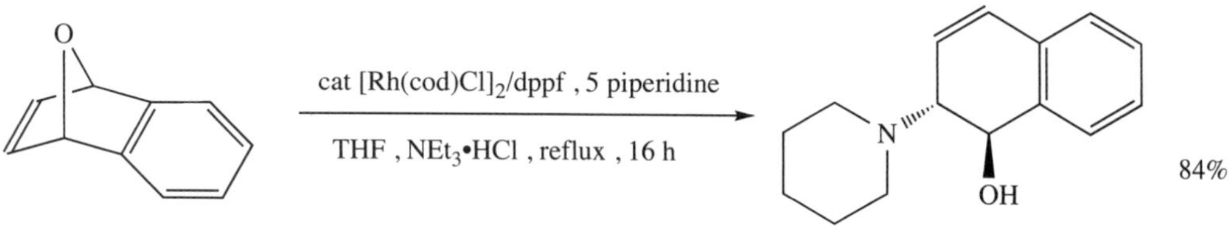
O
cat [Rh(cod)Cl]2/dppf , 5 piperidine
THF , NEt3•HCl , reflux , 16 h
N
OH
84%

Lautens, M.; Fagnou, K.; Yang, D. *J. Am. Chem. Soc.* ***2003***, *125*, 14884.

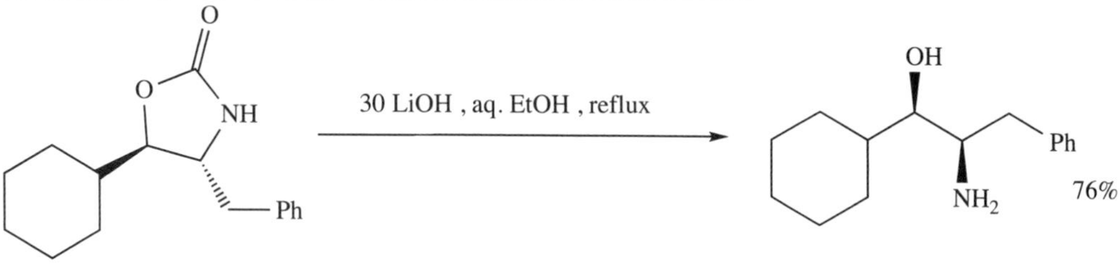
O
O
NH
Ph
30 LiOH , aq. EtOH , reflux
OH
Ph
NH2
76%

Katz, S.J.; Bergmeier, S.C. *Tetrahedron Lett.* ***2002***, *43*, 557.

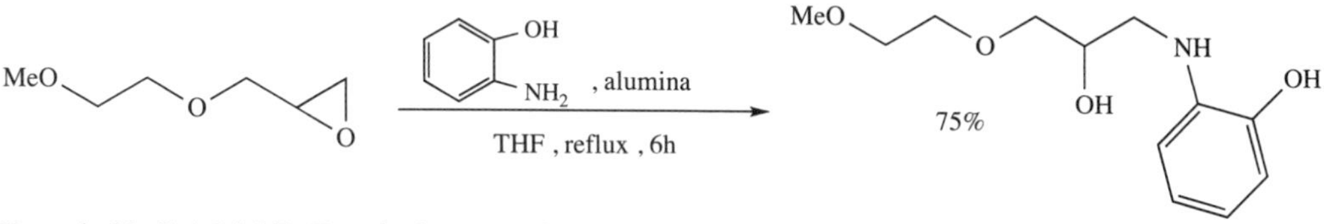
MeO
O
O
OH
NH2
, alumina
THF , reflux , 6h
MeO
O
NH
OH
OH
75%

Harrack, Y.; Pujol, M.D. *Tetrahedron Lett.* ***2002***, *43*, 819.

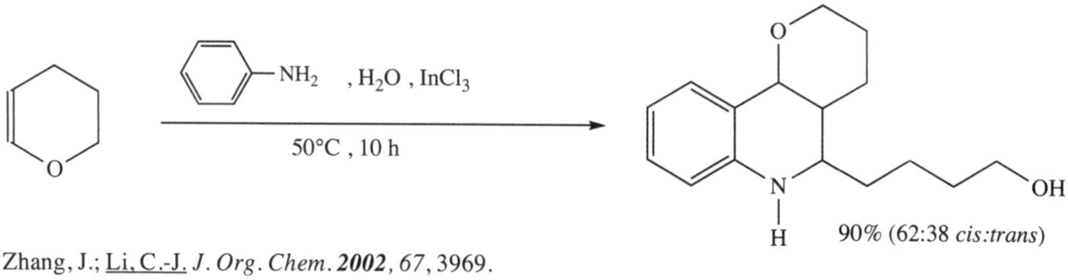
O
NH2
, H2O , InCl3
50°C , 10 h
O
N
H
OH
90% (62:38 cis:trans)

Zhang, J.; Li, C.-J. *J. Org. Chem.* ***2002***, *67*, 3969.

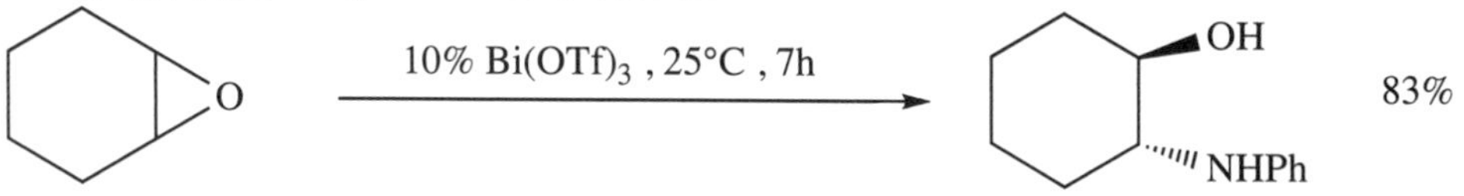
O
10% Bi(OTf)3 , 25°C , 7h
OH
NHPh
83%

Ollevier, T.; Lavie-Compin, G. *Tetrahedron Lett.* ***2004***, *45*, 49.

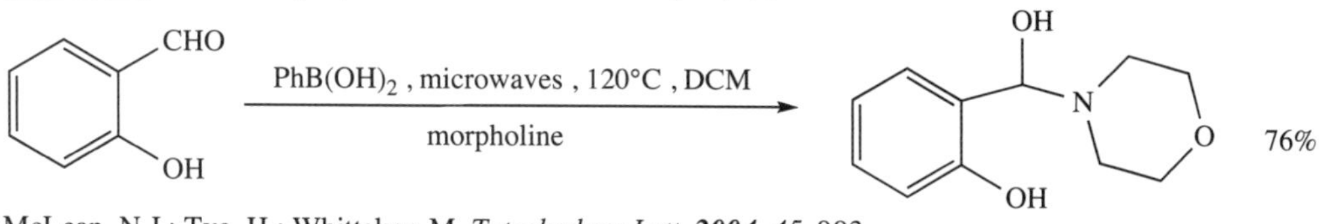
CHO
OH
PhB(OH)2 , microwaves , 120°C , DCM
morpholine
OH
N
O
OH
76%

McLean, N.J.; Tye, H.; Whittaker, M. *Tetrahedron Lett.* ***2004***, *45*, 993.

C_8H_{17} epoxide → t-Bu–C₆H₄–NH₂; cat $InBr_3$, DCM , rt → $C_8H_{17}CH(OH)CH_2NH$–C₆H₄–t-Bu 84%

Rodríguez, J.R.; Navarro, A. *Tetrahedron Lett.* ***2004***, *45*, 7495.

PhNH₂ , 10% $SmI_2(thf)_2$; DCM , rt 75%

Carrée, F.; Fil, R.; Collin, J. *Tetrahedron Lett.* ***2004***, *45*, 7749.

2,6-di-i-Pr-aniline; cat $CoCl_2$, MeCN , rt (12 : 88) 76%

Sundararajan, G.; Viyayakrishna, K.; Varghese, B. *Tetrahedron Lett.* ***2004***, *45*, 8253.

PhNH₂ , SiO_2 , neat , rt , 3h quant

Chakraborti, A.K.; Rudrawar, S.; Kondaskar, A. *Org. Biomol. Chem.* ***2004***, *2*, 1277.

chiral amino alcohol , H_2O ; pyrrolidine 90% (70% ee)

Steiner, D.; Sethofer, S.G.; Goralski, C.T.; Singaram, B. *Tetrahedron: Asymmetry* ***2002***, *13*, 1477.

1. Yb , THF-HMPA; 2. PhCH=N–C₆H₄–OMe, THF-HMPA no yield

Su, W.; Yang, B. *Synth. Commun.* ***2003***, *33*, 2613.

PhNH₂ , 1% $Sc(OTf)_3$; DCM , rt; 2 MeOH , 1% $Sc(OTf)_3$; DCM , rt 6%

Schneider, C.; Sreekanth, A.R.; Mai, E. *Angew. Chem. Int. Ed.* ***2004***, *43*, 5691.

PhCHO , 3 Zn-Cu , MeCN; 2 $BF_3{\bullet}OEt_2$, 4 TMSCl; 0°C → rt 66% (62:38 *syn:anti*) + 10%

Shimizu, M.; Iwata, A.; Makino, H. *Synlett* ***2002***, 1538.

$PhNH_2$, cat LiBr , neat , rt

OH, NHPh 70%

Chakraborti, A.K.; Rudrawar, S.; Kondaskar, A. *Eur. J. Org. Chem.* **2004**, 3597.

$PhNH_2$, 5% $BiCl_3$, 1 h / MeCN , rt

OH, NHPh 97%

Swamy, N.R.; Kondaji, G.; Nagaiah, K. *Synth. Commun.* **2002**, *32*, 2307.

Ph, CHO — $Ph_2C{=}NCH_2CO_2t$-Bu , toluene , 0°C , 2h / 1% aq NaOH , chiral ammonium salts

Ph, OH, CO_2t-Bu, NH_2 76% (3.3:1 *anti:syn*) 91%ee

Ooi, T.; Tankguchi, M.; Kameda, M.; Maruoka, K. *Angew. Chem. Int. Ed.* **2002**, *41*, 4542.

Bu, O — 5% $SnCl_4$•5 H_2O , $PhNH_2$, 50°C / solvent free , 12 h

OH, Bu, NHPh 89%

Zhao, P.-Q.; Xu, L.-W.; Xia, C.-G. *Synlett* **2004**, 846.

$PhNH_2$, DCM , 10% VCl_3 / rt , 2.5 h

OH, NHPh 87%

Sabitha, G.; Reddy, G.S.K.K.; Reddy, K.B.; Yadav, J.S. *Synthesis* **2003**, 2298.

EtO, O, OTMS, O + N-*p*-An, Ph, H — 2 TiI_4 , EtCN , rt

EtO, O, NH-*p*-An, Ph, OH 80% (64:36 *anti:syn*)

Shimizu, M.; Sahara, T. *Chem. Lett.* **2002**, *31*, 888.

Bi(OTf)$_3$, MeCN , $PhNH_2$ / microwaves , 105 sec

NHPh, OH 82%

Khosropour, A.R.; Khodaci, M.M.; Ghozati, K. *Chem. Lett.* **2004**, *33*, 304.

H, Me, N, Et, Ph, OH 65% (2:98 *syn:anti*) ← 1. TFA 2. Zn(BH$_4$)$_2$ — Boc, Me, N, Et, Ph, O — 1. $LiEt_3BH$ 2. TFA → H, Me, N, Et, Ph, OH quant (?99% *syn*)

Fraser, D.S.; Park, S.B.; Chong, J.M. *Can. J. Chem.* **2004**, *82*, 87.

PhO, O — $PhNH_2$, 5M $LiClO_4$/ether / rt , 30 min

PhO, OH, NHPh 97%

Heydari, A.; Mehrdad, M.; Malecki, A.; Ahmadi, N. *Synthesis* **2004**, 1563.

REVIEWS:

"The Sharpless Asymmetric Aminohydroxylation"
Bodkin, J.A.; McLeod, M.D. *J. Chem. Soc. Perkin Trans. 1* **2002**, 2733.

"Organolanthanide-Catalyzed Hydroamination"
Hong, S.; Marks, T.J. *Acc. Chem. Res.* **2004**, *37*, 673.

SECTION 327: ALCOHOL, THIOL - ESTER

cat. $CeCl_3$, Ac_2O
0.366 M , 0°C → rt
HO OH Ph Ph → HO OAc Ph Ph 95%

Clarke, P.A.; Kayaleh, N.E.; Smith, M.A.; Baker, J.R.; Bird, S.J.; Chan, C. *J. Org. Chem.* **2002**, *67*, 5226.

1. 2 eq *c*-hex$_2$BOTf , 2.4 NEt_3
−78°C , 2 h
2. EtCHO , −78°C → 0°C
Ph O N Bn SO_2Mes → Ph O OH Et N Bn SO_2Mes 90% (96:4 *anti:syn*)

Inoue, T.; Liu, J.-F.; Buske, D.C.; Abiko, A. *J. Org. Chem.* **2002**, *67*, 5250.

PhCHO
[Al_2O_3/$MeSO_3H$] , ethylene glycol
80°C , 4 h
O Ph O OH 80%

Sharghi, H.; Sarvari, M.H. *J. Org. Chem.* **2003**, *68*, 4096.

OHC O SEt
DBU , EtOH , DMAP , 78°C
DMAP•HCl , 1 h
OH O SEt 87%

Keck, G.E.; Welch, D.S. *Org. Lett.* **2002**, *4*, 3687.

O EtO Cl
i-PrCHO , Cp_2TiCl_2 , 6h
Mn , THF , rt
O OH EtO 88%

Parrish, J.D.; Shelton, D.R.; Little, R.D. *Org. Lett.* **2003**, *5*, 3615.

OH CO_2H
10% $B(OH)_3$
MeOH , 18 h , rt
OH CO_2Me 65%

Houston, T.A.; Wilkinson, B.L.; Blanchfield, J.T. *Org. Lett.* **2004**, *6*, 679.

O OPh
EtCHO , [(cod)RhCl]$_2$
6.5% *S*-BINAP
OH O OPh Et 76% (4.3:1 *syn:anti*) (88% ee *syn*)

Russell, A.E.; Fuller, N.O.; Taylor, S.J.; Aurriset, P.; Morken, J.P. *Org. Lett.* **2004**, *6*, 2309.

DABCO , DCM , 10 h , rt

62%

Krishna, P.R.; Kannan, V.; Sharma, G.V.M. *J. Org. Chem.* **2004**, *69*, 6467.

acetophenone , pyridine N-oxide

DCM , rt

quant

Denmark, S.E.; Fan, Y. *J. Am. Chem. Soc.* **2002**, *124*, 4233.

PhCHO , $SiCl_4$, DCM

5% phosphoramide , –78°C

97% (96.5:3.5 er)

Denmark, S.E.; Wynn, T.; Beutner, G.L. *J. Am. Chem. Soc.* **2002**, *124*, 13405.

PhCHO , 20% $Pr(OTf)_3$, aq. EtOH

24% chiral bis(piperidine), 18-crown-6
2,6-di-*t*-butylpyridine , 43 h

63% (95:5 *syn:anti*) (82% ee for *syn*)

Hamada, T.; Manabe, K.; Ishikawa, S.; Nagayama, S.; Shiro, M.; Kobayashi, S. *J. Am. Chem. Soc.* **2003**, *125*, 2989.

1. PhCHO , CuF•3 PPh_3•2 EtOH
$(EtO)_3SiF$, THF
2. 3 HF•NEt_3

95%

Oisaki, K.; Suto, Y.; Kanai, M.; Shibasaki, M. *J. Am. Chem. Soc.* **2003**, *125*, 5644.

3 BnOH , DCM , 30°C , 8% DIPEA , 15 h

10%

89% (?10:1 dr)

Chow, K.Y.-K.; Bode, J.W. *J. Am. Chem. Soc.* **2004**, *126*, 8126.

cat $Rh_2(cod)_2Cl_2$, H_2O

Ph_3Bi , 50°C

71%

Ding, R.; Ge, C.-S.; Chen, Y.-J.; Wang, D.; Li, C.-J. *Tetrahedron Lett.* **2002**, *43*, 7789.

10% $CeCl_3$, Ac_2O

23h

(85 : 15) 77%

Clarke, P.A. *Tetrahedron Lett.* **2002**, *43*, 4761.

CH_2=$CHCO_2t$-Bu → PhI , PhCHO , Me_2Zn , THF / 10% $Ni(cod)_2$, rt → Ph–CH(OH)–CH(Ph)–CO_2t-Bu 88% (86:14 dr)

Subburaj, K.; Montgomery, J. *J. Am. Chem. Soc.* **2003**, *125*, 11210.

HO(CH₂)₄OH → lipase from *Mucur mlehei* / CH_2=$CHCO_2Bz$, *t*-BuOMe / VB , 25°C , 4h → BzO(CH₂)₄OH + BzO(CH₂)₄OBz (82 : 18) 72%

Ciuffreda, P.; Casati, S.; Santaniello, E. *Tetrahedron Lett.* **2003**, *44*, 3663.

$HOCH_2CH_2OH$ → BzCl , cat $Cu(acac)_2$, reflux → $BnOCH_2CH_2OH$ 90%

Sirkecioglu, O.; Karliga, B.; Talinli, N. *Tetrahedron Lett.* **2003**, *44*, 8483.

Ph–CH₂CH₂–CHO → Me_2C=C(OTMS)OMe , (omim) Cl , rt , 12 h → Ph–CH₂CH₂–CH(OH)–C(Me)₂–CO_2Me 50%

Chen, S.-L.; Ji, S.-J.; Loh, T.-P. *Tetrahedron Lett.* **2004**, *45*, 375.

HO–CH₂–CH(OH)–CH₂CH₂–Ph → 1. 5% $Yb(OTf)_3$, 3 $MeC(OMe)_3$, MeCN , rt / 2. H_2O , –40°C → 0°C → AcO–CH₂–CH(OH)–CH₂CH₂–Ph + HO–CH₂–CH(OAc)–CH₂CH₂–Ph (17 : 83%) 96%

Ikejiri, M.; Miyashita, K.; Tsunemi, T.; Imanishi, T. *Tetrahedron Lett.* **2004**, *45*, 1243.

PhCHO → $BrCH_2CO_2Et$, Mn•THF , rt → Ph–CH(OH)–CH₂–CO_2Et 90%

Suh, Y.S.; Rieke, R.D. *Tetrahedron Lett.* **2004**, *45*, 1807.

$PhCO_2H$ → $HOCH_2CH_2OH$, Al_2O_3 , $MeSO_3H$, 80°C → $PhCO_2CH_2CH_2OH$ 94%

Sharghi, H.; Sarvari, M.H. *Tetrahedron* **2003**, *59*, 3627.

CH₃C(O)CH₂CO_2Me → Ru catalyst , bmim BF_4 , MeOH / H_2 → CH₃CH(OH)CH₂CO_2Me 94% ee

Ngo, H.L.; Hu, A.; Lin, W. *Chem. Commun.* **2003**, 1912.

CH₃C(O)CH₂COMe → MeOH , H_2 / mesoporous silcia-Ru catalyst → CH₃CH(OH)CH₂COMe 99% ee

Kesanli, B.; Lin, W. *Chem. Commun.* **2004**, 2284.

CH₃C(O)CH₂CO_2Me → H_2 , cat $[RuCl_2(benzene)]_2$, 15h / MeOH , 50°C → CH₃CH(OH)CH₂CO_2Me quant (99% ee)

Berthod, M.; Mignani, G.; Lemaire, M. *Tetrahedron: Asymmetry* **2004**, *15*, 1121.

PhC(O)CH₂CO_2Et → cat [Rh(L*)(benzene)Cl]Cl , H_2 / 65°C , EtOH , 20h / L* = biaryl diphosphine → PhCH(OH)CH₂CO_2Et 98% ee

Sun, Y.; Wan, X.; Guo, M.; Wang, D.; Dong, X.; Pan, Y.; Zhang, Z. *Tetrahedron: Asymmetry* **2004**, *15*, 2185.

Rhizopus arrhizus — 68% (99% ee)

Salvi, N.A.; Chattopadhyay, S. *Tetrahedron: Asymmetry* **2004**, *15*, 3397.

F_3C–CO–CO_2Et, $GaCl_3$ — 73%

Prakash, G.K.S.; Török, B.; Olah, G.A. *Synlett* **2003**, 527.

$NaBH_4$, EtOH — 65%

Padhi, S.K.; Chadha, A. *Synlett* **2003**, 639.

i-PrCHO, DCM, 0°C, 3 h; 10% Zr(O*t*-Bu)$_4$ / chiral diol — 86% (30% ee)

Schneider, C.; Hansch, M. *Synlett* **2003**, 837.

$HOCH_2CH_2OH$ —AcOH, $CHCl_3$, oxirane; HY-zeolite→ $HOCH_2CH_2OAc$ 90%

Srinkvas, K.V.N.S.; Mahender, I.; Das, B. *Synlett* **2003**, 2419.

$MeCO_2^-Na^+$, SDS solution; cat $Ce(OTf)_4$, rt — 87%

Iranpoor, N.; Firouzabadi, H.; Safavi, A.; Shekarriz, M. *Synth. Commun.* **2002**, *32*, 2287.

1. Bu_6CrLi_3, THF, –78°C, 1 h; 2. 1-phenyloxirane, Et_3Al, –78°C, 2 h — 69% + 23%

Hojo, M.; Sakata, K.; Maimaiti, X.; Ueno, J.; Nishikori, H.; Hosomi, A. *Chem. Lett.* **2002**, 142.

1. (OTMS, OMe ketene acetal), 25% Ph_2NH; 2. 20% MeLI, Py, 0°C, 7 h; 3. 1N HCl, THF, rt — 98%

Fujisawa, H.; Mukaiyama, T. *Chem. Lett.* **2002**, 182.

PhCHO — 1. ($OSiMe_3$, Me, OMe), 10% LiOAc, DCM, –45°C, 3 h; 2. 1N HCl, aq. THF, rt — 87% 1.8 :1 *syn:anti*

Nakagawa, T.; Fujisawa, H.; Mukaiyama, T. *Chem. Lett.* **2003**, *32*, 462.

O_2N–C$_6$H$_4$–CHO — ($OSiMe_3$, OMe), 10% LiOAc; 50:1 DMF:H_2O, –45°C, 3 h — 97%

Shintou, T.; Kikuchi, W.; Mukaiyama, T. *Chem. Lett.* **2003**, *32*, 696.

OTMS, OMe — 1. 1.5 piperidine; 2. 1.4 MeLi , THF , 0°C , 3 h → OH, O, Ph, OMe 67% (4:1 OH:OTMS)

Mukaiyama, T.; Fujisawa, H.; Nakagawa, T. *Helv. Chim. Acta* **2002**, *85*, 4518.

REVIEWS:

"The Reformatsky Reaction in Organic Synthesis: Recent Advances"
Ocampo, R.; Dolbier Jr. W.R. *Tetrahedron* **2004**, *60*, 9325.

"Boron-Mediated Aldol Reaction of Carboxylic Esters"
Abiko, A. *Acc. Chem. Res.* **2004**, *37*, 387.

Also via: Section 313 (Alcohol - Carboxylic Acid)

SECTION 328: ALCOHOL, THIOL - ETHER, EPOXIDE, THIOETHER

PhO, O — NaOPh , β-cyclodextrin , H_2O; 60°C → OH, PhO, OPh 94%

Surendra, K.; Krishnaveni, N.S.; Nageswar, Y.V.D.; Rao, K.R. *J. Org. Chem.* **2003**, *68*, 4994.

CF_3, O — 5 eq Oxone , aq. Na_2•EDTA; MeCN , 15 eq $NaHCO_3$, 1d; rt → O, OH, CF_3 74%

Wong, M.-K.; Chung, N.-W.; He, L.; Wang, X.-X.; Yan, Z.; Tang, Y.-C.; Yang, D. *J. Org. Chem.* **2003**, *68*, 6321.

Ph, O — PhSH , 10% $ZnCl_2$, 30°C; pH 7 , 20 min → OH, SPh, Ph quant (α:β = 84:16)

Fringuelli, F.; Pizzo, F.; Tortoioli, S.; Vaccaro, L. *J. Org. Chem.* **2003**, *68*, 8248.

OH — microencapsulated $VO(acac)_2$ (3.6%); 1.3 TBHP , hexane , rt , 2 h → O, OH 90%

Lattanzi, A.; Leadbeater, N.E. *Org. Lett.* **2002**, *4*, 1579.

O — 1% $Cu(BF_4)_2$•nH_2O; MeOH , DCM , rt → OH, OMe 94%

Barluenga, J.; Vázquez-Villa, H.; Ballesteros, A.; González, J.M. *Org. Lett.* **2002**, *4*, 2817.

OH — $Zr(Ot\text{-}Bu)_4$/L-DBTA → O, OH 92% (87% ee)

Okachi, T.; Murai, N.; Onaka, M. *Org. Lett.* **2003**, *5*, 85.

1. 2 eq SmI_2 , THF , CH_2I_2 2% NiI_2 , 0°C → rt
2. H_3O^+

HO HO 90%

Molander, G.A.; Brown, G.A.; de Graua, I.S. *J. Org. Chem.* **2002**, *67*, 3459.

OH

CH_2=CHCHO , PtO_2/H_2 DCM , 1 atm

Ru metathesis catalyst

OH O 60%

Cossy, J.; Bargiggia, F.; Bouz-Bouz, S. *Org. Lett.* **2003**, *5*, 459.

OH MeO

1. Na , NH_3 , *t*-BuOH , –30°C
2. PhCHO , TFA , DCM

O Ph OMe OH (93:7) 63%

Barry, C.St.J.; Crosby, S.R.; Harding, J.R.; Hughes, R.A.; King, C.D.; Parker, G.D.; Willis, C.L. *Org. Lett.* **2003**, *5*, 2429.

$C_{10}H_{21}$ O

1. *s*-BuLi , sparteine derivatives
2. PhCHO

$C_{10}H_{21}$ O Ph OH 75%

Hodgson, D.M.; Reynolds, N.J.; Coote, S.J. *Org. Lett.* **2004**, *6*, 4187.

Ph O Ph

1. TMEDA , BuLi , THF , –60°C
2. PhCHO

OH Ph O Ph Ph 56%

Florio, S.; Aggarwal, V.; Salomone, A. *Org. Lett.* **2004**, *6*, 4191.

O CO_2Me

Oxone , $NaHCO_3$, MeCN , H_2O

OH Me_2HCH_2C O CO_2Me 62% (1:10 *trans:cis*)

Wong, M.-K.; Chung, N.-W.; He, L.; Yang, D. *J. Am. Chem. Soc.* **2003**, *125*, 158.

CH_3CH_2CHO

1. $Zn(CH=CHEt)_2$, 3 Et_2Zn , 4% (N, OH, O ligand)
2. O_2 3. 20% $Ti(Oi\text{-}Pr)_4$

OH O 93% (60% ee)

Lurain, A.E.; Maestri, A.; Kelly, A.R.; Carroll, P.J.; Walsh, P.J. *J. Am. Chem. Soc.* **2004**, *126*, 13608.

MgI_2 , THF , 65 h

(2 : >98) 60%

Karikomi, M.; Watanabe, S.; Kimura, Y.; Uyehara, T. *Tetrahedron Lett.* **2002**, *43*, 1495.

$BnOCH_2CHO$, DCM
$TiBr_4$, 2,6-DTBP , –78°C

71% (ca. 1:1 OH)

Patterson, B.; Marumoto, S.; Rychnovsky, S.D. *Org. Lett.* **2003**, *5*, 3163.

1. chiral disulfonamide , $Ti(Oi\text{-}Pr)_4$, 40 h
$ZnMe_2$, hexane/toluene , rt
2. O_2 , 0°C → rt

81% (99% ee)

Jeon, S.-J.; Walsh, P.J. *J. Am. Chem. Soc.* **2003**, *125*, 9544.

MeO–C6H4–CHO , 5% phosphite , THF
5-20% BuLi , 1 h

quant (90% ee)

Linghu, X.; Potnick, J.R.; Johnson, J.S. *J. Am. Chem. Soc.* **2004**, *126*, 3070.

1. Cl–C6H4–CHO
$La(OTf)_3$•BINOL + BuLi , THF , rt
2. NaOMe , MeOH

Gnanadesikan, V.; Horiuchi, Y.; Ohshima, T.; Shibasaki, M. *J. Am. Chem. Soc.* **2004**, *126*, 7782.

1. BuLi
2. chiral bis(oxazoline) ligand , cumeme
–78°C
3. benzophenone

60% (85% ee)

Nakamura, S.; Aoki, T.; Ogura, T.; Wang, L.; Toru, T. *J. Am. Chem. Soc.* **2004**, *126*, 8916.

PhCHO , cat $Ce(OTf)_3$, H_2O
rt , 12h

60%

Keh, C.C.K.; Namboodiri, V.V.; Varma, R.S.; Li, C.-J. *Tetrahedron Lett.* **2002**, *43*, 4993.

cat 2-furyl-CMe_2OOH , DCM
Ti(O*i*-Pr)$_4$, L-DET , MS 4Å
(a renewable hydroperoxide)
45% (85% ee)

Lattanzi, A.; Tannece, P.; Scettri, A. *Tetrahedron Lett.* **2002**, *43*, 5629.

vanadyl akyl phosphonate modified silica
MeCN , *t*-BuOOH , 80°C , 6h
79%

Jurado-Gonzalez, M.; Sullivan, A.C. *Tetrahedron Lett.* **2004**, *45*, 4465.

Sn(IV)(tpp)(OTf)$_2$, MeOH
rt , 10 min
99%

Moghadam, M.; Tangestaninejad, S.; Mirkhani, V.; Shaibai, R. *Tetrahedron* **2004**, *60*, 6105.

t-BuOOH , toluene
cat VO(acac)$_2$, 3h
microwaves
(62 : 38) 63%

Torres, G.; Torres, W.; Prieto, J.A. *Tetrahedron* **2004**, *60*, 10245.

oxaborolidine catalyst/EtPhN(*i*-Pr)BH_3
THF , 25°C
94% (92% ee)

Cho, B.T.; Choi, O.K.; Kim, D.J. *Tetrahedron: Asymmetry* **2002**, *13*, 697.

Ti(O*i*-Pr)$_4$, *t*-BuOOH
39% (53:47 *syn:anti*)

Krasinski, A.; Jurczak, J. *Tetrahedron: Asymmetry* **2002**, *13*, 2075.

1. OsO_4 , DCM , TMEDA
2. H^+ , MeOH
60%

Donohoe, T.J.; Butterworth, S. *Angew. Chem. Int. Ed.* **2003**, *42*, 948.

cat Ru(III)-polymetalate , DCE , 20°C
2 eq 30% H_2O_2 , 6 h
95% (92:8 *threo:erythro*)

Adam, W.; Alsters, P.L.; Neumann, R.; Saha-Möller, C.; Sloboda-Rozner, D.; Zhang, R. *Synlett* **2002**, 2011.

5% $ZnCl_2$, PhSH , H_2O
pH 4.0 , 75 min
(49 : 51) 95%

Amantini, D.; Fringuelli, F.; Pizzo, F.; Tortioli, S.; Vaccaro, L. *Synlett* **2003**, 2292.

PhSH , cat. $InCl_3$, 3 h
90%

Yadav, J.S.; Reddy, B.V.S.; Baishya, G. *Chem. Lett.* **2002**, *31*, 906.

CHO, OH → ($Ph\text{-}CH{=}CH\text{-}NO_2$, 1.5 h; DABCO, 40°C) → OH, NO_2, O, Ph 98%

Yan, M.-C.; Jang, Y.-J.; Kuo, W.-Y.; Tu, Z.; Shen, K.-H.; Cuo, T.-S.; Ueng, C.-H.; Yao, C.-F. *Heterocycles* ***2002***. *57*, 1033.

SECTION 329: ALCOHOL, THIOL - HALIDE, SULFONATE

O, Ph, Me → ($BrCHFCO_2Et$, THF; $Zn/CeCl_3{\bullet}7\ H_2O$) → OH, O, Ph, Me, OEt, F 90% (42:58 *RR/SS:RS/SR*)

Ocampo, R.; Dolbier Jr. W.R.; Abboud, K.A.; Zuluaga, F. *J. Org. Chem.* ***2002***, *67*, 72.

MeO, CHO → (1. Cl, OTMS, OMe; chiral Ti catalyst, −78°C, toluene; 2. aq HCl) → MeO, Cl, HO, CO_2Me + MeO, Cl, HO, CO_2Me (6 : 94) 53%; 60% ee, 91%ee

Imashiro, R.; Kuroda, T. *J. Org. Chem.* ***2003***, *68*, 974.

Ph, C, CO_2H → (1. $2\ I_2$, $LiOAc{\bullet}2\ H_2O$, THF, 3 h; 2. THF-DMF, O_2, 40°C) → I, Ph, HO, O, O 87%

Ma, S.; Wu, B.; Shi, Z. *J. Org. Chem.* ***2004***, *69*, 1429.

Ph → ($NaIO_4$, LiBr, aq. MeCN; 30% aq. H_2SO_4) → OH, Ph, Br 91%

Dewkar, G.K.; Narina, S.V.; Sudalai, A. *Org. Lett.* ***2003***, *5*, 4501.

MeO_2C, Ph, O → ($POCl_3$, DMAP, rt, DCM) → MeO_2C, Ph, OH, Cl 73%

Sartillo-Piscil, F.; Quintero, L.; Villegas, C.; Santacruz-Juárez, E.; de Parrodi, C.A. *Tetrahedron Lett.* ***2002***, *43*, 15.

O, Ph, CHO → ($TiCl_4$, $BnNEt_3Cl$, DCM; 0°C) → O, OH, Ph, Cl 90%

Yagi, K.; Turitani, T.; Shinokubo, H.; Oshima, K. *Org. Lett.* ***2002***, *4*, 3111.

PhS allene, Me → 2 I_2, aq. acetone, 9h → PhS, I 61% (97:3 *Z:E*)

Ma, S.; Hao, X.; Huang, X. *Org. Lett.* ***2003***, *5*, 1217.

Ph, O, Ph → 1. $(TMNP)_2ZrCl_2$, $CDCl_3$, 6 h; 2. HCl, ether → OH, Ph, Ph, Cl (7 : 1) + OH, Ph, Ph, Cl 70%

Wang, L.-S.; Hollis, T.K. *Org. Lett.* ***2003***, *5*, 2543.

O, O, N, O, Ph → 1. Et_2AlI, DCM; 2. BnO–C6H4–CHO → HO, O, O, BnO, I, N, O, Ph 82%

Timmons, C.; Chen, D.; Cannon, J.F.; Headley, A.D.; Li, G. *Org. Lett.* ***2004***, *6*, 2075.

C_7H_{15}, O, O, $N(i\text{-}Pr)_2$ → SmI_2, THF, 25°C → I, O, C_7H_{15}, $N(i\text{-}Pr)_2$, OH 78%

Concellón, J.M.; Bardales, E.; Concellón, C.; García-Granda, S.; Díaz, M.R. *J. Org. Chem.* ***2004***, *69*, 6923.

Et_3SiOCH_2CHO → $SiMe_2Ph$, BF_3•OEt_2; 2,6-di-*t*-butyl-4-methylpyridine, DCM, –78°C → HO, O, $SiMe_2Ph$ 60% (5.4:1)

Angle, S.R.; El-Said, N.A. *J. Am. Chem. Soc.* ***2002***, *124*, 3608.

Ph, O → (bmim) PF_6, TMSCl, rt → Cl, Ph, OH 98%

Xu, L.-W.; Li, L.; Xia, C.-G.; Zhao, P.-Q. *Tetrahedron Lett.* ***2004***, *45*, 2435.

O_2N–C6H4–CHO → F_2C=C(OTMS)$SiPh_3$; $TiCl_4$, DCM, 0°C→rt → O_2N, OH, O, $SiPh_3$ 84%

Chung, W.J.; Ngo, S.C.; Higashiya, S.; Welch, J.T. *Tetrahedron Lett.* ***2004***, *45*, 5403.

Ph, O → cat. diamine compound, I_2, DCM; 0.17h, rt → OH, Ph, I 97%

Nikam, K.; Nashei, T. *Tetrahedron* ***2002***, *58*, 10259.

Ph, O → S, H_2N, NH_2, I_2, MeCN, 25°C; 0.83 h → OH, Ph, I 90%

Sharghi, H.; Eskandari, M.M. *Tetrahedron* ***2003***, *59*, 8509.

BH_3-SMe_2 , toluene , 110°C; diazaphospha-oxo ligand — 88% (91% ee)

Basavaiah, D.; Reddy, G.J.; Chandrashekar, V. *Tetrahedron: Asymmetry* ***2004**, 15*, 47.

naphthalene , $AlCl_3$, DCM; rt, 1 h — 47% (89:11 *syn:anti*)

Lin, J.; Kanazaki, S.; Kashino, S.; Tsuboi, S. *Synlett* **2002**, 899.

$Al(OtBu)_3$, 10% acetophenone; $PyHBr_3$, toluene , 60°C, 28h — 62%

Cami-Kobeci, G.; Williams, J.M.J. *Synlett* **2003**, 124.

KO*t*-Bu , THF/DMF — 95%

Jablonski, L.; Joubert, J.; Billard, T.; Langlois, B.R. *Synlett* **2003**, 230.

NaI , H_2O_2 , HBF_4 , aq. THF; 0°C → rt , 2 h — 94%

Barluenga, J.; Marco-Arias, M.; González-Bobes, F.; Ballesteros, A.; González, J.-M. *Chem. Eur. J.* ***2004**, 10,* 1677.

0.1 SmI_2 , 0.05 EtO_2CCH_2Br; 1.2 I_2 , THF , 25°C — 95%

Kwon, D.W.; Cho, M.S.; Kim, Y.H. *Synlett* **2003**, 959.

Bu_4NF_2-KHF_2 , 120°C — (9 : 10) 68%

Aikiyama, Y.; Fukihara, T.; Hara, S. *Synlett* **2003**, 1530.

polyvinylpyrrolidine/$SOCl_2$; DCM , 10 min , rt — 95%

Tamami, B.; Ghazi, I.; Mahdavi, H. *Synth. Commun.* ***2002**, 32*, 3725.

Br_2 , cat $PhNHNH_2$, DCM — 95%

Sharghi, H.; Eskadari, M.M. *Synthesis* **2002**, 1519.

NEt_3•3 HF; microwaves — 61%

Inagaki, T.; Fukuhara, T.; Hara, S. *Synthesis* **2003**, 1157.

SECTION 330: ALCOHOL, THIOL - KETONE

1. (morpholino enamine), hexane, rt
2. H^+

88%

Funabiki, K.; Matsunaga, K.; Nojiri, M.; Hashimoto, W.; Yamamoto, H.; Shibata, K.; Matsui, M. *J. Org. Chem.* **2003**, *68*, 2853.

$OSiCl_3$

PhCHO, DCM

cat chiral phosphoramidite
–78°C

(16 : 1) 85%
19:1 er

Denmark, S.E.; Pham, S.M. *J. Org. Chem.* **2003**, *68*, 5045.

$OSi(OMe)_3$

PhCHO, 5% KF, THF
5% 18-crown-6, –20°C

6% R-BINAP, 10% AgOTf, 6 h

78% (9 : 91 *syn:anti*)
93% ee

Wadamoto, M.; Ozasa, N.; Yanagisawa, A.; Yamamoto, H. *J. Org. Chem.* **2003**, *68*, 5593.

RuO_4, 2.5 $NaHCO_3$

5 Oxone, 10 min

94%

Plietker, B. *J. Org. Chem.* **2003**, *68*, 7123.

O_2, *t*-BuOH, 65°C, 1 d
5% $[Pd[(-)\text{-sparteine}]Cl_2$

50% Na_2CO_3, MS 3Å

73% (84.2% ee)

Mandal, S.K.; Sigman, M.S. *J. Org. Chem.* **2003**, *68*, 7535.

acetone, aq. β-cyclodextrin, 12h
IBX, rt

IBX = 2-iodoxybenzoic acid

90%

Surendra, K.; Krishnaveni, N.S.; Reddy, M.A.; Nageswar, Y.V.D.; Rao, K.R. *J. Org. Chem.* **2003**, *68*, 9119.

PhCHO

$OSiMe_3$, 5% $Sc(OTf)_3$, CO_2

$PEG(OMe)_2$, 50°C, 15 mPa, 3 h

72%

Komoto, I.; Kobayashi, S. *J. Org. Chem.* **2004**, *69*, 680.

PhCHO, cat $RuCl_2(Ph_3)_2$
$[bmim]PF_6$, 90°C

76 : 24 (73%)

Yang, X.-F.; Wang, M.; Varma, R.S.; Li, C.-J. *Org. Lett.* **2003**, *5*, 657.

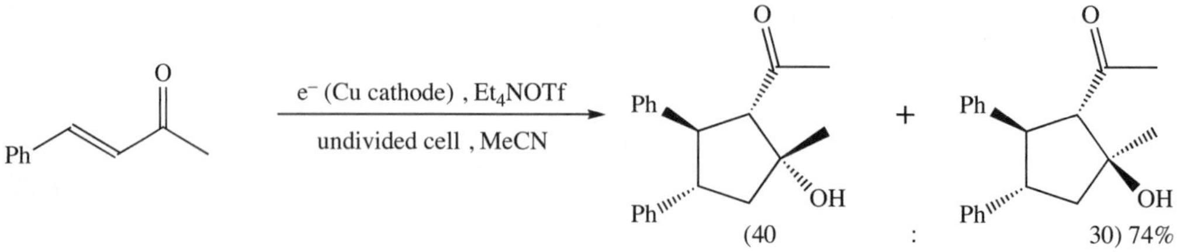

e⁻ (Cu cathode), Et_4NOTf; undivided cell, MeCN

Ph, Ph, OH (40 : 30) 74%

Kise, N.; Kitagishi, Y.; Ueda, N. *J. Org. Chem.* ***2004***, *69*, 959.

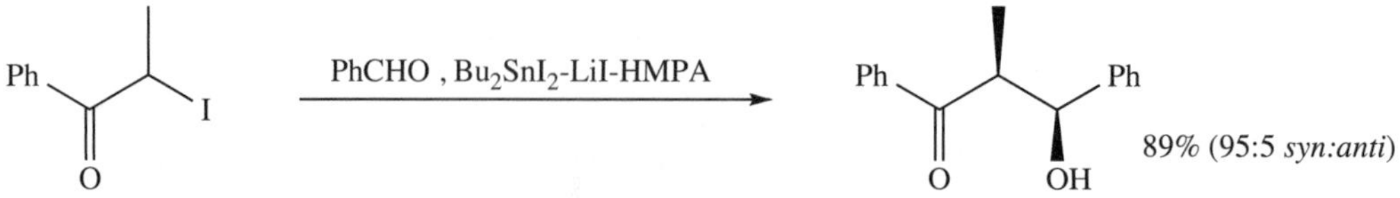

PhCHO, Bu_2SnI_2-LiI-HMPA

89% (95:5 *syn:anti*)

Shibata, I.; Suwa, T.; Sakakibara, H.; Baba, A. *Org. Lett.* ***2002***, *4*, 301.

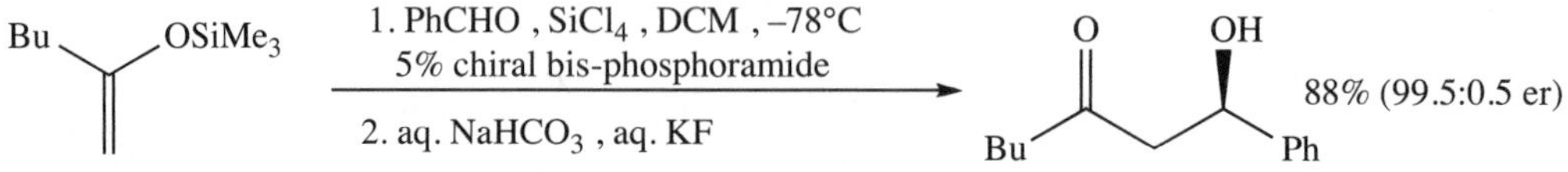

Bu, $OSiMe_3$ — 1. PhCHO, $SiCl_4$, DCM, −78°C; 5% chiral bis-phosphoramide; 2. aq. $NaHCO_3$, aq. KF

88% (99.5:0.5 er)

Denmark, S.E.; Heemstra Jr. J.R. *Org. Lett.* ***2003***, *5*, 2303.

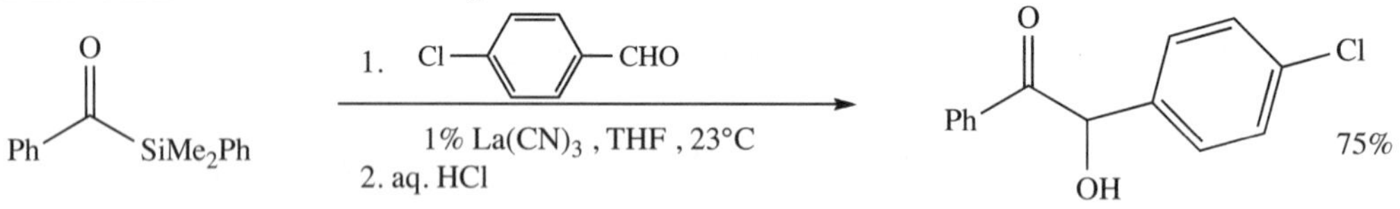

Ph, $SiMe_2Ph$ — 1. Cl–C₆H₄–CHO; 1% $La(CN)_3$, THF, 23°C; 2. aq. HCl

75%

Bausch, C.C.; Johnson, J.S. *J. Org. Chem.* ***2004***, *69*, 4283.

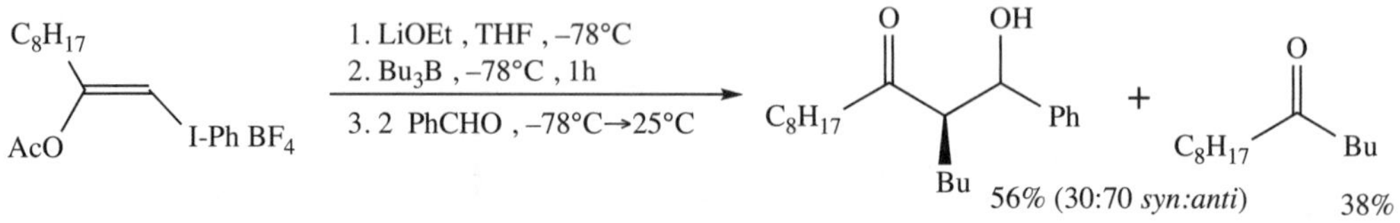

C_8H_{17}, AcO, I-Ph BF_4 — 1. LiOEt, THF, −78°C; 2. Bu_3B, −78°C, 1h; 3. 2 PhCHO, −78°C→25°C

C_8H_{17}, OH, Ph, Bu 56% (30:70 *syn:anti*) + C_8H_{17}, Bu 38%

Ochiai, M.; Tuchimoto, Y.; Higasiura, N. *Org. Lett.* ***2004***, *6*, 1505.

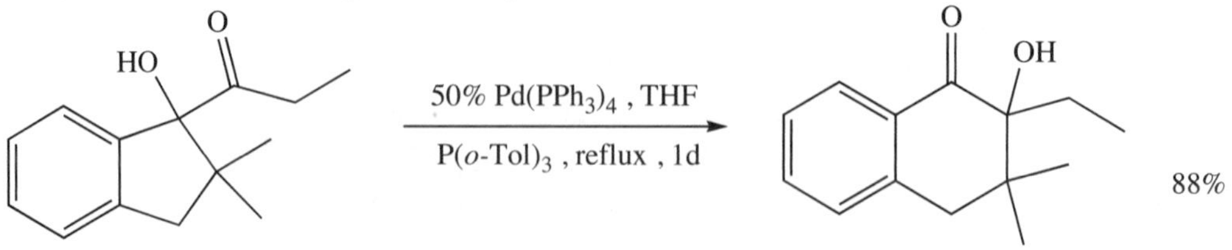

HO — 50% $Pd(PPh_3)_4$, THF; $P(o\text{-}Tol)_3$, reflux, 1d

OH 88%

Nagao, Y.; Tanaka, S.; Ueki, A.; Kumazawa, M.; Goto, S.; Ooi, T.; Sano, S.; Shiro, M. *Org. Lett.* ***2004***, *6*, 2133.

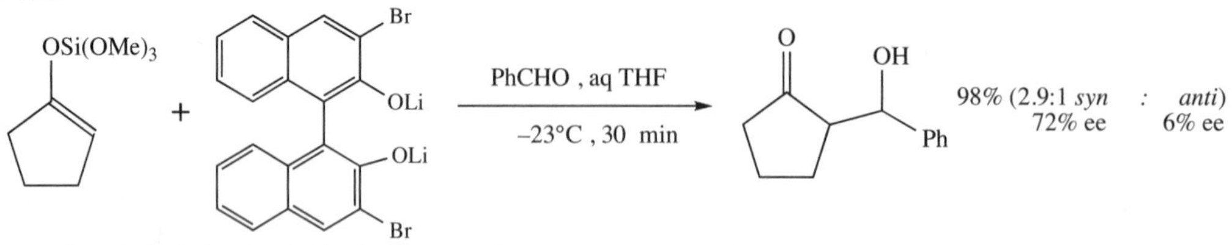

$OSi(OMe)_3$ + Br, OLi, OLi, Br — PhCHO, aq THF; −23°C, 30 min

OH, Ph 98% (2.9:1 *syn* : *anti*) 72% ee 6% ee

Nakajima, M.; Orito, Y.; Ishizuka, T.; Hashimoto, S. *Org. Lett.* ***2004***, *6*, 3763.

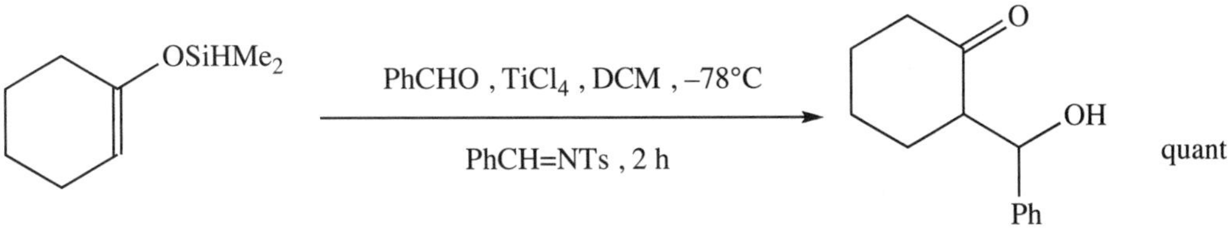

OSiHMe$_2$ — PhCHO , $TiCl_4$, DCM , –78°C / PhCH=NTs , 2 h → O, OH, Ph — quant

Miura, K.; Nakagawa, T.; Hosami, A. *J. Am. Chem. Soc.* ***2002**, 124*, 536.

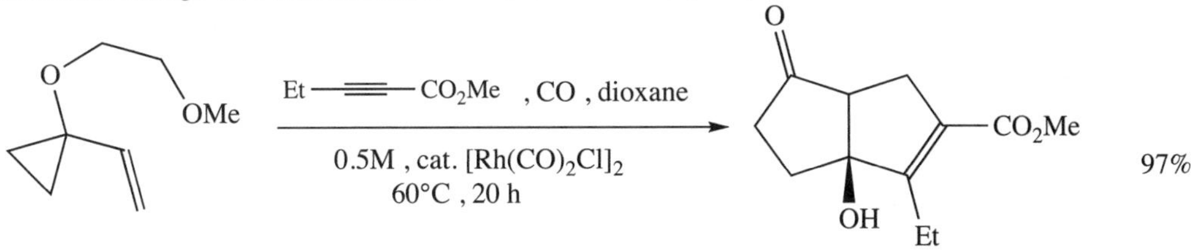

O, OMe — Et—≡—CO_2Me , CO , dioxane / 0.5M , cat. $[Rh(CO)_2Cl]_2$ 60°C , 20 h → O, CO_2Me, OH, Et — 97%

Wender, P.A.; Gamber, G.G.; Hubbard, R.D.; Zhang, L. *J. Am. Chem. Soc.* ***2002**, 124*, 2876.

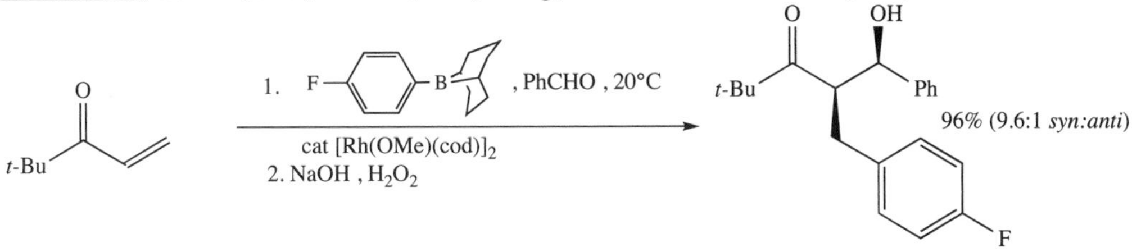

O, *t*-Bu — 1. F—C$_6$H$_4$—B(9-BBN) , PhCHO , 20°C / cat $[Rh(OMe)(cod)]_2$ 2. NaOH , H_2O_2 → O, OH, *t*-Bu, Ph, F — 96% (9.6:1 *syn:anti*)

Yoshida, K.; Ogasawara, M.; Hayashi, T. *J. Am. Chem. Soc.* ***2002**, 124*, 10984.

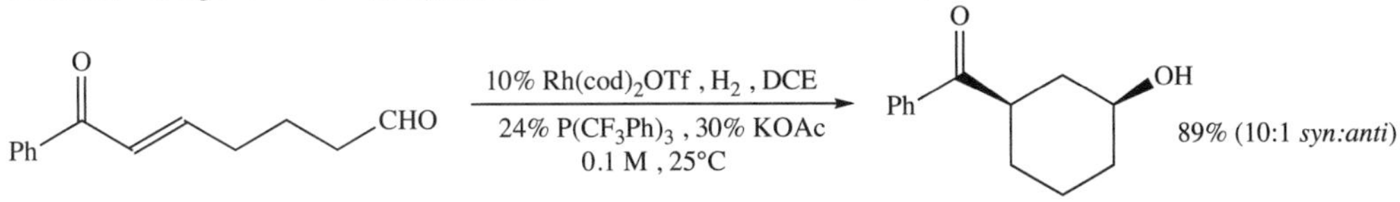

O, Ph, CHO — 10% $Rh(cod)_2OTf$, H_2 , DCE / 24% $P(CF_3Ph)_3$, 30% KOAc 0.1 M , 25°C → O, Ph, OH — 89% (10:1 *syn:anti*)

Jang, H.-Y.; Huddleston, R.R.; Krische, M.J. *J. Am. Chem. Soc.* ***2002**, 124*, 15156.

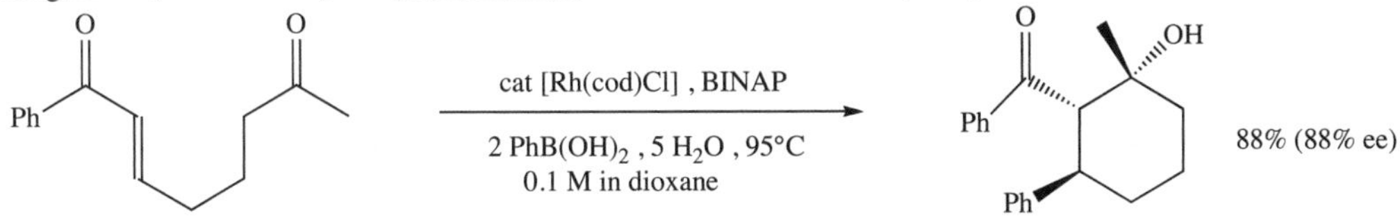

O, O, Ph — cat [Rh(cod)Cl] , BINAP / 2 $PhB(OH)_2$, 5 H_2O , 95°C 0.1 M in dioxane → O, OH, Ph, Ph — 88% (88% ee)

Cauble, D.F.; Gipson, J.D.; Krische, M.J. *J. Am. Chem. Soc.* ***2003**, 125*, 1110.

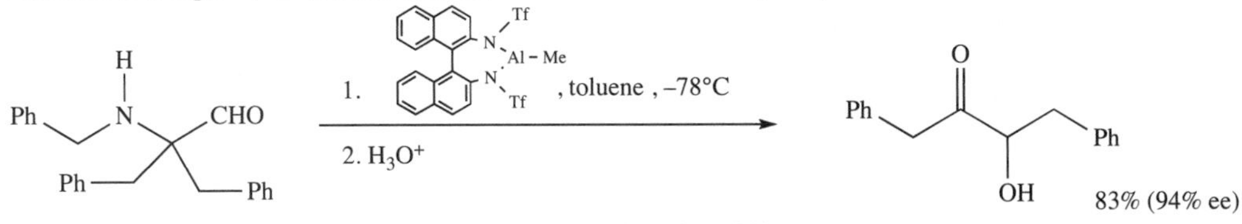

H, Ph, N, CHO, Ph, Ph — 1. (binaphthyl)(NTf)$_2$Al–Me , toluene , –78°C / 2. H_3O^+ → O, Ph, Ph, OH — 83% (94% ee)

Ooi, T.; Saito, A.; Maruoka, K. *J. Am. Chem. Soc.* ***2003**, 125*, 3220.

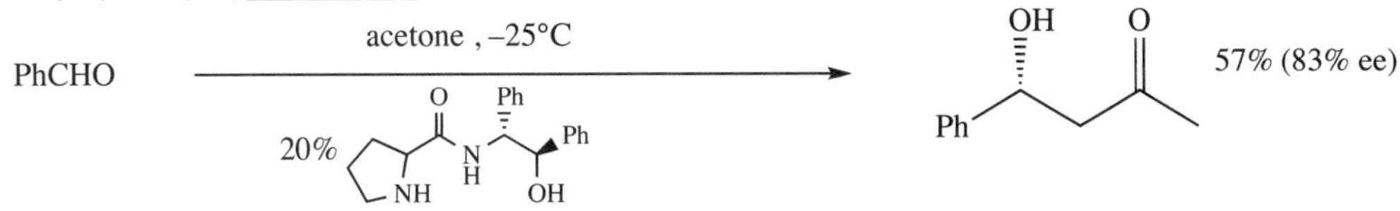

PhCHO — acetone , –25°C / 20% (prolinamide: O, NH, N H, Ph, Ph, OH) → OH, O, Ph — 57% (83% ee)

Tang, Z.; Jiang, F.; Yu, L.-T.; Cui, X.; Gong, L.-Z.; Mi, A.-Q.; Jiang, Y.-Z.; Wu, Y.-D. *J. Am. Chem. Soc.* ***2003**, 125*, 5262.

1% $RuCl_3$, 5 Oxone , –20°C
2.5 $NaHCO_3$, 90 min

OH
O
78%

Plietker, B. *J. Org. Chem.* **2004**, *69*, 8287.

OPMB O

1. Bu_2BOTf , *i*-Pr_2NEt , –78°C , ether
2. Ph CHO , –78°C

OPMB O OH Ph
89% (95:5 *anti:syn*)

Evans, D.A.; Côté, B.; Coleman, P.J.; Connell, B.T. *J. Am. Chem. Soc.* **2003**, *125*, 10893.

OEt
CHO
OEt

O
TES , THF
1% chiral diamine ligand , 0°C
10% Et_2Zn , MS 4Å , 17.5 h

EtO OEt
TES
OH O
76% (>98% ee)

Trost, B.M.; Fettes, A.; Shireman, B.T. *J. Am. Chem. Soc.* **2004**, *126*, 2660.

O
Ph
O

2.5% $Cu(OTf)_2$, 5% $P(OEt)_3$
1.5 Et_2Zn , DCM

O
Ph
OH
Me
Et
81% (>95:1 dr)

Agapiou, K.; Cauble, D.F.; Krische, M.J. *J. Am. Chem. Soc.* **2004**, *126*, 4528.

O O
t-Bu O

Ph CHO , $YbCl_3$, DCM
cat Pd_2dba_3/DIOP , rt

O OH
t-Bu Ph
93%

Lou, S.; Westbrook, J.A.; Schaus, S.E. *J. Am. Chem. Soc.* **2004**, *126*, 11440.

OTMS
Ph

aq. HCHO (5 eq) , 10% $Sc(OTf)_3$
aq. DME , –20°C , 1 d
12% bis(pyridyl) diol

O
HO Ph
80% (90% ee)

Ishikawa, S.; Hamada, T.; Manabe, K.; Kobayashi, S. *J. Am. Chem. Soc.* **2004**, *126*, 12236.

$BnMe_2Si$ O
Ph

1. 3 TBAF , THF , DMF , 0°C→ rt
2. 30% aq H_2O_2 , 80°C

OH O
Ph Ph
75%

Trost, B.M.; Ball, Z.T. *J. Am. Chem. Soc.* **2004**, *126*, 13942.

PhCHO

OH
, 3% $RuCl_2(PPh_3)_2$, 110°C
H_2O , toluene

OH O
76% (78:27 *syn:anti*)

Wang, M.; Li, C.-J. *Tetrahedron Lett.* **2002**, *43*, 3589.

PhCHO

1. MgI_2 , DCM , rt
2. HC≡C-CO_2Me , 0°C

OH O
Ph
I
86% (>96:4 *Z:E*)

Wei, H.-X.; Hu, J.; Purkiss, D.W.; Paré, P.W. *Tetrahedron Lett.* **2003**, *44*, 949.

PhCHO , BEt_3 , THF; rt , 6h

(20 : 80) 82%

Chandrasekhar, S.; Narsihmulu, Ch.; Reddy, N.R.; Reddy, M.S. *Tetrahedron Lett.* **2003**, *44*, 2583.

PhCHO , cat dendrimer-$Cu(OTf)_2$, 0°C; aq EtOH

90% (1.7:1 *syn:anti*) (50% ee)

Yang, B.-Y.; Chen, X.-M.; Deng, G.-J.; Zhang, Y.-L.; Fan, Q.-H. *Tetrahedron Lett.* **2003**, *44*, 3535.

acetone , 40% proline , 1d; sodium dodecyl sulfate

87%

Peng, Y.-Y.; Sing, Q.-P.; Li, Z.; Wang, P.G.; Cheng, J.-P. *Tetrahedron Lett.* **2003**, *44*, 3871.

PhCHO , cat $Bu_3Sn(OMe)_3$,THF; 5 MeOH , rt , 4h

91% (73:27 *syn:anti*)

Yanigisawa, A.; Sekiguchi, T. *Tetrahedron Lett.* **2003**, *44*, 7163.

$PhCO_2Et$ — 1. BrClCHLi 2. Zn , BuBr , Et_2AlCl — 2 SmI_2, THF

66% (>98%de)

Concellón, J.M.; Huerta, M. *Tetrahedron Lett.* **2003**, *44*, 1931.

PhCHO — 1. BrMgC≡CH 2. $LiAlH_4$, $AlCl_3$ 3. AD-mix

94x63x45% (88%ee)

Fleming, S.A.; Carroll, S.M.; Hirschi, J.; Liu, R.; Pace, J.L.; Redd, J.Ty. *Tetrahedron Lett.* **2004**, *45*, 3341.

PhCHO — , 35% L-prolinol , DMSO

82% (84% ee)

Zhong, G.; Fan, J.; Barbas III, C.F. *Tetrahedron Lett.* **2004**, *45*, 5681.

$MgBr_2$, MeCN; −10°C , 5h

96%

Ha, J.D.; Kim, S.Y.; Lee, S.J.; Kang, S.K.; Ahn, J.H.; Kim, S.S.; Choi, J.-K. *Tetrahedron Lett.* **2004**, *45*, 5969.

CH_2O , D-proline , rt / DMSO , 1h — 45% (>99% ee)

Casas, J.; Sundén, H.; Córdova, A. *Tetrahedron Lett.* ***2004****, 45*, 6117.

Ti(O*i*-Pr)$_4$, +DET , *t*-BuOOH / –20°C , 46h — OH, OH — 20% (86% ee)

Paju, A.; Kanger, T.; Pehk, T.; Lopp, M. *Tetrahedron* ***2002****, 58*, 7325.

PhCHO — OTMS, Ph , 0.1 Ph_2BOH , 0.01 $PhCO_2H$ / 0.1 SDS , H_2O , 80°C — OH, O, Ph, Ph — 90% 92:8 *syn:anti*)

Mori, Y.; Kobayashi, J.; Manabe, K.; Kobayashi, S. *Tetrahedron* ***2002****, 58*, 8263.

O, Ph — O, Ph , $TiCl_4$•NBu_3 — O, OH, Ph, Ph — 86%

Tanabe, Y.; Matsumoto, N.; Higashi, T.; Misaki, T.; Itoh, T.; Yamamoto, M.; Mitarai, K.; Nishii, Y. *Tetrahedron* ***2002****, 58*, 8269.

O — 1. Li, N, $(CH_2)_5$, –78°C / 2. PhCHO , –78°C → rt — O, OH, Ph — 92% (69:31 *syn:anti*)

Seki, A.; Ishiwata, F.; Takizawa, Y.; Asami, M. *Tetrahedron* ***2004****, 60*, 5001.

PhCHO — benzaldehyde lyase from *Pseudomonas fluoresces* / CH_2O — O, OH, Ph — 94%

Demir, A.S.; Ayhan, P.; Igdir, A.C.; Duygu, A.N. *Tetrahedron* ***2004****, 60*, 6509.

O, Ph — PhCHO , MgI_2 , DCM , piperidine — OH, O, Ph, Ph — 91% (95:5 *anti:syn*)

Wei, H.-X.; Jasoni, R.L.; Saho, H.; Hu, J.; Paré, P.W. *Tetrahedron* ***2004****, 60*, 11829.

O, Cl, Ph — PhCHO , Ba , THF , –78°C , 1h — O, OH, Ph, Ph — 59%

Yanagisawa, A.; Takahashi, H.; Arai, T. *Chem. Commun.* ***2004****,* 580.

CHO, O, O — 1. 1-octyne , BuLi , $(i\text{-PrO})_2TiCl$, –78°C / 2. $BnMe_2SiH$, acetone , 0°C→rt / cat $Cp^*Ru(MeCN)_3PF_6$ / 3. TBAF , H_2O_2 , MeOH , $KHCO_3$ — OH, O, O, O, C_6H_{13} — 92x7%

Trost, B.M.; Ball, Z.T.; Jöge, T. *Angew. Chem. Int. Ed.* ***2003****, 42*, 3415.

PhCHO → (but-3-en-2-ol, cat $RuCl_2(PPh_3)_3$, toluene/H_2O (4:1)) → 76% (3:37 *syn:anti*)

Wang, M.; Yang, X.-F.; Li, C.-J. *Eur. J. Org. Chem.* **2003**, 998.

O_2, 5% $CeCl_3$•7 H_2O, 23°C / *i*-PrOH, 16 h — CO_2Et → CO_2Et, OH — 99%

Christoffers, J.; Werner, T. *Synlett* **2002**, 119.

Ph–CO–CHN_2 → zeolite-H-Y, DCM, rt → Ph–CO–CH_2OH — 92%

Muthusamy, S.; Babu, S.A.; Gunanathan, C.; Jasra, R.V. *Synlett* **2002**, 407.

Ph–CO–CH=CH_2 → O_2N–C₆H₄–CHO, H_2, DCE / 5% $Rh(cod)_2OTf$, 0.5KOAc, rt → Ph, OH, NO_2 — 92%

Jang, H.-Y.; Krische, M.J. *Eur. J. Org. Chem.* **2004**, 3953.

PhCHO, 0.2 gPa, rt, 1 d / L-proline → OH, Ph — 90% (72% ee)

Sekiguchi, Y.; Sasaoka, A.; Shimomoto, A.; Fujioka, S.; Kotsuki, H. *Synlett* **2003**, 1655.

O_2N–C₆H₄–CHO → acetone, 30% Pro-Ala / DMSO, rt, 18 h → O_2N, OH, O — 70%

Martin, H.J.; List, B. *Synlett* **2003**, 1901.

OTMS → aq. HCHO, 5 TBAF, 1 h / THF, –20°C → O, OH — 71%

Ozasa, N.; Wadamoto, M.; Ishihara, K.; Yamamoto, H. *Synlett* **2003**, 2219.

O → PhI(OH)OTs, DMSO-H_2O / rt → O, OH — 59%

Xie, Y.-Y.; Chen, Z.-C. *Synth. Commun.* **2002**, *32*, 1875.

C_5H_{11}, OH → 3% $MgBr_2$, PhCHO, THF, 50 min / 3% (Ph_2P PPh_2 Cl–Ni–H), –50°C → rt → Ph, C_5H_{11}, OH, O — 99% (60:40 *syn:anti*)

Cuperly, D.; Crévisy, C.; Grée, R. *Synlett* **2004**, 93.

Ph(C=O)Ph → Sm , TMSCl , DMF , 90°C , 10 h → PhC(=O)CH(OH)Ph 91%

Liu, Y.; Xu, X.; Zhang, Y. *Synlett* ***2004***, 445.

PhC(=O)C(=O)Ph → $AlBr_3$, PPh_3 , H_2O , MeCN; rt , 1 d → PhC(=O)CH(OH)Ph 98%

Kikuchi, S.; Hashimoto, Y. *Synlett* ***2004***, 1267.

acetone + *i*-PrCHO → 10% L-proline , H_2O-DMF; rt , 9 d → 42% (92% ee)

Nyberg, A.I.; Usano, A.; Pihko, P.M. *Synlett* ***2004***, 1891.

→ PhCHO , $PhSiH_3$, 10% $In(OAc)_3$; EtOH , 0°C , 36 h → 84% (92:8 *syn:anti*)

Miura, K.; Yamada, Y.; Tomita, M.; Hosomi, A. *Synlett* ***2004***, 1985.

O_2N–C_6H_4–CHO → acetone , 20% Pro-Phe-OH , DMSO; NMM , rt , 30 h; [+ PGME 5000, 1 d, 0°C → 96%, 73% ee] → 95% (66% ee)

Shi, L.-X.; Sun, Q.; Ge, Z.-M.; Zhu, Y.-Q.; Cheng, T.-M.; Li, R.-T. *Synlett* ***2004***, 2215.

OTMS → PhCHO , $SiCl_4$, 5% chiral bis(phosphoramide); 20% *i*-Pr_2NEt , DCM , –50°C , 1 d → 80% (99:1 er)

Denmark, S.E.; Heemstra Jr. J.R. *Synlett* ***2004***, 2411.

CO_2Et → 1. Dibal , toluene , –78°C; 2. Ph(OTMS)C=CH_2 , THF; BF_3•OEt_2 , –78°C → rt → 90%

Sasaki, M.; Yudin, A.K. *Synlett* ***2004***, 2443.

→ KHMDS , 18-crown-6; THF , rt → 91%

Vilotijevic, I.; Yang, J.; Hilmey, D.; Paquette, L.A. *Synthesis* ***2003***, 1872.

microwaves, 65°C, 3 h

75%

Yadav, J.S.; Reddy, B.V.S.; Gupta, M.K. *Synthesis* **2004**, 1983.

PhCHO, DCM, $(o\text{-Tol})_3P/TiCl_4$

–78°C, 1 d

91% (96:4 *syn:anti*)

Hashimoto, Y.; Kikuchi, S. *Chem. Lett.* **2002**, 126.

1. 1.5 9-BBN, 1.5 2,6-lutidine THF, rt

2. PhCHO, –78°C, 5 h

95% (99:1 *syn:anti*)

Mukaiyama, T.; Imachi, S.; Yamane, K.; Mizuta, M. *Chem. Lett.* **2002**, *31*, 698.

1. C_6H_4-$(CH_2)_n$–N*i*-Pr(Li) –78°C → rt

2. PhCHO, –78°C

3. H_3O^+

n = 4 (71 : 29) 87%

Seki, A.; Takizawa, Y.; Ishiwata, F.; Asami, M. *Chem. Lett.* **2003**, *32*, 342.

acetone, 30% Co catalyst DMSO, MS 4Å

30% L-valine

30% *trans*-2,5-dimethylpiperazine

82% (68% ee)

Gao, M.Z.; Gao, J.; Lane, B.S.; Zingaro, R.A. *Chem. Lett.* **2003**, *32*, 524.

1. Chx_2BCl, ether

2. PhCHO, ether

(81 : 19) 58%

Zaidlewicz, M.; Sokól, W.; Wolan, A.; Cytarska, A.; Tafelska-Kaczmarek, C.A.; Dzieledziak, A.; Prewysz-Kwinto, A. *Pure Appl. Chem.* **2003**, *75*, 1349.

Bi, ZnF_2, 12 h, H_2O

PhCHO

95%

Lee, Y.J.; Chan, T.H. *Can. J. Chem.* **2003**, *81*, 1406.

REVIEWS:

"The Direct Catalytic Asymmetric Aldol Reaction"
Alcaide, B.; Almendros, P. *Eur. J. Org. Chem.* **2002**, 1595.

"Theory of Asymmetric Organocatalysis of Aldol and Related Reactions: Rationalizations and Predictions"
Allemann, C.; Gordillo, R.; Clemente, F.R.; Cheong, P.H.-Y.; Houk, K.N. *Acc. Chem. Res.* **2004**, *37*, 558.

"Design of Acid-Base Catalysis for the Asymmetric Direct Aldol Reaction"
Saito, S.; Yamamoto, H. *Acc. Chem. Res.* **2004**, *37*, 570.

"Enamine-Based Organocatalysts with Proline and Diamines: The Development of Direct Catalytic Asymmetric Aldol"
Notz, E.; Tanaka, F.; Barbas III, C.F. *Acc. Chem. Res.* **2004**, *37*, 580.

SECTION 331: ALCOHOL, THIOL - NITRILE

PhI , $PhB(OH)_2$, DMF; 5% $Pd(dba)_2$,CsF , 7 h — Ph, Ph — 89%

Huang, T.-H.; Chang, H.-M.; Wu, M.-Y.; Cheng, C.-H. *J. Org. Chem.* **2002**, *67*, 99.

CN — 1. $P(RNCH_2CH_2)_3N$/THF , –94°C 2. PhCHO , –94°C , 3 h 3. MeOH , –80°C , 5 min — OH, Ph, CN — 94%

Kisanga, P.B.; Verkade, J.G. *J. Org. Chem.* **2002**, *67*, 426.

PhCHO — 1. salen cat , Ti(O*i*-Pr)$_4$ DCM , TMSCN 2. HCN — OH, Ph, CN — 92% (97% ee)

Liang, S.; Bu, X.R. *J. Org. Chem.* **2002**, *67*, 2702.

CN, OH — 1. TMSCl , –78°C 2. PhMgCl , —78°C → rt 3. PhCHO — Ph, CN, OH, Ph, HO — 64%

Fleming, F.F.; Wang, Q.; Steward, O.W. *J. Org. Chem.* **2003**, *68*, 4235.

O, C_3H_7, OH — NaCN , DMF; $B(OMe)_3$ — OH, C_3H_7, OH, CN (85 + CN, C_3H_7, OH : OH 15) 97%

Sasaki, M.; Tanino, K.; Hirai, A.; Miyashita, M. *Org. Lett.* **2003**, *5*, 1789.

PhCHO — 10% CuO*t*-Bu , DMSO 15% dppe , 20 eq MeCN , rt , 2h — OH, Ph, CN — 95%

Suto, Y.; Kumagai, N.; Matsunaga, S.; Kanai, M.; Shibasaki, M. *Org. Lett.* **2003**, *5*, 3147.

PhCHO → [1. TMSCN, toluene, $Ph_3P{=}O$, 0°C, 40 h; chiral oxazaboroidinone salt catalyst; 2. H_3O^+] → Ph–CH(OH)–CN 94% (95% ee)

Ryu, D.H.; Corey, E.J. *J. Am. Chem. Soc.* ***2004**, 126*, 8106.

MeO–C_6H_4–CHO → [1. TMSCN, cat Pt NCN pincer complex; rt, 1d; 2. aq. HCl] → MeO–C_6H_4–CH(OH)–CN 92%

Fossey, J.S.; Richards, C.J. *Tetrahedron Lett.* ***2003**, 44*, 8773.

PhCHO → [KCN, Ac_2O, DCM/*t*-BuOH/H_2O, –20°C; cat Ti(*i*-Pr)$_4$/polymeric salen ligand] → Ph–CH(OH)–CN 93% (89% ee)

Huang, W.; Song, Y.; Bai, C.; Cao, G.; Zheng, Z. *Tetrahedron Lett.* ***2004**, 45*, 4673.

PhC(O)CH$_3$ → [Ti(O*i*-Pr)$_4$-salen, N-oxide] → Ph–C(OH)(CH$_3$)–CN 94% (81% ee)

He, B.; Chen, F.-X.; Li, Y.; Feng, X.; Zhang, G. *Tetrahedron Lett.* ***2004**, 45*, 5465.

Cyclohexene oxide → [2.5 LiCN-acetone, THF, 0.8h] → *trans*-2-hydroxycyclohexanecarbonitrile (OH, CN) 67%

Ciaccio, J.A.; Smrtka, M.; Maio, W.A. *Tetrahedron Lett.* ***2004**, 45*, 7201.

PhCHO → [1. TMSCN, TiO(*i*-Pr)$_4$, chiral diamide ligand; DCM, MS 4Å, 2d; 2. 1M HCl] → Ph–CH(OH)–CN 79% (94% ee)

Uang, B.-J.; Fu, I.-P.; Hwang, C.-D.; Chang, C.-W.; Yang, C.-T.; Hwang, D.-R. *Tetrahedron* ***2004**, 60*, 10479.

2-Phenyloxirane-2-carboxamide (Ph, $CONH_2$) → [Et_2AlCN, toluene, rt, 18h] → NC–CH$_2$–C(HO)(Ph)–C(O)NH$_2$ 90%

Ruano, J.L.G.; Fernández-Ibáñez, M. Á.; Castro, A.M.M.; Ramos, J.H.R.; Flamarique, A.C.R. *Tetrahedron: Asymmetry* ***2002**, 13*, 1321.

PhCHO → [phosphonodiamide ligand, TMSCN; DCM, cat Ti(O*i*-Pr)$_4$] → [H^+] → Ph–CH(OH)–CN 98% (43% ee)

Yang, Z.; Zhou, Z.; He, K.; Wang, L.; Zhao, G.; Zhou, Q.; Tang, C. *Tetrahedron: Asymmetry* ***2003**, 14*, 3937.

PhCHO → [HCN, (bmim) BF_4, aq. buffer; lyase from *Prunus amygdalus*] → Ph–CH(OH)–CN 95%

Gaisberger, R.P.; Fechter, M.H.; Griengl, H. *Tetrahedron: Asymmetry* ***2004**, 15*, 2959.

PhCHO → [1. TMSCN, cat. chiral sulfoxide; DCM, 12 h, –35°C; 2. HCl] → Ph–CH(OH)–CN >90% (40% ee)

Rowlands, G.J. *Synlett* ***2003***, 236.

REVIEW:

"Chemically Catalyzed Asymmetric Cyanohydrin Syntheses"
Brunel, J.-M.; Holmes, I.P. *Angew. Chem. Int. Ed.* **2004**, *43*, 2752.

SECTION 332: ALCOHOL, THIOL - ALKENE

Allylic and benzylic hydroxylation (C=C-C-H → C=C-C-OH, etc.) is listed in Section 41 (Alcohols and Thiols from Hydrides).

CHO, N; O, NH_2, DABCO; aq. dioxane, 1 d; OH, O, NH_2, N

Yu, C.; Hu, L. *J. Org. Chem.* **2002**, *67*, 219.

Ph; 20% Ni(0)–PPh_3; PhCHO, toluene; 1.5 eq i-Bu_2Al-acac; rt, 21 h; Ph, Ph, OH, 30%; +; Ph, Ph, OH, 42%

Sato, Y.; Sawaki, R.; Saito, N.; Mori, M. *J. Org. Chem.* **2002**, *67*, 656.

O; 1.5 LDA, 5 DBU, THF; 5% chiral diamine; 0°C, 1 d; OH; 95% (99% ee)

Bertilsson, S.K.; Södergren, M.J.; Andersson, P.G. *J. Org. Chem.* **2002**, *67*, 1567.

Ts—N, CHO; PhI, DMF, 80°C, 2 h; cat $PdCl_2$/dppf; Ph, Ts—N, OH, 80%; +; Ph, Ts—N, OH, 13%

Kang, S.-K.; Lee, S.-W.; Jung, J.; Lim, Y. *J. Org. Chem.* **2002**, *67*, 4376.

N, O, O, Ph, Ph; LDA, ether; –98°C → rt; N, O, OH, Ph, Ph; 76%

Perna, F.M.; Capriati, V.; Florio, S.; Luisis, R. *J. Org. Chem.* **2002**, *67*, 8351.

Ph—≡—SO_2Tol; O_2N–C_6H_4–CHO; PhSeMgBr, THF/DCM; –20°C, 50 min; HO, Ph, NO_2, PhSe, SO_2Tol; 80% (>96:14 *Z*:*E*)

Huang, X.; Xie, M. *J. Org. Chem.* **2002**, *67*, 8895.

PhCHO , $La(OTf)_3$-chiral ligand, 10 h — 97% (75% ee)

Yang, K.-S.; Lee, W.-D.; Pan, J.-F.; Chen, K. *J. Org. Chem.* ***2003***, *68*, 915.

3.5 eq SmI_2 , THF , rt; 18 eq HMPA , 2 eq *i*-PrOH — 75%

Ohno, H.; Wakayama, R.; Maeda, S.-i.; Iwasaki, H.; Okumura, M.; Iwata, C.; Mikamiyama, H.; Yanaka, T. *J. Org. Chem.* ***2003***, *68*, 5909.

Ph—≡—Ph; $Ni(dppe)Br_2$, Zn , MeCN , 80°C — 87%

Rayabarapu, D.K.; Yang, C.-H.; Cheng, C.-H. *J. Org. Chem.* ***2003***, *68*, 6726.

PhCHO , aq. $NaHCO_3$, THF; an imidazole , rt , 14 h — 83%

Luo, S.; Wang, P.G.; Cheng, J.-P. *J. Org. Chem.* ***2004***, *69*, 555.

LHMDS , THF — 69%

Greatrex, B.W.; Taylor, D.K.; Tiekink, E.R.T. *Org. Lett.* ***2002***, *4*, 221.

C_6H_{13}CH=CH$ZrClCp_2$; $Ni(cod)_2$, $ZnCl_2$ — 56%

Ni, Y.; Amarasinghe, K.K.D.; Montgomery, J. *Org. Lett.* ***2002***, *4*, 1743.

1. LDA , THF , –78°C; 2. EtO_2CCH($PO(OEt)_2$)CH_2CHO , –78°C → –10°C — 53%

Kraus, G.A.; Choudhury, P.K. *Org. Lett.* ***2002***, *4*, 2033.

$C_8H_{17}CHO$, $CrCl_2$, H_2O, cat. $NiCl_2$, cat. PPh_3; DMF, 25°C

Ph, C_8H_{17}, OH, 95 : 5 (82%)

Takai, K.; Sakamoto, S.; Isshiki, T. *Org. Lett.* ***2003***, *5*, 653.

1% $Hf(OTf)_2$, 5 H_2O

O, Ph, OH, 85%

Nishizawa, M.; Yadav, V.K.; Skwarczynski, M.; Takao, H.; Imagawa, H.; Sugihara, T. *Org. Lett.* ***2003***, *5*, 1609.

CHO, NO_2; methyl vinyl ketone, proline; peptide cat, $THF/CHCl_3$

OH, O, NO_2, 81% (78% ee)

Imbriglio, J.E.; Vasbinder, M.M.; Miller, S.J. *Org. Lett.* ***2003***, *5*, 3741.

I, CHO; Ph-C≡C-Ph, $[Co(dppe)]_2$; Zn, MeCN, 82°C, 3h

Ph, Ph, HO, 85%

Chang, K.-J.; Rayabarapu, D.K.; Cheng, C.-H. *Org. Lett.* ***2003***, *5*, 3963.

Br, CO_2Et, C; O, cat $Pd(PPh_3)_4$; dppp, THF, N_2, 1d

Br, EtO_2C, OH, 80% (39:61 *trans:cis*)

Nanayakkara, P.; Alper, H. *J. Org. Chem.* ***2004***, *69*, 4686.

PhS, C, Me; 2 Br_2, Na_2CO_3, aq. MeCN, 0°C, 3 h

PhS, Br, HO, 60% (72:1 *Z*:*E*)

Ma, S.; Hao, X.; Meng, X.; Huang, X. *J. Org. Chem.* ***2004***, *69*, 5720.

Me, Me, Si, O, I; $[(\pi\text{-allyl})PdCl]_2$, 10 TBAF, THF (0.01M), slow addition; rt

OH, 75%

Denmark, S.E.; Yang, S.-M. *J. Am. Chem. Soc.* ***2002***, *124*, 2012.

$NaBH_4$, MeOH; –34°C → 0°C — 70%

Meta, C.T.; Koide, K. *Org. Lett.* ***2004***, *6*, 1785.

1. Cp_2TiCl_2, Zn, THF
2. add epoxide by cannula, THF

OAc — 72%

Bermejo, F.; Sandoval, C. *J. Org. Chem.* ***2004***, *69*, 5275.

Ph—≡—Br; $C_5H_{11}CHO$, THF-H_2O, heat; β-SnO[Rh(cod)Cl]$_2$, 5 h

Banerjee, M.; Roy, S. *Org. Lett.* ***2004***, *6*, 2137.

MeO–C6H4–B(OH)$_2$; =C=/C_6H_{13}, PhCHO; chiral π-allyl Pd catalyst; CsF, THF, rt, 1 d; 0.05 [HPPh(*t*-Bu)$_2$]BF_4

76% (5.3:1 dr *syn:anti*)

Hopkins, C.D.; Malinakova, H.C. *Org. Lett.* ***2004***, *6*, 2221.

OTs; 1. Li_2Te, rt; 2. PhCHO, –78°C

1 min	(17	:	1) 68%
10 min	(12	:	1) 91%

Avilov, D.V.; Malasare, M.G.; Arslancan, E.; Dittmer, D.L. *Org. Lett.* ***2004***, *6*, 2225.

NO_2, CHO + NH_2; DABCO, phenol; 55°C, aq. *t*-BuOH, 18 h — 91%

Faltin, C.; Fleming, E.M.; Connow, S.J. *J. Org. Chem.* ***2004***, *69*, 6496.

Et_2Zn, Ni(cod)$_2$, PBu_3 — 85%

Lozanov, M.; Motgomery, J. *J. Am. Chem. Soc.* ***2002***, *124*, 2106.

$CH_2CMe(OMe)$, BaO
5% Cr complex , acetone
4°C

97% (98% ee)

Ruck, R.T.; Jacobsen, E.N. *J. Am. Chem. Soc.* **2002**, *124*, 2882.

[$Cy_2BH/BuC{\equiv}CH$] , Et_2Zn
toluene/hexane , 0°C , 20 h

77% (>20:1 dr)

García, C.; Libra, E.R.; Carroll, P.J.; Walsh, P.J. *J. Am. Chem. Soc.* **2003**, *125*, 3210.

i-PrCHO , 10% $Ni(cod)_2$, 2 BEt_3
20% chiral phosphine , EtOAc/DMI

95% (>95:5; 90% ee)

Miller, K.M.; Huang, W.-S.; Jamison, T.F. *J. Am. Chem. Soc.* **2003**, *125*, 3442.

1. $Ti(O\textit{i}\text{-}Pr)_4$, 2 *i*-PrMgCl
−78°C → 0°C
2. H^+

76%

Urabe, H.; Suzuki, D.; Sasaki, M.; Sato, F. *J. Am. Chem. Soc.* **2003**, *125*, 4036.

1. PhCHO , ATPH/toluene , −78°C
2. LDA , THF

84% (>99:1 *E:Z*)

Saito, S.; Nagahara, T.; Shiozawa, M.; Nakadai, M.; Yamamoto, H. *J. Am. Chem. Soc.* **2003**, *125*, 6200.

O_2N–C_6H_4–CHO , MeOH
0.5 NaOMe , 2 h

62%

Luo, S.; Mi, X.; Xu, H.; Wang, P.G.; Cheng, J.-P. *J. Org. Chem.* **2004**, *69*, 8413.

(2-methyloxirane) , 10% $Ni(cod)_2$, PBu_3
20% PBu_3 , 2 BEt_3

71%

Molinaro, C.; Jamison, T.F. *J. Am. Chem. Soc.* **2003**, *125*, 8076.

PhCOCHO , PhH , H_2
chiral La catalyst , 25°C

74% (91% ee)

Huddleston, R.R.; Jang, H.-Y.; Krische, M.J. *J. Am. Chem. Soc.* **2003**, *125*, 11488.

Ph–CH2CH2–CHO → (2-cyclohexen-1-one, PEt_3, THF; 10% chiral biaryl compound, –10°C) → 88% (90% ee)

McDougal, N.T.; Schaus, S.E. *J. Am. Chem. Soc.* ***2003***, *125*, 12094.

TIPS–C≡C–CH2–Br → ($C_8H_{17}CHO$, In, aq THF) → allenyl alcohol (TIPS, C_8H_{17}, HO) + homopropargyl alcohol (TIPS, OH, C_8H_{17}) (95 : 5) 52%

Lin, M.-J.; Loh, T.-P. *J. Am. Chem. Soc.* ***2003***, *125*, 13042.

α-methylstyrene → (HCHO encapsulated in zeolite; cyclohexane, 20°C) → Ph–C(=CH2)–CH2CH2–OH 94%

Okachi, T.; Onaka, M. *J. Am. Chem. Soc.* ***2004***, *126*, 2306.

Bu–C≡C–Br → (1. Et_2BH, 0°C; 2. Et_2Zn, –78°C; 3. isovaleraldehyde–CHO) → allylic alcohol (Et, OH, Bu) 61%

Chen, Y.K.; Walsh, P.J. *J. Am. Chem. Soc.* ***2004***, *126*, 3702.

CH2=CH–C≡C–Ph → (5% Ni(cod)$_2$, 10% P(*c*-hexyl)$_3$, 0°C; BEt_3, *i*-PrCHO) → dienyl alcohol (OH, *i*-Pr, Ph) 71% (95:5)

Miller, K.M.; Luanphaisarnnont, T.; Moinaaro, C.; Jamison, T.F. *J. Am. Chem. Soc.* ***2004***, *126*, 4130.

CH2=CH–C≡C–Ph → (OHC–C(=O)–Ph, DCE, H_2; cat $Rh(cod)_2BF_4$/BIPHEP, 25°C) → (Ph, O, Ph, OH) 87%

Jang, H.-Y.; Huddleston, R.R.; Krische, M.J. *J. Am. Chem. Soc.* ***2004***, *126*, 4664.

Bu–C≡CH → (1. $[Cp_2ZrHCl]_n$-$Ti(Oi\text{-}Pr)_4$, chiral bis(sulfoxamide) acetophenone, rt; 2. aq. $NaHCO_3$) → (OH, Ph, Bu) 85% (93% ee)

Li, H.; Walsh, P.J. *J. Am. Chem. Soc.* ***2004***, *126*, 6538.

CH2=CH–CH2–Et → (benzoyl chloride, 0.33 Cp_2ZrCl_2, PhH; 0.66 cyclopentylmagnesium bromide, THF) → (OH, Ph, Et) 76% (75:25 *anti:syn*)

Fujita, K.; Yorimitsu, H.; Shinokubo, H.; Oshima, K. *J. Am. Chem. Soc.* ***2004***, *126*, 6776.

MOMO–CH2–C(F_3C)=CBr2 → (1. BuLi, THF, –78°C; 2. PhCHO) → (OH, F_3C, Ph, MOMO, Br) 93%

Li, Y.; Lu, L. *Org. Lett.* ***2004***, *6*, 4467.

Ph, Me–CH–CHO; 1. $CH_2{=}C(CH_2OH)_2$, 10% $Pd(OAc)_2$, 20% PPh_3 , BEt_3 , NEt_3 , LiCl; 2. 10% $Pd(OAc)_2$, 20% PPh_3 , 2.4 BEt_3, THF , 39 h → OH, Me, Ph 93x82%

Mukai, R.; Horino, Y.; Tanaka, S.; Tamaru, Y.; Kimura, M. *J. Am. Chem. Soc.* ***2004***, *126*, 11138.

$(CH_2)_3$ — C_6H_{13}; MeCHO , cat $Ni(cod)_2$; BEt_3 , EtOAc → HO, $(CH_2)_3$, C_6H_{13} 69%

Miller, K.M.; Jamison, T.F. *J. Am. Chem. Soc.* ***2004***, *126*, 15342.

O, Ph; 1% $[Ph_3PAu)_3O]$ BF_4 , DCM , rt; $NaBH_4$, MeOH , 5 h → Ph, C, OH 78%

Sherry, B.D.; Toste, F.D. *J. Am. Chem. Soc.* ***2004***, *126*, 15978.

OH, Ph, CCl_3; 1. 4 $CrCl_2$; 2. PhCHO → OH, Ph, Ph, Cl 78%

Baati, R.; Barma, D.K.; Falck, J.R.; Mioskowski, C. *Tetrahedron Lett.* ***2002***, *43*, 2183.

O + CHO, NO_2; imidazole , aq. THF; rt , 36h → O, OH, NO_2 69%

Luo, S.; Zhang, B.; He, J.; Janczuk, A.; Wang, P.G.; Cheng, J.-P. *Tetrahedron Lett.* ***2002***, *43*, 7369.

O; PhCHO , cat imidazole, aq. THF, rt , 65d → O, OH, Ph 69%

Gatri, R.; El Gaïed, M.M. *Tetrahedron Lett.* ***2002***, *43*, 7835.

O; 1. PhCHO , BF_3•OEt_2 , tetrahydrothiophene; DCM , 0°C; 2. NEt_3 → OH, O, Ph 52%

Walsh, L.M.; Winn, C.L.; Goodman, J.M. *Tetrahedron Lett.* ***2002***, *43*, 8219.

O, CO_2Me; $NaBH_4$, MeOH; –72°C → 0°C → OH, CO_2Me 70%

Naka, T.; Koide, K. *Tetrahedron Lett* ***2003***, *44*, 443.

1. $ScCl_3$, THF , 0°C
2. PhCHO , –78°C

MgCl

OH Ph 89% (4:94 *cis:trans*) + OH Ph 88% (1:99 *cis:trans*)

Matsukawa, S.; Funabashi, Y.; Imamoto, T. *Tetrahedron Lett.* **2003**, *44*, 1007.

O Br , In THF , 1h OH 87%

Oh, B.K.; Cha, J.H.; Cho, Y.S.; Choi, K.I.; Koh, H.Y.; Chang, M.H.; Pe, A.N. *Tetrahedron Lett.* **2003**, *44*, 2911.

O O PhCHO , 1.2 Et_2AlI , DCM , 0°C OH O Ph O 65%

Karur, S.; Hardin, J.; Headley, A.; Li, G. *Tetrahedron Lett.* **2003**, *44*, 2991.

OAc Ts—N CHO cat $Ru_3(CO)_{12}$, 15 atm CO , 3 NEt_3 THF , 90°C , 16h Ts N OH 89%

Yu, C.-M.; Lee, S.; Hong, Y.-T.; Yoon, S.-K. *Tetrahedron Lett.* **2004**, *45*, 6557.

O Ph Br $InCl_3$, $NaBH_4$, MeCN OH Ph 82%

Ranu, B.C.; Banerjee, S.; Das, A. *Tetrahedron Lett.* **2004**, *45*, 8579.

O 5% NH N , 1.5 LDA , DBU THF , 0°C OH 98% (95% ee)

Bertilsson, S.K.; Andersson, P.G. *Tetrahedron* **2002**, *58*, 4665.

BnO O 3 Me Ph LiHN OLi , THF/hexane –78°C → rt , 16h BnO HO 91% (86% ee)

Brookes, P.C.; Milne, D.J.; Murphy, P.J.; Spolaore, B. *Tetrahedron* **2002**, *58*, 4675.

O O PhCHO , MgI_2 , 0°C DCM , 16h OH O Ph O I 86% (44% ee)

Wei, H.-X.; Chen, D.; Xu, X.; Li, G.; Paré, P.W. *Tetrahedron: Asymmetry* **2003**, *14*, 971.

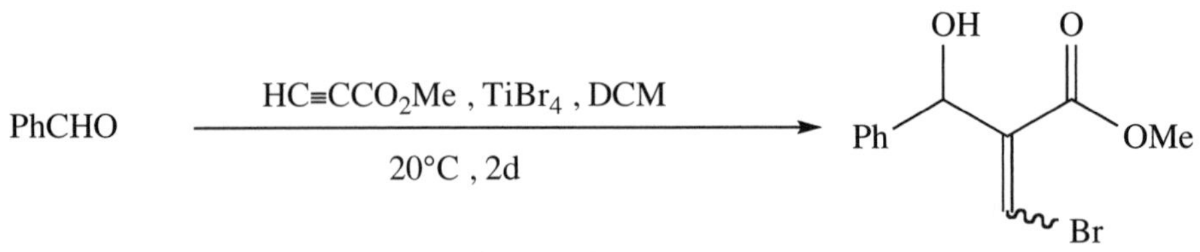

PhCHO → ($HC{\equiv}CCO_2Me$, $TiBr_4$, DCM; 20°C, 2d) 1:2 *E:Z*

Shi, M.; Wang, C.-J. *Tetrahedron* ***2002**, 58*, 9063.

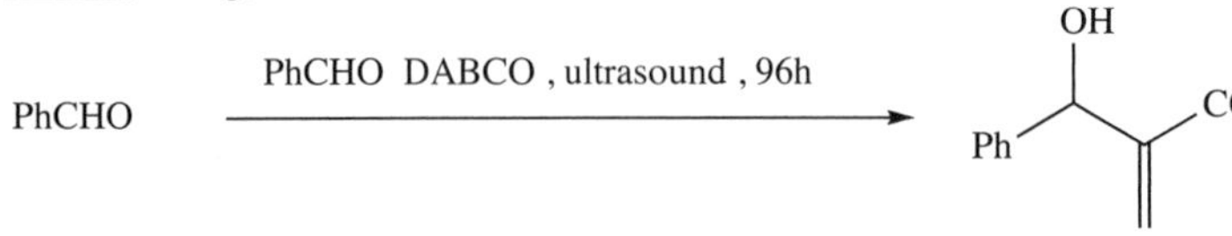

PhCHO → (PhCHO DABCO, ultrasound, 96h) 74% (only 5% without ultrasound)

Coelho, F.; Almeida, W.P.; Veronese, D.; Mateus, C.R.; Lopes, E.C.S.; Rossi, R.C.; Silveira, G.P.C.; Pavam, C.H. *Tetrahedron* ***2002**, 58*, 7437.

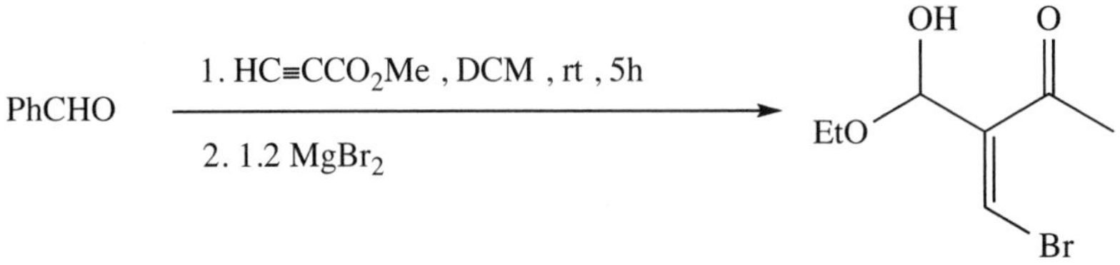

PhCHO → (1. $HC{\equiv}CCO_2Me$, DCM, rt, 5h; 2. 1.2 $MgBr_2$) 82% (87:13 *Z:E*)

Wei, H.-X.; Jasoni, R.L.; Hu, J.; Li, G.; Paré, P.W. *Tetrahedron* ***2004**, 60*, 10233.

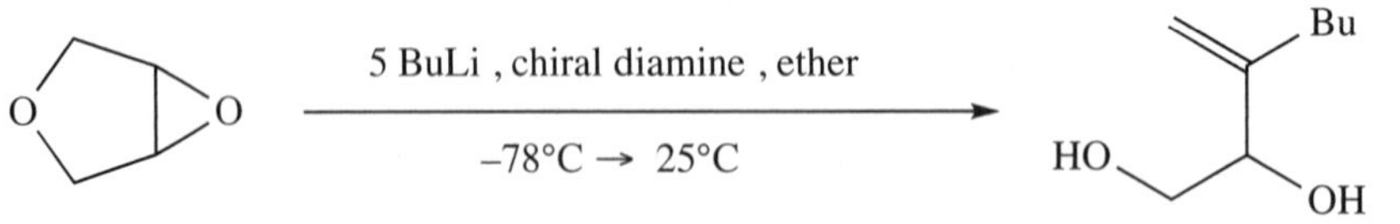

→ (5 BuLi, chiral diamine, ether; −78°C → 25°C) 63% (20% ee)

Hodgson, D.M.; Stent, M.A.H.; Stefane, B.; Wilson, F.X. *Org. Biomol. Chem.* ***2003**, 1*, 1139.

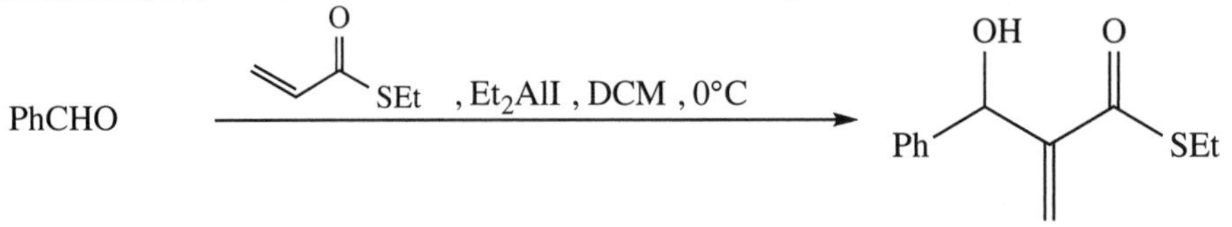

PhCHO → (SEt, Et_2AlI, DCM, 0°C) 64%

Pei, W.; Wei, H.-X.; Li, G. *Chem. Commun.* ***2002**,* 1856.

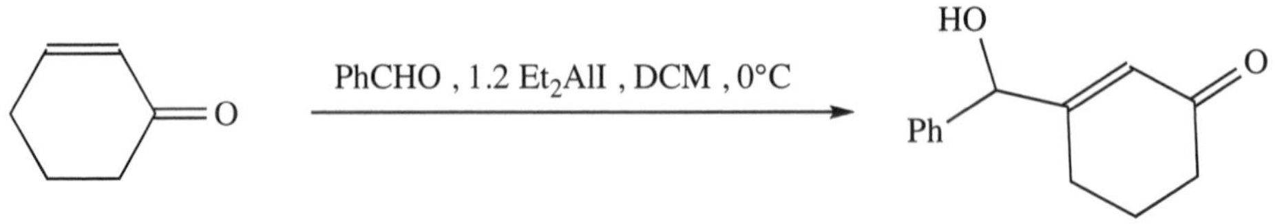

→ (PhCHO, 1.2 Et_2AlI, DCM, 0°C) 65%

Pei, W.; Wei, H.X.; Li, G. *Chem. Commun.* ***2002**,* 2412.

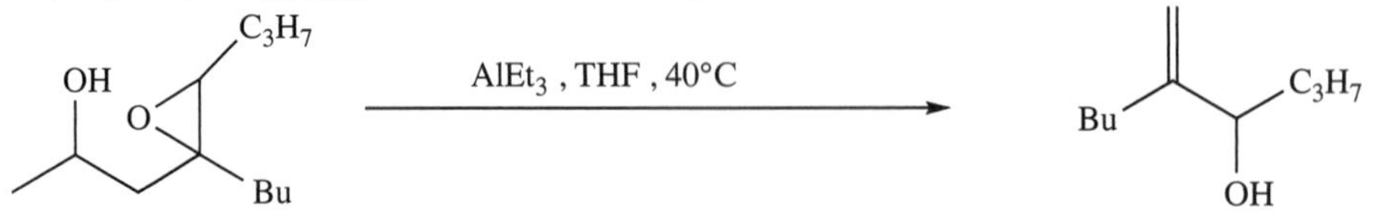

→ ($AlEt_3$, THF, 40°C) 96%

Wang, F.; Wang, S.H.; Tu, Y.Q.; Ren, S.K. *Tetrahedron: Asymmetry* ***2003**, 14*, 2189.

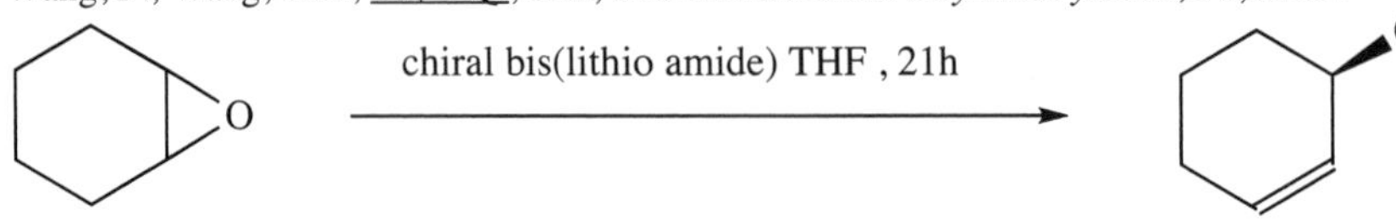

→ (chiral bis(lithio amide) THF, 21h) 94% (82% ee)

Equey, O.; Alexakis, A. *Tetrahedron: Asymmetry* ***2004**, 15*, 1069.

Ts—N(CH₂CH=C=CH₂)(CH₂CHO) → $Me_3SiSnBu_3$, cat $(\pi\text{-allyl})_2Pd_2Cl_2$ / THF , rt , 10 min → Ts—N (pyrrolidine, TMS, OH) 71%

Kang, S.-K.; Ha, Y.-H.; Ko, B.-S.; Lim, Y.; Jung, J. *Angew. Chem. Int. Ed.* ***2002**, 41*, 343.

AcO—CH₂CH=CHCH₂—OAc + (pinacol)B—CH₂CH=CH₂ → 1. DCM , Grubbs' II , 40°C / 2. PhCHO → OH, Ph, OAc 75% (4.5:1 *anti:syn*)

Goldberg, S.D.; Grubbs, R.H. *Angew. Chem. Int. Ed.* ***2002**, 41*, 807.

O_2N—C₆H₄—CHO → cyclohexenone (=O) , $TiCl_4$, DCM , rt / proazaphosphatrane → O_2N—C₆H₄—CH(OH)—(2-cyclohexenone) 94%

You, J.; Xu, J.; Verkade, J.G. *Angew. Chem. Int. Ed.* ***2003**, 42*, 5054.

2-Ph-cyclohexenone → cat $[RuCl_2(p\text{-cymene})]_2$ / cat amino amide , $HCOOH/NEt_3$ / overnight , 28°C → OH, Ph 47% (72% ee)

Hannedouche, J.; Kenny, J.A.; Walsgrove, T.; Wills, M. *Synlett* ***2002***, 263.

cyclohexene oxide (O) → 1.5 $LiZnBu_3$, THF , rt , 2 h → OH, Bu 86%

Equey, O.; Vrancken, E.; Alexakis, A. *Eur. J. Org. Chem.* ***2004***, 2151.

cyclopentanone (=O) → Ph—CH=C=CH—Ph , 2.2 SmI_2 , 18 HMPA / 2 *t*-BuOH , THF , overnight , rt → OH, Ph, Ph 69% (97:3 *E:Z*)

Hölemann, A.; Reißig, H.-U. *Chem. Eur. J.* ***2004**, 10*, 5493.

cyclohexenone (=O) → (CH₂=CHCH₂)₄Sn , MeOH , 12 h / reflux → OH 65%

Leitch, S.K.; McCluskey, A. *Synlett* ***2003***, 699.

CH₂=C(ZnCl)CH₂—SiMe₃ → 1. PhCHO , ether , –20°C → rt / 2. aq. HCl → Ph—CH(OH)—CH₂—C(=CH₂)SiMe₃ (OH, $SiMe_3$, Ph) 92%

Viseux, E.M.E.; Parsons, P.J.; Pavey, J.B.J. *Synlett* ***2003***, 861.

PhMe$_2$Si — heptanal , Me$_2$AlCl → C_6H_{13} OH 80%

Braddock, D.C.; Badine, D.M.; Gottschalk, T.; Matsuno, A.; Rodriguez-Lens, M. *Synlett* ***2003***, 345.

O-(l-menthyl) — 1. (Ph, Ph, O, Al)$_3$, toluene PhCHO; 2. LTMP , –78°C → OH, O, Ph, O-(l-menthyl) 88% (60% ee)

Takikawa, H.; Ishihara, K.; Saito, S.; Yamamoto, H. *Synlett* ***2004***, 732.

O — Br , 20% $InCl_3$, In; H_2O , rt , 16 h → OH, HO 83%

Juan, S.; Hua, Z.-H.; Qi, S.; Ji, S.-J.; Loh, T.-P. *Synlett* ***2004***, 829.

H, O, CHO — 1. 4 TMSCl , 4 Mn , DMF 0.2 $CrCl_2$; 2. 4 TBAF , THF → OH, OH 60%

Groth, U.; Jung, M.; Vogel, T. *Synlett* ***2004***, 1054.

Ph, CHO — Br , In , THF , rt , 8 h; ultrasound , 1 drop H_2O → OH, C, Ph 53%

Miao, W.; Chan, T.H. *Synthesis* ***2003***, 785.

O — Br , In , DMF , 22°C → OH 30%

Villalva-Servín, N.P.; Melekov, A.; Fallis, A.G. *Synthesis* ***2003***, 790.

O, Ph, Ph — I , In/$InCl_3$; DMF , 1 h → HO, Ph, Ph 96%

Kim, H.Y.; Choi, K.I.; Pae, A.N.; Koh, H.Y.; Choi, J.H.; Cho, Y.S. *Synth. Commun.* ***2003***, *33*, 1899.

O_2N, CHO — O , $TiBr_4$, DCM; –78°C → OH, CO_2Me, O_2N 90% (19:1 *E:Z*)

Shi, M.; Wang, C.-J. *Helv. Chim. Acta.* ***2002***, *85*, 841.

CO_2Me — 1. TMSI , DCM , rt , 3 h; 2. PhCHO , 10% MgI_2 , rt , 10 h → OH, O, Ph, OMe, I 85% (>98% *Z:E*)

Deng, G.-H.; Hu, H.; Wei, H.-X.; Paré, P.W. *Helv. Chim. Acta* ***2003***, *86*, 3510.
Wei, H.-X.; Li, K.; Zhang, Q.; Jasoni, R.L.; Hu, J.; Paré, P.W. *Helv. Chim. Acta* ***2004***, *87*, 2354.

Also via: Section 302 (Alkyne - Alcohol)

SECTION 333: ALDEHYDE - ALDEHYDE

NO ADDITIONAL EXAMPLES

SECTION 334: ALDEHYDE - AMIDE

NHOMe

$MeCH(OMe)_2$, $BF_3 \bullet OEt_2$

MeCN , –40°C

CHO N OMe 48%

Marson, C.M.; Pucci, S. *Tetrahedron Lett.* ***2004***, *45*, 9007.

SECTION 335: ALDEHYDE - AMINE

O H C_5H_{11}

PMP–N=CH–CO_2Et , *S*-proline

DMSO , rt , 3 h

PMP = *p*-methoxyphenyl

O NHPMP H CO_2Et C_5H_{11}

88%, 32:1 *syn* :*anti*
syn = >99% ee

Notz, W.; Tanaka, F.; Watanabe, S.; Chowdari, N.S.; Turner, J.M.; Thayumanavan, R.; Barbas III, C.F. *J. Org. Chem.* ***2003***, *68*, 9624.

N Me

CHO

10% chiral amine , DCM , 19 h
i-PrOH , –87°C

CHO N Me 82% (92% ee)

Austin, J.F.; MacMillan, D.W.C. *J. Am. Chem. Soc.* ***2002***, *124*, 1172.

N — PMP
EtO_2C

$C_5H_{11}CH_2CHO$, 10% L-proline
aq. THF , rt

NHPMP CHO EtO_2C C_5H_{11}

88% (>19:1 dr)

Córdova, A.; Barbas III, C.F. *Tetrahedron Lett.* ***2003***, *44*, 1923.

SECTION 336: ALDEHYDE - ESTER

$SiMePh_2$

1. ICH_2CO_2Bn , BEt_3
2. Py , H_3PO_2 , BEt_3 , PhH

O H CO_2Bn 68% + Ph_2MeSi CO_2Bn 25%

Kondo, J.; Shinokubo, H.; Oshima, K. *Angew. Chem. Int. Ed.* ***2003***, *42*, 825.

TMSO-furan + CHO (crotonaldehyde), 20% chiral amine-DNBA, DCM-H_2O → CHO 87% (8:1 *syn:anti*) (90% ee)

Brown, S.P.; Goodwin, N.C.; MacMillan, D.W.C. *J. Am. Chem. Soc.* **2003**, *125*, 1192.

SECTION 337: ALDEHYDE - ETHER, EPOXIDE, THIOETHER

PhCHO → PhS, Cl, In, DMF/H_2O, rt → OH, Ph, SPh 83% (20:80 *syn:anti*)

Mitzel, T.M.; Palomo, C.; Jendza, K. *J. Org. Chem.* **2002**, *67*, 136.

PhCHO → 10 eq BEt_3, 6 eq *t*-BuOOH, THF, 0°C → rt → O, Ph, OH + O, Ph, OH (86 : 14) 82%

Yoshimitsu, T.; Arano, Y.; Nagaoka H. *J. Org. Chem.* **2003**, *68*, 625.

O_2N-C6H4-CHO → O, imidazole, 1d, 30% proline, DMF → O_2N, OH, O 91%

Shi, M.; Jiang, J.-K.; Li, C.-Q. *Tetrahedron Lett.* **2002**, *43*, 127.

O, N—SePh, O + O, H → 20% L-prolinamide, DCM, rt, 10 min → O, H, SePh 81%

Wang, W.; Wang, J.; Li, H. *Org. Lett.* **2004**, *6*, 2817.

EtCHO + Ph-N=O → 5% L-proline, $CHCl_3$, 4°C, 4 h → OHC, O-NHPh 88% (97% ee)

Brown, S.P.; Brochu, M.P.; Sinz, C.J.; MacMillan, D.W.C. *J. Am. Chem. Soc.* **2003**, *125*, 10808.

SECTION 338: ALDEHYDE - HALIDE, SULFONATE

O, H, C_6H_{13} + O, Cl, Cl, Cl, Cl, Cl, Cl → $CHCl_3$, 5% L-proline, –30°C, 8 h → O, H, C_6H_{13}, Cl 91% (91% ee)

Brochu, M.P.; Brown, S.P.; MacMillan, D.W.C. *J. Am. Chem. Soc.* **2004**, *126*, 4108.

cat (2,5-diphenylpyrrolidine); NCS , DCM , rt — 90% (97% ee)

Halland, N.; Braunton, A.; Bachmann, S.; Marigo, M.; Jørgensen, K.A. *J. Am. Chem. Soc.* **2004**, *126*, 4790.

CHO; 2% TEAC , DCM , Cl_2 — Cl, CHO — 93%

Bellesia, F.; DeBuyck, L.; Ghelfi, F.; Pagnoni, U.M.; Parsons, A.F.; Pinetti, A. *Synthesis* **2003**, 2173.

SECTION 339: ALDEHYDE - KETONE

1. TMSI , DCM , –78°C→0°C; 2. PhCHO , BF_3•OEt_2 , –15°C — 76%

Wei, H.-X.; Kim, S.H.; Li, G. *Org. Lett.* **2002**, *4*, 3691.

C_6H_{13} CHO; , mesoporous silica FSM-16; toluene , reflux — CHO, C_6H_{13}, O — 59%

Shimizu, K.; Suzuki, H.; Hayashi, E.; Kodama, T.; Tsuchiya, Y.; Hagiwara, H.; Kitayama, Y. *Chem. Commun.* **2002**, 1068.

SECTION 340: ALDEHYDE - NITRILE

NO ADDITIONAL EXAMPLES

SECTION 341: ALDEHYDE - ALKENE

For the oxidation of allylic alcohols to alkene aldehydes, see also Section 48 (Aldehydes from Alcohols and Thiols)

MeO–C6H4–I; 1. CH_2=$CH(OEt)_2$, cat. $Pd(OAc)_2$, DMF 90°C , Bu_4NOAc , K_2CO_3 , KCl; 2. hydrolysis — MeO–C6H4–CH=CH–CHO 88%

Battistuzzi, G.; Cacchi, S.; Fabrizi, G. *Org. Lett.* **2003**, *5*, 777.

+ CHO; (oxazaborolidine: Ph, Ph, H, O, N–B, 2-MeC_6H_4) , TfOH; DCM , –95°C — CHO — 91% (89:11 *exo:endo*) 90%ee

Corey, E.J.; Shibata, T.; Lee, T.W. *J. Am. Chem. Soc.* **2002**, *124*, 3808.

$EtO_2C-CO-CO_2Et$, DCE, 90°C; cat Cp*Ru(cod)Cl, 2 h — 46%

Yamamoto, Y.; Takagishi, H.; Itoh, K. *J. Am. Chem. Soc.* ***2002***, *124*, 6844.

10% CuI, 20%, $C_3H_7CH=CHCH_2OH$; 3 Cs_2CO_3, 120°C, o-xylene, 2d, Ar; 1,10-phenantrholine derv. — 55% (92:8 dr)

Nordmann, G.; Buchwald, S.L. *J. Am. Chem. Soc.* ***2003***, *125*, 4978.

1. NaHMDS, BnBr, THF, –80°C → –60°C; 2. TBAF, EtOH, –80°C → –70°C — 85%

Sasaki, M.; Takeda, K. *Org. Lett.* ***2004***, *6*, 4849.

$PhCO_2H$, cat $RuCpCl(PMe_3)_2$, H_2O; *i*-PrOH, 100°C, sealed tube, 12h — 96%

Suzuki, T.; Tokunaga, M.; Wakatsuki, Y. *Tetrahedron Lett.* ***2002***, *43*, 7531.

$C_5H_{11}CHO$ → 10% Ph_2BClO_4, $EtNO_2$; rt, 1h — 97%

Kiyooka, S.-i.; Fujimoto, H.; Mishima, M.; Kobayashi, S.; Uddin, K.Md.; Fujio, M. *Tetrahedron Lett* ***2003***, *44*, 927.

cat Rh(BINAP) BF_4, DCM; rt — 50%

Tanaka, K.; Fu, G.C. *Chem. Commun.* ***2002***, 684.

$PhCH_2OH$ → $Br^- Ph_3P^+$-CH₂-(1,3-dioxolane), MnO_2, heat, 2d; then H^+ → Ph-CH=CH-CHO — 86%

Reid, M.; Rowe, D.J.; Taylor, R.J.K. *Chem. Commun.* ***2003***, 2284.

i-PrCHO, TsOH; toluene, Dean-Stark — 57% (2:1 *E:Z*)

Freiría, M.; Whitehead, A.J.; Motherwell, W.B. *Synlett* ***2003***, 805.

0.1 $Hg(OAc)_2$, NaOAc, 10 h; vinyl ether, reflux — 79%

Tokuyama, H.; Makido, T.; Ueda, T.; Fukuyama, T. *Synth. Commun.* ***2002***, *32*, 869.

REVIEWS:

"The Thio-Claisen Rearrangement, 1980-2001"
Mujumdar, K.C.; Ghosh, S.; Ghosh, M. *Tetrahedron* ***2003***, *59*, 7251.

"Catalysis of the Claisen Rearrangement of Aliphatic Allyl Vinyl Ethers"
Hiersemann, M.; Abraham, L. *Eur. J. Org. Chem.* **2002**, 1461.

"Claisen Rearrangement Over the Last Nine Decades"
Castro, A.M.M. *Chem. Rev.* **2004**, *104*, 2939.

Also via β-hydroxyaldehydes: Section 324 (Alcohol - Aldehyde)

SECTION 342: AMIDE - AMIDE

Ts-N=CH–C$_6$H$_4$–Br; MgI_2 , THF , reflux , 0.1M

NHTS, Br, O, N, H

Lautens, M.; Han, W.; Liu, J.-H.C. *J. Am. Chem. Soc.* **2003**, *125*, 4028.

, THF , cat NaH; 1% $[(C_3H_5)PdCl]_2$, 2% Ph-BINAP

Bz, Bn, NHBz, O, N, H, + (97 : 3) 95%

Cook, G.R.; Yu, H.; Sankaranarayanan, S.; Shanker, P.S. *J. Am. Chem. Soc.* **2003**, *125*, 5115.

2 SmI_2 , THF , 2 HMPA

Cl, N—S, *t*-Bu, O, *t*-Bu(O=)SHN, NHS(=O)*t*-Bu, (4-Cl)C_6H_4, (4-Cl)C_6H_4 99%

Zhong, Y.-W.; Izumi, K.; Xu, M.-H.; Lin, G.-Q. *Org. Lett.* **2004**, *6*, 4747.

$C_6H_{13}CHO$, morpholine; toluene , NH_4Cl , rt

NC, Bn, CO_2K, O, N, Bn, H, C_6H_{13} 90%

Bonne, D.; Dekhane, M.; Zhu, J. *Org. Lett.* **2004**, *6*, 4771.

$HNEt_2$ → 100 atm CO , 0.25 DCM , rt; 0.005 $[PdCl_2(PPh_3)_2]$, THF → Et_2N–C(=O)–C(=O)–NEt_2 95%

Hiwtari, K.; Kayaki, Y.; Okita, K.; Uki, T.; Shimizu, I.; Yamamoto, A. *Bull. Chem. Soc. Jpn.* **2004**, *77*, 2237.

Also via Dicarboxylic Acids: Section 312 (Carboxylic Acid - Carboxylic Acid)
Diamines Section 350 (Amine - Amine)

SECTION 343: AMIDE - AMINE

AcHN NHEt O → Me_2NH , H_2O , rt , 36h → AcHN NHEt O NMe_2 90%

Naidu, B.N.; Sorenson, M.E.; Connolly, T.P.; Ueda, Y. *J. Org. Chem.* ***2003***, *68*, 10098.

CO_2H Bn NH_2 → *t*-BuNC , $TiCl_4$, MeOH, OHC Cl N → CO_2H CONH*t*-Bu Bn N H Cl N 92% (*SS*/*SR* = 6), 26% chromatographed

Godet, T.; Bonvin, Y.; Vincent, G.; Merle, D.; Thozet, A.; Ciufolini, M.A. *Org. Lett.* ***2004***, *6*, 3281.

O S N S → 1. Ph N OMe , DCM , 0°C; 3 $TiCl_4$, 2.5 (–)-sparteine; 2. benzoyl chloride , 4 H_2O → NHBz O S Ph N S 65x78%

Ambhaikar, N.B.; Snyder, J.P.; Liotta, D.C. *J. Am. Chem. Soc.* ***2003***, *125*, 3690.

O O N N Bn → Me +N – O Ph , Cu(OTf)$_2$, DCM; chiral ligand , rt → O Z Ph N Me 85% (66:34 *exo*:*endo*) (99% ee for *exo*)

Sibi, M.P.; Ma, Z.; Jasperse, C.P. *J. Am. Chem. Soc.* ***2004***, *126*, 718.

NH_2 Br → H N O , 5% CUI , toluene , 1 d; 10% dimethylethylenediamine; 2 K_2CO_3 , 110°C → NH O N H 96%

Klapars, A.; Parris, S.; Anderson, K.W.; Buchwald, S.L. *J. Am. Chem. Soc.* ***2004***, *126*, 3529.

MeO N Ts → BF_3•OEt_2 , MeCN , DCM , rt → MeO N Me N Ts 47%

Prasad, B.A.B.; Bisai, A.; Singh, V.K. *Org. Lett.* ***2004***, *6*, 4829.

Ts N Ph → pyrrole , 10% $InCl_3$, DCM; 2.5 h , rt → NHTs N Ph + NHTs N Ph (87 : 13) 90%

Yadav, J.S.; Reddy, B.V.S.; Abraham, S.; Sabitha, G. *Tetrahedron Lett.* ***2002***, *43*, 1565.

1. BF_3 , *i*-Pr_2NEt , –78°C→reflux; THF , 13h; 2. H_2O — 57%

Blid, J.; Somfai, P. *Tetrahedron Lett.* ***2003***, *44*, 3159.

$PhCH_2NH_2$, MeCN , heat — NHTs, NHBn — 62%

Scheuermann, J.E.W.; Ilyashenko, G.; Griffiths, D.V.; Watkinson, M. *Tetrahedron: Asymmetry* ***2002***, *13*, 269.

N—Ts; $PhNH_2$, $LiClO_4$; MeCN — NHTs, NHPh — 90%

Yadav, J.S.; Reddy, B.V.S.; Jyothirmai, B.; Murty, M.S.R. *Synlett* ***2002***, 53.
Yadav, J.S.; Reddy, B.V.S.; Rao, K.; Raj, K.S.; Prasad, A.R. *Synthesis* ***2002***, 1061.

N—Ts; $PhNH_2$, 10% $BiCl_3$, rt; MeCN , 1.5 h — NHTs, NHPh — 96%

Swamy, N.R.; Venkateswarlu, Y. *Synth. Commun.* ***2003***, *33*, 547.

N—Ts; 2.5% $TaCl_5/SiO_2$, DC; PhNHMe , rt — NHTs, N(Me)Ph — 90%

Chandrasekhar, S.; Prakash, S.J.; Shyamsunder, T.; Ramachandar, T. *Synth. Commun.* ***2004***, *34*, 3865.

N-Ph, Ph; 1. Yb , THF-HMPA, rt , 3 h; 2. Me_2CH-N=C=O, rt , 1 h , MeI — NHPh, $NHCHMe_2$, Ph, O — 60%

Ueno, R.; Yano, K.; Makioka, Y.; Fujiwara, Y.; Kitamura, T. *Chem. Lett.* ***2002***, *31*, 790.

Ts, N, Ph, OBn; 10 $BnNH_2$, MeCN , 4 h; 0°C → rt — NHBn, Ph, OBn, NHTs — 98%

Chakraborty, T.K.; Ghosh, A.; Raju, T.V. *Chem. Lett.* ***2003***, *32*, 82.

H, O, N, Bn, N, H; PhCHO , microwaves; 5 min — Bn, N, O, Ph, N — 89%

Pospísil, J.; Potácek, M. *Heterocycles* ***2004***, *63*, 1165.

SECTION 344: AMIDE - ESTER

MeO_2C–N=CH–CO_2Me → PhNMe$_2$, $MeO_2CN=PBnPh_2$ toluene-THF , –78°C , 1 d; 10% *R*-Tol-BINAP-CuPF$_6$ → 4-Me_2N-C_6H_4-CH(NHCO_2Me)CO_2Me 80% (98% ee)

Saaby, S.; Bayón, P.; Aburel, P.S.; Jørgensen, K.A. *J. Org. Chem.* ***2002**, 67*, 4352.

PhCH$_2$COCl → 1. $EtO_2CH(Cl)NHBz$, 10% benzoquinone 3 eq Proton Sponge , toluene; 2. MeOH , reflux → EtO_2C–CH(NHBz)–CH(Ph)–CO_2Me 62% (95% ee , 12:1 dr)

Hafez, A.M.; Dudding, T.; Wagerle, Ty.R.; Shah, M.H.; Taggi, A.E.; Lectka, T. *J. Org. Chem.* ***2003**, 68*, 5819.

PhCH$_2CO_2Et$ → 1. KO*t*-Bu , THF , –78°C; 2. DCM , –78°C, bis(4-acetylphenyl)P(O)NH$_2$; 3. Ac_2O , NEt_3 → PhCH(NHAc)CO_2Et 70%

Smulik, J.A.; Vedejs, E. *Org. Lett.* ***2003**, 5*, 4187.

Ph(CH$_2$)$_2$CH=N–NHBz → $CH_2=C(SMe)_2$, PrOH , toluene; chiral Zr catalyst , 0°C , 18 h → 1-Bz-3,3-bis(SMe)-5-(2-phenylethyl)pyrazolidine (HN–N–Bz) 87% (97% ee)

Yamashita, Y.; Kobayashi, S. *J. Am. Chem. Soc.* ***2004**, 126*, 11279.

PhCH(N(Boc)$_2$)CO_2Me → Zn , MeOH , heat , 15 h → PhCH(NHBoc)CO_2Me 90% with In (20 h) , 92%

Yadav, J.S.; Reddy, B.V.S.; Reddy, K.S.; Reddy, K.B. *Tetrahedron Lett.* ***2002**, 43*, 1549.

MeO(OTMS)C=CHCH$_3$ → 1. SO_2 , cat TBSOTf; 2. Br_2; 3. $NHEt_2$ → MeOC(O)CH(CH$_3$)SO_2NEt_2 65%

Bouchez, L.C.; Dubbaka, S.R.; Turks, M.; Vogel, P. *J. Org. Chem.* ***2004**, 69*, 6413.

2-(MeO_2CC≡C)-C$_6H_4$-NHMs → 3% Pd(PPh$_3$)$_4$, *i*-Pr$_2$NEt; HC≡CCO_2Me , THF , reflux → 1-Ms-indole-2-CO_2Me 55%

Horoya, K.; Matsumoto, S.; Sakamoto, T. *Org. Lett.* ***2004**, 6*, 2953.

Ph CHO + Ph Ph N CO_2t-Bu → 1. chiral ammonium salt , –50°C 2.5 BTTP , toluene-$CHCl_3$ 2. 0.25N HCl , THF 3. $NaHCO_3$, PhCOCl → Ph OH CO_2H HN Ph O 64% (1:1.3 *anti* ; *syn*) 33% ee 80% ee

Mettath, S.; Srikanth, G.S.C.; Dangerfield, B.S.; Castle, S.L. *J. Org. Chem.* ***2004***, *69*, 6489.

MeO N—Boc → 2% chiral phosphonic acid, acac , DCM , rt , 1 h → NHBoc MeO Ac Ac 93% (90% ee)

Uraguchi, D.; Terada, M. *J. Am. Chem. Soc.* ***2004***, *126*, 5356.

CO_2t-Bu Ph CN → Boc-N=N-Boc , –78°C, chiral Cinchona base → Boc N N-Boc Ph CO_2t-Bu CN 99% (>98% ee)

Saaby, S.; Bella, M.; Jørgensen, K.A. *J. Am. Chem. Soc.* ***2004***, *126*, 8120.

N—Ts → AcOH , $In(OTf)_3$, DCM, rt , 3 h → NHTs OAc 90%

Yadav, J.S.; Reedy, B.V.S.; Sadashiv, K.; Harikishan, K. *Tetrahedron Lett.* ***2002***, *43*, 2099.

N—Ts → 1. Ac_2O , toluene , 10% PBu_3 , reflux , 1d 2. H_2O , rt → NHTs OAc 85%

Fan, R.-H.; Hou, X.-L. *Tetrahedron Lett.* ***2003***, *44*, 4411.

O CO_2Et → EtO_2C-N=N-CO_2Et , $InCl_3/SiO_2$, microwaves , 4 min → O CO_2Et N $NHCO_2Et$ EtO_2C 90%

Yadav, J.S.; Reddy, B.V.S.; Venugopal, Ch.; Padmavani, B. *Tetrahedron Lett.* ***2004***, *45*, 7507.

O CO_2Et → BnO_2C-N=N-CO_2Bn , DCM , 2h, [Ph-BOX-Cu$(OTf)_2$] , Schlenk tube → O $NHCO_2Bn$ N CO_2Bn CO_2Et 98% (98%ee)

Marigo, M.; Juhl, K.; Jørgensen, K.A. *Angew. Chem. Int. Ed.* **2003**, *42*, 1367.

BocHN CO_2Et → cat Rh(acac)$(C_2H_4)_2$, *R*-BINAP, (*p*-Cl)$C_6H_4B(OH)_2$, 2 Na_2CO_3 , 2 d, 100°C , aq. dioxane → BocHN CO_2Et C_6H_4(*p*-Cl) 62% (82% ee)

Meyer, O.; Becht, J.-M.; Helmchen, G. *Synlett* ***2003***, 1539.

0.5 $TsSO_2H$, reflux; toluene, 6 h — 84%

Knight, D.W.; Sharland, C.M. *Synlett* **2003**, 2258.

1. 4-$TSNCl_2$, CuOTf, MeCN, MS 4Å; 2. Na_2SO_3, NEt_3 — 66%

Chen, D.; Timmons, C.; Guo, L.; Xu, X.; Li, G. *Synthesis* **2004**, 2479.

PhCHO, $(NH_2)_2C{=}O$, LiBr; THF, heat, 4 h — 90%

Baruah, P.P.; Gadhwal, S.; Prajapati, D.; Sandu, J.S. *Chem. Lett.* **2002**, *31*, 1038.

1.4 [$OSiMe_3$ / OMe ketene silyl acetal], DMF, rt, 6 h; 10% PhC(=O)NHLi — 95%

Fujisawa, H.; Takahashi, E.; Nakagawa, T.; Mukaiyama, T. *Chem. Lett.* **2003**, *32*, 1036.

1. [$OSiMe_3$ / OMe ketene silyl acetal], DMF, –45°C, 1 h; 2. 1N aq HCl, THF, rt — 99%

Nakagawa, T.; Fujisawa, H.; Mukaiyama, T. *Chem. Lett.* **2004**, *33*, 92.

$MeSCH_2CO_2Et$, PIFA; DCE — 87%

Kang, L.-J.; Wang, H.-M.; Su, C.-H.; Chen, L.-C. *Heterocycles* **2002**, *57*, 1.

Related Methods: Section 315 (Carboxylic Acid - Amide)
Section 316 (Carboxylic Acid - Amine)
Section 351 (Amine - Ester)

SECTION 345: AMIDE - ETHER, EPOXIDE, THIOETHER

NTs → PhOH , 10% PBu_3 / toluene , reflux → NHTs, OPh 95%

Hou, X.-L.; Fan, R.-H.; Dai, L.-X. *J. Org. Chem.* **2002**, *67*, 5295.

Ph, Ph, N, CO_2Me → PhCH=NTs , NEt_3 , –20°C / cat chiral phosphine/$CuClO_4$ / THF , MS 4Å → N=CPh_2, Ph, CO_2Me, NHTs

Bernardi, L.; Gothef, A.S.; Hazell, R.G.; Jørgensen, K.A. *J. Org. Chem.* **2003**, *68*, 2583.

Ph, N, Ts → tetrahydropyran , Me_2Zn / 1 d → O, H, NHTs, H, Ph 76%

Yamada, K.; Fujihara, H.; Yamamoto, Y.; Miwa, Y.; Taga, T.; Tomioka, K. *Org. Lett.* **2002**, *4*, 3509.

SMe, OMe → 1. Br, O, NEt_2 / 2. PhCHO , KOH , EtOH , –5°C → O, Ph, NEt_2, O 93% (97% ee)

Aggarwal, V.K.; Hynd, G.; Picoul, W.; Vasse, J.-L. *J. Am. Chem. Soc.* **2002**, *124*, 9964.

MeO, N, CO_2Me → 1. e⁻ , TMS-Cl , TEA / 2. BzCl , TEA → TMSO, OMe, MeO, N, Bz 60%

Kise, N.; Ozaki, H.; Moriyama, N.; Kitagishi, Y.; Ueda, N. *J. Am. Chem. Soc.* **2003**, *125*, 11591.

TsHN, CO_2Me → $BnOCH=C=CH_2$, TEA , MeCN / cat $Pd(OAc)_2$/dppe , rt , 2 h → BnO, N, CO_2Me, Ts 85%

Kinderman, S.S.; de Gelder, R.; van Maarseveen, J.H.; Schoemaker, H.E.; Hiemstra, H.; Rutjes, F.P.J.T. *J. Am. Chem. Soc.* **2004**, *126*, 4100.

Ph, O, NHMe → 10% Sm-(S)-BINOL-Ph_3As=O / TBHP , THF , MS 4Å , rt → Ph, O, O, NHMe 99% (>99% ee)

Nemoto, T.; Kakei, H.; Gnanadesikan, V.; Tosaki, S.-y.; Ohshima, T.; Shibasaki, M. *J. Am. Chem. Soc.* **2002**, *124*, 14544.

Ph, CHO → furan , TsN=S-O , $ZnCl_2$ / THF → NHTs, Ph, O 81%

Padwa, A.; Zanka, A.; Cassidy, M.P.; Harris, J.M. *Tetrahedron* **2003**, *59*, 4939.

EtCHO → 1. Li-C(=O)-N(Me)-CH₂-OMe, THF, −78°C; 2. H_2O → OTMS, Me, N, OMe, O — 77%

Cunico, R.F. *Tetrahedron Lett.* **2002**, *43*, 355.

1. $TiCl_4$, *i*-Pr_2NEt, DCM
2. PhCHOBn)$_2$, carbene palladacycle

OBn, Ph — 92% (87:13 *anti:syn*)

Cosp, A.; Larrosa, I.; Vilasís, I.; Romea, P.; Urpí, F.; Vilarrasa, J. *Synlett* **2003**, 1109.

N—Ts → 5% Bi(OTf)$_3$, MeCN; rt, 4.5 h → NHTS, SPh — 92%

Yadav, J.S.; Reddy, B.V.S.; Baishya, G.; Reddy, V.; Harshavardhan, S.J. *Synthesis* **2004**, 1854.

SECTION 346: AMIDE - HALIDE, SULFONATE

Ts, N, I → OMe, BEt_3, air; PhH → Ts—N, OMe, I — 71% (1:1.7 *cis:trans*)

Kitagawa, O.; Miyaji, S.; Yamada, Y.; Fujiwara, H.; Taguchi, T. *J. Org. Chem.* **2003**, *68*, 3184.

N—Ts → KF•2 H_2O, MeCN; Bu_4NHSO_4, 45°C → NHTs, F — 96%

Fan, R.-H.; Zhou, Y.-G.; Zhang, W.-X.; Hou, X.-L.; Dai, L.-X. *J. Org. Chem.* **2004**, *69*, 335.

Bu, I, O, NH_2 → 3 eq *t*-BuOCl, 2 eq I_2; DCM, rt, 10 h → O, N, Bu, I — 72%

Tang, Y.; Li, C. *Org. Lett.* **2004**, *6*, 3229.

1. BuLi, 0°C
2. I_2

60% (59% ee)

Shen, M.; Li, C. *J. Org. Chem.* **2004**, *69*, 7906.

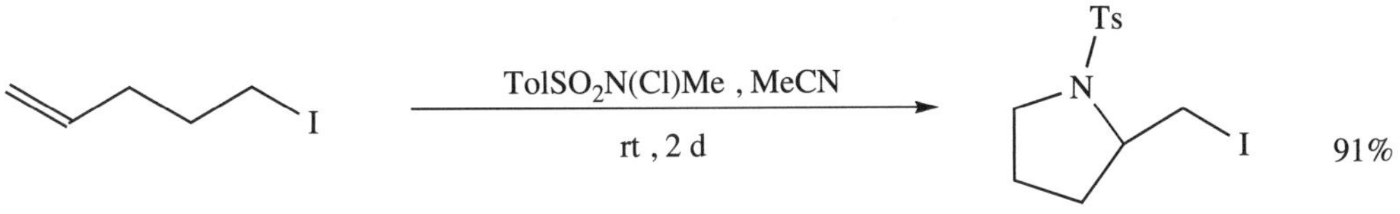

$TolSO_2N(Cl)Me$, MeCN; rt , 2 d — 91%

Minakata, S.; Kano, D.; Oderaotoshi, Y.; Komatsu, M. *Org. Lett.* **2002**, *4*, 2097.

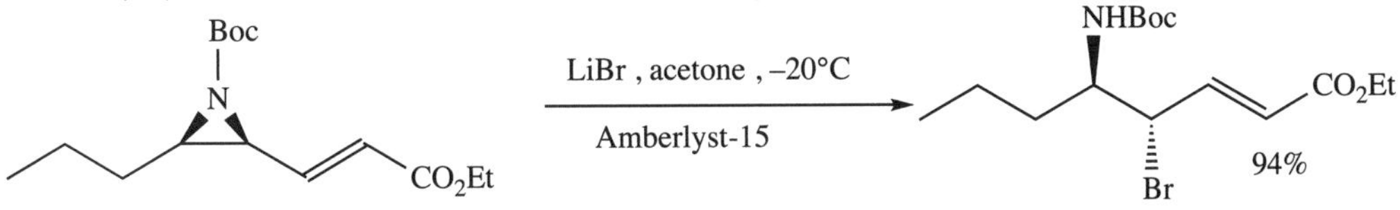

LiBr , acetone , –20°C; Amberlyst-15 — 94%

Righi, G.; Poini, C.; Bovicelli, P. *Tetrahedron Lett.* **2002**, *43*, 5867.

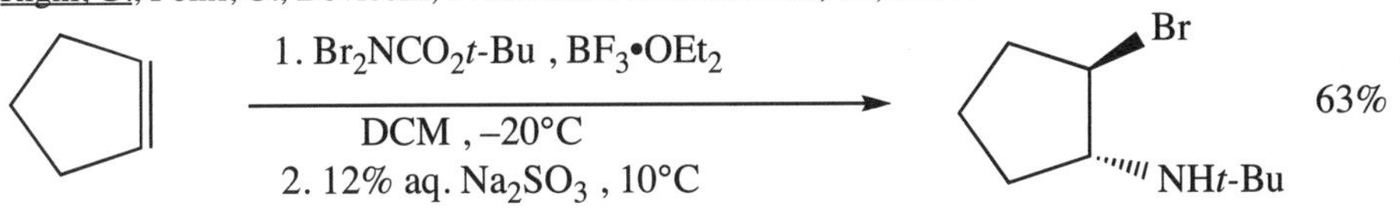

1. Br_2NCO_2t-Bu , BF_3•OEt_2; DCM , –20°C; 2. 12% aq. Na_2SO_3 , 10°C — 63%

Siwnnska, A.; Zwierzak, A. *Tetrahedron Lett.* **2003**, *44*, 9323.

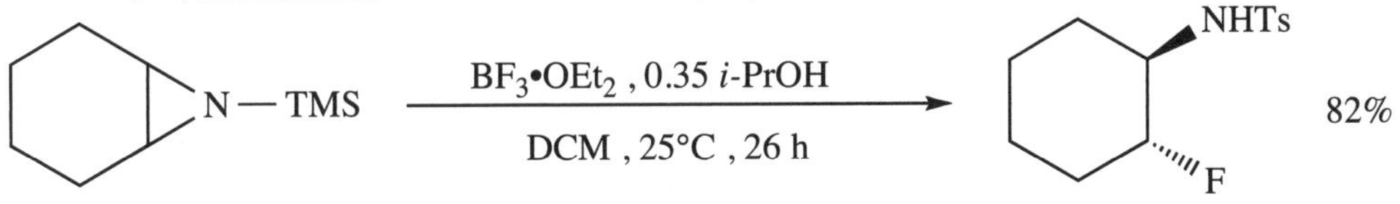

BF_3•OEt_2 , 0.35 *i*-PrOH; DCM , 25°C , 26 h — 82%

Ding, C.-H.; Dai, L.-X.; Hou, X.-L. *Synlett* **2004**, 2218.

SECTION 347: AMIDE - KETONE

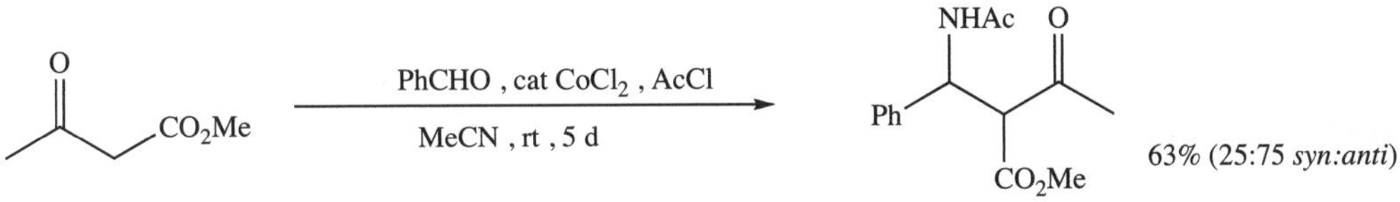

PhCHO , cat $CoCl_2$, AcCl; MeCN , rt , 5 d — 63% (25:75 *syn:anti*)

Rao, I.N.; Prabhakaran, E.N.; Das, S.K.; UIqbal, J. *J. Org. Chem.* **2003**, *68*, 4079.

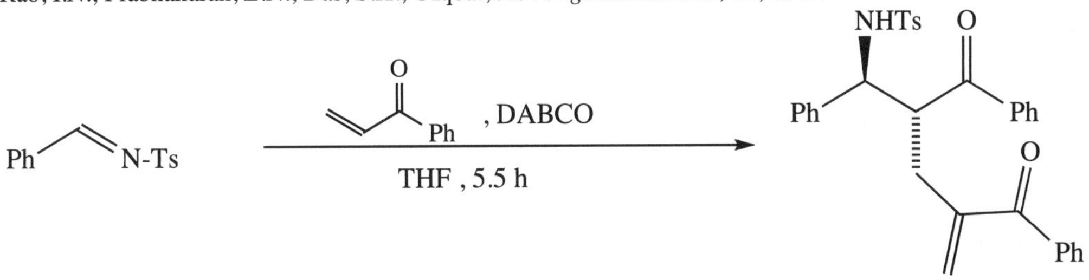

, DABCO; THF , 5.5 h

Shi, M.; Xu, Y.-M. *J. Org. Chem.* **2003**, *68*, 4784.

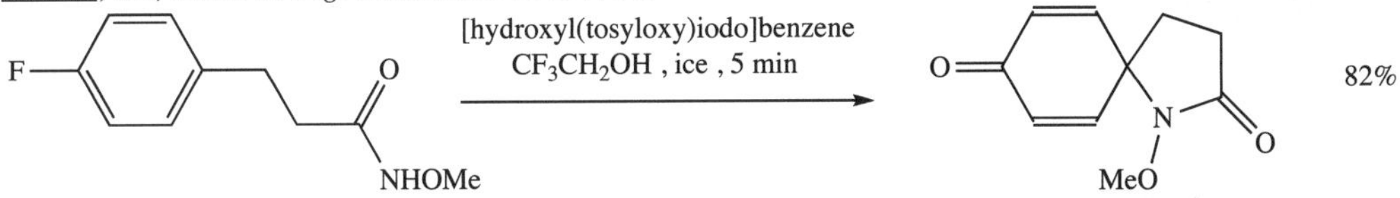

[hydroxyl(tosyloxy)iodo]benzene; CF_3CH_2OH , ice , 5 min — 82%

Miyazawa, E.; Sakamoto, T.; Kikugawa, Y. *J. Org. Chem.* **2003**, *68*, 5429.

PhCHO → (acetophenone, AcCl, MeCN, montmorillonite K-10, 7 h, 70°C) Ph, Ph, NHAc, O — 80%

Buhulayan, D.; Das, S.K.; Iqbal, J. *J. Org. Chem.* **2003**, *68*, 5735.

HO, N—Ts → (SmI_2, DCM, rt, 40 min; other Lewis acids can also be used) TsHN, O — 93%

Wang, B.M.; Song, Z.L.; Fan, C.A.; Tu, Y.Q.; Shi, Y. *Org. Lett.* **2002**, *4*, 363.

MeO, HO, C, N—Me, O → (5% $Pd(PPh_3)_4$, THF; K_2CO_3, reflux) O, OMe, N, Me, O

Nagao, Y.; Ueki, A.; Asano, K.; Tanaka, S.; Sano, S.; Shiro, M. *Org. Lett.* **2002**, *4*, 455.

→ (O, N, Troc; CF_3SO_3H, DCE 0°C → rt; Troc = $-CO_2CH_2CCl_3$) O, NHTroc — 84%

Anderson, K.W.; Tepe, J.J. *Org. Lett.* **2002**, *4*, 459.

N, OMe, O → (1. Et_2NCH_2CN, THF, 2.5 NaHMDS; 2. NaOCl) N, O, NEt_2, O — 73%

Yang, Z.; Zhang, Z.; Meanwell, N.A.; Kadow, J.F.; Wang, T. *Org. Lett.* **2002**, *4*, 1103.

O, Ph → (O, MeHN, OBn, DCM, 1 d; $PtCl_4$•5 H_2O) O, NMeCbz, Ph — quant

Kobayashi, S.; Kukumoto, K.; Sugiura, M. *Org. Lett.* **2002**, *4*, 1319.

O, Ph → (1.5 H_2NCbz, 0.1 Tf_2NH; MeCN, –20°C, 10 min) O, NHCbz, Ph — 98%

Wabintz, T.C.; Spencer, J.B. *Org. Lett.* **2003**, *5*, 2141.

Bn, N; $MeCO_2H$, 5% $Pd_2(dba)_3$•$CHCl_3$ / 10% PPh_3, THF, 100°C; O, Bn, N, O; 75%

Oh, B.H.; Nakamura, I.; Yamamoto, Y. *J. Org. Chem.* **2004**, *69*, 2856.

O, TMS, NMe_2; acetyl chloride, THF, 11 h; O, NMe_2, O; 72%

Chen, J.; Cunico, R.F. *J. Org. Chem.* **2004**, *69*, 5509.

MeO, CHO, N, O, OMe; TMST, DCM, 0°C; TMST = 2-(trimethylsilyl)thiazole; OMe, O, O, Thz, HN, OMe; 58%; + OTMS, MeO, Thz, N, O, OMe; 31%

Alcaide, B.; Almendros, P.; Redondo, M.C. *Org. Lett.* **2004**, *6*, 1765.

Ph, +, N, O, O; 1. PhC≡CH, –25°C; 2. H_2O, NEt_3; Ph, O, N, O, Ph, O; 63%

Suga, S.; Kageyama, Y.; Babu, G.; Itami, K.; Yoshida, J.-i. *Org. Lett.* **2004**, *6*, 2709.

Bn, CbzHN, S, O, N; O, Phe(OMe); SmI_2, THF, –78°C; Bn, O, CbzHN, Phe(OMe), O; 63%

Blakskjær, P.; Høj, B.; Riber, D.; Skrydstrup, T. *J. Am. Chem. Soc.* **2003**, *125*, 4030.

PhCHO; benzophenone, 1.5 NH_2CO_2Et / 5% $AuCl_3$-PPh_3, MeCN, rt, 1 d; O, $NHCO_2Me$, Ph, Ph; 82%

Xu, L.-W.; Xia, C.-G.; Li, L. *J. Org. Chem.* **2004**, *69*, 8482.

O, Ph, Ph, N, O, O; 1. bmim BF_4, cat CuOTf, MS 4Å, rt / $TsSO_2NCl_2$, 12 h; 2. aq. Na_2SO_3; Cl, O, Ph, Ph, N, O, NHTs, O; 63% (3:1 dr)

Xu, X.; Kotti, S.R.S.S.; Liu, J.; Cannon, J.F.; Headley, A.D.; Li, G. *Org. Lett.* **2004**, *6*, 4881.

OMe

TMSO

1. PhCH=N-SO_2Tol , 10% $AgClO_4$, DCM
chiral ferrocene-Ca complex , –20°C
2. 5 TFA

N SO_2Tol
O Ph

90% (97% ee)

Mancheño, O.G.; Arrayás, R.G.; Carretero, J.C. *J. Am. Chem. Soc.* ***2004,*** *126*, 456.

O
N S *t*-Bu
C_6H_{13}

1. Ph(Me)N-MnMe•4 LiBr , THF , rt
2. Boc-N=NBoc , –30°C → rt
3. HCl , rt

O
C_6H_{13}
N
Boc NHBoc

65%

Dessole, G.; Bernardi, L.; Bonin, B.F.; Capitò, E.; Fochi, M.; Herrera, R.P.; Ricci, A.; Cahiez, G. *J. Org. Chem.* ***2004,*** *69*, 8525.

O Ph
HO

O
H_2N OBn , DCM , rt
chiral catalyst , 71 h

O O
HO HN OBn
Ph

86% (96% ee)

Palomo, C.; Oiarbide, M.; Halder, R.; Kelso, M.; Gómez-Bengoa, E.; García, J.M. *J. Am. Chem. Soc.* ***2004,*** *126*, 9188.

O

BnO_2CNH_2 , 10% $Cu(OTf)_2$, MeCN , rt

O
NHCbz

81%

Watanabe, T.C.; Spencer, J.B. *Tetrahedron Lett.* ***2002,*** *43*, 3891.

O O

R* = 3,5-diMeC_6H_3 N SO_2Ph OH

$NsONHCO_2R^*$, CaO , DCM , 0°C

O O
N — CO_2R^*

82% (72:28)

Fioravanti, S.; Morreale, A.; Lellacani, L.; Tardella, P.A. *Tetrahedron Lett.* ***2003,*** *44*, 3031.

CN

1. O Ph Cl , MeCN microwaves
2. aq. $CaCO_3$, microwaves

O H
N
Ph
O

77%

Chen, J.J.; Deshpande, S.V. *Tetrahedron Lett.* ***2003,*** *44*, 8873.

O

10% proline , EtO_2C-N=N-CO_2Et , MeCN

O $NHCO_2Et$
CO_2Et

73% (93%ee)

Duthaler, R.O. *Angew. Chem. Int. Ed.* ***2003,*** *42*, 975.

OH
Ph
C

PhCH=N-CO_2Me , DCM
5% $VO(PSiPh_3)_3$

O $NHCO_2Me$
Ph Ph

85% (*anti/syn* = 9)

Trost, B.M.; Jonasson, C. *Angew. Chem. Int. Ed.* ***2003,*** *42*, 2063.

$CbzNH_2$, MeCN , 12 h

polymer supported acid catayst

NHCbz 53%

Wabnitz, T.C.; Yu, J.-Q.; Spencer, J.B. *Synlett* ***2003***, 1070.

NH_2CO_2Et , 10% TMSCl , Bu_4NBr , rt

20% BF_3•OEt_2 , 1 d

$NHCO_2Et$ 95% (25% without Bu_4NBr

Xu, L.-W.; Li, L.; Xia, C.-G.; Zhou, S.-L.; Li, J.-W.; Hu, X.-X. *Synlett* ***2003***, 2337.

$BzNH_2$, 1% $Pd(PhCN)_2Cl_2$

60°C , neat

NHBz 62%

Takasu, K.; Nishida, N.; Ihara, M. *Synlett* ***2004***, 1844.

HO ... N

1. 2 BuLi

2. butanoyl chloride

74%

Chen, Y.; Sieburth, S.Mc.N. *Synthesis* ***2002***, 2191.

$BrCH_2SO_2NHCH_2CH{=}CH_2$ — $ClCH_2COMe$, K_2CO_3 , DMF

Ac, O=S—N, O 59%

Barton, W.R.; Paquette, L.A. *Can. J. Chem.* ***2004***, *82*, 113.

SECTION 348: AMIDE - NITRILE

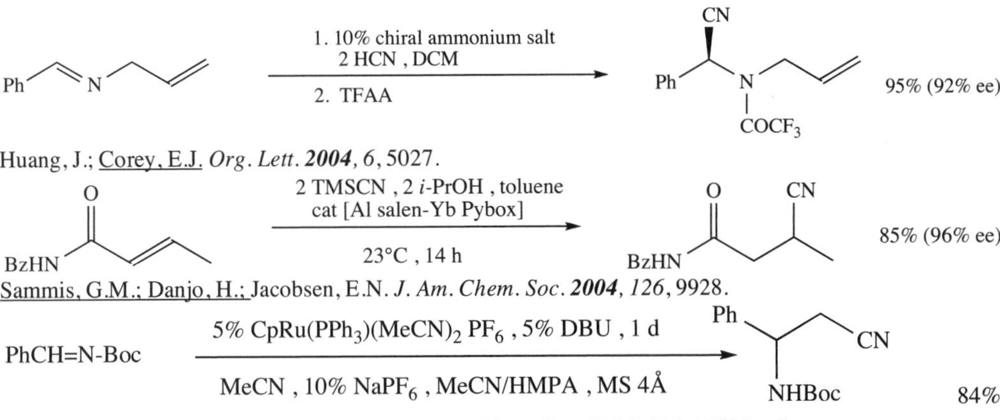

Huang, J.; Corey, E.J. *Org. Lett.* ***2004***, *6*, 5027.

Sammis, G.M.; Danjo, H.; Jacobsen, E.N. *J. Am. Chem. Soc.* ***2004***, *126*, 9928.

Kumagai, N.; Matsunaga, S.; Shibasaki, M. *J. Am. Chem. Soc.* ***2004***, *126*, 13632.

1. t-BuLi , THF , –70°C , 1h
2. PhOCN

N(i-Pr)$_2$ → N(i-Pr)$_2$, CN 98%

Sato, N. *Tetrahedron Lett* ***2002****, 43*, 6403.

Ph N-Ts → TMSCN , BF_3•OEt_2 , 8h; MeCN , rt → CN, Ph NHTs 92%

Prasad, B.A.B.; Bisai, A.; Singh, V.K. *Tetrahedron Lett.* ***2004****, 45*, 9565.

SECTION 349: AMIDE - ALKENE

NH_2 → CO_2 , Pd(OAc)$_2$, toluene; 20°C , 1 d → O NH O 85%

Shi, M.; Shen, Y.-M. *J. Org. Chem.* ***2002****, 67*, 16.

AcO OH → 1. TsNCO , THF; 2. Pd(OAc)$_2$, LiBr , reflux → O N Ts O 88% (93:7 *E:Z*)

Lei, A.; Liu, G.; Lu, X. *J. Org. Chem.* ***2002****, 67*, 974.

O N → CH_2=CH_2 Grubbs' catalyst I → N O 80%

Arjona, O.; Csákÿ, A.G.; Medel, R.; Plumet, J. *J. Org. Chem.* ***2002****, 67*, 1380.

SO_2NH_2 → CH_2=$CHCO_2Me$, PhCHO; 3-HQD , Ti(Oi-Pr)$_4$, rt; i-PrOH , MS 4Å → Tol—SO_2 NH CO_2Me 85%

Balan, D.; Adolfsson, H. *J. Org. Chem.* ***2002****, 67*, 2329.

Ph_3Si O NBn_2 → 1. KHMDS , THF , –78°C; 2. PhCHO → Ph O NBn_2 88% (>97:3 *Z:E*)

Kojima, S.; Inai, H.; Hidaka, T.; Fukuzaki, T.; Ohkata, K. *J. Org. Chem.* ***2002****, 67*, 4093.

Me–N(Boc)–CH$_2$–C(=CH–*i*-Pr)–CH$_2$–CO_2Et → 1. *sec*-BuLi, sparteine, THF, –78°C; 2. CuCN, 2 LiCl, –55°C; 3. 30 eq PhOH, TMSCl, DCM, 25°C, 2 h → *i*-Pr; O; N; Me 60%

Lu, K. *J. Org. Chem.* **2002**, *67*, 847.

N; Me; O; N H; O → hν, MeOH → Me; N; O; NH; O 33%

Zhu, M.; Qiu, Z.; Hiel, G.P.; Sieburth, S.McN. *J. Org. Chem.* **2002**, *67*, 3487.

O → CN; $CONH_2$; *t*-BuOK, DMSO; no oxygen → O; HN 60%

Carles, L.; Narkunan, K.; Penlou, S.; Rousset, L.; Couchu, D.; Ciufolini, M.A. *J. Org. Chem.* **2002**, *67*, 4304.

Ph–CH=CH–CH=N–*t*-Bu → 10 atm CH_2=CH_2, 160°C; cat $Ru_3(CO)_{12}$, 10 atm CO; stainless steel autoclave, 20 h → Ph; N; *t*-Bu; O 73%

Chatani, N.; Kamitani, A.; Murai, S. *J. Org. Chem.* **2002**, *67*, 7014.

Br; N; Me; O; N; CO_2Me; TMS → Bu_3SnH/AIBN; $NaBH_3CN$, *t*-BuOH; 80°C → TMS; N; CO_2Me; O 90%

Andrukiewicz, R.; Loska, R.; Prisyahnyuk, V.; Stalinski, K. *J. Org. Chem.* **2003**, *68*, 1552.

PhCH=NTs → Ph, 5% Yb(OTf)$_3$; TMSOTf, DCE-THF, reflux, 2 h → NHTs; Ph; Ph + NHTs; Ph; Ph (<1 : 10) 84:16 *E*:*Z* 76%

Yamanaka, M.; Nishida, A.; Nakagawa, M. *J. Org. Chem.* **2003**, *68*, 3112.

Me_3Si; Se; N → 1. BuLi, 0°C; 2. CHO → Se; N 93%

Murai, T.; Fujishima, A.; Iwamoto, C.; Kato, S. *J. Org. Chem.* **2003**, *68*, 7979.

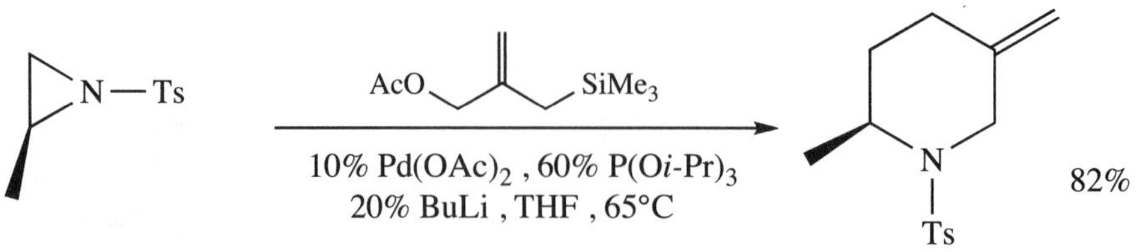

N—Ts

AcO SiMe$_3$

10% Pd(OAc)$_2$, 60% P(O*i*-Pr)$_3$
20% BuLi , THF , 65°C

N
Ts

82%

Hedley, S.J.; Moran, W.J.; Price, D.A.; Harrity, J.P.A. *J. Org. Chem.* ***2003***, *68*, 4286.

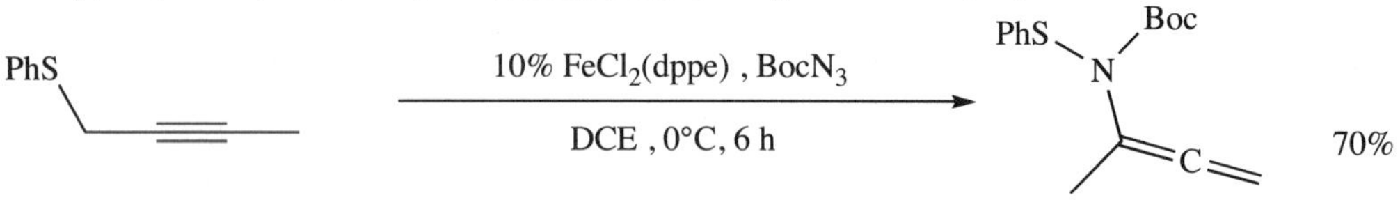

PhS

10% FeCl$_2$(dppe) , BocN$_3$

DCE , 0°C, 6 h

PhS Boc
N
C

70%

Bacci, J.P.; Greenman, K.L.; Van Vranken, D.L. *J. Org. Chem.* ***2003***, *68*, 4955.

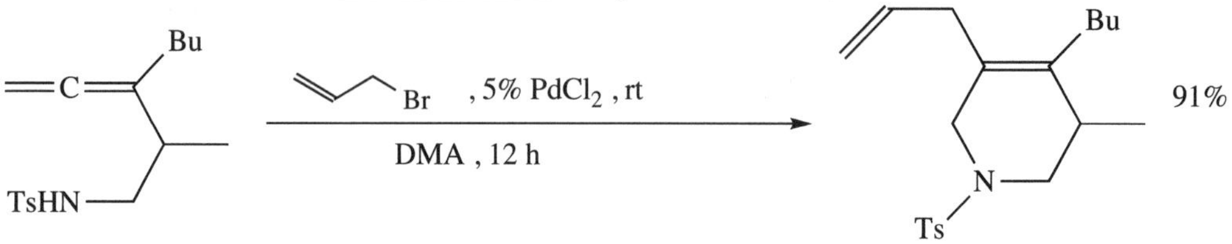

Bu
C
TsHN

Br , 5% PdCl$_2$, rt

DMA , 12 h

Bu
N
Ts

91%

Ma, S.; Yu, F.; Gao, W. *J. Org. Chem.* ***2003***, *68*, 5943.

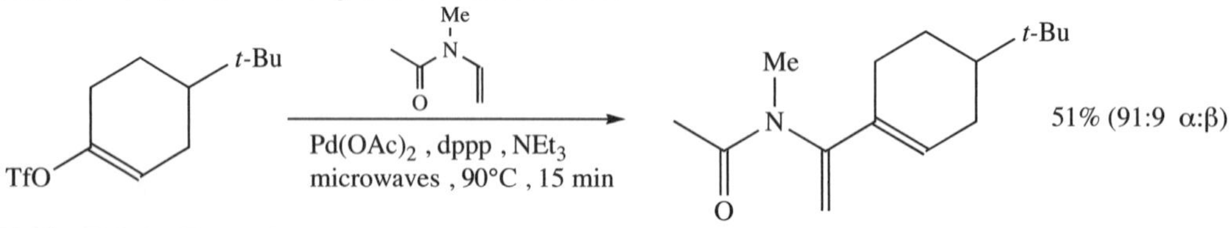

t-Bu
TfO

Me
N
O

Pd(OAc)$_2$, dppp , NEt$_3$
microwaves , 90°C , 15 min

Me
N
O
t-Bu

51% (91:9 α:β)

Vallin, K.S.A.; Zhang, Q.; Larhed, M.; Curran, D.P. *J. Org. Chem.* ***2003***, *68*, 6639.

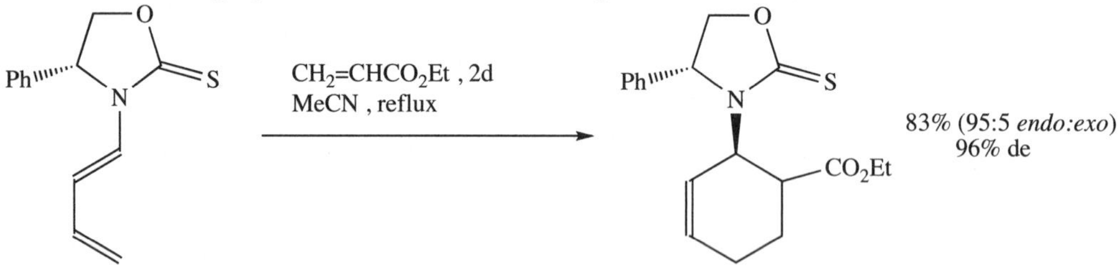

O
Ph
N
S

CH$_2$=CHCO$_2$Et , 2d
MeCN , reflux

O
Ph
N
S
CO$_2$Et

83% (95:5 *endo:exo*)
96% de

Robiette, R.; Cheboub-Benchaba, K.; Peeters, D.; Marchand-Brynaert, J. *J. Org. Chem.* ***2003***, *68*, 9809.

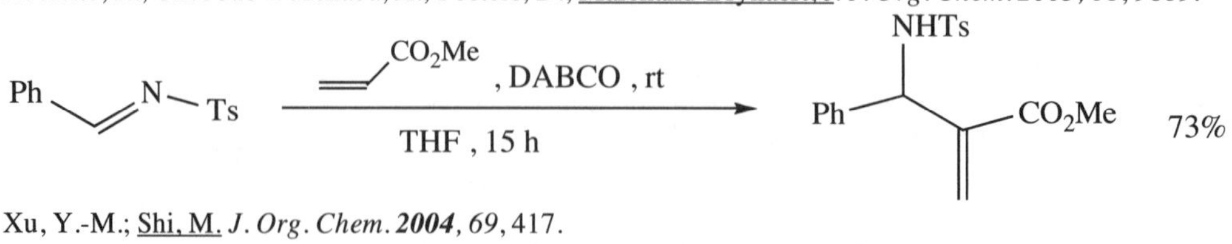

Ph N Ts

CO$_2$Me , DABCO , rt

THF , 15 h

NHTs
Ph CO$_2$Me

73%

Xu, Y.-M.; Shi, M. *J. Org. Chem.* ***2004***, *69*, 417.

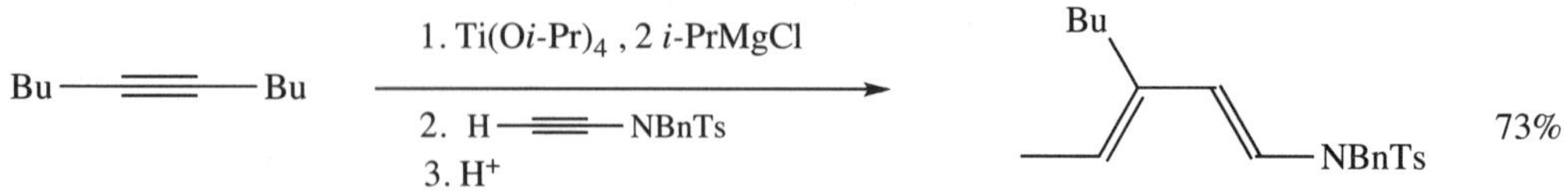

Bu—≡—Bu

1. Ti(O*i*-Pr)$_4$, 2 *i*-PrMgCl

2. H—≡—NBnTs
3. H$^+$

Bu
NBnTs

73%

Tanaka, R.; Hirano, S.; Urabe, H.; Sato, F. *Org. Lett.* ***2003***, *5*, 67.

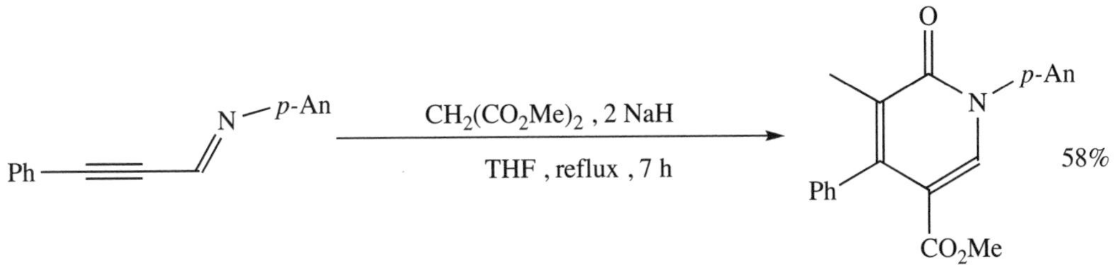

$CH_2(CO_2Me)_2$, 2 NaH

THF , reflux , 7 h

58%

Hachiya, I.; Ogura, K.; Shimizu, M. *Org. Lett.* **2002**, *4*, 2755.

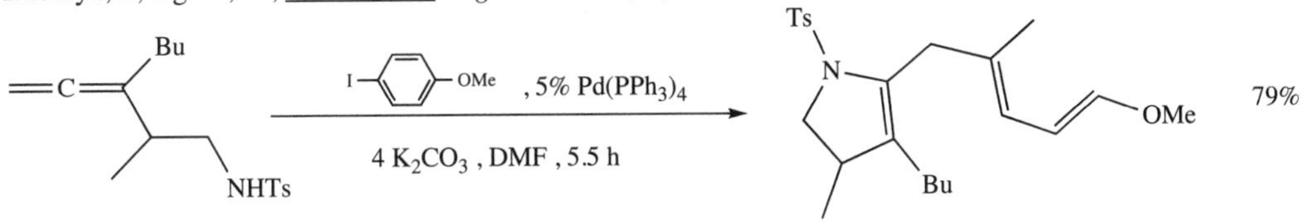

, 5% $Pd(PPh_3)_4$

4 K_2CO_3 , DMF , 5.5 h

79%

Ma, S.; Gao, W. *Org. Lett.* **2002**, *4*, 2989.

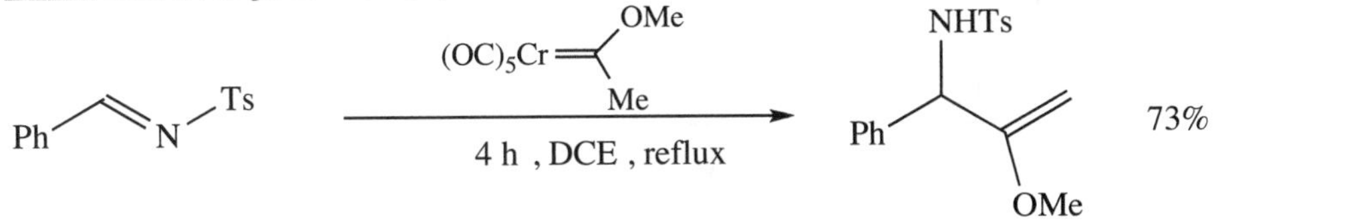

4 h , DCE , reflux

73%

Sangu, K.; Kagoshima, H.; Fuchibe, K.; Akiyama, T. *Org. Lett,* **2002**, *4*, 3967.

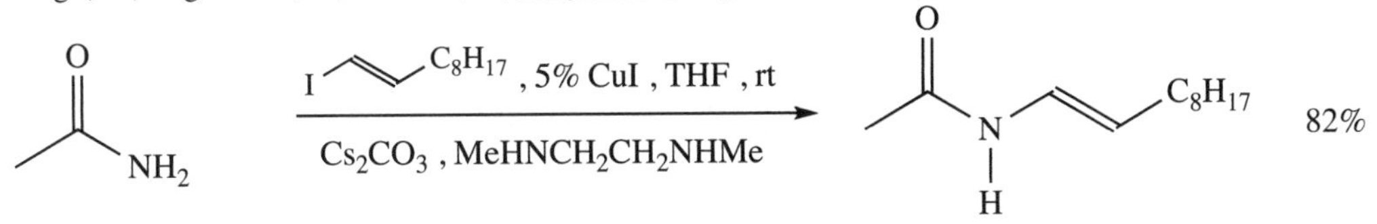

, 5% CuI , THF , rt

Cs_2CO_3 , $MeHNCH_2CH_2NHMe$

82%

Jiang, L.; Job, G.E.; Klapars, A.; Buchwald, S.L. *Org. Lett.* **2003**, *5*, 3667.

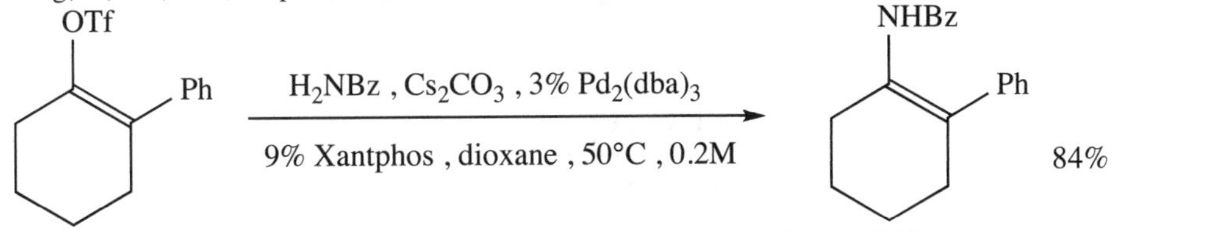

H_2NBz , Cs_2CO_3 , 3% $Pd_2(dba)_3$

9% Xantphos , dioxane , 50°C , 0.2M

84%

Wallace, D.J.; Klauber, D.J.; Chen, C.-y.; Volante, R.P. *Org. Lett.* **2003**, *5*, 4749.

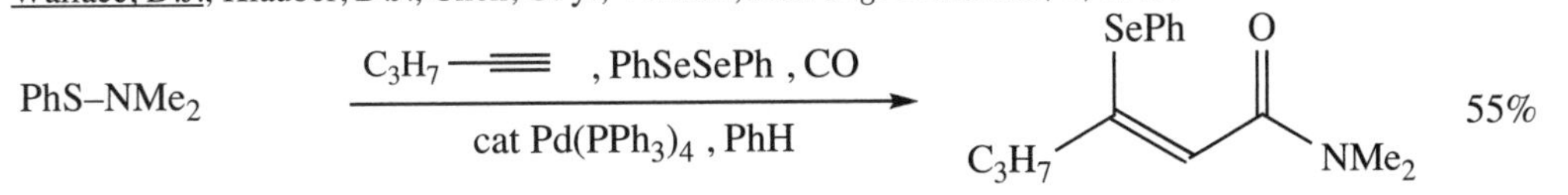

$PhS–NMe_2$

, PhSeSePh , CO

cat $Pd(PPh_3)_4$, PhH

55%

Knapton, D.J.; Meyer, T.Y. *Org. Lett.* **2004**, *6*, 687.

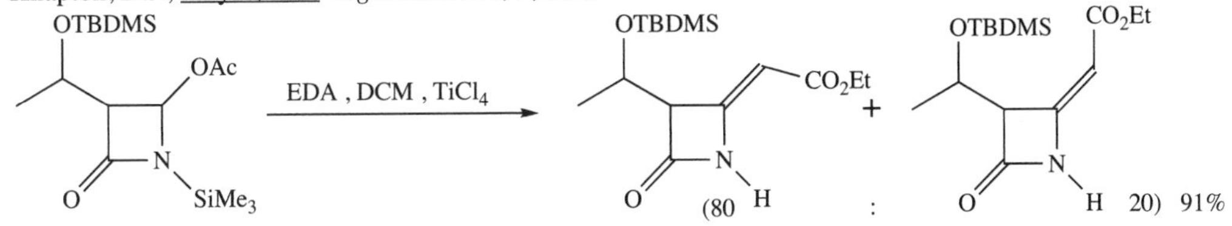

EDA , DCM , $TiCl_4$

(80 : 20) 91%

Cainelli, G.; Galletti, P.; Gazzano, M.; Giacomini, D.; Quintavalla, A. *Tetrahedron Lett.* **2002**, *43*, 233.

10% $Pd(PPh_3)_4$, 2 PhI
2 K_2CO_3
MTS = trimethylphenylsulfonyl
46%

Ohno, H.; Takeoka, Y.; Kadoh, Y.; Miyamura, K.; Tanaka, T. *J. Org. Chem.* ***2004**, 69*, 4541.

5% $Pd(PPh_3)_2Cl_2$, CO , DMA
NEt_3 , 65°C, 0.03M
78%

Trost, B.M.; Ameriks, M.K. *Org. Lett.* ***2004**, 6*, 1745.

10 $BuCH{=}CH_2$, air , 75°C
5% Pd(dppp) $(OTf)_2$
89%

Brice, J.L.; Meerdink, J.E.; Stahl, S.S. *Org. Lett.* ***2004**, 6*, 1845.

5% $[Rh(CO)_2Cl]_2$
DCE , 90°C , N_2 , 2h
72%

Brummond, K.M.; Chen, H.; Mitasev, B.; Casarez, A.D. *Org. Lett.* ***2004**, 6*, 2161.

2.5 eq. LDA , THF
–30°C → rt
81%

Reed, M.A; Chang, M.T.; Snieckus, V. *Org. Lett.* ***2004**, 6*, 2297.

2 In , Py-HBr_3 , DMF
1d
75% *E*

Yanada, R.; Obika, S.; Oyama, M.; Takemoto, Y. *Org. Lett.* ***2004**, 6*, 2825.

1. $Hg(ClO_4)_2$, THF
2. KBr

HgBr 90%

Singh, O.V.; Han, H. *Org. Lett.* ***2004****, 6*, 3067.

Grubbs II catalyst
DCM, reflux
CH_2=CH_2, 11 h

69%

Ma, S.; Ni, B.; Liang, Z. *J. Org. Chem.* ***2004****, 69*, 6305.

5 PrC≡CPr, CO, Bu_4NCl
5% $Pd(OAc)_2$, 2 pyridine
DMF, 100°C

75%

Kadnikov, D.V.; Larock, R.C. *J. Org. Chem.* ***2004****, 69*, 6772.

1. *t*-BuLi, –78°C
2. DMF, –78°C
3. 2M HCl, THF

75%

Kessler, A.; Coleman, C.M.; Chaoenying, P.; O'Shea, D. *J. Org. Chem.* ***2004****, 69*, 7836.

5% cobalt oxazoline palladacycle
hfacac, THF, 0°C

93% (91%ee)

Kirsch, S.F.; Overman, L.E.; Watson, M.P. *J. Org. Chem.* ***2004****, 69*, 8101.

1. Cp_2ZrEt_2
2. PhCN
3. C_3H_7—≡—C_3H_7

61%

Takahashi, T.; Tsai, F.-Y.; Li, Y.; Wang, H.; Kondo, Y.; Yamanaka, M.; Nakajima, K.; Kotora, M. *J. Am. Chem. Soc.* ***2002****, 124*, 5059.

Ts—N(Li)—CH$_2$C≡CH ; allyl–OCO_2Me , AgOTf , heat ; cat $Rh(PPh_3)_3Cl$, toluene , rt ; 1,3-butadiene → Ts—N (bicyclic) 83%

Evans, P.A.; Robinson, J.E.; Baum, E.W.; Fazal, A.N. *J. Am. Chem. Soc.* ***2002**, 124*, 8782.

Ts–NH–CH$_2$C≡CH , cat $[Rh(nbd)Cl]_2$; phosphine , $AgSbF_6$, DCM-EtOAc ; 60°C , 12 h → NHTs ; Ts—N ; H 49%

Gilbertson, S.R.; DeBoef, B. *J. Am. Chem. Soc.* ***2002**, 124*, 8784.

N—Ts ; Ph ; =C= ; CN ; CO_2Et ; 20% PBu_3 , DCM , rt → Ts ; Ph ; N ; CN ; CO_2Et 99% (982 dr)

Zhu, X.-F.; Lan, J.; Kwon, O. *J. Am. Chem. Soc.* ***2003**, 125*, 4716.

N ; Me_3Si—≡—Ph , 150°C , 7 h ; 20% $CpRu(PPh_3)_2Cl$, 22% $NaPF_6$ → N ; Ph 87%

Murakami, M.; Hori, S. *J. Am. Chem. Soc.* ***2003**, 125*, 4720.

N ; Me_3SiCl , MeOH , Bu_3SnH , AIBN ; CO , PhH → O ; N quant

Ryu, I.; Miyazato, H.; Kuriyama, H.; Matsu, K.; Tojino, M.; Fukuyama, T.; Minakata, S.; Komatsu, M. *J. Am. Chem. Soc.* ***2003**, 125*, 5632.

OTf ; O ; OMe ; N ; Br ; Ph ; cat $Pd(OAc)_2$, BINAP ; PMP , THF , 80°C → Ph ; OMe ; O ; N ; Br 86% (84% ee)

Dounay, A.B.; Hatanaka, K.; Kodanko, J.J.; Oestreich, M.; Overman, L.E.; Pfeifer, L.A.; Weiss, M.M. *J. Am. Chem. Soc.* ***2003**, 125*, 6261.

Ts ; N ; 5% $(MeCN)_4Pd\ BF_4$, HCOOH ; 10% S-BINAP , DMSO ; 100°C , 1 h → N ; Ts 53% (86% ee)

Hatano, M.; Mikami, K. *J. Am. Chem. Soc.* ***2003**, 125*, 4704.

CHO, CO_2H — $MeNH_3Cl$, AcOH , KCN , MeOH / reflux — NC, NH_2, N—Me, O — 53% (3:1 *Z*:*E*)

Opatz, T.; Ferenc, D. *J. Org. Chem.* ***2004***, *69*, 8496.

CCl_3, HN, O, C_3H_7 — chiral Pd catalyst , DCM , 38°C / 18 h — CCl_3, HN, O, C_3H_7 — 99% (94% ee)

Anderson, C.E.; Overman, L.E. *J. Am. Chem. Soc.* ***2003***, *125*, 12412.

Ph—≡—CH_3 — $BsNCl_2$, 10% $Pd(OAc)_2$ / MeCN , 80°C — Ph, Cl, Cl, NHBs — 61%

Karur, S.; Kotti, S.R.S.S.; Xu, X.; Cannon, J.F.; Headley, A.; Li, G. *J. Am. Chem. Soc.* ***2003***, *125*, 13340.

Ph, NHTf — 5% $Pd_2(dba)_3$•$CHCl_3$, 10% $PhCO_2H$, 72 h / 25% (R,R)-RENOR-PhTOS , PhH , 100°C — N, Tf, Ph — 65% (82% ee)

Lutete, L.M.; Kadota, I.; Yamamoto, Y. *J. Am. Chem. Soc.* ***2004***, *126*, 1622.

O, N — 10% Ir complex , C_6H_{12} , 150°C , 13 h — N, O — 66% + 2 other products

DeBoef, B.; Pastine, S.J.; Sames, D. *J. Am. Chem. Soc.* ***2004***, *126*, 6556.

OHC—N — PPh_3 , PhCHO , toluene / 100°C , 2 h — Ph, O, N — 97% (>20:1 *E*:*Z*)

Matsunaga, S.; Kinoshita, T.; Okada, S.; Harada, S.; Shibasaki, M. *J. Am. Chem. Soc.* ***2004***, *126*, 7559.

Cl, $P(=O)(OEt)_2$ — cat Pybox derivative , DCM / $[IrCl(cod)]_2$, –20°C , 20 h — Cl, BnO(Bz)N (70 : 30) 75%, 87% ee + Cl, N(Bz)OBn

Miyabe, H.; Matsumura, A.; Moriyama, K.; Takemoto, Y. *Org. Lett.* ***2004***, *6*, 4631.

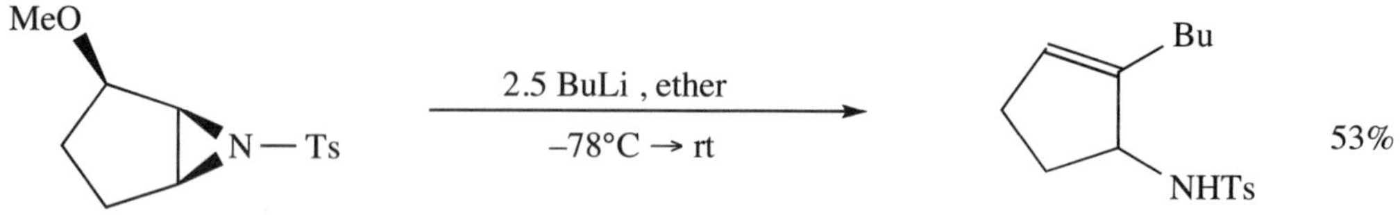

MeO
N — Ts
2.5 BuLi , ether
–78°C → rt
Bu
NHTs
53%

Rosser, C.M.; Coote, S.C.; Kirby, J.P.; O'Brien, P.; Caine, D. *Org. Lett.* **2004**, *6*, 4817.

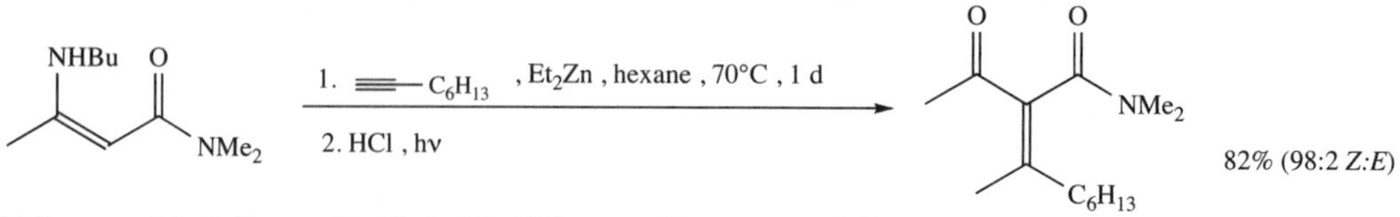

NHBu O
NMe_2
1. ≡—C_6H_{13} , Et_2Zn , hexane , 70°C , 1 d
2. HCl , hν
O O
NMe_2
C_6H_{13}
82% (98:2 *Z:E*)

Nakamura, M.; Fujimoto, T.; Endo, K.; Nakamura, E. *Org. Lett.* **2004**, *6*, 4837.

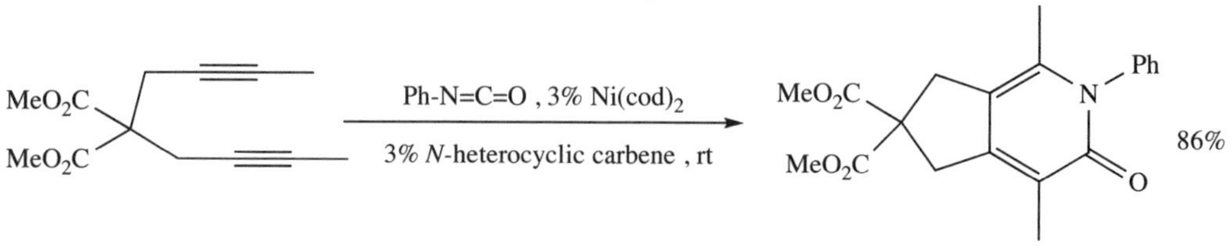

MeO_2C
MeO_2C
Ph-N=C=O , 3% $Ni(cod)_2$
3% *N*-heterocyclic carbene , rt
MeO_2C
MeO_2C
N Ph
O
86%

Dung, H.A.; Cross, M.J.; Louie, J. *J. Am. Chem. Soc.* **2004**, *126*, 11438.

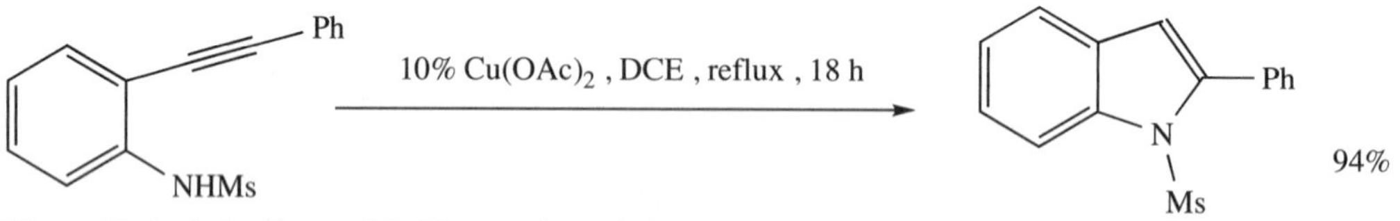

Ph
NHMs
10% $Cu(OAc)_2$, DCE , reflux , 18 h
Ph
N
Ms
94%

Hiroya, K.; Itoh, S.; Ozawa, M.; Kanamori, Y.; Sakamoto, T. *Tetrahedron Lett.* **2002**, *43*, 1277.

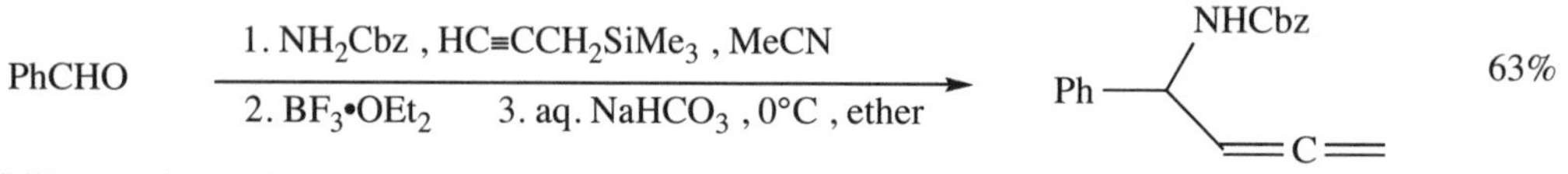

PhCHO
1. NH_2Cbz , $HC{\equiv}CCH_2SiMe_3$, MeCN
2. $BF_3{\bullet}OEt_2$ 3. aq. $NaHCO_3$, 0°C , ether
NHCbz
Ph
=C=
63%

Billet, M.; Schoenfelder, A.; Klotz, P.; Mann, A. *Tetrahedron Lett.* **2002**, *43*, 1453.

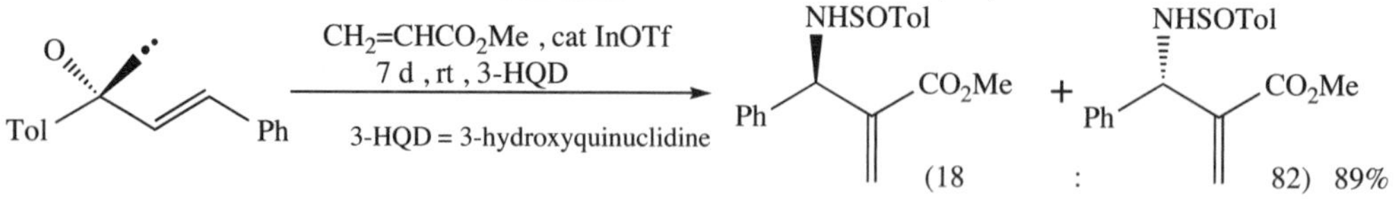

O
Tol
Ph
CH_2=$CHCO_2Me$, cat InOTf
7 d , rt , 3-HQD
3-HQD = 3-hydroxyquinuclidine
NHSOTol
Ph
CO_2Me
+
NHSOTol
Ph
CO_2Me
(18 : 82) 89%

Aggarwal, V.K.; Castro, A.M.M.; Mereu, A.; Adams, H. *Tetrahedron Lett.* **2002**, *43*, 1577.

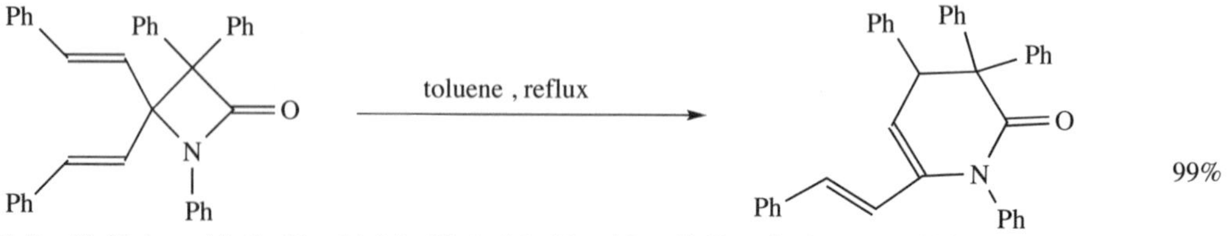

Ph Ph Ph
O
N
Ph
Ph
toluene , reflux
Ph Ph
Ph
O
N
Ph
Ph
99%

Saito, T.; Kobayashi, S.; Ohgaki, M.; Wada, M.; Nagahiro, C. *Tetrahedron Lett.* **2002**, *43*, 2627.

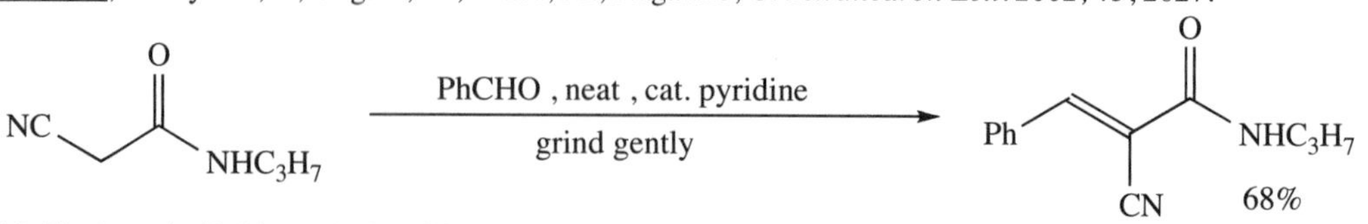

O
NC
NHC_3H_7
PhCHO , neat , cat. pyridine
grind gently
O
Ph
NHC_3H_7
CN
68%

McCluskey, A.; Robinson, P.J.; Hill, T.; Scott, J.L.; Edwards, J.K. *Tetrahedron Lett.* **2002**, *43*, 3117.

PhCHO → (CH_2=$CHCO_2Me$, $TsNH_2$, 2% Ti(O*i*-Pr)$_4$ / 1% chiral quinuclidine , MS 4Å , THF , 2d , rt) → NHTs, Ph, CO_2Me 78% (68% ee)

Balan, D.; Adolfsson, H. *Tetrahedron Lett.* ***2003***, *44*, 2521.

O, P, Ph, O, OEt, OEt → (cat Pd(OAc)$_2$, P*t*-Bu$_3$, CO , K_2CO_3 , rt / [ClTi=NTNS , $Cl_2TiN(TMS)_2$ toluene , HMPA , overnight) → Ph, NH_2, O 59%

Ueda, K.; Mori, M. *Tetrahedron Lett.* ***2004***, *45*, 2907.

AcHN → (cat RuCl(CO)(PPh$_3$)$_3$ / PhH , 80°C) → AcHN 85% (59:41 *E:Z*)

Krompiec, S.; Pigulla, M.; Krompiec, M.; Baj, S.; Mrowiec-Bialon, J.; Kasperczk, J. *Tetrahedron Lett.* ***2004***, *45*, 5257.

Et—≡—Et → (1. PhN=C=O , CuCl , Cp_2ZrEt_2 / 2. 2 NBS , rt) → Et, Et, Br, NHPh, O 58%

Li, Y.; Matsumura, H.; Yamanaka, M.; Takahashi, T. *Tetrahedron* ***2004***, *60*, 1393.

O, HN → (Grubbs' I catalyst / DCM , rt) → O, HN 86%

Brimble, M.A.; Trzoss, M. *Tetrahedron* ***2004***, *60*, 5613.

PhCHO → (chiral bis(hydroxy)sulfonamide ligand / Et_2Zn , Ti(O*i*-Pr)$_4$, toluene , 0°C , 8h) → OH, Ph, Et >95% (90% ee)

Yus, M.; Ramón, D.J.; Prieto, O. *Tetrahedron: Asymmetry* ***202***, *13*, 1573.

NHTs → (Py , O_2 , xylenes , 80°C , 2h / cat Pd(OAc)$_2$) → Ts, N 87%

Fix, S.R.; Brice, J.L.; Stahl, S.S. *Angew. Chem. Int. Ed.* ***2002***, *41*, 164.

Et, N—Ts → (O, , DMF , –40°C , 1d / chiral quinoline derivative) → Et, NHTS, O 50% 96%ee)

Shi, M.; Xu, Y.-M. *Angew. Chem. Int. Ed.* ***2002***, *41*, 4507.

N–Et → (benzoyl chloride, CH_2=$CHSnBu_3$, rt / cat Pd$_2$(bda)$_3$•$CHCl_3$, MeCN , DCM , 16h) → O, N, Ph, Et 82%

Davis, J.L.; Dhawan, R.; Arndtsen, B.A. *Angew. Chem. Int. Ed.* ***2004***, *43*, 590.

OH

PhH , heat

89%

Eradl, S.N.; Kennedy-Smith, J.J.; Kim, J.; Trauner, D. *Synlett* ***2002***, 411.

$Re(CO)_5$, 100°C , neat

95% (61:39 *E:Z*)

Seregeyev, S.; Hesse, M. *Synlett* ***2002***, 1313.

4 SmI_2 , 5 HMPA

70%

Concellón, J.M.; Bardales, E. *Eur. J. Org. Chem.* ***2004***, 1523.

TsOH

$(Me_2CHO)_2CH_2$

42%

Ledneczki, I.; Agócs, P.M.; Molnár, Á. *Synlett* ***2003***, 2255.

Also via alkenyl acids: Section 322 (Carboxylic Acid -Alkene)

SECTION 350: AMINE - AMINE

$PhNH_2$, 10% PBu_3 , H_2O

also with ArOH , ArSH, RNH_2

98%

Fan, R.-H.; Hou, X.-L. *J. Org. Chem.* ***2003***, *68*, 726.

$PhNH_2$, SiO_2(activated)

1h

91%

Anand, R.V.; Pandey, G.; Singh, V.K. *Tetrahedron Lett.* ***2002***, *43*, 3975.

130°C , microwaves , 3h

92%

Naryan, S.; Seelhammer, T.; Gawley, R.E. *Tetrahedron Lett.* ***2004***, *45*, 757.

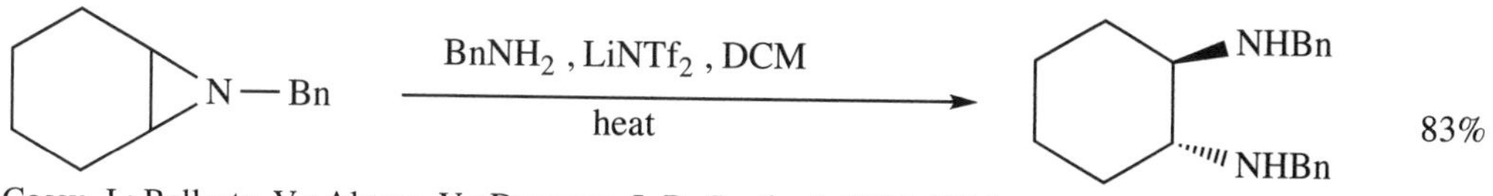

$BnNH_2$, $LiNTf_2$, DCM, heat — 83%

Cossy, J.; Bellosta, V.; Alauze, V.; Desmurs, J.-R. *Synthesis* **2002**, 2211.

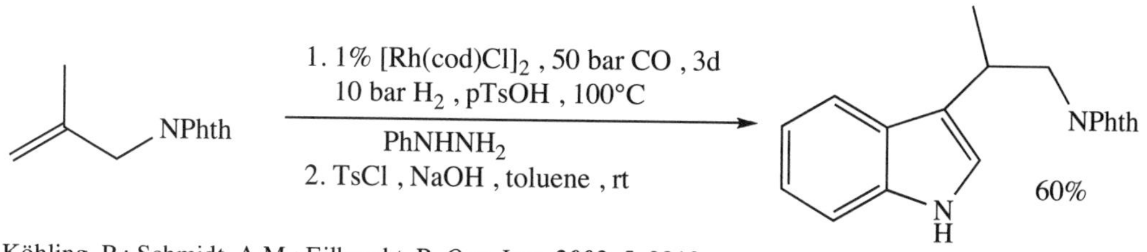

1. 1% $[Rh(cod)Cl]_2$, 50 bar CO , 3d, 10 bar H_2 , pTsOH , 100°C, $PhNHNH_2$
2. TsCl , NaOH , toluene , rt

NPhth — 60%

Köhling, P.; Schmidt, A.M.; Eilbracht, P. *Org. Lett.* **2003**, *5*, 3213.

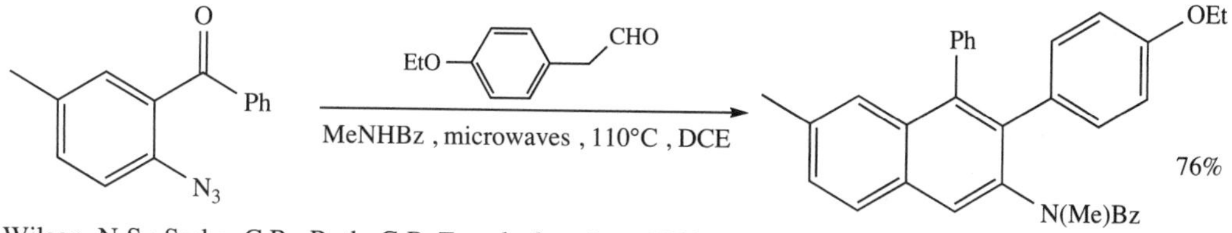

N_3, Ph, O; EtO, CHO; MeNHBz , microwaves , 110°C , DCE — Ph, OEt, N(Me)Bz — 76%

Wilson, N.S.; Sarko, C.R.; Roth, G.P. *Tetrahedron Lett.* **2002**, *43*, 581.

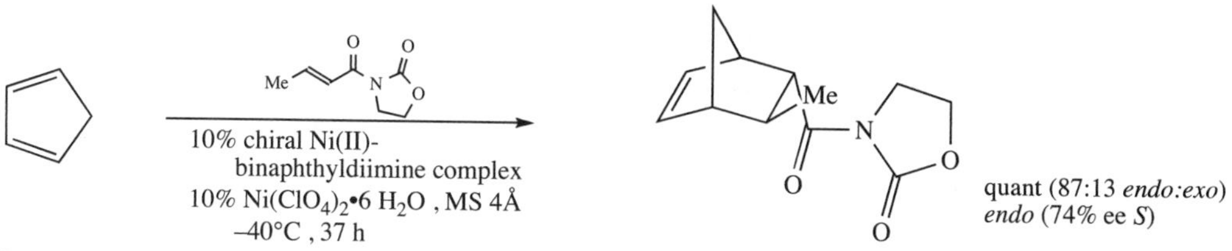

10% chiral Ni(II)-binaphthyldiimine complex, 10% $Ni(ClO_4)_2$•6 H_2O , MS 4Å, –40°C , 37 h

quant (87:13 *endo:exo*) *endo* (74% ee *S*)

Suga, H.; Kakehi, A.; Mitsuda, M. *Chem Lett.* **2002**, *31*, 900.

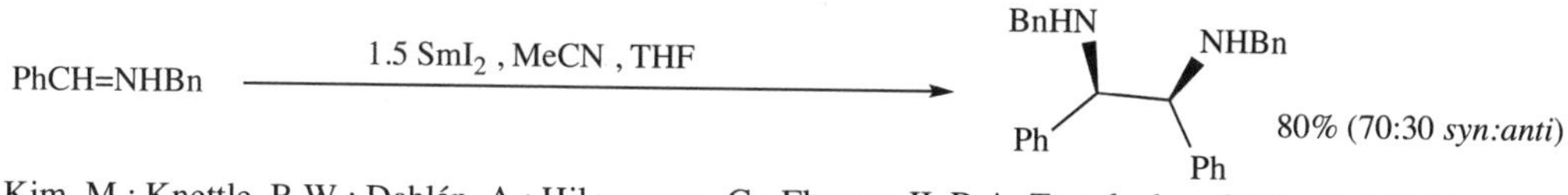

PhCH=NHBn — 1.5 SmI_2 , MeCN , THF — BnHN, NHBn, Ph, Ph — 80% (70:30 *syn:anti*)

Kim, M.; Knettle, B.W.; Dahlén, A.; Hilmersson, G.; Flowers II, R.A. *Tetrahedron* **2003**, *59*, 10397.

SECTION 351: AMINE - ESTER

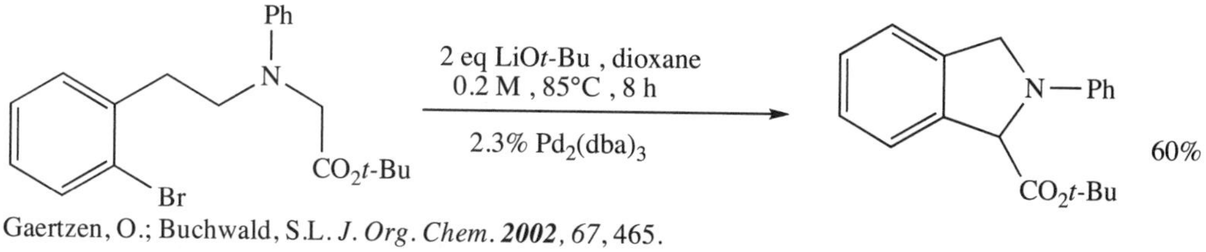

Ph, N, CO_2t-Bu, Br — 2 eq LiO*t*-Bu , dioxane, 0.2 M , 85°C , 8 h, 2.3% $Pd_2(dba)_3$ — N—Ph, CO_2t-Bu — 60%

Gaertzen, O.; Buchwald, S.L. *J. Org. Chem.* **2002**, *67*, 465.

SmI_2 , t-BuOH , THF
20°C
75%

Jacobsen, M.F.; Turks, M.; Hazell, R.; Skrydstrup, T. *J. Org. Chem.* ***2002***, *67*, 2411.

$N_2C(CO_2Me)_2$, DCM , rt
cat $Rh_2(OAc)_4$, 1.5 h
96%

Gibe, R.; Kerr, M.A. *J. Org. Chem.* ***2002***, *67*, 6247.

$MeCO_2Me$, LDA , THF
2 eq $ClTi(Oi\text{-}Pr)_3$, –78°C
94% (99:1 dr)

Tang, T.P.; Ellman, J.A. *J. Org. Chem.* ***2002***, *67*, 7819.

L-proline (30%) , DMSO
rt , 6 h
66% (85:15 *syn:anti*)
(86:25 ee (*syn:anti*)

Chowdari, N.S.; Suri, J.T.; Barbas III, C.F. *Org. Lett.* ***2004***, *6*, 2507.

1. MeO_2CCH_2CN , THF
t-BuOK , 0°C , 1 h
2. H_2O
55%

Bullington, J.L.; Wolff, P.R.; Jackson, P.F. *J. Org. Chem.* ***2002***, *67*, 9439.

, Me_2Zn
$BrCH_2CO_2Me$, DCM , rt
cat $NiCl(PPh_3)_3$
96%

Adrian Jr. J.C.; Snapper, M.L. *J. Org. Chem.* ***2003***, *68*, 2143.

cat $Cu(acac)_2$, N_2CHCO_2Et
xylene , reflux

N Me → N Me CO_2Et 49%

Heath, P.; Roberts, E.; Sweeney, J.B.; Wessel, H.P.; Workman, J.A. *J. Org. Chem.* ***2003***, *68*, 4083.

CO_2t-Bu N

1. a chiral cinchonidinium bromide
KOH , BnBr , KOH , 0°C
toluene , 2 h
2. 1N HCl , THF

H_2N CO_2t-Bu Ph
93% (88% ee)

Jew, S.-i.; Jeong, B.-S.; Le, J.-H.; Yoo, M.-S.; Lee, Y.-J.; Park, B.-i.; Kim, M.G.; Park, H.-g. *J. Org. Chem.* ***2003***, *68*, 4514.

Ph Ph N O Bu Me

$Co_2(CO)_8$, MeCN
75°C , 30 min

Me Ph N O Ph OBu H H 64%

Ishikawa, T.; Kudoh, T.; Yoshida, J.; Yasuhara, A.; Manabe, S.; Saito, S. *Org. Lett.* ***2002***, *4*, 1907.

BnO Cl O N

4 NaOMe , 2;N in MeOH
heat , 4 h

O MeO OBn N H 80%

Dejaegher, Y.; De Kimpe, N. *J. Org. Chem.* ***2004***, *69*, 5974.

Me N CO_2Et

t-Bu O
t-BuO N P – OEt OEt
t-Bu
0.5M , *t*-BuOH , 120°C

Me N CO_2Et *t*-Bu 54%

Leroi, C.; Bertin, D.; Dufils, P.-E.; Gigmes, D.; Marque, S.; Tordo, P.; Couturier, J.-L.; Guerret, O.; Ciufolini, M.A. *Org. Lett.* ***2003***, *5*, 4943.

MeO_2C NH_2

Cl–C6H4–CHO , OTMS OMe
20% $InCl_3$, MeOH , rt
overnight

H N MeO_2C CO_2Me Cl 57% (99:1 *R*:*S*)

Loh, T.-P.; Chen, S.-L. *Org. Lett.* ***2002***, *4*, 3647.

Bu_3Sn–$CH_2C(=CH_2)CO_2Et$, 5% Pd cat; 120h, rt, THF, MeOH — NHBn, Ph, CO_2Et — 81% (82% ee)

Fernandes, R.A.; Yamamoto, Y. *J. Org. Chem.* ***2004***, *69*, 3562.

OTMS, OEt, , % $Sc(OTf)_3$; MeCN, rt — O, N, NH, Bn, CO_2Et, EtO_2C — 53% (9:1 dr)

N, N, O, O, Bn, EtO_2C

Jacobsen, M.F.; Ionita, L.; Skrydstrup, T. *J. Org. Chem.* ***2004***, *69*, 4792.

AcO — 1. 2.25 LDA, THF, –78°C; 2. Ph–CH=N–C$_6$H$_4$–OMe, –40°C → –20°C — MeO, NH, O, Ph, O — 76% (96:4)

Hata, S.; Iguchi, M.; Iwsawa, T.; Yamada, K.-i.; Tomioka, K. *Org. Lett.* ***2004***, *6*, 1721.

Bn-I + BnO-N=$CHCO_2Me$ — BEt_3, hexane; aq. MeOH — Bn, CO_2Me, NHOBn — 62%

McNabb, S.B.; Ueda, M.; Naito, T. *Org. Lett.* ***2004***, *6*, 1911.

H, O, CO_2Et — acetone, 20% L-proline; DMSO — O, NHPMP, CO_2Et — 86% (99% ee)

Córdova, A.; Notz, W.; Zhong, G.; Betancort, J.M.; Barbas III, C.F. *J. Am. Chem. Soc.* ***2002***, *124*, 1842.
Córdova, A.; Watanabe, S.-i.; Tanaka, F.; Notz, W.; Barbas III, C.F. *J. Am. Chem. Soc.* ***2002***, *124*, 1866.

Bn, +, N, Ph, O– — $OSiMe_2t$-Bu, OMe; cat Ti(BINOL), –78°C — Ph, Bn, N, CO_2Me, $OSiMe_2t$-Bu — 99% (92% ee)

Murahashi, S.-I.; Imada, Y.; Kawakami, T.; Harada, K.; Yonemushi, Y.; Tomita, N. *J. Am. Chem. Soc.* ***2002***, *124*, 2888.

CO_2t-Bu — LiN(TMS)(Piv), toluene; –78°C, 15 min — CO_2t-Bu, NHBn — 92% (97% ee)

Doi, H.; Sakai, T.; Iguchi, M.; Yamada, K.-i.; Tomioka, K. *J. Am. Chem. Soc.* ***2003***, *125*, 2886.

N-*p*-An, Ph, CO_2Et — 1. $(BzO)_2$, EtCN, 2 TMS_2AlCl . allyl-$SnBu_3$, –20°C → 50°C, 23 h; 2. aq. KF — NH*p*-An, Ph, CO_2Et — 93%

Niwa, Y.; Shimizu, M. *J. Am. Chem. Soc.* ***2003***, *125*, 3720.

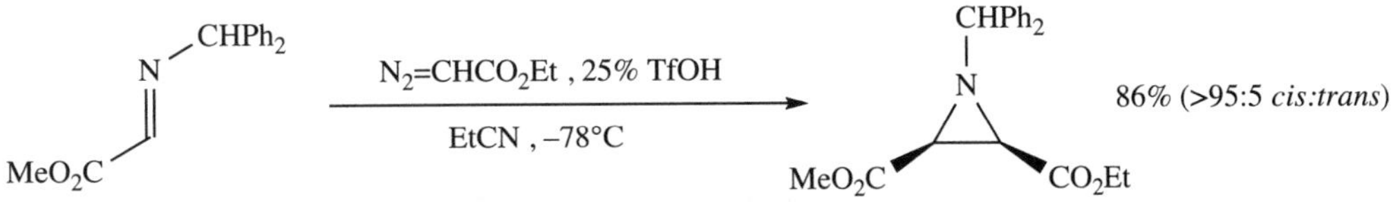
CHPh2
N
MeO2C
N2=CHCO2Et , 25% TfOH
EtCN , –78°C
CHPh2
N
MeO2C
CO2Et
86% (>95:5 cis:trans)

Williams, A.L.; Johnston, J.N. *J. Am. Chem. Soc.* ***2004***, *126*, 1612.

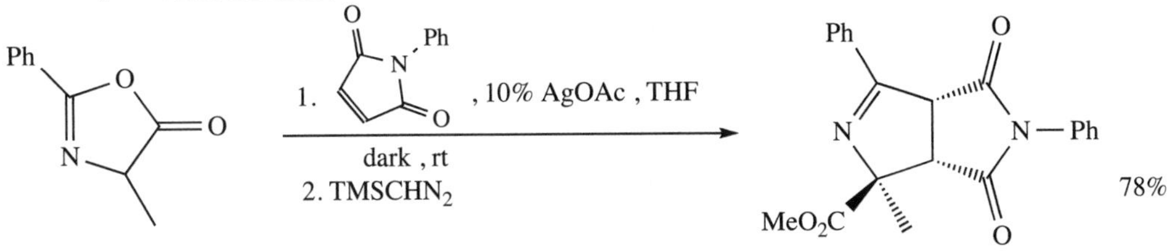
Ph
O
N
O
1.
, 10% AgOAc , THF
dark , rt
2. TMSCHN2
Ph
O
N
N — Ph
MeO2C
O
78%

Peddibhotla, S.; Tepe, J.J. *J. Am. Chem. Soc.* ***2004***, *126*, 12776.

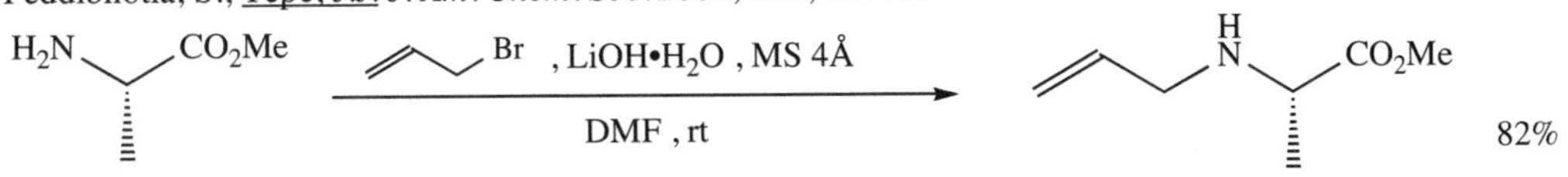
H2N
CO2Me
Br
, LiOH•H2O , MS 4Å
DMF , rt
H
N
CO2Me
82%

Cho, J.H.; Kim, B.M. *Tetrahedron Lett.* ***2002***, *43*, 1273.

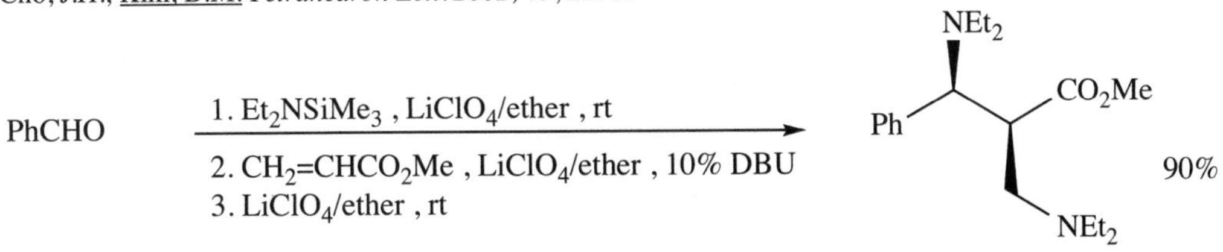
PhCHO
1. Et2NSiMe3 , LiClO4/ether , rt
2. CH2=CHCO2Me , LiClO4/ether , 10% DBU
3. LiClO4/ether , rt
NEt2
CO2Me
Ph
NEt2
90%

Azizi N.; Saidi, M.R. *Tetrahedron Lett.* ***2002***, *43*, 4305.

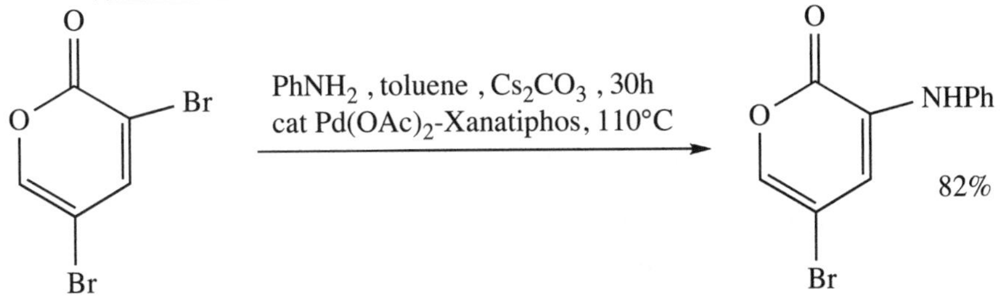
O
O
Br
Br
PhNH2 , toluene , Cs2CO3 , 30h
cat Pd(OAc)2-Xanatiphos, 110°C
O
O
NHPh
82%
Br

Lee, J.H.; Cho, C.-G. *Tetrahedron Lett.* ***2003***, *44*, 65.

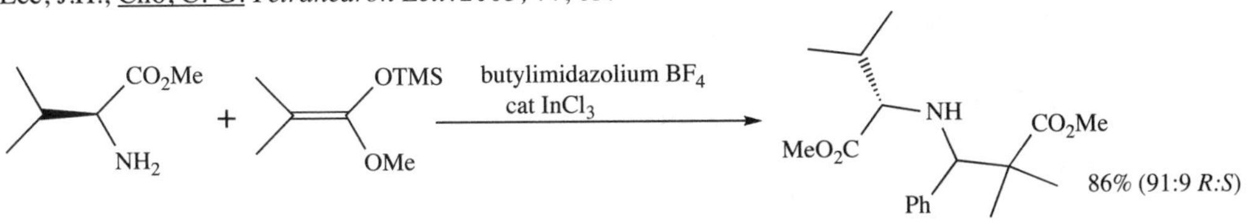
CO2Me
NH2
+
OTMS
OMe
butylimidazolium BF4
cat InCl3
NH
MeO2C
CO2Me
Ph
86% (91:9 R:S)

Chen, S.-L.; Ji, S.-J.; Loh, T.-P. *Tetrahedron Lett.* ***2003***, *44*, 2405.

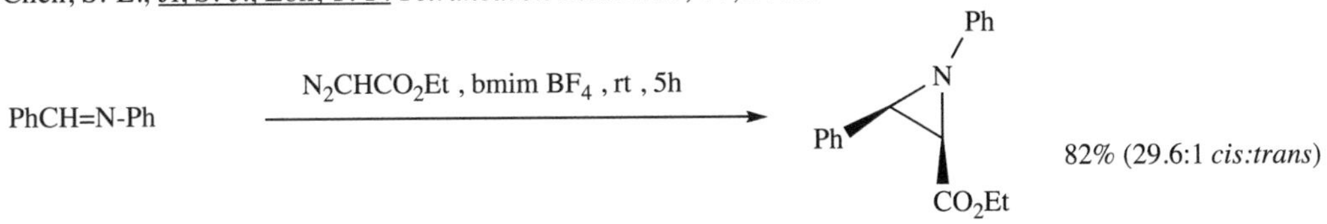
PhCH=N-Ph
N2CHCO2Et , bmim BF4 , rt , 5h
Ph
N
Ph
CO2Et
82% (29.6:1 cis:trans)

Sun, W.; Xia, C.-G.; Wang, H.-W. *Tetrahedron Lett.* ***2003***, *44*, 2409.

OTMS → $CF_3C{=}N\text{-}Bn$, 0.5 $BF_3{\bullet}OEt_2$, 0.5h / DCM , −78°C → F_3C, BnHN, O, O 90%

Spanedda, M.V.; Ourévitch, M.; Crousse, B.; Bégué, J.-P.; Bonnet-Delpon, D. *Tetrahedron Lett.* ***2004***, *45*, 5023.

$Ph_2C{=}N$, CO_2t-Bu → MeI , [bmim] PF_6 , aq KOH , rt / 40 min → $Ph_2C{=}N$, CO_2t-Bu 82%

Loureco, N.M.T.; Afonso, C.A.M. *Tetrahedron* ***2003***, *59*, 789.

CO_2Me → piperidine , $LiClO_4$, heat / rt , 1h → N, CO_2Me 86%

Azizi, N.; Saidi, M.R. *Tetrahedron* ***2004***, *60*, 383.

CO_2Me, MeO_2C, N H, CO_2Et → O, CO_2Et / CAN , MeOH → Me, EtO_2C, MeO_2C, N, CO_2Et, CO_2Me 60%

Chuang, C.-P.; Wu, Y.-L. *Tetrahedron* ***2004***, *60*, 1841.

MeO_2C, $N\text{–}NPh_2$ → Me_2CHI , In , H_2O-MeOH / 20°C , 1h → MeO_2C, $NHNPh_2$, $CHMe_2$ 98%

Miyabe, H.; Ueda, M.; Nishimura, A.; Naito, T. *Tetrahedron* ***2004***, *60*, 4227.

Ph, N, CO_2Et → 1. $ZnBr_2$, ether , 20°C / 2. *t*-BuZnBr , ether , 20°C / 3. NH_4Cl → Ph, N H, CO_2Et 64% (80:20 dr)

Chiev, K.P.; Roland, S.; Mangeney, P. *Tetrahedron: Asymmetry* ***2002***, *13*, 2205.

O^-, N^+, Bn → $CH_2{=}CHCO_2Et$, 2 SmI_2 , 10h / THF , −78°C → HO, N, Bn, CO_2Et 76%

Masson, G.; Cividino, P.; Py, S.; Vallée, Y. *Angew. Chem. Int. Ed.* ***2003***, *42*, 2265.

Ph, N, OMe → 3 OTMS, OMe / HBF_4 , 0.4 SDS H_2O , rt → HN, OMe, Ph, CO_2Me quant

Akiyama, T.; Itoh, J.; Fuchibe, K. *Synlett* ***2002***, 1269.

NH_2, CO_2Et + Ph, OH → DMSO-AcOH , 65°C , DMSO / CO_2H, IO → EtO_2C, N, Ph 70%

Bagley, M.C.; Hughes, D.D.; Sabo, H.M.; Taylor, P.H.; Xiong, X. *Synlett* ***2003***, 1443.

CH_2=CO_2Et + $PhCH_2NH_2$ → (5% $Cu(OTf)_2$, H_2O; rt, 12 h) → $BnNHCH_2CH_2CO_2Et$ 98%

Xu, L.-W.; Wei, J.-W.; Xia, C.-G.; Zhou, S.-L.; Hu, X.-X. *Synlett* ***2003***, 2425.

O O | Ot-Bu → ($PhNH_2$, DCM, 30% $MgSO_4$; 30% $Zn(ClO_4)_2$•6 H_2O, 21 h) → NHPh O | Ot-Bu 98%

Bartoli, G.; Bosco, M.; Locatelli, M.; Marcantoni, E.; Melchiorre, P.; Sambri, L. *Synlett* ***2004***, 239.

N—H → (CH_2=$CHCO_2Et$, silica gel; 40°C, 1 h) → N — CO_2Et 93%

Basu, B.; Das, P.; Hossain, I. *Synlett* ***2004***, 2630.

CH_2=$CHCO_2Et$ + Me_2C=N-OH → (MeCN, 0.2 PPh_3, 65°C) → N O CO_2Et 85%

Bhuniya, D.; Mohan, S.; Narayanan, S. *Synthesis* ***2003***, 1018.

O | CO_2Et → ($BnNH_2$, 0.1 AcOH, 0.2 h; ultrasound) → NHBn | CO_2Et 98%

Brandt, C.A.; da Silva, A.C.M.P.; Pancote, C.G.; Brito, C.L.; da Silveira, M.A.B. *Synthesis* ***2004***, 1557.

CO_2Et → (Ph · N N · H; [bmim] PF_6, rt, 8min) → N N Ph CO_2Et 89%

Yadav, J.S.; Reddy, B.V.S.; Basak, A.K.; Narsaiah, A.V. *Chem. Lett.* ***2003***, *32*, 988.

CO_2Et → (piperidine, $Co(OAc)_2$) → N CO_2Et quant

Xu, L.-W.; Li, L.; Xia, C.-G. *Helv. Chim. Acta* ***2004***, *87*, 1522.

O | H_2N OBn → (PhCHO, acetophenone, TMSCl; $FeCl_3$•6 H_2O, rt) → NHBn O | Ph OEt 72%

Xu, L.-W.; Wang, Z.-T.; Xia, C.-G.; Li, L.; Zhao, P.-Q. *Helv. Chim. Acta* ***2004***, *87*, 2608.

$BrCH_2CO_2Et$ + PhCH=N-Ph → ($RhCl(PPh_3)_3$, Et_2Zn; THF, 0°C) → NHPh | Ph CO_2Et [Ph N O Ph (in toluene at 40°C, 65%)]

Kanai, K.; Wakabayashi, H.; Honda, T. *Heterocycles* ***2002***. *58*, 47.

Me_2HC CO_2Me NH_2 → (EtCHO, MeOH; AcOH) → Me_2HC CO_2Me N= H Et → ($NaBH_4$; 0°C) → Me_2HC CO_2Me HN H Et 82%

Verardo, G.; Geatti, P.; Pol, E.; Giumanini, A.G. *Can. J. Chem.* ***2002***, *80*, 779.

REVIEW:

"Catalytic Enantioselective Strecker Reactions and Analogous Syntheses"
Gröger, H. *Chem. Rev.* **2003**, *103*, 2795.

Related Methods: Section 315 (Carboxylic Acid - Amide)
Section 316 (Carboxylic Acid - Amine)
Section 344 (Amide - Ester)

SECTION 352: AMINE - ETHER, EPOXIDE, THIOETHER

PhI, toluene, 70°C
1% $Pd(PPh_3)_4$, 5% TBAB
2 K_2CO_3, 47 h
C_3H_7 NHBn O Ph C_3H_7 O NBn 95%

Ma, S.; Xie, H. *J. Org. Chem.* **2002**, *67*, 6575.

cat CuCl, MeOH, 2.5h
toluene, 100°C
C_3H_7 C_3H_7 N MeO 70%

Kamijo, S.; Sasaki, Y.; Yamamoto, Y. *Tetrahedron Lett.* **2004**, *45*, 35.

O + N≡C
cat $GaCl_3$, toluene, 60°C
methylcyclohexane, 18 hg
O N 94%

Chatani, N.; Oshita, M.; Tobisu, M.; Ishii, Y.; Murai, S. *J. Am. Chem. Soc.* **2003**, *125*, 7812.

OMe
PhCOOH + N N MeO N Cl
DCM, rt, 3h
OH NH_2
O Ph N 78%

Bandgar, B.P.; Pandit, S.S. *Tetrahedron Lett.* **2003**, *44*, 2331.

Ph N *t*-Bu
$Zn(Et_2CHCO_2)_2$, hν, Pyrex
i-PrOH
Ph NH*t*-Bu Ph NH*t*-Bu 40% + Ph NH*t*-Bu Ph NH*t*-Bu 35%

Ortega, M.; Rodríguez, M.A.; Campos, P.J. *Tetrahedron* **2004**, *60*, 6475.

SECTION 353: AMINE - HALIDE, SULFONATE

O SO_2Ph F N H C_7H_{15}
1. $NaBH_4$, dioxane, reflux
AcOH, 3 h
2. H_2O
F N H C_7H_{15} 86%

Mataloni, M.; Petrini, M.; Profeta, R. *Synlett* **2003**, 1129.

diphosgene , MeCN
130°C (sealed tube)
12 h
75%

Lee, B.S.; Lee, J.H.; Chi, D.Y. *J. Org. Chem.* **2002**, *67*, 7884.

PhC≡CH , $ZnCl_2$
NEt_3 , 25°C → 50°C
60%

Jiang, B.; Si, Y.-G. *J. Org. Chem.* **2002**, *67*, 9449.

F_3C—≡—C_6H_4—Cl
10% $Pd_2(dba)_3$•$CHCl_3$, 80% P(*o*-Tol)$_3$
Net_3 , DMF , 80°C , 1 h
(34 : 66) 85%

Konno, T.; Chae, J.; Ishihara, T.; Yamanaka, H. *J. Org. Chem.* **2004**, *69*, 8258.

1. 1.5 PhSeBr , Na_2CO_3 , MeCN , rt , 16h
2. aq. Na_2CO_3
78%

Outurquin, F.; Pannecoucke, X.; Berthe, B.; Paulmier, C. *Eur. J. Org. Chem.* **2002**, 1007.

4 HBr , 3.1 H_2O_2 , MeOH , 8h
0°C → rt
91%

Vyas, P.V.; Bhatt, A.K.; Ramachandraiah, G.; Bedekar, A.V. *Tetrahedron Lett.* **2003**, *44*, 4085.

aq. HBr , hν , 0.025M , 4h
96%

McIntyre, B.P.; Coleman, B.D.; Wubbels, G.G. *Tetrahedron Lett.* **2004**, *45*, 7709.

HCl , −78°C , DCM , 1 h
99%

Noack, M.; Kalsow, S.; Göttlich, R. *Synlett* **2004**, 1110.

$Me_2CHCH_2NH_2$ —trichloroisocyanuric acid / DCM, 0.5 h, 15°C→ $Me_2CHCH_2NCl_2$ quant

DeLuca, L.; Giacomelli, G. *Synlett* ***2004***, 2180.

O, N, H —$Ca(OCl)_2$, moist Al_2O_3 / DCM, 10 h→ O, N, Cl 95%

Larionov, O.V.; Kozhushkov, S.I; de Meijere, A. *Synthesis* ***2003***, 1916.

SECTION 354: AMINE - KETONE

Cl, Cl, Ph, N, *i*-Pr —1. 2 NaH, DMSO, 8°C, 3 h 2. aq. H^+→ O, Ph, N, *i*-Pr 46%

Djaegher, Y.; Mangelinckx, S.; De Kimpe, N. *J. Org. Chem.* ***2002***, *67*, 2075.

N*t*-Bu, Ph —MeO–C₆H₄–I / cat $Pd(PPh_3)_4$, DMF rt, 5 min→ N, Ph, O, MeO 74%

Dai, G.; Larock, R.C. *J. Org. Chem.* ***2002***, *67*, 7042.

O —piperidine, DCM, rt, 12 h $Bi(NO)_3$→ O, N 65%

Srivastava, N.; Banik, B.K. *J. Org. Chem.* ***2003***, *68*, 2109.

OMe, TMSO —PhCHO, $PhNH_2$, MeOH / rt, 2 h→ Ph, N, O, Ph >99%

Yuan, Y.; Li, X.; Ding, K. *Org. Lett.* ***2002***, *4*, 3309.

PhNO , THF , –20°C
2 h
92%

Momiyama, N.; Yamamoto, H. *Org. Lett.* **2002**, *4*, 3579.

18-crown-6 , DMF
4 KO_2 , 16 h , 25°C
52%

Jiang, W.; Zhang, X.; Sui, Z. *Org. Lett.* **2003**, *5*, 43.

acetone , 30% proline
90% (93% ee)

List, B.; Pojarliev, P.; Biller, W.T.; Martin, H.J. *J. Am. Chem. Soc.* **2002**, *124*, 827.

2.2 *t*-BuLi , –78°C
3 h
87%

Ruiz, J.; Sotomayor, N.; Lete, E. *Org. Lett.* **2003**, *5*, 1115.

PhCHO , DMF , 50°C
72%

Suginome, M.; Uehlin, L.; Yamamoto, A.; Murakami, M. *Org. Lett.* **2004**, *6*, 1167.

PhN=O , 30% L-proline
DMF , rt , 5.5h
79% (>99% ee)

Hayashi, Y.; Yamaguchi, J.; Sumiya, T.; Hibino, K.; Shoji, M. *J. Org. Chem.* **2004**, *69*, 5966.

$EtO_2C-N=N-CO_2Et$
chiral Cu complex
DCM
55% (68% ee)

Juhl, K.; Jørgensen, K.A. *J. Am. Chem. Soc.* **2002**, *124*, 2420.

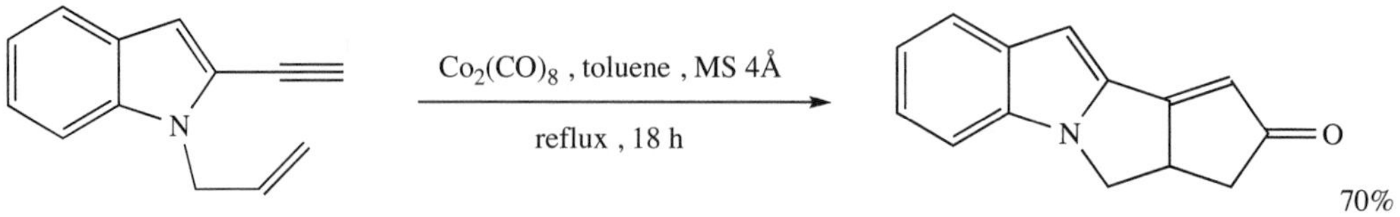

$Co_2(CO)_8$, toluene , MS 4Å; reflux , 18 h — 70%

Pérez-Serrano, L.; Domínguez, G.; Pérez-Castells, J. *J. Org. Chem.* **2004**, *69*, 5413.

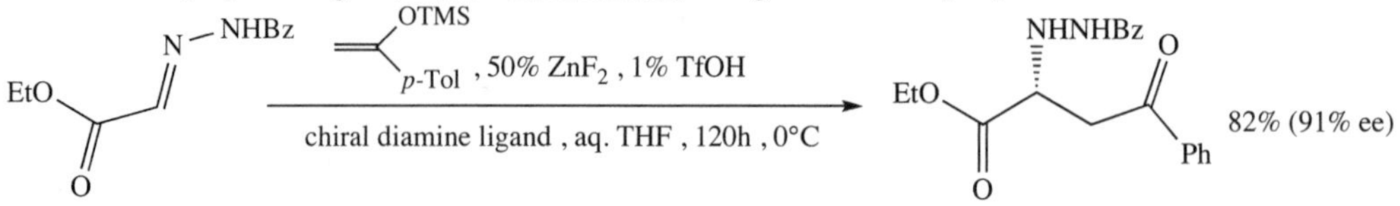

OTMS, *p*-Tol, 50% ZnF_2 , 1% TfOH; chiral diamine ligand , aq. THF , 120h , 0°C — 82% (91% ee)

Kobayashi, S.; Hamada, T.; Manabe, K. *J. Am. Chem. Soc.* **2002**, *124*, 5640.

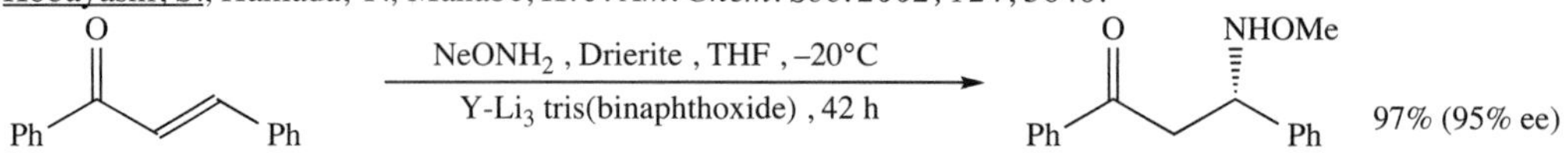

$NeONH_2$, Drierite , THF , −20°C; Y-Li_3 tris(binaphthoxide) , 42 h — 97% (95% ee)

Yamagiwa, N.; Matsunaga, S.; Shibasaki, M. *J. Am. Chem. Soc.* **2003**, *125*, 16178.

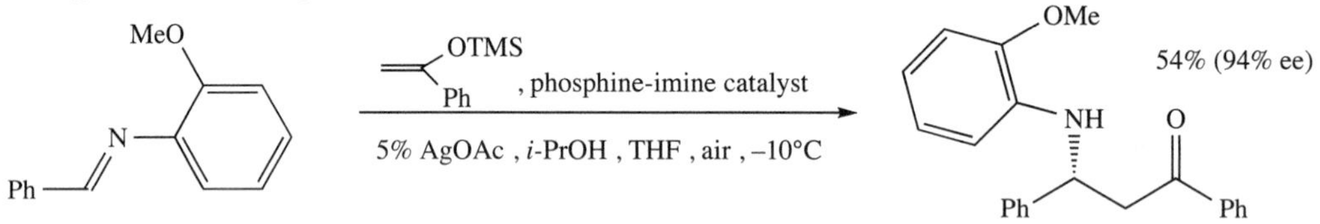

OTMS, Ph, phosphine-imine catalyst; 5% AgOAc , *i*-PrOH , THF , air , −10°C — 54% (94% ee)

Josephsohn, N.S.; Snapper, M.L.; Hoveyda, A.H. *J. Am. Chem. Soc.* **2004**, *126*, 3734.

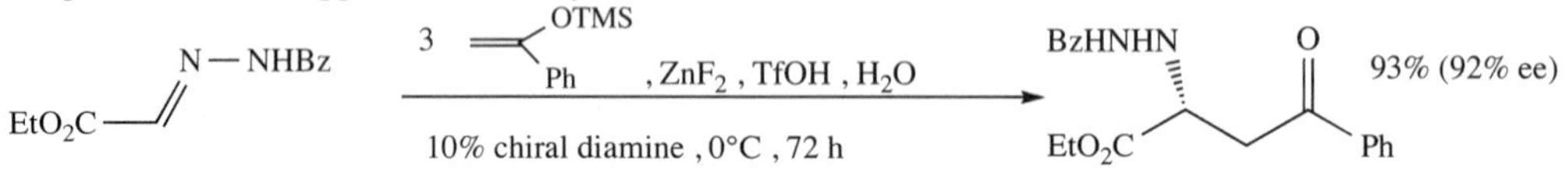

3 OTMS, Ph, ZnF_2 , TfOH , H_2O; 10% chiral diamine , 0°C , 72 h — 93% (92% ee)

Hamada, T.; Manabe, K.; Kobayashi, S. *J. Am. Chem. Soc.* **2004**, *126*, 7768.

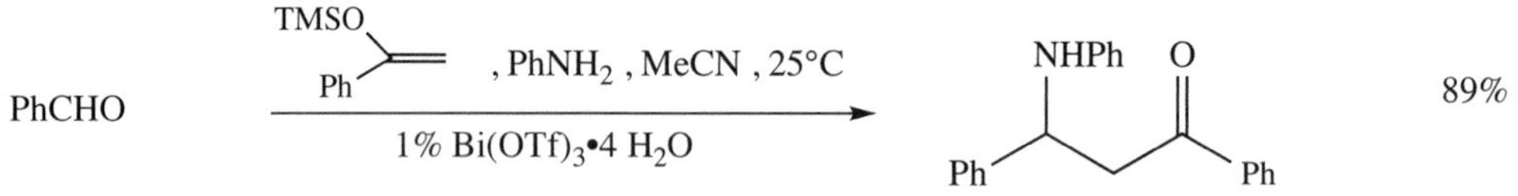

PhCHO — TMSO, Ph, , $PhNH_2$, MeCN , 25°C; 1% $Bi(OTf)_3$•4 H_2O — 89%

Ollevier, T.; Nadeau, E. *J. Org. Chem.* **2004**, *69*, 9293.

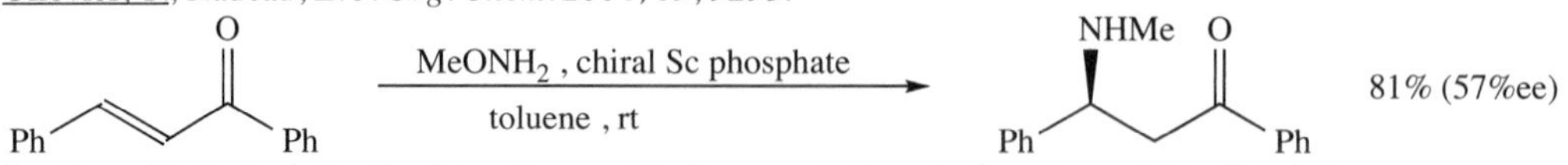

$MeONH_2$, chiral Sc phosphate; toluene , rt — 81% (57%ee)

Sugihara, H.; Daikai, K.; Jin, X.L.; Furuno, H.; Inanaga, J. *Tetrahedron Lett.* **2002**, *43*, 2735.

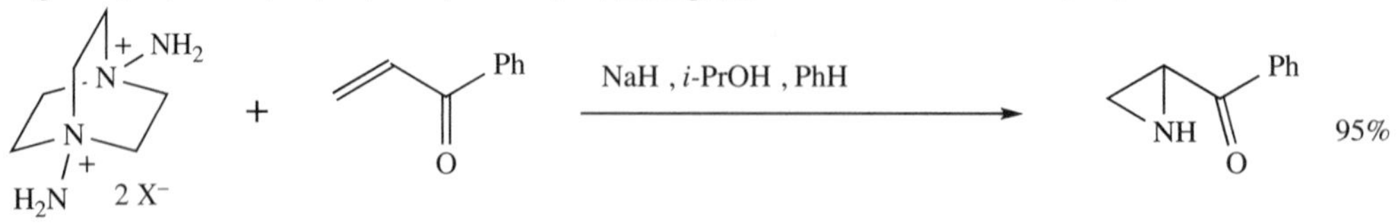

2 X^- + ... NaH , *i*-PrOH , PhH — 95%

Xu, J.; Jiao, P. *J. Chem. Soc. Perkin Trans. 1* **2002**, 1491.

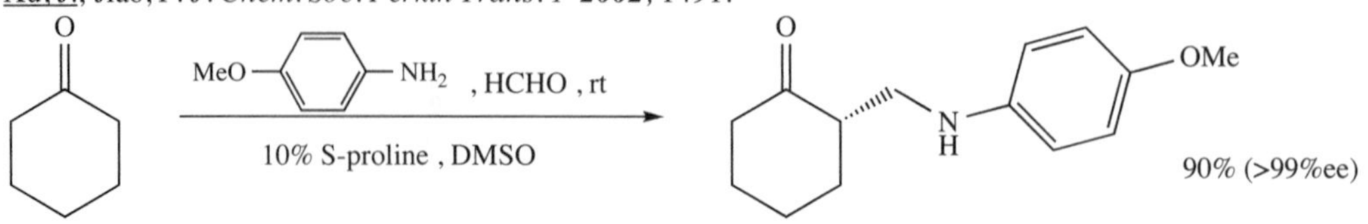

MeO–C6H4–NH_2 , HCHO , rt; 10% S-proline , DMSO — 90% (>99%ee)

Ibrahem, I.; Casas, J.; Córdova, A. *Angew. Chem. Int. Ed.* **2004**, *43*, 6528.

5% L-proline, 30 min
[bmim] BF_4

PMP = *p*-anisidine

99% (>99% ee)
>19:1 dr

Chowdari, N.S.; Ramachary, D.B.; Barbas III, C.F. *Synlett* ***2003***, 1906.

4-Cl-PhCHO, $PhNH_2$, perfluorinated naphthalene
2% $Sc(OSO_2C_8H_{17})_3$, hexane, 60°C

73%

Shi, M.; Cui, S.-C. *New J. Chem.* ***2004***, *28*, 1286

1. MeCN, 2 TiI_4, –40°C → rt, 21 h
2. 5 Ac_2O, rt, 6 h

52%

Shimizu, M.; Manabe, N.; Goto, H. *Chem Lett.* ***2003***, *32*, 1088.

1. $PhNH_2$, microwaves
2. 255°C

93x65%

Cernuchová, P.; Vo-Thanh, G.; Milata, V.; Loupy, A. *Heterocycles* ***2004***, *64*, 177.

$PhNMe_2$, HCHO
5 AcOH, 80°C
30 min

63% + 7%

Takahashi, H.; Kashiwa, N.; Hashimoto, Y.; Nagasawa, K. *Tetrahedron Lett.* ***2002***, *43*, 2935.

REVIEWS:

"Electrophilic α-Amination of Carbonyl Compounds"
Erdik, E. *Tetrahedron* ***2004***, *60*, 8747.

"The Direct Catalytic Asymmetric Mannich Reaction"
Córdova, A. *Acc. Chem. Res.* ***2004***, *37*, 102.

SECTION 355: AMINE - NITRILE

1. Bu_3SnH , AIBN
2. Bu_3SnH , H_2O

NC CN N3 → CN HN N H 60%

Benati, L.; Bencivenni, G.; Leardini, R.; Minozzi, M.; Nanni, D.; Scialpi, R.; Spagnolo, P.; Zanardi, G.; Rizzoli, C. *Org. Lett.* **2004**, *6*, 417.

H_2N OMe Me

1. PhCHO , PhH , reflux
2. $NaBH_4$, EtOH
3. malonitrile , 2 TsOH

→ CN NH_2 N Bn 19% overall

Chien, T.-C.; Meade, E.A.; Hinkley, J.M.; Townsend, L.B. *Org. Lett.* **2004**, *6*, 2857.

MeO—C6H4—N≡C

OCO_2Me , toluene
$TMSN_3$, cat $Pd_2(dba)\cdot CHCl_3$
dppe , toluene , 60°C

→ MeO—C6H4—N(CN) 99%

Kamijo, S.; Yamamoto, Y. *J. Am. Chem. Soc.* **2002**, *124*, 11940.

N CH-CN

3 BHT , toluene , 0.05M , 120°C
sealed Pyrex tube

→ H N CN 70%

Amos, D.T.; Renslo, A.R.; Danheiser, R.L. *J. Am. Chem. Soc.* **2003**, *125*, 4970.

$PhNMe_2$

5% $RuCl_3\cdot n\text{-}H_2O$, O_2 , NaCN
MeOH , AcOH

→ Me N Ph CH_2CN 88%

Murahashi, S.-I.; Komiya, N.; Terai, H.; Nakae, T. *J. Am. Chem. Soc.* **2003**, *125*, 15312.

CN

$(t\text{-Bu})_2PH$, rt
cat (PigiPhos)Ni(thf) $(ClO_4)_2$

→ $(t\text{-Bu})_2P$ CN 65%

Sadow, A.D.; Haller, I.; Fadini, L.; Togni, A. *J. Am. Chem. Soc.* **2004**, *126*, 14704.

PhCHO

$PhNH_2$, TMSCN , cat $BiCl_3$, MeCN , rt , 10h

→ NHPh Ph CN 84%

De, S.K.; Gibbs, R.A. *Tetrahedron Lett.* **2004**, *45*, 7407.

PhCHO

$PhNH_2$, KCN , $InCl_3$, THF , 6h

→ NHPh Ph CN 75%

Ranu, B.C.; Dey, S.S.; Hajra, A. *Tetrahedron* **2002**, *58*, 2529.

PhCHO

$(Et_2N)_2BCN$, THF , rt , 20h

→ CN Ph NEt_2 92%

Suginome, M.; Yamamoto, A.; Ito, Y. *Chem. Commun.* **2002**, 1392.

PhCH=N-Ph $\xrightarrow[\text{BINOL derivative}]{1.5\ Et_2AlCN\text{ , toluene}}$ PhCH(NHPh)CN 96% (61% ee)

Nakamura, S.; Sato, N.; Sugimoto, M.; Toru, T. *Tetrahedron: Asymmetry* ***2004***, *15*, 1513.

$PhNMe_2$ $\xrightarrow{O_2\text{ , NaCN , MeOH , 5\% }RuCl_3\text{ ,AcOH}}$ $Ph(Me)NCH_2CN$ 88%

North, M. *Angew. Chem. Int. Ed.* ***2004***, *43*, 4126.

$Me_2NCH(CN)SiMe_3$ $\xrightarrow{\text{1. }sec\text{-BuLi , –78°C; 2. PhCHO , THF , –78°C → 20°C}}$ $(NC)(Me_2N)C=CHPh$ 91% (64:36 *E:Z*)

Adam, W.; Ortega-Schulte, C.M. *Synlett* ***2003***, 414.

PhCH=CH-CH=N-*p*-An $\xrightarrow[\text{0.5 }AlCl_3\text{ , DCM , –80°C → rt}]{Me_2C=C(OTMS)(OEt)\text{ , 1.5 TMSCN}}$ EtO_2C-C(Me)$_2$-CH(Ph)-CH$_2$-CH(NH*p*-An)CN 81% (47:53 *syn:anti*)

Shimizu, M.; Kamiya, M.; Hachiya, I. *Chem. Lett.* ***2003***, *32*, 606

Et_2NH + CH_2=CHCN $\xrightarrow{\text{cat bmim }BF_4\text{ , }H_2O\text{ , rt}}$ $Et_2NCH_2CH_2CN$ 97%

Xu, L.-W.; Li, J.-W.; Zhou, S.-L.; Xia, C.-G. *New J. Chem.* ***2004***, *28*, 183.

SECTION 356: AMINE - ALKENE

I-CH=CH-CH$_2$-Br $\xrightarrow[\text{2. PhI , LiCl , cat }Pd(PPh_3)_4]{\text{1. In , PhCH=NPh}}$ PhCH(NHPh)CH$_2$CH=CHPh 57%

Hirashita, T.; Hayashi, Y.; Mitsui, K.; Araki, S. *J. Org. Chem.* ***2003***, *68*, 1309.

Ph—≡—Ph $\xrightarrow[\text{cat }Yb(PPh_2)(thf)_4]{Ph_2PH\text{ , THF , rt}}$ PhCH=C(Ph)PPh$_2$ quant (95:5)

Takai, K.; Koshoji, G.; Komeyama, K.; Takeda, M.; Shishido, T.; Kitani, A.; Takhira, K. *J. Org. Chem.* ***2003***, *68*, 6554.

cyclohexyl-C(Me)=C=CH-Me $\xrightarrow[\text{BF}_3\text{•OEt}_2\text{ , DCM, rt , 1.5 d}]{\text{PhCH=N-Bn}}$ N-Bn octahydroquinoline (4-Me, 3-ethylidene, 2-Ph) 54%

Regás, D.; Afonso, M.M.; Rodríguez, M.L.; Palenzuela, J.A. *J. Org. Chem.* ***2003***, *68*, 7845.

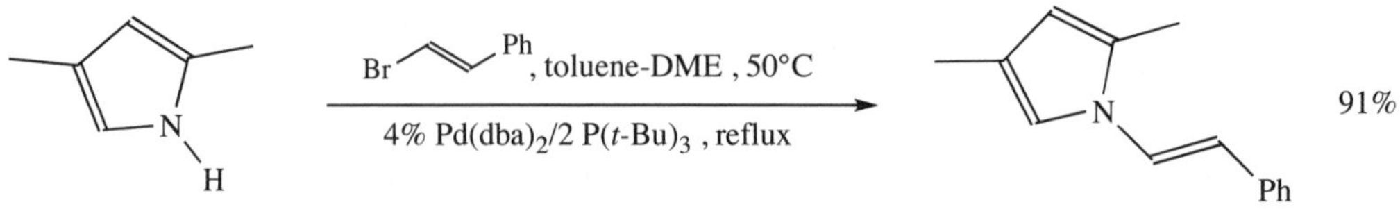

Br–CH=CH–Ph, toluene-DME, 50°C; 4% $Pd(dba)_2$/2 $P(t\text{-}Bu)_3$, reflux — 91%

Lebedev, A.Y.; Izmer, V.V.; Kazyul'kin, D.N.; Beletskaya, I.P.; Voskoboynikov, A.Z. *Org. Lett.* ***2002**, 4*, 623.

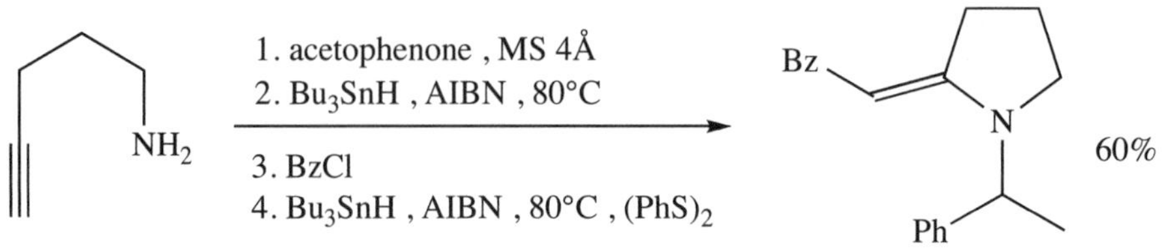

NH_2

1. acetophenone, MS 4Å
2. Bu_3SnH, AIBN, 80°C
3. BzCl
4. Bu_3SnH, AIBN, 80°C, $(PhS)_2$

Bz, N, Ph — 60%

Prabhakaran, E.N.; Nugent, B.M.; Williams, A.; Nailor, K.E.; Johnston, J.N. *Org. Lett.* ***2002**, 4*, 4197.

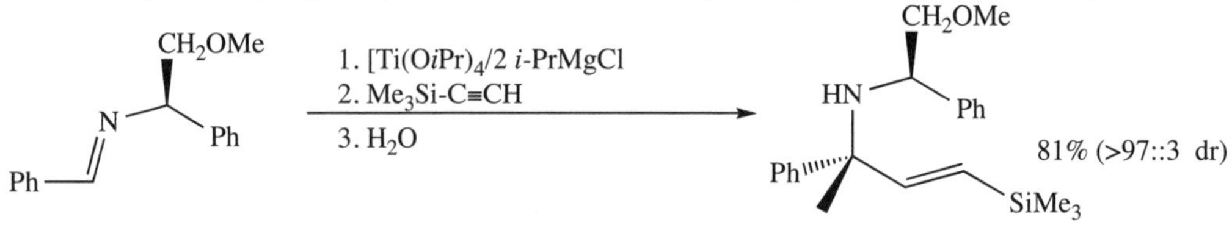

CH_2OMe, N, Ph, Ph

1. $[Ti(OiPr)_4/2\ i\text{-PrMgCl}$
2. $Me_3Si\text{-C}{\equiv}CH$
3. H_2O

CH_2OMe, HN, Ph, Ph, $SiMe_3$ — 81% (>97::3 dr)

Fukuhara, K.; Okamoto, S.; Sato, F. *Org. Lett.* ***2003**, 5*, 2145.

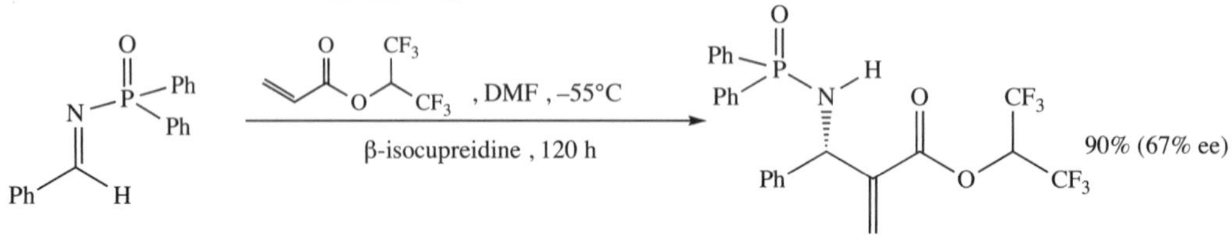

O, N, P, Ph, Ph, Ph, H

O, CF_3, O, CF_3, DMF, –55°C; β-isocupreidine, 120 h

O, Ph, P, Ph, N, H, O, CF_3, Ph, O, CF_3 — 90% (67% ee)

Kawahara, S.; Nakano, A.; Esumi, T.; Iwabuchi, Y.; Hatakeyama, S. *Org. Lett.* ***2003**, 5*, 3013.

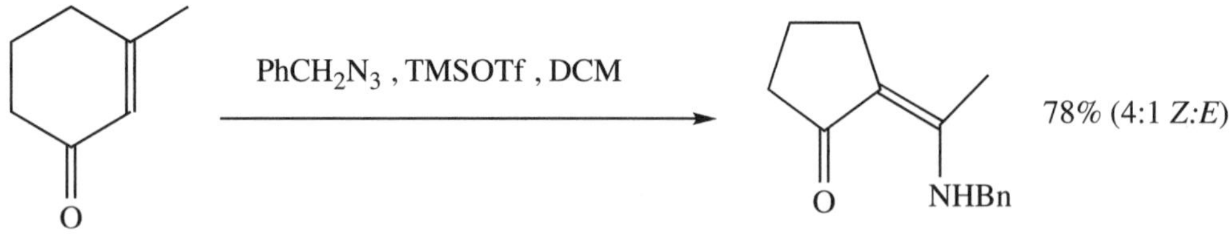

O

$PhCH_2N_3$, TMSOTf, DCM

O, NHBn — 78% (4:1 *Z:E*)

Reddy, D.S.; Judd, W.R.; Aubé, J. *Org. Lett.* ***2003**, 5*, 3871.

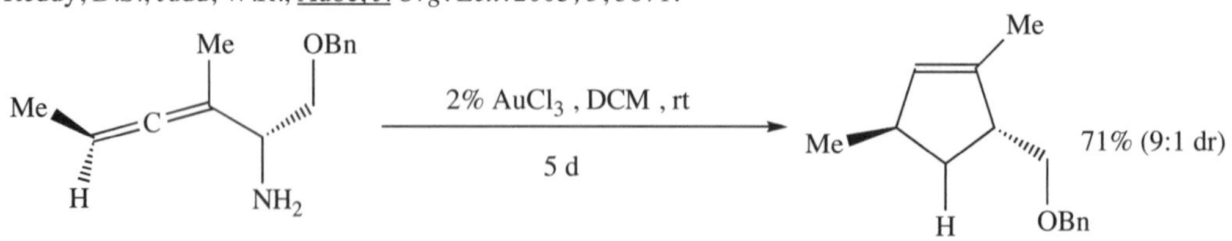

Me, OBn, Me, C, H, NH_2

2% $AuCl_3$, DCM, rt; 5 d

Me, Me, H, OBn — 71% (9:1 dr)

Morita, N.; Krause, N. *Org. Lett.* ***2004**, 6*, 4121.

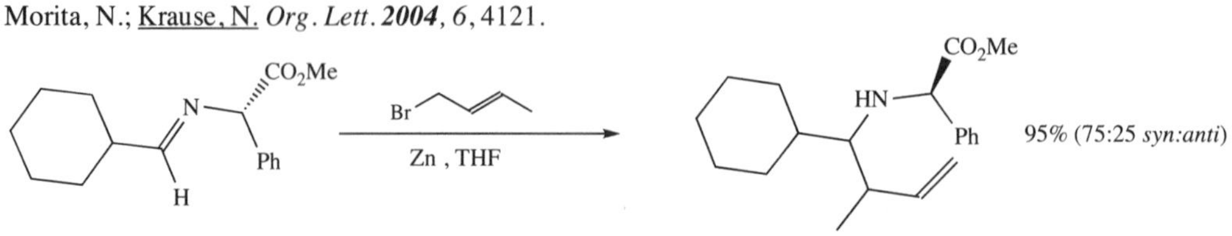

CO_2Me, N, Ph, H

Br–CH₂CH=CHCH₃; Zn, THF

CO_2Me, HN, Ph — 95% (75:25 *syn:anti*)

Lee, C.-L.K.; Ling, H.-Y.; Loh, T.-P. *J. Org. Chem.* ***2004**, 69*, 7787.

Co(III)-salen catalyst; DCM, 16 h — 98% (98% ee)

Huang, Y.; Iwama, T.; Rawal, V.H. *J. Am. Chem. Soc.* **2002**, *124*, 5950.

5% Mo carbene catalyst; PhH, 22°C — 78% (98% ee)

Dolman, S.J.; Sattely, E.S.; Hoveyda, A.H.; Shrock, R.R. *J. Am. Chem. Soc.* **2002**, *124*, 6991.

cat (OHF*)•SmN(TMS)$_2$; C_6D_6 — >95% (93:7 *E:Z*)

OHF = S-Me$_2$Si(η^5-octahydrofluorenyl)(CpR*), R* = (–)-menthyl

Hong, S.; Marks, T.J. *J. Am. Chem. Soc.* **2002**, *124*, 7886.

MeCH=C=CH-CO$_2$Bn, DCM; 10% Zn(OTf)$_2$, 23°C — 95% (>98:2 *syn:anti*)

Lambert, T.H.; MacMillan, D.W.C. *J. Am. Chem. Soc.* **2002**, *124*, 13646.

MeO$_2$C—≡—CO$_2$Me, 0.1 M, 60°C; 5% [Rh(CO)$_2$Cl]$_2$, toluene, 12. 5 h

Wender, P.A.; Pedersen, T.M.; Scanio, M.J.C. *J. Am. Chem. Soc.* **2002**, *124*, 15154.

PhCH=CHCH$_2$OCO$_2$Me; cat Ir-phospharmidite, THF, rt — 80% (94% ee)

Ohmura, T.; Hartwig, J.F. *J. Am. Chem. Soc.* **2002**, *124*, 15164.

Et—≡—Et: 1. 1.5 Cp$_2$ZrHCl, DCM, rt; 2. 1.5 Me$_2$Zn, toluene, –78°C; 3. PhCH=N-P(=O)Ph$_2$, toluene, rt — 72%

Wipf, P.; Kendall, C.; Stephenson, C.R.J. *J. Am. Chem. Soc.* **2003**, *125*, 761.

PhNH$_2$: Bu-C≡CH, *t*-BuN≡C, toluene, 100°C; 10% Ti(NMe$_2$)$_2$(dpma), 2 d — 77%

Cao, C.; Shi, Y.; Odom, A.L. *J. Am. Chem. Soc.* **2003**, *125*, 2880.

Ph, CO_2Me, N_2 — Ph–CH=N–C_6H_4–NO_2, DCM; cat $Rh_2(OAc)_4$-cu$(OTf)_2$, reflux → Ph, Ph, N, O_2N, CO_2Me 66% (<98:2 *trans:cis*)

Doyle, M.P.; Yan, M.; Hun, W.; Gronenberg, L.S. *J. Am. Chem. Soc.* ***2003***, *125*, 4692.

N—NBz (+ −) — 1. MeMgBr, DCM, rt; 2. $NaBH_4$, MeOH → N—NHBz, Me 87%

Legault, C.; Charette, A.B. *J. Am. Chem. Soc.* ***2003***, *125*, 6360.

Me, N — 2% $PtCl_2$, dioxane, 60°C, trace HCl → Me, N 92%

Liu, C.; Han, X.; Wang, X.; Widenhofer, R.A. *J. Am. Chem. Soc.* ***2004***, *126*, 3700.

Ph, O, Si, Cl, N, Me + N-NHBz, Ph — $CHCl_3$, 40°C, 1 d → NHNHBz, Ph 86% (90% ee)

Berger, R.; Duff, K.; Leighton, J.L. *J. Am. Chem. Soc.* ***2004***, *126*, 5686.

Ph—≡—Me — PhNHMe, 5% $Pd(PPh_3)_4$, 100°C; 10% $PhCO_2H$, dioxane, 4 h → Me, N, Ph, Ph 93%

Patil, N.T.; Wu, H.; Kadota, I.; Yamamoto, Y. *J. Am. Chem. Soc.* ***2004***, *126*, 8745.

MeO, N—Ac, Me — 5% $PtCl_2$, anisole; 80°C → O, N, Me 89%

Shimada, T.; Nakamura, I.; Yamamoto, Y. *J. Am. Chem. Soc.* ***2004***, *126*, 10546.

Cl, NH_2, Cl — Ph, Ph (cyclopropylidene), $Sn(OTf)_2$; DCE, 85°C, 1d → Ph, Ph, ( )$_2$N, Cl, Cl 85%

Shi, M.; Chen, Y.; Xu, B.; Tang, J. *Tetrahedron Lett.* ***2002***, *43*, 8019.

PhCH=N-Ph , cat $[RhCl(cod)]_2$ / dioxane , 1h — 99%

Kakuuchi, A.; Taaguchi, T.; Hanazawa, Y. *Tetrahedron Lett* **2003**, *44*, 923.

, PhH , Ms 4Å / 1% $Pd(OAc)_2$, reflux , 5h / 25% Ti(O-*i*-Pr)$_4$ — 38% + 10%

Shue, Y.-J.; Yang, S.-C.; Lai, H.-C. *Tetrahedron Lett.* **2003**, *44*, 1481.

0.2 $Pd(PPh_3)_4$, 2.5 *t*-BuOK , THF / reflux , 0.5h — 49%

Solé, D.; Urbaneja, X.; Bonjoch, J. *Tetrahedron Lett.* **2004**, *45*, 3131.

MeO–C$_6$H$_4$–NHLi / toluene , –78°C → –10°C — 44%

Satoh, T.; Ogino, Y.; Nakamura, M. *Tetrahedron Lett.* **2004**, *45*, 5785.

benzophenone , Sm/SmI_2 / THF , reflux , 3h — 80%

Xu, X.; Zhang, Y. *Tetrahedron* **2002**, *58*, 503.

morpholine , cat Pd_2dba_3/P(*o*-Tol)$_3$ / NaO*t*-Bu , toluene , 90°C , 6h — 78%

Barluenga, J.; Fernández, M.A.; Aznar, F.; Valdés, C. *Chem. Commun.* **2002**, 2362.

(*i*-Pr)$_3$N — $[\eta^3\text{-2,6-}(t\text{-}Bu_2)(PCH_2)_2C_6H_3]IrH_2$ / d_{10}-*p*-xylene , 90°C , 5h , 1d → (*i*-Pr)$_2$NCH=CH$_2$ 98%

Zhang, X.; Fried, A.; Knapp, S.; Goldman, A.S. *Chem. Commun.* **2003**, 2060.

Ph–CH=N–Ph + TMS–CH=CH–CH$_2$–TeBu$_2$ Br → (LHMDS, HMPA; toluene, –78°C) → N-Ph aziridine (Ph, CH=CH–TMS) 84% (98:2 *trans:cis*)

Liao, W.-W.; Deng, X.-M.; Tang, Y. *Chem. Commun.* **2004**, 1516.

CH_2=C(Ph)Cl → (morpholine, cat Pd_2dba_3/Xantphos; toluene, NaO*t*-Bu, 6h) → CH_2=C(Ph)–N(morpholine) 96%

Barluenga, J.; Fernández, M.A.; Aznar, F.; Valdés, C. *Chem. Commun.* **2004**, 1400.

1-ethynylcyclohexene → ($PhNH_2$, cat Ti(dmp)$_2$(NMe$_2$)$_2$; toluene, 50°C, 16h) → cyclohexenyl–C(=N–Ph)CH$_3$ 88%

Cao, C.; Li, Y.; Shi, Y.; Odom, A.L. *Chem. Commun.* **2004**, 2002.

Ph–C≡C–Me → (PhCH=N-Me, cat [Ni(cod)]$_2$; 3 BEt$_3$, MeOAc, MeOH; PBu$_3$, 50°C) → PhC(Et)=C(Me)–CH(NHMe)Ph + PhCH=C(Me)–CH(NHMe)Ph (94 : 6) 80%

Patel, S.J.; Jamison, T.F. *Angew. Chem. Int. Ed.* **2003**, *42*, 1364.

1-butene → (piperidine, CO/H$_2$, 100°C, toluene, 12h; NAPHOS ligand, cat [Rh(CO)$_2$(acac)]$_2$) → $CH_3(CH_2)_3$CH=CH–N(piperidine) 75%

Ahmed, M.; Seayad, A.M.; Jackstell, R.; Beller, M. *Angew. Chem. Int. Ed.* **2003**, *42*, 5615.

PhC(=O)–C≡C–Ph → (Me$_3$Si–P(pyrrolidine), THF; 0°C → 20°C) → PhC(=O)–C(SiMe$_3$)=C(Ph)–P(pyrrolidine) 70%

Reisser, M.; Maier, A.; Maas, G. *Synlett* **2002**, 1459.

(E)-BuCH=CH–B(pin) → (1. $^-$O–N$^+$(Bn)=CHPh, Me$_2$Zn, DMF, 60°C, 3.5 h; 2. H$_2$O) → HO–N(Bn)–CH(Ph)–CH=CH–Bu 90%

Pandya, S.U.; Pinet, S.; Chavant, P.Y.; Vallée, Y. *Eur. J. Org. Chem.* **2003**, 3621.

PhNHCH$_3$ → (1. *t*-BuLi, –78°C; 2. [Cl(Me)ZrCp$_2$], –50°C → –20°C; 3. CH$_2$=CHOMe, 80°C) → PhNHCH$_2$CH=CH$_2$ 83%

Barluenga, J.; Rodríguez, F.; Álvarez-Rodrigo, L.; Fañanás, F.J. *Chem. Eur. J.* **2004**, *10*, 109.

Ph_2CHNH_2 , MeCN , 0°C , 2 h
cat $[IrCl(cod)]_2$
$CHPh_2$ N 60%

Miyabe, H.; Yoshida, K.; Kobayashi, Y.; Matsumura, A.; Takemoto, Y. *Synlett* ***2003***, 1031.

$NaBH_4$, NaOEt
EtOH , 0°C , 1 d
OBn N^+ Br–
Me_2N OBn 73% (94:6 *E:Z*)

Honda, K.; Yasui, H.; Inoue, S. *Synlett* ***2003***, 2380.

N-Bn Cl
5% $Pd_2dba_3{\bullet}CHCl_3$, THF
rt, 140 h
N—Bn 56% (61:39 *cis:trans*)

Bao, M.; Nakamura, H.; Inoue, A.; Yamamoto, Y. *Chem. Lett.* ***2002***, *31*, 158.

$Ph\text{-}N_3$
Br , In , NaI , DMF
rt , 2 h
NHPh 90%

Yadav, J.S.; Madhuri, Ch.; Reddy, B.V.S.; Reddy, G.S.K.K.; Sabitha, G. *Synth. Commun.* ***2002***, *32*, 2771.

C_7H_{15}
THF , rt , 6 h
2.5% $RhH(PPh_3)_4$
PPh_3 , CF_3SO_3H
PPh_3 O_3SCF_3 C_7H_{15} 31% + PPh_3 O_3SCF_3 C_7H_{15} 56%

Arisawa, M.; Momozuka, R.; Yamaguchi, M. *Chem. Lett.* ***2002***, *31*, 272.

HO $CO_2i\text{-}Pr$
HO $CO_2i\text{-}Pr$
1. PrZnBr 2. $i\text{-}Pr_2Zn$
3. OH
4. PhN=O , $CHCl_3$, 20 h , 0°C
Ph N O OH H 86% (72% ee)

Ding, X.; Ukaji, Y.; Fujinami, S.; Inomata, K. *Chem. Lett.* ***2003***, *32*, 582.

Ph
1. 1.2 Ph_2SO , 1.2 Tf_2O
DCM , –78°C → 0°C
2. $NbNH_2$, neat , rt , 2 h
NHBn Ph 79%

Yamanaka, H.; Matsuo, J.; Kawana, A.; Mukaiyama, T. *Chem. Lett.* ***2003***, *32*, 626.

$N\text{-}CHPh_2$ C_3H_7
0.5(Sn)4 5 HN_3 , 0.5 $AlCl_3$
DCM , –78°C → rt , 12 h
N_3 $NHCHPh_2$ C_3H_7 76%

Shimizu, M.; Yamauchi, C.; Ogawa, T. *Chem. Lett.* ***2004***, *33*, 606.

C NHBn
PhI , 2 $i\text{-}Pr_2NEt$, 5% $Pd(PPh_3)_4$
toluene , 60°C , 15 h
Ph N Bn 82%

Shibata, T.; Kadowaki, S.; Takagi, K. *Heterocycles* ***2002***. *57*, 2261.

REVIEWS:

"Recent Advances in Synthetic Applications of Azadienes"
Jayakumar, S.; Ishar, M.P.S.; Mahajan, M.P. *Tetrahedron* ***2002**, 58*, 379.

"Advances and Adventures in Amination Reactions of Olefins and Akynes"
Beller, M.; Breindl, C.; Eichberger, M.; Hartung, C.G.; Seayad, J.; Thiel, O.R.; Tillack, A.; Trauthwen, H. *Synlett* ***2002***, 1579.

SECTION 357: ESTER - ESTER

4 $KHSO_5$, rt , 18h; MeOH → MeO_2C–$CH_2CH_2CH_2$–CO_2Me 85%

Yan, J.; Travis, B.R.; Borhan, B. *J. Org. Chem.* ***2004**, 69*, 9299.

5 $CH_2(CO_2H)_2$, 2.2 $Mn(OAc)_3$; AcOH → 50%

Méou, A.; Lamarque, L.; Brun, P. *Tetrahedron Lett.* ***2002**, 43*, 5301.

Ac_2O , 2% $Bi(OTf)_3$, MeCN , rt 45 min → 91%

Yadav, J.S.; Reddy, B.V.S.; Swamy, T.; Rao, K.R. *Tetrahedron Lett.* ***2004**, 45*, 6037.

PhCHO → 2 Ac_2O , Zn , DME , 2 imidazole , 80°C , 1d → 97%

Hirao, T.; Santhitikul, S.; Takeuchi, H.; Ogawa, A.; Sakurai, H. *Tetrahedron* ***2003**, 59*, 10147.

2.5% $PdCl_2$/2.5% thiourea ligand , 20h; 20% CuCl , CO/O_2 , MeOH , 50°C → 90%

Dai, M.; Wang, C.; Dong, G.; Xiang, J.; Luo, T.; Liang, B.; Chen, J.; Yang, Z. *Eur. J. Org. Chem.* ***2003***, 4346.

Also via Dicarboxylic Acids: Section 312 (Carboxylic Acid - Carboxylic Acid)
Hydroxy-esters Section 327 (Alcohol - Ester)
Diols Section 323 (Alcohol - Alcohol)

SECTION 358: ESTER - ETHER, EPOXIDE, THIOETHER

TBSO; $MeO_2C(=N_2)Ph$, 2,2-dimethyltuane; cat $Rh_2(DOSP)_4$/2.2-dmb, 23°C → MeO_2C, Ph, Me, OTBS; 45% (>98% de/96% ee)

Davies, H.M.L.; Beckwith, R.E.J.; Antoulinakis, E.G.; Jin, Q. *J. Org. Chem.* **2003**, *68*, 6126.

TFEA = trifluoroethyl trifluoroacetate
NBSA = 2-(NO_2)-$C_6H_4SO_3N_3$

1. $LN(TMS)_2$, TFEA, TFA, –78°C
2. 2% $Rh_2(oct)_4$, *o*-NBSA, DCM
Cs_2CO_3, MS 3Å, 4 BnOH → BnO; 68%

Brodsky, B.H.; DuBois, J. *Org. Lett.* **2004**, *6*, 2619.

OH; 1. KHMDS, 18-crown-6, THF, –78°C; 2. S, CO_2Me, CO_2Me; 3. 2 AcOH → O, S, CO_2Me, CO_2Me

Buchanan, D.J.; Dixon, D.J.; Hernandez-Juan, F.A. *Org. Lett.* **2004**, *6*, 1357.

HO, BnO_2C, CO_2Bn; 1. oxalyl chlloride; 2. DMAP, toluene, hv; S, N, HO, –15°C → N, S, CO_2Bn, CO_2Bn; 51% (84:16)

Daglard, J.E.; Rychnovsky, S.D. *Org. Lett.* **2004**, *6*, 2713.

t-Bu, CHO; $ClCH_2CO_2t$-Bu, 10% THAB; KOH, rt, 25 h; THAB = tetrahexylammonium bromide → O, CO_2t-Bu, *t*-Bu; 61% (cis only)

Aria, S.; Suzuki, Y.; Tokumaru, K.; Shioiri, T. *Tetrahedron Lett.* **2002**, *43*, 833.

Ph, OTMS, $SiMe_3$; $PhCH(OMe)_2$, $SnCl_4$, DCM; –78°C → OMe, O, Ph, $SiMe_3$, Ph; 96% (81:19 *anti:syn*)

Honda, M.; Oguchi, W.; Segi, M.; Nakakima, T. *Tetrahedron* **2002**, *58*, 6815.

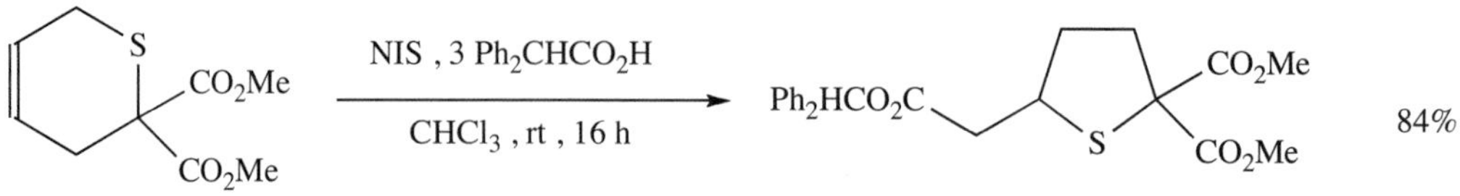

NIS , 3 Ph_2CHCO_2H

$CHCl_3$, rt , 16 h

CO_2Me CO_2Me S Ph_2HCO_2C CO_2Me CO_2Me S 84%

Lucassen, A.C.B.; Zwanenburg, B. *Eur. J. Org. Chem.* **2004**, 74.

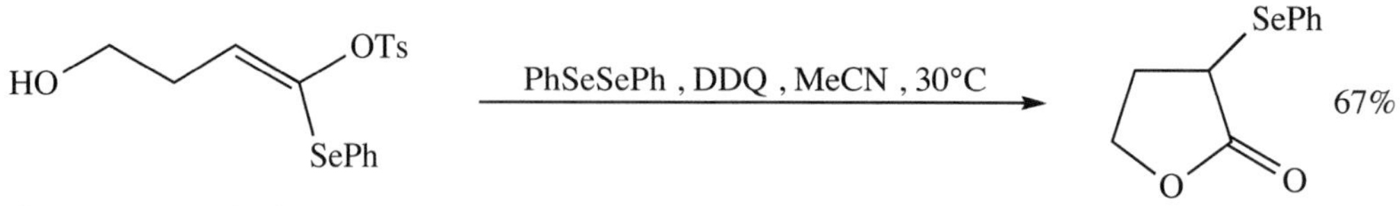

HO OTs SePh PhSeSePh , DDQ , MeCN , 30°C SePh O O 67%

Tiecco, M.; Testaferri, L.; Temperini, A.; Bagnoli, L.; Marini, F.; Santi, C. *Synlett* **2003**, 655.

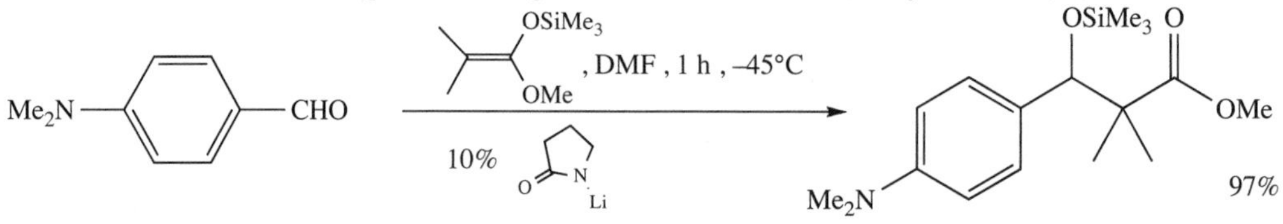

Me_2N CHO $OSiMe_3$ OMe , DMF , 1 h , –45°C 10% O N Li $OSiMe_3$ O OMe Me_2N 97%

Fujisawa, H.; Mukaiyama, T. *Chem. Lett.* **2002**, *31*, 858.

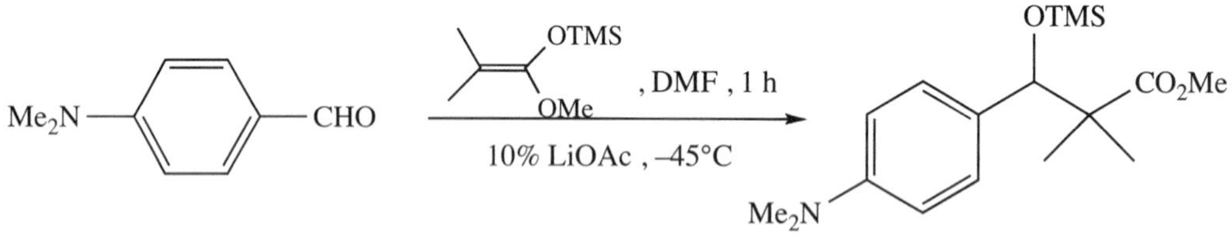

Me_2N CHO OTMS OMe , DMF , 1 h 10% LiOAc , –45°C OTMS CO_2Me Me_2N

Nakagawa, T.; Fujisawa, H.; Nagata, Y.; Mukaiyama, T. *Bull. Chem. Soc. Jpn.* **2004**, *77*, 1555.

SECTION 359: ESTER - HALIDE, SULFONATE

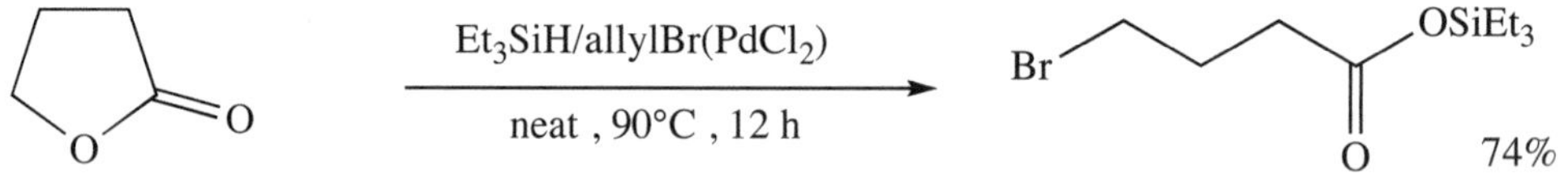

O O Et_3SiH/allylBr($PdCl_2$) neat , 90°C , 12 h Br $OSiEt_3$ O 74%

Iwata, A.; Ohshita, J.; Tang, H.; Kunai, A.; Yamamoto, Y.; Matui, C. *J. Org. Chem.* **2002**, *67*, 3927.

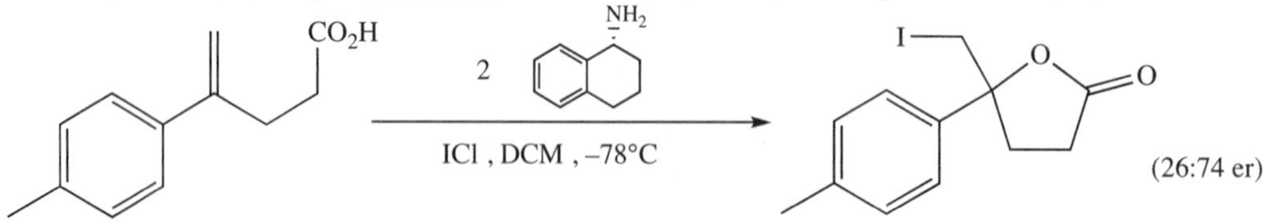

CO_2H 2 NH_2 ICl , DCM , –78°C I O O (26:74 er)

Haas, J.; Piguel, S.; Wirth, T. *Org. Lett.* **2002**, *4*, 297.

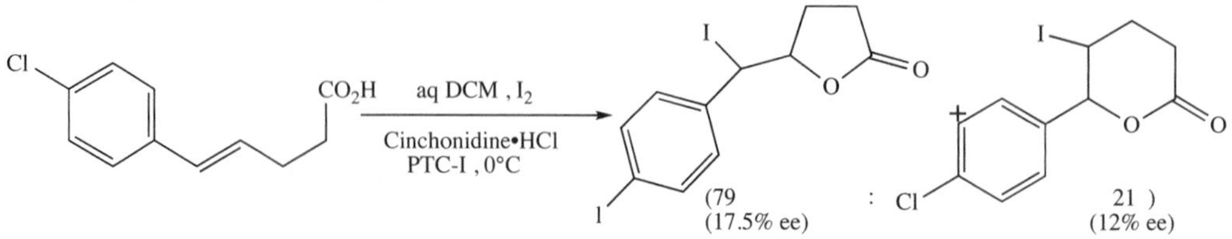

Cl CO_2H aq DCM , I_2 Cinchonidine•HCl PTC-I , 0°C I O O I (79 (17.5% ee) + I O O Cl : 21) (12% ee)

Wang, M.; Gao, L.X.; Mai, W.P.; Xia, A.X.; Wang, F.; Zhang, S.B. *J. Org. Chem.* **2004**, *69*, 2874.

1. BF_3 catalyst, O_2, DCM
2. cat H_2SO_4, rt, 4 h

92%

Yu, H.; Wu, T.; Li, C. *J. Am. Chem. Soc.* ***2002***, *124*, 10302.

PhH, hν, reflux, 1h

$SnBu_2$ $SnBu_2$

quant

Hernán, A.G.; Kilburn, J.D. *Tetrahedron Lett.* ***2004***, *45*, 831.

, 10% benzoquinone

THF, 4h, −78°C → 25°C

OC_6Cl_5

80% (99%ee)

France, S.; Wack, H.; Taggi, A.E.; Hafez, A.M.; Waagerle, Ty.R.; Shah, M.H.; Susich, C.L.; Lectka, T. *J. Am. Chem. Soc.* ***2004***, *126*, 4245.

AcCl, 2 eq SmI_2

92%

Kwon, D.W.; Kim, Y.H. *J. Org. Chem.* ***2002***, *67*, 9488.

Et CO_2H NBS, cat PhSeSePh, MeCN

52%

Mellegaard, S.R.; Tunge, J.A. *J. Am. Chem. Soc.* ***2004***, *126*, 8979.

e^-, $Et_3N{\bullet}5$ HF

65%

Hasegawa, M.; Ishii, H.; Fuchigami, T. *Tetrahedron Lett.* ***2002***, *43*, 1503.

C_6H_{13} $ClCO_2Et$, 110°C, toluene, 20h

cat $RhCl(CO)(PPh_3)_2$

CO_2Et

64%

Hua, R.; Tanaka, M. *Tetrahedron Lett.* ***2004***, *45*, 2367.

NCS, cat $Cu(Tf)_2$, ether, rt

chiral bis(oxazoline) ligand

98% (77% ee)

Marigo, M.; Kumaragurubaran, N.; Jørgensen, K.-A. *Chem. Eur. J.* ***2004***, *10*, 2133.

$BzO^{-+}PPh_3I$, 10% SmI_3

1.2 h

BzO I

94%

Liu, Y.; Zhang, Y. *Org. Prep. Proceed. Int.* ***2002***, *34*, 213.

Bu—C(CO_2Me)(CO_2Me) —[e^- , NaCl , MeOH]→ Bu—C(Cl)(CO_2Me)(CO_2Me) 83%

Okimoto, M.; Takahashi, Y. *Synthesis* **2002**, 2215.

cyclohexene —[I_2 , NH_4OAc , AcOH , 10 min]→ trans-2-iodocyclohexyl acetate (I, OAc) 94%

Myint, Y.Y.; Pasha, M.A. *Synth. Commun.* **2004**, *34*, 4477.

SECTION 360: ESTER - KETONE

Ph(Bu$_3$SnO)C=CH_2 —[CH_2=CHCO_2Me , Bu_4NBr , THF , 12 h]→ PhCO$CH_2CH_2CH_2CO_2Me$ >99%

Yasuda, M.; Chiba, K.; Ohigashi, N.; Katoh, Y.; Baba, A. *J. Am. Chem. Soc.* **2003**, *125*, 7291.

PhCH(OH)CH_3 —[3 eq PhIO , 2.5 eq TsOH•H_2O ; MeCN , heat , 1.5 h]→ PhCH(OH)CH_2OTs 96%

Ueno, M.; Nabana, T.; Togo, H. *J. Org. Chem.* **2003**, *68*, 6424.

O, CO_2Me —[CH_2=CHCH$_2$Br , KOH , toluene , rt ; chiral phase transfer catalyst]→ O, CO_2Me 94% (68% ee)

Park, E.J.; Kim, M.H.; Kim, D.Y. *J. Org. Chem.* **2004**, *69*, 6897.

H, O, O, O, H —[F_3C-C$_6$H$_4$-CH=CH_2 , 10% Ni(cod)$_2$; 12% PhPHOS , Me_2Zn , THF]→ H, CO_2H, H, O 86%

Bercot, E.A.; Rovis, T. *J. Am. Chem. Soc.* **2002**, *124*, 174.

I —[BuOH , 40 atm CO , NEt_3 ; PhH , 12 h]→ O, CO_2Bu 82%

Ryu, I.; Kreimerman, S.; Araki, S.; Oderaotoshi, Y.; Minakata, S.; Komaatsu, M. *J. Am. Chem. Soc.* **2002**, *124*, 3812.

Ph, O —[St-Bu, OSiHMe$_2$, *t*-BuOMe , DCM ; chiral oxazaborolidine , –78°C ; 2,6-diisopropylphenol]→ *t*-BuS, O, Ph, O 83% (95% ee)

Harada, T.; Adachi, S.; Wang, X. *Org. Lett.* **2004**, *6*, 4877.

Ac_2O , CD_2Cl_2 , rt
chiral ferrocenyl derivative

84% (90% ee)

Mermerian, A.H.; Fu, G.C. *J. Am. Chem. Soc.* ***2003***, *125*, 4050.

PhCHO → EtCOCl, Mg . DMF , 15°C

61%

Nishiguchi, I.; Sakai, M.; Mackawa, H.; Ohno, T.; Yamamoto, Y.; Ishino, Y. *Tetrahedron Lett.* ***2002***, *43*, 635.

Ph—≡—Br → $AgNO_3$, NBS , rt → PhCOCO$_2$Me 93%

Li, L.-S.; Wu, Y.-L. *Tetrahedron Lett.* ***2002***, *43*, 2427.

PhCHO → 1. NaCN , BTAC 2. $ClCO_2Et$ 3. LDA → PhCOCO$_2$Et 85x64%

Thasana, N.; Prachyawarakorn, V.; Tontoolarug, S.; Ruchirawat, S. *Tetrahedron Lett.* ***2003***, *44*, 1019.

PhCHO → 1. THF , 12 Me_2Zn , air , 50°C , 2d 2. CrO_3 , H_2SO_4 , acetone , 0°C , 15 min

54%

Yamamoto, Y.; Yamada, K.-i.; Tomioka, K. *Tetrahedron Lett.* ***2004***, *45*, 795.

, DCM
3% Cu(hfacac)$_2$

85%

Yan, M.; Zhao, W.-J.; Huang, D.; Ji, S.-J. *Tetrahedron Lett.* ***2004***, *45*, 6365.

$BuNO_2$ → 1. EtO_2CHO , Amberlyst A-21 2. NEt_3 , TsCl , DCM , 0°C 3. $TiCl_3$, NH_4OAc , aq. DME → $C_3H_7COCH_2CO_2Et$ 70x87x61%

Ballini, R.; Fiorini, D.; Palmieri, A. *Tetrahedron Lett.* ***2004***, *45*, 7027.

AIBN , $Ph_4Si_2H_2$, toluene
reflux

66%

Sugi, M.; Togo, H. *Tetrahedron* ***2002***, *58*, 3171.

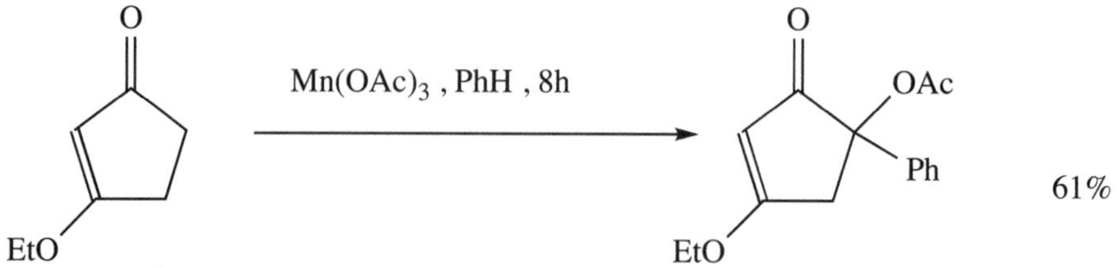

$Mn(OAc)_3$, PhH , 8h

EtO O OAc Ph

61%

Tanyeli, C.; Özdemîrhan, D.; Sezen, B. *Tetrahedron* ***2002***, *58*, 9983.

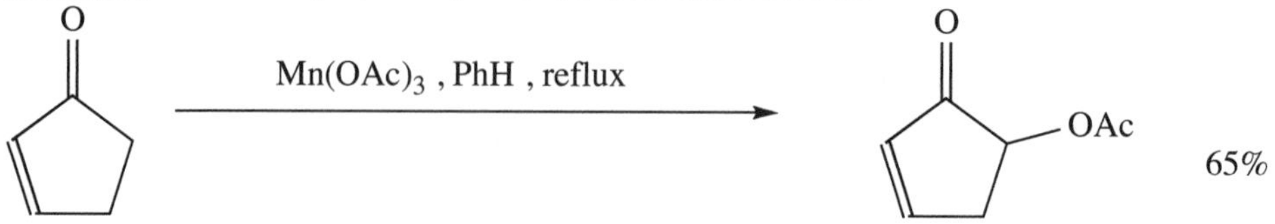

$Mn(OAc)_3$, PhH , reflux

O OAc

65%

Tanyeli, C.; Tosun, A.; Turkut, E.; Sezen, B. *Tetrahedron* ***2003***, *59*, 1055.

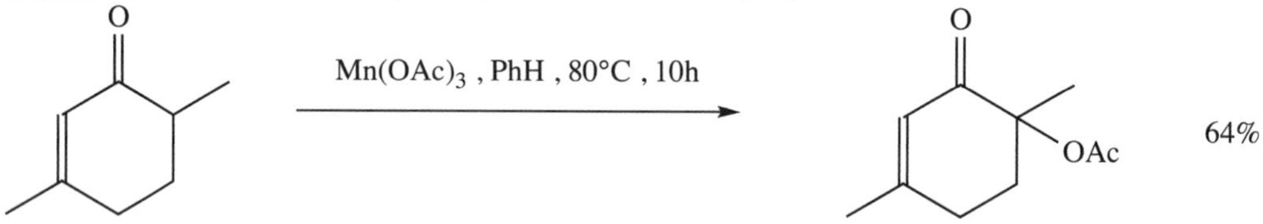

$Mn(OAc)_3$, PhH , 80°C , 10h

O OAc

64%

Tanyeli, C.; Iyigün, C. *Tetrahedron* ***2003***, *59*, 7135.

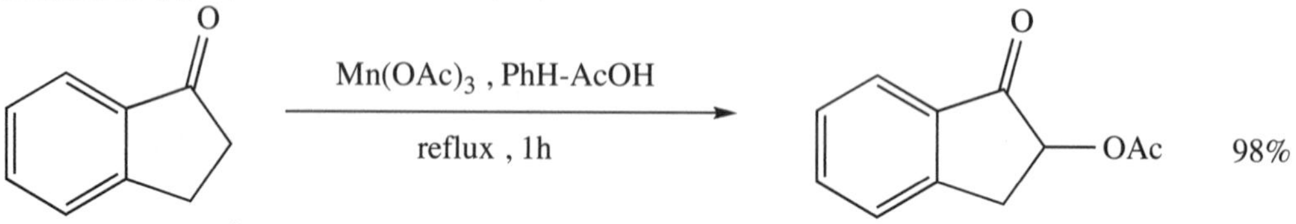

$Mn(OAc)_3$, PhH-AcOH

reflux , 1h

O OAc

98%

Demir, A.S.; Reis, Ö.; Igdir, A.C. *Tetrahedron* ***2004***, *60*, 3427.

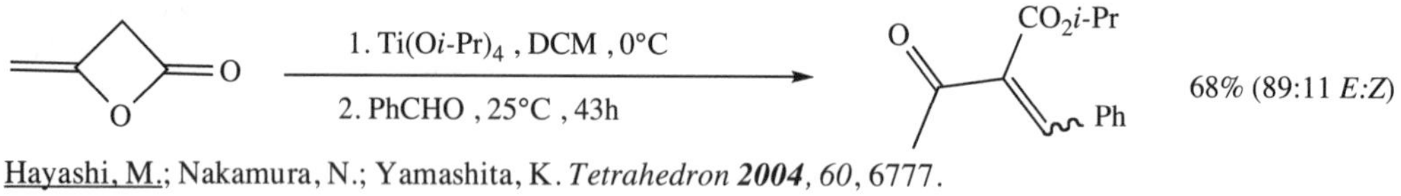

1. $Ti(Oi\text{-}Pr)_4$, DCM , 0°C

2. PhCHO , 25°C , 43h

O CO_2i-Pr Ph

68% (89:11 *E:Z*)

Hayashi, M.; Nakamura, N.; Yamashita, K. *Tetrahedron* ***2004***, *60*, 6777.

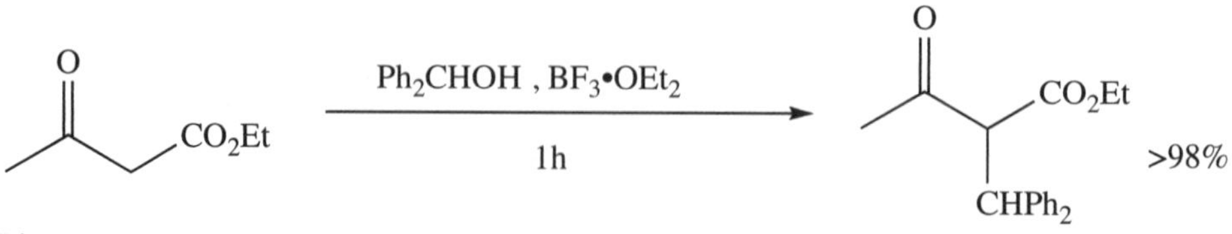

O CO_2Et

Ph_2CHOH , BF_3•OEt_2

1h

O CO_2Et $CHPh_2$

>98%

Bisaro, F.; Prestat, G.; Vitale, M.; Poli, G. *Synlett* ***2002***, 1823.

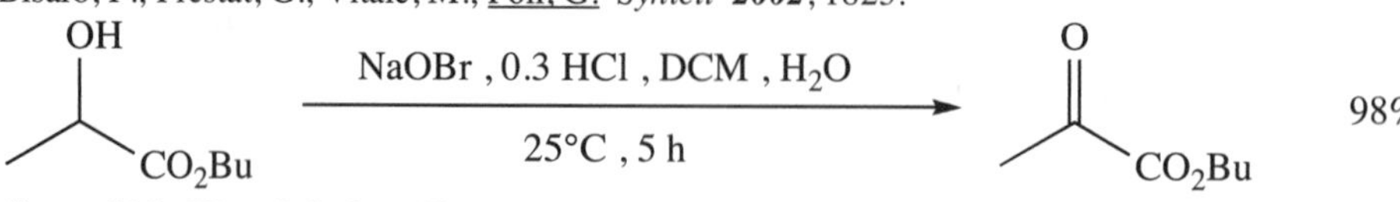

OH CO_2Bu

NaOBr , 0.3 HCl , DCM , H_2O

25°C , 5 h

O CO_2Bu

98%

Chang, H.S.; Woo, J.C.; Lee, K.M.; Ko, Y.K.; Moon, S.-S.; Kim, D.-W. *Synth. Commun.* ***2002***, *32*, 31.

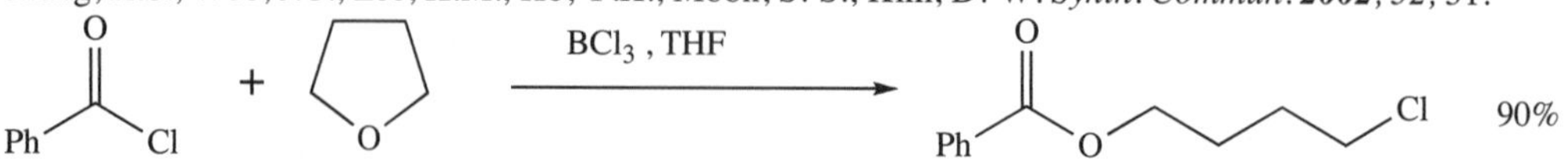

O Ph Cl + O

BCl_3 , THF

O Ph O Cl

90%

Malladi, R.R.; Kabalka, G.W. *Synth. Commun.* ***2002***, *32*, 1997.

$TiCl_4$, PPh_3, DCM/DCE; 0°C, 3 h — 50%

Hashimoto, Y.; Konishi, H.; Kikuchi, S. *Synlett* **2004**, 1264.

CH_2=$CHCO_2Me$, 10% $InCl_3$; rt, 6 h — 90%

Yadav, J.S.; Geetha, V.; Reddy, B.V.S. *Synth. Commun.* **2002**, *32*, 3519.

CO, MeOH, NBu_3, 110°C, 2 h; cat $PdCl_2(PPh_3)_2$ — 80%

Lapidus, A.L.; Eliseev, O.L.; Bondarenko, T.N.; Sizan, O.E.; Ostaenko, A.G.; Beletskaya, I.P. *Synthesis* **2002**, 317.

$PhCO_2K$, bmim BF_4, 20°C — 88%

Liu, Z.; Chen, Z.-C.; Zheng, Q.-G. *Synthesis* **2004**, 33.

10% $PdCl_2(MeCN)_2$; dioxane, 25°C — 76%

Widenhoefer, R.A. *Pure Appl. Chem.* **2004**, *76*, 671.

$P(NMe_2)_3$ — 60%

Bellee, C.; Gaurat, O. *Can. J. Chem.* **2004**, *82*, 1289.

Also via Ketoacids Section 320 (Carboxylic Acid - Ketone)
Hydroxyketones Section 330 (Alcohol - Ketone)

SECTION 361: ESTER - NITRILE

PhBr, *t*-BuOK, toluene, 90°C, 5 h; 2% $Pd(dba)_2$, 4% $P(i\text{-}BuNCH_2CH_2)_3N$ — 98%

Yu, J.; Verkade, J.G. *J. Org. Chem.* **2003**, *68*, 8003.

Ph(C=O)SiEt$_3$ → EtOC(=O)CN , cat KCN , ether , 12 h; 5% 18-crown-6 , 25°C → NC, CO$_2$Et, Ph, OSiEt$_3$ 81%

Lighu, X.; Nicewicz, D.A.; Johnson, J.S. *Org. Lett.* **2002**, *4*, 2957.

PHCHO → 2 EtOC(=O)CN , 18 h; cat [(salen)TiO]$_2$ → OCO$_2$Et, Ph, CN 90% (95% ee)

Belokon', Y.N.; Blacker, A.J.; Clutterbuck, L.A.; North, M. *Org. Lett.* **2003**, *5*, 4505.

CN, CN, SePh → EtOCO$_2$Bn , CHCl$_3$; sunlamp (hν) → NC, CN, CO$_2$Et 69%

Roepel, M.G. *Tetrahedron Lett.* **2002**, *43*, 1973.

PhCHO → KCN , Ac$_2$O , DCM/*t*-BuOH; bis(Ti-salen) catalyst , –42°C , 7h → Ph, OAc, CN 93% (90% ee)

Belokon, Y.N.; Gutnov, A.V.; Moskalenko, M.A.; Yashkina, L.V.; Lesovoy, D.E.; Ikonnikov, N.S.; Larichev, V.S.; North, M. *Chem. Commun.* **2002**, 244.

NC, NC, O, O*t*-Bu → ClCH$_2$COOH , DCM; EtCHO , 0°C → rt → O, O, Cl, O, O*t*-Bu, O, CN 27%

Müller, S.; Neidlein, R. *Helv. Chim. Acta* **2002**, *85*, 2222.

2 S, N-CN, N-CN → NaCH(CO$_2$Me)$_2$, THF; rt , 5 h → NC, CO$_2$Me, CO$_2$Me 81%

Maezaki, N.; Furusawa, A.; Uchida, S.; Tanaka, T. *Heterocycles* **2003**. *59*, 161.

SECTION 362: ESTER - ALKENE

This section contains syntheses of enol esters and esters of unsaturated acids as well as ester molecules bearing a remote alkenyl unit.

O, O → hydroquinone , toluene; reflux → O, O, H 78% (95:5)

Richter, F.; Bauer, M.; Perez, C.; Maichle-Mössmer, C.; Maier, M.E. *J. Org. Chem.* **2002**, *67*, 2474.

Cr(CO)$_5$ / OMe; PPh$_3$, dioxane, reflux, 1.5 h — 72%

Zhang, Y.; Herndon, J.W. *J. Org. Chem.* **2002**, *67*, 4177.

Bu_3Sn … $OSnBu_3$; 5% Pd(PPh$_3$)$_4$, dioxane, 50°C — 84%

Thibonnet, J.; Abarbri, M.; Parrain, J.-L.; Duchêne, A. *J. Org. Chem.* **2002**, *67*, 3941.

1. Meldrums acid, DCC, DMAP, DCM
2. NaBH$_4$, AcOH, DCM, –5°C
3. $^{+}Me_2N{=}CH_2\ I^{-}$, MeOH, 65°C — 75x95%

Hin, B.; Majer, P.; Tsukamoto, T. *J. Org. Chem.* **2002**, *67*, 7365.

1. Et$_2$Zn, CH$_2$I$_2$
2. I$_2$
3. satd Na$_2$S$_2$O$_3$, DBU — 86%

Ronsheim, M.D.; Zercher, C.K. *J. Org. Chem.* **2003**, *68*, 4535.

Br, DMA, 50°C; 5% PdCl$_2$, 16 h — 85%

Ma, S.; Yu, Z. *J. Org. Chem.* **2003**, *68*, 6149.

1% Ru$_3$(CO)$_{12}$, dioxane; 10 atm CO, 100°C, 8h; NEt$_3$ — 94%

Yoneda, E.; Zhang, S.-W.; Zhou, D.-Y.; Onitsuka, K.; Takahashi, S. *J. Org. Chem.* **2003**, *68*, 8571.

C$_3$H$_7$—≡—C$_3$H$_7$; CO, Py, Bu$_4$NCl, 1 d; 5% Pd(OAc)$_2$, DMF, 120°C — 63%

Kadnikov, D.V.; Larock, R.C. *J. Org. Chem.* **2003**, *68*, 9423.

CO$_2$Me, DCM; 5% Grubbs II catalyst — MeO$_2$C… 83%

Morgan, J.P.; Morrill, C.; Grubbs, R.H. *Org. Lett.* **2002**, *4*, 67.

$CH_2(CO_2Me)_2$, DMF , 4 h
10% K_2CO_3 , 80°C
88%

Ma, S.; Yin, S.; Li, L.; Tao, F. *Org. Lett.* ***2002***, *4*, 505.

1. $HInCl_2$, BEt_3 , THF , –78°C
. I_2
99% (7:93 *E:Z*)

Takami, K.; Yorimitsu, H.; Oshima, K. *Org. Lett.* ***2002***, *4*, 2993.

$EtO_2C\text{-}CHN_2$, 0.5 $PhCO_2H$
1.5% Fe(TPP)Cl , toluene
PPh_3 , 50 h
84% (59:41 *Z:E*)

Chen Y.; Huang, L.; Zhang, X.P. *Org. Lett.* ***2003***, *5*, 2493.

1. CO_2 , $Ni(cod)_2$, THF , 2 h
2 eq DBU , 0°C
2. 2 eq PhCHO , rt , 6 h
3. TsOH , PhH , reflux
57%

Takimoto, K.; Kawamura, M.; Mori, M. *Org. Lett.* ***2003***, *5*, 2599.

10% $PdCl_2$, 5 eq $CuCl_2$
20 atm CO , THF , 30°C
89% (92% ee)

Ma, S.; Wu, B.; Zhao, S. *Org. Lett.* ***2003***, *5*, 4429.

AcOH , acetone , 5% $Pd(OAc)_2$
25% benzoquinone
53% (>98% *trans*)

Verboom, R.C.; Persson, B.A.; Bäckvall, J.-E. *J. Org. Chem.* ***2004***, *69*, 3102.

1. acetophenone , rt , 30 min
2. MeI , HMPA
99%

Shindo, M.; Sato, Y.; Yoshikawa, T.; Koretsune, R.; Shishido, K. *J. Org. Chem.* ***2004***, *69*, 3912.

5% Grubbs' II , 19°C , DCM
$CH_2{=}CHCO_2Me$
73%

Lautens, M. ; Maddess, M.L. *Org. Lett.* ***2004***, *6*, 1883.

$MeCO_2H$, 1% $ReBr(CO)_5$, 110°C
toluene , 18h
73% (28:72 *E:Z*)

Hua, R.; Tian, X. *J. Org. Chem.* ***2004***, *69*, 5782.

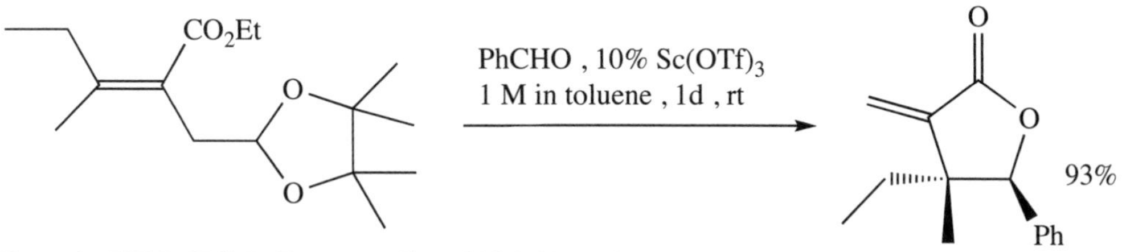

CO_2Et

PhCHO , 10% $Sc(OTf)_3$
1 M in toluene , 1d , rt

O O O Ph 93%

Kennedy, J.W.J.; Hall, D.G. *J. Org. Chem.* **2004**, *69*, 4412.

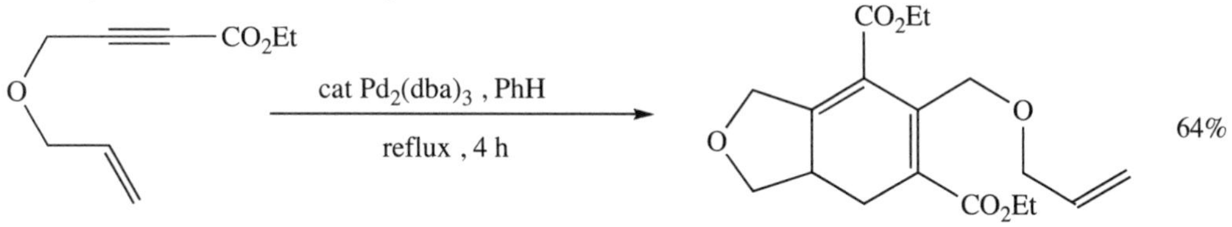

CO_2Et O

cat $Pd_2(dba)_3$, PhH
reflux , 4 h

CO_2Et O O CO_2Et 64%

Yamamoto, Y.; Kuwabara, S. ; Ando, Y.; Nagata, H.; Nishiyama, H.; Itoh, K. *J. Org. Chem.* **2004**, *69*, 6697.

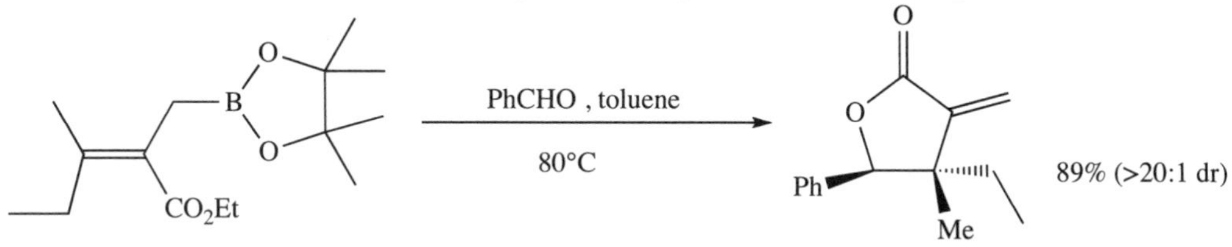

O B O CO_2Et

PhCHO , toluene
80°C

O O Ph Me 89% (>20:1 dr)

Kennedy, J.W.J.; Hall, D.G. *J. Am. Chem. Soc.* **2002**, *124*, 898.

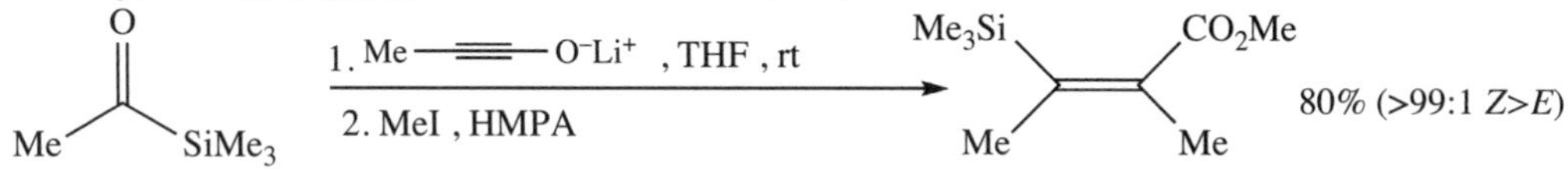

O Me $SiMe_3$

1. Me—≡—O^-Li^+ , THF , rt
2. MeI , HMPA

Me_3Si CO_2Me Me Me 80% (>99:1 *Z*>*E*)

Shindo, M.; Matsumoto, K.; Mori, S.; Shishido, K. *J. Am. Chem. Soc.* **2002**, *124*, 6840.

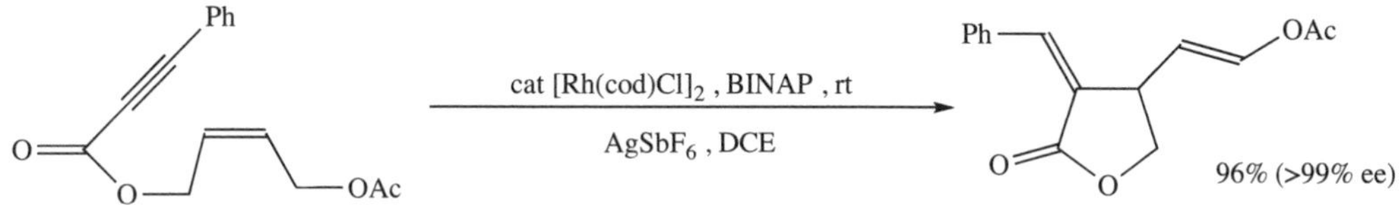

Ph O O OAc

cat $[Rh(cod)Cl]_2$, BINAP , rt
$AgSbF_6$, DCE

Ph OAc O O 96% (>99% ee)

Lei, A.; He, M.; Zhang, X. *J. Am. Chem. Soc.* **2002**, *124*, 8198.

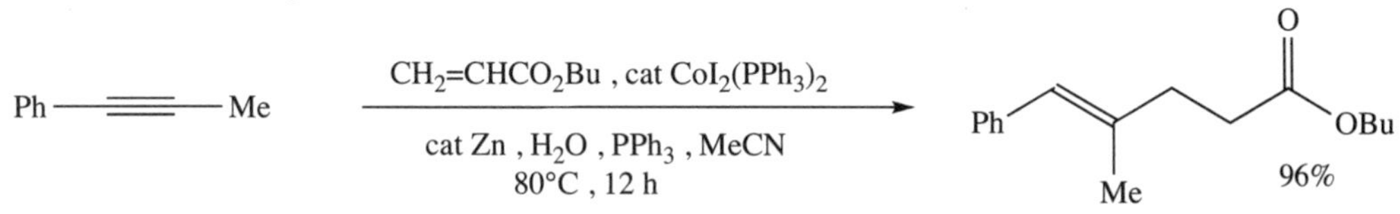

Ph—≡—Me

CH_2=$CHCO_2Bu$, cat $CoI_2(PPh_3)_2$
cat Zn , H_2O , PPh_3 , MeCN
80°C , 12 h

O Ph OBu Me 96%

Wang, C.-C.; Lin, P.-S.; Cheng, C.-H. *J. Am. Chem. Soc.* **2002**, *124*, 9696.

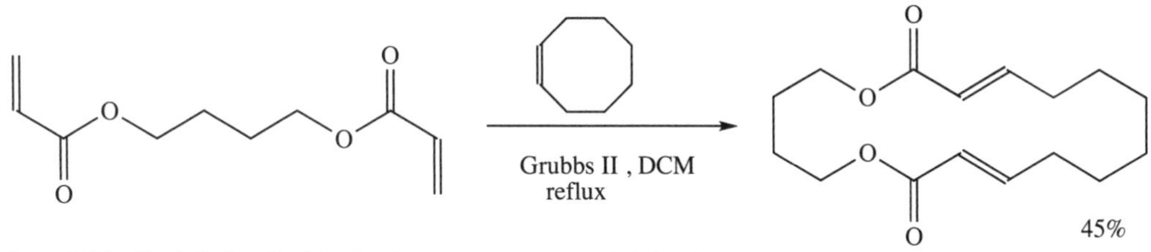

O O O O

Grubbs II , DCM
reflux

O O O O 45%

Lee, C.W.; Choi, T.-L.; Grubbs, R.H. *J. Am. Chem. Soc.* **2002**, *124*, 3224.

3 $CH_2(CO_2Me)_2$, 0.25% $[PdCl(\eta^3\text{-}C_3H_5)]_2$
6.25% P,N-ligand, Cs_2CO_3, DCM, 0°C

OAc, Ph, Ph → $CH(CO_2Me)_2$, Ph, Ph 91% (90% ee)

Lyle, M.P.A.; Navine, A.A.; Wilson, P.D. *J. Org. Chem.* ***2004**, 69*, 5060.

cat $Ni(acac)_2/PPh_3$, excess Me_2Zn
CO_2, THF, 19 h, rt

Ts—N → CO_2Me, Ts—N 94%

Takimoto, M.; Mori, M. *J. Am. Chem. Soc.* ***2002**, 124*, 10008.

PhCHO —N_2CHCO_2Et, PPh_3, Fe(II)(TTP) / toluene, rt, 2h→ $PhCH=CHCO_2Et$ 90% (24:1 *trans;cis*)

Mirafzal, G.A.; Cheng, G.; Woo, L.K. *J. Am. Chem. Soc.* ***2002**, 124*, 176.

NHAc —$CH_2=CHCO_2Bu$ / cat $Pd(OAc)_2$, 80°C AcOH→ NHAc, CO_2Bu 85%

Boele, M.D.K.; van Strijdonck, G.P.F.; de Vries, A.H.M.; Kamer, P.C.J.; de Vries, J.G.; van Leeuwen, P.W.N.M. *J. Am. Chem. Soc.* ***2002**, 124*, 1586.

I, OH —C_5H_{11}—≡—CO_2Me / Ni(dppe)Br, Zn, MeCN, 80°C→ C_5H_{11}, O, O 81%

Rayabarapu, D.K.; Cheng, C.-H. *J. Am. Chem. Soc.* ***2002**, 124*, 5630.

2 O, Et, Et —2 CO (at atm), cat $Ru_3(CO)_{12}$ / NEt_3, THF, 140°C, 20h→ Et, Et, O, O, O, O, Et, Et 70%

Kondo, T.; Kaneko, Y.; Taguchi, Y.; Nakamura, A.; Okada, T.; Shiotsuki, M.; Ura, Y.; Wada, K.; Mitsudo, T.-a. *J. Am. Chem. Soc.* ***2002**, 124*, 6824.

CO_2Et, O, B, O —PhCHO, $Sc(OTf)_3$, toluene→ O, O, Ph 72%

Kennedy, J.W.J.; Hall, D.G. *J. Am. Chem. Soc.* ***2002**, 124*, 11586.

MeO_2C / MeO_2C — 5% $Ni(cod)_2$, toluene , 60°C; 10% imdazolylidene derivative → MeO_2C / MeO_2C … O … O 93%

Louie, J.; Gibby, J.E.; Farnsworth, M.V.; Tekavec, T.N. *J. Am. Chem. Soc.* ***2002***, *124*, 15188.

PhCHO — EtO_2C-CH=N_2 , $P(OMe)_3$, LiBr; 1% ClFe(TPP) , dioxane , 85°C , 12 h → Ph … CO_2Et 86% (96:4 *E:Z*)

TPP = tetraphenylporphyrin

Aggarwal, V.K.; Fulton, J.R.; Sheldon, C.G.; de Vincente, J. *J. Am. Chem. Soc.* ***2003***, *125*, 6034.

OH, MeO, OMe — HC≡C-CO_2Et , HCOOH , 50°C; cat $Pd_2(dba)_3$, 20% NaOAc → MeO … O … O, OMe 88%

Trost, B.M.; Toste, F.D.; Greenman, K. *J. Am. Chem. Soc.* ***2003***, *125*, 4518.

Ts—N … C … O — 1.2 $Mo(CO)_3$, DMSO , 10°C; toluene → H, Ts—N … O … O 80%

Yu, C.-M.; Hong, Y.-T.; Lee, J.-H. *J. Org. Chem.* ***2004***, *69*, 8506.

Ph— — HO_2CCH_2CN , cat $RuCl(C_5Me_5((cod)$; dioxane , rt , 17.5 h → Ph … O … CN, Ph, O 85%

LePaih, J.; Monnier, F.; Dérien, S.; Dixneuf, P.H.; Clot, E.; Eisenstein, O. *J. Am. Chem. Soc.* ***2003***, *125*, 11964.

C_8H_{17} … CHO, O — 1. $SnBu_2Cl$, THF; 2. PhN=C=O , 0°C → 60°C; 3. H_2O → O, PhHN … O, C_8H_{17} … O 99%

Shibata, I.; Kato, H.; Kanazawa, N.; Yasuda, M.; Baba, A. *J. Am. Chem. Soc.* ***2004***, *126*, 466.

Ts—N — 1. 10% $Ni(acac)_2$, 4.5 Et_2Zn , 0°C , 8 h; 20% 15-MeO-MOP , THF , CO_2; 2. CH_2N_2 , ether → H, CO_2Me, Ts—N, Et, H 57% (94% ee)

Takimoto, M.; Nakamra, Y.; Kimura, K.; Mori, M. *J. Am. Chem. Soc.* ***2004***, *126*, 5956.

Me … Cl, O, O — 10% $[Rh(cod)Cl]_2$, dppb/$AgSbF_6$; DCE , rt , 3 h → Cl, Me, O, O 87%

Tong, X.; Li, D.; Zhang, Z.; Zhang, X. *J. Am. Chem. Soc.* ***2004***, *126*, 7601.

Ph–CH=CH–Br (82:18 *E:Z*) → 1. *t*-BuLi; 2. $InCl_3$; 3. cat Et_3B–C(CO_2Bn), ether, 25°C → Ph–CH=CH–CH_2–C(O)OBn 93% (90:10 *E:Z*)

Takami, K.; Yorimitsu, H.; Oshima, K. *Org. Lett.* ***2004**, 6*, 4555.

PhCH=CH–CH(OAc)Ph → $CH_2(CO_2Me)_2$, BSA-THF, KOAc, 0.4h, 20h; $[Pd(\eta^3\text{-}C_3H_7)Cl]_2$- (bis-N-Me benzimidazoline-pyridine ligand) → PhCH=CH–CH($CH(CO_2Me)_2$)Ph 97% (97% ee)

Jin, M.-J.; Kim, S.-H.; Lee, S.-J.; Kim, Y.-M. *Tetrahedron Lett.* ***2002**, 43*, 7409.

CH_2=$CHCO_2Me$ → CH_2=CHOAc, AcOH, O_2, 90°C, 5 h; cat $Pd(OAc)_2/H_4PMo_{11}VO_{40}$/NaOAc → AcO–CH=CH–CH=CH–CO_2Me 62% (60:40 *E:Z*)

Hatamoto, Y.; Sakaguchi, S.; Ishii, Y. *Org. Lett.* ***2004**, 6*, 4623.

HC≡C–CH_2–O–C(O)–β-naphth → cyclooctadiene, 5% Grubbs II, DCM, rt → cyclooctatriene–CH_2–O–C(O)–β-naphth 89%

Kulkarni, A.A.; Diver, S.T. *J. Am. Chem. Soc.* ***2004**, 126*, 8110.

PhCH(OH)CH_2OH → Ph_3P=$CHCO_2Me$, $NaIO_4$, SiO_2 → PhCH=CHCO_2Me 85% (8:1 *E:Z*)

Dunlap, N.K.; Mergo, W.; Jones, J.M.; Carrick, J.D. *Tetrahedron Lett.* ***2002**, 43*, 3923.

C_6H_{13}–C≡CH → $PhCO_2H$, cat $[Ir(cod)Cl]_2$, 2 Na_2CO_3; cat $P(OMe)_3$ → PhC(O)O–C(C_6H_{13})=CH_2 79% + PhC(O)O–CH=CH–C_6H_{13} 17% (41:53 *E:Z*)

Nakagawa, H.; Okimoto, Y.; Sakaguchi, S.; Ishii, Y. *Tetrahedron Lett.* ***2003**, 44*, 103.

1-OTMS-3-$SiMe_3$-cyclohexene → 1. MeLi, THF, 0°C; 2. BF_3•OEt_2, –78°C, (2-phenylethyl)oxirane; 3. $PhI(OAc)_2$, I_2, DCM, 0°C → 2-phenylethyl-substituted unsaturated 10-membered lactone 54%

Hatcher, M.A.; Borstnik, K.; Posner, G.H. *Tetrahedron Lett.* ***2003**, 44*, 5407.

Resin–CH_2–O–C(O)–C(CH_2Br)=CH–4-CF_3-C_6H_4 → 1. DMF, $CrCl_2$; 2. THF, HCl → 2-methylene-3-(4-CF_3-C_6H_4)-4-C_3H_7-cyclopentanone

Breitenstein, K.; Llebaria, A.; Delgado, A. *Tetrahedron Lett.* ***2004**, 45*, 1511.

C_7H_{15}CH(OH)–CH(Cl)CO_2Me → 3 $CrCl_2$, THF → C_7H_{15}CH=CHCO_2Me 65%

Concellón, J.M.; Rodríguez-Solla, H.; Méjica, C. *Tetrahedron Lett.* **2004**, *45*, 2977.

Me—≡ → cat $Pd(OH)_2$, Ph_2Py , MeOH , CO / 60°C , cat TsOH , tdtbpp 40h → $MeO_2C(Me)C{=}CH_2$ >98%

Clarke, M.L. *Tetrahedron Lett.* ***2004****, 45*, 4043.

cyclohexanone → $Ph_3P{=}CHCO_2Et$, MeCN , 190°C / microwaves , 20 min → cyclohexylidene=$CHCO_2Et$ 62%

Wu, J.; Wu, H.; Wei, S.; Dai, W.-M. *Tetrahedron Lett.* ***2004****, 45*, 4401.

$MeO_2CCH_2P(O)(OCH_2CF_3)_2$ → *t*-BuOK , PhCHO , THF → Ph–CH=CH–CO_2Me quant (98% Z)

Touchard, F.P. *Tetrahedron Lett.* ***2004****, 45*, 5519.

acetophenone → $CH_2(CO_2Et)_2$, cat $[Ir(cod)Cl]_2$/dppe / THF , reflux , 3h → $CH_2{=}C{=}CH–C(Me)(Ph)CH(CO_2Et)_2$ 74%

Kezuka, S.; Kanemoto, K.; Takeuchi, R. *Tetrahedron Lett.* ***2004****, 45*, 6403.

resorcinol (HO, OH) → $CH_3COCH_2CO_2Et$ / 10% $Sm(NO_3)_3$•6 H_2O → 7-hydroxy-4-methylcoumarin (HO, O, O) 98%

Bahekar, S.S.; Shinde, B.D. *Tetrahedron Lett.* ***2004****, 45*, 7999.

Ph—≡—Me → $CH_2(CO_2Et)_2$, microwaves , 15 min / cat $Pd(PPh_3)_4$, AcOH → Ph–CH=CH–CH_2–$CH(CO_2Et)_2$ 88%

Patil, N.T.; Lkan, F.N.; Yamamoto, Y. *Tetrahedron Lett.* ***2004****, 45*, 8497.

C_6H_{13}—≡ → PhSH , CO , MeCN , cat $Pt(PPh_3)_4$ / 120°C , 5h → $C_6H_{13}C(=CH_2)C(O)SPh$ 60%

Kawakami, J.-i.; Miharaq, M.; Kamiya, I.; Takeba M.; Ogawa, A.; Sonoda, N. *Tetrahedron* ***2003****, 59*, 3521.

$CH_2{=}C(Br)CH_2OCH_2CH{=}CH_2$ → CO , NEt_3, cat $Cl_2Pd(PPh_3)_2/PPh_3$ / MeOH , DMF , H_2O , 85°C , 1d → 3-methylene-4-(CH_2CO_2Me)tetrahydrofuran (O) 69%

Aggarwal, V.K.; Davies, P.W.; Moss, W.O. *Chem. Commun.* ***2002****,* 972.

Ts—N($CH_2CH{=}C{=}CH_2$)(CH_2CH_2CHO) → cat $[Ru_3(CO)_{12}]$, CO , dioxane / 120°C , 12h → Ts–N bicyclic α-methylene lactone (H, H, O, O) 60%

Kwang, S.-K.; Kim, K.-J.; Hong, Y.-T. *Angew. Chem. Int. Ed.* ***2002****, 41*, 1584.

PhCHO → $MoC(CO_2Et)_3$, cat PhOH , 16h / 130°C → $PhCH{=}CHCO_2Et$ 85%

Kumar, H.M.S.; Rao, M.S.; Joyawal, S.; Yadav, J.S. *Tetrahedron Lett.* ***2003***, *44*, 4287.

1. 10% $AcNMe_2$, B-H (catecholborane) 2. CO_2Et, SO_2Ph 3% *t*-BuON-NO*t*-Bu → CO_2Et 50%

Schaffner, A.-P.; Renaud, P. *Angew. Chem. Int. Ed.* ***2003***, *42*, 2658.

O, Ph, OH → $Ph_3P{=}C(Me)CO_2Et$, MnO_2 , DCM / heat , 1h → O, Ph, CO_2Et 78%

Runcie, K.A.; Taylor, R.J.K. *Chem. Commun.* ***2002***, 974.

$CH_2{=}C{=}CH_2$ → cat $Ru_3(CO)_{12}$, 100°C , CO , MeOH → CO_2Me 84%

Zhou, D.-Y.; Yoneda, E.; Onitsuka, K.; Takahashi, S. *Chem. Commun.* ***2002***, 2868.

OH, C_7H_{15}, CO_2Me, Cl → 1. Sm / 2. CH_2I_2 → C_7H_{15}, CO_2Me 87%

Concellón, J.M.; Rodríguez-Solla, H.; Huerta, M.; Pérez-Andrés, J.A. *Eur. J. Org. Chem.* ***2002***, 1839.

O → Ac_2O , Montmorillonite KSF / 23°C , 3 h → OAc 88%

Kalita, B.; Bezbarua, M.S.; Barua, N.C. *Synth. Commun.* ***2002***, *32*, 3181.

I, OH → Ph-C≡C-Ph , cat Co_2Rh_2 , THF pyridine , CO , 120°C , 18 h → Ph, Ph, O, O 87%

Park, K.H.; Jung, I.G.; Chung, Y.K. *Synlett* ***2004***, 2541.

PhCHO → EtO_2CH_2CN , bmim BF_4 , rt / 20% EDDA , 1 h → CO_2Et, Ph, CN 95%

Su, C.; Chen, Z.-C.; Zheng, Q.G. *Synthesis* ***2003***, 555.

Ph—≡ → (OH)OTs / I_2 , DCM , rt → Ph, I, TsO 95%

Chen, J.-M.; Huang, X. *Synthesis* ***2004***, 1577.

Ph, =C= → $PhCO_2H$, $CH_2(CO_2Et)_2$ / cat $Pd(PPh_3)_4$ → Ph, CO_2Et, CO_2Et 90%

Patil, N.T.; Pahadi, N.K.; Yamamoto, Y. *Synthesis* ***2004***, 2186.

5% Grubbs' II , 40°C

89%

Chatterjee, A.K.; Toste, F.D.; Goldberg, S.D.; Grubbs, R.H. *Pure Appl. Chem.* ***2003****, 75,* 421.

Related Methods:	Section 60A (Protection of Aldehydes)
	Section 180A (Protection of Ketones)
Also via Acetylenic Esters:	Section 306 (Alkyne - Ester)
Alkenyl Acids:	Section 322 (Carboxylic Acid - Alkene)
β-Hydroxy-esters:	Section 327 (Alcohol - Ester)

SECTION 363: ETHER, EPOXIDE, THIOETHER - ETHER, EPOXIDE, THIOETHER

H SeBr N–OH

Bu Bu → OMe Bu Bu SeR

DCM , MeOH , –78°C

73%

Back, T.G.; Moussa, Z.; Parvez, M. *J. Org. Chem.* ***2002****, 67,* 499.

See Section 60A (Protection of Aldehydes) and Section 180A (Protection of Ketones) for reactions involving formation of acetals and ketals.

SECTION 364: ETHER, EPOXIDE, THIOETHER - HALIDE, SULFONATE

CO_2Me Ph

I_2 , LiBr , AcOH , 1 h

O O Ph I

90%

Yao, T.; Larock, R.C. *J. Org. Chem.* ***2003****, 68,* 5936.

OAc; I_2 , $NaHCO_3$; O; Ph; 76%; Ph; I

Arcadi, A.; Cacchi, S.; Giusepe, S.D.; Fabrizi, G.; Marinelli, F. *Org. Lett.* ***2002**, 4*, 2409.

EtO_2C; O; C_6H_{13}; $TiCl_4$, DCM; rt , 3 d; EtO_2C; O; C_6H_{13}; Cl; (20 : 1); EtO_2C; O; C_6H_{13}; Cl; 63%

Hart, D.J.; Bennett, C.E. *Org. Lett.* ***2003**, 5*, 1499.

HO; 1. bis(oxazolidine) ligand , DCM; K_2CO_3 , MeOH , $Hg(TFA)_2$, –78°C; 2. aq. KBr , $LiBH_4$, BEt_3; 3. THF , –78°C; I; O; 91% (89% ee)

Kang, S.H.; Kim, M. *J. Am. Chem. Soc.* ***2003**, 125*, 4684.

HO; 1. [Co salen complex] + NCS , toluene; 2. I_2 , –78°C; I; O; 94% (84% ee)

Kang, S.H.; Lee, S.B.; Park, C.M. *J. Am. Chem. Soc.* ***2003**, 125*, 15748.

Ph; OH; [Ti(O*i*-Pr)$_4$/BINOL derivative , *t*-BuOMe , rt]; NIS , 0°C; I; O; Ph; 93% (65% ee)

Kang, S.H.; Park, C.M.; Lee, S.B.; Kim, M. *Synlett* **2004**, 1279.

O; IPy_3 BF_4 , HBF_4 , DCM; –40°C , 30 h; tetrahydroquinolines can also be prepared; O; I; 92%

Barluenga, J.; Trincado, M.; Rubio, E.; González J.M. *J. Am. Chem. Soc.* ***2004**, 126*, 3416.

O; e^- , $Et_4NF{\bullet}4$ HF; O; I; 56%

Hasegawa, M.; Ishii, H.; Fuchigami, T. *Tetrahedron Lett.* ***2002**, 43*, 1503.

Br; Ph; O; 10 LiBr , MeOH , 50°C; Br; Ph; O; 63% (93:7 *cis:trans*)

Karikomi, M.; Takayama, T.; Haga, K. *Tetrahedron Lett.* ***2002**, 43*, 4487.

Ph; Ph; PhSe-SePh , 4 (tol)IF_2 , DCM , rt; Ph; Ph; F; SePh; 58%

Panunzi, B.; Picardi, A.; Tingoli, M. *Synlett* **2004**, 2339.

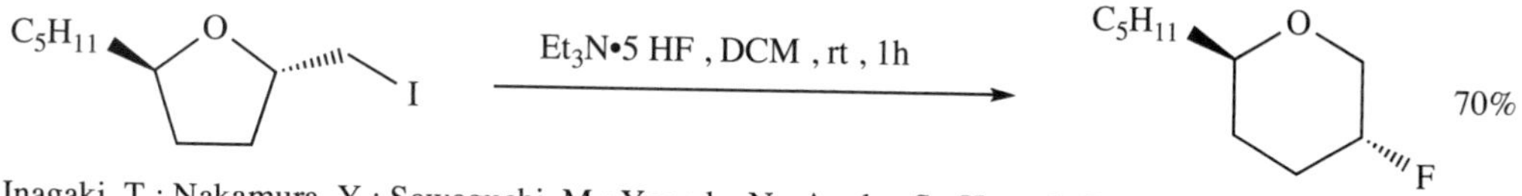

C_5H_{11} ... I → $Et_3N{\bullet}5\ HF$, DCM, rt, 1h → C_5H_{11} ... F 70%

Inagaki, T.; Nakamura, Y.; Sawaguchi, M.; Yoneda, N.; Ayuba, S.; Hara, S. *Tetrahedron Lett.* **2003**, *44*, 4117.

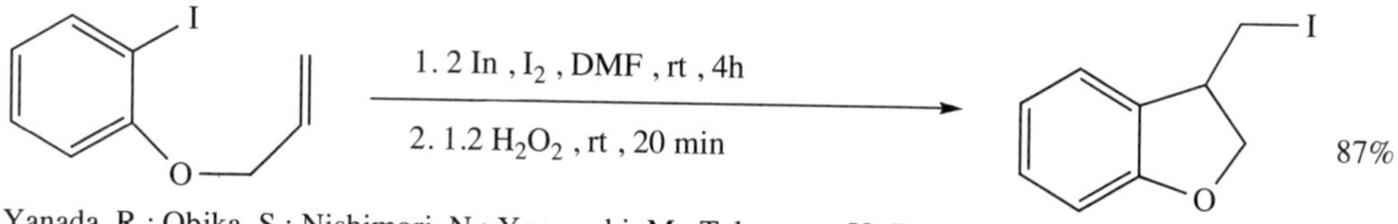

1. 2 In, I_2, DMF, rt, 4h
2. 1.2 H_2O_2, rt, 20 min

87%

Yanada, R.; Obika, S.; Nishimori, N.; Yamauchi, M.; Takemoto, Y. *Tetrahedron Lett.* **2004**, *45*, 2331.

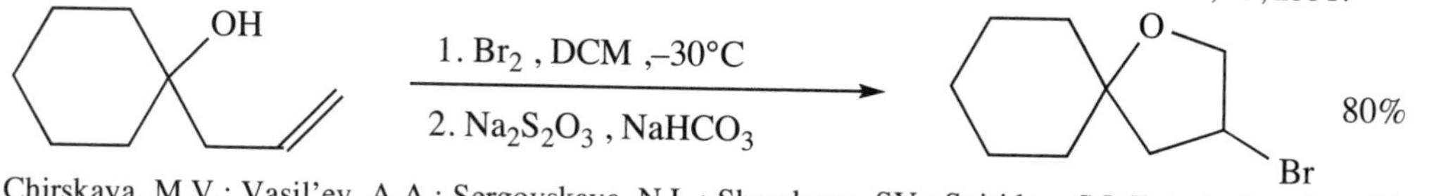

OH

1. Br_2, DCM, –30°C
2. $Na_2S_2O_3$, $NaHCO_3$

Br 80%

Chirskaya, M.V.; Vasil'ev, A.A.; Sergovskaya, N.L.; Shorshnev, SV.; Sviridov, S.I. *Tetrahedron Lett.* **2004**, *45*, 8811.

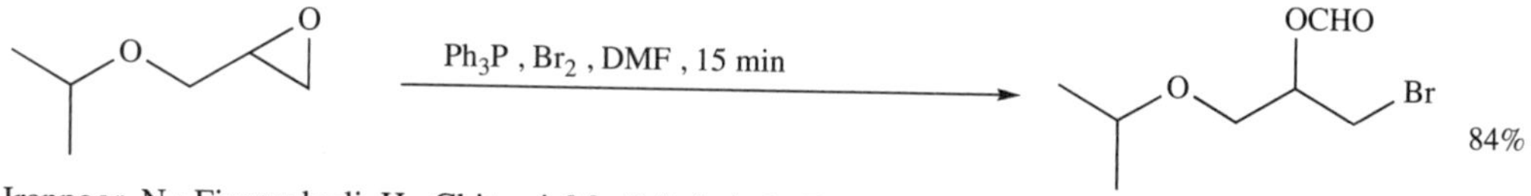

Ph_3P, Br_2, DMF, 15 min

OCHO Br 84%

Iranpoor, N.; Firouzabadi, H.; Chitsazi, M.; Jafari, A.A. *Tetrahedron* **2002**, *58*, 7037.

SECTION 365: ETHER, EPOXIDE, THIOETHER - KETONE

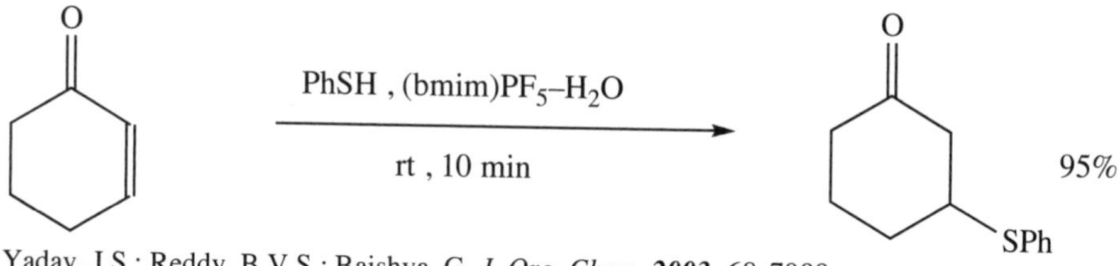

PhSH, (bmim)PF_5–H_2O
rt, 10 min

SPh 95%

Yadav, J.S.; Reddy, B.V.S.; Baishya, G. *J. Org. Chem.* **2003**, *68*, 7098.

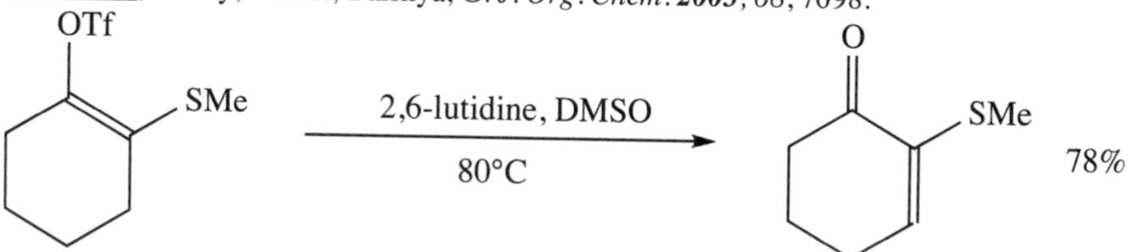

OTf SMe

2,6-lutidine, DMSO
80°C

SMe 78%

Hynes Jr. J.; Nasser, T.; Overman, L.E.; Watson, D.A. *Org. Lett.* **2002**, *4*, 929.

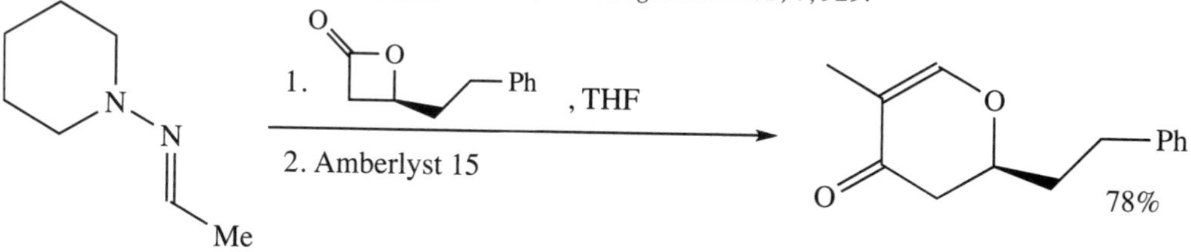

N N Me

1. (β-lactone, Ph), THF
2. Amberlyst 15

O Ph 78%

Zipp, G.G.; Hilfiker, M.A.; Nelson, S.G. *Org. Lett.* **2002**, *4*, 1823.

chiral ketone, Bu_4NHSO_4, Oxone, 1 d; $NaHCO_3$, MeCNaq Na_2(edta), 0°C → rt — 73% (96% ee)

Wu, X.-Y.; She, X.; Shi, Y. *J. Am. Chem. Soc.* **2002**, *124*, 8792.

2.4% $RuCl_3$•nH_2O, 4.1 eq $NaIO_4$; $H_2O/CCl_4/MeCN$, rt, 75 min — 56%

Ferraz, H.M.C.; Longo Jr. L.S. *Org. Lett.* **2003**, *5*, 1337.

cat $PdCl_2$, CuCl, DME; Na_2HPO_4, 50°C, 4h — 79%

Reiter, M.; Ropp, S.; Gouverneur, V. *Org. Lett.* **2004**, *6*, 91.

Bu_2BOTf/*i*-Pr_2NEt; DCM, −78°C — 94%

Li, L.-S.; Das, S.; Sinha, S.C. *Org. Lett.* **2004**, *6*, 127.

1. DCM, rt, mol sieve; 2. DDQ, $LiClO_4$; 3. OTMS, Ph — 84%

Ying, B.-P.; Trogden, B.G.; Kohlman, D.T.; Liang, S.X.; Xu, Y.-C. *Org. Lett.* **2004**, *6*, 1523.

1. methoxycyclopentane, $NaTMS_2$; 2. Me_3Si, $SiMe_2t$-Bu, −98°C → −50°C — 83%

Sawada, Y.; Sasaki, M.; Takeda, K. *Org. Lett.* **2004**, *6*, 2277.

PhSH, MeOH, $Na_2CaP_2O_7$, rt; 20 min — 94%

Zahouily, M.; Abrouki, Y.; Rayadh, A. *Tetrahedron Lett.* **2002**, *43*, 7729.

BnOH, toluene, 100°C; cat Rh(cod)$(OMe)_2$, 0.6 Na_2CO_3 — 72%

Farnsworth, M.V.; Cross, M.J.; Louie, J. *Tetrahedron Lett.* **2004**, *45*, 7441.

$Ph_3P{=}O$, CMHP, MS 4Å, THF; cat La-BINOL complex, 20h, rt — 96% (98% ee)

Jayaprakash, D.; Kobayashi, Y.; Watanabe, S.; Arai, T.; Sasai, H. *Tetrahedron: Asymmetry* **2003**, *14*, 1587.

OTIPS O N O ~OMe — e^- , 2.6-lutidine , MeOH / 0.1M Et_4NOTs → O O N O $CH(OMe)_2$ 74% (2:1)

Huang, Y.-t.; Moeller, K.D. *Org. Lett.* ***2004***, *6*, 4199.

$OSnBu_3$ — PhN=O , cat BINAP-AgOTf / THF , –78°C → O O-NHPh 95% (95% ee)

Momiyama, N.; Yamamoto, H. *J. Am. Chem. Soc.* ***2003***, *125*, 6038.

O — 5% PMe_3 , MeOH , 45°C , 6 h → O OMe 95%

Stewart, I.C.; Bergman, R.G.; Toste, F.D. *J. Am. Chem. Soc.* ***2003***, *125*, 8696.

O N_2 O OMe H — 10% $Cu(hfacac)_2$, DCM / reflux → O O OMe H 82%

Marmsäter, F.P.; Murphy, G.K.; West, F.G. *J. Am. Chem. Soc.* ***2003***, *125*, 14724.

O Ph — 1. PhCHO , toluene , 110°C , nanocrystalline MgO / 2. TBHP , (+)-DET , 25°C → Ph O O Ph 70% (90% ee)

Choudary, B.M.; Kantam, M.L.; Ranganath, K.V.S.; Mahendar, K.; Shedhar, B. *J. Am. Chem. Soc.* ***2004***, *126*, 3396.

O Ph CHN_2 — cat $In(OTf)_3$, BuOH , rt / 2 min → O Ph OBu 93%

Muthusamy, S.; Arulananda, S.; Babu, A.; Gunanathan, C. *Tetrahedron Lett.* ***2002***, *43*, 3133.

O — a hydrotalcite , 1.5 of 5% aq. H_2O_2 → O O 94%

Honma, T.; Nakajo, M.; Mizugaki, T.; Ebitani, K.; Kaneda, K. *Tetrahedron Lett.* ***2002***, *43*, 6229.

N-(phenylthio)phthalimide, cat (pyrrolidinyl)-NHTf, MeCN; rt, 4h — 83%

Wang, W.; Li, H.; Wang, J.; Liao, L. *Tetrahedron Lett.* ***2004***, *45*, 8229.

chiral phase transfer cat, LiOH, 4°C; 30% H_2O_2, Bu_2O, 26 h — 72% (73% ee)

Arai, S.; Tsuge, H.; Oku, M.; Miura, M.; Shiori, T. *Tetrahedron* ***2002***, *58*, 1623.

chiral amino ether, toluene, 0°C; $Me_2C(Pr)OOLi$ — 99% (40% ee)

Tanaka, Y.; Nishimura, K.; Tomioka, K. *Tetrahedron* ***2003***, *59*, 4549.

chiral phase transfer catalyst, 1d; 15% aq. NaOCl, toluene, 25°C — >90% (84% ee)

Lygo, B.; To, D.C.M. *Chem. Commun.* ***2002***, 2360.

chiral phase transfer catlayst; toluene, 50% KOH, 8h — quant (81% ee)

Ye, J.; Wang, Y.; Liu, R.; Zhang, G.; Zhang, Q.; Chen, J.; Liang, X. *Chem. Commun.* ***2003***, 2714.

cat $(DHDQ)_2PYR$, 17h; –60°C — 77% (94% ee)

McDaid, P.; Chen, Y.; Deng, L. *Angew. Chem. Int. Ed.* ***2002***, *41*, 338.

Ph-N=O, DMSO, 20% S-proline, rt — ONHPh — 70% (>99%ee)

Bøgevig, A.; Sundén, H.; Córdova, A. *Angew. Chem. Int. Ed.* ***2004***, *43*, 1109.

Ph-N=O, DMF, cat L-proline; 0°C — ONHPh — 79% (>99%ee)

Hayashi, Y.; Yamaguchi, J.; Sumiya, T.; Shoihi, M. *Angew. Chem. Int. Ed.* ***2004***, *43*, 1112.

1.5 BF_3•OEt_2, MeOH, 2 h — OMe — 76%

Lee, A.S.-Y.; Wang, S.-H.; Chang, Y.-T.; Chu, S.-F. *Synlett* ***2003***, 2359.

BuS-SnBu , In-I , THF , 3 h

BuS 85%

Ranu, B.C.; Mandal, T. *Synlett* ***2004***, 1239.

EtO OEt OEt — EtMgBr , THF , –78°C , 2 h → Et OEt OEt 66%

Adamczyk, M.; Johnson, D.D.; Mattingly, P.G.; Pan, Y.; Reddy, R.E. *Synth. Commun.* ***2002***, *32*, 3199.

Ph Ph — 1. 20% (R)-BINOL , ether; 2. 36% Et_2Zn , 0°C; 3. *t*-BuOOH → Ph Ph 99% (68% ee)

Minalti, A.; Dötz, K.H. *Synlett* ***2004***, 1634.

MeO OH — *p*-TsOH , EtOH , 70°C , 5 h → MeO OEt 94%

Olivi, N.; Thomas, E.; Reyrat, J.-F.; Alami, M.; Brion, J.-D. *Synlett* ***2004***, 2175.

Ph Ph — NaOCl , toluene , 20°C; chiral quaternary ammonium salt → Ph Ph 91% (60% ee)

Kim, D.Y.; Choi, Y.J.; Park, H.Y.; Joung, C.U.; Koh, K.O.; Mang, J.Y.; Jung, K.-Y. *Synth. Commun.* ***2003***, *33*, 435.

1. acetone , 0°C , 30% H_2O_2; 2. NaOEt , EtOH → 92%

Patra, A.; Bandyopadhyay, M.; Ghorai, S.K.; Mal, D. *Org. Prep. Proceed. Int.* ***2003***, *35*, 515.

$CH_3CH_2C(=O)Me$, TMSCl; LDA , CuBr , PPh_3 , ether; reflux , 4 h → OTMS Me Me 75%

Mitani, M.; Ishimoto, K.; Koyam, R. *Chem. Lett.* ***2002***, *31*, 1142

SECTION 366: ETHER, EPOXIDE, THIOETHER - NITRILE

PhCHO — TMSCN , DCM , –20°C; β-amino alcohol–Ti complex → Ph OTMS H CN 94% (85% ee)

Li, Y.; He, B.; Qin, B.; Feng, X.; Zhang, G. *J. Org. Chem.* ***2004***, *69*, 7910.

PhCOMe → (4 TMSCN, 10% $(DHQ)_2AQN$; $CHCl_3$, –24°C, 16 h) → Ph(Me)C(CN)(OTMS) 91% (31% ee)

Tian, S.-K.; Hong, R.; Deng, L. *J. Am. Chem. Soc.* ***2003***, *125*, 9900.

PhCHO → (cat $(TfO)_2Sc$ complex; 1.2 TMSCN, DCM, rt, 3 h) → PhCH(CN)OTMS 95%

Karimi, B.; Ma'Mani, L. *Org. Lett.* ***2004***, *6*, 4813.

PhCHO → (VO(salen), TMSCN, 3h; [bmim] PF_6) → PhCH(OTMS)CN 94%

Baleizão, C.; Gigante, B.; Garcia, H.; Corma, A. *Tetrahedron Lett.* ***2003***, *44*, 6813.

PhCOMe → (TMSCN, $Ti(Oi\text{-}Pr)_4$/salen complex) → Ph(Me)C(OTMS)CN 95% (67% ee)

Chen, F.-X.; Qin, B.; Feng, X.; Zhang, G.; Jiang, Y. *Tetrahedron* ***2004***, *60*, 10449.

PhCOMe → (TMSCN, cat Bu_4NCN, DCM, rt) → Ph(Me)C(OTMS)CN 95%

Amurrio, I.; Córdoba, R.; Csákÿ, A.G.; Plumet, J. *Tetrahedron* ***2004***, *60*, 10521.

PhCHO → (chiral Schiff base, $Ti(Oi\text{-}Pr)_4$, 36 h; DCM, –78°C) → PhCH(OTMS)CN 69% (85% ee)

Gama, A.; Flores-López, L.-Z.; Aguirre, G.; Parra-Hake, M.; Somanathan, R.; Walsh, P.J. *Tetrahedron: Asymmetry* ***2002***, *13*, 149.

PhCOMe → ($Al(OMe)_3$, TMSCN, MEOH, MS 3Å; peptide ligand, –78°C, 2d) → Ph(Me)C(OTMS)CN 93% (88% ee)

Deng, H.; Isler, M.P.; Snapper, M.L.; Hoveyda, A.H. *Angew. Chem. Int. Ed.* ***2002***, *41*, 1009.

PhCOMe → (TMSCN, DCM, 96 h; $T(Oi\text{-}Pr)_4$ - chiral N-oxide) → Ph(Me)C(CN)OTMS 78% (54% ee)

Shen, Y.; Feng, X.; Zhang, G.; Jiang, Y. *Synlett* ***2002***, 1353.
Shen, Y.; Feng, X.; Li, Y.; Zhang, G.; Jiang, Y. *Eur. J. Org. Chem.* ***2004***, 129.
Shen, Y.; Feng, X.; Li, Y.; Zhang, G.; Jiang, Y. *Tetrahedron* ***2003***, *59*, 5667.
Shen, Y.; Feng, X.; Li, Y.; Zhang, G.; Jiang, Y. *Synlett* ***2002***, 793.

PhCHO + 10% (1,3-dibenzylimidazolidin-2-yl dithiocarboxylate, Bn, N, S^-, S) → (DCM, rt, 16 h) → PhCH(OTMS)CN 99%

Blanrue, A.; Wilhelm, R. *Synlett* ***2004***, 2621.

TMSCN , 20% Bn_2NMe_2 Br

20% *N,N*-methylaniline *N*-oxide

68%

Zhou, H.; Chen, F.-X.; Qin, B.; Feng, X.; Zhang, G. *Synlett* **2004**, 1077
Chen, F.; Feng, X.; Qin, B.; Zhang, G.; Jiang, Y. *Synlett* **2003**, 558.
He, B.; Li, Y.; Feng, X.; Zhang, G. *Synlett* **2004**, 1776.

PhCHO → 10% hydroxynaphthylphosphonodiamide; 20% *i*-PrOH , Me_3SiCN , DCM

84% (40% ee)

He, K.; Zhou, Z.; Wang, L.; Li, K.; Zhao, G.; Zhou, Q.; Tang, C. *Synlett* **2004**, 1525.

5% , 2 TMSCN , ether; 6 h , 23°C

99%

Li, Y.; He, B.; Fen, X.; Zhang, G. *Synlett* **2004**, 1598.

SECTION 367: ETHER, EPOXIDE, THIOETHER - ALKENE

Enol ethers are found in this section as well as alkenyl ethers.

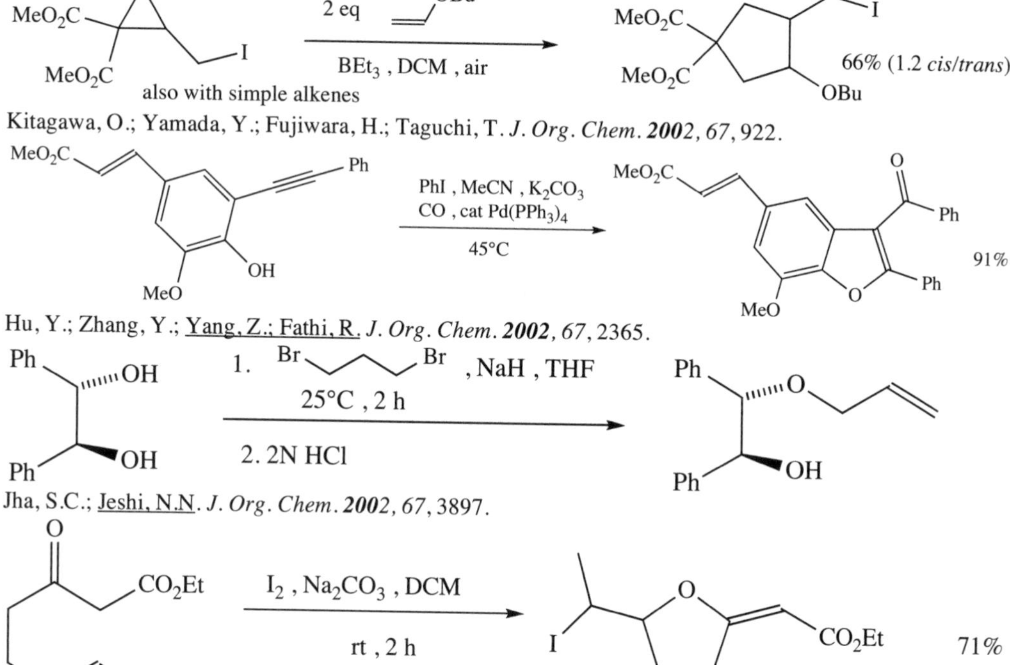

Kitagawa, O.; Yamada, Y.; Fujiwara, H.; Taguchi, T. *J. Org. Chem.* **2002**, *67*, 922.

Hu, Y.; Zhang, Y.; Yang, Z.; Fathi, R. *J. Org. Chem.* **2002**, *67*, 2365.

Jha, S.C.; Jeshi, N.N. *J. Org. Chem.* **2002**, *67*, 3897.

Ferraz, H.M.C.; Sano, M.K.; Nunes, M.R.S.; Bianco, G.G. *J. Org. Chem.* **2002**, *67*, 4122.

1. BINOL derivative–$SnCl_4$ DCM , –97°C
2. 2,4,6-collidine

$SiMe_3$ 97x85% (81% ee)

Nakamura, H.; Ishikara, K.; Yamamoto, H. *J. Org. Chem.* **2002**, *67*, 5124.

Bu

Br , 18.5 h

C

OH

5% $PdCl_2$, DMA , rt

Bu

O

76%

Ma, S.; Gao, W. *J. Org. Chem.* **2002**, *67*, 6014.

2 N S O

PhCHO , $[bpy^+]BF_4^-$

2.5 eq K_2CO_3 , rt , 1.5 h

O O Ph S-Btz

93% (97:3 *E:Z*)

Calò, V.; Scordari, F.; Nacci, A.; Schingaro, E.; D'Accolti, L.; Monopoli, A. *J. Org. Chem.* **2003**, *68*, 4406.

O O N_2 CO_2t-Bu

cat $Rh_2(DOSP)_4$, 25°C
hexane

O CO_2t-Bu O

83% (57% ee)

Hodgson, D.M.; Labande, A.H.; Pierard, F.Y.T.M.; Castro, M.A.E. *J. Org. Chem.* **2003**, *68*, 6153.

SMe O

1. 2 $HC(OMe)_3$, 3 BF_3•OEt_2 MeCN , –40°C , 13 h
2. quench

SMe O OMe OMe

89%

Kinoshita, H.; Osamura, T.; Kinoshita, S.; Iwamura, T.; Watanabe, S.; Kataoka, T.; Tanabe, G. *J. Org. Chem.* **2003**, *68*, 7532.

HS $SiMe_3$

$PhCH_2CHO$, $InCl_3$
DCM , reflux

S Ph

51%

Dobbs, A.P.; Guesné, S.J.J.; Martinovic, S.; Coles, S.J.; Hursthouse, M.B. *J. Org. Chem.* **2003**, *68*, 7880.

OH OH

$W(CO)_6$, hv
DABCO , THF
50-60°C , 14h

HO O (1 + OH O : 1) 49%

Wipf, P.; Graham, T.H. *J. Org. Chem.* **2003**, *68*, 8798.

C_7H_{15} SMe ZnCl + RO Br

cat $Pd(PPh_3)_4$, THF
reflux , 2 d

MeS OR C_7H_{15}

86%

Su, M.; Kang, Y.; Yu, W.; Hua, Z.; Jin, Z. *Org. Lett.* **2002**, *4*, 691.

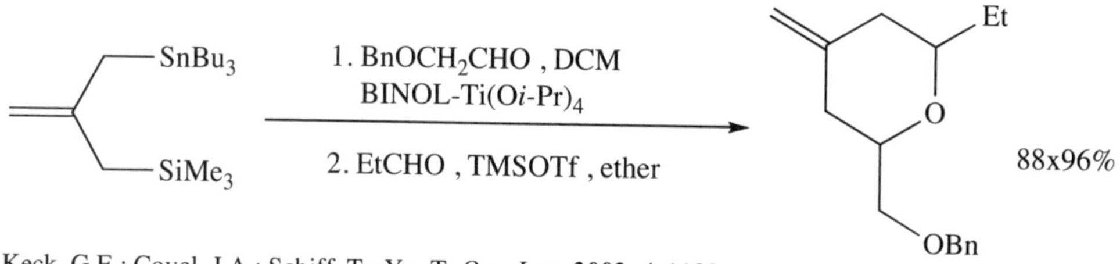

SnBu$_3$ / SiMe$_3$ → 1. BnOCH$_2$CHO , DCM, BINOL-Ti(O*i*-Pr)$_4$; 2. EtCHO , TMSOTf , ether → Et, O, OBn 88x96%

Keck, G.E.; Covel, J.A.; Schiff, T.; Yu, T. *Org. Lett.* ***2002****, 4*, 1189.

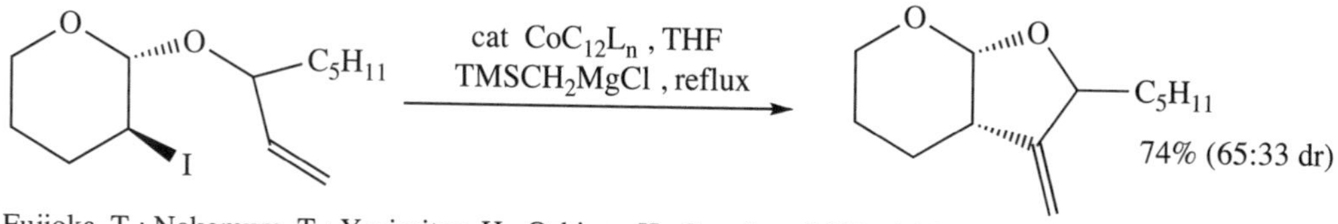

O, O, C_5H_{11}, I → cat $CoC_{12}L_n$, THF, $TMSCH_2MgCl$, reflux → O, O, C_5H_{11} 74% (65:33 dr)

Fujioka, T.; Nakamura, T.; Yorimitsu, H.; Oshima, K. *Org. Lett.* ***2002****, 4*, 2257.

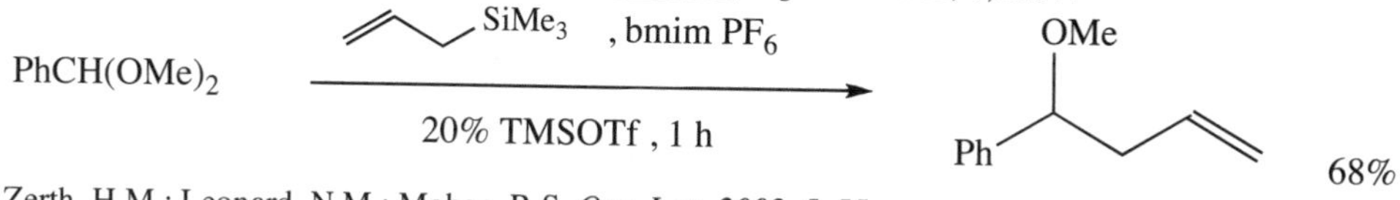

$PhCH(OMe)_2$ → SiMe$_3$, bmim PF_6; 20% TMSOTf , 1 h → OMe, Ph 68%

Zerth, H.M.; Leonard, N.M.; Mohan, R.S. *Org. Lett.* ***2003****, 5*, 55.

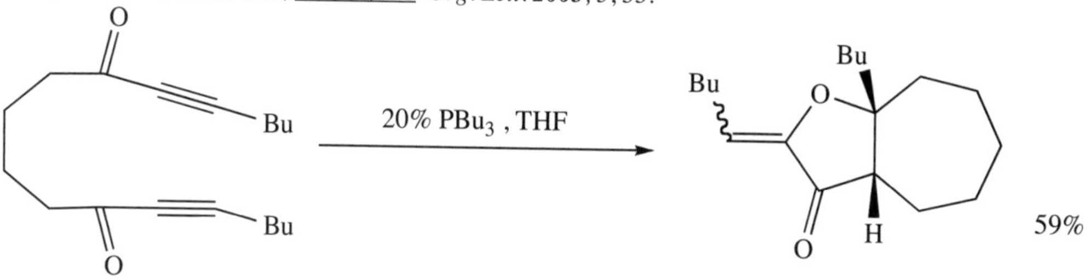

O, Bu, Bu, O → 20% PBu_3 , THF → Bu, Bu, O, O, H 59%

Kuroda, H.; Tomita, I.; Endo, T. *Org. Lett.* ***2003****, 5*, 129.

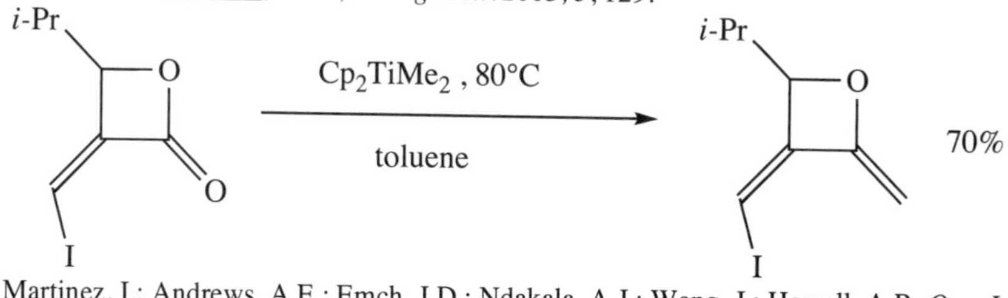

i-Pr, O, O, I → Cp_2TiMe_2 , 80°C; toluene → *i*-Pr, O, I 70%

Martinez, I.; Andrews, A.E.; Emch, J.D.; Ndakala, A.J.; Wang, J.; Howell, A.R. *Org. Lett.* ***2003****, 5*, 399.

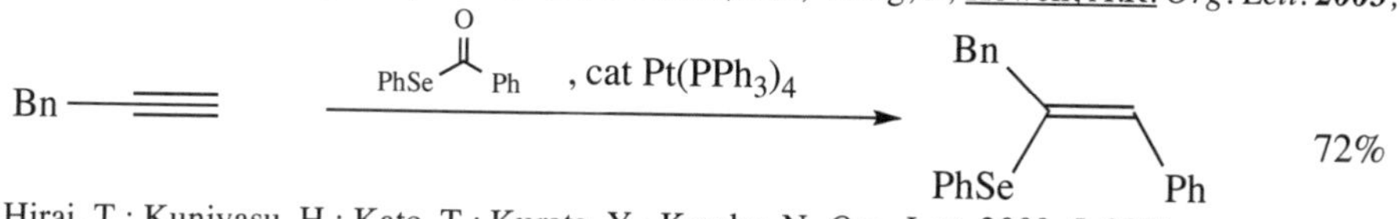

Bn → PhSe(C=O)Ph , cat $Pt(PPh_3)_4$ → Bn, PhSe, Ph 72%

Hirai, T.; Kuniyasu, H.; Kato, T.; Kurata, Y.; Kambe, N. *Org. Lett.* ***2003****, 5*, 3871.

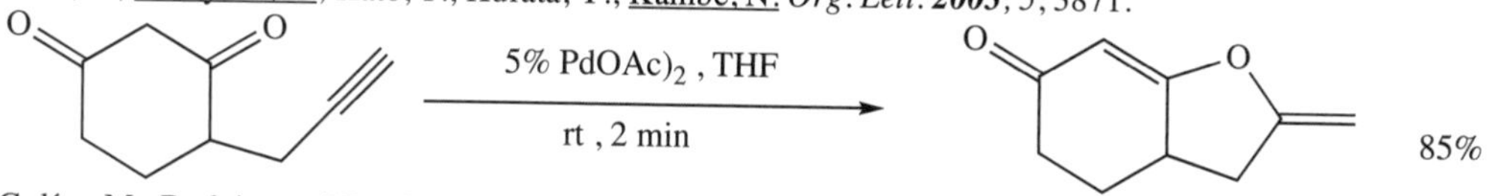

O, O → 5% $PdOAc)_2$, THF; rt , 2 min → O, O 85%

Gulías, M.; Rodríguez, J.R.; Castedo, L.; Mascareñas, J.L. *Org. Lett.* ***2003****, 5*, 1975.

2.5 CH_2=CHCH(OEt)$_2$, e$^-$
10% $NiBr_2$•3 H_2O , 70°C
DMF , Py

59% (59:41 *Z*:*E*)

Condon, S.; Dupré, D.; Nédélec, J.Y. *Org. Lett.* **2003**, *5*, 4701.

, 5% $Pd_2(dba)_3$
10% bpy , K_2CO_3 , 50°C , 5h

64%

Hu, Y.; Nawoschik, K.J.; Liao, Y.; Ma, J.; Fathi, R.; Yang, Z. *J. Org. Chem.* **2004**, *69*, 2235.

1% $PtCl_4$, dioxane , rt

86%

Pastine, S.J.; Youn, S.W.; Sames, D. *Org. Lett.* **2003**, *5*, 1055.

$Ce(Bu_4N)_6(NO_3)_3$, DCM

72%

Zhang, Y.; Raines, A.J.; Flowers II, R.A. *Org. Lett.* **2003**, *5*, 2363.

$PhB(OH)_2$, KF , toluene
20% $Pd(PPh_3)_4$, 60°C

84%

Zhu, G.; Zhang, Z. *Org. Lett.* **2003**, *5*, 3645.

$C_6H_{11}CHO$, $FeCl_3$
DCM

80%

Miranda, P.O.; Díaz, D.D.; Padrón, J.I.; Bermejo, J.; Martín, V.S. *Org. Lett.* **2003**, *5*, 1979.

–80°C→–10°C , THF

45%

Takeda, K.; Karauchi, H.; Okamoto, Y. *Org. Lett.* **2003**, *5*, 3705.

1. $MeO_2C-C_6H_4-C\equiv CH$, 4% CuI, 4% $PdCl_2(PPh_3)_2$, MeCN-NEt_3, 1 d
2. $MeO_2C-C_6H_4-I$, 60°C, 2 d (one pot)

83%

Bossharth, E.; Desbordes, P.; Monteiro, N.; Balme, G. *Org. Lett.* ***2003***, *5*, 2441.

$BnOCH_2SO_2BT$, THF, LiHMDS, 0°C, 1.5h

BT =

87% (1:1 *E:Z*)

Concellón, J.M.; Bardales, E. *Org. Lett.* ***2003***, *5*, 4783.

PhCHO, 5 *t*-$BuMe_2SiH$, 20% $Ni(cod)_2$, 40% PPh_3, imidazolium salt, reflux, 40% Cs_2CO_3, toluene

$PhMe_2Si$ — $OSiEt_3$ — Ph 71%

Sawaki, R.; Sato, Y.; Mori, M. *Org. Lett.* ***2004***, *6*, 1131.

5% $[Cp^*RuCl(\mu_2SMe)_2Cp^*RuCl$, DCE, 60°C, 1h

+ H_2O 93%

Nishibayashi, Y.; Yoshikawa, M.; Inada, Y.; Hidai, M.; Uemura, S. *J. Org. Chem.* ***2004***, *69*, 3408.

$Cu(OAc)_2$, Na_2CO_3, DCM, rt, 1d

99%

McKinley, N.F.; O'Shea, D.F. *J. Org. Chem.* ***2004***, *69*, 5087.

BuOH, 10% CuI, DMF, 70°C

87%

Patil, N.T.; Yamamoto, Y. *J. Org. Chem.* ***2004***, *69*, 5139.

DCM, MeOH, 0.3 I_2, 0.25 K_2CO_3

93%

Yue, D.; Cà, N.D.; Larock, R.C. *Org. Lett.* ***2004***, *6*, 1581.

Ph—≡—Ph → BuSeLi, THF, EtOH → (Ph)CH=C(SeBu)(Ph) 60%

Zeni, G.; Stracke, M.P.; Nogueira, C.W.; Braga, A.L.; Menezes, P.H.; Stefani, H.A. *Org. Lett.* ***2004***, *6*, 1135.

5% $W(CO)_6$, DABCO, THF; hν, 55°C, 6 h → NHBoc 88%

Davidson, M.H.; McDonald, F.E. *Org. Lett.* ***2004***, *6*, 1601.

Ph, CO_2Et, CO_2Et; HO; 20% $Zn(OTf)_2$, NEt_3, neat, 25°C → EtO_2C, EtO_2C, Ph, O 93%

Nakamura, M.; Liang, C.; Nakamura, E. *Org. Lett.* ***2004***, *6*, 2015.

OH, HO, O, O; 1. 15% $W(CO)_6$, NEt_3 THF, hν, 55°C; 2. Ac_2O, cat DMAP → O, O, O, AcO 82%

Alcázar, E.; Pletcher, J.M.; McDonald, F.E. *Org. Lett.* ***2004***, *6*, 3877.

$POPh_2$, C, OH; *t*-BuOK, *t*-BuOH, 30°C → $POPh_2$, Me, O 85%

Mukai, C.; Ohta, M.; Yamashita, H.; Kitagaki, S. *J. Org. Chem.* ***2004***, *69*, 6867.

OMe, O; 1. 5% Grubbs I, toluene, 25°C; 2. $EtOCH=CH_2$ (C=C isomerization) → OMe, O 77%

Kang, B.; Lee, J.M.; Kwak, J.; Lei, Y.S.; Chang, S. *J. Org. Chem.* ***2004***, *69*, 7661.

Cp, Cp, Zr, Et, Et, Et, Et; BCl_3; 2 equiv. EtO_2C–C(=O)–CO_2Et → O, Et, Et, Et, Et, EtO_2C, EtO_2C 80%

Takahashi, T.; Li, Y.; Ito, T.; Xu, F.; Nakajima, K.; Liu, Y. *J. Am. Chem. Soc.* ***2002***, *124*, 1144.

OMe, Ph, OMe — OTMS, CuBr / microwaves → OMe, Ph — 92%

Jung, M.E.; Maderna, A. *J. Org. Chem.* **2004**, *69*, 7755.

TMSO, OMe — 1. PhCHO, chiral Ti catalyst, 1d; 2. CF_3CO_2H → O, O, Ph — >99% (99%ee)

Long, J.; Hu, J.; Shen, X.; Ji, B.; Ding, K. *J. Am. Chem. Soc.* **2002**, *124*, 10.

O, O — BnOH, toluene, Na_2CO_3 / cat $[Ir(cod)Cl]_2$, 100°C, 2 h → Ph, O — 94%

Okimoto, Y.; Sakaguchi, S.; Ishii, Y. *J. Am. Chem. Soc.* **2002**, *124*, 1590.

OTBS, Me_2N + O — 1. 2-butanol; 2. AcCl, –78°C → O, O — 78%

Huang, Y.; Rawal, V.H. *J. Am. Chem. Soc.* **2002**, *124*, 9662.

Ph, O — Grubbs I catalyst / DCM → Ph, O — 61%

Sutton, A.E.; Seigal, B.A.; Finnegan, D.F.; Snapper, M.L. *J. Am. Chem. Soc.* **2002**, *124*, 13390.

OTMS, O*t*-Bu — 1. PhCHO, chiral Zr catalyst, toluene; 2. H^+ → O, Ph, O — 99% (124 *cis:trans*; 97% ee)

Yamashita, Y.; Saito, S.; Ishitani, H.; Kobayashi, S. *J. Am. Chem. Soc.* **2003**, *125*, 3793.

BnO, O, BnO, OAc — BnOH, 0.5 Et_2Zn, ligand / 10% $Pd(OAc)_2$ → BnO, O, OBn, BnO

+ 15% dppf — 95% (1:5 α:β)
+ 30% $P(OMe)_3$ — 90% (7:1 α:β)

Kim, H.; Men, H.; Lee, C. *J. Am. Chem. Soc.* **2004**, *126*, 1336.

Ph, Ph, C — OMe, $(OC)_5Cr$, Ph / 10% Rh complex, 25°C → MeO, Ph, Ph, Ph — 84%

Barluenga, J.; Vicente, R.; Barrio, P.; López, L.A.; Tómas, M. *J. Am. Chem. Soc.* **2004**, *126*, 5974.

O, Ph, OEt — $TiCl_4$/Mg (1:8), DCM → Ph, OEt — 96%

Yan, T.-H.; Chien, C.-T.; Tsai, C.-C.; Lin, K.-W.; Wu, Y.-H. *Org. Lett.* **2004**, *6*, 4965.

PhSH , 1.5 K_3PO_4 , toluene
5% $[Cu(phen)(PPh_3)_2]NO_3$
110°C , 4 h
93%

Bates, C.G.; Saejung, P.; Doherty, M.Q.; Venkataraman, D. *Org. Lett.* ***2004***, *6*, 5005.

OTIPS
10% Tf_2NH , DCM
20°C , 10 min
OTIPS
86%

Zhang, L.; Kozmin, S.A. *J. Am. Chem. Soc.* ***2004***, *126*, 10204.

OH
2 $PhCH{=}CH_2$, THF , 22°C , 1 h
2% chiral Ru carbene complex
OH
69% (96% ee)

Gillingham, D.G.; Kataoka, O.; Garber, S.B.; Hoveyda, A.H. *J. Am. Chem. Soc.* ***2004***, *126*, 12288.

C_3H_7 $B(OH)_2$
PhSeCl , bmim BF_4 , rt , 2h
C_3H_7 SePh
82%

Kabalka, G.W.; Venkataiah, B. *Tetrahedron Lett.* ***2002***, *43*, 3703.

$SiMe_3$
$PhCH(OMe)_2$, 1% $Bi(OTf)_3{\bullet}n\ H_2O$
DCM , rt , 15 min
OMe
Ph
84%

Wieland, C.; Zerth, H.M.; Mohan, R.S. *Tetrahedron Lett.* ***2002***, *43*, 4597.

MeO_2C
5% $PdCl_2(MeCN)_2$, CO , MOH
1.1 *p*-benzoquinone , 1h , 0°C → rt
MeO_2C MeO CO_2Me
82%

Kato, K.; Yamamoto, Y.; Akita, H. *Tetrahedron Lett.* ***2002***, *43*, 4915.

Me_3Si OH
$PhCH_2CHO$, $InCl_3$, rt
DCM
Ph
81%

Dobbs, A.P.; Martinovic, S. *Tetrahedron Lett.* ***2002***, *43*, 7055.

Ph Ph
EtO_2C CO_2Et , DCE , 40°C , 1d
5% Yb(OTfO$_4$
Ph Ph EtO_2C CO_2Et
84%

Shi, M.; Xu, B. *Tetrahedron Lett.* ***2003***, *44*, 3839.

Ph P Ph Ph
Cl CHO
[BuLi/S] . THF , rt
Cl Ph BuS
80%

Huang, X.; Xiong, Z.-C. *Tetrahedron Lett.* ***2003***, *44*, 5913.

Br–CH₂CH=CH–SiMe₃ , Cs_2CO_3 ; t-BuOH , Bu_2Te^+–CH₂CH=CH–SiMe₃ Br^-

Cl–C₆H₄–CHO → Cl–C₆H₄–epoxide–CH=CH–SiMe₃

82% (49:51 *cis:trans*)

Li, K.; Huang, Z.-Z.; Tang, Y. *Tetrahedron Lett.* ***2003***, *44*, 4137.

HC≡C-Ts , THF , CF_3CH_2OH ; 0.5% NEt_3 , 15°C , 5h

PhCH₂SH → PhCH₂S–CH=CH–Ts quant (95:5 *Z:E*)

Arjona, O.; Medel, R.; Rojas, J.; Costa, A.M.; Vilarrasa, J. *Tetrahedron Lett.* ***2003***, *44*, 6369.

1. $Cp_2Ti[P(OEt)_3]_2$; 2. PhCH₂CH₂C(O)CH₂CH₂Ph

PhCH₂OCH₂Cl → PhCH₂O–CH=C(CH₂CH₂Ph)₂ 66%

Takeda, T.; Shono, T.; Ito, K.; Sasak, H. *Tetrahedron Lett.* ***2003***, *44*, 7897.

$ClPh_3P$–CH₂–O–CH(Ph)Cl , 2.5 t-BuOK ; t-BuOH , THF , 0°C

Ar₂C=O → O(CH=C(Ph)Ar)... 66%

Ar = 3,5-dimethoxyphenyl

Kulkarni, M.G.; Doke, A.K.; Davawala, S.I.; Doke, A.V. *Tetrahedron Lett.* ***2003***, *44*, 4913.

MeO–C₆H₄–OH , CuCl , Cs_2CO_3 , toluene ; piperidinyl–CH₂CH₂–O–C₆H₄–OMe , reflux 5h

Ph(Br)C=CPh₂ → Ph(ArO)C=CPh₂ (Ar = 4-MeOC₆H₄) 94%

Wan, Z.; Jones, C.D.; Koenig, T.M.; Pu, Y.J.; Mitchell, D. *Tetrahedron Lett.* ***2003***, *44*, 8257.

Mg , TMSCl DMF

cyclopentenone → 3,3'-bi(1-TMSO-cyclopentene) (TMSO, OTMS) 61%

Maekawa, H.; Sakai, M.; Uchida, T.; Kita, Y.; Nishiguchi, I. *Tetrahedron Lett.* ***2004***, *45*, 607.

MeOH , 3% I_2 , rt , 3 min

1,3-cyclohexanedione → 3-methoxy-2-cyclohexenone (O, OMe) 94%

Bhosale, R.S.; Bhosale, S.V.; Bhosale, S.V.; Wang, T.; Zubaidha, P.K. *Tetrahedron Lett.* ***2004***, *45*, 7187.

THF , 200°C , microwaves , 40 min

Ph–C≡CH → 2-(CH=CH–Ph)tetrahydrofuran 60% (53:47 *cis:trans*)

Zhang, Y.; Li, C.-J. *Tetrahedron Lett.* ***2004***, *45*, 7581.

1. (a) $PhSCH_2CH_2CH(OMe)_2$, BuLi , THF/HMPA (b) CuBr , 5 min ; 2. $TiCl_4$, DCM , MS 4Å

cyclohexenone → bicyclic enone with SPh 87x81%

Ding, P.; Ghosez, L. *Tetrahedron* ***2002***, *58*, 1565.

Ph_2MeSiH , cat $B(C_6F_5)_3$ / toluene — 81%

Blackwell, J.M.; Morrison, D.J.; Piers, W.E. *Tetrahedron* **2002**, *58*, 8247.

Ph_2S_2 , neat , 80°C , 8h / cat Pd)$(PPh_3)_4$, PPh_3 — Bu, SPh — 97%

Ananikov, V.P.; Beletskaya, I.P. *Org. Biomol. Chem.* **2004**, *2*, 284.

MeCON(Me)TMS , cat NaOH / DMF , 20°C, 1h — OTMS + OTMS (18 : 82) 8%

Tanabe, Y.; Misaki, T.; Kurihara, M.; Iida, A.; Nishii, Y. *Chem. Commun.* **2002**, 1628.

$PhCO_2H$ — BuC≡CH , P(fur)$_3$, $PhCO_2Na$, toluene / 60°C , 16h — Ph, O, Bu — 97%

Goossen, L.J.; Paetzold, J.; Koley, D. *Chem. Commun.* **2003**, 706.

PhC≡CH , $Me_3SiTePh$, 100°C / sealed tube , 12h — OTMS, Ph, TePh — 93% 96:4 *E:Z*)

Bu_3SnH/AIBN reduces CTePh to CH

Yamago, S.; Miyoshi, M.; Miyazoe, H.; Soshida, J. *Angew. Chem. Int. Ed.* **2002**, *41*, 1407.

5% chiral Cr catalyst , CH_2=CHOEt , 1d / neat , MS 4Å — O, OEt — 75% (94% ee)

Gademann, K.; Chavez, D.E.; Jacobsen, E.N. *Angew. Chem. Int. Ed.* **2002**, *41*, 3059.

cat $[Rh(cod)Cl)]_2$, $AgSbF_6$, cat BINAP / cat biaryl diphosphine ligand — quant (99% ee)

Lei, A.; He, M.; Wu, S.; Zhang, X. *Angew. Chem. Int. Ed.* **2002**, *41*, 3457.

cat $Pd(OAc)_2$, bpy , 80°C , 10h — C_7H_{15}, AcO, OAc — 50%

Zhao, L.; Lu, X. *Angew. Chem. Int. Ed.* **2002**, *41*, 4343.

1. Sm / 2. CH_2I_2 — C_7H_{15}, CO_2Me — 87%

Kim, J.T.; Kel'in, A.V.; Gevorgyan, V. *Angew. Chem. Int. Ed.* **2003**, *42*, 98.

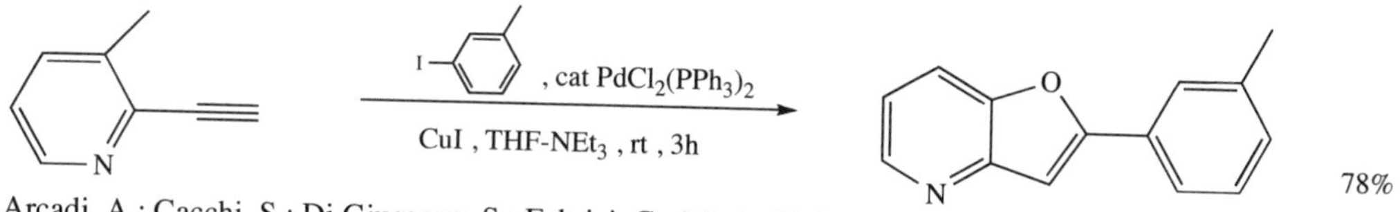

, cat $PdCl_2(PPh_3)_2$
CuI , $THF-NEt_3$, rt , 3h
78%

Arcadi, A.; Cacchi, S.; Di Giuseppe, S.; Fabrizi, G.; Marinelli, F. *Synlett* **2002**, 457.

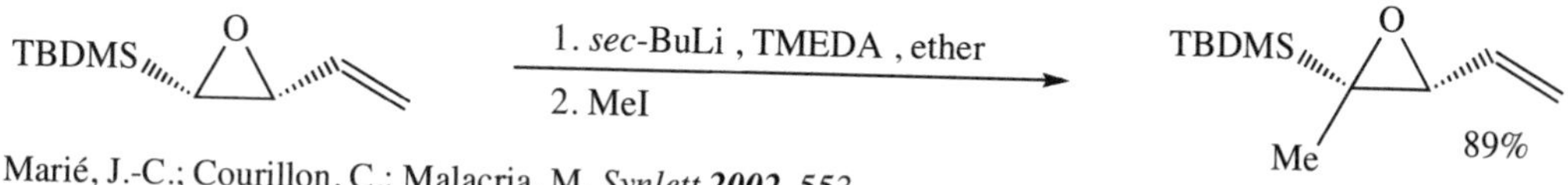

TBDMS
1. *sec*-BuLi , TMEDA , ether
2. MeI
TBDMS
Me
89%

Marié, J.-C.; Courillon, C.; Malacria, M. *Synlett* **2002**, 553.

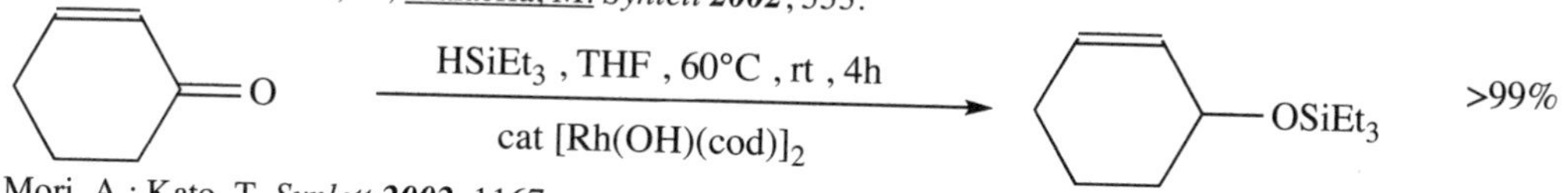

O
$HSiEt_3$, THF , 60°C , rt , 4h
cat $[Rh(OH)(cod)]_2$
$OSiEt_3$
>99%

Mori, A.; Kato, T. *Synlett* **2002**, 1167.

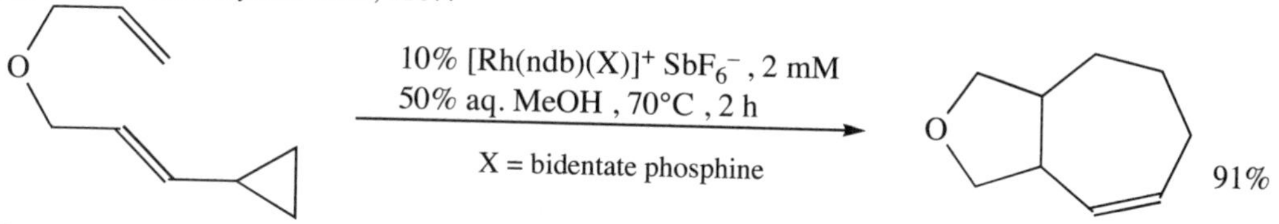

O
10% $[Rh(ndb)(X)]^+ SbF_6^-$, 2 mM
50% aq. MeOH , 70°C , 2 h
X = bidentate phosphine
O
91%

Wender, P.A.; Love, J.A.; Williams, T.J. *Synlett* **2003**, 1295.

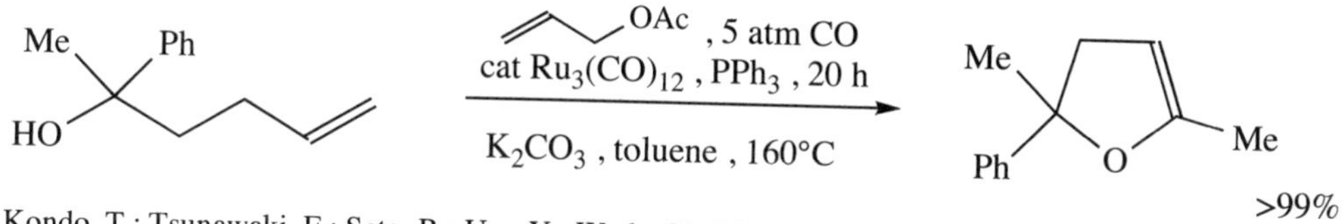

Me Ph
HO
OAc , 5 atm CO
cat $Ru_3(CO)_{12}$, PPh_3 , 20 h
K_2CO_3 , toluene , 160°C
Me
Ph O Me
>99%

Kondo, T.; Tsunawaki, F.; Sato, R.; Ura, Y.; Wada, K.; Mitsuda, T. *Chem. Lett.* **2003**, *32*, 24.

PhSSPh
Sm , allyl bromide , MeOH
rt , 30 min
$PhSCH_2CH{=}CH_2$ 76%

Zhan, Z.-P.; Lang, K. *Chem. Lett.* **2004**, *33*, 1370.

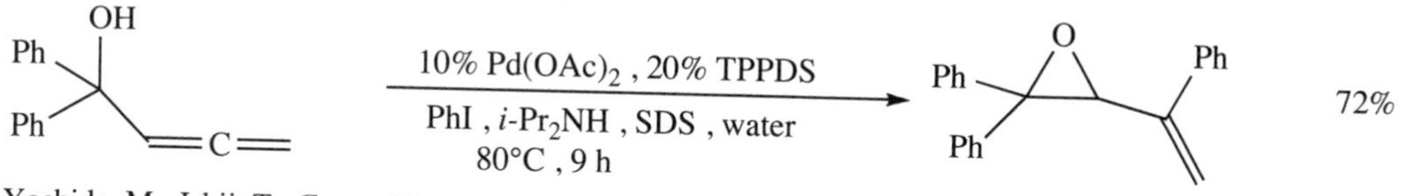

OH
Ph
Ph
C
10% $Pd(OAc)_2$, 20% TPPDS
PhI , i-Pr_2NH , SDS , water
80°C , 9 h
O
Ph
Ph
Ph
72%

Yoshida, M.; Ishii, T.; Gotou, T.; Ihara, M. *Heterocycles* **2004**, *64*, 41.

Related Method: Section 180A (Protection of Ketones)

SECTION 368: HALIDE, SULFONATE - HALIDE, SULFONATE

Halocyclopropanations are given in Section 74F (Cyclopropanations, including Halocyclopropanations).

Ph —≡— Me
Ph OH , H^+
cat. $Pd(PPh_4)_4$
Ph O Ph
89%

Zhang, W.; Haight, A.R.; Hsu, M.C. *Tetrahedron Lett.* **2002**, *43*, 6575.

InI_2 , MeOH , 5 h

60% (1:3.6 *E:Z*)

Yanada, R.; Koh, Y.; Nishimori, N.; Matsumura, A.; Obika, S.; Mitsuya, H.; Fujii, N.; Tekemoto, Y. *J. Org. Chem.* ***2004***, *69*, 2417.

Selecfluor , KBr , MeCN/MeOH

98%

Ye, C.; Shreeve, J.M. *J. Org. Chem.* ***2004***, *69*, 8561.

N_2=$CHCO_2Et$ silver-(pyrazoyl)borate ligand CH_2Br_2

65%

Dias, H.V.R.; Browning, R.G.; Polach, S.A.; Diyabalanage, H.V.K.; Lovely, C.J. *J. Am. Chem. Soc.* ***2003***, *125*, 9270.

$ZnBr_2$, $Pb(OAc)_4$, $CHCl_3$ rt , 5 min

91%

Muathen, H.A. *Synth. Commun.* ***2004***, *34*, 3545.

SECTION 369: HALIDE, SULFONATE - KETONE

Ph-I(OH)OTs , MeCN reflux

93%

Nabana, T.; Togo, H. *J. Org. Chem.* ***2002***, *67*, 4362.

NCS , MeCN , 0.3 $Mg(ClO_4)_2$

99%

Yang, D.; Yan, Y.-L.; Lui, B. *J. Org. Chem.* ***2002***, *67*, 7429.

NaOCl , AcOH , acetone 0°C

72%

Van Brunt, M.P.; Ambenge, R.O.; Weinreb, S.M. *J. Org. Chem.* ***2003***, *68*, 3323.

CF_3I , Et_2Zn , THF cat $RhCl(PPh_3)_3$, 0.5 h

77%

Sato, K.; Omote, M.; Ando, A.; Kumadaki, I. *Org. Lett.* ***2004***, *6*, 4359.

OTBDMS

$ZrCl_4$, DCM , –78°C

O O O O Cl Cl 1-naphth 1-naphth

O Cl 88% ee

Zhang, Y.; Shibatomi, K.; Yamamoto, H. *J. Am. Chem. Soc.* **2004**, *126*, 15038.

OH

$Ag/AgClO_3$, e^- , Et_3N-5 HF

1.7V , 6h

O F F 89%

Fukuhara, T.; Akiyama, Y.; Yoneda, N.; Tada, T.; Hara, S. *Tetrahedron Lett* **2002**, *43*, 6583.

O CO_2Me

$NaBH_4$, MeOH

–72°C → 0°C

OH CO_2Me 70%

Paul, S.; Gupta, V.; Guta, R.; Loupy, A. *Tetrahedron Lett.* **2003**, *44*, 439.

O O

NF_3O , MeCN , Bu_4NOH

rt , 12h

O O F 80%

Gupta, O.D.; Chreeve, J.M. *Tetrahedron Lett.* **2003**, *44*, 2779.

O Ph

1. PhI(OH)OTs , microwaves , 80 sec

2. $MgBr_2$, microwaves , 120 sec

O Ph Br 85%

Lee, J.C.; Park, J.Y.; Yoon, S.Y.; Bae, Y.H.; Lee, S.J. *Tetrahedron Lett.* **2004**, *45*, 191.

t-Bu O

NBS , NH_4OAc , ether

25°C , 30 min

Br *t*-Bu O 92%

Tanemura, K.; Suzuki, T.; Nishida, Y.; Satsumabayashi, K.; Horaguchi, T. *Chem. Commun.* **2004**, 470.

O CO_2t-Bu

cat $Cu(OTf)_2$-chiral bis(oxazoline)

$(PhSO_2)_2NF$ ether , 0°C

O CO_2t-Bu F 94% (82% ee)

Ma. J.-A.; Cahard, D. *Tetrahedron: Asymmetry* **2004**, *15*, 1007.

O Ph

NIS , TsOH , microwaves , neat

O Ph I 84%

Lee, J.C.; Bae, Y.H. *Synlett* **2003**, 507.

ClH_2C-N^+ N^+-F $(BF_4)_2$

I_2 , MeOH , 1d

86%

Jereb, M.; Stavber, S.; Zupan, M. *Tetrahedron* ***2003***, *59*, 5935.

Ph OH Ph

Chloramine-T , $ZnBr_2$

MeCN , rt

Ph O Ph Br

94%

Wang, B.M.; Song, Z.L.; Fan, C.A.; Tu, Y.Q.; Chen, W.M. *Synlett* *2003*, 1497.

HO N Ts

, NBS , *i*-PrOH

–78°C → 25°C

Br N Ts O

85%

Hurley, P.B.; Dake, G.R. *Synlett* ***2003***, 2131.

Ph O

4 $CuCl_2$, DMF

LiCl , 4 h

Cl Ph Cl O

76%

Nobrega, J.A.; Goncalves, S.M.C.; Reppe, C. *Synth. Commun.* ***2002***, *32*, 3711.

MeO O

I_2 , MeOH , 20 h

HO N N-F $(BF_4^-)_2$

MeO O I

86%

Jereb, M.; Stavber, S.; Zupan, M. *Synthesis* ***2003***, 853.

O Ph CH_3

HBr , P_2O_5

dioxane

O Ph CH_2Br

95%

Tillu, V.H.; Shinde, P.D.; Bedekar, A.V.; Wakharkar, R.D. *Synth. Commun.* ***2003***, *33*, 1399.

trichloroisocyanuric acid

H_2O , acetone

O Cl Cl

61%

Hiegel, G.A.; Bayne, C.D.; Ridley, B. *Synth. Commun.* ***2003***, *33*, 1997.

O Ph

H_2O_2 , HCl , MeOH

O Ph $CHCl_2$

89%

Terent'ev, A.O.; Khodykin, S.V.; Troitskii, N.A.; Ogibin, Y.N.; Nikishin, G.I. *Synthesis* ***2004***, 2845.

SECTION 370: HALIDE, SULFONATE - NITRILE

NO ADDITIONAL EXAMPLES

SECTION 371: HALIDE, SULFONATE - ALKENE

2-Naphthaldehyde → (2 eq $Ph_2P(O)CH_2CF_3$; 10 eq TBAF, THF, rt) → 2-naphthyl-CH=CH-CF_3 98% (85:15 *E:Z*)

Kobayashi, T.; Eda, T.; Tamura, O.; Ishibashi, H. *J. Org. Chem.* ***2002***, *67*, 3156.

Ph-C≡C-CH(OH)Bu → (ICl, DCM, DCE, rt) → Ph(Cl)C=C(I)-CH(OH)Bu 70%

Bellina, F.; Colzi, F.; Mannina, L.; Rossi, R.; Viel, S. *J. Org. Chem.* ***2003***, *68*, 10175.

PhCHO → (2 PhC≡CH, $TiCl_4$, 10 h; DCM, 10°C) → Ph(Cl)C=CH-CH(Ph)-CH=C(Ph)Cl 56%

Kabalka, G.W.; Wu, Z.; Ju, Y. *Org. Lett.* ***2002***, *4*, 3415.

EtO_2C, CO_2Et diene with Cl → (Grubbs' II, PhH, 65°C) → chlorocyclohexene diester 99%

Cha, W.; Weinreb, S.M. *Org. Lett.* ***2003***, *5*, 2505.

EtO_2C-CH=C(I)CF_3 → (1. *i*-PrMgBr, –78°C, THF; 2. CuCN•2 LiCl; 3. allyl bromide, THF, –78°C → –20°C) → F_3C(allyl)C=CH-CO_2Et 95%

Thibonnet, J.; Duchêne, A.; Parrain, J.-L.; Abarbri, M. *J. Org. Chem.* ***2004***, *69*, 4262.

$PhCH_2CH_2CHO$ (also for ketones) → (CBr_3F, Ph_3, Et_2Zn; THF, rt) → $PhCH_2CH_2CH$=C(F)Br 88% (1:1.05 *Z:E*)

Lei, X.; Dutheuil, G.; Pannecoucke, X.; Quirion, J.-C. *Org. Lett.* ***2004***, *6*, 2101.

2-(4-Methoxyphenylethynyl)biphenyl → (1.2 ICl, DCM, 1 h, N_2, –78°C) → 9-iodo-10-(4-methoxyphenyl)phenanthrene 99%

Yao, T.; Campo, M.A.; Larock, R.C. *Org. Lett.* ***2004***, *6*, 2677.

O I I O $h\nu$, $(Bu_3Sn)_2$, PhH C 85% (1:1.3)

Lin, H.-H.; Chang, W.-S.; Luo, S.-Y.; Sha, C.-K. *Org. Lett.* ***2004***, *6*, 3289.

TIPS TIPS F MeO_2C cat $Mo(CO)_6$ F F MeO_2C MeO_2C 72% F MeO_2C

Shen, Q.; Hammond, G.B. *J. Am. Chem. Soc.* ***2002***, *124*, 6534.

O methyl vinyl ketone, Me_4NCl , DMF C_6H_{13} C_6H_{13} cat $CpRu(NCMe)_3$ PF_6 72% (>15:1 *E:Z*) Cl

Trost, B.M.; Pinkerton, A.B. *J. Am. Chem. Soc.* ***2002***, *124*, 7376.

O Cl_3CCO_2Me , $CrCl_2$ CO_2Me Ph Ph 60% (>5:1 *Z:E*) Cl

Barma, D.K.; Kundu, A.; Zhang, H.; Mioskowski, C.; Falck, J.R. *J. Am. Chem. Soc.* ***2003***, *125*, 3218.

Cl 10% $RhCl(PPh_3)_3$, DCE 92% O reflux , 1 h O O O Cl

Tong, X.; Zhang, Z.; Zhang, X. *J. Am. Chem. Soc.* ***2003***, *125*, 6370.

CO_2Me I CO_2Me 10% NaI , 0.5 Na_2CO_3 , BnI CO_2Me 81% acetone , reflux CO_2Me Bn

Ma, S.; Zhang, J.; Cai, Y.; Lu, L. *J. Am. Chem. Soc.* ***2003***, *125*, 13954.

HO Cl $TiCl_4$, DCM , –78°C HO OMe 74% OMe OMe

Kim, Y.-H.; Lee, X.-Y.; Oh, C.-Y.; Yang, J.-G.; Ham, W.-H. *Tetrahedron Lett.* ***2002***, *43*, 837.

$CrCl_2$, THF , rt , 12 h — 93%

Baati, R.; Barma, D.K.; Krishna, U.M.; Mioskowski, C.; Falck, J.R. *Tetrahedron Lett.* **2002**, *43*, 959.

NBS , DCM-THF , rt , 5% TMSCl / 5% $Yb(OTf)_3$ — 84%

Yamanaka, M.; Arisawa, M.; Nishida, A.; Nakagawa, M. *Tetrahedron Lett.* **2002**, *43*, 2403.

In , DMF , reflux , 5h — 76%

Ranu, B.C.; Samanta, S.; Das, A. *Tetrahedron Lett.* **2002**, *43*, 5945.

HCl , dioxane , 120°C — 99%

Siriwardana, A.I.; Nakamura, I.; Yamamoto, Y. *Tetrahedron Lett.* **2003**, *44*, 985.

I_2 , NEt_3 , DCM , rt , <5 min — 95%

Kim, J.M.; Na, J.E.; Kim, J.N. *Tetrahedron Lett.* **2003**, *44*, 6317.

$CeCl_3$•7 H_2O , NaClO / DCM , H_2O — 98%

Moreno-Dorado, F.J.; Guerra, F.M.; Manzano, F.L.; Aladro, F.J.; Jorge, Z.D.; Massanet, G.M. *Tetrahedron Lett.* **2003**, *44*, 6691.

Zn-Cu , AcOH , THF-MeOH , 0°C — (83 : 17) 74%

Kadota, I.; Ueno, H.; Ohno, A.; Yamamoto, Y. *Tetrahedron Lett.* **2003**, *44*, 8645.

SmI_2 , THF , rt , 5 min — 85%

Li, J.; Xu, X.; Zhang, Y. *Tetrahedron Lett.* **2003**, *44*, 9349.

Chloramine-T , aq. THF , NaI / rt , 10 min — 95%

Kabalka, G.W.; Mereddy, A.R. *Tetrahedron Lett.* **2004**, *45*, 1417.

Dess-Martin periodinane , DCM

NEt_4Br , 1 d

O

O

Br

68%

Ramanarayanan, G.V.; Shukla, V.G.; Akamanchi, K.G. *Synlett* ***2002***, 2059.

Ph

Ph

$TiCl_4$, DEAD , MeCN

MS 4Å

Ph

Ph

Cl

89%

Shao, L.-X.; Zhao, L.-J.; Shi, M. *Eur. J. Org. Chem.* ***2004***, 4894.

Br

2 $CuCl_2$, 65°C

Cl

Cl

Br

89%

Zhou, H.; Huang, X.; Chen, W. *Synlett* ***2003***, 2080.

$PhCH_2OH$

MnO_2 , Ph_3PCHBr_2 Br , DCM

MS 4Å , reflux , 17 h

Ph

Br

Br

73%

Raw, S.A.; Reid, M.; Roman, E.; Taylor, R.J.K. *Synlett* ***2004***, 819.

Me—≡—CO_2H

1. $SOCl_2$, DMF

2. H_2O

Me

CO_2H

Cl

88% (80:20 *E:Z*)

Urdaneta, N.A.; Herrera, J.C.; Salazar, J.; López, S.E. *Synth. Commun.* ***2002***, *32*, 3003.

OH

Ph

C

$InCl_3$, PhH , 3°C → rt

Ph

Cl

64%

Cho, Y.S.; Jun, B.K.; Pae, A.N.; Cha, J.H.; Koh, J.H.; Koh, H.Y.; Chang, M.H.; Han, S.-Y. *Synthesis* ***2004***, 2620.

Cl

Cl

O

O

Ph

NaN_3 , DMF , rt

also with other nucleophiles

Cl

O

O

Ph

N_3

Hamura, T.; Kakinuma, M.; Tsuji, S.; Matsumoto, T.; Suzuki, K. *Chem. Lett.* ***2002***, *31*, 748.

Cl

O

Bu

Me

Me

Li

, ether

5 h

AcO

Me

Cl

Bu

Me

88%

Hamura, T.; Kawano, N.; Tsuji, S.; Matsumoto, T.; Suzuki, K. *Chem. Lett.* ***2002***, *31*, 1042.

SECTION 372: KETONE - KETONE

Ph—≡—Ph, cat $Pd(dba)_2$, NEt_3, $DMF-H_2O$, 130°C — 96%

Aetnev, A.A.; Tian, Q.; Larock, R.C. *J. Org. Chem.* ***2002**, 67*, 9276.

BnZnCl, CO, 3.5 TMSCl, 5 LiCl, 1.5% $Pd(PPh_3)_4$, 31°C — Ph — 59%

Yuguchi, M.; Tokuda, M.; Orito, K. *J. Org. Chem.* ***2004**, 69*, 908.

OH, Na_2CO_3, $EtOAc/H_2O$, cat $[PdCl(\eta^3\text{-}C_3H_7)]_2$-TppTs, rt, 14 h — 84%

Kinoshita, H.; Shinokubo, H.; Oshima, K. *Org. Lett.* ***2004**, 6*, 4085.

OEt — 1. $PhCO(B_t)$, NaH; 2. NH_4Cl, NH_4OH — Ph — 55%

Katritzky, A.R.; Wang, Z.; Wang, M.; Wilkerson, C.R.; Hall, C.D.; Akhmedov, N.G. *J. Org. Chem.* ***2004**, 69*, 6617.

Ph, $OSnBu_3$ + Cl, Ph — cat $ZnCl_2$, THF, 40°C, 2 h — Ph, Ph — >99%

Yasuda, M.; Tsuji, S.; Shigeyoshi, Y.; Baba, A. *J. Am. Chem. Soc.* ***2002**, 124*, 7440.

Ph, $SiMe_3$ — 1. Ph, Cl, DBU, *i*-PrOH, THF; 30% S, N, Et, Br^-, HO; 2. H_2O — Ph, Ph, Cl — 82%

Mattson, A.E.; Bharadwaj, A.R.; Scheidt, K.A. *J. Am. Chem. Soc.* ***2004**, 126*, 2314.

Ph — 1. NO_2, O_2N, S, O_2, O, I, Ph, OH; microwaves, 2 min; 2. pyridine *N*-oxide, microwaves, 30 sec — Ph — 83%

Lee, J.C.; Park, H.-J.; Park, J.Y. *Tetrahedron Lett.* ***2002**, 43*, 5661.

2.2 SmI_2 , THF , rt , 10 min

68%

Wang, X.; Zhang, Y. *Tetrahedron Lett.* ***2002**, 43*, 5431.

YbI_2 , THF , 5h

80%

Saikia, P.; Laskar, D.D.; Prajapati, D.; Sandhu, J.S. *Tetrahedron Lett.* ***2002**, 43*, 7525.

NBS , CCl_4 , Py , 85°C , 2.5h

94%

Khurana, J.M.; Kandpal, B.M. *Tetrahedron Lett.* ***2003**, 44*, 4909.

DBU , AL_2O_3 , microwaves , 8 min

thiazolium salt , Me_2C=CHCHO

90%

Yadav, J.S.; Anuradha, K.; Reddy, B.V.S.; Eshwaraiah, B. *Tetrahedron Lett.* ***2003**, 44*, 8959.

$MeReO_3$/aq. H_2O_2 , MeCN , $MgSO_4$

reflux , 7.5h

92%

Jain, S.L.; Sharma, V.B.; Sain, B. *Tetrahedron Lett.* ***2004**, 45*, 1233.

1. BuLi
2. I(CH₂)₄I
3. HBF_4•OEt_2
4. TBAF , $KHCO_3$, aq. H_2O_2

59%

Inoue, A.; Kondo, J.; Shinokubo, H.; Oshima, K. *Chem. Commun.* ***2002***, 114.

2% Fe(tpp)OTf , dioxane ,1.5h

reflux

95%

Suda, K.; Baba, K.; Nakajima, S.; Takanami, T. *Chem. Commun.* ***2002***, 2570.

lauroyl peroxide , DCE , heat

81%

Ouvry, G.; Zard, S.Z. *Chem. Commun.* ***2003***, 778.

$C_8H_{17}C(O)Zr(Cl)Cp_2$

1% $Pd(OAc)_2$, BF_3•OEt_2 , 0°C

ether-THF , 12h

70%

Hanzawa, Y.; Tabuchi, N.; Narita, K.; Kakuuchi, A.; Yabe, M.; Taguchi, T. *Tetrahedron* ***2002**, 58*, 7559.

$PhB(OH)_2$, cat $RhH(CO)(PPh_3)_3$

CO , MeOH , 80°C , 15h

78%

Sauthier, M.; Castanet, Y.; Mortreux, A. *Chem. Commun.* ***2004**,* 1520.

cat $RuCl_2$(*p*-cymene)$_2$, O_2

Cs_2CO_3 , PhH , 80°C

83%

Chang, S.; Lee, M.; Ko, S.; Lee, P.H. *Synth. Commun.* ***2002**, 32,* 1279.

$OSiHMe_2$ +

1. $MgCl_2$, DMF , 1d
2. H^+

quant (98:2 *anti:syn*)

Miura, K.; Nkagawa, T.; Hosomi, A. *Synlett* ***2003**,* 2068.

Fe(III)-EDTA , O_2 , pH 9.5

aq. MeOH , 3.5 h

98%

Rao, T.V.; Dongre, R.S.; Jain, S.L.; Sain, B. *Synth. Commun.* ***2002**, 32,* 2637.

2 InBr , THF , 20°C , 1 d

90%

Peppe, C.; das Chagas, R.P. *Synlett* ***2004**,* 1187.

3 $NaNO_2$, HCl , THF

0°C , 6 h

85%

Rüedi, G.; Oberli, M.A.; Nagel, M.; Weymuth, C.; Hansen, H.-J. *Synlett* ***2004**,* 2315.

$PhMe_3N\ Br_3$, $SbBr_3$, 19 h

pyridine , MeOH

91%

Sayama, S.; Onami, T. *Synlett* ***2004**,* 2369.

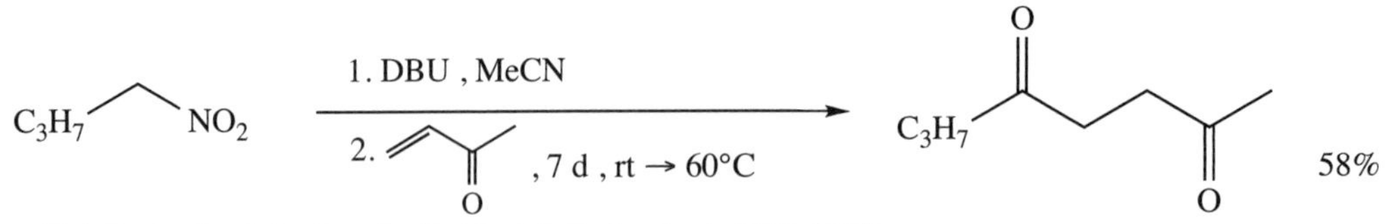

1. DBU , MeCN

2. , 7 d , rt → 60°C

58%

Ballini, R.; Barboni, L.; Boscia, G.; Fiorini, D. *Synthesis* **2002**, 2725.

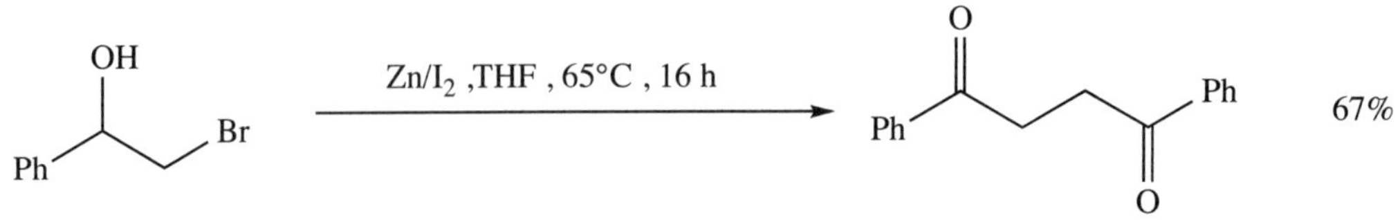

Zn/I_2 ,THF , 65°C , 16 h

67%

Ceylan, M.; Gürdere, M.B.; Budak, Y.; Kazaz, C.; Secen, H. *Synthesis* **2004**, 1750.

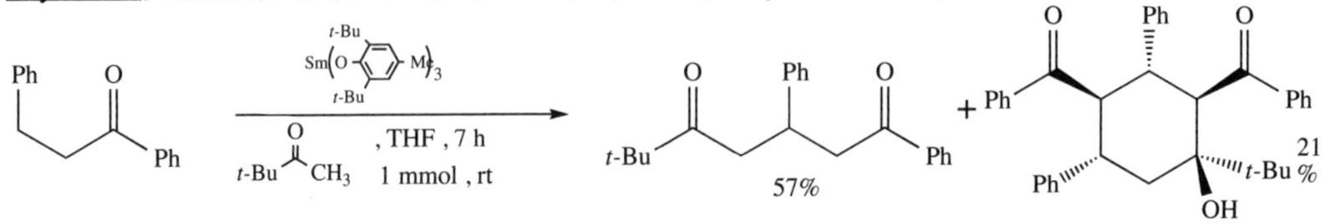

, THF , 7 h

1 mmol , rt

57%

21 %

Katagiri, K.; Kameoka, M.; Nishiura, M.; Imamoto, T. *Chem. Lett.* **2002**, *31*, 426.

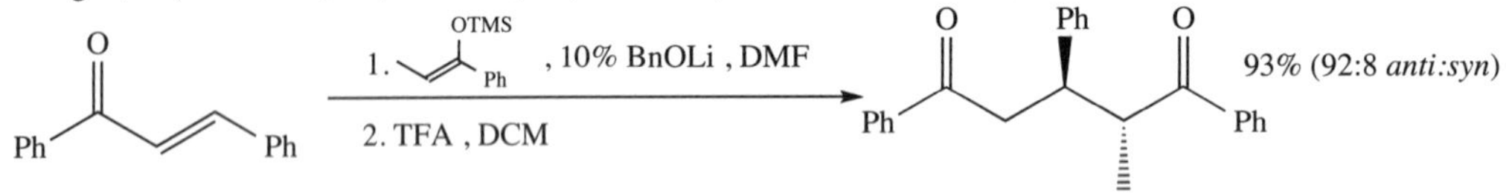

1. , 10% BnOLi , DMF

2. TFA , DCM

93% (92:8 *anti:syn*)

Mukaiyama, T.; Tozawa, T.; Fujisawa, H. *Chem. Lett.* **2004**, *33*, 1410.

SECTION 373: KETONE - NITRILE

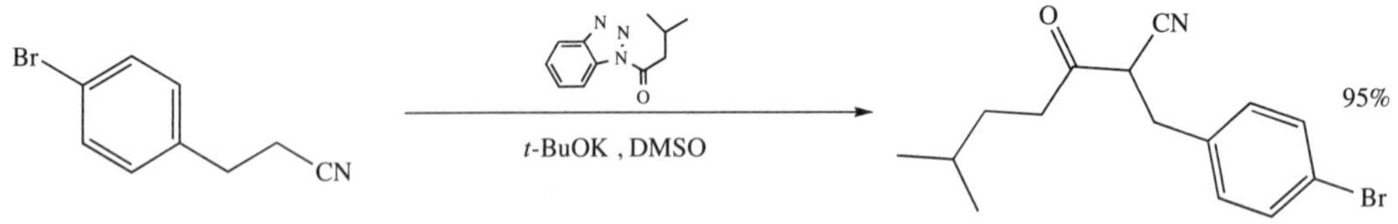

t-BuOK , DMSO

95%

Katritzky, A.R.; Abdel-Fattach, A.A.A.; Wang, M. *J. Org. Chem.* **2003**, *68*, 4932.

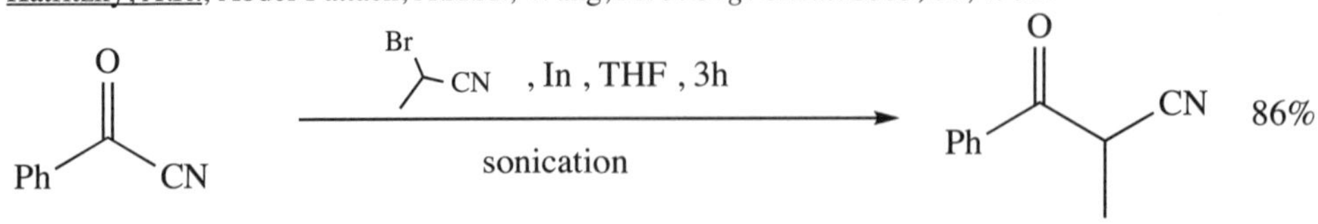

, In , THF , 3h

sonication

86%

Yoo, B.W.; Hwang, S.K.; Kim, D.Y.; Choi, J.W.; Ko, J.J.; Choi, K.I.; Kim, J.H. *Tetrahedron Lett.* **2002**, *43*, 4813.

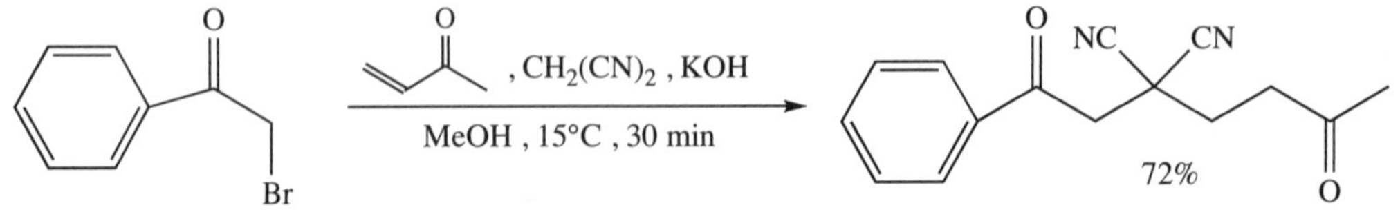

, $CH_2(CN)_2$, KOH

MeOH , 15°C , 30 min

72%

Saikia, A.; Chetia, A.; Bora, U.; Boruah, R.C. *Synlett* 2003, 1506.

SECTION 374: KETONE - ALKENE

For the oxidation of allylic alcohols to alkene ketones, see Section 168 (Ketones from Alcohols and Phenols)

For the oxidation of allylic methylene groups (C=C-CH_2 → C=C-C=O), see Section 170 (Ketones from Alkyls, Methylenes, and aryls).

For the alkylation of alkene ketones, also see Section 177 (Ketones from Ketones), and for conjugate alkylations, see Section 74E (Conjugate Alkylations).

2 PhI, 2 Bu_4NCl, DMF; 10% $Pd(OAc)_2$, 20% PPh_3, 2 i-Pr_2NEt, 80°C, 12 h — 70%

Larock, R.C.; Reddy, Ch.K. *J. Org. Chem.* **2002**, *67*, 2027.

DMSO, O_2 (1 atm), 60°C, 10% $Pd(OAc)_2$ — 42%

Toyota, M.; Rudyanto, M.; Ihara, M. *J. Org. Chem.* **2002**, *67*, 3374.

6% $Co_2(CO)_8$, 120°C, 10% chiral aryl biphosphite; toluene, CO, 1 d — 75% (75% ee)

Sturla, S.J.; Buchwald, S.L. *J. Org. Chem.* **2002**, *67*, 3398.

toluene, 110°C, 1 h — 92%

Padwa, A.; Ginn, J.D.; Bur, S.K.; Eidell, C.K.; Lynch, S.M. *J. Org. Chem.* **2002**, *67*, 3412.

1. BuCHO, pH 7.8, H_2O; 2. pH 1, 70°C — 80x85% (one-pot procedure)

Kourouli, T.; Kefalas, P.; Ragoussis, N.; Ragoussis, V. *J. Org. Chem.* **2002**, *67*, 4615.

Ph–CH=CH–CHO, 2 h; 5% $Rh(dppp)_2Cl$, 120°C — 98%

Shibata, T.; Toshida, N.; Takagi, K. *J. Org. Chem.* **2002**, *67*, 7446.

$SnBu_3$ / Me_3Si — 5% $Pd(PPh_3)_4$, LiCl, CO, THF, 65°C — OTf — O — 60%

Mazzola Jr. R.D.; Giese, S.; Benson, C.L.; West, F.G. *J. Org. Chem.* **2004**, *69*, 220.

EtO_2C, EtO_2C — Co nanoparticles, 20 atm CO, 12 h, 130°C — EtO_2C, EtO_2C, O (3 : 1) 95% + EtO_2C, EtO_2C, O

Son, S.U.; Lee, S.I.; Chung, Y.K.; Kim, S.-W.; Hyeon, T. *Org. Lett.* **2002**, *4*, 277.

CHO, C_6H_{13} — Ph—≡—CO_2Et, DCM, rt; 10% [Rh(dppe)] BF_4 — O, Ph, CO_2Et, C_6H_{13} (6 : 1) 79% + O, CO_2Et, Ph, C_6H_{13}

Tanaka, K.; Fu, G.C. *Org. Lett.* **2002**, *4*, 933.

Ph, O — Ph/CHO, 2 h; 5% $Rh(dppp)_2Cl$, 120°C — Ph, O, O — 98% + styrene

Shibata, T.; Toshida, N.; Takagi, K. *Org. Lett.* **2002**, *4*, 1619.

Ph, C, $SiMe_3$ — 5% $[Rh(CO)_2Cl]_2$, toluene; CO — Ph, O, $SiMe_3$ — 64%

Brummond, K.M.; Chen, H.; Fisher, K.D.; Kerekes, A.D.; Rickards, B.; Sill, P.C. *Org. Lett*, **2002**, *4*, 1931.

Li, H, C_3H_7, C_3H_7, C_3H_7, C_3H_7 — 1. CO, –78°C; 2. H_2O — C_3H_7, C_3H_7, O, C_3H_7, C_3H_7 — 93%

Song, Q.; Li, Z.; Chen, J.; Wang, C.; Xi, Z. *J. Org. Chem.* **2002**, *67*, 4627.

Me_3Si, C=O, Ph + N=N, OMe, OMe, O — Me_3Si, O, OMe, OMe, Ph — 63%

Rigby, J.H.; Wang, Z. *Org. Lett.* **2003**, *5*, 263.

CO , *i*-PrOH; $Fe(CO)_5$, hν → (5.9 : 1) 80%

Taber, D.F.; Joshi, P.V.; Kanai, K. *J. Org. Chem.* ***2004***, *69*, 2268.

10% $[CpRu(MeCN)_3]\ PF_6$, 3 h; 2% H_2O , acetone , 40°C; malonic acid → 80% (96% ee)

Trost, B.M.; Rudd, M.T. *Org. Lett.* ***2003***, *5*, 1467.

PhCHO , $Me_2NCO_2^-NH_2Me_2^+$; 50°C , 5.25 h → 76%

Kreher, U.P; Rosamillia, A.E.; Raston, C.L.; Scott, J.; Strauss, C.R. *Org. Lett.* ***2003***, *5*, 3107.

600°C → $(CH_2)_8$ 75% (1.1 *E/Z*)

Rüedi, G.; Nagel, M.; Hansen, H.-J. *Org. Lett.* ***2003***, *5*, 4211.

$AlCl_3$; DCM → 91%

Liang, G.; Gradl, S.N.; Trauner, D. *Org. Lett.* ***2003***, *5*, 4931.

chiral bis(oxazoline) , $CuBr_2$; $AgSbF_6$, DCM , rt → 73% (76% ee)

Aggarwal, V.K.; Belfield, A.J. *Org. Lett.* ***2003***, *5*, 5075.

Ph—≡ , Co_2Rh_2 , 130°C; THF , 18h → 48%

Park, K.H.; Jung, I.G.; Chung, Y.K. *Org. Lett.* ***2004***, *6*, 1183.

cat $TRuPPh_3(MeCN)_2\ PF_6$, DCE

80°C , 12 h

89% (E/Z = 11.1)

Yeh, K.-L.; Liu, B.; Lai, Y.-T.; Li, C.-W.; Liu, R.-S. *J. Org. Chem.* ***2004***, *69*, 4692.

1. nitrosobenzene , 20% L-proline, DMSO , rt

2. Cs_2CO_3 , $(EtO)_2P(O)CH_2CH_2CH_3$

O-NHPh

74% (98% ee)

Zhong, G.; Yu, Y. *Org. Lett.* ***2004***, *6*, 1637.

PhI , cat. $Pd_2(dba)_3$•$CHCl_3$

20% P(o-Tol)$_3$, 2 Ag_2CO_3

toluene , 45°C

89%

Yoshida, M.; Sugimoto, K.; Thara, M. *Org. Lett.* ***2004***, *6*, 1979.

2.5% $[Cp^*RuCl]_4$, 0.2M

10% bpy , 25°C , DCM

1.5 h

(>19 : 1) quant

Burger, E.C.; Tunge, J.A. *Org. Lett.* ***2004***, *6*, 2603.

2 MAD , NEt_3 , 60°C

toluene

76%

Magomedev, N.A.; Ruggiero, P.L.; Tang, Y. *Org. Lett.* ***2004***, *6*, 3373.

1. 2.2 eq BuLi

2. HCl-H_2O

60%

Denichoux, A.; Ferreira, F.; Chemia, F. *Org. Lett.* ***2004***, *6*, 3509.

cat $[Rn(cod)Cl]_2$, 1 atm CO

DCE , 50°C , 1 d

93%

Bennacer, B.; Fujiwara, M.; Ojima, I. *Org. Lett.* ***2004***, *6*, 3589.

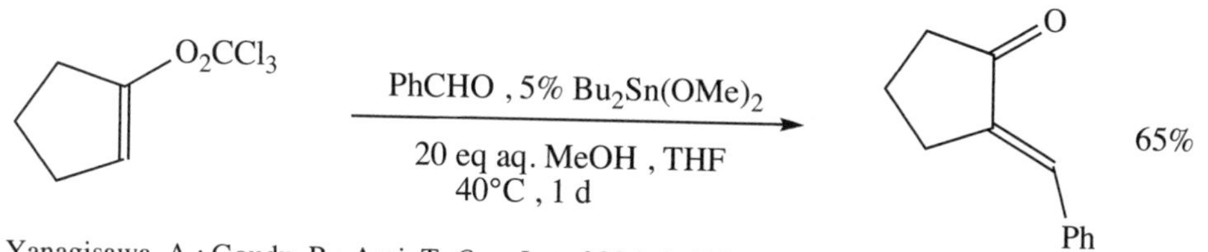

O_2CCl_3

PhCHO , 5% $Bu_2Sn(OMe)_2$

20 eq aq. MeOH , THF
40°C , 1 d

O

65%

Ph

Yanagisawa, A.; Goudu, R.; Arai, T. *Org. Lett.* ***2004***, *6*, 4281.

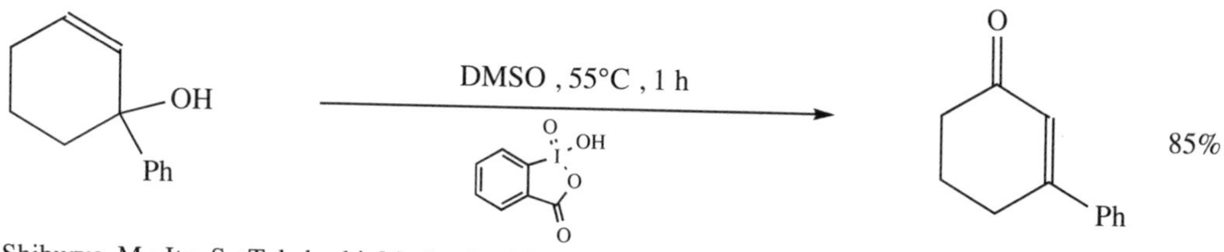

OH

Ph

DMSO , 55°C , 1 h

O OH I O O

O

85%

Ph

Shiburya, M.; Ito, S.; Takahashi, M.; Iwabuchi, Y. *Org. Lett.* ***2004***, *6*, 4303.

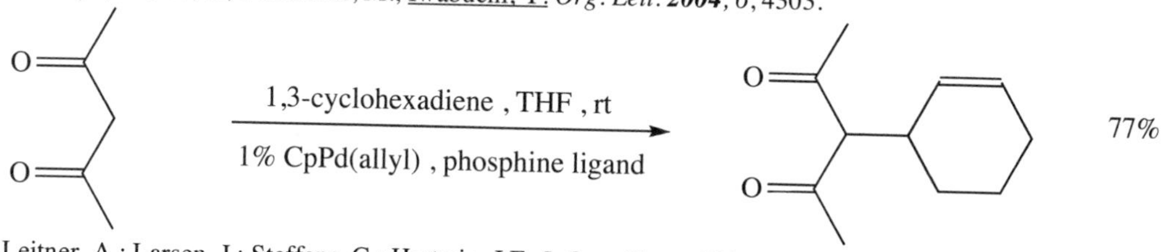

O

O

1,3-cyclohexadiene , THF , rt

1% CpPd(allyl) , phosphine ligand

O

O

77%

Leitner, A.; Larsen, J.; Steffens, C.; Hartwig, J.F. *J. Org. Chem.* ***2004***, *69*, 7552.

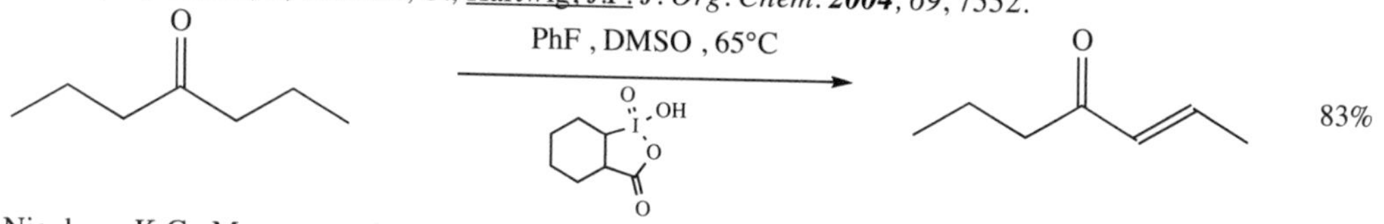

O

PhF , DMSO , 65°C

O OH I O O

O

83%

Nicolaou, K.C.; Montagnon, T.; Baran, P.S.; Zhong, Y.-L. *J. Am. Chem. Soc.* ***2002***, *124*, 2245.

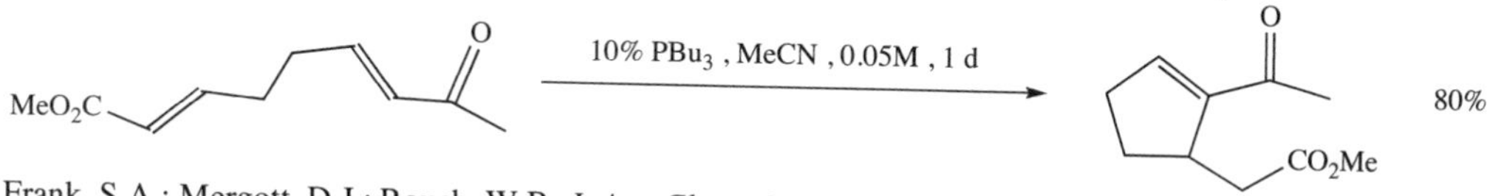

MeO_2C

O

10% PBu_3 , MeCN , 0.05M , 1 d

O

CO_2Me

80%

Frank, S.A.; Mergott, D.J.; Roush, W.R. *J. Am. Chem. Soc.* ***2002***, *124*, 2404.

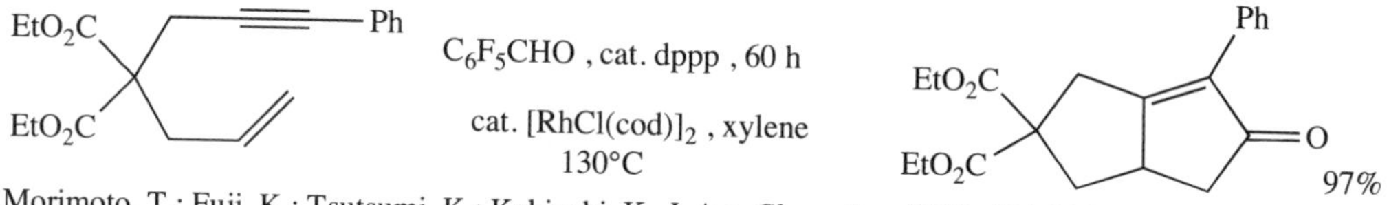

EtO_2C

EtO_2C

Ph

C_6F_5CHO , cat. dppp , 60 h

cat. $[RhCl(cod)]_2$, xylene
130°C

Ph

EtO_2C

EtO_2C

O

97%

Morimoto, T.; Fuji, K.; Tsutsumi, K.; Kakiuchi, K. *J. Am. Chem. Soc.* ***2002***, *124*, 3806.

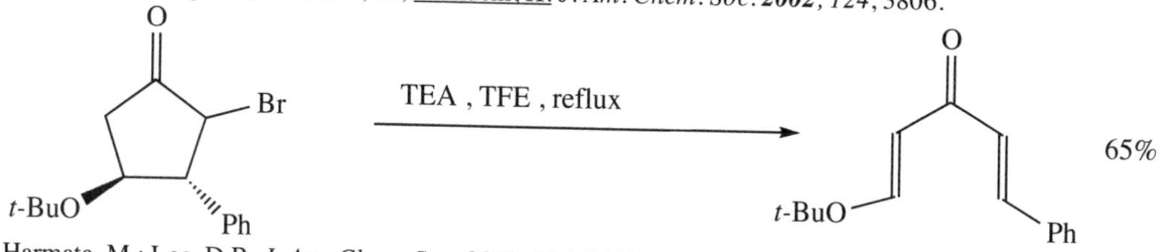

O

Br

t-BuO

Ph

TEA , TFE , reflux

O

t-BuO

Ph

65%

Harmata, M.; Lee, D.R. *J. Am. Chem. Soc.* ***2002***, *124*, 14328.

C_6H_{13}—≡ → Ge(2-furyl)$_3$, quinoline, CO; cat Pd(η^3-C_3H_7)$_2$, phosphite ligand; toluene → C_6H_{13}CH=CHC(O)Ge 78%

Kinoshita, H.; Shinokubo, H.; Oshima, K. *J. Am. Chem. Soc.* ***2002***, *124*, 4220.

Ph—≡—CH(OMe)CH$_2$CHO → 5% [Rh(ligand)] BF_4, DCM; (*S,S*)-*i*-Pr-DUPHOS. 30°C → 2-phenyl-4-methoxycyclopent-2-enone

kinetic resolution
56% conversion (93% ee)

Tanaka, K.; Fu, G.C. *J. Am. Chem. Soc.* ***2002***, *124*, 10296.

TMS—≡—CO_2Me → PhCHO, DABCO; PhH, reflux → PhC(O)CH=C(CO_2Me)CH(OTMS)Ph (74 : 26) + PhC(O)CH$_2$CH(CO_2Me)C(O)Ph 73%

Matsuya, Y.; Heyashi, K.; Nemoto, H. *J. Am. Chem. Soc.* ***2003***, *125*, 646.

cat $PdCl_2(MeCN)_2$, DCE; 2.5 $CuCl_2$, rt 97%

Pei, T.; Wang, X.; Widenhoefer, R.A. *J. Am. Chem. Soc.* ***2003***, *125*, 648.

OHC, OMe, Ph → 5% [Rh(R-Tol-BINAP)], DCM; rt → O, OMe, Ph 47% + O, Ph, OMe 45%

Tanaka, K.; Fu, G.C. *J. Am. Chem. Soc.* ***2003***, *125*, 8078.

O, CO_2Me, OMe, O, MeO, OMe → 2% $Cu(OTf)_2$, 0.06 M, DCE; 25°C, 5 min → O, CO_2Me, OMe, O, MeO, OMe >99%

He, W.; Sun, X.; Frontier, A.J. *J. Am. Chem. Soc.* ***2003***, *125*, 14278.

H—≡—CO_2Me → (OC)$_5$Cr=C(OMe)CH=CHPh, DCM, 25°C; 10% Rh(naphth)(cod) SbF_6 → O, Ph, CO_2Me 75%

Barluenga, J.; Vicente, R.; López, L.A.; Rubio, E.; Tomás, A.; Álvarez-Rúa, C. *J. Am. Chem. Soc.* ***2004***, *126*, 470.

2.5 [cyclopentadiene], $AlCl_3$, DCM, –78°C → 0°C

CHO

H

Me

O

90% (96% de)

Davies, H.M.L.; Dai, X. *J. Am. Chem. Soc.* ***2004**, 126*, 2692.

cat $Rh_4(CO)_{12}$, 2 h

CO, toluene, 130°C

H, Et, O, H, O

85%

Lee, S.I.; Park, J.H.; Chung, Y.K.; Lee, S.-G. *J. Am. Chem. Soc.* ***2004**, 126*, 2714.

1. 5% $[DHQ]_2$ PHAL

2. [2-acetylcyclopentanone], PBu_3

Ph

99% (1:1 *E:Z*) (80% ee *E*)

Bella, M.; Jørgensen, K.A. *J. Am. Chem. Soc.* ***2004**, 126*, 5672.

MeO_2C

MeO_2C

CO, DCE, 80°C, 3.8 h

5% $[RhCl(CO)]_2$

H

H

84%

Wender, P.A.; Croatt, M.P.; Deschamps, N.M. *J. Am. Chem. Soc.* ***2004**, 126*, 5948.

Ph, Br

1. LDA, CuI

2. OCO_2Me, cat $RhCl(PPh_3)_3/P(OAc)_3$

Ph

OBn

97% (24:1 dr)

Evans, P.A.; Lawler, M.J. *J. Am. Chem. Soc.* ***2004**, 126*, 8642.

chiral Sc catalyst

THF

H

H

53% (61% ee)

Liang, G.; Trauner, D. *J. Am. Chem. Soc.* ***2004**, 126*, 9544.

N

Si

Me Me

1. 5% $Ru_3(CO)_{12}$, CO, toluene, 100°C

2. Ph-C≡C-H, 1 d

Ph

O

55%

Itami, K.; Mitsudo, K.; Fujita, K.; Ohashi, Y.; Yoshida, J.-i. *J. Am. Chem. Soc.* ***2004**, 126*, 11058.

Ph—≡—Ph → pentanal, DCM, cat $GaCl_3$ 59%

Viswanathan, G.S.; Li, C.-J. *Tetrahedron Lett.* **2002**, *43*, 1613.

5 benzophenone, neat, 100°C, 4 d; 5% $Pd(PPh_3)_4$ 79%

Camacho, D.H.; Nakamura, I.; Oh, B.H.; Saito, S.; Yamamoto, Y. *Tetrahedron Lett.* **2002**, *43*, 2903.

1. Et_2Zn, DCM; 2. TFA-DCM; 3. CH_2I_2 58%

Gautier, A.; Garipova, G.; Deléens, R. *Tetrahedron Lett.* **2002**, *43*, 4959.

10% $Co_2(CO)_8$, toluene, air; 65°C, MS-pretreated with CO 65%

Blanco-Urgotti, J.; Casarrubios, L.; Domínguez, G.; Pérez-Castells, J. *Tetrahedron Lett.* **2002**, *43*, 5763.

PhI, cat $Pd(OAc)_2$; MeCN, NEt_3 68%

Wei, L.-M.; Wei, L.-L.; Pan, W.-B.; Wu, M.-J. *Tetrahedron Lett.* **2003**, *44*, 595.

TBHP, $CHCl_3$, 40°C, 1d; cat $Mo(CO)_6$

Massa, A.; Acocella, M.R.; De Rosa, M.; Soriente, A.; Villano, R.; Scettri, A. *Tetrahedron Lett.* **2003**, *44*, 835.

chiral Co catalyst/CH_2=CHPh; toluene, CO, 80°C 75% (36% ee)

Konya, D.; Robert, F.; Gimbert, Y.; Greene, A.E. *Tetrahedron Lett.* **2004**, *45*, 6975.

48%

HCHO, SDS, H_2O, 110°C; cat $[RhCl(cod)]_2$, toluene; BINAP/TPPTs, 6h 83% (81% ee)

Fuji, K.; Morimoto, T.; Tsutsumi, K.; Kakiuchi, K. *Tetrahedron Lett.* **2004**, *45*, 9163.

BnO_2C / BnO_2C (diyne, Ph) → 5% $IrCl(CO)(PPh_3)_2$, CO; xylene, 120°C, 5h → 86%

Shibata, T.; Yamashita, K.; Katayama, E.; Takagi, K. *Tetrahedron* **2002**, *58*, 8661.

Cl_3CCHO, 1.5 K_2CO_3, THF; rt, 2d → Cl_3C enone 74%

Nakatsu, S.; Gubaidullin, A.T.; Mamedov, V.A. *Tetrahedron* **2004**, *60*, 2337.

NHBoc → 1. Ti(IV), PPh_3; 2. *i*-PrCHO 3. K_2CO_3 → $CHMe_2$, NHBoc 39% (8.3:1 *E:Z*)

Dauvergne, J.; Happe, A.M.; Roberts, S.M. *Tetrahedron* **2004**, *60*, 2551.

EtO_2C / EtO_2C → CO, dendritic Co catalyst; THF, 70°C → EtO_2C / EtO_2C 68%

Dahan, A.; Portnoy, M. *Chem. Commun.* **2002**, 2700.

DMSO, DCM, 40 min; MeO, N - O, I - OH → 94%

Nicolaou, K.C.; Gray, D.L.F.; Montagnon, T.; Harrison, S.T. *Angew. Chem. Int. Ed.* **2002**, *41*, 996.

Ph, C_6H_{13} → HIO_5, I_2O_5, DMSO, 50°C, 18h → Ph, C_6H_{13} 88%

Nicolaou, K.C.; Montagnon, T.; Baran, P.S. *Angew. Chem. Int. Ed.* **2002**, *41*, 1386.

PhCHO → BuC≡CH, cat [$Rh(PPh_3)_3Cl$], $PhCO_2H$; 2-amino-3-picoline, toluene, 80°C, 12h → Ph, Bu 92%

Jun, C.-H.; Lee, H.; Hong, J.-B.; Kwon, B.-I. *Angew. Chem. Int. Ed.* **2002**, *41*, 2146.

EtO_2C / EtO_2C → CO, 1% $RhCl(CO)(PPh_3)_2$, 1% $AgSbF_6$, 0.1M, DCE; CO, rt → EtO_2C / EtO_2C 89%

Wender, P.A.; Deschamps, N.M.; Gamber, G.C. *Angew. Chem. Int. Ed.* **2003**, *42*, 1853.

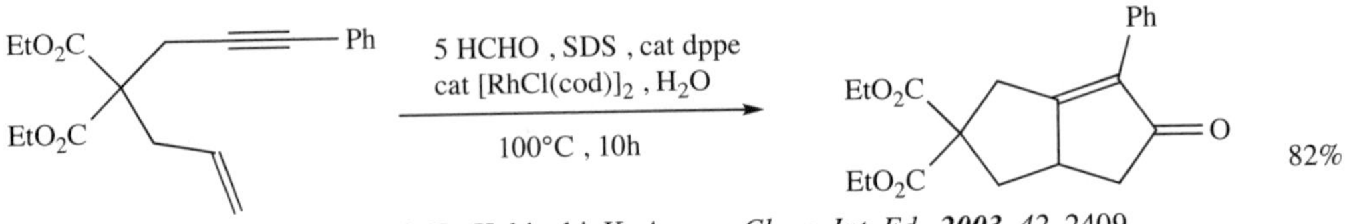

EtO_2C EtO_2C Ph — 5 HCHO, SDS, cat dppe, cat $[RhCl(cod)]_2$, H_2O, 100°C, 10h → Ph EtO_2C EtO_2C O 82%

Fuji, K.; Morimoto, T.; Tsutsumi, K.; Kakiuchi, K. *Angew. Chem. Int. Ed.* ***2003***, *42*, 2409.

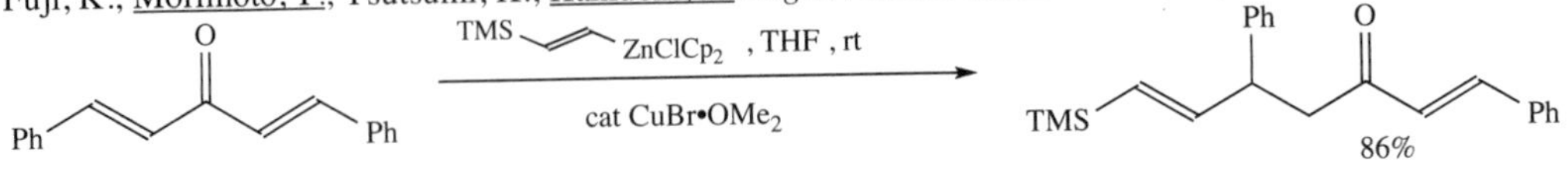

O Ph Ph — TMS ZnClCp$_2$, THF, rt; cat CuBr•OMe$_2$ → Ph O TMS Ph 86%

Huang, X.; Pi, J. *Synlett* ***2003***, 481.

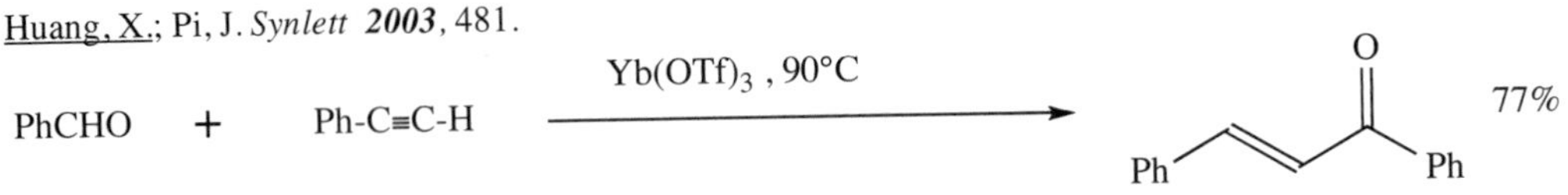

PhCHO + Ph-C≡C-H — Yb(OTf)$_3$, 90°C → O Ph Ph 77%

Curini, M.; Epifano, F.; Maltese, F.; Rosati, O. *Synlett* ***2003***, 552.

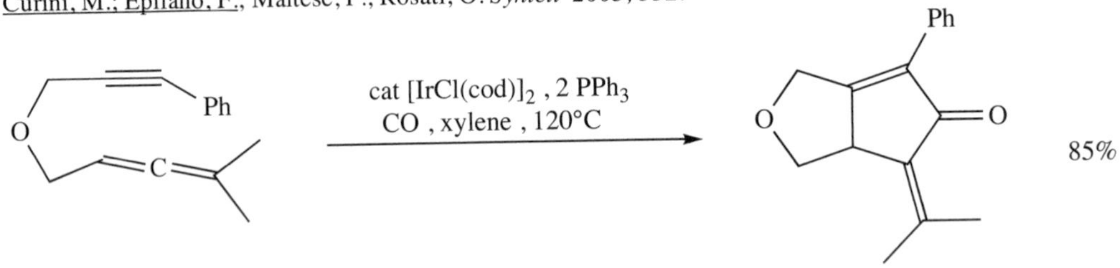

O Ph C — cat $[IrCl(cod)]_2$, 2 PPh$_3$; CO, xylene, 120°C → Ph O O 85%

Shibata, T.; Kadowaki, S.; Hirase, M.; Takagi, K. *Synlett* ***2003***, 573.

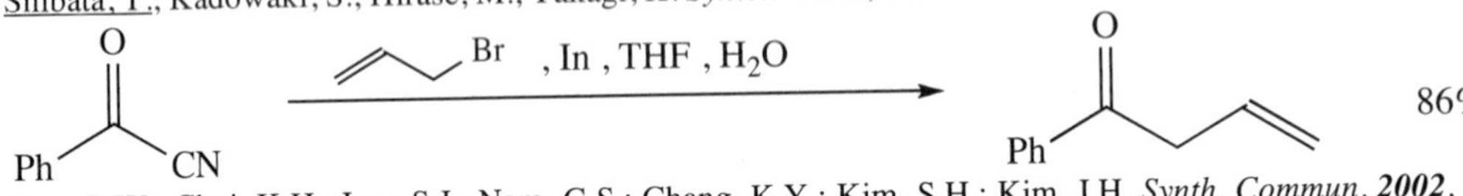

O Ph CN — Br, In, THF, H_2O → O Ph 86%

Yoo, B.W.; Choi, K.H.; Lee, S.J.; Nam, G.S.; Chang, K.Y.; Kim, S.H.; Kim, J.H. *Synth. Commun.* ***2002***, *32*, 839.

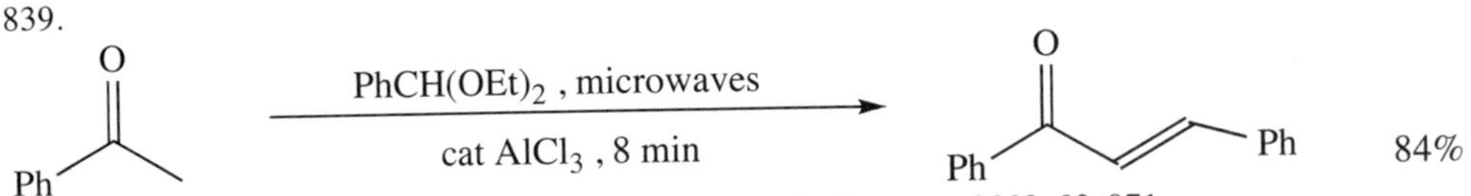

O Ph — PhCH(OEt)$_2$, microwaves; cat AlCl$_3$, 8 min → O Ph Ph 84%

Huang, D.; Wang, J.-X.; Hu, Y.; Zhang, Y.; Tang, J. *Synth. Commun.* ***2002***, *32*, 971.

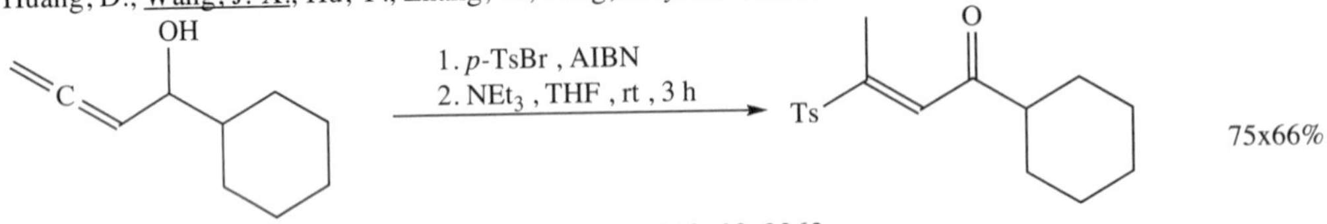

OH C — 1. *p*-TsBr, AIBN; 2. NEt$_3$, THF, rt, 3 h → O Ts 75x66%

Kang, S.-K.; Ko, B.-S.; Lee, D.M. *Synth. Commun.* ***2002***, *32*, 3263.

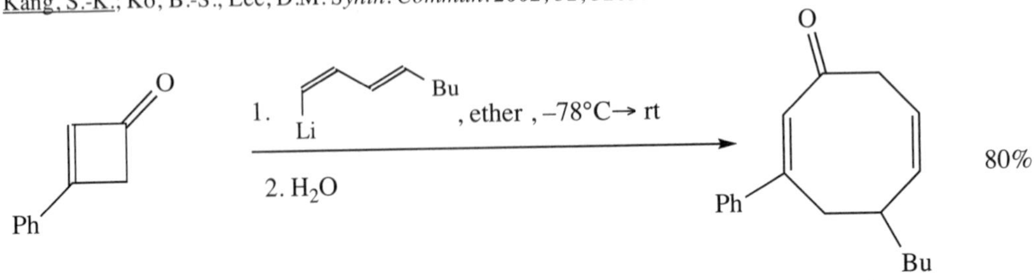

O Ph — 1. Li Bu, ether, −78°C→ rt; 2. H_2O → O Ph Bu 80%

Hamura, T.; Kakinuma, M.; Tsuji, S.; Matsumoto, T.; Suzuki, K. *Chem. Lett.* ***2002***, *31*, 750.

2 PhCHO , 20% SmI_3 , rt , 3 h; bmim BF_4 — 95%

Zheng, X.; Zhang, Y. *Synth. Commun.* **2003**, *33*, 161.

2 PhCHO , 10% $InCl_3$•4 H_2O; 110°C , 6h — Ph, Ph — 95%

Deng, G.; Ren, T. *Synth. Commun.* **2003**, *33*, 2995.

70°C , Na_2CO_3/H_2O , 1 h; CHO — 94%

Zhang, Z.; Dong, Y.-W.; Wang, G.-W. *Chem. Lett.* **2003**, *32*, 966.

PhCHO , Pd/C , DMF; TMSCl , 5 h — CHPh — 83%

Zhu, Y.; Pan, Y. *Chem. Lett.* **2004**, *33*, 668.

PhCHO + PhC≡CH — HBr , bmim OTs , 100°C; 12 h — Ph, Ph — 90%

Xu, L.-W.; Li, L.; Xia, C.-G.; Zhao, P.-Q. *Helv. Chim. Acta* **2004**, *87*, 3080.

$Co_2(CO)_8$, DCM , NMO; rt , 2 h — 57%

Ishizaki, M.; Masamoto, M.; Hoshino, O. *Heterocycles* **2002**. *57*, 1409.

MeO_2C, MeO_2C, Ph — $Co_2(CO)_8$, BuSMe , DCE; 83°C — MeO_2C, MeO_2C, Ph — 73%

Ishizaki, M.; Satoh, H.; Hoshino, O.; Nishitani, K.; Hara, H. *Heterocycles* **2004**, *63*, 827.

PhC≡CH , $Co_2(CO)_8$, *i*-PrOH; KOH , heat — Ph — 99%

Hätzelt, A.; Laschat, S. *Can. J. Chem.* **2002**, *80*, 1327.

REVIEWS:

"The Pauson-Khand Reaction: The Catalytic Age Is Here!"
Gibson, S.E.; Stevenazzi, A. *Angew. Chem. Int. Ed.* **2003**, *42*, 1800.

"Recent Advances in the Baylis-Hillman Reaction and Applications"
Basavaiah, D.; Rao, A.J.; Satyanarayana, T. *Chem. Rev.* **2003**, *103*, 811.

SECTION 375: NITRILE - NITRILE

NO ADDITIONAL EXAMPLES

SECTION 376: NITRILE - ALKENE

Me_2t-BuSi-cyclopropyl-CH=CH-CH(OSit-BuMe$_2$)CN —(LDA; MeI / THF, –78°C)→ Me_2t-BuSi-CH=CH-CH=CH-C(OSit-BuMe$_2$)(Me)CN 87% (9.8 *E/Z*)

Sasaki, M.; Kawanishi, E.; Nakai, Y.; Matsumoto, T.; Yamaguchi, K.; Takeda, K. *J. Org. Chem.* **2003**, *68*, 9330.

$Ph_2(t\text{-BuO})SiCH_2CN$ —(1. KHMDS, THF, –78°C; 2. MeO-C$_6$H$_4$-CHO)→ MeO-C$_6$H$_4$-CH=CH-CN 94% (96:4 *Z:E*)

Kojima, S.; Fukuzaki, T.; Yamakawa, A.; Murai, Y. *Org. Lett.* **2004**, *6*, 3917.

F_3C-C$_6$H$_4$-CN —(Pr-C≡C-Pr, 100°C, toluene, 1 d / cat Ni(cod)$_2$PMe$_3$)→ F_3C-C$_6$H$_4$-C(Pr)=C(Pr)CN 84%

Nakao, Y.; Oda, S.; Hiyama, T. *J. Am. Chem. Soc.* **2004**, *126*, 13904.

(EtO)$_2$P(O)CH(CH$_3$)CN —(PhCHO, LiOH, THF, rt, 1h)→ PhCH=C(CH$_3$)CN 85% (77:23 *E:Z*)

Lattanzi, A.; Orelli, L.R.; Barone, P.; Massa, A.; Iannece, P.; Scettri, A. *Tetrahedron Lett.* **2003**, *44*, 1333.

PhCHO —(EtO_2CCH_2CN, TPP, microwaves / 3 min)→ PhCH=C(CO$_2$Et)CN 90%

Yadav, J.S.; Reddy, B.V.S.; Basak, A.K.; Visali, B.; Narsaiah, A.V.; Nagaiah, K. *Eur. J. Org. Chem.* **2004**, 546.

PhCHO —(EtO_2CCH_2CN, Zr(KPO$_4$)$_2$ / 60°C, 3 h)→ PhCH=C(CN)CO$_2$Et 88%

Curini, M.; Epifano, F.; Marcotullio, M.C.; Rosati, O.; Tsadjout, A. *Synth. Commun.* **2002**, *32*, 355.

PhCHO + $CH_2(CN)_2$ —(grinding, 1 h / no solvent)→ $PhCH{=}C(CN)_2$ 68%

Ren, Z.; Cao, W.; Tong, W. *Synth. Commun.* **2002**, *32*, 3475.

2 KCN, DMSO, rt — 65%

SO$_2$Ph, Cl, CN

Temmem, O.; Uguen, D.; DeCian, A.; Gruber, N. *Tetrahedron Lett.* ***2002**, 43*, 3175.

PhCHO → $CH_2(CN)_2$, NaF, solvent free / microwaves → Ph, CN, CN — 96%

Mogilaiah, K.; Prashanthi, M.; Reddy, G.R.; Reddy, Ch.S.; Reddy, N.V. *Synth. Commun.* ***2003**, 33*, 2309.

PhCHO → $CH_2(CN)_2$, 0.5% $ReBr(CO)_5$, 110°C / no solvent → PhCH=C(CN)$_2$ 90%

Zuo, W.-X.; Hua, R.; Qiu, X. *Synth. Commun.* ***2004**, 34*, 3219.

PhCHO + NC–CH$_2$–CN → KI_3, 15 min → Ph, CN, CN — 98%

Thakur, A.J.; Prajapati, D.; Gogoi, B.J.; Sandhu, J.S. *Chem. Lett.* ***2003**, 32*, 258.

SECTION 377: ALKENE - ALKENE

Ph, CHO → 1. NH_2NH_2, EtOH 2. MnO_2, DCM 3. $Rh_2(OAc)_4$, DCM, 0°C → Ph, Ph (98:2 *trans:cis*) 52%

Doyle, M.P.; Yan, M. *J. Org. Chem.* ***2002**, 67*, 602.

MeO$_2$C, MeO$_2$C → $HSiMe_2Bn$, $B(C_6F_5)_3$, toluene, 110°C, 15 min; Ph, Me, N, Pt, Me, Me, N, Me, Ph → MeO$_2$C, MeO$_2$C, SiMe$_2$Bn — 75% (>30:1 *Z:E*)

Wang, X.; Chakrapani, H.; Madine, J.W.; Keyerleber, M.A.; Widenhoefer, R.A. *J. Org. Chem.* ***2002**, 67*, 2778.

BzO → H_2C=CHOEt / Grubb II → OBz, OEt — 99%

Giessert, A.J.; Snyder, L.; Markham, J.; Diver, S.T. *Org. Lett.* ***2003**, 5*, 1793.

BnO → H_2C=CHC$_6H_{13}$, CH_2=CH_2, DCM / 10% Grubbs' II, rt → BnO, C$_6$H$_{13}$ — 82% (100% *E*)

Lee, H-Y.; Kim, B.G.; Snapper, M.L. *Org. Lett.* ***2003**, 5*, 1855.

OPiv, BnMe$_2$Si → H_2C=CHC$_6H_{13}$, acetone, rt, 4h / 10% $CpRu(MeCN)_3PF_6$ → BnMe$_2$Si, PivO, C$_6$H$_{13}$ — 79%

Trost, B.M.; Machecek, M.R.; Ball, Z.T. *Org. Lett.* ***2003**, 5*, 1895.

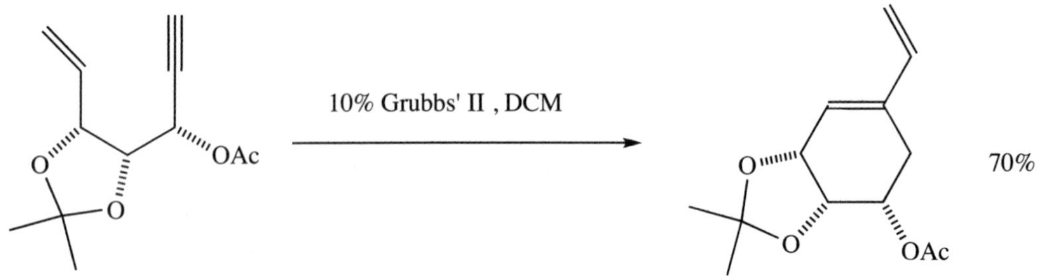
OAc
O
O
10% Grubbs' II , DCM
O
O
OAc
70%

Poulsen, C.S.; Madsen, R. *J. Org. Chem.* ***2002****, 67*, 4441.

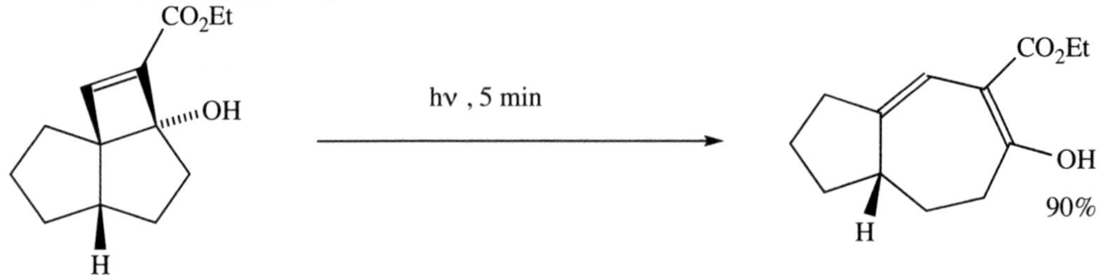
CO_2Et
OH
H
hν , 5 min
CO_2Et
OH
H
90%

Mislin, G.L.; Miesch, M. *J. Org. Chem.* ***2003****, 68*, 433.

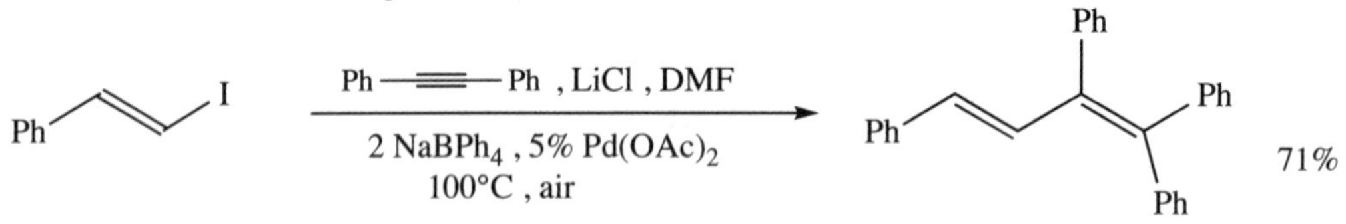
Ph
I
Ph—≡—Ph , LiCl , DMF
2 $NaBPh_4$, 5% $Pd(OAc)_2$
100°C , air
Ph
Ph
Ph
Ph
71%

Zhang, X.; Larock, R.C. *Org. Lett.* ***2003****, 5*, 2993.

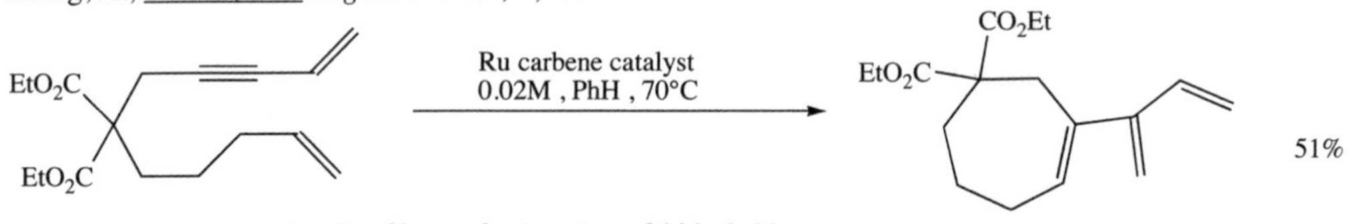
EtO_2C
EtO_2C
Ru carbene catalyst
0.02M , PhH , 70°C
CO_2Et
EtO_2C
51%

Kang, B.; Kim, D.-L.; Do, Y.; Chang, S. *Org. Lett.* ***2003****, 5*, 3041.

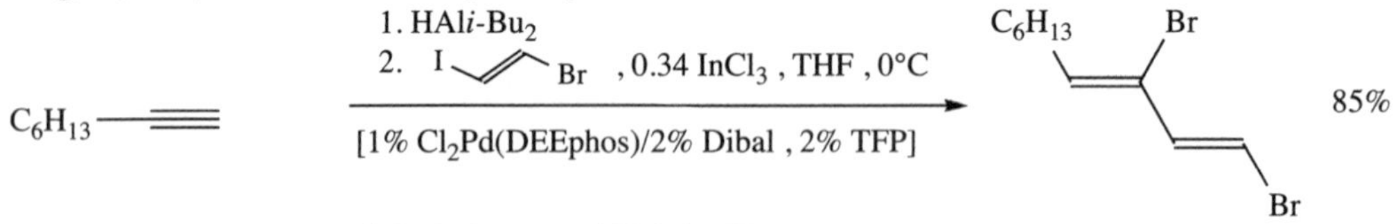
C_6H_{13}
1. HAl*i*-Bu_2
2. I Br , 0.34 $InCl_3$, THF , 0°C
[1% Cl_2Pd(DEEphos)/2% Dibal , 2% TFP]
C_6H_{13}
Br
Br
85%

Qian, M.; Huang, Z.; Negishi, E.-i. *Org. Lett.* ***2004****, 6*, 1531.

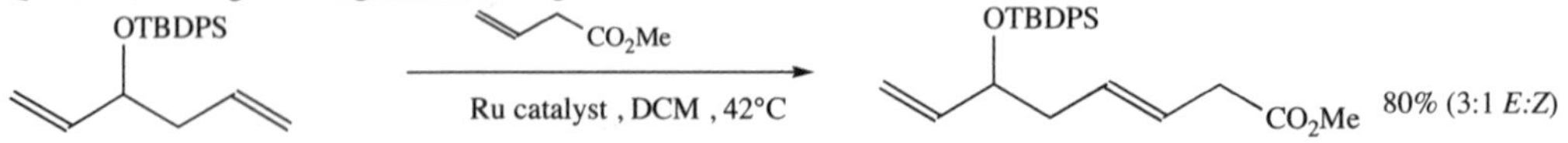
OTBDPS
CO_2Me
Ru catalyst , DCM , 42°C
OTBDPS
CO_2Me
80% (3:1 *E:Z*)

Bouz-Bouz, S.; Simmons, R.; Cossy, J. *Org. Lett.* ***2004****, 6*, 3465.

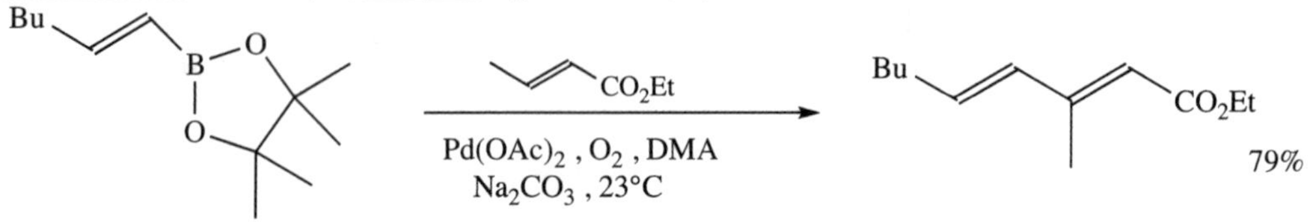
Bu
B
O
O
CO_2Et
$Pd(OAc)_2$, O_2 , DMA
Na_2CO_3 , 23°C
Bu
CO_2Et
79%

Yoon, C.H.; Yoo, K.S.; Yi, S.W.; Mishra, R.K.; Jung, K.W. *Org. Lett.* ***2004****, 6*, 4037.

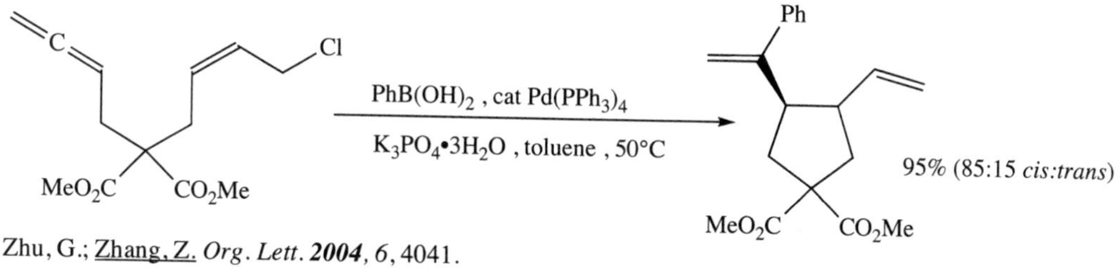
Cl
MeO2C
CO2Me
PhB(OH)2 , cat Pd(PPh3)4
K3PO4•3H2O , toluene , 50°C
Ph
MeO2C
CO2Me
95% (85:15 cis:trans)

Zhu, G.; Zhang, Z. *Org. Lett.* ***2004***, *6*, 4041.

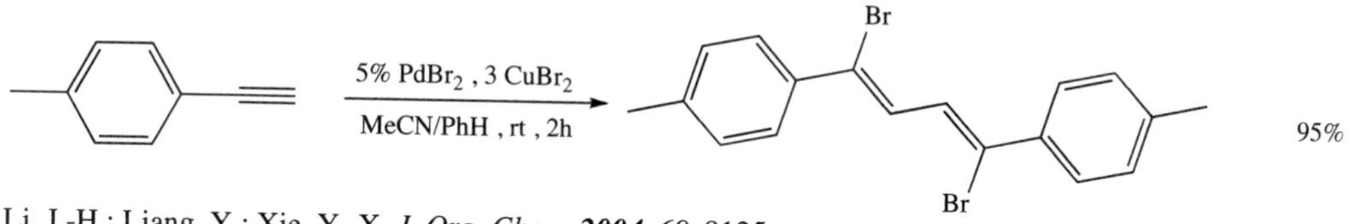
5% PdBr2 , 3 CuBr2
MeCN/PhH , rt , 2h
Br
Br
95%

Li, J.-H.; Liang, Y.; Xie, Y.-X. *J. Org. Chem.* ***2004***, *69*, 8125.

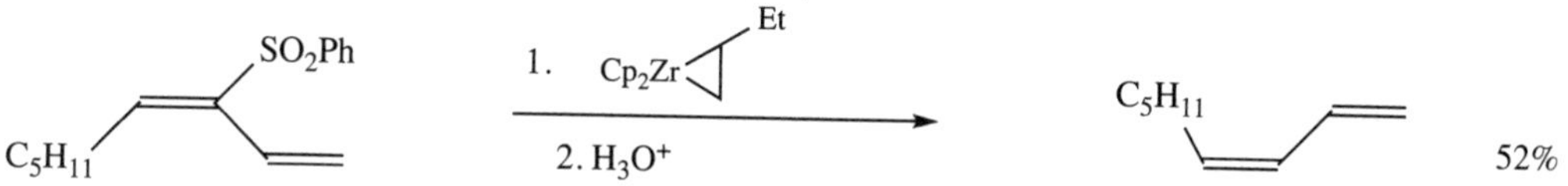
SO2Ph
C5H11
1. Cp2Zr
Et
2. H3O+
C5H11
52%

Chinkov, N.; Mujumdar, S.; Marek, I. *J. Am. Chem. Soc.* ***2002***, *124*, 10282.

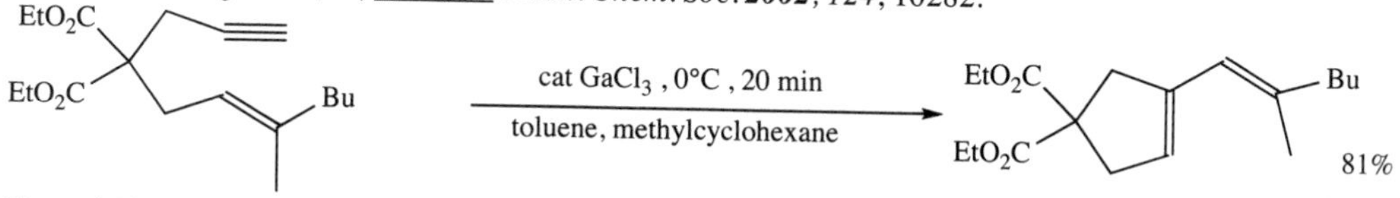
EtO2C
EtO2C
Bu
cat GaCl3 , 0°C , 20 min
toluene, methylcyclohexane
EtO2C
EtO2C
Bu
81%

Chatani, N.; Inoue, H.; Kotsuma, T.; Murai, S. *J. Am. Chem. Soc.* ***2002***, *124*, 10294.

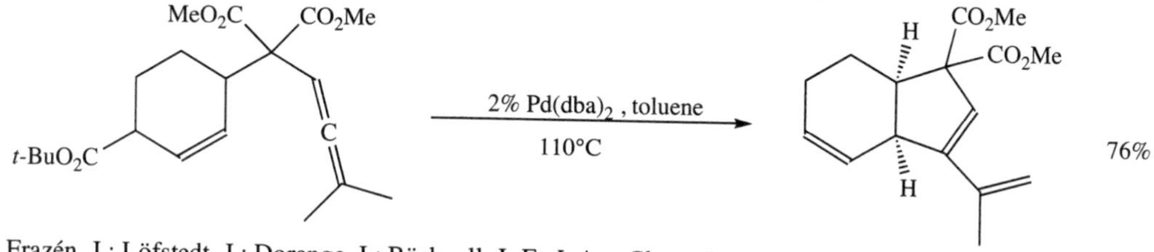
MeO2C
CO2Me
t-BuO2C
C
2% Pd(dba)2 , toluene
110°C
H
CO2Me
CO2Me
H
76%

Frazén, J.; Löfstedt, J.; Dorange, I.; Bäckvall, J.-E. *J. Am. Chem. Soc.* ***2002***, *124*, 11246.

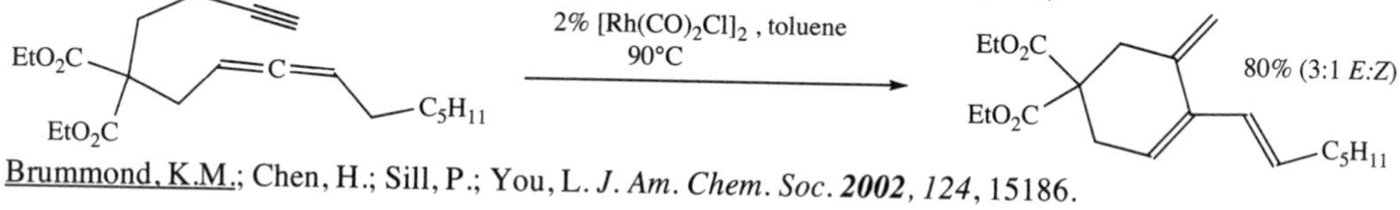
EtO2C
EtO2C
C
C5H11
2% [Rh(CO)2Cl]2 , toluene
90°C
EtO2C
EtO2C
C5H11
80% (3:1 E:Z)

Brummond, K.M.; Chen, H.; Sill, P.; You, L. *J. Am. Chem. Soc.* ***2002***, *124*, 15186.

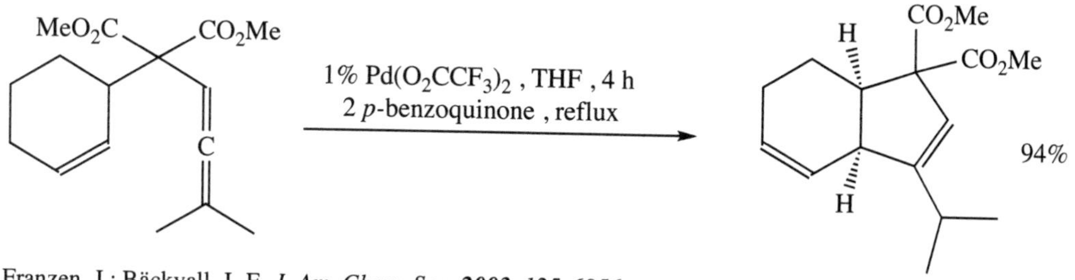
MeO2C
CO2Me
C
1% Pd(O2CCF3)2 , THF , 4 h
2 p-benzoquinone , reflux
H
CO2Me
CO2Me
H
94%

Franzen, J.; Bäckvall, J.-E. *J. Am. Chem. Soc.* ***2003***, *125*, 6056.

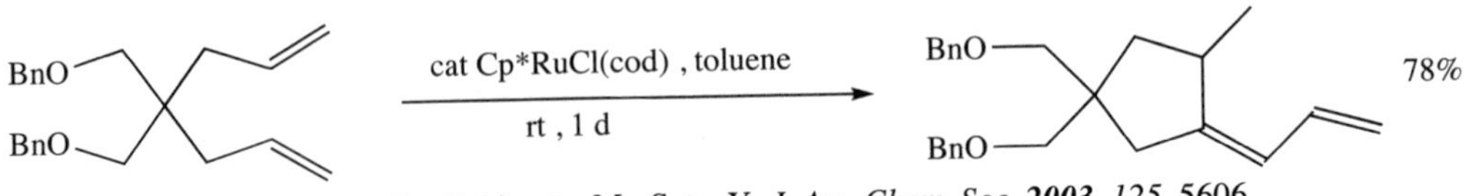

BnO, BnO — cat Cp*RuCl(cod), toluene; rt, 1 d — BnO, BnO 78%

Mori, M.; Saito, N.; Tanaka, D.; Takimoto, M.; Sato, Y. *J. Am. Chem. Soc.* **2003**, *125*, 5606.

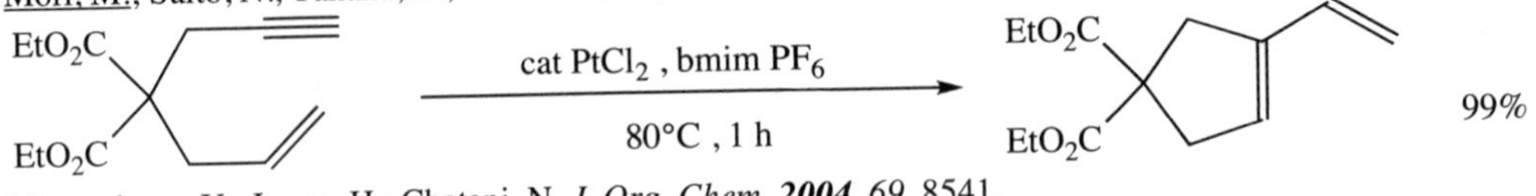

EtO_2C, EtO_2C — cat $PtCl_2$, bmim PF_6; 80°C, 1 h — EtO_2C, EtO_2C 99%

Miyanohana, Y.; Inoue, H.; Chatani, N. *J. Org. Chem.* **2004**, *69*, 8541.

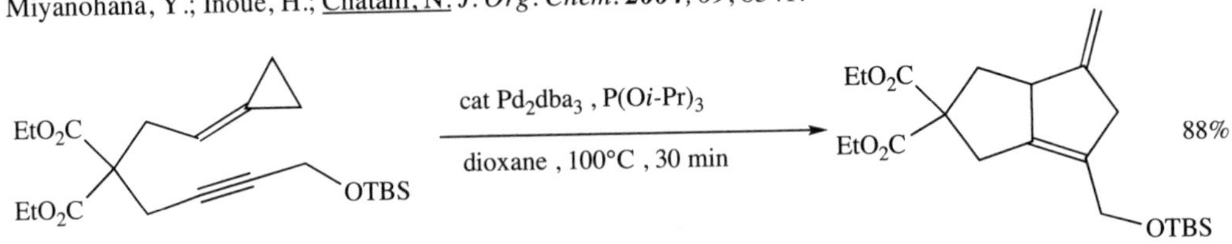

EtO_2C, EtO_2C, OTBS — cat Pd_2dba_3, P(O*i*-Pr)$_3$; dioxane, 100°C, 30 min — EtO_2C, EtO_2C, OTBS 88%

Delgado, A.; Rodríguez, J.R.; Castedo, L.; Mascareñas, J.L. *J. Am. Chem. Soc.* **2003**, *125*, 9282.

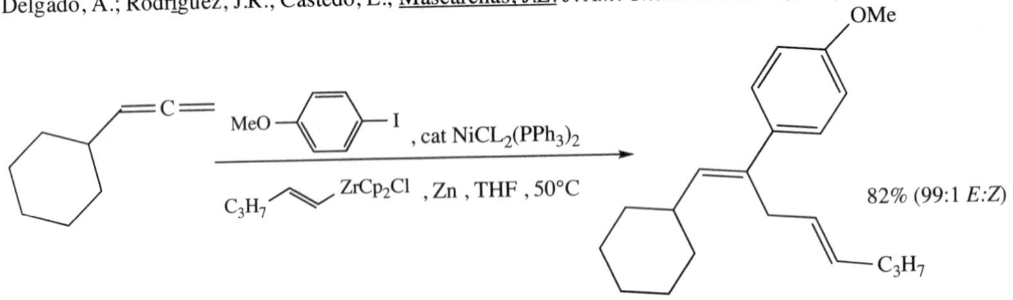

=C= ; MeO–C$_6$H$_4$–I, cat NiCL$_2$(PPh$_3$)$_2$; C_3H_7–CH=CH–ZrCp$_2$Cl, Zn, THF, 50°C — OMe, C_3H_7 82% (99:1 *E:Z*)

Wu, M.-S.; Rayabarapu, D.K.; Cheng, C.-H. *J. Am. Chem. Soc.* **2003**, *125*, 12426.

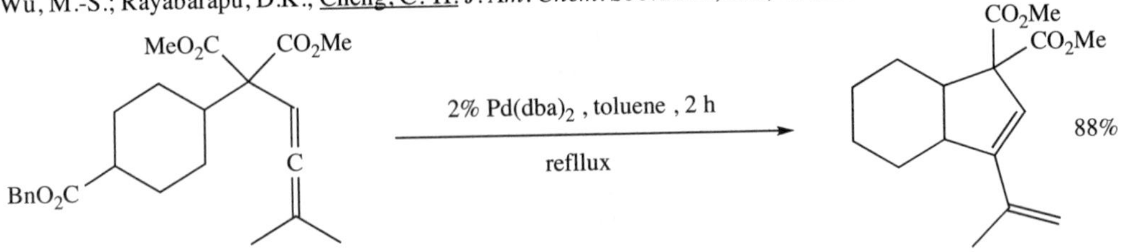

MeO_2C, CO_2Me, C, BnO_2C — 2% Pd(dba)$_2$, toluene, 2 h; refllux — CO_2Me, CO_2Me 88%

Franzén, J.; Löfstedt, J.; Falk, J.; Bäckvall, J.-E. *J. Am. Chem. Soc.* **2003**, *125*, 14140.

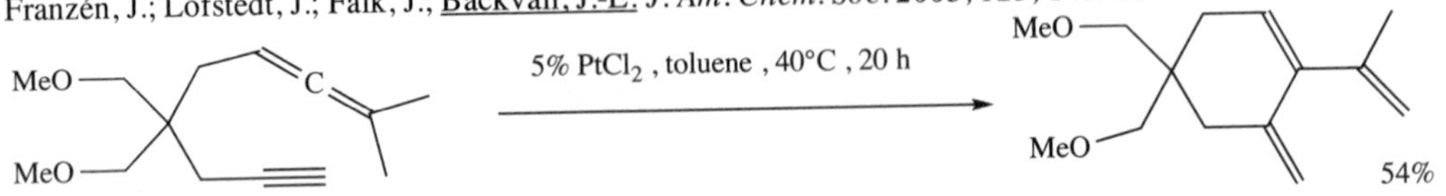

MeO, MeO, C — 5% $PtCl_2$, toluene, 40°C, 20 h — MeO, MeO 54%

Chandran, N.; Cariou, K.; Hervé, G.; Aubert, C.; Fensterbank, L.; Malacria, M. *J. Am. Chem. Soc.* **2004**, *126*, 3408.

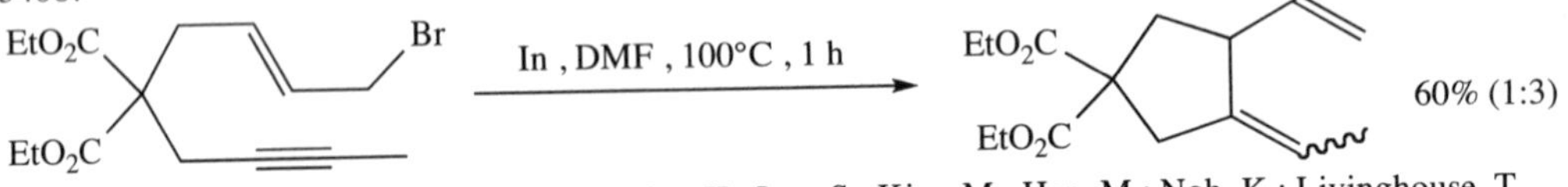

EtO_2C, EtO_2C, Br — In, DMF, 100°C, 1 h — EtO_2C, EtO_2C 60% (1:3)

Lee, P.H.; Kim, S.; Lee, K.; Seomoon, D.; Kim, H.; Lee, S.; Kim, M.; Han, M.; Noh, K.; Livinghouse, T. *Org. Lett.* **2004**, *6*, 4825.

5% $PtCl_2$, PhH , reflux , 20 h

OMe

OMe

42%

Harrack, Y.; Blaszykowski, C.; Bernard, M.; Cariou, K.; Mainetti, E.; Mouriès, V.; Dhimane, A.-L.; Fensterbank, L.; Malacria, M. *J. Am. Chem. Soc.* ***2004****, 126*, 8656.

AcO

$SiMe_3$

Br , DCM

cat $H_2(I_{MES})(PCy_3)Cl_2Ru$=CHPh

AcO

Br

$SiMe_3$

79%

Kim, M.; Park, S.; Maifeld, S.V.; Lee, D. *J. Am. Chem. Soc.* ***2004****, 126*, 10242.

EtO_2C

EtO_2C

Grubbs I , toluene , 110°C

45 min

EtO_2C

EtO_2C

78%

López, F.; Delgado, A.; Rodríguez, J.R.; Castedo, L.; Mascareñas, J.L. *J. Am. Chem. Soc.* ***2004****, 126*, 10262.

CO_2Et

CO_2Et

+

O

cat $Ni(cod)_2$, MeCN , rt

2 $ZnCl_2$, 1 d

O

EtO_2C

CO_2Et

54% (96:4 *E:Z*)

Ikeda, S.-i.; Sanuki, R.; Miyachi, H.; Miyashita, H.; Tankguchi, M.; Odashima, L. *J. Am. Chem. Soc.* ***2004****, 126*, 10331.

CO_2Et

TMS-C≡C-H , 10% $Ni(cod)_2$

20% PPh_3 , toluene

EtO_2C

TMS

TMS

70%

Saito, S.; Masuda, M.; Komagawa, S. *J. Am. Chem. Soc.* ***2004****, 126*, 10540.

MeO_2C

MeO_2C

$N_2CHSiMe_3$, cat $Ni(cod)_2$

THF , 60°C

$SiMe_3$

MeO_2C

MeO_2C

H

76% (13:1)

Ni, Y.; Montgomery, J. *J. Am. Chem. Soc.* ***2004****, 126*, 11162.

Ph

OTIPS

1% AuCl DCM , 20°C

Ph

73%

OTIPS

Zhang, L.; Kozmin, S.A. *J. Am. Chem. Soc.* ***2004****, 126*, 11806.

1. p-TsCl , TEA , toluene
2. CH_2=$CHCO_2Me$, 1% $Pd(OAc)_2/PPh_3$ DMA/DMF/TEA , 105°C, 32 h

CO_2Me 90%

Fu, X.; Zhang, S.; Yin, J.; McAllister, T.L.; Jiang, S.A.; Tann, C.-H.; Thiruvengadam, T.K.; Zhang, F. *Tetrahedron Lett.* **2002**, *43*, 573.

MeO, OAc

1 atm CH_2=CH_2 , toluene
10% Grubbs II , 80°C , 30 min

MeO, OAc quant

Tonogaki, K.; Mori, M. *Tetrahedron Lett.* **2002**, *43*, 2235.

Ts—N, C

cat. $RuClH(CO)(PPh_3)$, toluene
reflux , 8h

Ts—N 60%

Kang, S.-K.; Ko, B.-S.; Lee, D.-M. *Tetrahedron Lett.* **2002**, *43*, 6693.

EtO_2C, EtO_2C, Br

3% $Pd(PPh_3)_4$, 2 eq Cs_2CO_3 , DMF
2 eq Et_3SiH , 80°C , 4h

EtO_2C, EtO_2C 72%

Oh, C.H.; Park, S.J. *Tetrahedron Lett.* **2003**, *44*, 3785.

C, Ts—N

cat $[RhCl(cod)]_2$, $P(OTol)_3$, dioxane
110°C , 3h

Ts—N 93%

Makino, T.; Itoh, K. *Tetrahedron Lett.* **2003**, *44*, 6335.

PhCHO

1. cyclopropyl–MgBr , 0°C
2. AcCl , 60°C 3. t-BuOK , reflux

Ph 81x63%

Wong, K.-T.; Hung, Y.-Y. *Tetrahedron Lett.* **2003**, *44*, 8033.

Me_3Si, $ZrCp_2Cl$

1. Br–C_6H_4–CHO , DCM , –78°C
2. aq. $NaHCO_3$

Br 64% (89:11 *E:Z*)

Pi, J.-H.; Huang, X. *Tetrahedron Lett.* **2004**, *45*, 2215.

CHO, N, N, Ph

allyl bromide (Br) , In , BF_3•OEt_2 , THF
30°C

N, N, Ph 63%

Kumar, V.; Chimni, S.S.; Kumar, S. *Tetrahedron Lett.* **2004**, *45*, 3409.

OH ; B(OH)$_2$, cat Pd(PPh$_3$)$_4$, 2h ; dioxane , 80°C ; 95%

Yoshida, M.; Gotou, T.; Ihara, M. *Chem. Commun.* **2004**, 1124.

PhCHO ; HC≡CCH$_2$Br , In , DMF , 100°C ; cat Pd(PPh$_3$)$_4$, 3 LiI ; Ph, Ph ; 80%

Lee, K.; Seomoon, D.; Lee, P.H. *Angew. Chem. Int. Ed.* **2002**, *41*, 3901.

Ph, I ; 1. *t*-BuLi , –78°C ; 2. [Cl(Me)ZrCp$_2$] , –78°C → –60°C ; 3. CH$_2$=CHBr , –6°C 4. H$_2$O ; Ph ; 69%

Barluenga, J.; Rodríguez, F.; Álvarez-Rodrigo, L.; Fañanás, F.J. *Chem. Eur. J.* **2004**, *10*, 101.

Ph, Br ; CH$_2$=CHCO$_2$Bu , DMF , K$_2$CO$_3$; cat [Pd(C$_3$H$_7$)Cl]$_2$, tedicyp ; 130°C , 20 h ; Ph, CO$_2$Bu ; 95%

Berthiol, F.; Doucet, H.; Santelli, M. *Synlett* **2003**, 841.

O ; TMS-Cl , NEt$_3$, DMF , 90°C ; 18 h ; OTMS ; 58%

Uyeda, R.T.; Vu, P.; Holsworth, D.D. *Org. Prep. Proceed. Int.* **2002**, *34*, 540.

C$_6$H$_{13}$—≡—Ph ; 4 Br , 5% CoCl$_2$, THF , rt , 2 d ; C$_6$H$_{13}$, Ph ; 64%

Nishikawa, T.; Yorimitsu, H.; Oshima, K. *Synlett* **2004**, 1573.

Ph, O ; [IrCl(cod)]$_2$, bmim BF$_4$; 60°C , 1 h ; Ph, O ; 95%

Shibata, T.; Yamasaki, M.; Kadowaki, S.; Takagi, K. *Synlett* **2004**, 2812.

O, P, OEt, OEt ; BuLi , THF , PhCHO , HMPA ; –78°C → rt ; Ph ; 60% (40:1 *E:Z*)

Wang, Y.; West, F.G. *Synthesis* **2002**, 99.

O, C$_6$H$_{13}$; 1. Ti(O*i*-Pr)$_4$, 2 *i*-PrMgCl , ether ; –50°C → rt ; 2. 1N HCl ; HO, C$_6$H$_{13}$; 91%

Nakajima, R.; Urabe, H.; Sato, F. *Chem. Lett.* **2002**, 4.

1. Ga , cat In , THF , 10°C
2. $C_{10}H_{21}C{\equiv}CH$, *i*-Pr_2NEt THF , 25°C , 4 h

Br → $C_{10}H_{21}$ 96%

Takai, K.; Ikawa, Y.; Ishii, K.; Kumanda, M. *Chem. Lett.* ***2002***, 172.

$PhCH{=}CH_2$, Pyridine , 12 h
10% $CpRu(PPh_3)_2Cl$, 100°C
12% $NaPF_6$

H Ph H 56%

Murakami, M.; Ubukata, M.; Ito, Y. *Chem Lett.* ***2002***, *31*, 294.

MeO_2C MeO_2C $PtCl_2$ → MeO_2C MeO_2C 98%

Echavarren, A.M.; Méndez, M.; Muñoz, M.P.; Nevado, C.; Martín-Matute, B.; Nieto-Oberhuber, C.; Cárdenas, D.J. *Pure Appl. Chem.* ***2004***. *76*, 453.

REVIEWS:

"The Reverse Cope Cyclisation: A Classical Reaction Goes Backwards"
Cooper, N.J.; Knight, D.W. *Tetrahedron* ***2004***, *60*, 243.

"The Mechanisms of the Stille Reaction"
Espinet, P.; Echavarren, A.M. *Angew. Chem. Int. Ed.* ***2004***, *43*, 4704.

SECTION 378: OXIDES - ALKYNES

Ph—≡—S—Ph + (Cl, N, O, Cl, N, O, N, O, Cl) Py , DCM , MeCN , H_2O → Ph—≡—S(=O)—Ph 86%

Zhong, P.; Guo, M.-P.; Huang, N.-P. *Synth. Commun.* ***2002***, *32*, 175.

SECTION 379: OXIDES - ACID DERIVATIVES

NO ADDITIONAL EXAMPLES

SECTION 380: OXIDES - ALCOHOLS, THIOLS

CH_3CH_2CHO tetramethylquanidinium lactate (an ionic liquid) / $MeNO_2$, 20h → OH NO_2 73%

Jiang, T.; Gao, H.; Han, B.; Zhao, G.; Chang, Y.; Wu, W.; Gao, L.; Yang, G. *Tetrahedron Lett.* ***2004***, *45*, 2699.

$MeNO_2$, 20% cu(OTf)$_2$; 20% NEt_3 , rt — 95% (92% ee)

Christensen, C.; Juhl, K.; Hazell, R.G.; Jørgensen, K.A. *J. Org. Chem.* **2002**, *67*, 4875.

10% (hydroxyacetone), $CHCl_3$; 15% (pyrrolidine catalyst, N-*i*-Pr), rt, 7d — 79% (83:17 *anti:syn*, 97% ee)

Andrey, O.; Alexakis, A.; Bernaardinelli, G. *Org. Lett.* **2003**, *5*, 2555.

PhCHO — Me(nitronate)$OSiMe_3$, cat chiral ammonium salt, THF; 1N HCl — 92% (92:8 *anti:syn*) (95% ee)

Ooi, T.; Doda, K.; Maruoka, K. *J. Am. Chem. Soc.* **2003**, *125*, 2054.

I_2 , LiOAc•2 H_2O , rt , aq. MeCN — 87%

Ma, S.; Ren, H.; Wei, Q. *J. Am. Chem. Soc.* **2003**, *125*, 4817.

PhCHO — $MeNO_2$, EtOH , rt , 22 h ; [bis(oxazolone) ligand] (OAc)$_2$ — 76% (94% ee)

Evans, D.A.; Seidel, D.; Reuping, M.; Lam, H.W.; Shaw, J.T.; Downey, C.W. *J. Am. Chem. Soc.* **2003**, *125*, 12692.

i-PrCHO , SmI_2 , THF — 79%

Orsini, F.; Caselli, A. *Tetrahedron Lett.* **2002**, *43*, 7255.

NO (trace O_2) , DCM , rt , 15h — 91%

Liu, Z.; Li. R.; Yang, D.; Wu, L. *Tetrahedron Lett.* **2004**, *45*, 1565.

PhCHO , Cu(OT)$_2$, THF ; chiral bis(oxazline) — 70% (66% threo)

Risgaard, T.; Gothelf, K.V.; Jorgensen, K.A. *Org. Biomol. Chem.* **2003**, *1*, 153.

TsCl , H_2O , THF , 50°C ; 20% sulfonic acid — 75%

Xi, C.; Lai, C.; Chen, C.; Wang, R. *Synlett* **2004**, 1595.

$NaBH_4$, TMSCl , THF

cat

98% (94% ee)

Zhao, G.; Hu, J.-b.; Qian, Z.-s.; Yin, X.-x. *Tetrahedron: Asymmetry* **2002**, *13*, 2095.

cat Zn dimer complex , MS 4Å

$MeNO_2$, THF , 1d , –78°C→–35°C

75% (85% ee)

Trost, B.M.; Yeh, V.S.C. *Angew. Chem. Int. Ed.* **2002**, *41*, 861.

10 $MeNO_2$, MeCN , 8 kbar

rt , 12h

81% (>99% ee)

Misumi, Y.; Matsumoto, K. *Angew. Chem. Int. Ed.* **2002**, *41*, 1031.

$MeNO_2$, MeCN , K_3PO_4 , 15 min

95%

Desai, U.V.; Pore, D.M.; Mane, R.B.; Solabannavar, S.B.; Wadgaonkar, P. *Synth. Commun.* **2004**, *34*, 19.

SECTION 381: OXIDES - ALDEHYDES

, *i*-PrOH , 4°C , 2 d

87% (80% ee)

Mase, N.; Thayumanavan, R.; Tanaka, F.; Barbas III, C.F. *Org. Lett.* **2004**, *6*, 2527.

Me_2CHCH_2CHO , (bmim) PF_6

cat *S*-proline

83% (9:1 *syn:anti*)

Kotreusz, P.; Toma, S.; Schmalz, H.-G.; Adler, A. *Eur. J. Org. Chem.* **2004**, 1577.

SeO_2 , 70% TBHP

dioxane , reflux , 3 h

55%

Tagawa, Y.; Yamashita, K.; Higuchi, Y.; Goto, Y. *Heterocycles* **2003**. *60*, 953

SECTION 382: OXIDES - AMIDES

cat $Rh_2(OAc)_4$, DCM

reflux

65%

Yoon, C.H.; Nagle, A.; Chen, C.; Gandhi, D.; Jung, K.W. *Org. Lett.* ***2003***, *5*, 2259.

$EtNO_2$, chiral diamine catalyst , –20°C

69% (14:1 dr) 59% ee

Nugent, B.M.; Yoder, R.A.; Johnston, J.N. *J. Am. Chem. Soc.* ***2004***, *126*, 3418.

1. LHMDS , –78°C , THF

2. AcOH

71%

Hernandez, N.M.; Sedano, M.J.; Jacobs, H.K.; Gopalan, A.S. *Tetrahedron Lett.* ***2003***, *44*, 4035.

SECTION 383: OXIDES - AMINES

1. LHMDS , THF , –78°C , 1 h

2. $F_4B^- Me_2Se^+$ ... THF , –78°C

78% (83% de)

Watanabe, S.-i.; Ikeda, T.; Kataoka, T.; Tanabe, G.; Muraoka, O. *Org. Lett.* ***2003***, *5*, 565.

, toluene , 23°C , 1 h

chiral thiourea catalyst

quant (80% ee)

Joly, G.D.; Jacobsen, E.N. *J. Am. Chem. Soc.* ***2004***, *126*, 4102.

, 2 HFIP , MS 4Å

$Cu(OTf)_2$ /cat. diamine , DCM , 0°C

86% (91% ee)

Kobashi, S.; Kiyohara, H.; Nakamura, Y.; Maatsubara, R. *J. Am. Chem. Soc.* ***2004***, *126*, 6558.

$PhNH_2/Al_2O_3$, microwaves

5 min

55%

Koboudin, B. *Tetrahedron Lett.* ***2003***, *44*, 1051.

$PhSO_2Cl$, $InBr_3$, DCE; reflux , 2.5h — SO_2Ph 87%

Yadav, J.S.; Reddy, B.V.S.; Krishna, A.D.; Swamy, T. *Tetrahedron Lett.* ***2003***, *44*, 6055.

Ph—≡—Ph; 1. MeO–C_6H_4–NH_2 , 110°C; 3% Cp_2TiMe_2 , toluene , 3d; 2. $HP(=O)(OEt)_2$, 5% Me_2AlCl , rt , 2h — Ph, Ph, MeO, N H, P, OEt, OEt, O 68%

Haak, E.; Bytschkov, I.; Doye, S. *Eur. J. Org. Chem.* ***2002***, 457.

PhCHO; $PhNH_2$, $P(OMe)_3$, TMSCl; $LiClO_4$ (ether) , rt — NHPh, Ph, P, OMe, OMe, O 98%

Saidi, M.R.; Azizi, N. *Synlett* ***2002***, 1347.

PhCHO + N—H; $P(OMe)_3$, $LiClO_4$, rt; solvent free , 40 min — O, P, OMe, OMe, N, Ph 90%

Azizi, N.; Saidi, M.R. *Eur. J. Org. Chem.* ***2003***, 4630.

PhCHO + $PhNH_2$; $(EtO)_2P$-OH , 10% SmI_2 , 1d; MeCN , MS 4Å , 80°C — O, P, OMe, OMe, PhHN, Ph 60%

Xu, F.; Luo, Y.; Deng, M.; Shen, Q. *Eur. J. Org. Chem.* ***2003***, 4728.

PhCHO; 1. $PhNH_2$, $P(OEt)_3$, neat , 15 min; 2. SiO_2 — NHPh, Ph, P, OEt, OEt, O 95%

Chandrasekhar, S.; Narsihmulu, Ch.; Sultana, S.S.; Saritha, B.; Prakash, S.J. *Synlett* ***2003***, 505.

NO_2, Cl + H_2N; micrwaves , 90°C; 8 min — NO_2, H, N 93%

Xu, Z.-B.; Lu, Y.; Guo, Z.-R. *Synlett* ***2003***, 564.

O; $BnNH_2$, DCM , $HP(=)(OEt)_2$; phthalocyanine — NHBn, P, OEt, OEt, O 85%

Maatveeva, E.D.; Podrugina, T.A.; Tishkoskaya, E.V.; Tomilova, L.G.; Zefirov, N.S. *Synlett* ***2003***, 2321.

MeO–C6H4–NH2 → PhCHO , $(ErO)_2P(=O)OH$, 1d / 0.25 CF_3CO_2H , solvent free , rt → Ph–CH(NH–C6H4–OMe)–P(=O)(OEt)2 96%

Akiyama, T.; Sanada, M.; Fuchibe, K. *Synlett* *2003*, 1463.

SECTION 384: OXIDES - ESTERS

Ph–CH=CH–NO_2 → 2 $CH_2(CO_2Et)_2$, toluene , rt / cat thiourea derivative , 1 d → Ph–CH($CH(CO_2Et)_2$)–CH_2NO_2 86% (93% ee)

Okino, T.; Hoashi, Y.; Takemoto, Y. *J. Am. Chem. Soc.* ***2003***, *125*, 12672.

MeO–C(OTMS)=CH–CH_3 → 1. SO_2 , cat TBSOTf / 2.TBAF , MeI → MeO–C(=O)–CH(CH_3)–CO_2Me 52%

Bouchez, L.; Vogel, P. *Synthesis* ***2002***, 225.

SECTION 385: OXIDES - ETHERS, EPOXIDES, THIOETHERS

CH_2=CH–P(=O)(OEt)2 → cyclohexanone monooxygenase / 25°C , 0.05M TMS•HCl , pH 8.6 / NaDPH , glucose-6-phosphite , 2 d → (oxiranyl)–P(=O)(OEt)2 40% >98% ee)

Colonna, S.; Gaggero, N.; Carrea, G.; Ottolina, G.; Pasta, P.; Zambianchi, F. *Tetrahedron Lett.* ***2002***, *43*, 1797.

cyclohexanone → $ClCH_2SO_2Me$, 1.2 *t*-BuOK , THF / –78°C , 30 min → spiro epoxide with SO_2Me 98%

Makosza, M.; Urbanska, N.; Chesnokov, A.A. *Tetrahedron Lett.* ***2003***, *44*, 1473.

PhCHO → $ClCH_2SO_2Ph$, 10% chiral phase transfer cat → Ph–epoxide–SO_2Ph 85% (69% ee)

Arai, S.; Shioiri, T. *Tetrahedron* ***2002***, *58*, 1407.

CH_2=CH–OBu → Ph_2P-I , 2% $Ni(PPh_3)_2Br_2$ / PhH 80°C , 2 h → CH_3–CH(OBu)–P(=O)Ph2 83%

Kazankova, M.A.; Shulyupin, M.O.; Beletskaya, I.P. *Synlett* ***2003***, 2155.

SECTION 386: OXIDES - HALIDES, SULFONATES

Me-S-CH_2Cl —[5% H_2O_2, ether, AcOH, rt; $MgSO_4$, 30 h, reflux]→ $MeSO_2CH_2Cl$ 89%

Makosza, M.; Surowiec, M. *Org. Prep. Proceed. Int.* **2003**, *35*, 412.

SECTION 387: OXIDES - KETONES

C_8H_{17}, AcO, $I(Ph)BF_4$ —[2 eq $Ph_2P(O)OH$, MeOH; DCM, NEt_3]→ C_8H_{17}, O, P, Ph, Ph, O 88%

Ochiai, M.; Nishitani, J.; Nishi, Y. *J. Org. Chem.* **2002**, *67*, 4407.

O, Ph —[*i*-$PrNO_2$, neat, 240 h; cat Me, N, Bn, N, H, CO_2H]→ NO_2, O, Ph quant, 79% ee

Halland, N.; Hazell, R.G.; Jørgensen, K.A. *J. Org. Chem.* **2002**, *67*, 8331.

O, Ph —[OTMS, S*t*-Bu, cat oxazaborolidinone; *i*-Pr, OH, *i*-Pr, DCM, –78°C, $MeCO_2$*t*-Bu]→ O, Ph, O, *t*-BuS 88% (89% ee)

Wang, X.; Adachi, S.; Iwai, H.; Takatsuki, H.; Fujita, K.; Kubo, M.; Oku, A.; Harada, T. *J. Org. Chem.* **2003**, *68*, 10046.

O —[Ph, NO_2, rt, 15 h; chiral diamine cat., 15% HCl]→ O, Ph, NO_2 82% (25% ee)

Alexakis, A.; Andrey, O. *Org. Lett.* **2002**, *4*, 3611.

S, SO_2Ph —[1. BuLi, THF, –78°C; 2. THF, –78°C → rt, O, Ph]→ Ph, S, SO_2Ph, O quant (>99% ee)

Magomedov, N.A.; Ruggiero, P.L.; Tang, Y. *J. Am. Chem. Soc.* **2004**, *126*, 1624.

O, Ph —[Me_3SiCl, *i*-AmONO, DCM; –20°C, 1h]→ O, Ph, N–OH 90%

Mohammed, A.H.A.; Nagendrappa, G. *Tetrahedron Lett.* **2003**, *44*, 2753.

$PhC(=O)OMe$ → [$ICH_2P(=O)(OEt)_2$, cat $Co(PMe_3)_4$] → $PhC(=O)CH_2P(=O)(OEt)_2$ 63%

Orsini, F.; Diteodoro, E.; Ferrari, M. *Synthesis* **2002**, 1683.

SECTION 388: OXIDES - NITRILES

$PhC(=N\text{-}P(=O)Ph_2)CH_3$ → [cat La complex, 1.5 TMSCN; EtCN, –48°C, 1 d] → $PhC(CH_3)(CN)NH\text{-}P(=O)Ph_2$ 94% (95% ee)

Masumoto, S.; Usuda, H.; Suzuki, M.; Kanai, M.; Shibasaki, M. *J. Am. Chem. Soc.* **2003**, *125*, 5634.

1. 5 TMSCN, 0.5 urea, DCM, –78°C
2. HCl, MeOH

79%

Okino, T.; Hashi, Y.; Takemoto, Y. *Tetrahedron Lett.* **2003**, *44*, 2817.

SECTION 389: OXIDES - ALKENES

Bu_3SnH, AIBN, PhH; $P(OMe)_3$, reflux

P—OMe, OMe + P—OMe, OMe

(94 : 6) 83%

Jiao, X.-Y.; Bentrude, W.G. *J. Org. Chem.* **2003**, *68*, 3303.

Ph—≡ → [THF, rt, $=C=CH\text{-}P^+(O^-)Ph_3$; 10% $Pf(OAc)_2$, 20% TDMPP] → 90%

Rubin, M.; Markov, J.; Chuprakov, S.; Wink, D.J.; Gevorgyan, V. *J. Org. Chem.* **2003**, *68*, 6251.

NO_2 … → [Pb, $Et_4N^+ HCO_2^-$, MeOH; 58°C] → N-OH indole 92%

Wong, A.; Kuethe, J.T.; Davies, I.W. *J. Org. Chem.* **2003**, *68*, 9865.

$S(O_2)$ diallyl → [2% Grubbs I, DCM, 1 h; 0.05 M] → SO_2 95%

Yao, Q. *Org. Lett.* **2002**, *4*, 427.

Bu—≡—P(O)(OEt)(OEt) → 1. Cp_2ZrCl_2–2 BuLi, THF; 2. EtCHO → 78%

Quntar, A.A.A.; Melman, A.; Srebnik, M. *J. Org. Chem.* ***2002***, *67*, 3769.

PhB(OH)$_2$, cat. Pd(PPh$_3$)$_4$ → 93%

Kobayashi, Y.; Williams, A.D. *Org. Lett.* ***2002***, *4*, 4241.

Me$_2$C=CHMe, neat, 16 h; 5% Ru carbene complex; reflux → 98%

Demchuk, O.M.; Pietrusiewicz, K.M.; Michrowska, A.; Grela, K. *Org. Lett.* ***2003***, *5*, 3217.

LiO*t*-Bu, THF, rt → (29 : 71) 90%

Morton, D.; Pearson, D.; Field, R.A.; Stockman, R.A. *Org. Lett.* ***2004***, *6*, 2377.

2.5 eq LDA, THF; −78°C → 60% (97:3 *S*:*R*)

Maezaki, N.; Sawamoto, H.; Yuyama, S.; Yoshigami, R.; Suzuki, T.; Izumi, M.; Ohishi, H.; Tanaka, T. *J. Org. Chem.* ***2004***, *69*, 6335.

$Cl(F_2C)_4O_2S$CH$_2$C(O)Ph → PhCHO, AcOH, toluene; piperidine, reflux, 4 h → 92%

Xing, C.; Zhu, S. *J. Org. Chem.* ***2004***, *69*, 6486.

C_6H_{13}—≡ → menthyl—P(O)(H)Ph, toluene, 70°C; 5% Me$_2$Pd(PPhMe$_2$)$_2$, 10% Ph$_2$P(O)OH → 96%

Han, L.-B.; Zhao, C.-Q.; Onozawa, S.-y.; Goto, M.; Tanakta, M. *J. Am. Chem. Soc.* ***2002***, *124*, 3842.

Bu–C≡C–S(O)Tol → 1. EtCu , $MgBr_2$ 2. PhCHO 3. Zn $(CH_2I)_2$, THF → 78% (>99:1 dr)

Sklute, G.; Amsallem, D.; Shabli, A.; Varghese, J.P.; Marek, I. *J. Am. Chem. Soc.* ***2003***, *125*, 11776.

$PhCH=CH_2$, Grubbs' II catalyst , DCM, 40°C → 51% (70% ee)

He, A.; Yan, B.; Thanavaro, A.; Spilling, C.D.; Rath, N.P. *J. Am. Chem. Soc.* ***2004***, *126*, 8643.

EtO_2C, Br , DBU , MeCN, 30°C , rt → 63%

Ballini, R.; Barboni, L.; Bosica, G.; Fiorini, D.; Mignini, E.; Palmieri, A. *Tetrahedron* ***2004***, *60*, 4995.

PhCHO → $PhSO_2CH_2CN$, MeOH , rt, $Na_2CaP_2O_7$ → 62%

Zahouily, M.; Salah, M.; Bennazha, J.; Rayadh, A.; Sebti, S. *Tetrahedron Lett.* ***2003***, *44*, 3217.

Ph–C≡CH → (MeO)$_2$P(O)H , cat Ni(cod)$_2$, $PPhMe_2$, 67°C , 3 h , THF → 79%

Han, L.-B.; Zhang, C.; Yazawa, H.; Shimada, S. *J. Am. Chem. Soc.* ***2004***, *126*, 5080.

$PhB(OH)_2$ → CH_2=CHP(=O)$(OEt)_2$, 10% $Pd(OAc)_2$, 2 Na_2CO_3 , DMF , O_2 , 60°C , 18h → 82%

Kabalka, G.W.; Guchhait, S.K.; Naravane, A. *Tetrahedron Lett.* ***2004***, *45*, 4685.

Ph$CH_2SO_3R^*$ → 1.BuLi , THF , –95°C 2. Br , –95°C → –78°C , 15 h → 80% (90% de)

Enders, D.; Harnying, W.; Vignola, N. *Eur. J. Org. Chem.* ***2003***, 3939.

→ Grubbs' II, PhH , 70°C → 99%

LeFlohic, A.; Meyer, C.; Cossy, J.; Desmurs, J.-R.; Galland, J.-C. *Synlett* ***2003***, 667.

C_8H_{19}–CH=CH–CH_2OH → 1. NBS , PPh_3 , MeCN 2. Bu_4NI , $TolSO_2Me$ → C_8H_{19}–CH=CH–SO_2Tol 84%

Murakami, T.; Furusawa, K. *Synthesis* ***2002***, 479.

Ph—≡ + Bn—P(=O)(OH)—OEt —[10% $Hg(OAc)_2$, toluene / 2% BF_3•OEt_2, 80°C]→ Ph–C(=CH_2)–O–P(=O)(Bn)–OEt 55%

Peng, A.; Ding, Y. *Synthesis* **2003**, 205.

SECTION 390: OXIDES - OXIDES

CH_2=CH–C(Me)=C=CH–S(=O)–Ph —[SO_2, DCE, reflux / 2.5 h]→ 3-methyl-2,5-dihydrothiophene 1,1-dioxide, 2-(=CH–S(=O)–Ph) 72%

Christov, V.Ch.; Ivanov, I.K. *Heterocycles* **2004**, *63*, 2203.

AUTHOR INDEX